INVESTIGATION OF LATE QUATERNARY PALEOCEANOGRAPHY AND PALEOCLIMATOLOGY

The Geological Society of America, Inc.
Memoir 145

Investigation of Late Quaternary Paleoceanography and Paleoclimatology

Edited by
R. M. CLINE
AND
J. D. HAYS

1976

To the memory of
RICHARD FOSTER FLINT
whose work guided and inspired the
CLIMAP *program*

Copyright 1976 by The Geological Society of America, Inc.
Copyright is not claimed on any material prepared by
U.S. Government employees within the scope of their employment.
Library of Congress Catalog Card Number 75-40899
I.S.B.N. 0-8137-1145-2

Published by
THE GEOLOGICAL SOCIETY OF AMERICA, INC.
3300 Penrose Place
Boulder, Colorado 80301

*The Memoir series was originally made possible
through the bequest of
Richard Alexander Fullerton Penrose, Jr.*

Printed in the United States of America

Contents

Preface ix

ATLANTIC

New transfer function for estimating past sea-surface conditions from sea-bed distribution of planktonic foraminiferal assemblages in the North Atlantic
. . . . *Nilva G. Kipp* 3

Glacial North Atlantic 18,000 years ago: a CLIMAP reconstruction
. . . . *Andrew McIntyre and Nilva G. Kipp with Allen W. H. Bé, Thomas Crowley, Thomas Kellogg, James V. Gardner, Warren Prell, and William F. Ruddiman* 43

Late Quaternary climatic changes: Evidence from deep-sea cores of Norwegian and Greenland Seas . . . . *Thomas B. Kellogg* 77

Northeast Atlantic paleoclimatic changes over the past 600,000 years
. . . . *W. F. Ruddiman and A. McIntyre* 111

O^{18} record of the Atlantic Ocean for the entire Pleistocene Epoch
. . . . *Jan van Donk* 147

Late Quaternary climatic record in western equatorial Atlantic sediment
. . . *Allan W. H. Bé, John E. Damuth, Leroy Lott, and Rosemary Free* 165

Late Pleistocene faunal and temperature patterns of the Colombia Basin, Caribbean Sea . . . . *Warren L. Prell and James D. Hays* 201

Responses of sea-surface temperature and circulation to global climatic change during the past 200,000 years in the eastern equatorial Atlantic Ocean
. . . . *James V. Gardner and James D. Hays* 221

Equatorial Atlantic and Caribbean foraminiferal assemblages, temperatures, and circulation: Interglacial and glacial comparisons
. . . . *Warren L. Prell, James V. Gardner, Allan W. H. Bé, and James D. Hays* 247

Corresponding patterns of contemporary pollen and vegetation in central North America . . . . *T. Webb III and J. H. McAndrews* 267

ANTARCTIC

Relationship of radiolarian assemblages to sediment types and physical oceanography in the Atlantic and western Indian Ocean sectors of the Antarctic Ocean . . . . *Jose A. Lozano and James D. Hays* 303

Reconstruction of the Atlantic and western Indian Ocean sectors of the 18,000 B.P. Antarctic Ocean . . . . *James D. Hays, Jose A. Lozano, Nicholas Shackleton, and Grace Irving* 337

PACIFIC

Late Quaternary sediment of the Panama Basin: Sedimentation rates, periodicities, and controls of carbonate and opal accumulation
.... *Nicklas G. Pisias*   375

Late Quaternary accumulation rates of opal, quartz, organic carbon, and calcium carbonate in the Cascadia Basin area, northeast Pacific
.... *G. Ross Heath, Ted C. Moore, Jr., and J. Paul Dauphin*   393

Glacial advance in the Gulf of Alaska area implied by ice-rafted material
.... *Roland von Huene, Jim Crouch, and Edwin Larson*   411

Modern Pacific coccolith assemblages: Derivation and application to late Pleistocene paleotemperature analysis
..... *Kurt R. Geitzenauer, Michael B. Roche, and Andrew McIntyre*   423

Oxygen-isotope and paleomagnetic stratigraphy of Pacific core V28-239 late Pliocene to latest Pleistocene . . . *N. J. Shackleton and N. D. Opdyke*   449

Microfiche .... in pocket inside back cover

Kipp Appendixes:
 I. Species identification and North Atlantic planktonic foraminiferal counts
 II. Locations and descriptions of core-top samples
 III. Varimax factor matrix (B)
 IV. Estimated temperatures and salinities for core-top locations

Kellogg Appendixes:
 I. Raw coarse-fraction and $CaCO_3$ data T.W. cores
 II. Raw faunal counts T.W. cores
 III. Raw faunal counts
 IV. Raw coarse-fraction and $CaCO_3$ data for cores

Ruddiman and McIntyre Appendixes:
 I. Coccolith downcore data
 II. Foraminifera downcore data
 III. Coarse-fraction carbonate data

Bé and others Appendix:
 I. Western equatorial Atlantic cores: Species expressed in percent of total planktonic foraminifera

Prell and Hays Appendixes:
 I. V18-357 Caribbean raw foraminifera data
 II. V28-127 Caribbean raw foraminifera data

Gardner and Hays Appendixes:
 I. Raw counts of planktonic foraminifera
 II. Raw counts of planktonic foraminifera

Webb and McAndrews Table:
 1. Location of unpublished samples

Lozano and Hays and Hays and others Appendix:
 I. Core tops and 18,000 B.P. level species counts

Pisias Appendixes:
 II. Sediment composition data Y69-106P
 III. δO^{18} for core Y69-106P (with respect to PDB)

Heath and others Appendixes:
  I. Analytical data core Y6604-10, Y6609-5, and Y6910-2
  II. Sample depths and bulk density (mass of dry material per cubic centimetre of wet sediment) of Cascadia Basin cores
Geitzenauer and others Appendixes:
  I. Coccolith counts from Pacific Ocean surface sediment
  II. Surface sediment varimax matrix
  III. Coccolith counts from cores V28-238, V28-239, and V19-55
  IV. Downcore varimax matrix

Note: Hard copy is on file with Lamont-Doherty Geological Observatory.

Preface

The mission of the CLIMAP (Climate/Long Range Investigation Mapping and Predictions) project (operational since 1971) is to document global climatic change over the past million years for the purpose of understanding both the underlying factors that motivate climatic change and the response of the ocean atmosphere ice system that amplifies these factors. The program has as its central thrust the monitoring of paleoceanographic changes through the study of a global distribution of deep-sea cores.

Deep-sea cores have extraordinary advantages over other kinds of proxy paleoclimatic data because comparable data on a global scale can be extracted from them; they contain continuous or nearly continuous records, and since sedimentation rates vary regionally, they can be used to study long time series (hundreds of thousands to millions of years; see Ruddiman and McIntyre; Pisias) or details of the past few tens of thousands of years. Deep-sea sediments contain information that can be used to monitor a variety of aspects of the ocean-ice-atmosphere system. For example, the shells of myriads of planktonic species, each adapted to its own peculiar set of oceanographic conditions, are preserved in the sediments. Through appropriate counting and multivariate statistical techniques, these fossil populations can be grouped and quantitative estimates of water properties can be obtained (see Kipp; McIntyre and others; Prell and Hays; Gardner and Hays; Prell and others; Lozano and Hays; Hays and others; Geitzenauer and others). Changes in global ice volume are reflected in the oxygen isotope ratios of foraminiferal skeletons (see van Donk; Shackleton and Opdyke). The mineral content of the sediments reflect variations in density of drifting icebergs (see von Huene and others) and variation of continental erosion and oceanic productivity (see Heath and others).

Although these data provide a treasury of paleoclimatic and paleoceanographic information, there are other critical sources of information that lie on the continents. Consequently, it was decided early in the CLIMAP Program that to be truly effective we must work with those scientists probing the climatic records above the strandline such as paleo-sea levels, vegetational patterns (see Webb and McAndrews), ice sheets, and fluctuation of alpine glaciers.

As an aid in synthesizing this mass of proxy climatic data, we have encouraged various global numerical atmospheric modeling groups to use our data to establish the boundary conditions to reconstruct atmospheric circulation during various past climatic regimes. The first CLIMAP reconstruction selected was 18,000 yr B.P., the maximum of the latest glacial. A number of papers in this volume present 18K data that currently are being used in global atmospheric modeling experiments (see McIntyre and others; Kellogg; Bé and others; Prell and Hays; Gardner and Hays; Prell and others; Hays and others; Geitzenauer and others).

All papers in the volume represent work of the CLIMAP Program funded by the Office of the International Decade of Ocean Exploration, except the paper by von Huene and others that contains important data on fluctuations of ice-rafted detritus in the northeast Pacific which is closely related to CLIMAP work.

We gratefully acknowledge not only the encouragement and support of the International Decade of Ocean Exploration but also the encouragement and help of a host of dedicated scientists throughout the world who are working or who have worked with us.

Only through the co-operation of our national and international members has CLIMAP been able to accumulate its present data base. We are fortunate that these members have so willingly contributed both their knowledge and their friendship.

ATLANTIC

New Transfer Function for Estimating Past Sea-Surface Conditions from Sea-Bed Distribution of Planktonic Foraminiferal Assemblages in the North Atlantic

NILVA G. KIPP
Department of Geological Sciences
Brown University
Providence, Rhode Island 02912

ABSTRACT

This paper presents maps of the distribution of 29 planktonic foraminiferal species in 191 North Atlantic Ocean core tops and describes and tests a transfer function (F13) derived from these data by factor analysis and regression techniques. This transfer function relates six assemblages of planktonic foraminifera to seasonal temperatures and salinities at the sea surface and at a depth of 100 m. The data represent revision and expansion of a data base used previously to derive another transfer function (F3). The new set of data comes mostly (81%) from trigger-weight cores, has greater geographical coverage and depth range, includes faunas from shallow and upwelling waters, and incorporates samples with moderately high dissolution. The transfer function derived from these data yields estimates that have 80% confidence intervals ranging from ±1.5° to ±2.8°C and from ±0.36‰ to ±0.64‰, depending on season and depth. Estimates that fall outside the 80% confidence intervals are randomly distributed in latitude, distance from continents, depth of water, and dissolution effects.

Two independent sets of data were used to test the equations: 60 core tops from the South Atlantic and five short cores from the North Atlantic. For the sea-surface temperatures estimated from the core tops, 78% to 82% of the derived temperatures fall within the previously calculated 80% confidence intervals. Application of F13 to the five short cores located between 0° and 54°N yields reasonable temperature and salinity ranges for the past 20,000 yr.

INTRODUCTION

Planktonic foraminifera have been used as qualitative indicators of marine paleotemperatures since the pioneer work of Schott (1935). This effort was continued by Phleger and others (1953) and Ericson and others (1964). More recently, quantitative temperature estimates have been published by Imbrie and Kipp (1971), Hecht (1973), and Lynts and others (1973).

The quantitative method introduced by Imbrie and Kipp (1971) and revised in Imbrie and others (1973) relates planktonic foraminiferal assemblages on the sea bed to oceanographic parameters of the sea surface. The relations are expressed as a transfer function that may be used on downcore samples to obtain estimates of late Pleistocene salinity and temperature.

The main purpose of this paper is to describe a new transfer function (here designated F13) that reflects significant improvements in the size and quality of the previously used data base, and to describe the data set used to derive this new transfer function. Imbrie and Kipp (1971) used 34 piston core tops from the North Atlantic Ocean, but their data also included tops from the South Atlantic and Indian Oceans. The new data comprise 191 tops taken exclusively from the North Atlantic. Eighty-one percent of the samples are from trigger-weight cores, the tops of which are more representative of the present sea bed than piston-core tops. In addition, a more accurate set of oceanographic data has been assembled, and a deliberate effort has been made to sample a wider range of oceanographic and postdepositional environments. The new set of data spans a greater depth range than the old set; introduces shallow-water, nearshore faunas; includes samples with significant differential dissolution; and incorporates more samples from upwelling areas.

The transfer function F13 was derived from factor analysis and regression analysis of this new set of data. This transfer function has been used by CLIMAP workers to provide maps of sea-surface conditions for the North Atlantic 18,000 yr ago (see McIntyre and others, this volume). Any investigator following the same laboratory procedures and recognizing the same taxonomic categories can use this transfer function, provided the sample is from Quaternary sediment of the North Atlantic Ocean and has a faunal composition that falls within the environmental range of the F13 data base. As discussed later in this paper, applications of F13 are also possible with selected samples from the South Atlantic.

MATHEMATICAL METHOD

The method used here to derive the transfer function is fully described in Imbrie and Kipp (1971) and Imbrie and others (1973). As in the latter paper, calculations here are based on the relative abundances of species without a percent-range transformation.

As summarized in Figure 1, the method can be described in terms of three independent starting points.

1. The raw paleontological data from core tops are factor-analyzed into varimax assemblages. Start 1 represents the acquisition of taxonomic data on the sea bed (matrix X_{ct}). These data are presented graphically in the accompanying figures as species distribution maps. The line of calculation beginning at Start 1 produces a varimax matrix B and an assemblage description matrix F'. The columns of the B matrix are presented in the figures as distribution maps of the assemblages.

2. A least-squares technique is used to write a set of paleoecological equations

relating the varimax assemblages to observed oceanographic parameters. Start 2 represents the acquisition of oceanographic data. Regression techniques combine these data with the faunal information (matrix **B**) to produce a set of equations **K**.

3. The fossil data from a core are described in terms of the core-top varimax assemblages. These assemblages are used with the paleoecological equations to obtain estimates of paleoenvironments. Start 3 represents the application of the matrices **F**′ and **K** to any set of samples (X_c). The final result is a set of environmental estimates (**Y**). The procedure originating at Start 3 is described (in the section titled "Testing the Transfer Function") in an application of the transfer function to 60 South Atlantic tops, and to a set of short cores in the North Atlantic. McIntyre and others (this volume) have used the same procedure to produce environmental estimates for the North Atlantic 18,000 yr ago.

DATA

Data Storage

Foraminiferal counts and oceanographic parameters for 191 locations in the North Atlantic are deposited with the National Geophysical and Solar-Terrestrial Data Center, Washington, D.C., and are presented in Appendix I.[1]

Sea-Bed Samples

Of 229 core-top samples examined, 38 were deleted after applying criteria discussed below. The remaining 191 samples form the data base for this study. They consist of 155 samples from trigger-weight cores, 35 from piston cores and one from a Kasten core. Although the tops of trigger-weight cores are less likely to be disturbed than piston core tops, no consistent difference in quality between the two was found in this study. The sample locations are shown in Figure 2 and listed in Appendix II (see footnote 1).

Twenty-four of the 229 core-top samples examined contained fewer than 300 planktonic foraminifera in the >149-μm sieve fraction. Imbrie and others (1973) suggested that a count of fewer than 300 specimens is too small for a reliable estimate. Seven of the samples with fewer than 300 specimens were retained, however; one of them had fewer than 250 specimens. All these small samples represented single-assemblage faunas for which a reliable description could be obtained from smaller counts. These samples were retained in order to describe transition areas not well-represented by other samples. The average number of specimens counted per sample was 427 and the largest sample counted contained 886 specimens.

Twenty of the samples examined showed evidence of postdepositional mixing or loss of material from the top of the core, based on the shipboard core description, the presence of turbidites or of pre-Pleistocene species, or a faunal composition anomalous compared to several nearby samples. No sample without close neighbors on several sides was rejected on this latter basis. For example, sample number 128 was retained, even though it contains a much colder fauna than nearby sites, because it has no close neighbors to the north and west. It is similar, however, to distant neighbors situated directly to the north.

---

[1] Appendixes I-IV on microfiche in pocket inside back cover.

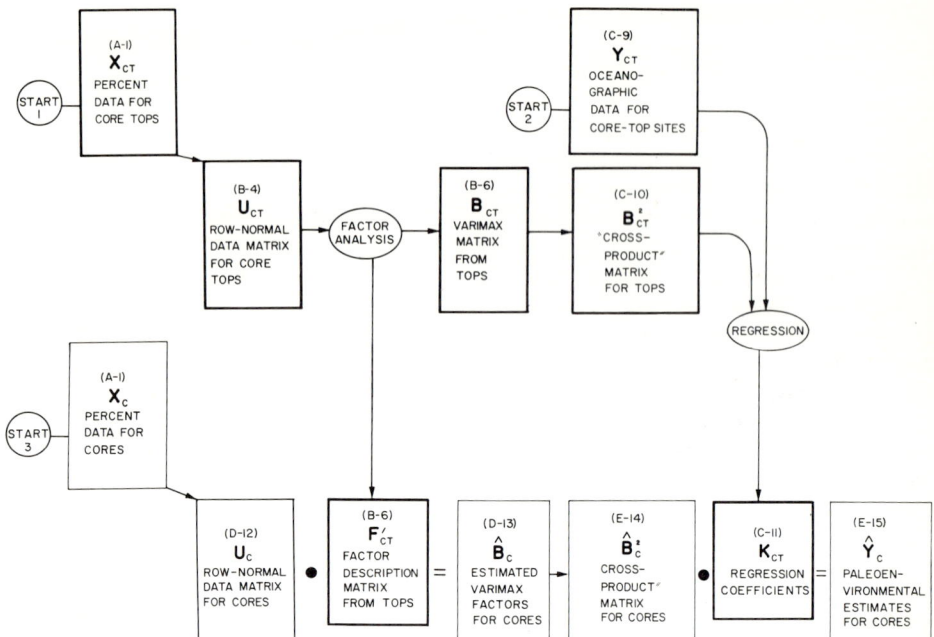

Figure 1. Summary of a mathematical method for estimating oceanographic parameters from faunal data. Matrices used to derive transfer function represented as boldly outlined rectangles. Matrices used in applying transfer function indicated by lightly outlined rectangles. Alphanumeric designations at top of each box refer to step numbers described by Imbrie and Kipp (1971). Bold capital letters stand for matrices defined in same paper. Arrows indicate simple and ovals complex mathematical operations. Matrix operation of multiplication symbolized by bold dots and equal signs.

Oceanographic Data

Oceanographic parameters for the 191 locations were read from monthly or seasonal maps in the Oceanographic Atlas of the North Atlantic Ocean and the Oceanographic Atlas of the Polar Seas (U. S. Naval Oceanographic Office, 1958, 1967). Because these maps are on a larger scale than those of Defant (1961), which were used by Imbrie and Kipp (1971), we consider the data used here to be more accurate. However, the density of observations on which the variable maps were based varies considerably. For example, for much of the North Atlantic the contours given for seasonal parameters at 100-m depth are controlled by no more than a single observation for each 1° quadrangle. However, the contours for August sea-surface temperature are generally based on more than 25 observations for each 1° quadrangle.

To arrive at an estimate of the error involved in reading the maps, two records of winter temperature at the surface were made from separate plots of the locations. The same procedure was used to obtain two records of summer temperature at 100 m from a smaller scale map. The average difference between the two records for winter temperature at the surface was 0.16°C and the average difference for summer temperature at 100 m was 0.36°C. In both cases, deviations were largest in areas of steep thermal gradients. These estimates of error do not, of course, include any error inherent in the maps themselves due to insufficient or inaccurate data.

Definition of Seasons

Equations were written for 16 variables: temperature and salinity at the sea surface and at 100-m depth for each of the four seasons. Equations for five of the variables are reported in this paper. These are summer sea-surface temperature, winter sea-surface temperature, summer temperature at 100 m, winter sea-surface salinity and winter salinity at 100 m. Summer is defined as the warmest month for each location and winter as the coldest month. For most of the North Atlantic, the warmest month is August and the coldest is February. However, a small area in the eastern equatorial Atlantic (south of the dashed line in Figure 2) is warmer in February than in August. The cool August temperatures in this area are linked to the increased intensity of the eastern boundary currents of the South Atlantic. The winter temperature and salinity for this area were read from the August maps and summer values from the February maps.

Multiple correlation coefficients for equations with a calendar definition based on all 191 sample locations are not significantly lower than those with a thermal definition. However, estimates for half the samples in the eastern equatorial Atlantic incorrectly indicated that August is warmer than February when a calendar-based equation was used. In addition, estimates in this area made with a calendar-based equation exceeded the standard error of estimate more frequently than did estimates made with a temperature-based equation.

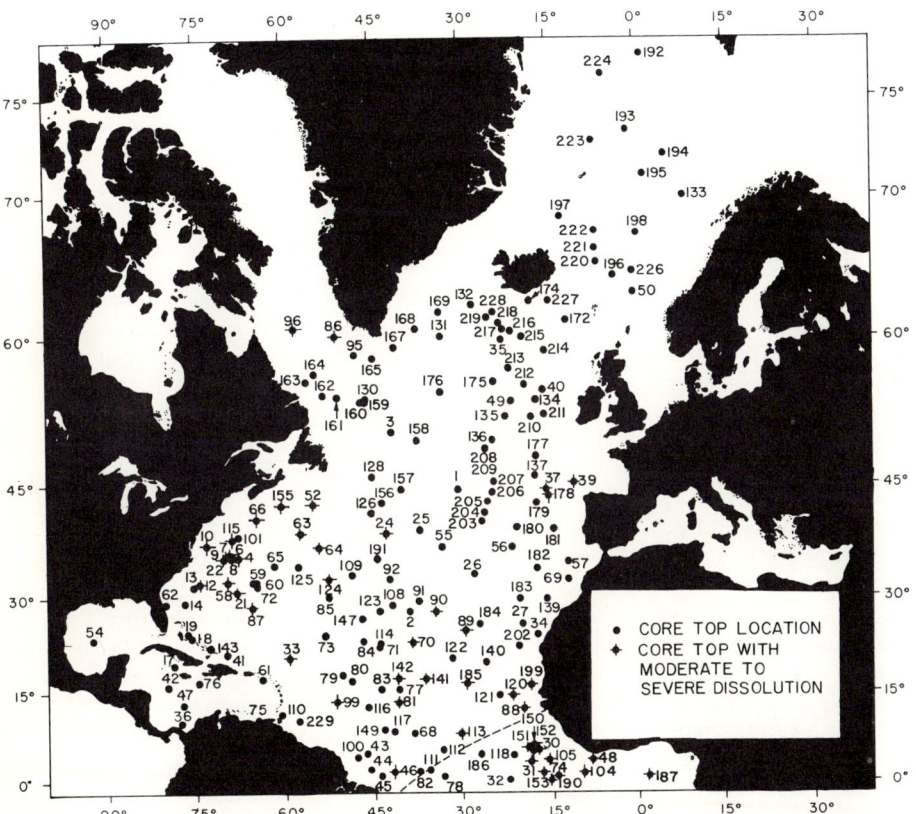

Figure 2. Locations of 191 North Atlantic core-top samples. February sea-surface temperatures are warmer than August temperatures for area south of dashed line.

Two other tests were used to determine if the differences in the equations are significant. First, comparisons were made between the residuals of the two sets of equations for the eastern equatorial area. Student's t tests for the difference between 17 paired samples showed that the temperature- and calendar-based equations are not significantly different for salinity and 100-m-depth temperature estimates. However, winter sea-surface temperature estimates from the calendar-based equation are significantly higher ($\bar{X} = -0.74°C$; $t = -3.14$) and summer sea-surface temperature estimates are significantly lower ($\bar{X} = +0.75°C$; $t = +3.21$) at the 0.005 level (one-tailed test).

Second, the mean of the residuals for the eastern equatorial section was compared with the mean of the residuals for the remaining 174 locations. For the problem area, the mean of winter temperature estimates from the calendar-based equation is significantly higher ($\bar{X}_1 - \bar{X}_2 = +0.06°C$; $t = -4.16$) at the 0.005 level and the mean of summer temperature estimates is significantly lower ($\bar{X}_1 - \bar{X}_2 = -0.07°C$; $t = +2.20$) at the 0.025 level (one-tailed tests). The means of estimates from temperature-based equations for the eastern equatorial section are not significantly different from the means of estimates for the remaining 174 samples at the 0.30 level (two-tailed test). Formulas for the t tests were taken from Simpson and others (1960, Chap. 10).

Faunal Data

Samples were processed in the manner described in Imbrie and Kipp (1971), with the following exceptions: samples were not cleaned ultrasonically; H_2O_2 was used only on samples that did not disaggregate in water; and all half-specimens were included in counts. The half-specimens were either spiral or umbilical halves, or included half of each side. Counts were made from splits of the fraction retained on a 149-μm sieve.

The taxonomic categories conform to the taxonomy of Parker (1962, 1967), but use the nomenclature of Bé (1967) and Bé and Hamlin (1967). These are the categories used in Imbrie and others (1973), with the following exceptions:

1. Pink and white varieties of *Globigerinoides ruber* (d'Orbigny) were distinguished. This split was considered advantageous because maximum occurrence of *G. ruber* (pink) is in the Caribbean Sea, significantly south of the maximum occurrence of *G. ruber* (white).

2. *Globigerinoides sacculifer* (Brady) was divided into two varieties, those with and those without a final saclike chamber. Although the distribution of *G. sacculifer* (with sac) is somewhat patchy, its highest abundances with respect to total *G. sacculifer* do tend to occur only in the gyre margin, especially in samples with moderate dissolution.

3. *Globorotalia menardii* (d'Orbigny) and *G. tumida* (Brady) were counted separately. As in the case of *G. sacculifer*, dissolution causes patchy occurrence of the two species; *G. tumida* has its highest percentages in the most severely dissolved samples, and *G. menardii* in moderately dissolved samples. The distribution of the sum of the two species percentages exhibits a simple geographic pattern.

4. A separate category for "intergrades" between *Globoquadrina dutertrei* (d'Orbigny) and right-coiling *Globigerina pachyderma* (Ehrenberg) was established to reduce inconsistencies in species counts made by a large number of workers with varied micropaleontological experience. This category includes the following varieties: (a) specimens of right-coiling *G. pachyderma* that have more than four chambers per whorl when viewed from the umbilical side, such as that figured by Parker (1962, Pl. 1, fig. 35); (b) immature specimens of *G. dutertrei* without tooth, of

the type shown by Boltovskoy (1969, Pl. 1, figs. 7, 8); and (c) right-coiling specimens of *Globigerina incompta* Cifelli with more that four chambers per whorl such as those illustrated by Cifelli (1973, Pl. 2, figs. 8-12), and specimens of *Neogloboquadrina pachyderma incompta* (Cifelli) such as those figured by Rögl and Bolli (1973, Pl. 10, figs. 12-20 and Pl. 11, fig. 1). Relative distribution of the three categories, *G. dutertrei*, *G. pachyderma* (right-coiling), and "P-D intergrade," is shown in Figure 3. Samples with fewer than ten specimens were omitted from consideration. The highest relative abundance of "P-D intergrade" occurs in areas separating the maximum relative abundances of *G. pachyderma* (right-coiling) and *G. dutertrei*. Future restriction and simplification of "P-D intergrade" may emphasize this marginal aspect of it.

Of the 42 taxonomic categories identified, 13 made up less than 2% of any sample and were eliminated from calculations. The remaining 29 categories on which calculations are based are listed in Table 1 along with the maximum occurrence of each.

Dissolution

Differential dissolution affects the percentage composition of planktonic foraminifera in deep-sea sediment samples (see, for example, Berger, 1968, 1970; Olausson,

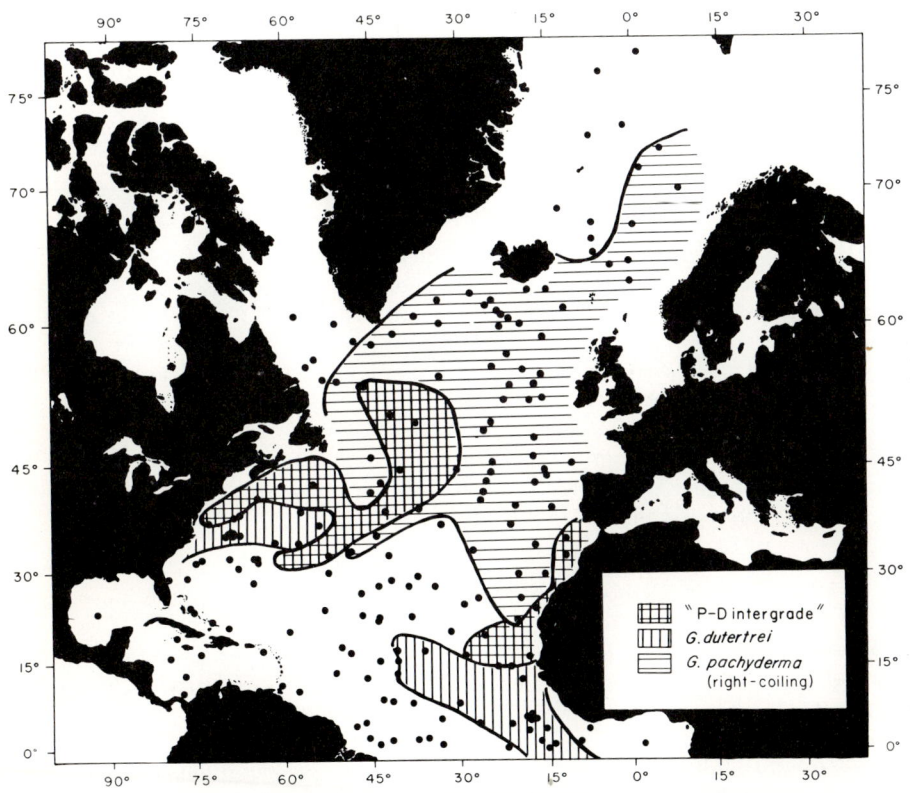

Figure 3. Distribution of areas of relative dominance of three taxonomic categories in North Atlantic sea bed: *Globoquadrina dutertrei* (d'Orbigny), *Globigerina pachyderma* (Ehrenberg) (right-coiling) and "P-D intergrade." Blank areas on map represent samples in which the three categories are either absent or very rare.

TABLE 1. TAXONOMIC CATEGORIES

Species	Maximum occurrence (% of sample)
Orbulina universa	18.432
Globigerinoides conglobatus	10.000
G. ruber (pink)	21.782
G. ruber (white)	78.170
G. tenellus	11.370
G. sacculifer (without saclike final chamber)	23.822
G. sacculifer (with saclike final chamber)	24.096
Sphaeroidinella dehiscens	5.357
Globigerinella aequilateralis	12.849
Globigerina calida	4.401
G. bulloides	53.191
G. falconensis	19.749
G. digitata	2.679
G. rubescens	18.944
G. quinqueloba	45.623
G. pachyderma (left-coiling)	98.421
G. pachyderma (right-coiling)	32.177
Globoquadrina dutertrei	31.944
Pulleniatina obliquiloculata	25.595
Globorotalia inflata	39.560
G. truncatulinoides (left-coiling)	14.430
G. truncatulinoides (right-coiling)	8.211
G. crassaformis	7.335
"P-D intergrade"	28.637
G. hirsuta	8.333
G. scitula	6.551
G. menardii	30.925
G. tumida	62.829
Globigerinita glutinata	29.102

1971). It is important to include this effect in writing transfer functions so that the equations may be used on downcore samples exhibiting dissolution. Estimates made on such samples are judged to be valid provided the dissolution intensity does not exceed the range included in the core-top data base. Therefore, samples showing evidence of moderate to severe dissolution were not deleted from the data base for F13. The locations of these samples are shown on Figure 2 and in Appendix II (see footnote 1).

Dissolution criteria used in this study are large numbers of fragments and benthic foraminifera, absence of fragile forms such as pteropods, pitted tests, and eroded apertures or internal walls.

SPECIES DISTRIBUTIONS

Abundance maps of 20 of the 29 species and morphotypes considered in this study show simple gradients of abundance on the sea bed. Although each of these species has a distributional pattern that is distinctly different in detail from the others (Figs. 4-21), it is nevertheless possible to recognize groups of species that have generally similar distributions. These groups are discussed below. Abundance maps of nine of the species are not presented, however, either because the general level of abundance is too low or the patterns exhibited are too patchy. These species are *Orbulina universa* d'Orbigny, *Globigerinoides conglobatus* (Brady), *G. tenellus* Parker, *Sphaeroidinella dehiscens* (Parker and Jones), *Globigerinella aequila-*

teralis (Brady), *Globigerina calida* Parker, *G. digitata* Brady, *G. rubescens* Hofker, and *Globorotalia crassaformis* (Galloway and Wissler). If larger samples were counted, it is possible that systematic gradients could be established on sea-bed samples for these species.

Polar Species

Only two varieties have distributional maxima in latitudes north of 60°N: left-coiling *Globigerina pachyderma* and *G. quinqueloba* Natland. The maximum of *G. quinqueloba* is south of that for left-coiling *G. pachyderma* (Figs. 4, 5), and the maximum for both species lies north of the influence of the Gulf Stream. Left-coiling *G. pachyderma* is essentially the only species found in the sediments of the Labrador and Greenland Seas.

Subpolar Species

Four forms have distributional maxima in the central and eastern parts of the North Atlantic just south of 60°N: right-coiling *G. pachyderma*, "P-D intergrade," *G. bulloides* d'Orbigny and *Globigerinita glutinata* (Egger). The first three (Figs. 6, 7, 8) form one narrow band of intermediate levels of relative abundance that extends southwestward from the maximum into the area of the Gulf Stream, and

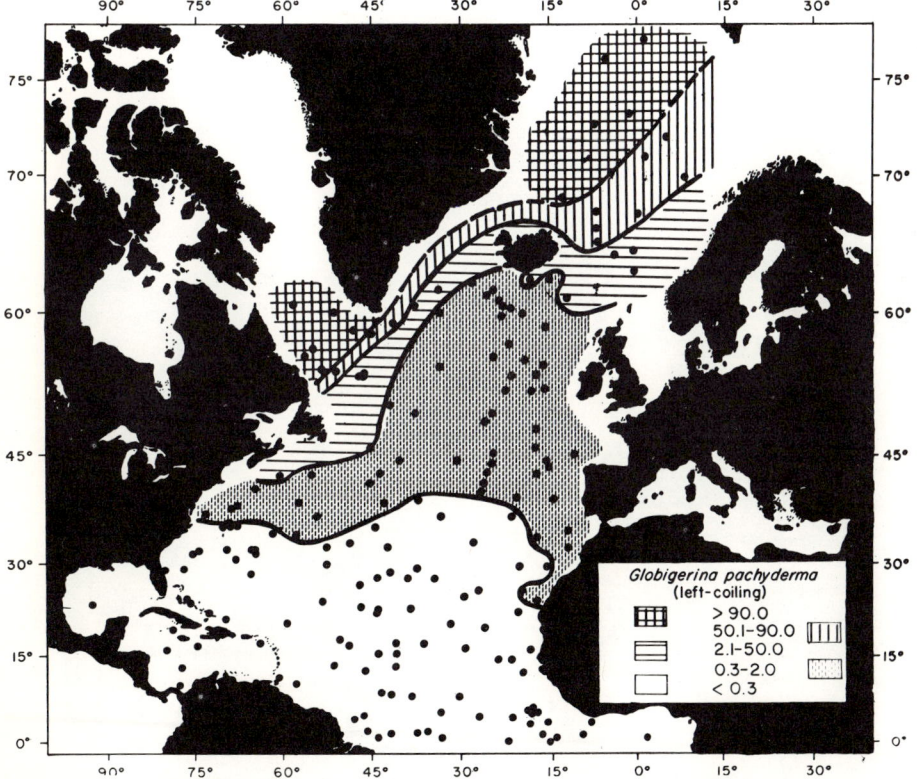

Figure 4. Relative abundance (%) of *Globigerina pachyderma* (Ehrenberg) (left-coiling) on North Atlantic sea bed.

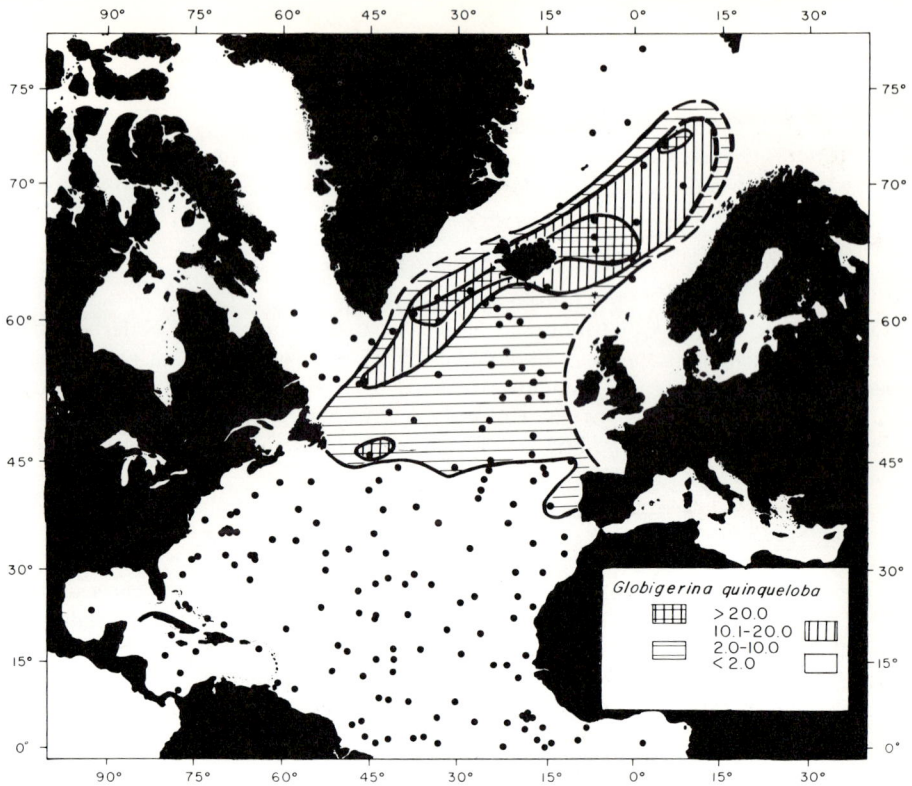

Figure 5. Relative abundance (%) of *Globigerina quinqueloba* Natland on North Atlantic sea bed.

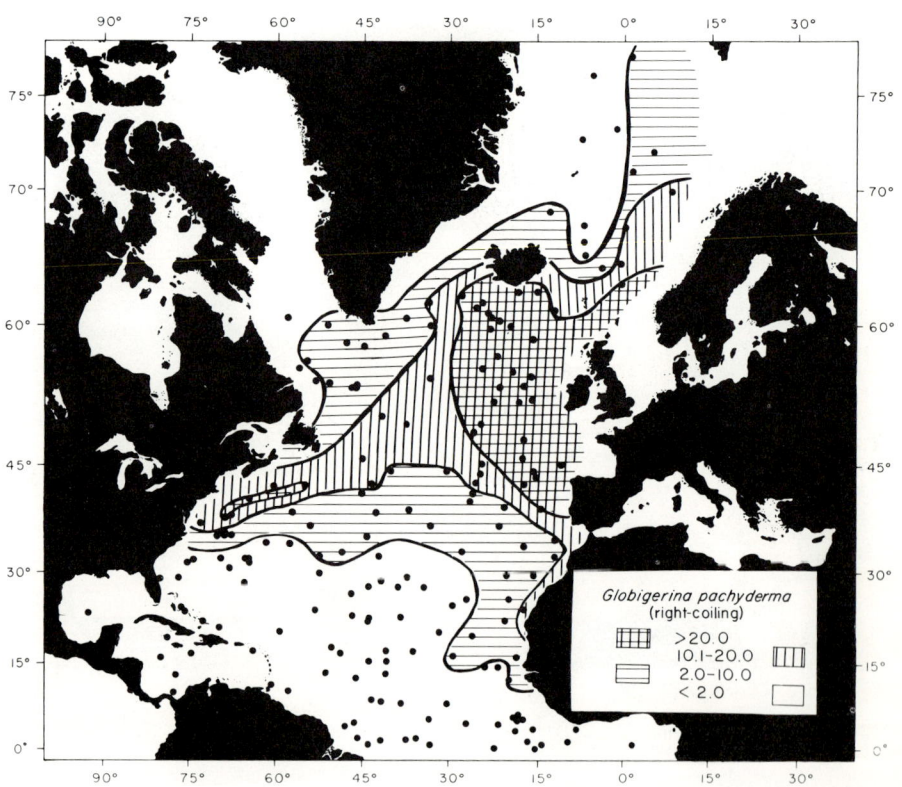

Figure 6. Relative abundance (%) of *Globigerina pachyderma* (Ehrenberg) (right-coiling) on North Atlantic sea bed.

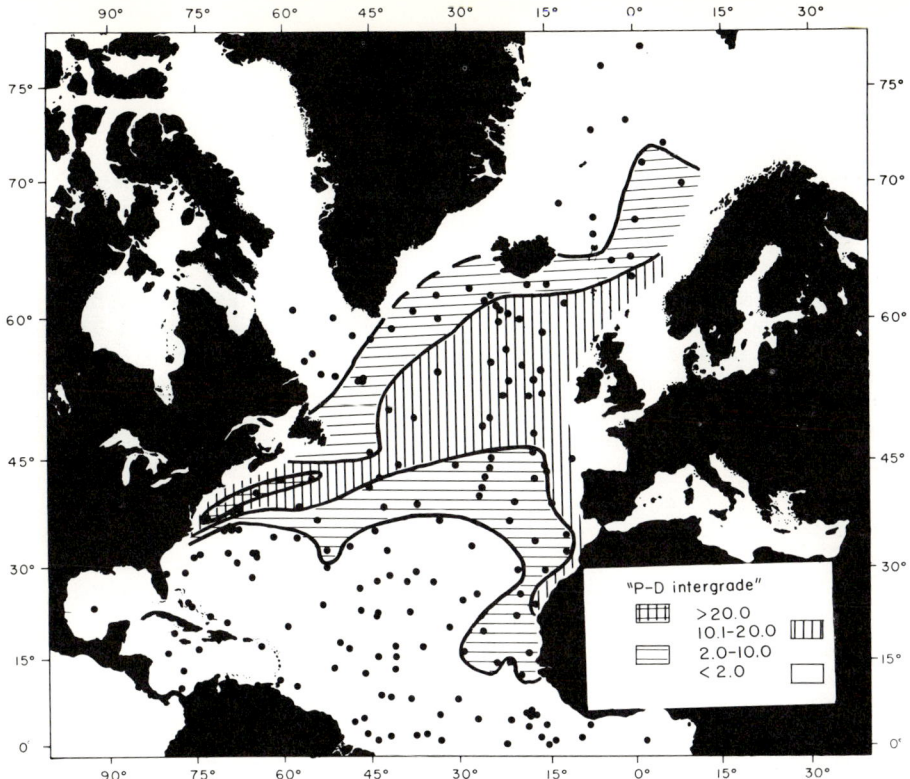

Figure 7. Relative abundance (%) of "P-D intergrade" on North Atlantic sea bed.

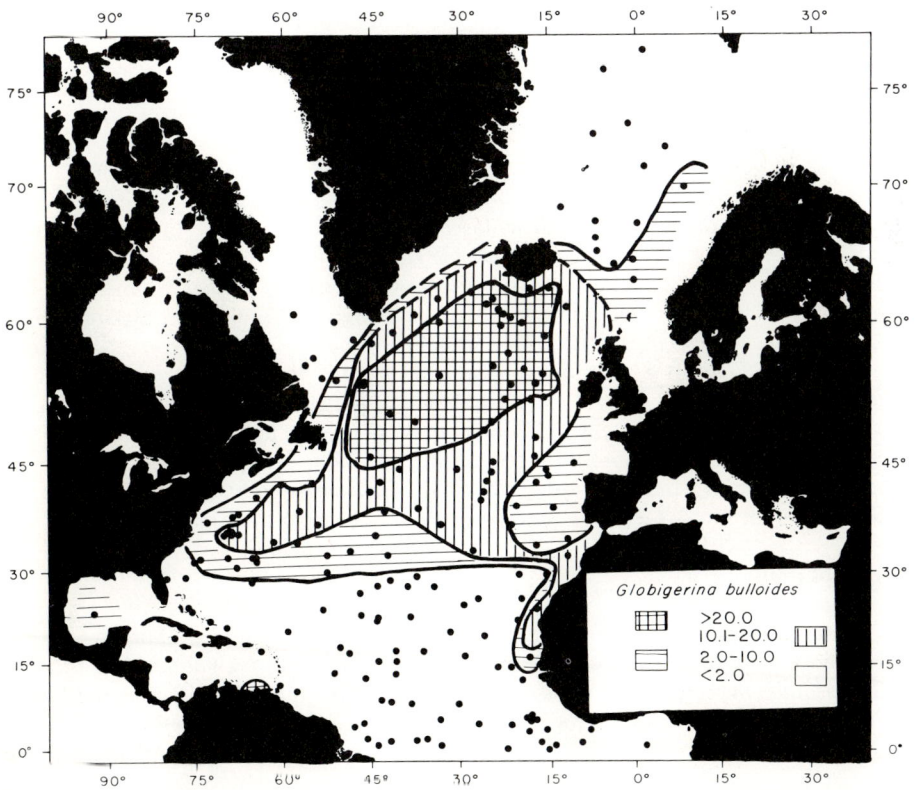

Figure 8. Relative abundance (%) of *Globigerina bulloides* d'Orbigny on North Atlantic sea bed.

another band which extends southward as far as the African coast along the eastern side of the Atlantic. Maximum abundances of *G. glutinata* (Fig. 9), unlike the other members of this group, occur in two geographically distinct areas.

Transitional Species

A single species, *Globorotalia inflata* (d'Orbigny), characterizes a transition zone between subpolar and subtropical waters (see Bé and Tolderlund, 1971). The major area of maximum occurrence is in the western part of the Atlantic north of 30°N. (Fig. 10) and in a small area of the eastern Atlantic. The high values in the latter area are a reflection of differential dissolution, which the condition of the tests confirms. A band of intermediate abundance extends southward along the eastern side of the Atlantic.

Subtropical Species

Five forms have maximum abundances near 30°N: *Globigerina falconensis* Blow, *Globorotalia scitula* (Brady), right-coiling *G. truncatulinoides* (d'Orbigny), left-coiling *G. truncatulinoides* and *G. hirsuta* (d'Orbigny). The maximum abundances of *G. falconensis* and *G. scitula* (Figs. 11, 12) occur in the eastern Atlantic, and the two forms of *G. truncatulinoides* (Figs. 13, 14) have maxima both to the east and to the west of the central area. The maximum abundance of *G. hirsuta* (Fig. 15) occurs in the west-central Atlantic. *G. falconensis* (Fig. 11) occurs in greater relative abundance than the other subtropical forms and exhibits areas of high and intermediate abundance levels extending tongue-like westward from the Strait of Gibraltar. The western areas of maximum abundance of *G. scitula* (Fig. 12) and both varieties of *G. truncatulinoides* (Figs. 13, 14) are marginal to those of *G. falconensis* (Fig. 11).

Tropical Species

Three varieties have maximum abundances in the tropical region of the North Atlantic: *Globigerinoides ruber* (white), *G. ruber* (pink) and *G. sacculifer*. Each of the three dominates in a particular area: the white variety of *G. ruber* (Fig. 16) in the central Sargasso Sea; pink *G. ruber* (Fig. 17) in the Gulf of Mexico, the northern Caribbean Sea and adjacent waters of the North Atlantic; and *G. sacculifer* (Fig. 18) in the southern Caribbean Sea and equatorial Atlantic. Both varieties of *G. ruber* have a wide distributional range. They also show strong latitudinal gradients, with one exception: the fragile tests of *G. ruber* are readily dissolved, creating an area of low abundance in the southeastern equatorial region. Occasional anomalously low percentages of *G. ruber* in samples with evidence of dissolution were ignored in contouring. Although the pattern of maximum occurrence of *G. sacculifer* qualifies it for inclusion in the tropical group, its distribution at moderately low abundance levels forms a gyre-margin pattern.

Gyre-Margin Species

Four other species besides *G. sacculifer* occur in moderate abundance along the southern and northwestern margins of the central gyre: *Globoquadrina dutertrei*, *Globorotalia menardii*, *G. tumida* and *Pulleniatina obliquiloculata* (Parker and Jones). These four species (Figs. 19, 20, 21) occur in maximum abundance in the southeastern area off the coast of Africa. All the samples in this area contain evidence of moderate to severe dissolution. Occasional samples with anomalously high percentages of resistant species generally were ignored in contouring. However, most of the samples that contain more than 5% *P. obliquiloculata* are here classified as moderately to severely dissolved; no samples were ignored in contouring this species.

The species distributions shown in Figures 4 through 21 are remarkably similar to those reported for the >125-μm fraction by Ruddiman (1969). In addition to the difference in sieve size, different taxonomic categories were used in the two studies. For example, Ruddiman included all specimens of *G. falconensis* in the *G. bulloides* category and did not use the "P-D intergrade" classification. When taxonomic categories used in this study are adjusted to those used by Ruddiman, similar contour intervals produce similar distributional patterns and maxima. Exceptions can be explained in one of two ways. (1) An area of high abundance of a species in one study may be shown as two smaller areas in the other study due to a difference in sample distribution. (2) Small species, especially *G. rubescens* and *G. tenellus*, occur in higher percentages in Ruddiman's data. For example, in one sample used in both studies (RC10-22), the two species account for 20% of the specimens greater than 125 μm, but only 5% of the specimens greater than 149 μm. The 15% difference is balanced by increased percentages of *G. ruber* and *G. sacculifer* in the larger fraction.

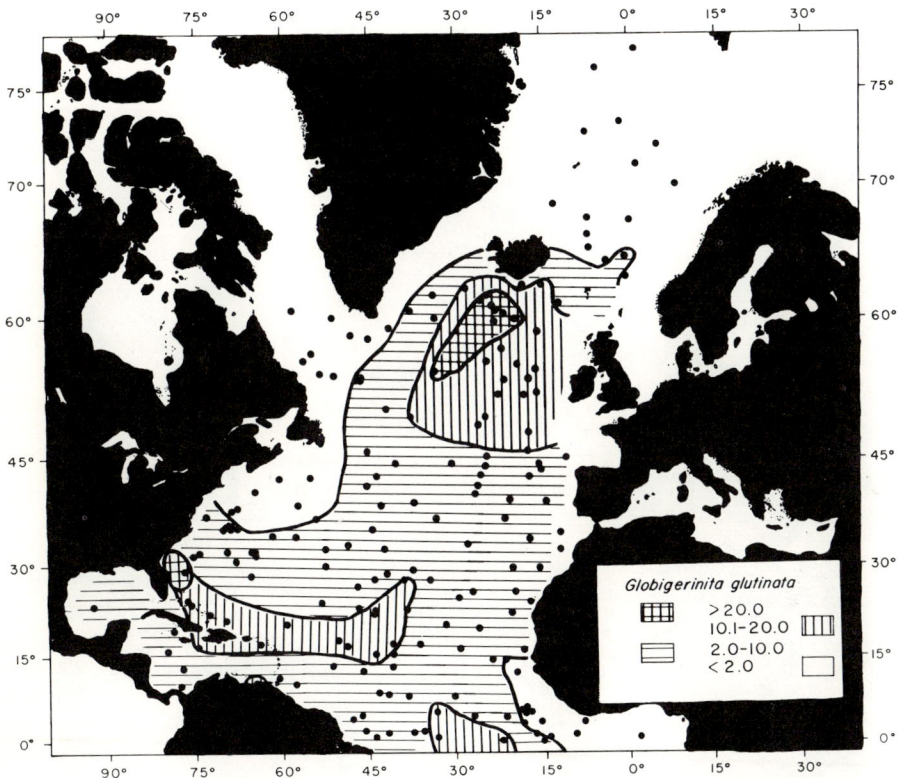

Figure 9. Relative abundance (%) of *Globigerinita glutinata* (Egger) on North Atlantic sea bed.

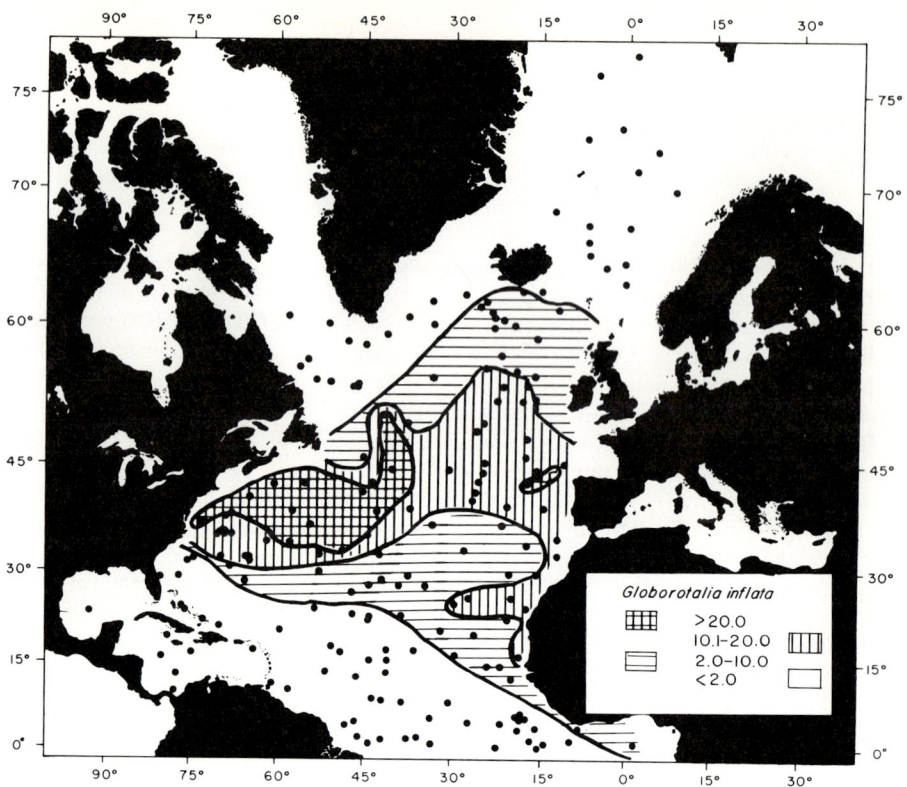

Figure 10. Relative abundance (%) of *Globorotalia inflata* (d'Orbigny) on North Atlantic sea bed.

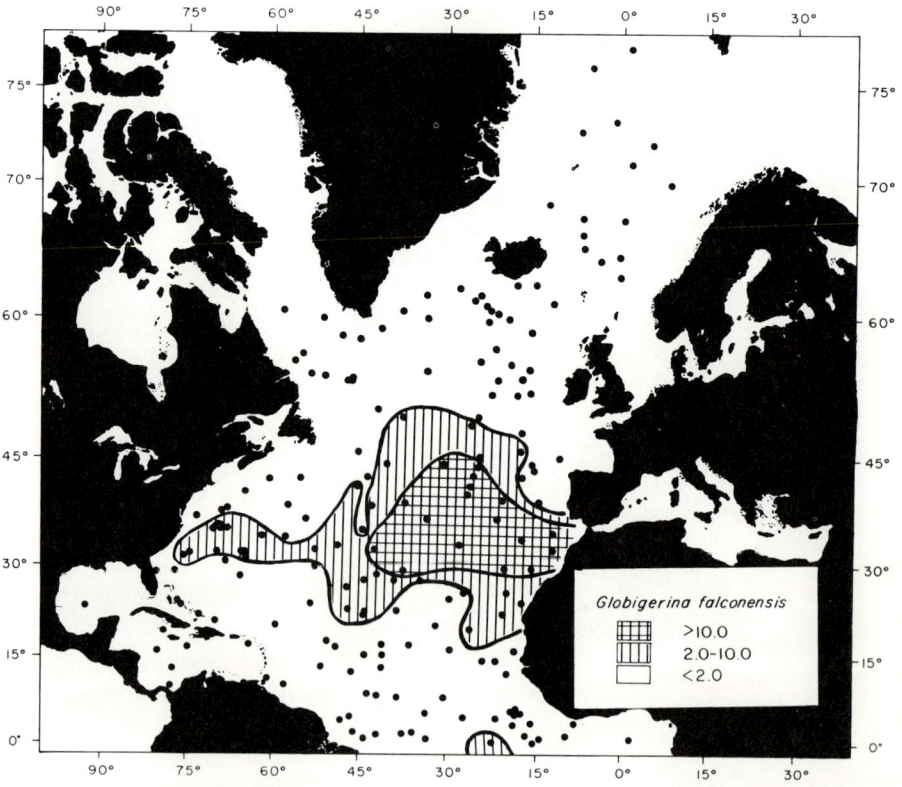

Figure 11. Relative abundance (%) of *Globigerina falconensis* Blow on North Atlantic sea bed.

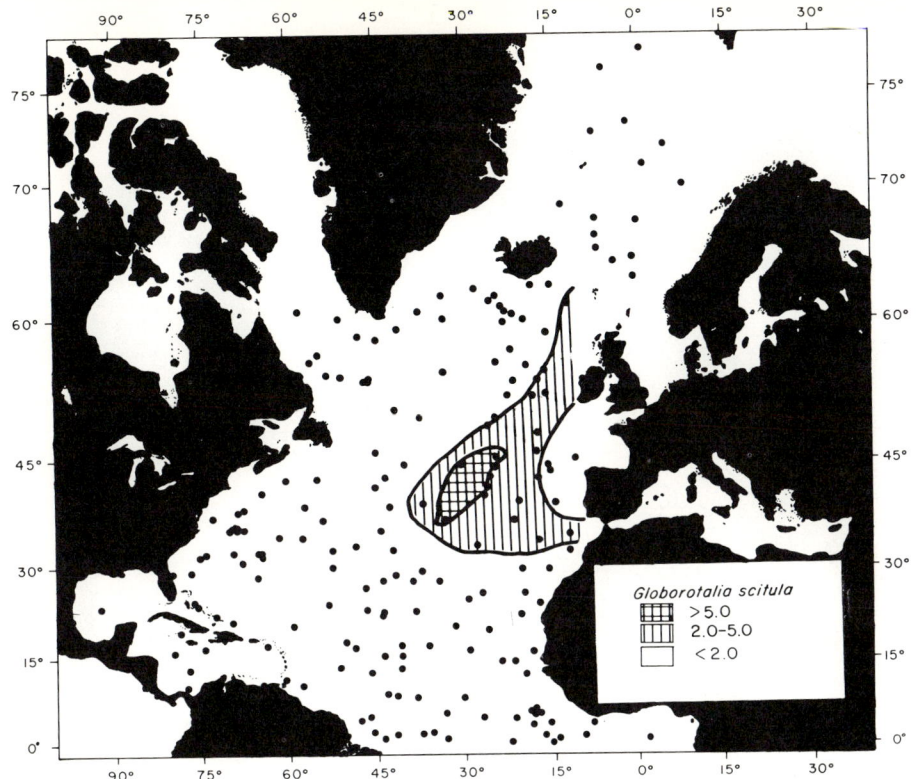

Figure 12. Relative abundance (%) of *Globorotalia scitula* (Brady) on North Atlantic sea bed.

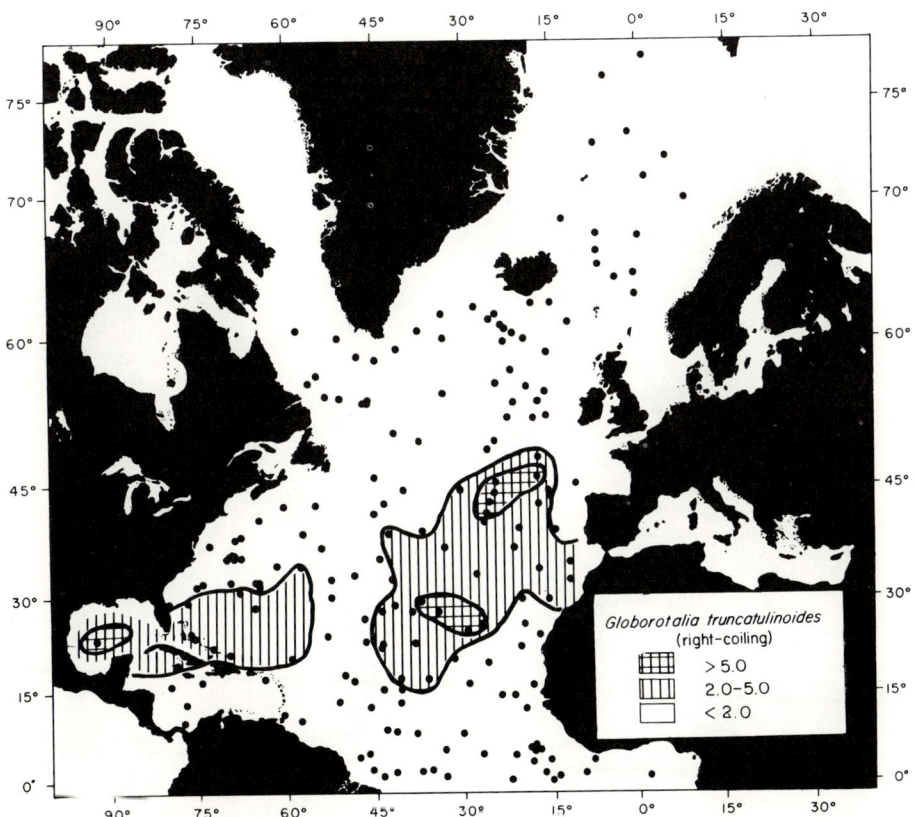

Figure 13. Relative abundance (%) of *Globorotalia truncatulinoides* (d'Orbigny) (right-coiling) on North Atlantic sea bed.

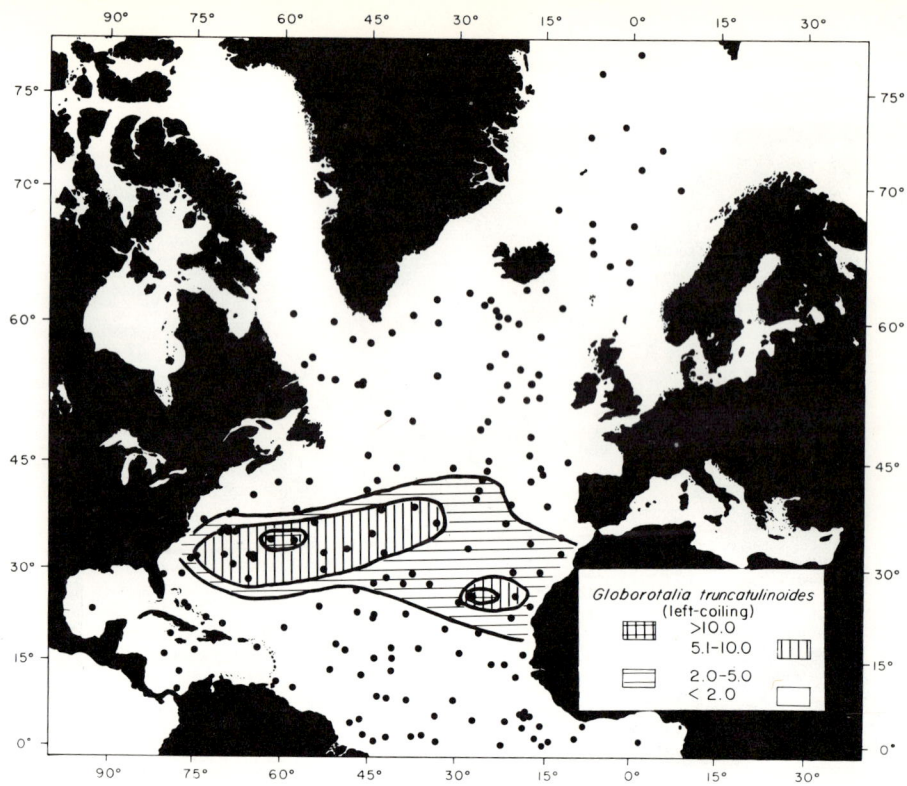

Figure 14. Relative abundance (%) of *Globorotalia truncatulinoides* (d'Orbigny) (left-coiling) on North Atlantic sea bed.

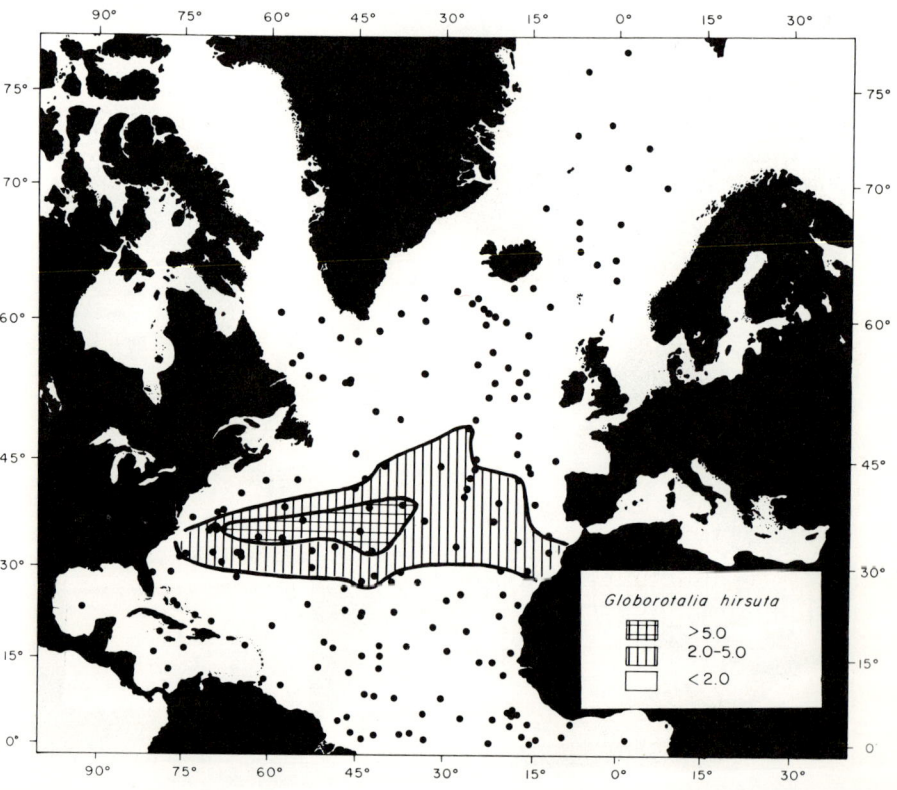

Figure 15. Relative abundance (%) of *Globorotalia hirsuta* (d'Orbigny) on North Atlantic sea bed.

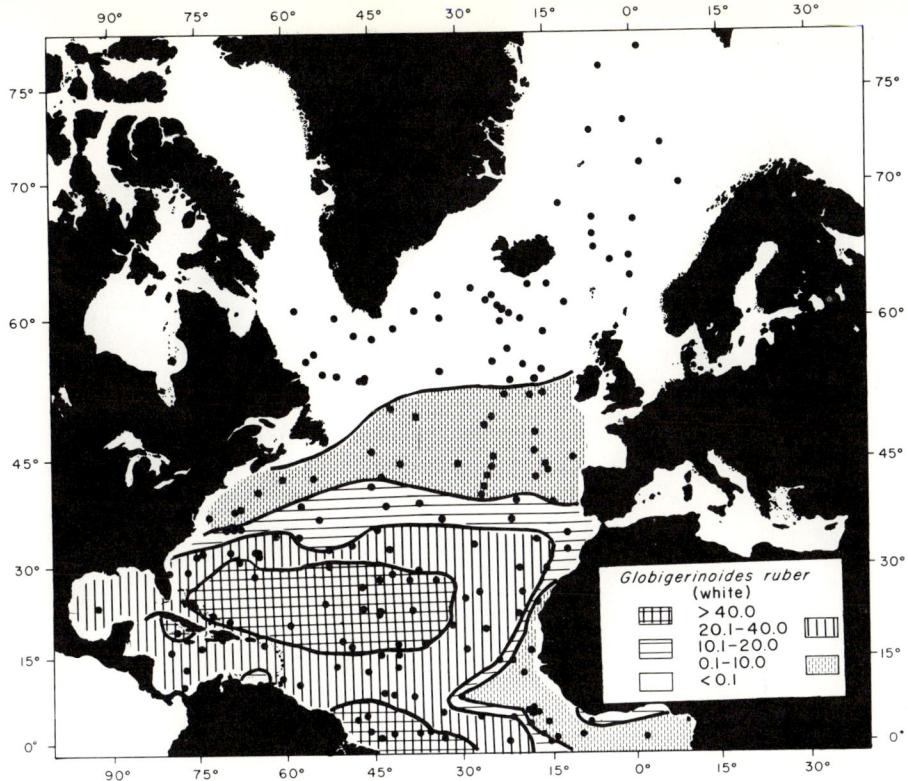

Figure 16. Relative abundance (%) of *Globigerinoides ruber* (d'Orbigny) (white) on North Atlantic sea bed.

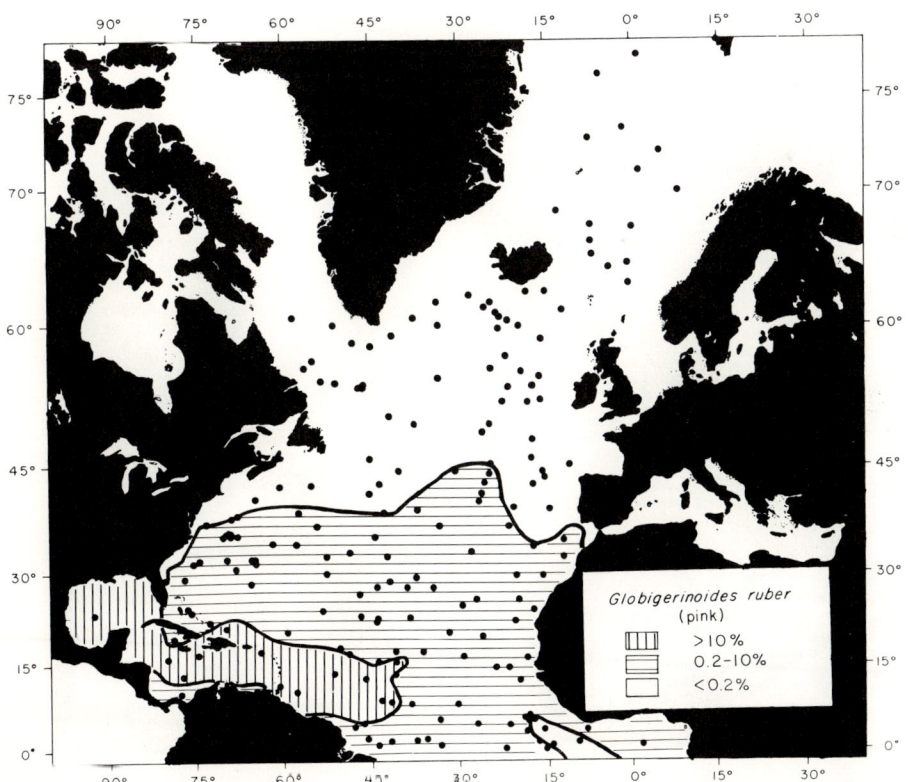

Figure 17. Relative abundance (%) of *Globigerinoides ruber* (d'Orbigny) (pink) on North Atlantic sea bed.

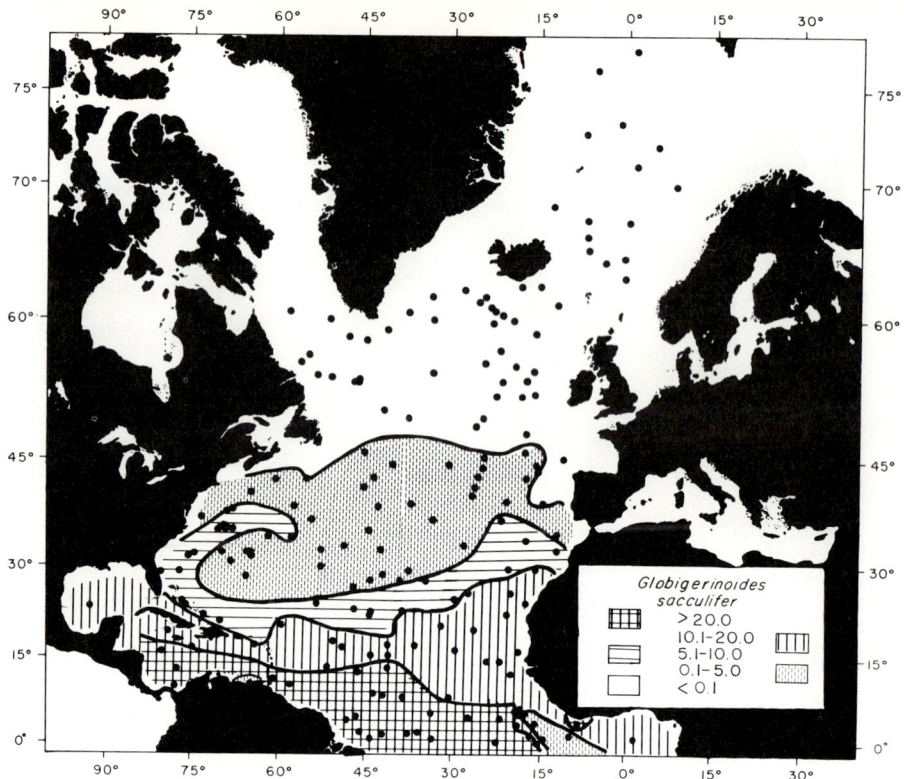

Figure 18. Relative abundance (%) of *Globigerinoides sacculifer* (Brady) on North Atlantic sea bed.

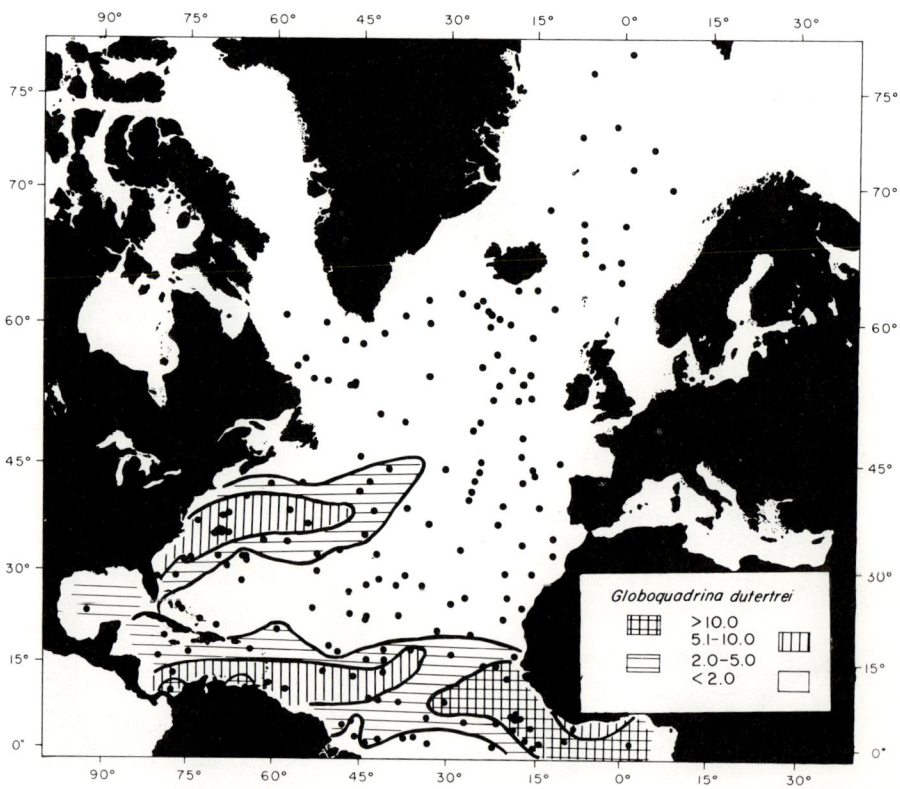

Figure 19. Relative abundance (%) of *Globoquadrina dutertrei* (d'Orbigny) on North Atlantic sea bed.

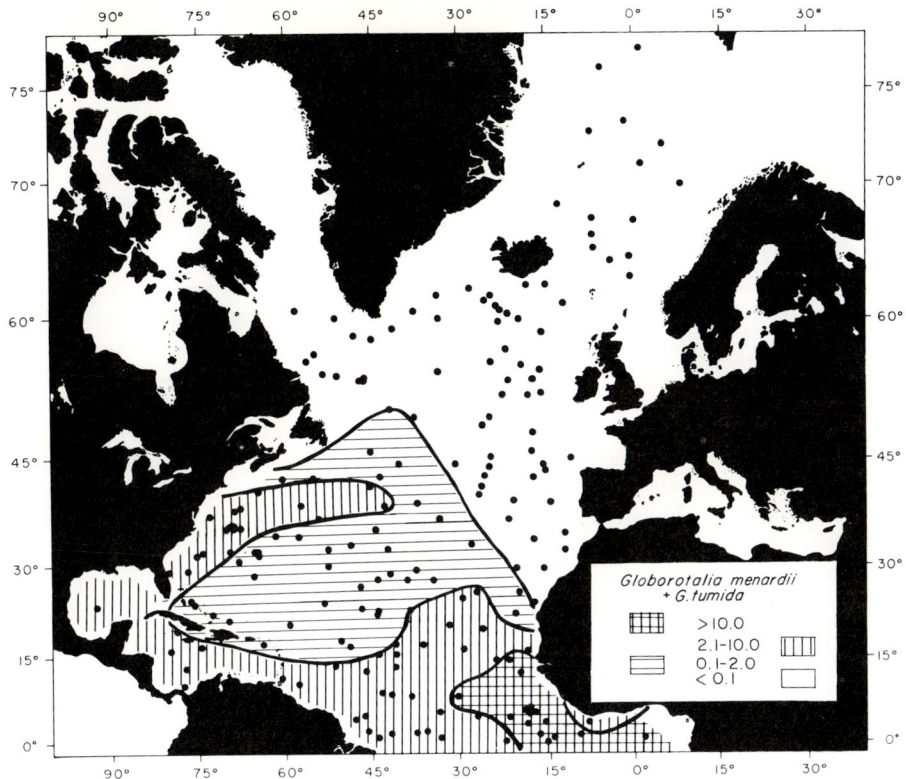

Figure 20. Relative abundance (%) of *Globorotalia menardii* (d'Orbigny) plus *G. tumida* (Brady) on North Atlantic sea bed.

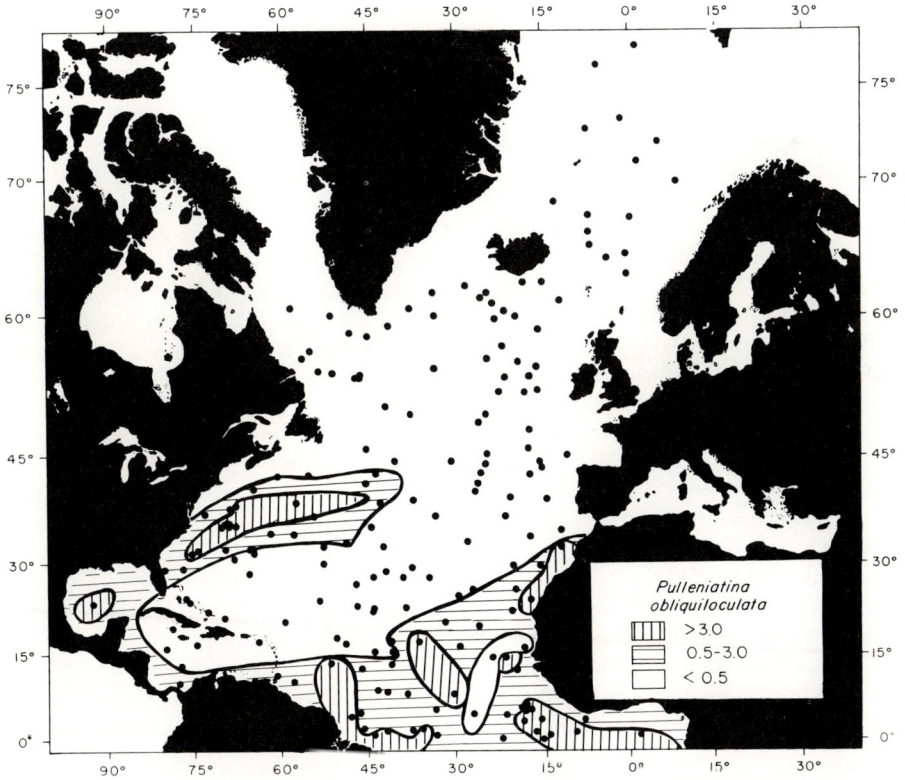

Figure 21. Relative abundance (%) of *Pulleniatina obliquiloculata* (Parker and Jones) on North Atlantic sea bed.

Comparison of Sea-Bed Data with Plankton-Tow Data

Species distributions on the North Atlantic sea bed reported here are generally similar to distributions of living planktonic foraminiferal species reported by Bé and Tolderlund (1971). Most of the differences can be explained in one of six ways.

1. Both studies lack control in some areas. The present study, for example, lacks control in several small areas, especially east of Newfoundland, and in deep basins. Bé and Tolderlund (1971) did not have samples for an area of the central water north of 30°N. Figures 9 and 13 show maximum abundances of *G. glutinata* and right-coiling *G. truncatulinoides* in the sediment of this area. Some nearshore areas are not represented in either of the data sets.

2. The 200-μm mesh size used in the collection of surface plankton tow samples results in higher percentages of large species than are found in the 149-μm fraction of sea-bed samples. Examples of these large species are *G. sacculifer*, *G. dutertrei*, *Globigerinoides conglobatus*, and *Orbulina universa*.

3. Extremely fragile species are rare in sediment samples. One such species, *Hastigerina pelagica* (d'Orbigny), occurs in part of the northwestern Sargasso Sea plankton in abundances greater than 20%.

4. Discrepancies in some areas are caused by dissolution of fragile species and consequent enrichment of sea-bed samples in the relatively resistant and generally large species such as *G. tumida*, *G. dutertrei* and *P. obliquiloculata*. The most conspicuous example of this phenomenon occurs in the eastern equatorial area.

5. Deep-living species that are not found in surface plankton tows are found in sea-bed samples. *G. scitula*, for example, is rare in surface plankton tows, but shows a well-defined distributional pattern in quantities of 2% to 6.5% in the sediment (Fig. 12).

6. Although bottom-transport processes can be expected to produce differences between plankton and sea-bed distributions (see, for example, Ruddiman, 1968), the density of control points in this study seems to be insufficient to document the effect.

A combination of the above explanations accounts for differences between sea-bed and plankton distributions of two species. The sea-bed distribution pattern for the total population of *G. ruber* is essentially that shown for the white variety (Fig. 16); it has a single abundance maximum in the central Sargasso Sea. The plankton data for the same species has a significantly different pattern, characterized by an area of relatively low abundance in the central Atlantic that includes part of the region of the sea-bed maximum. This difference can be attributed to the difference in size fractions examined; plankton was collected with a 200-μm mesh net but sediment samples were sieved through a 149-μm screen. Large species, notably *G. sacculifer*, *H. pelagica* and *G. conglobatus*, are relatively more abundant in the plankton of the central area. Thus the percentage map of the comparatively small species *G. ruber* shows a region of lower values in the plankton.

In quantities greater than 20%, *G. quinqueloba* shows a more restricted distribution in the 149-μm sediment fraction than in plankton-tow data. Conversely, the larger species, *G. bulloides* and *G. pachyderma*, are found in smaller percentages in this area in the plankton tows than in sediment samples. Several factors can be suggested for these differences: (1) misidentification of *G. quinqueloba* in the sediment, where spines are seldom preserved; (2) dissolution of the more fragile *G. quinqueloba* tests in the sediment; (3) *G. quinqueloba*'s preference for the shallow depths sampled by the plankton tows in contrast to the deeper habitat of right-coiling *G. pachyderma*; (4) retention of a disproportionately large number of the smaller

G. quinqueloba tests in the plankton nets in an area of extremely high productivity. All of these factors are discussed by Bé and Tolderlund (1971) or suggested by their data.

FACTOR ANALYSIS

Definition of Foraminiferal Assemblages

Six assemblages derived from factor analysis and varimax rotation of the sea-bed data are described by the F' matrix given in Table 2 (see Imbrie and Kipp, 1971, p. 81,92, for discussion of derivation). These assemblages correspond to the species groups discussed above. Factor 1 is the tropical assemblage, in which *G. ruber* and *G. sacculifer* are the most important species. Dominant forms in factor 2, the subpolar assemblage, are *G. pachyderma* (right-coiling), *G. bulloides, G. glutinata* and "P-D intergrade." A single variety, *G. pachyderma* (left), dominates factor 3, the polar assemblage, with *G. quinqueloba* making a small but significant contribution. Factor 4 is the gyre margin assemblage, in which *G. dutertrei, G. tumida, G. menardii, P. obliquiloculata,* and *G. sacculifer* are all important. Factor 5, the transitional assemblage, is dominated by *G. inflata*; "P-D intergrade" also makes a substantial contribution. *G. bulloides, G. glutinata* and *G. quinqueloba*

TABLE 2. VARIMAX ASSEMBLAGE DESCRIPTION MATRIX (F')

Species	Tropical	Subpolar	Polar	Gyre margin	Transitional	Subtropical
Orbulina universa	0.024	0.020	−0.003	0.053	−0.094	−0.029
Globigerinoides conglobatus	0.019	−0.005	0.001	0.038	−0.013	0.007
G. ruber (pink)	0.141	0.023	0.001	0.021	0.024	−0.118
G. ruber (white)	0.922	−0.054	0.030	−0.098	−0.013	0.112
G. tenellus	0.037	−0.001	0.000	−0.027	0.010	0.022
G. sacculifer (without saclike final chamber)	0.255	0.024	0.001	0.272	0.003	−0.213
G. sacculifer (with saclike final chamber)	0.109	0.007	−0.001	0.192	−0.009	−0.092
Sphaeroidinella dehiscens	−0.003	−0.002	−0.001	0.057	0.005	0.004
Globigerinella aequilateralis	0.094	0.007	−0.002	0.038	0.017	0.017
Globigerina calida	0.019	0.005	−0.002	0.006	0.014	0.031
G. bulloides	−0.052	0.574	0.016	0.072	0.472	0.435
G. falconensis	0.026	−0.008	−0.015	−0.066	0.033	0.491
G. digitata	0.003	0.002	−0.001	0.030	−0.005	0.001
G. rubescens	0.035	0.003	0.000	−0.015	0.013	0.000
G. quinqueloba	−0.006	0.150	0.150	0.005	0.180	−0.028
G. pachyderma (left-coiling)	−0.018	−0.027	0.987	0.011	−0.029	0.016
G. pachyderma (right-coiling)	0.002	0.604	0.017	−0.045	−0.285	−0.307
Globoquadrina dutertrei	0.005	−0.012	−0.009	0.587	−0.031	0.057
Pulleniatina obliquiloculata	0.000	−0.009	−0.004	0.277	−0.005	0.034
Globorotalia inflata	−0.010	0.118	−0.015	0.049	−0.708	0.480
G. truncatulinoides (left-coiling)	0.029	−0.033	−0.002	−0.008	−0.034	0.238
G. truncatulinoides (right-coiling)	0.024	0.008	−0.005	−0.015	−0.021	0.096
G. crassaformis	0.004	0.000	−0.001	0.070	0.010	0.011
"P-D intergrade"	0.004	0.342	0.011	−0.006	−0.330	−0.182
G. hirsuta	0.001	−0.010	−0.003	0.004	−0.025	0.198
G. scitula	0.000	0.037	−0.009	−0.010	0.016	0.074
G. menardii	0.053	−0.001	−0.005	0.405	−0.030	−0.059
G. tumida	−0.039	−0.024	−0.004	0.513	0.054	0.067
Globigerinita glutinata	0.176	0.378	−0.042	−0.028	0.183	−0.070

tend to be absent in factor 5. Five species are important in factor 6, the subtropical assemblage: *G. falconensis, G. bulloides, G. truncatulinoides* (left) and *G. hirsuta*. Also important is the tendency of *G. pachyderma* (right) to be absent.

Distribution of Assemblages

The geographic distributions of these assemblages are mapped in Figures 22 through 27 from values in the columns of the varimax matrix (Appendix III; see footnote 1). The distribution of five of the assemblages (tropical, subtropical, subpolar, polar and gyre margin) is similar to that of the five assemblages in Imbrie and Kipp (1971). One exception is the high abundance of the gyre margin assemblage in the dissolution region of the eastern equatorial Atlantic. In the present study, a fivefold increase in sampling density for the North Atlantic also permits more detailed mapping and definition of a sixth assemblage, the transitional (Fig. 24). This assemblage, characterized by *G. inflata*, corresponds to the Transition Zone of Bé and Tolderlund (1971) for living planktonic foraminifera. A decrease in the size of the area occupied by the subtropical assemblage arises from the recognition of the transitional assemblage.

In general, each of the six assemblages has a maximum beneath a specific hydrographic province. The polar assemblage (Fig. 22) occurs north of the 4.5°C winter isotherm, and the subpolar assemblage (Fig. 23) occurs in the northeastern

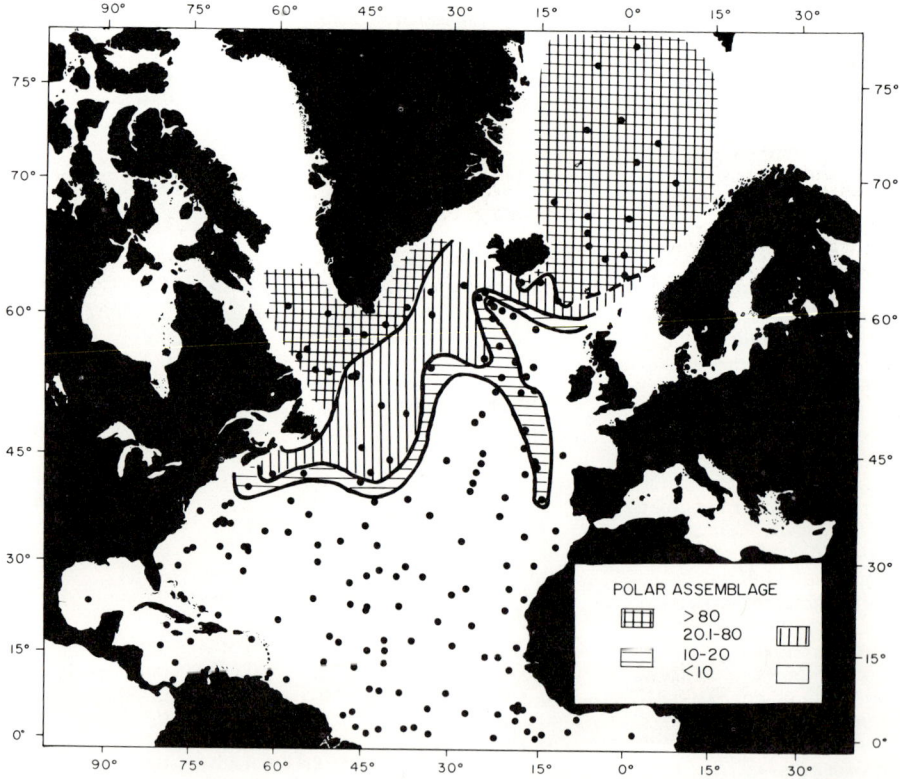

Figure 22. Distribution of polar assemblage on North Atlantic sea bed. Values plotted are those given in varimax matrix (App. III) times 100.

Atlantic between the Arctic Convergence and the northern Sargasso Sea in the region of the North Atlantic Drift. The subtropical assemblage (Fig. 25) has its maximum in the sediments of the northern Sargasso Sea. The transitional assemblage (Fig. 24) has a maximum in the western Atlantic along the Arctic Convergence. The tropical assemblage (Fig. 26) is dominant south of approximately 30°N with the exception of the Guinea region, which is the maximum for the gyre margin assemblage (Fig. 27) and a small area of mixed assemblages in the area of the Canary Current. The major components of the tropical assemblage are *G. ruber* in the southern Sargasso Sea and *G. sacculifer* in the equatorial region. The latter also contributes to the gyre margin assemblage, which occurs in moderate abundance in equatorial water and its extension, the Gulf Stream.

Although the general match between the faunal distributions and the hydrographic characteristics of the North Atlantic is good, there are a number of patterns on the assemblage maps that are puzzling. For example, what is the significance of the tongue of low abundance of the polar assemblage which extends southward along the eastern side of the Atlantic (Fig. 22)? Although the rough spatial correlation with an extension of high-density water during the autumn at 100 m (U. S. Naval Oceanographic Office, 1967, Fig. II-81) is suggested, no conclusions should be drawn without a much more detailed analysis. When making such analyses it is important to remember that the samples used represent an average faunal distribution over a time interval of about 500 yr. Such an interval would be characteristic of a typical North Atlantic core with a sedimentation rate of 2 cm/1,000 yr. However,

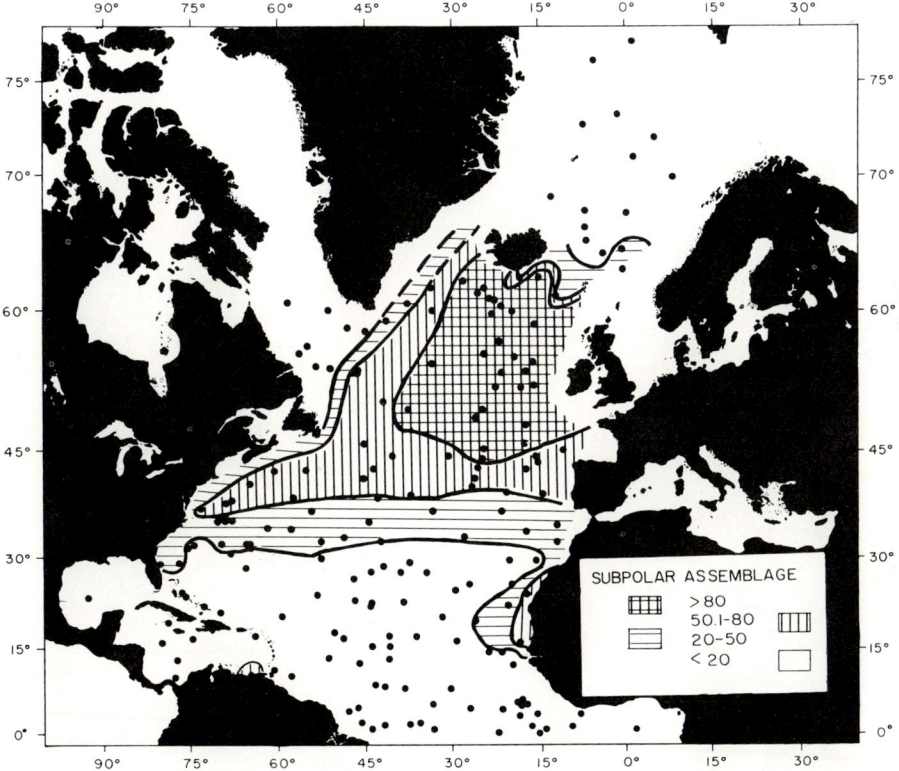

Figure 23. Distribution of subpolar assemblage on North Atlantic sea bed. Values plotted are those given in varimax matrix (App. III) times 100.

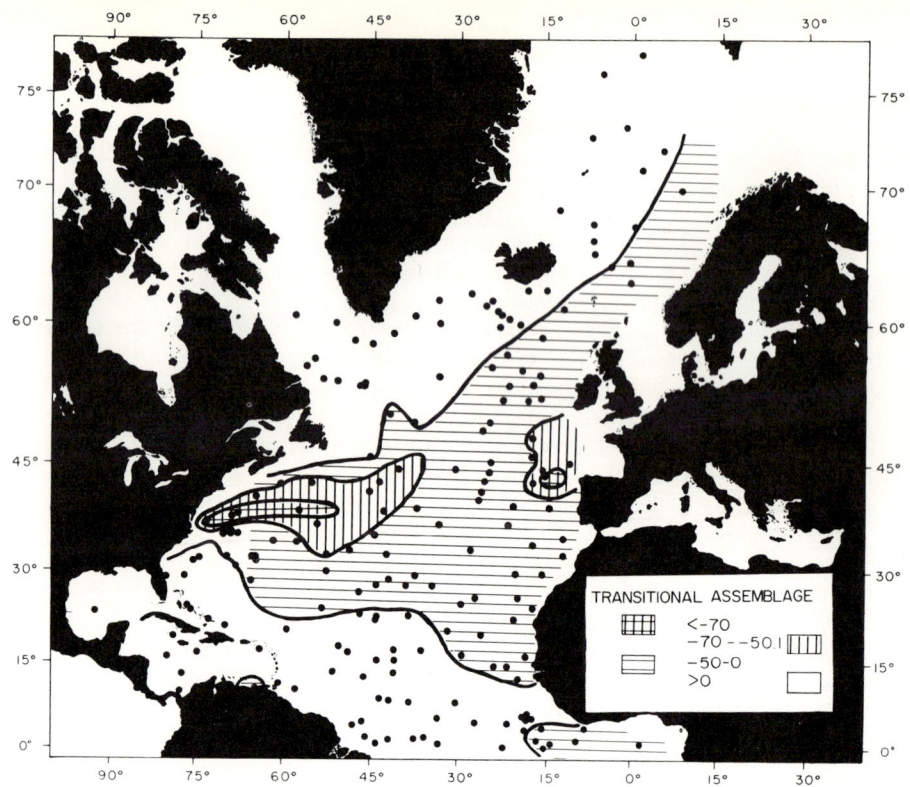

Figure 24. Distribution of transitional assemblage on North Atlantic sea bed. Values plotted are those given in varimax matrix (App. III) times 100.

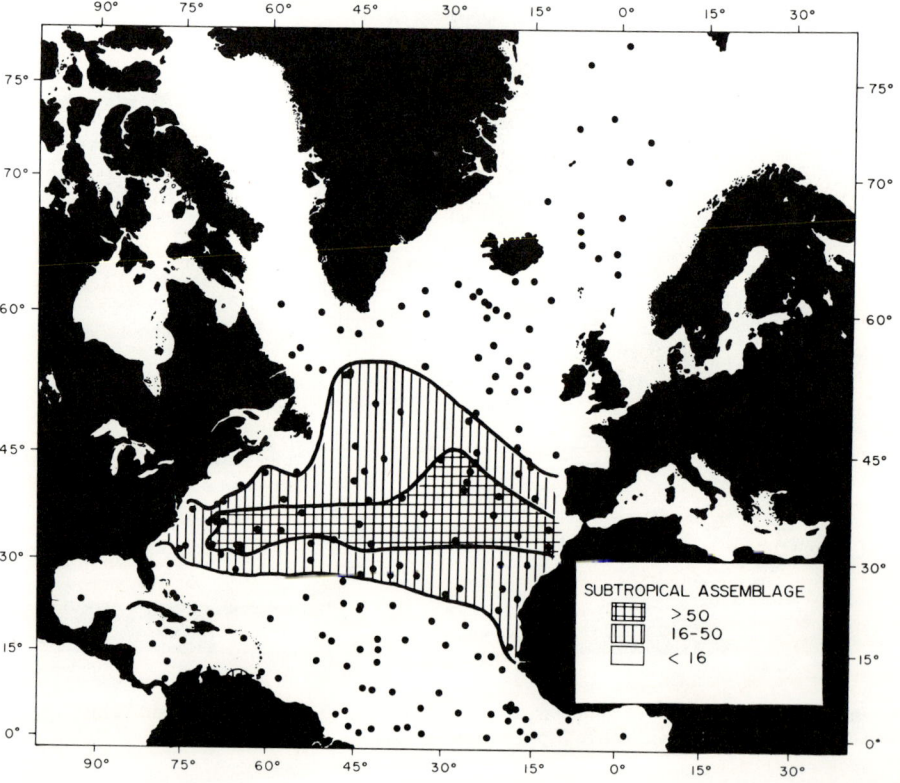

Figure 25. Distribution of subtropical assemblage on North Atlantic sea bed. Values plotted are those given in varimax matrix (App. III) times 100.

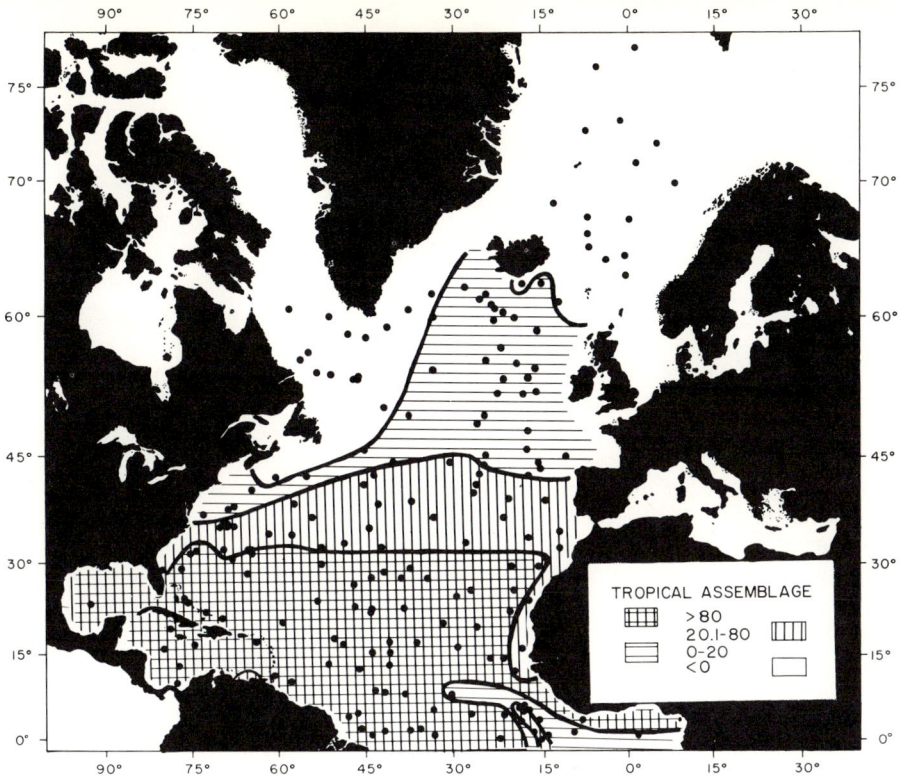

Figure 26. Distribution of tropical assemblage on North Atlantic sea bed. Values plotted are those given in varimax matrix (App. III) times 100.

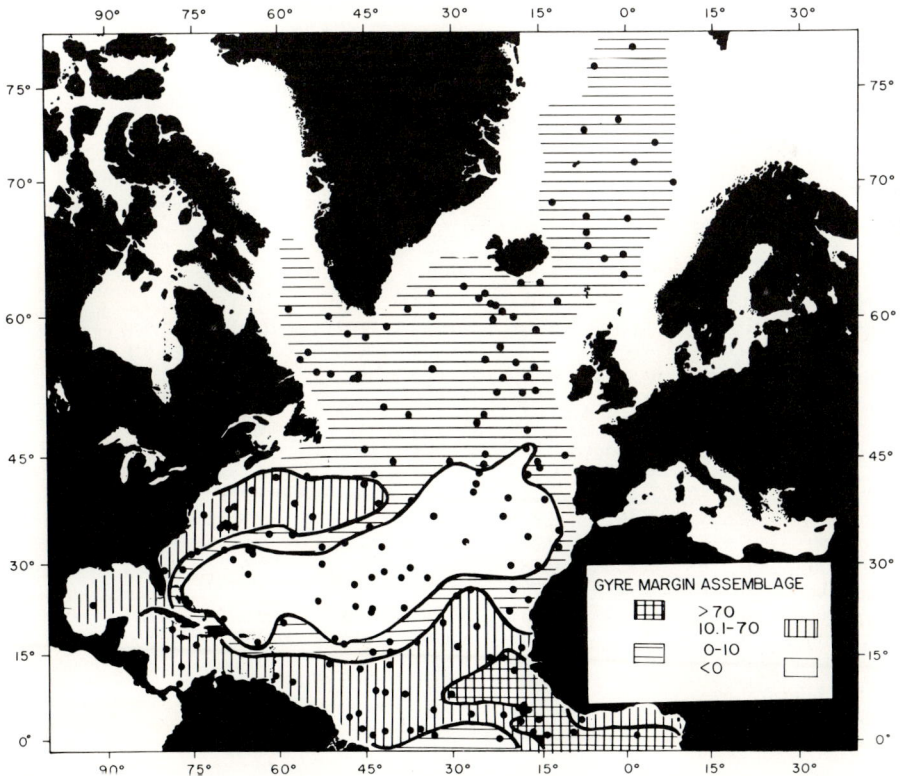

Figure 27. Distribution of gyre margin assemblage on North Atlantic sea bed. Values plotted are those given in varimax matrix (App. III) times 100.

tops used in this study are from cores with sedimentation rates ranging from 40 cm/1,000 yr to 0.5 cm/1,000 yr.

TRANSFER FUNCTION F13

Derivation

As the term is used here, a transfer function consists of a set of equations relating a set of oceanographic parameters to faunal data. Each equation is derived by standard multivariate regression techniques and is a nonlinear equation of the second degree. The technique is fully described in Imbrie and Kipp (1971).

Several options for deriving the set of equations **K** were explored. One group of options represents choices among alternate taxonomic schemes. Originally, the equations were written using the taxonomic categories of Imbrie and others (1973). Then various combinations of the four taxonomic divisions discussed in the section "Faunal Data" were tried. In all, eight taxonomic options were explored. The equation that gave the most consistent and accurate results, as determined by comparison of multiple correlation coefficients, employed the taxonomic categories listed in Table 1.

Other options investigated included defining season on a thermal or a calendar basis. As discussed above, the thermal definition improved the accuracy of the equations. A total of ten sets of equations were written. Transfer function F13B-4SE, which produced the most consistent and accurate results, is described below and given the abbreviated designation F13.

In all, the transfer function F13 contains 16 equations: surface temperature and salinity for winter, spring, summer and autumn, and temperature and salinity at 100 m for the same seasons. In order to determine which parameters were statistically independent, linear correlation coefficients between all pairs of variables were calculated. Table 3 gives the correlation matrix. These values range from a high

TABLE 3. CORRELATION COEFFICIENTS BETWEEN

	T-winter	T-spring	T-summer	T-autumn	T_{win}-100	T_{spr}-100	T_{sum}-100	T_{aut}-100
T-winter*	1.000	0.995	0.955	0.991	0.911	0.911	0.919	0.912
T-spring	0.995	1.000	0.974	0.995	0.916	0.915	0.928	0.921
T-summer	0.955	0.974	1.000	0.979	0.925	0.924	0.938	0.930
T-autumn	0.991	0.995	0.979	1.000	0.915	0.916	0.921	0.912
T_{win}-100	0.911	0.916	0.925	0.915	1.000	0.988	0.982	0.968
T_{spr}-100	0.911	0.915	0.924	0.916	0.988	1.000	0.974	0.952
T_{sum}-100	0.919	0.928	0.938	0.921	0.982	0.974	1.000	0.980
T_{aut}-100	0.912	0.921	0.930	0.912	0.968	0.952	0.980	1.000
S-winter†	0.561	0.565	0.621	0.579	0.757	0.725	0.708	0.695
S-spring	0.606	0.602	0.645	0.620	0.769	0.742	0.718	0.706
S-summer	0.531	0.526	0.577	0.541	0.717	0.691	0.664	0.659
S-autumn	0.511	0.520	0.582	0.527	0.721	0.679	0.694	0.692
S_{win}-100	0.741	0.741	0.770	0.750	0.904	0.886	0.866	0.846
S_{spr}-100	0.763	0.769	0.811	0.783	0.913	0.903	0.881	0.863
S_{sum}-100	0.738	0.750	0.802	0.762	0.898	0.877	0.876	0.864
S_{aut}-100	0.752	0.759	0.803	0.771	0.906	0.894	0.882	0.871

*T = temperature.
†S = salinity.

of 0.995 (summer versus winter surface temperatures) to a low of 0.511 (winter surface temperature versus autumn surface salinity). The variables are arrayed in four covariance groups: surface temperatures, 100-m temperatures, surface salinities, and 100-m salinities. Within each group correlations are high; between groups the correlations are lower. The essential statistics of only five of the 16 regression equations are listed in Table 4. Equations for summer and winter surface temperatures are presented because estimates from these equations are essential parts of the CLIMAP research. One equation from each of the remaining covariance groups is summarized, with preference given to equations having the best multiple correlation coefficients. Estimated temperatures and salinities for core-top locations are presented in Appendix IV (see footnote 1).

Accuracy

The accuracy of these equations may be judged in part by scanning the multiple correlation coefficients, R, and the standard errors of estimate (Table 4). Because each of these equations has 27 terms, it is important to note that, both in the table and in the discussions that follow, these statistics have been adjusted for the number of degrees of freedom lost. The R values range from 0.884 to 0.991, and the standard errors from 1.165° to 1.698°C and from 0.308‰ to 0.389‰. Scatter diagrams of observed versus estimated variables are plotted and the geographical distribution of the residuals is mapped in Figures 28 through 32. In general the distribution of errors is random in relation to geographic position. No sample exceeds the standard error of estimate for all five parameters.

As noted earlier, samples in this study display a considerable range of dissolution intensity, ranging from samples with no evidence of dissolution to tropical samples in which the species diversity is substantially decreased by dissolution. Figure 2 shows the distribution of dissolved as well as undissolved samples. By comparing this distribution with the anomaly maps, it can be seen that the equation anomalies are random in relation to dissolution. To test the significance of the difference between the means of the 50 samples with dissolution and the remaining 141 samples,

ALL PAIRS OF OCEANOGRAPHIC PARAMETERS

S-winter	S-spring	S-summer	S-autumn	S_{win}-100	S_{spr}-100	S_{sum}-100	S_{aut}-100	
0.561	0.606	0.531	0.511	0.741	0.763	0.738	0.752	T-winter
0.565	0.602	0.526	0.520	0.741	0.769	0.750	0.759	T-spring
0.621	0.645	0.577	0.582	0.770	0.811	0.802	0.803	T-summer
0.579	0.620	0.541	0.527	0.750	0.783	0.762	0.771	T-autumn
0.757	0.769	0.717	0.721	0.904	0.913	0.898	0.906	T_{win}-100
0.725	0.742	0.691	0.679	0.886	0.903	0.877	0.894	T_{spr}-100
0.708	0.718	0.664	0.694	0.866	0.881	0.876	0.882	T_{sum}-100
0.695	0.706	0.659	0.692	0.846	0.863	0.864	0.871	T_{aut}-100
1.000	0.953	0.910	0.922	0.890	0.886	0.891	0.880	S-winter
0.953	1.000	0.928	0.901	0.882	0.890	0.884	0.884	S-spring
0.910	0.928	1.000	0.926	0.825	0.824	0.820	0.815	S-summer
0.922	0.901	0.926	1.000	0.839	0.832	0.860	0.844	S-autumn
0.890	0.882	0.825	0.839	1.000	0.973	0.958	0.974	S_{win}-100
0.886	0.890	0.824	0.832	0.973	1.000	0.975	0.980	S_{spr}-100
0.891	0.884	0.820	0.860	0.958	0.975	1.000	0.979	S_{sum}-100
0.880	0.884	0.815	0.844	0.974	0.980	0.979	1.000	S_{aut}-100

Student's t test was applied. The t ratios for the five equations range from 0.106 to 0.804. These ratios, with 189 degrees of freedom, indicate that the differences between the two data sets are not significant at the 0.40 level. The statistics are summarized in Table 5.

One important limitation on the accuracy of the transfer function should be stressed. For each season a single temperature estimate is obtained for all areas of the sea bed dominated by the monospecific polar assemblage (samples containing more than 80% left-coiling *G. pachyderma*). This temperature is the average for all polar sites for a given season, as is shown particularly well by the distribution of polar estimates for summer temperature (Figure 29a). The measured summer temperatures for 13 samples containing more than 80% left-coiling *G. pachyderma* range from 3.1° to 9.3°C. The range of temperature estimates for these sites is only 0.8°C, from 6.2° to 7.0°C.

Samples that contain intermediate abundances of left-coiling *G. pachyderma* (30% to 80%), with significant percentages of the less resistant species *G. bulloides* and *G. quinqueloba*, occur at sites with summer sea-surface temperatures ranging from approximately 8° to 11°C. The resistant species, right-coiling *G. pachyderma* and *G. inflata*, are relatively abundant with left-coiling *G. pachyderma* at locations

TABLE 4. STATISTICS OF TRANSFER FUNCTION F13B-4SE

	T-winter	T-summer	T_{sum}-100	S-winter	S_{win}-100
Multiple correlation coefficient*	0.991	0.985	0.973	0.884	0.919
Standard error of estimate*	1.165°C	1.380°C	1.698°C	0.389‰	0.308‰
80% confidence interval	±1.5°C	±1.8°C	±2.2°C	±0.50‰	±0.40‰
Tropical	35.696	29.562	30.273	−1.167	4.104
Subpolar	4.822	1.758	−2.532	−2.997	−0.226
Polar	−22.520	−18.650	−28.249	−6.521	−1.402
Gyre margin	41.114	25.646	9.493	−6.044	−1.371
Transitional	8.070	5.336	14.982	2.263	−0.446
Subtropical	6.758	10.389	−8.451	2.326	1.901
Tropical-subpolar	−15.748	−10.599	−17.459	1.976	−1.850
Tropical-polar	27.906	24.685	40.502	11.272	3.399
Tropical-gyre margin	−32.273	−23.619	−14.218	3.787	−0.161
Tropical-transitional	−0.074	8.066	−0.269	−3.542	−0.306
Tropical-subtropical	−17.195	−8.132	2.762	−0.657	−0.948
Subpolar-polar	14.779	11.933	16.613	4.560	0.780
Subpolar-gyre margin	−4.341	6.229	−9.056	3.650	−0.730
Subpolar-transitional	−12.539	−14.289	−17.527	−2.247	−0.199
Subpolar-subtropical	−0.615	−0.987	11.465	−0.492	−0.320
Polar-gyre margin	−65.617	60.248	10.749	−24.220	−15.781
Polar-transitional	−6.917	−3.223	−12.765	−1.286	0.240
Polar-subtropical	−13.522	−23.019	−2.048	−0.433	−0.029
Gyre margin-transitional	18.374	4.546	2.314	−1.070	0.757
Gyre margin-subtropical	−8.524	−6.404	28.165	2.932	4.306
Transitional-subtropical	−1.869	1.576	−4.663	1.163	1.823
Tropical, squared	−18.067	−19.186	−15.608	0.175	−2.897
Subpolar, squared	−3.974	−6.507	3.326	1.039	0.097
Polar, squared	17.788	8.962	23.196	4.549	1.157
Gyre margin, squared	−23.658	−15.558	−3.927	3.519	1.018
Transitional, squared	4.331	2.890	17.274	1.507	0.379
Subtropical, squared	−0.397	−6.283	7.464	−1.903	−1.729
Intercept	6.874	16.384	8.789	37.238	35.343

*Adjusted for degrees of freedom.

with summer temperatures above 11°C. Dissolution at sites with summer temperatures of less than 11°C tends to reduce the fauna in the sediments to the monospecific polar assemblage. Therefore, inclusion in the data base of samples with intermediate abundances of left-coiling *G. pachyderma* and evidence of dissolution could tend to increase the average, and hence the estimated, temperatures for true polar samples. Consequently, temperature estimates for downcore polar samples that exhibit evidence of severe dissolution are suspect.

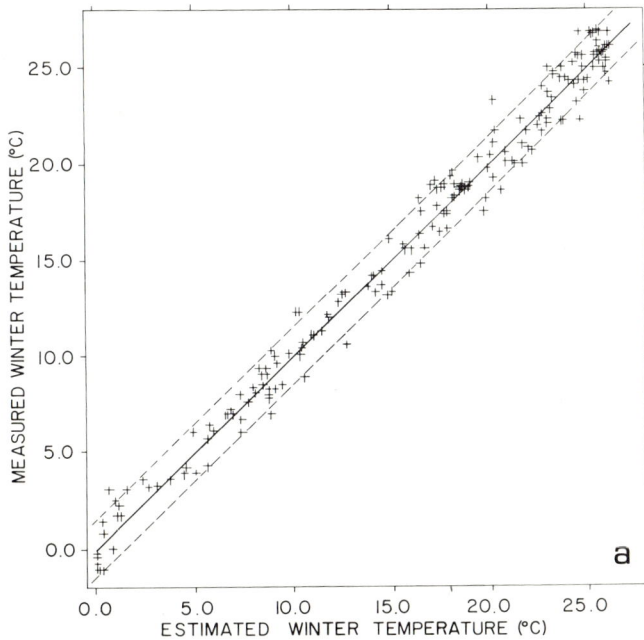

Figure 28. Observed values of winter sea-surface temperature versus estimates calculated by transfer function F13 for 191 North Atlantic seabed samples. (a) Scatter diagram with 80% confidence intervals indicated by dashed lines. (b) Geographic distribution of residuals. T-win is observed winter temperature; T̂-win is estimated winter temperature. |Δ| is absolute value of difference between observed and estimated temperatures. Standard error is 1.165°C. Mean deviation indicated for areas with more than two points in which more than half of residuals are of same sign.

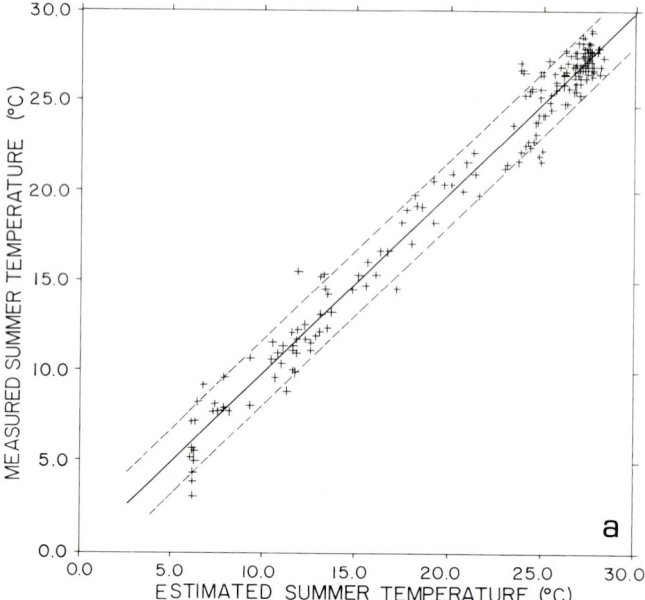

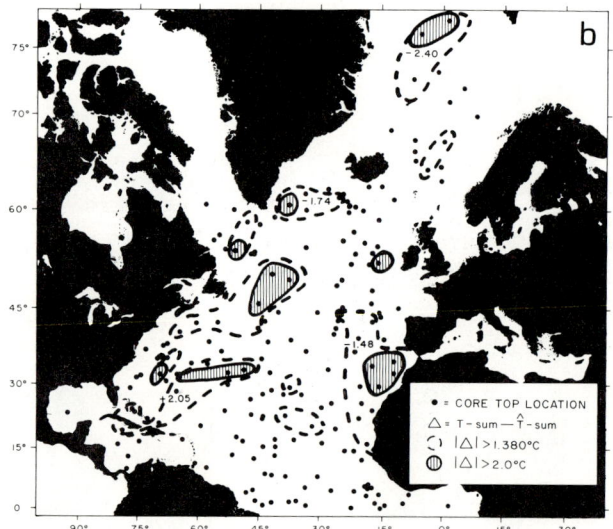

Figure 29. Observed values of summer sea-surface temperature versus estimates calculated by transfer function F13 for 191 North Atlantic sea-bed samples. (a) Scatter diagram with 80% confidence intervals indicated by dashed lines. (b) Geographic distribution of residuals. Standard error is 1.380°C. Other symbols explained in caption for Figure 28.

Precision

Analytical precision, or laboratory error, was estimated by applying transfer function F13 to counts of five replicate samples from each of seven core tops. One location was selected from the area of maximum occurrence for each of the six assemblages; the seventh location was selected from an area of assemblage overlap. Three sets of samples (V12-122, V4-32, and SP9-3) are the same as those for which 80% confidence intervals were calculated by Imbrie and others (1973,

Figure 30. Observed values of summer temperature at 100-m depth versus estimates calculated by transfer function F13 for 191 North Atlantic sea-bed samples. (a) Scatter diagram with 80% confidence intervals indicated by dashed lines. (b) Geographic distribution of residuals. Standard error is 1.698°C. Other symbols explained in caption for Figure 28.

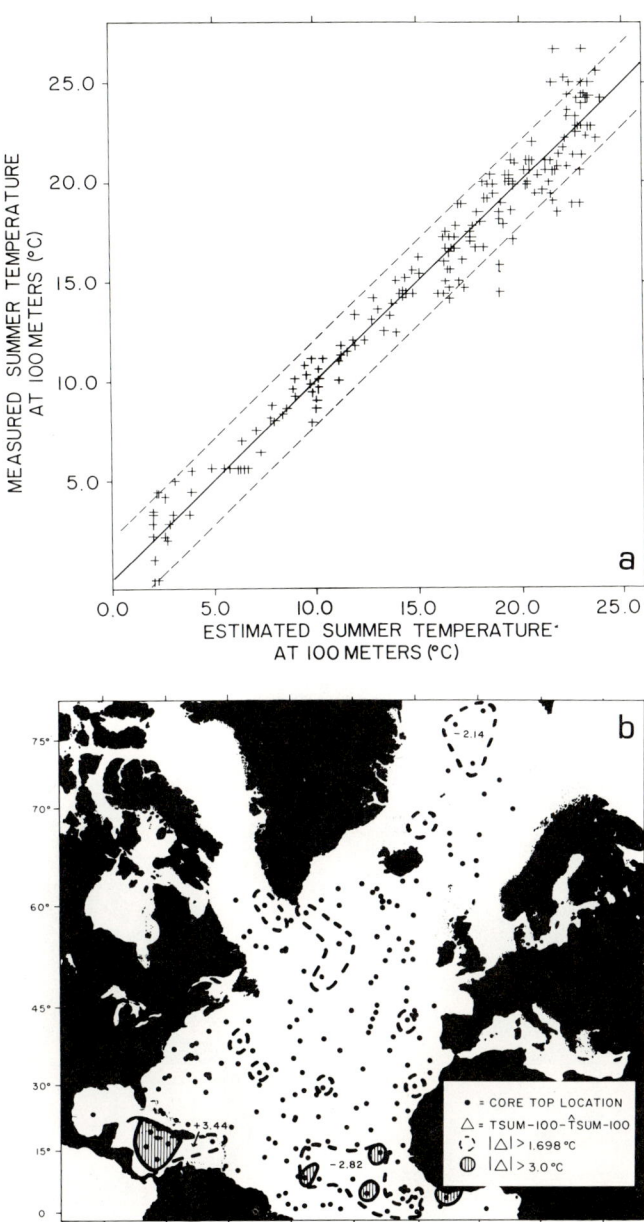

their Table 7) from transfer function F3. The same method of calculation was used to obtain 80% confidence intervals for transfer function F13. These confidence intervals are presented in Table 6. Comparison of the confidence intervals for estimates obtained from both transfer functions on samples V12-122, V4-32, and SP9-3 shows somewhat larger confidence intervals for the new equation. This is attributed to the splitting of four taxonomic categories for F13, and the resulting increase in the counting error associated with smaller percentages.

Those locations that contain essentially a single-assemblage fauna (tropical sample

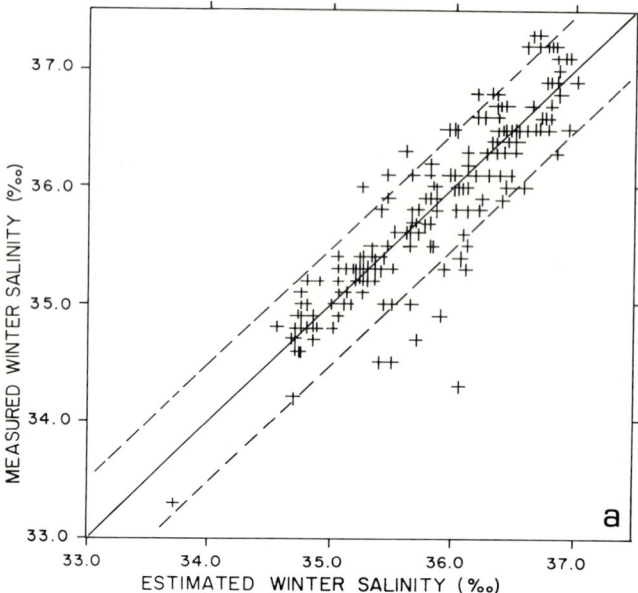
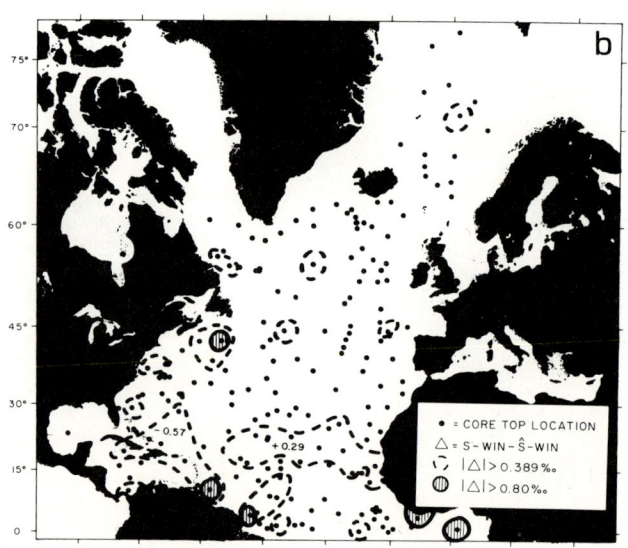

Figure 31. Observed values of winter sea-surface salinity versus estimates calculated by transfer function F13 for 191 North Atlantic seabed samples. (a) Scatter diagram with 80% confidence intervals indicated by dashed lines. (b) Geographic distribution of residuals. Standard error is 0.389‰. Other symbols explained in caption for Figure 28.

V12-122, polar sample V29-211, subpolar sample SP9-3) are in general characterized by the highest precision. Although Sample V4-32 is located in the area of the subtropical maximum, tropical and subpolar faunas occur in this sample in amounts approximately equal to that of the subtropical fauna, and this mixture of assemblages results in confidence intervals of intermediate range. The gyre margin and transitional assemblages are themselves mixtures; the former is a warm, low-salinity fauna intensified by dissolution, along with elements of the tropical fauna, and the latter is a restricted subpolar fauna with one additional characteristic species. Samples dominated by these assemblages (V26-53 and V18-373), therefore, also have

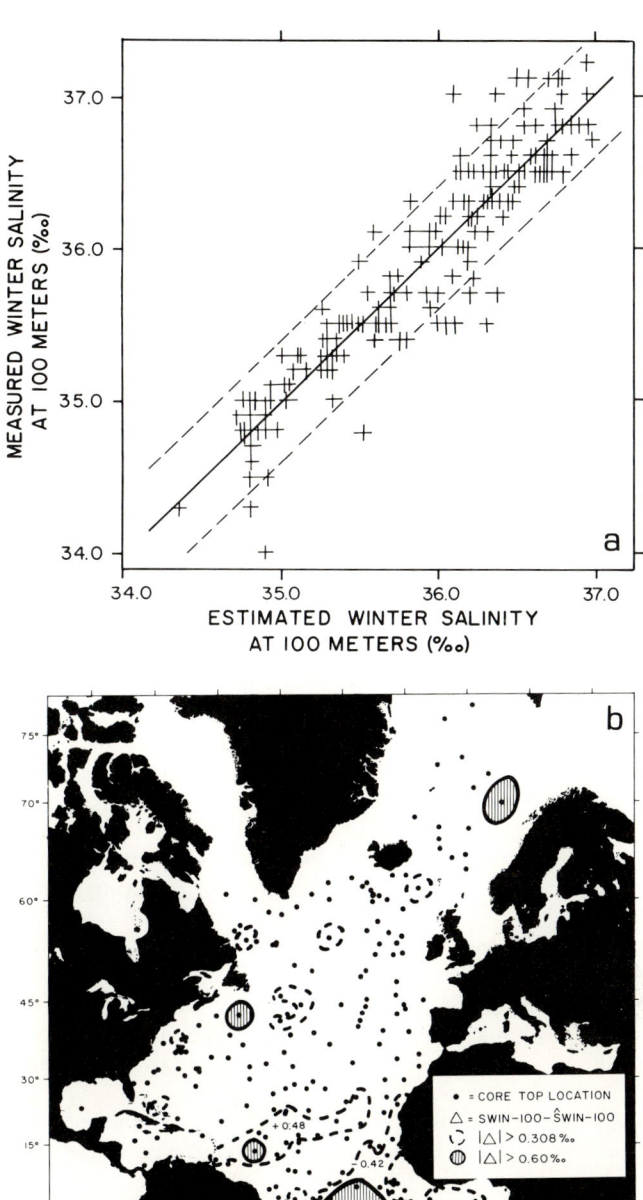

Figure 32. Observed values of winter salinity at 100 m versus estimates calculated by transfer function F13 for 191 North Atlantic seabed samples. (a) Scatter diagram with 80% confidence intervals indicated by dashed lines. (b) Geographic distribution of residuals. Standard error is 0.308‰. Other symbols explained in caption for Figure 28.

confidence intervals of intermediate range. Sample V27-16, from an area of mixing, contains approximately equal contributions from the transitional, subpolar and subtropical assemblages. This complex combination of the mixed transitional fauna with single-assemblage faunas results in the lowest precision.

The confidence intervals given in the section "Accuracy" include uncertainties originating from laboratory error (precision) as well as inaccuracies from all other sources. Of the 175 estimates made for the above seven locations, only 25, or approximately 14%, exceeded the 80% confidence intervals describing the accuracy of transfer function F13. Of these 25, 10 were estimates of summer temperature

TABLE 5. STATISTICS FOR STUDENT'S T TEST

	T-winter (°C)	T-summer (°C)	T_{sum}-100 (°C)	S-winter (‰)	S_{win}-100 (‰)
No dissolution					
Mean ($\bar{X}_1$)	+0.042	−0.041	+0.007	−0.002	−0.003
Standard deviation (s_1)	0.983	1.238	1.431	0.313	0.258
Number of samples (n_1) = 141					
Dissolution					
Mean ($\bar{X}_2$)	−0.052	+0.116	−0.019	+0.009	+0.012
Standard deviation (s_2)	1.053	1.034	1.556	0.394	0.284
Number of samples (n_2) = 50					
t ratio	+0.571	+0.804	+0.106	−0.201	−0.332
Degrees of freedom = $n_1 + n_2 - 2 = 189$					

Note: Student's t test is for significance of difference between two sample means: samples with little or no dissolution and samples with moderate to severe dissolution.

at 100 m for samples V12-122 and V26-53. The average error for the former was 4.4°C, and for the latter, 3.0°C. It is possible that the unusually large error for sample V12-122 is due to the lack of oceanographic data for this location.

The confidence intervals given in Table 4 are the appropriate ones to apply when the purpose of the investigation is the reconstruction of past sea-surface conditions. One reason for measuring and presenting separately the laboratory or counting error is to identify and control any comparatively large errors from this source. For example, in this study it was determined that counts of more than 300 specimens should be made for samples that contain mixed assemblages. Other reasons were given by Imbrie and others (1973), who argued that for stratigraphic and other purposes it is useful to consider the estimates made by transfer functions simply as faunal indices, and that in evaluating such plots it is important to know the magnitude of the counting errors.

Testing the Transfer Function

The equations for sea-surface temperatures were tested on an independent data set by applying them to foraminiferal counts from 60 samples from the South Atlantic Ocean. Twenty additional samples were not included in this set because they were either polar samples with evidence of severe dissolution, or samples that contained species or assemblages with abundance levels exceeding the corresponding maxima in the North Atlantic data set. The results of the test are encouraging: for the 60 locations, 78% of the winter temperature estimates and

TABLE 6. ANALYTICAL PRECISION OF ESTIMATES AT REPRESENTATIVE SITES

Map no.	Core	Avg. sample size	Province	T-winter (°C)	80% Confidence intervals T-summer (°C)	T_{sum}-100 (°C)	S-winter (‰)	S_{win}-100 (‰)
221	V29-211	320	Polar	±0.400	±0.535	±0.317	±0.061	±0.042
49	SP9-3	333	Subpolar	±0.403	±0.752	±0.737	±0.131	±0.104
101	V18-373	383	Transitional	±1.191	±0.806	±0.662	±0.162	±0.196
57	V4-32	337	Subtropical	±0.838	±1.057	±0.662	±0.174	±0.156
76	V12-122	330	Tropical	±0.923	±0.184	±0.737	±0.274	±0.193
153	V26-53	346	Gyre margin	±1.188	±1.267	±1.013	±0.145	±0.154
157	V27-16	347	Mixed	±2.003	±3.345	±1.649	±0.622	±0.347

82% of the summer temperature estimates fall within the 80% confidence intervals calculated for the North Atlantic core tops. We conclude that North Atlantic equations may be used on carefully selected samples from the South Atlantic with reasonable confidence.

Five short cores were selected for additional testing of the equations. The locations of these cores (Fig. 33) were chosen to insure a wide geographic and temperature range, as well as a variety of faunal compositions.

Core SP9-3 contains a polar fauna that changes suddenly upward to a subpolar fauna. Throughout core V17-165 there are significant contributions of tropical, subtropical and transitional assemblages. Core A180-47 shows significant fluctuations in faunal composition; its contributing assemblages range from tropical to subpolar. Core A179-15 is tropical throughout. Core A180-76 is tropical throughout with small but significant contributions from the subpolar and gyre margin assemblages. (As noted in the discussion of definition of seasons for this latter area, winter temperature estimates are *August* temperature estimates.)

Winter temperature estimates (Fig. 34) for two of the cores—a tropical, nearshore core from the western side of the Atlantic (A179-15) and a core from central waters (V17-165)—display a range of less than 2°C. From the work of McIntyre and others (1972; this volume), it is apparent that an oceanic front, with accompanying

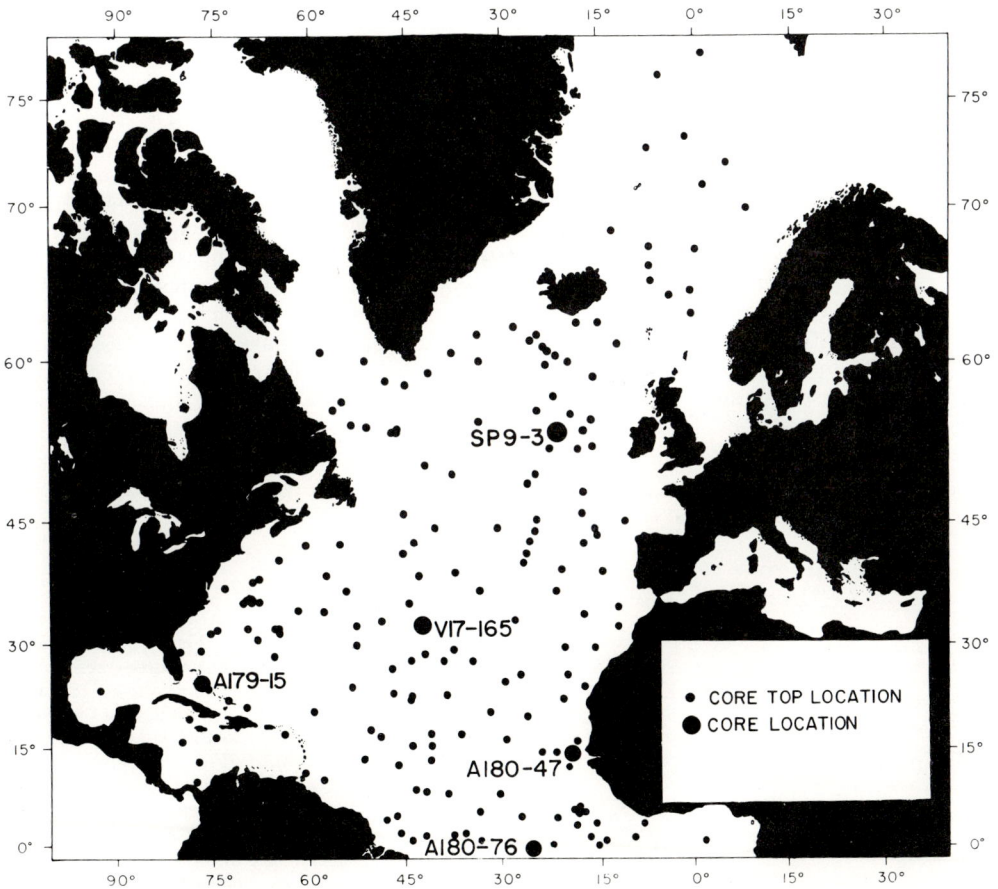

Figure 33. Locations of five short North Atlantic cores used in testing transfer function F13.

large temperature changes across the front, did not traverse this area within the last 20,000 yr. The calculated temperature ranges are consistent with this conclusion. One core (SP9-3) displays an estimated temperature range of more than 10°C. From evidence presented in McIntyre and others (1972; this volume), it is clear that the oceanic polar front migrated across the core site at the end of Wisconsin glaciation. The calculated temperature range, as well as the absolute temperature levels calculated, is consistent with this conclusion.

Cores A180-47 and A180-76, both located in the eastern equatorial Atlantic, show calculated temperature ranges of about 10° and 5°C, respectively. Both cores are in areas of anomalously cool water. Off Africa, this effect is due to seasonal upwelling. In the equatorial region, it is due to the occasional surfacing of the equatorial undercurrent (Neumann and Pierson, 1966). Abundance of *G. dutertrei* has been correlated with upwelling (Gardner, 1973) and the Atlantic equatorial undercurrent (Jones, 1967). The increase in abundance of *G. dutertrei* down these cores might indicate more permanent year-around upwelling, because the magnitude of temperature change correlates with the magnitude of the change in upwelling intensity.

A third test of the equations is made by using them to prepare synoptic maps of the North Atlantic 18,000 yr ago (McIntyre and others, this volume). The distribution of environmental estimates displayed on these maps is characterized by systematic gradients and can be explained by principles of physical oceanography. This result is an additional confirmation of the validity of transfer function F13.

CAUTIONS

Certain cautions should be observed in using this or any transfer function.

1. Laboratory procedure, size fraction, sample size and taxonomy must be consistent with those used for the data base of the transfer function.

2. The equations should not be applied to any sample with species or assemblage abundances exceeding their maxima in the calibration data described in this paper. In addition, estimates from samples containing percentages of species that are near the maximum in the data base of the transfer function should be viewed with reservations. For example, the data base for F13 contains one sample with 62% *G. tumida*. This is the only sample in the data base that contains more than 38% of the species; only six samples contain more than 18%. Estimates on samples with an abundance of *G. tumida* in this upper range should be used with care.

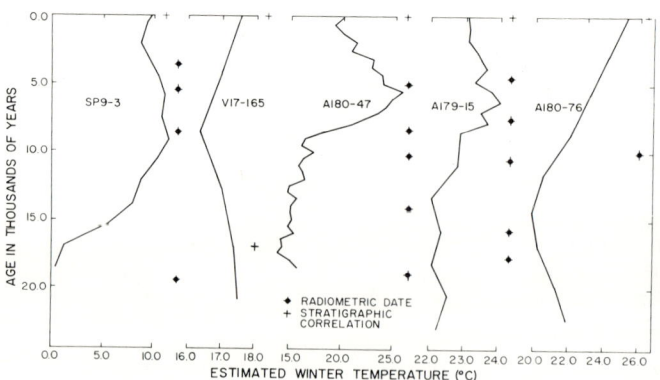

Figure 34. Estimates of winter sea-surface temperature plotted versus age for five North Atlantic cores shown in Fig. 33. Solid circle with cross indicates level dated by C^{14}. Cross indicates level was dated stratigraphically. For discussion of stratigraphy see McIntyre and others (this volume).

In general, this precaution should eliminate from consideration estimates on samples containing faunas that are not well represented in the data base, including those affected by extreme dissolution.

SUMMARY AND CONCLUSIONS

1. Imbrie and Kipp (1971) and Imbrie and others (1973) used planktonic foraminiferal counts to derive transfer functions (F2 and F3) for estimating past sea-surface temperatures and salinity. The same methods were used to derive transfer function F13, described in this paper. However, the data base has been expanded and improved and more accurate oceanographic data has been employed.

2. Six species groups can be recognized by visual inspection of sea-bed species distributions: polar, subpolar, transitional, subtropical, tropical and gyre margin. The distributions of these groups are similar to the distribution of the living plankton reported by Bé and Tolderlund (1971). Major differences may be explained by lack of control in small areas in both data sets, difference in mesh or sieve size, differential dissolution, and depth habitats sampled.

3. Six assemblages derived from factor analysis correspond approximately to the groups described above. Five of the assemblages are similar to those described by Imbrie and Kipp (1971). However, increased sampling density allows more detailed mapping and definition of a sixth assemblage, the transitional.

4. Maps of the 20 species and six assemblages display systematic regional gradients and are clearly correlated with physical and chemical gradients in the ocean.

5. Transfer function F13 represents the optimum result obtained by experimenting with ten different taxonomic and counting options.

6. Distribution of errors for the five parameters estimated by F13 is generally random with respect to geographic location and dissolution intensity. Eighty percent confidence intervals for temperature estimates range from 1.5° to 2.2°C and for salinity estimates from 0.40‰ to 0.50‰.

7. Results of use of the sea-surface temperature equations on 60 South Atlantic core tops indicate that the North Atlantic equations may be used on carefully selected South Atlantic samples with reasonable confidence.

8. Estimates for five North Atlantic cores with a wide geographic and temperature distribution exhibit reasonable ranges of temperature and salinity for the past 20,000 yr.

9. Care must be exercised in the use of transfer functions; laboratory procedure, size fraction, sample size, and taxonomy must be consistent with those used for the data base of the transfer function. In addition, the maximum percentage for any species or assemblage employed in the calibration data must not be exceeded.

ACKNOWLEDGMENTS

I specially thank J. Imbrie, who provided continual support and encouragement, and who critically reviewed the manuscript. I also gratefully acknowledge constructive criticism by T. Webb and T. J. Crowley and stimulating discussions with A. McIntyre, M. B. Roche and other members of the CLIMAP community.

Unwashed samples were provided by R. Capo at the Lamont-Doherty Geological Observatory Core Laboratory, supported by National Science Foundation Grant GA-35454 and Office of Naval Research Grant N00014-67-A-0108-0004. W. L. Balsam, D. P. Towner, C. Krause and the staff of the Brown University micropaleontology laboratory assisted in sample preparation and computations. In addition, washed samples were received from D. B. Ericson, W. F. Ruddiman, M. B. Roche, B. Molfino, W. L. Prell, J. V. Gardner, and T. B. Kellogg. Data for cores A180-76 and V17-165 were supplied by R. Z. Poore and M. Bauer.

This study has been supported by National Science Foundation Grant GX-28672 (CLIMAP) and by Advanced Research Projects Agency and the Air Force Office of Scientific Research Contract F44620-73-C-0021.

REFERENCES CITED

Bé, A.W.H., 1967, Foraminifera, families: Globigerinidae and Globorotaliidae, Fiche no. 108, in Fraser, J. H., ed., Fiches d'identification du zooplancton: Charlottenlund, Denmark, Conseil International pour l'Exploration de la Mer, sheet 108.

Bé, A.W.H., and Hamlin, W. H., 1967, Ecology of Recent planktonic foraminifera, Part 3: Distribution in the North Atlantic during the summer of 1962: Micropaleontology, v. 13, p.87-106.

Bé, A.W.H., and Tolderlund, D. S., 1971, Distribution and ecology of living planktonic foraminifera in the surface waters of the Atlantic and Indian oceans, in Funnell, B. M., and Riedel, W. R., eds., The micropaleontology of oceans: Cambridge, Cambridge Univ. Press, p. 105-149.

Berger, W. H., 1968, Planktonic foraminifera: Selective solution and paleoclimatic interpretation: Deep-Sea Research, v. 15, p. 31-43.

——1970, Planktonic foraminifera: Selective solution and the lysocline: Marine Geology, v. 8, p. 111-138.

Boltovskoy, E., 1969, Living planktonic foraminifera at 90°E. meridian from the Equator to the Antarctic: Micropaleontology, v. 15, p. 237-255.

Cifelli, R., 1973, Observations on *Globigerina pachyderma* (Ehrenberg) and *Globigerina incompta* Cifelli from the North Atlantic: Jour. Foram. Research, v. 3, p. 157-166.

Defant, A., 1961, Physical oceanography, Vol. 1: New York, Pergamon Press, 729 p.

Ericson, D. B., Ewing, M., and Wollin, G., 1964, The Pleistocene epoch in deep-sea sediments: Science, v. 146, p. 723-732.

Gardner, J. V., 1973, The eastern Equatorial Atlantic: Sedimentation, faunal and sea-surface temperature responses to global climatic changes during the past 200,000 years [Ph.D. dissert.]: New York, Columbia Univ., 387 p.

Hecht, A. D., 1973, A model for determining Pleistocene paleotemperatures from planktonic foraminiferal assemblages: Micropaleontology, v. 19, p. 68-77.

Imbrie, J., and Kipp, N. G., 1971, A new micropaleontological method for quantitative paleoclimatology: Application to a late Pleistocene Caribbean core, in Turekian, K. K., ed., The late Cenozoic glacial ages: New Haven, Yale Univ. Press, p. 71-181.

Imbrie, J., van Donk, J., and Kipp, N. G., 1973, Paleoclimatic investigation of a late Pleistocene Caribbean deep-sea core: Comparison of isotopic and faunal methods: Quaternary Research, v. 3, p. 10-38.

Jones, J. I., 1967, Significance of distribution of planktonic foraminifera in the equatorial Atlantic undercurrent: Micropaleontology, v. 13, p. 489-501.

Lynts, G. W., Judd, J. B., and Stehman, C. F., 1973, Late Pleistocene history of Tongue of the Ocean, Bahamas: Geol. Soc. America Bull., v. 84, p. 2665-2683.

McIntyre, A., Ruddiman, W. F., and Jantzen, R., 1972, Southward penetrations of the North Atlantic Polar Front; faunal and floral evidence of large-scale surface water mass movements over the last 225,000 years: Deep-Sea Research, v. 19, p. 61-77.

McIntyre, A., and others, 1976, Glacial North Atlantic 18,000 years ago: A CLIMAP reconstruction, in Cline, R. M., and Hays, J. D., eds., Investigation of late Quaternary paleoceanography and paleoclimatology: Geol. Soc. America Mem. 145 (this volume).

Neumann, G., and Pierson, W. J., Jr., 1966, Principles of physical oceanography: Englewood Cliffs, N.J., Prentice-Hall Inc., p. 422-478.

Olausson, E., 1971, Quaternary correlations and the geochemistry of oozes, *in* Funnell, B. M., and Riedel, W. R., eds., The micropaleontology of oceans: Cambridge, Cambridge Univ. Press, p. 375-398.

Parker, F. L., 1962, Planktonic foraminiferal species in Pacific sediments: Micropaleontology, v. 8, p. 219-254.

———1967, Late Tertiary biostratigraphy (planktonic foraminifera) of tropical Indo-Pacific deep-sea cores: Bull. Am. Paleontology, v. 52, p. 115-208.

Phleger, F. B., Parker, F. L., and Peirson, J. F., 1953, North Atlantic core foraminifera: Repts. Swedish Deep-Sea Exped., v. 7, 122 p.

Rögl, F., and Bolli, H. M., 1973, Holocene to Pleistocene planktonic foraminifera of Leg 15, Site 147 (Cariaco Basin [Trench], Caribbean Sea) and their climatic interpretation, *in* Edgar, N. T., Kaneps, A. G., and Herring, J. R., eds.: Initial reports of the Deep Sea Drilling Project, v. XV, p. 553-615.

Ruddiman, W. F., 1968, Historical stability of the Gulf Stream meander belt: Foraminiferal evidence: Deep-Sea Research, v. 15, p. 137-148.

———1969, Planktonic Foraminifera of the subtropical north Atlantic gyre [Ph.D. dissert.]: New York, Columbia Univ., 295 p.

Schott, W., 1935, Die Foraminiferen in den Aquatorialen Teil des Atlantischen Ozeans: Deutsche Atlantische Exped., 11, heft 6, p. 43-134.

Simpson, G. G., Roe, A., and Lewontin, R. C., 1960, Quantitative zoology: New York, Harcourt, Brace and Co., 440 p.

U. S. Naval Oceanographic Office, 1958, Oceanographic Atlas of the Polar Seas, Part II, Arctic: Washington, U. S. Naval Oceanographic Office, H. O. Pub. No. 705, 149 p.

U. S. Naval Oceanographic Office, 1967, Oceanographic Atlas of the North Atlantic Ocean, Section II, Physical Properties: Washington, U. S. Naval Oceanographic Office, Pub. No. 700, 300 p.

MANUSCRIPT RECEIVED BY THE SOCIETY FEBRUARY 4, 1974
REVISED MANUSCRIPT RECEIVED SEPTEMBER 23, 1974
MANUSCRIPT ACCEPTED OCTOBER 7, 1974

Printed in U.S.A.

Geological Society of America
Memoir 145
© 1976

Glacial North Atlantic 18,000 Years Ago: A CLIMAP Reconstruction

ANDREW MCINTYRE
Lamont-Doherty Geological Observatory
Columbia University, Palisades, New York 10964
AND
NILVA G. KIPP
Department of Geological Sciences
Brown University, Providence, Rhode Island 02912

WITH

ALLEN W. H. BÉ
Lamont-Doherty Geological Observatory
Columbia University, Palisades, New York 10964

THOMAS CROWLEY AND THOMAS KELLOGG
Department of Geological Sciences
Brown University, Providence, Rhode Island 02912

JAMES V. GARDNER* AND WARREN PRELL*
Lamont-Doherty Geological Observatory
Columbia University, Palisades, New York 10964

WILLIAM F. RUDDIMAN
U.S. Naval Oceanographic Office
Code 6120, NRL/CBD, Chesapeake Beach, Maryland 20732

ABSTRACT

Temperature maps of surface water in the North Atlantic for 18,000 B.P. have been reconstructed for the four seasons. Temperatures were estimated by transfer-function analysis of foraminiferal assemblages, and geometric patterns of surface waters were derived from water-mass-related assemblages of Coccolithophorida and Foraminifera.

---

*Present address: (Gardner) U.S. Geological Survey, Menlo Park, California 94025. (Prell) Department of Geological Sciences, Brown University, Providence, Rhode Island 02912.

At 18,000 B.P. the Arctic Polar Front, which was characterized by a steep thermal gradient parallel and centered on lat 42°N, marked the fundamental dividing line for all climatic regimens between a northern dynamic zone and a southern area of relative stability.

North of lat 42°N the glacial Atlantic was polar in character with wide areas of seasonal pack ice. The Norwegian-Greenland and Labrador Seas had year-round ice cover. The polar sea was dominated by a counterclockwise gyre.

Subpolar and transitional surface water masses were squeezed into a narrow band between the polar front and the subtropical gyre, whose geometry differed only slightly from today. Increased upwelling occurred off the west coast of Africa, while the equatorial divergence–Benguela flow increased during Southern Hemisphere winter at 18,000 B.P.

The greatest temperature differences between today and 18,000 B.P. are found in a latitudinal band from 42°N to 60°N, with differences in some areas exceeding 10°C. North and south of this maximum anomaly band, temperature differences are smaller. The upwelling region off Africa shows a 6°C anomaly. The subtropical gyre shows no statistically significant anomaly.

INTRODUCTION

The central mission of the CLIMAP project (Climate Long Range Investigation, Mapping and Predictions) of the International Decade of Ocean Exploration is to document the history of climate over the past million years using the deep-sea sedimentary record as a primary data source. One facet of this project is global climate reconstruction which seeks to understand the dynamics of past quasi-equilibrium states of the global climate system. One of the first stages in this research program was the reconstruction of surface-water temperature and salinity maps of the North Atlantic during the youngest glacial period 18,000 yr ago.

The technique of paleoceanographic reconstruction comprises seven steps: First, a suite of deep-sea cores is obtained with a broad geographic coverage to provide both surface and downcore samples. Second, the relative abundance of the species in the surface samples is used in Q-mode factor analysis (CABFAC; Klovan and Imbrie, 1971) to give a quantitative description of species assemblages (factors) that form coherent distribution patterns that can be related to water masses in the modern ocean. This analysis provides a quantitative statement of both the relative importance of each species in each assemblage and the relative importance of each assemblage in each surface sample. The third step, transfer-function analysis, is to establish by means of the regression equation the relationship between the average monthly sea-surface temperature and the numerical value of the assemblage at each sample site (Imbrie and Kipp, 1971). This equation expresses sea-surface temperature as a function of various assemblages. Fourth, the stratigraphy of the time interval under study is determined using paleontologic, geochemical, and paleomagnetic techniques. Fifth, a suite of isochronous samples that represent the sediment deposited at the time we wish to reconstruct is chosen. Sixth, the microfossils in these sediments are examined and counted. Each sample is described in terms of the assemblages defined in the surface sediment. Using this numerical description of the ice-age samples in terms of modern assemblages and the regression equations relating modern assemblages to temperature, an estimate of the 18,000 B.P. surface temperatures and salinities is obtained. Seventh, the results are plotted and contoured to yield paleoisotherm and paleoisohaline maps.

Choices: Ocean, Time, and Cores

The North Atlantic was chosen for our first effort because it is characterized by (1) a wide geographic distribution of well-preserved carbonate sediments; (2) a relatively high average accumulation rate; (3) the best-documented physical oceanography of any ocean; (4) continental ice sheets on three sides from lat 40°N on the west to lat 50°N on the east during the glacial interval; and (5) polar front migration between glacial and interglacial periods (Manley, 1951; McIntyre, 1967; McIntyre and others, 1972).

The reasons for choosing 18,000 B.P. were both climatic and practical. First, we wished to model for purposes of comparison the climatic antithesis of today—a glacial maximum. Pleistocene land records indicate that, for Europe and North America, the interval centering upon approximately 18,000 B.P. represents the youngest glacial maximum (Bloom, 1971; Dreimanis and Karrow, 1972). In addition, since this was our first attempt at paleoceanographic reconstruction, a time markedly different from today was preferred, because at first we were unsure whether our transfer-function analyses could handle subtle deviations from modern conditions.

Perhaps the most difficult task in this project has been to establish a unified stratigraphy for an entire ocean basin to permit isochronous sampling. In most of the Atlantic basin, calcium carbonate concentrations reflect climatic change (McIntyre and others, 1972; Gardner and Hays, this volume; Bé and others, this volume). Plots of downcore carbonate yield curves that parallel O^{18}/O^{16} variations. Calcium carbonate curves for selected cores from each region were dated by C^{14} analysis. The use by CLIMAP of a 6 × 6 in. Kasten corer (Kögler, 1963) allowed dating of 1-cm-thick layers in regions of low calcium carbonate concentration. Thus, our second and equally pertinent reason for choosing this level was that 18,000 B.P. coincides with the youngest distinct carbonate minimum prior to termination I in most of the regions under study. This carbonate minimum is of short duration and yields a date with an error of ±1,500 yr, well within the acceptable limits for climatic similarity for this last glacial maximum. Not all regions, however, show a carbonate low at 18,000 B.P. Carbonate curves for cores between lat 40°N to 25°N and long 40°W to 80°W, underlying the North Atlantic central gyre, are qualitatively different and cannot be directly compared with curves from higher or lower latitudes (Crowley, 1975, personal commun.). Within this region, the 18,000 B.P. level has been dated by C^{14} and O^{18}/O^{16} techniques in both piston and Kasten cores to provide interregional correlation.

Regionally, floral and faunal curves and stratigraphic boundaries proved useful for corroborative time control within the period 0 to 130,000 B.P. The coccolith species *Emiliania huxleyi* and *Gephyrocapsa caribbeanica* were not utilized for CABFAC analysis, because they exhibit adaptational changes over the past 100,000 yr. *E. huxleyi* replaces *G. caribbeanica* in middle and high latitudes within the youngest glacial interval as the dominant species. The ratio of *E. huxleyi* to *G. caribbeanica* is thus useful as a stratigraphic tool. Today *E. huxleyi* is the dominant species, and the ratio is greater than five. As this ratio is followed back through time, the point at which it first falls to the value one is approximately at the 18,000-B.P. level. The *Emiliani* O^{18}/O^{16} zone 4 (carbonate minimum 2 in McIntyre and others, 1972) is clearly picked out by this ratio when it falls below 0.2, at approximately 73,000 B.P. In low latitudes in the North Atlantic, the presence or absence of the foraminifera *Globorotalia menardii* is used for glacial-interglacial stratigraphy (Prell and Hays, this volume; Gardner and Hays, this volume).

The time frame used in this study, that of Broecker and van Donk (1970), was applied by McIntyre and others (1972) and Ruddiman and McIntyre (this volume)

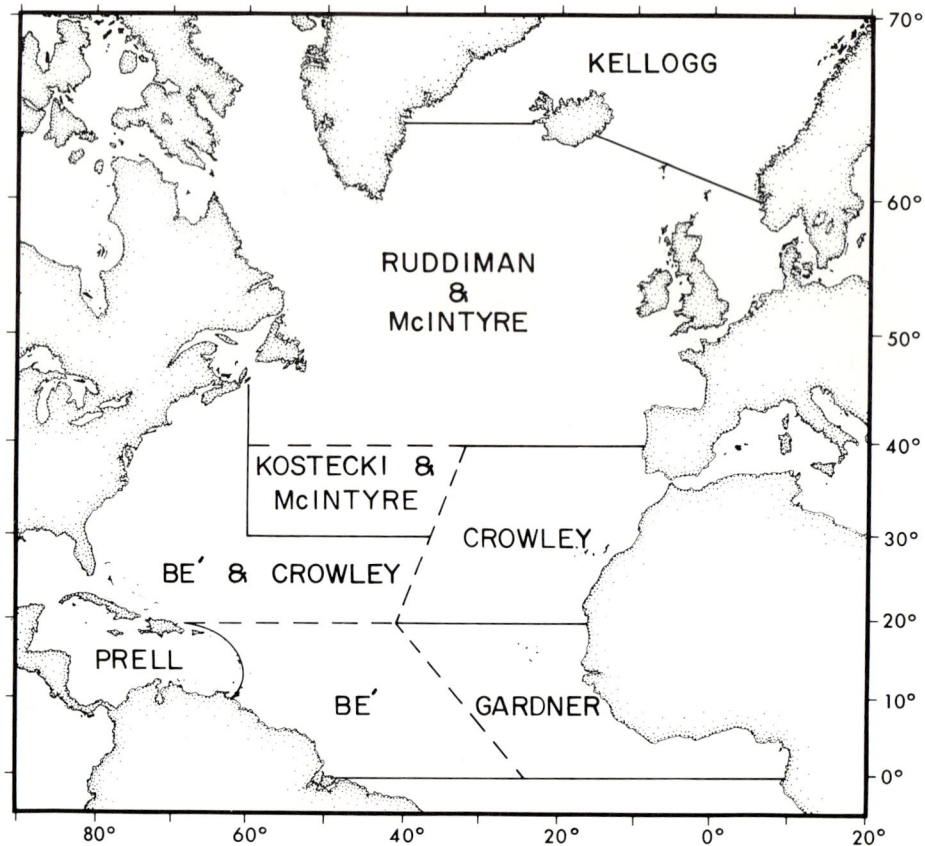

Figure 1. Map of each CLIMAP author's area of responsibility for stratigraphic and microfossil data.

to the North Atlantic. Corroborative evidence for the legitimacy of this frame comes from the Th^{230}-Th^{232} analysis of two cores in the North Atlantic by Ku and others (1972) and $0^{18}/0^{16}$ paleomagnetic studies (Shackleton and Opdyke, 1973, and this volume).

A project of this magnitude requires the concerted effort of numerous workers. Core selection necessitates a detailed knowledge of the stratigraphy. This could not be done by one person for such a large oceanic area within the time limit of this project. In the North Atlantic eight people working in different areas (Fig. 1) have produced a basic stratigraphy of the most recent glacial-interglacial cycle (Emiliani's zones 2 through 5). Each of the workers attempted to obtain a broad geographic distribution of selected cores in his area in which the 18,000 B.P. level was represented by well-preserved microfossils. $CaCO_3$ curves were used in six of the seven areas for initial delineation of climatic stratigraphy. This climatic signature was then placed in a time frame, and the 18,000 B.P. level in each core was selected. Where the $CaCO_3$ curves proved inadequate, C^{14} analysis, floral and faunal temperature curves, and $0^{18}/0^{16}$ curves were used for locating the 18,000 B.P. level.[1] From this reconnaissance and after analysis of each 18,000

---

[1] For exact details of how the 18,000 B.P. level was chosen for each area, the reader is referred to the following papers: Bé and others (this volume), Gardner and Hays (this volume), Prell and Hays (this volume), and Ruddiman and McIntyre (1973).

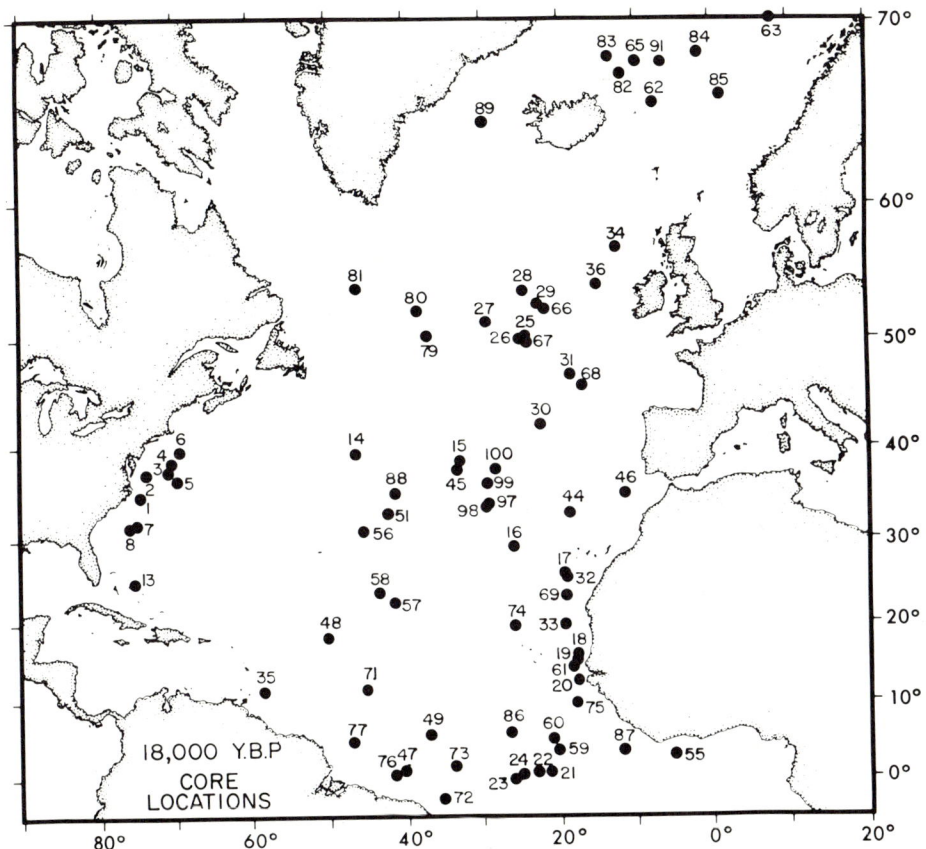

Figure 2. Core location map. Code numbers refer to cores in Table 1 of this paper.

B.P. sample, 100 cores proved useful (Table 1; Fig. 2). Because of geographic density, cores from the Caribbean are not plotted in Figure 2. The reader is referred to Prell and Hays (this volume) for position of the Caribbean cores.

Whereas geographic distribution is adequate, there remains a gap that could not be filled in the western North Atlantic basin. The sole exception is the Bermuda high. This is an area of red-clay deposition where carbonate constituents are dissolved. Attempts to obtain usable cores from this higher standing area have failed because of prevalent winnowing, turbidites, and hiatuses. The other gaps of minor importance in the areal distribution are the Labrador Basin and the slope and rise off the North American continent from Cape Cod to the Grand Banks. Suitable cores have yet to be found in these areas: the former because of dissolution, and the latter because of extreme variation in and great volume of glacial sediments.

RECONSTRUCTIONS

Data Base: Floral and Faunal Assemblages

The basic data to produce 18,000 B.P. surface-water paleoisotherm and paleo-isohaline maps comes from the floral and faunal assemblages. Calibration of the discrete faunal and floral assemblages from surface-sediment samples against modern

TABLE 1. LOCATION OF 18,000 B.P. MAP CORES

Code no.	Core no.	Sample depth (cm)	Lat	Long
1	A156-4	750	34°49'N	74°41'W
2	A156-5	126	37°07'N	73°37'W
3	A164-5	171	37°46'N	71°14.5'W
4	A164-6	109	38°08'N	69°51.5'W
5	A164-24	75	36°29'N	69°00'W
6	A164-61	100	39°32.5'N	68°47'W
7	A167-13	100	31°39'N	75°21'W
8	A167-14	80	31°28'N	76°28'W
9	A172-1	40	17°14'N	73°28'W
10	A172-2	60	16°12'N	72°19'W
11	A172-6	50	14°59'N	68°51'W
12	A179-4	40	16°36'N	74°48'W
13	A179-15	138	24°48'N	75°55.5'W
14	A180-9	60	39°27'N	45°57'W
15	A180-16	70	38°21'N	32°29'W
16	A180-32	30	29°07'N	26°15'W
17	A180-39	30	25°50'N	19°18'W
18	A180-47	311.5	15°19.5'N	17°55.5'W
19	A180-48	500	15°19'N	18°06'W
20	A180-56	100	12°14.5'N	17°46'W
21	A180-72	40	00°35.5'N	21°47'W
22	A180-73	38	00°10'N	23°00'W
23	A180-74	50	00°03'S	24°10'W
24	A180-76	40	00°46'S	26°02'W
25	K708-1	134	50°00'N	23°44.5'W
26	K708-4	90	49°59.3'N	35°01.3'W
27	K708-6	60	51°33.6'N	29°34.0'W
28	K708-7	63	53°56.0'N	24°05.0'W
29	K708-8	106	52°44.8'N	22°33.1'W
30	R5-34	40	42°33'N	21°58'W
31	R5-36	93	46°55'N	18°35'W
32	R5-54	10	25°52'N	19°03'W
33	R5-57	100	19°39'N	19°06'W
34	R10-2	60	56°59.3'N	12°28'W
35	RC9-49	30	11°11'N	58°35'W
36	RC9-225	135	54°58.6'N	15°23.5'W
37	RC10-50	100	11°46'N	81°05.8'W
38	RC13-151	60	15°51.4'N	74°47.1'W
39	RC13-152	60	16°42.5'N	75°26.5'W
40	RC13-153	60	15°03.9'N	75°57.2'W
41	RC13-154	60	14°52.7'N	78°44.6'W
42	RC13-158	80	13°10.7'N	79°49.6'W
43	RC13-159	83	14°41.7'N	77°10.2'W
44	SP8-4	30	32°49.5'N	18°31.5'W
45	V4-8	60	37°14'N	33°08'W
46	V4-32	30	35°03'N	11°37'W
47	V15-168	100	00°12'N	39°54'W
48	V16-20	50	17°56'N	50°21'W
49	V16-25	35	05°04'N	36°48'W
50	V17-39	100	11°45'N	80°26'W

TABLE 1. (*Continued*)

Code no.	Core no.	Sample depth (cm)	Lat	Long
51	V17-165	40	32°45'N	41°54'W
52	V18-357	50	15°02'N	80°14'W
53	V19-19	80	13°14'N	78°22'W
54	V19-21	60	10°36'N	79°21'W
55	V19-291	80	02°00'N	05°15'W
56	V19-309	31	31°10'N	45°08'W
57	V20-241	30	22°08'N	41°30'W
58	V20-242	29	23°22'N	43°39'W
59	V22-186	30	03°23'N	20°07'W
60	V22-188	20	04°40'N	20°55'W
61	V22-197	72	14°10'N	18°35'W
62	V23-58	40	65°46'N	07°07'W
63	V23-60	70	70°03'N	08°19'W
64	V23-63	20	77°57.4'N	00°12.2'W
65	V23-74	60	68°11.2'N	09°36'W
66	V23-82	106	52°35.1'N	21°56'W
67	V23-83	99	49°52.3'N	24°15'W
68	V23-84	90	46°00.3'N	16°55'W
69	V23-98	80	23°07'N	19°18'W
70	V24-28	60	15°19'N	77°57'W
71	V25-44	30	11°30.4'N	45°09.3'W
72	V25-56	60	03°32.8'S	35°13.8'W
73	V25-59	40	01°22.4'N	33°28.9'W
74	V26-41	43	19°19.9'N	26°06.8'W
75	V26-46	40	09°33.8'N	18°10.8'W
76	V26-104	150	00°05'S	41°47'W
77	V26-107	100	04°41'N	46°52'W
78	V26-124	60	16°08'N	74°27'W
79	V27-17	35	50°05.5'N	37°18.1'W
80	V27-19	39	52°05.6'N	38°48'W
81	V27-20	85	54°00'N	46°12'W
82	V27-46	30	67°35'N	11°31.2'W
83	V27-47	20	68°27.7'N	13°32.5'W
84	V27-84	40	68°37.7'N	01°35.7'W
85	V27-86	20	66°36.4'N	01°07.1'E
86	V27-178	55	05°06.1'N	26°39'W
87	V27-248	82	03°03.1'N	11°48.9'W
88	V27-263	20	35°00.5'N	40°55.2'W
89	V28-14	140	64°47'N	29°34'W
90	V28-25	20	76°49'N	01°20'W
91	V28-56	40	68°02'N	06°07'W
92	V28-119	80	17°02'N	76°36'W
93	V28-122	100	11°56'N	78°41'W
94	V28-127	60	11°39'N	80°08'W
95	V28-128	40	11°36'N	80°45'W
96	V28-129	80	11°07'N	80°24'W
97	V29-172	50	33°40'N	29°23'W
98	V29-173K	25	33°40'N	29°23'W
99	V29-174	40	36°18'N	29°22'W
100	V29-175	40	37°30'N	28°17'W

surface temperature and salinity allows quantitative distributional data to be transformed into estimated temperature and salinity values. For the modern surface-water temperatures and salinities used in this calibration, we have relied on the *Oceanographic Atlas of the North Atlantic Ocean* (U.S. Naval Oceanographic Office, 1967).

For this synthesis, only the Foraminifera and Coccolithophorida taxonomy and distribution had reached the level of sophistication necessary for paleoenvironmental analysis; the Diatomacea and Radiolaria require more study.

CABFAC analysis of foraminiferal counts from surface sediments and 18,000 B.P. samples yielded six factoral assemblages. The biogeography of an assemblage encompasses its area of viability. Thus, the assemblages are indicative of a set of environmental parameters that constitute an ecologically defined water mass. Since, in all cases examined to date, these are equivalent to surface water masses described by physical oceanographic means in the North Atlantic, we will refer to them throughout the paper simply as surface water masses.

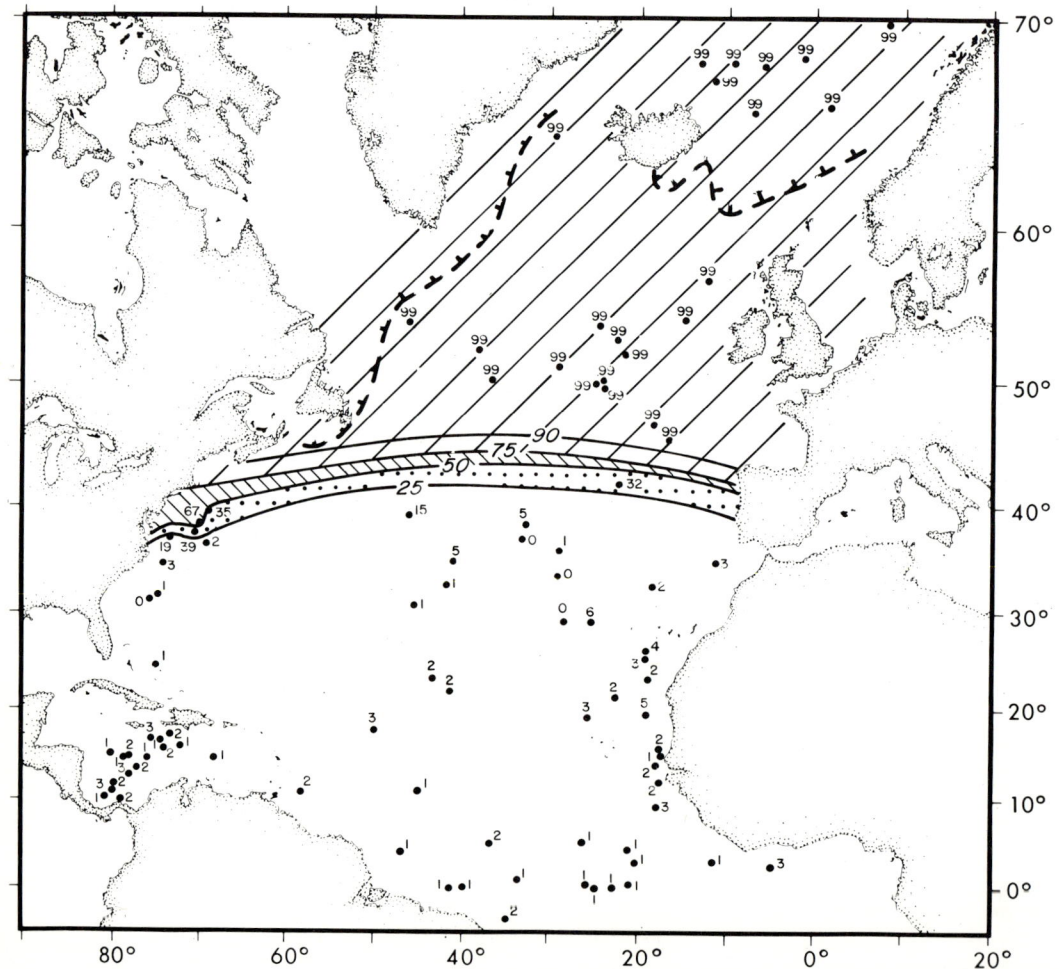

Figure 3. Polar foraminiferal assemblage distribution at 18,000 B.P. In Figures 3 to 8, the centers of concentration of each assemblage in the modern surface sediments have been marked by a line of truncated "T," the shank of the "T" pointing inward toward maximum concentrations.

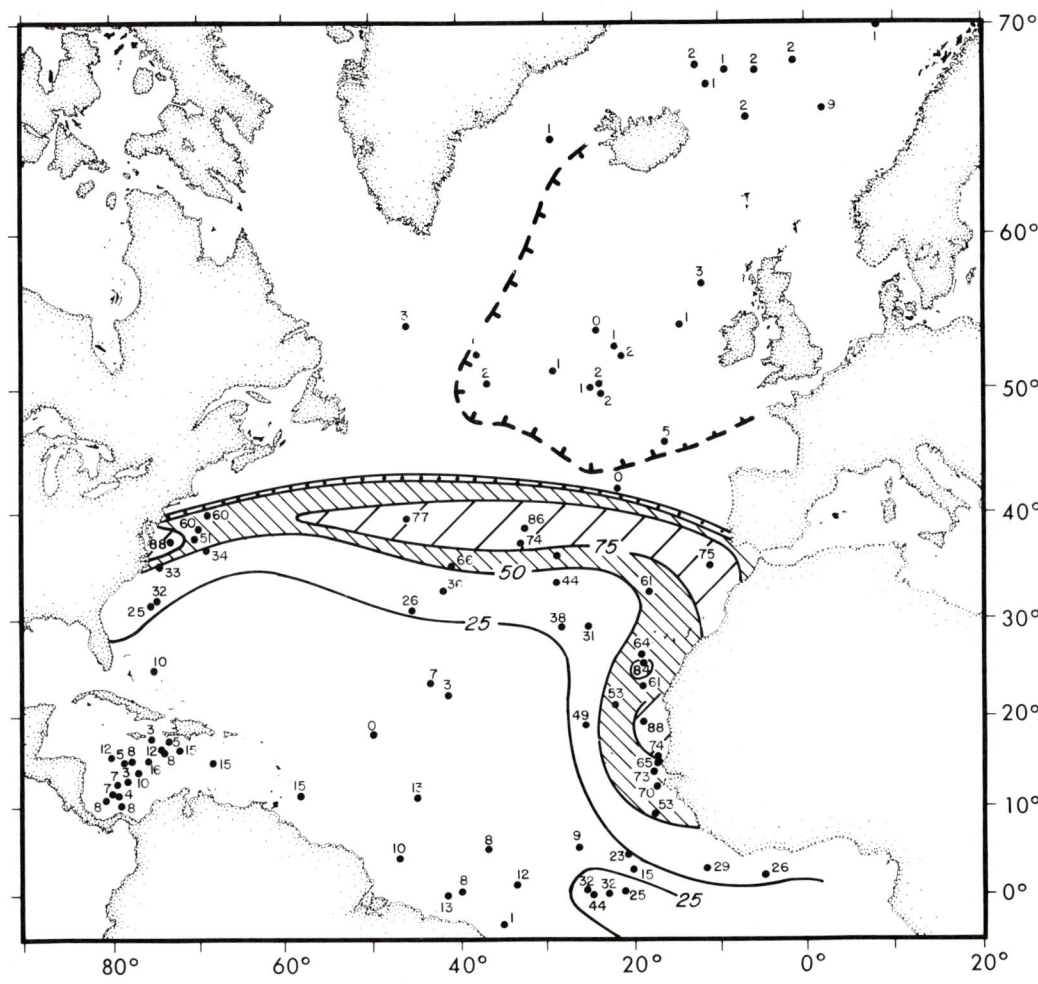

Figure 4. Subpolar foraminiferal assemblage distribution at 18,000 B.P.

The geographic patterns of the foraminiferal assemblages show minor to major variations between 18,000 B.P. and today (Kipp, this volume, Figs. 22 to 27; Figs. 3 to 8). The polar assemblage shows a sweeping southward extension from a small area off Greenland and Labrador to a dominant position throughout the oceanic area north of lat 42°N in the glacial period. Today, the subpolar assemblage shares dominance with the transitional assemblage in waters between Greenland and the northern boundary of the subtropical gyre. At 18,000 B.P. it was restricted to a narrow belt centered on lat 40°N (Fig. 4). Directly to the south, the transitional assemblage (Fig. 5) also shows a marked change in its areal distribution between glacial and interglacial periods. When the modern and 18,000 B.P. distributions of the three are compared (Kipp, this volume, Figs. 22, 23, 24 versus this paper, Figs. 3, 4, 5), it is apparent that in the modern interglacial period the boundaries between these three assemblages are gradational, indicative of wide-scale mixing in time and space by eddies, and that the transitional assemblage, which represents a mixed water mass, shares dominance with the subpolar, whereas the polar assemblage is spatially restricted in the North Atlantic. During a glaciation like the one encompassing 18,000 B.P., however, boundary conditions were the antithesis of today. There was little overlap between the three assemblages, while the dominance of the North Atlantic by polar surface waters was complete north of lat 45°N. The general similarity of factorial values for the polar assemblage throughout this area indicates the homogeneity of this water mass at 18,000 B.P.

The subtropical assemblage (Fig. 6) differed little from the present. The greatest change occurred in the northern boundary, which is characterized today by gradational conditions (Kipp, this volume, Fig. 25). At 18,000 B.P. this gradational zone was absent, its position taken by the subpolar and transitional assemblages. These were squeezed by the advancing polar water against the stable subtropical gyre to form two thin discrete belts reaching from east to west across the Atlantic Basin. The disappearance of the gradational zone suggests that during the glacial period there was no analogue to modern large-scale mixing of Gulf Stream-North Atlantic Drift water, with both cooler and warmer water masses mixing via the formation of cyclonic and anticyclonic eddies. On the contrary, boundaries were sharply defined with surface-water mixing restricted across them.

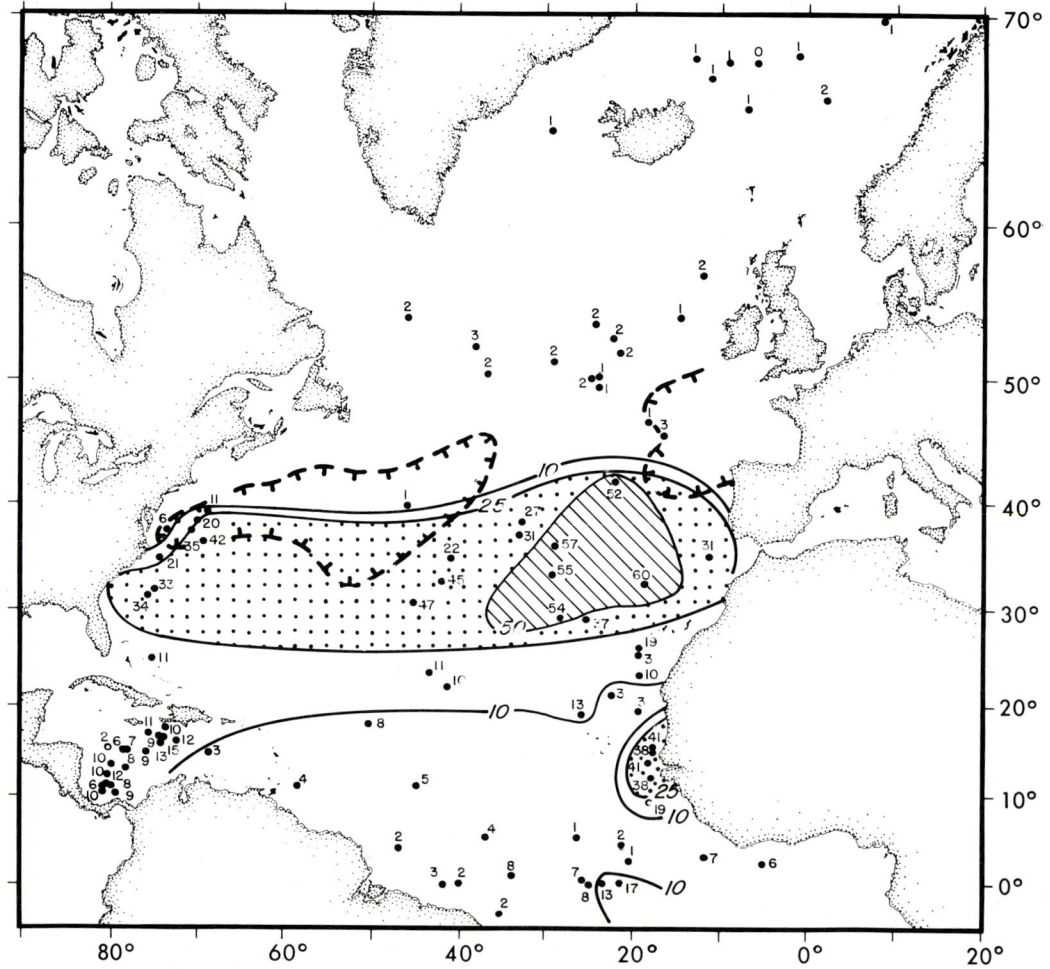

Figure 5. **Transitional foraminiferal assemblage distribution at 18,000 B.P.**

The tropical assemblage (Fig. 7), though showing little change from today, was more dominant in the Caribbean than it is at present, which suggests higher salinities and lower productivity (Prell and Hays, this volume).

A gyre margin assemblage is the sixth factor defined by the foraminiferal data. Like the transitional assemblage, this is representative of mixed surface waters and, at 18,000 B.P. (Fig. 8), was limited to the equatorial region with a concentration in what is roughly the area of the South Equatorial and Benguela Currents today.

CABFAC analysis (Imbrie and Kipp, 1971) of 91 North Atlantic trigger-weight core tops distinguished four coccolith assemblages related to subpolar, transitional, subtropical, and tropical surface water masses (Roche and others, 1975; Fig. 9a). The Coccolithophorida in the 18,000 B.P. samples have been described in terms of these same factoral units (Fig. 9b). Coccolithophorida do not live in polar waters, and thus, there is no analogue to the polar foraminiferal assemblage. Comparison of the two sets depicts important changes in assemblage patterns between today and the youngest glacial interval (McIntyre, 1967; McIntyre and others, 1972). The subpolar and transitional assemblages show a marked diminution in areal

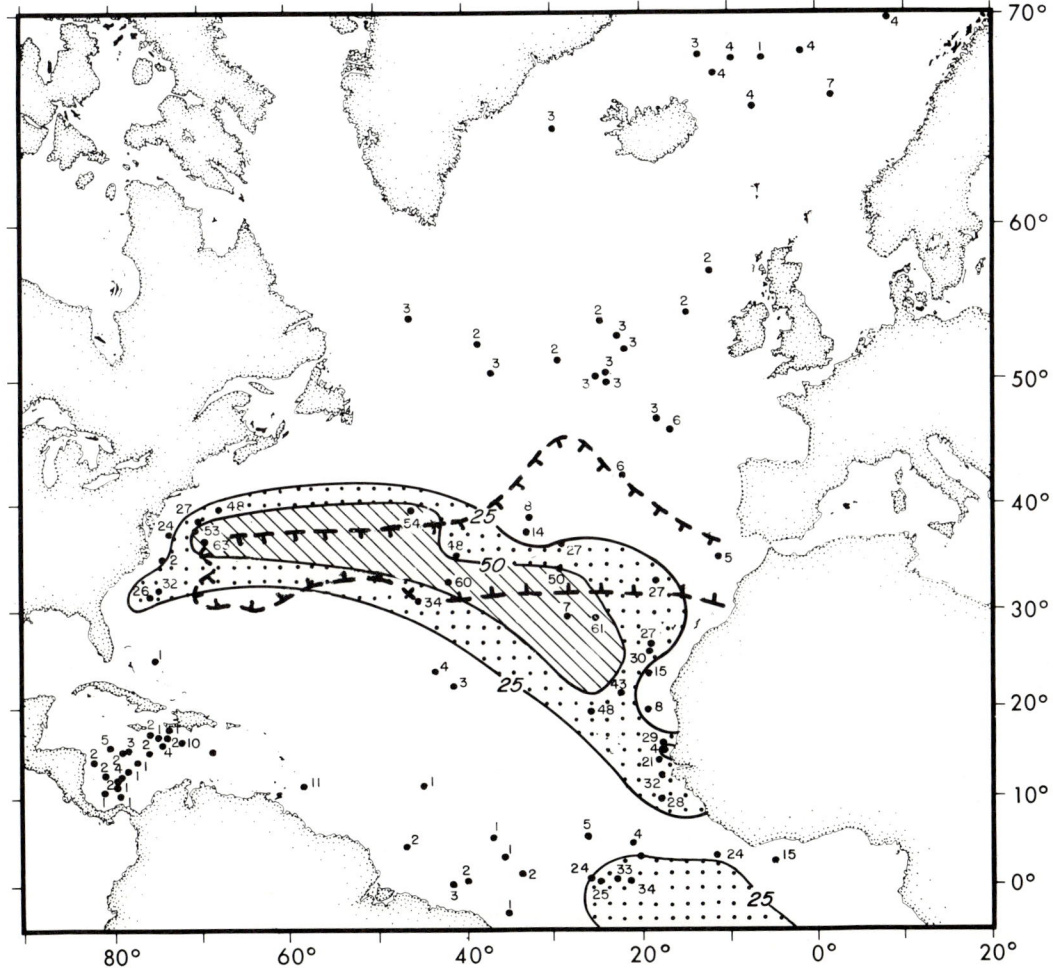

Figure 6. Subtropical foraminiferal assemblage distribution at 18,000 B.P.

distribution and a migration to the south as the polar front moved toward the Equator. The subtropical and tropical assemblages varied only slightly from today in their distribution. Since these factoral assemblages are found beneath surface water masses today, the presumption is that the same relation existed at 18,000 B.P. as well. Although the ecology of a species may vary through time, a wholesale change of the total biota through this short period of time is highly unlikely, because the assemblages are the same in both surface and 18,000 B.P. samples. We presume that ecological stability has been maintained, and thus, the patterns reflect a change in the position of North Atlantic surface water masses.

The surface temperatures used in our maps were not derived from the Coccolithophorida, because at the time of writing, refinement of the regression equations was still in progress. Comparison of modern and 18,000 B.P. foraminiferal and coccolith assemblages indicate that there has been a major change in surface-water patterns between 18,000 B.P. and today and that the boundary conditions along the polar frontal system have both altered and migrated large distances since the last glaciation.

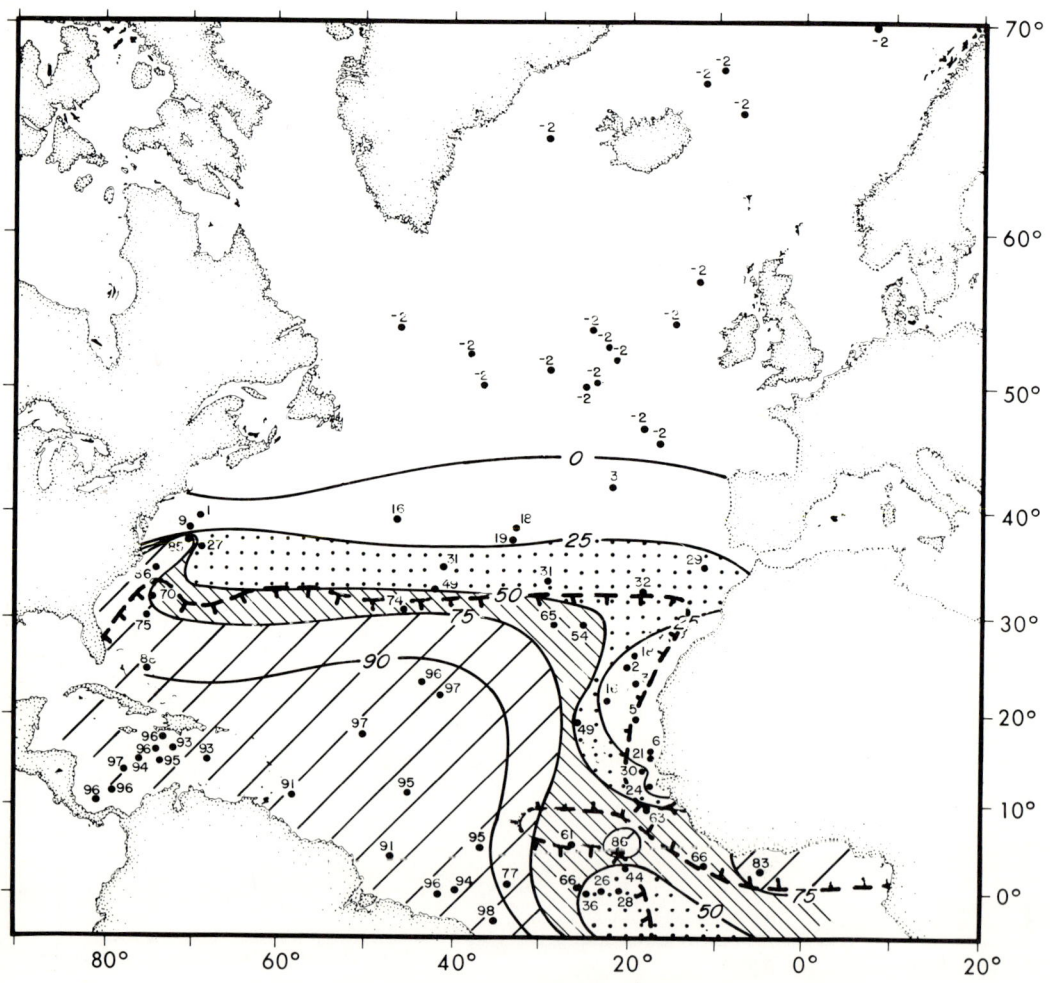

Figure 7. Tropical foraminiferal assemblage distribution at 18,000 B.P.

Map Projection

Although there are more sophisticated projections that would give a visual representation closer to the actual spherical surface of the globe, the Mercator projection is used. The choice is dictated by the fact that our first 18,000 B.P. map is the preparatory stage for global atmospheric modeling experiments. The first model to be used by CLIMAP is the Minz-Arakawa two-level model as modified by Gates (1975), which utilizes a Mercator projection with linear spacing of parallels; thus, we can digitize our parameters directly from the projection for computer input. The Mercator's drawback is that it visually overemphasizes the polar sea north of lat 45°.

Edge of the Sea, 18,000 B.P.

The transposition of water from ocean to continental ice cap is known to have lowered sea level. The maximum continental ice volume and lowest sea-level stand

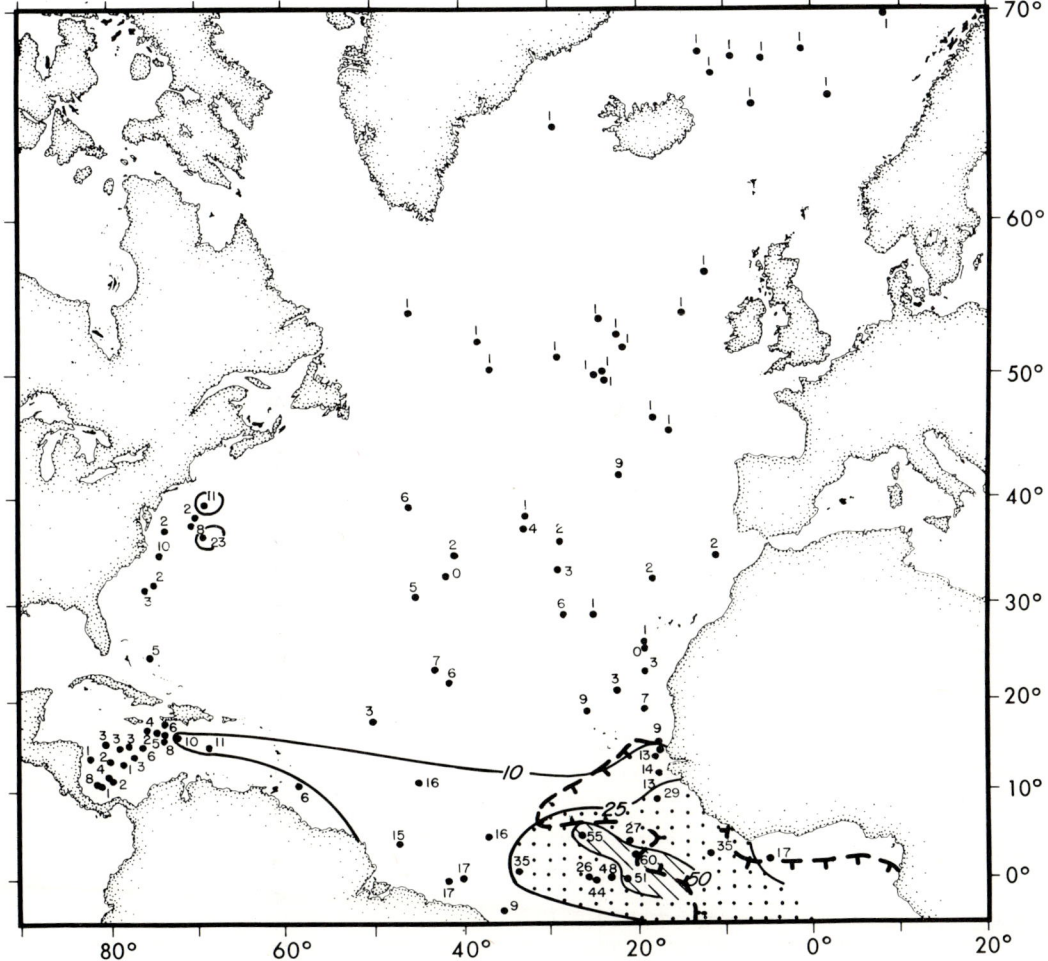

Figure 8. Gyre-margin foraminiferal assemblage distribution at 18,000 B.P.

occurred between 18,000 and 15,000 B.P. (Bloom, 1971). Estimates of the glacial sea level range from less than 80 to over 130 m below the present level (Emery, 1967; Bloom, 1971; Bloom and others, 1974); the consensus appears to be at least 85 m. Because of the topography of most of the Atlantic continental shelf margin and the scale of our map, any variation over 100 m has little effect on geometry, the glacial shoreline, or the total North Atlantic surface area. Thus, where no local studies have been made, the 100-m isobath has been used as the 18,000 B.P. shoreline. Local information can, however, help in delineating the geometry of paleoshorelines. The extent of exposed land is in part delineated by paleoshorelines and iceberg plow marks (Belderson and others, 1972). Similar data has been used to draw in the Faeroe Island ice cap and the northern European and American shorelines (Emery, 1967, Fig. 8; King, 1967; Flint, 1971; Manley, 1951).

North Atlantic Ice Margin

In the construction of the 18,000 B.P. map, one of the most critical features to depict is the sea-ice margin. Clearly, there are three regimens: (1) permanent sea-ice cover, (2) semipermanent sea-ice cover of varying seasonal extents, and (3) no cover. The problem lies in choosing the criteria for distinguishing solid pack ice from transient sea ice and the seasonal location of the margin between sea ice and open waters.

Examination of our data north of lat 42°N indicates that attempts to use foraminiferal barren zones (Kellogg, 1975) are inadequate, because the pattern and distribution at 18,000 B.P. do not present a cohesive picture. Today, even in the Arctic Ocean, Foraminifera are present in surface sediments, albeit in exceedingly low concentrations. The use of ice-rafted sediments is also unsatisfactory, since all samples north of lat 42°N contain them. However, an attempt to define the ice margin must be made, since the limit of the cryosphere is critical to our ultimate goal: a numerical model of the glacial climate. Permanent ice cover would markedly lower organic productivity and possibly detrital sediment input, depending on the duration of the ice cover.

Kellogg (1975) has shown that high planktonic abundance and resulting high

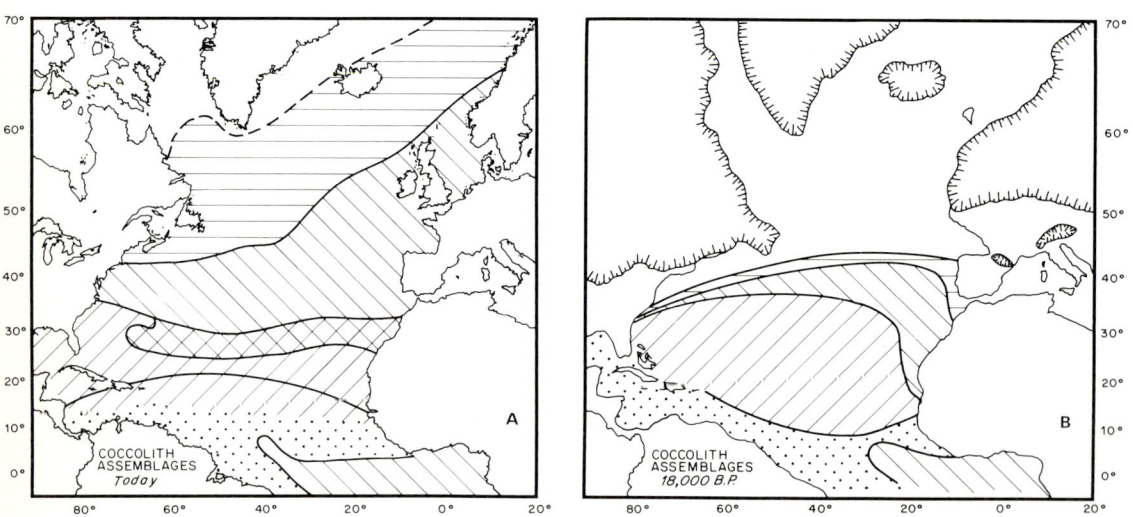

Figure 9. Coccolith assemblages: A, modern surface sediment distribution; B, 18,000 B.P.

carbonate values combined with low values of dextral *Globoquadrina pachyderma* and "warm species" are found today at the location of core V28-25 in the northern Greenland Sea. This core has ice cover nine or more months of the year. Thus, similar sediments in Norwegian Sea cores are considered to represent transient sea ice. Such conditions are found in most samples from the penultimate high-carbonate zone (equivalent to zone 5 of Emiliani or the early glaciation of van der Hammen and others, 1971), excepting only the 5E peak of Emiliani.

Throughout the last low-carbonate period (about 70,000 to 18,000 B.P.) planktonic abundance and carbonate values were low, as were dextral *G. pachyderma* and "warm fauna" percentages. These conditions do not exist in any Norwegian Sea core tops. Therefore, core samples containing these indicators represent conditions more severe than in any North Atlantic core top in the Lamont-Doherty collection— that is, solid year-round pack ice.

Barren zones are prevalent at approximately 18,000 and 70,000 B.P. in the Norwegian Sea cores. These represent times of very low fertility as well as possible sediment dilution due to ice-rafted input. Permanent pack-ice cover, regimen 1, characterizes the Labrador, Greenland, and Norwegian Seas, as well as nearshore areas adjacent to continental glaciers at high latitudes. Because permanent ice cover would lower the sedimentation rate and productivity, the opposite effect might be expected for semipermanent ice cover, regimen 2, in which spring and summer melting would deposit large quantities of detrital material, and there would be higher productivity near the pack ice. This basic model has been used where cores are available to determine the edge of the permanent pack ice. For example, comparison within a series of cores running due south from Greenland shows a marked break in sediment accumulation rate and composition. Core V23-23 (lat 56°4.5′N, long 44°33′W) has a peak glacial sedimentation rate of approximately 1.6 cm/1,000 yr, very low carbonate content, and no coccoliths, whereas to the south, core V27-20 (lat 54°00′N, long 46°12′W) has a peak glacial sedimentation rate of approximately 4.3 cm/1,000 yr, higher carbonate content, and rare coccoliths, as do all cores south of site V27-20. The permanent pack-ice line is, therefore, drawn between these two cores. Similar core series and comparisons have been used to continue the line across the Atlantic.

18,000 B.P. Surface-Water Isotherm Maps

Man tends to characterize climate, weather, or the condition of any natural phenomena by its extremes and a mean: for example, glaciation versus interglaciation, summer versus winter. However, man is directly involved with climatics in terms that require a fuller knowledge of seasonality. The length of the growing season, duration of winter in higher latitudes, duration and areal extent of sea ice and continental snow cover, and more are controlling factors in the evolution of civilization. Since the long-range goal of CLIMAP is to help predict future climate, we must attempt to reconstruct climatic patterns for all seasons. The four maps chosen to portray 18,000 B.P. represent the seasons: February T_w (winter temperature), May T_{sp} (spring temperature), August T_s (summer temperature), and November T_f (fall temperature).

Sea-surface temperatures for these four maps were estimated by transfer-function analysis of foraminiferal assemblages. These values were plotted on the 18,000 B.P. base map, and the results were examined to determine if a cohesive pattern was present. Contours were then drawn and the maps compared with the surface-water geometric distributions of both foraminiferal and coccolith assemblages.

Before applying the transfer-function equations to the 18,000 B.P. samples, it was necessary to determine the average position of the caloric or thermal equator. The thermal equator is defined as the line that bisects the zone of maximum surface temperature for whatever month or season is being mapped. The modern position of the Atlantic's thermal equator varies seasonally from a winter position coincident with lat 1°N to its summer position, where it reaches from lat 10°N off South America to almost lat 18°N along the African coast. This displacement of the thermal equator north of the geographic Equator is due to the northward transport of cool water by the Benguela Current. This effect is most pronounced during Southern Hemisphere winter (northern summer).

The Coccolithophorida species *Gephyrocapsa oceanica* has its maximum concentration coincident with the warmest surface waters in the world ocean. A line that bisects the maximum concentration plot of *G. oceanica* at 18,000 B.P. is our best estimate of the mean position of the thermal equator. Although there is no foraminifera analogous to *G. oceanica*, there are two foraminiferal assemblages that bracket this warmest surface water mass in the Atlantic (Gardner and Hays, this volume). An equipotential line between these two assemblages is correlative with the *G. oceanica* line, thus fixing the position of the 18,000 B.P. thermal equator. All samples south of this equator must be treated for Southern Hemisphere seasonality by reversing T_s for T_w and T_{sp} for T_f values. Six cores—code numbers 19, 20, 21, 22, 55, and 82—were corrected in this manner.

The 18,000 B.P. maps of surface-water temperature are our interpretations, using transfer-function techniques and value judgments, based on comparisons between modern and glacial biogeography. It is, therefore, open to question both on the judgments made by the authors in those areas where data were minimal and to the fact that transfer-function analysis yields a value of at least ±1-1/2°C. In examining the maps, however, we wish to stress that it is the pattern that is the important feature.

Seasonal Extremes

February 18,000 B.P. The North Atlantic surface thermal patterns during the glacial maximum show some similarity to the Southern Ocean today (Fig. 10). The dominant feature was a narrow steep thermal gradient nearly coincident with lat 42°N. Its northern boundary was the glacial polar front. North of this front, surface temperatures were below 2°C. There are no thermal indications of the North Atlantic Drift or the Norwegian or Irminger Currents.

South of this thermal slope the pattern is similar to the modern ocean. The isotherms were deflected southward along the African coast both by drift of cooler water toward the Equator and upwelling along the coast. Gardner and Hays (this volume) have inferred that the upwelling was stronger during the glaciation than at present, and this is mirrored in the contour pattern. On the western side of the basin, warm water moved north along the coast but was deflected sharply eastward off Cape Hatteras, presumably by both continental topography and the glacial westerlies.

August 18,000 B.P. The glacial summer at 18,000 B.P. retained the winter geometric pattern with the polar front and attendant steep thermal gradient in approximately the same positions (Fig. 11)—a marked change from today (Fig. 12). However, along the European continent there is an indication of some slight northward advective flow of warmer water at least to lat 45°N, since surface temperatures north of the front in this area were higher than in winter. Unfortunately, since transfer functions for the diatoms and Radiolaria have not as yet been completed for the North Atlantic at 18,000 B.P., our discrimination of isotherms north of the polar

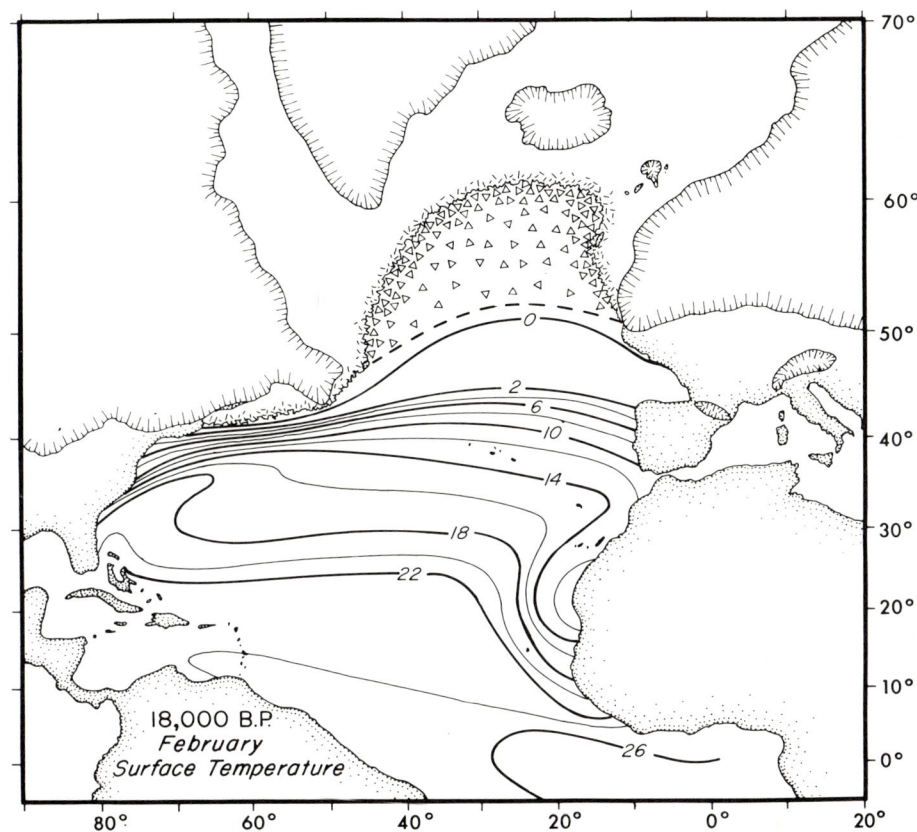

Figure 10. Surface-water isotherm map for February 18,000 B.P. in degrees Celsius. Contour interval is 2°C. Dashed isotherms are interpretive. Major continental ice masses are delineated by hachured borders, permanent pack ice by granulate borders, loose pack ice by triangles. Glacial shoreline is drawn using present bathymetry to a lowered sea level of 100 m.

front is poor. The input for this first map comes from Foraminifera factor I and an examination of the sedimentary data, both organic and inorganic. The dotted isotherms represent our best estimate, based on this imprecise data and analogy with similar areas near ice margins today. The position of the dotted isotherms, particularly the bulge eastward to the northeast of the emergent Grand Banks, is derived in part by analogy with data from Ruddiman and Glover (1975) that show the presence of an eastward-protruding low-temperature lobe in a counterclockwise gyre in the subpolar sea at 9300 B.P.

Ruddiman and Glover (1975) noted that 9300 B.P. was a time when the subpolar gyre was displaced roughly 3° to the south of its present position and probably to the east as well. The oceanic gyre and Icelandic low were compressed in such a way that the cellular nature in effect was less marked than it is today in the lower atmosphere. Instead, there was a greater tendency toward zonal flow in a broadened belt of relatively weak easterlies and a narrowed and southward displaced belt of strong westerlies. This tendency is much more evident at 18,000 B.P. with the disappearance of the cellular Icelandic low (Gates, 1975). At 18,000 B.P. an enlarged belt of weak surface-water flow to the southwest and a narrow "return" flow to the east along the polar front dominated the subpolar ocean. Because of the continental margins existing as oceanic flow obstructions, this flow had more of a cell-like or "gyre" form than did the lower air flow; yet it was only in a limited sense a cyclonic gyre.

The part of Rockall Bank emergent at 18,000 B.P. was not ice covered (D. Roberts, 1974, oral commun.). Lack of ice on this oceanic island indicates that for at least part of the year, sea-surface temperatures were above freezing. This single point of information corroborates our placement of the 2°C summer isotherm in the east.

South of the polar front the August pattern, like January 18,000 B.P., was similar to that of today except for the area south of the 18,000 B.P. thermal equator. Here, increased flow of the Benguela Current, coupled with increased upwelling along the divergence, caused a greater development of lower surface-water temperatures extending toward South America.

Transitional Seasons

May 18,000 B.P. The May T_{sp} values yield a geometric distribution similar to those already discussed (Fig. 13). The polar frontal system was present with slight surface warming in the southeastern part of the polar gyre. Temperatures were cooler than today except for the eastern part of the subtropical gyre, where a pod of water, centered on lat 21°N, long 35°W, was some 2°C warmer than today.

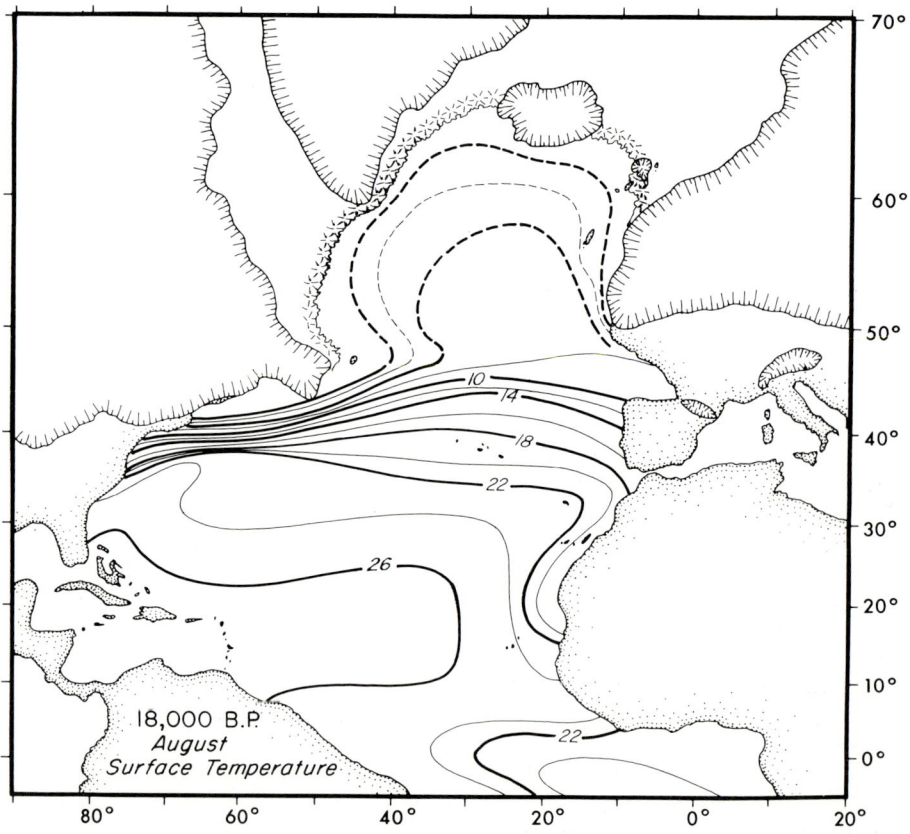

Figure 11. Surface-water isotherm map for August 18,000 B.P. See Figure 10 legend for explanation.

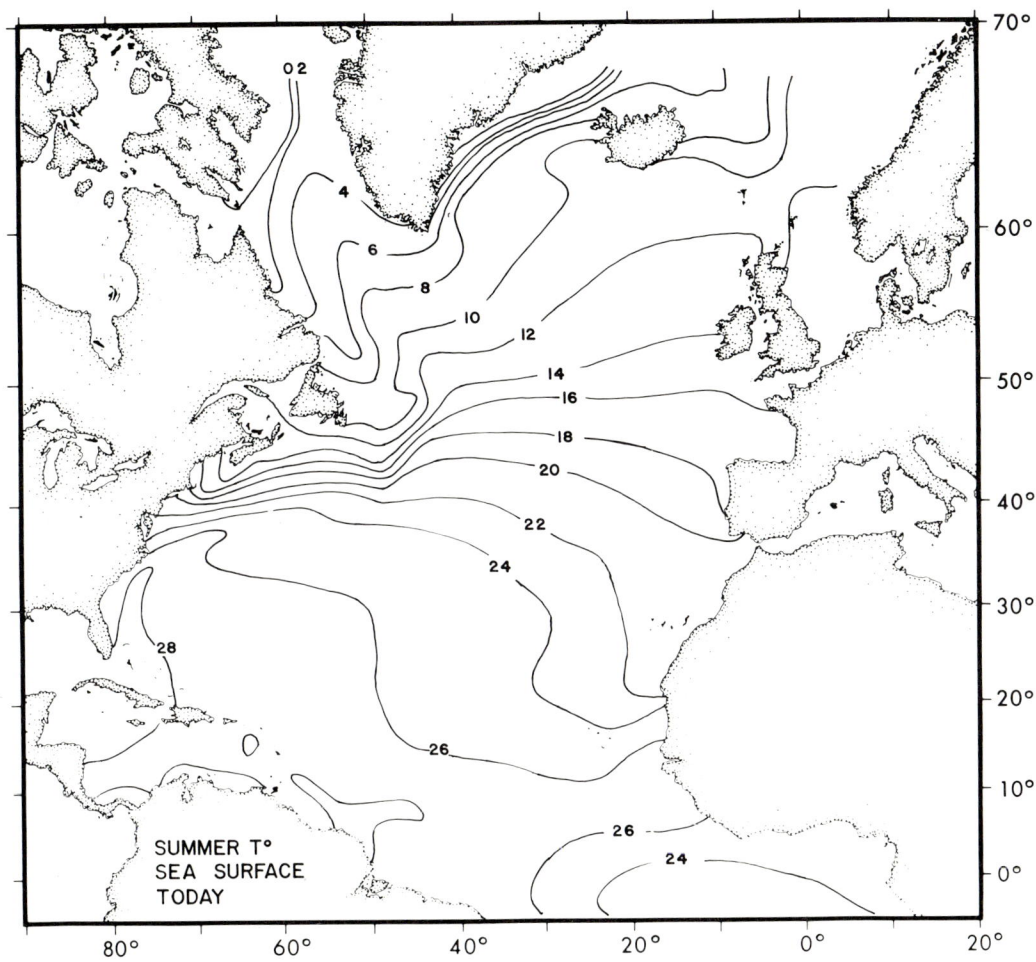

Figure 12. Surface-water isotherm map for August today.

Generally, the isothermal structure south of lat 40°N resembled today's March conditions. In the upwelling region off the west coast of Africa, temperatures were still cold, differing only slightly from February 18,000 B.P.

November 18,000 B.P. South of the polar front the 18,000 B.P. autumn in the North Atlantic was the season that most closely resembled its counterpart today (Fig. 14). The only exception is in the southeast where the glacial Benguela Current–equatorial divergence continued to pump cooler water into the South Equatorial Current. The 26°C isotherm, instead of continuing eastward along lat 12°N to touch Africa, bent south and then recurved at long 30°E. The result was an east to west temperature gradient between long 30°W and 18°W, for which no analogue exists today. Finally, the subtropical gyre was again slightly warmer than today in the eastern part of the gyre.

Salinity at 18,000 B.P.

If we are to produce valid reconstructions of past oceanic conditions, salinity values must also be determined. Unfortunately, it appears that our salinity and temperature estimates are not independent of our temperature estimates. This is so because salinity in the oceans is interdependent on solar energy and atmospheric conditions that control surface-water temperature via evaporation and precipitation. Consequently, the modern temperature-salinity calibration data set has a built-in interdependence. Yet the surface-water planktonic organisms theoretically are adapted to specific temperature and salinity ranges and to nutrients and other abiotic and biotic parameters that are poorly understood. Thus, the biogeography of single species, and perhaps entire assemblages, may not be primarily dependent on the parameters we wish to derive from them. The Imbrie-Kipp (1971) method obviates this latter problem by directly calibrating microfossil data against physical parameters. This method does not separate the interdependence of salinity and temperature in the ocean system. Its use in paleoceanography simply assumes that this interdependence does not change significantly with time.

The surface waters are the least consistent in salinity values. Examination of salinity variation at 0 and 100 m from NODC data tapes indicates that at 0 m, salinity has a mean variation 0.1‰ greater than at 100 m. It is not surprising that our derived salinity values at 100 m yield more cohesive patterns than surface salinities. Thus, we have mapped only the salinity at 100 m for 18,000 B.P. (Fig. 15). There are no values for waters of polar character, since coccoliths are absent and foraminifers are present as a monospecific fauna.

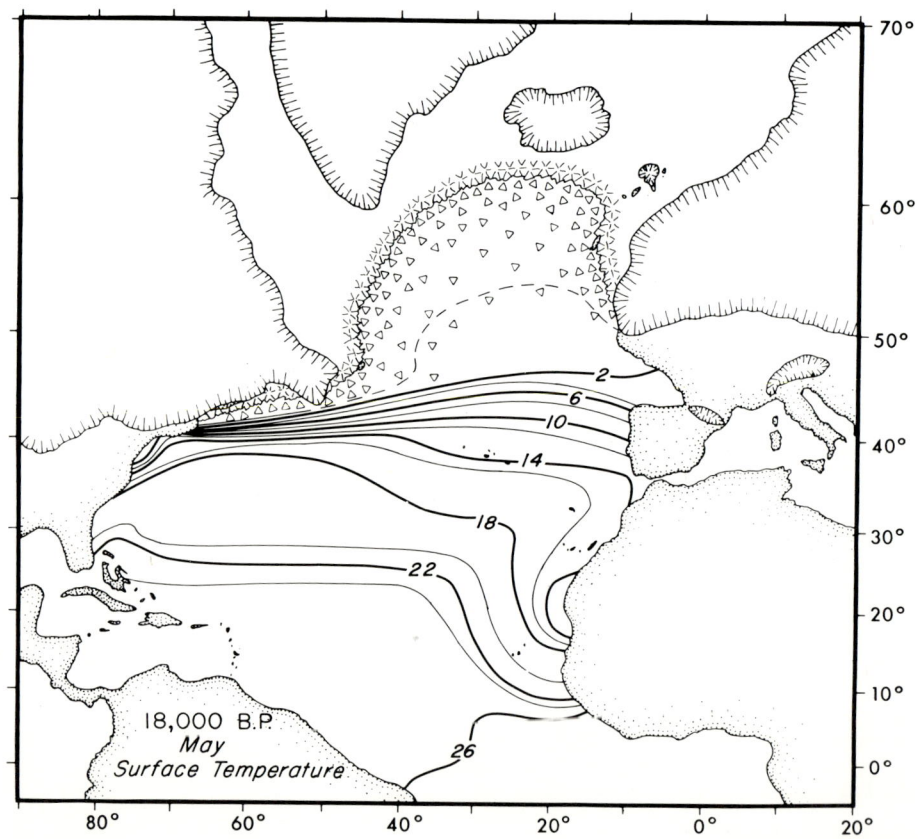

Figure 13. Surface-water isotherm map for May 18,000 B.P. See Figure 10 legend for explanation.

Compared to today, the pattern shows a marked southwest shift of the salinity high of the central water mass (Fig. 15; note 36.5‰ isohaline). The center has moved from approximately lat 27°N, long 53°W at 18,000 B.P. to lat 28°N, long 35°W today. The areal extent of the Atlantic's salinity high (or the waters encompassed by the 36.5‰ isohaline) was some 12% less than it is today. However, Prell and Hays (this volume) showed that at 18,000 B.P., the Caribbean Sea had surface waters like the central Sargasso Sea today. If this is indicative of salinity as well as temperature, then the high salinity area was actually increased by roughly 8%.

DISCUSSION

18,000 B.P. Circulation and Water Masses

The geometrical distribution of water masses and current systems undergoes a major change between glacial and interglacial modes. Basically, the glacial North Atlantic can be divided into two areas at approximately lat 42°N—the southern one stable, the northern one not. South of lat 42°N there were minor changes in surface temperature and slight shifts in the salinity high of the North Atlantic

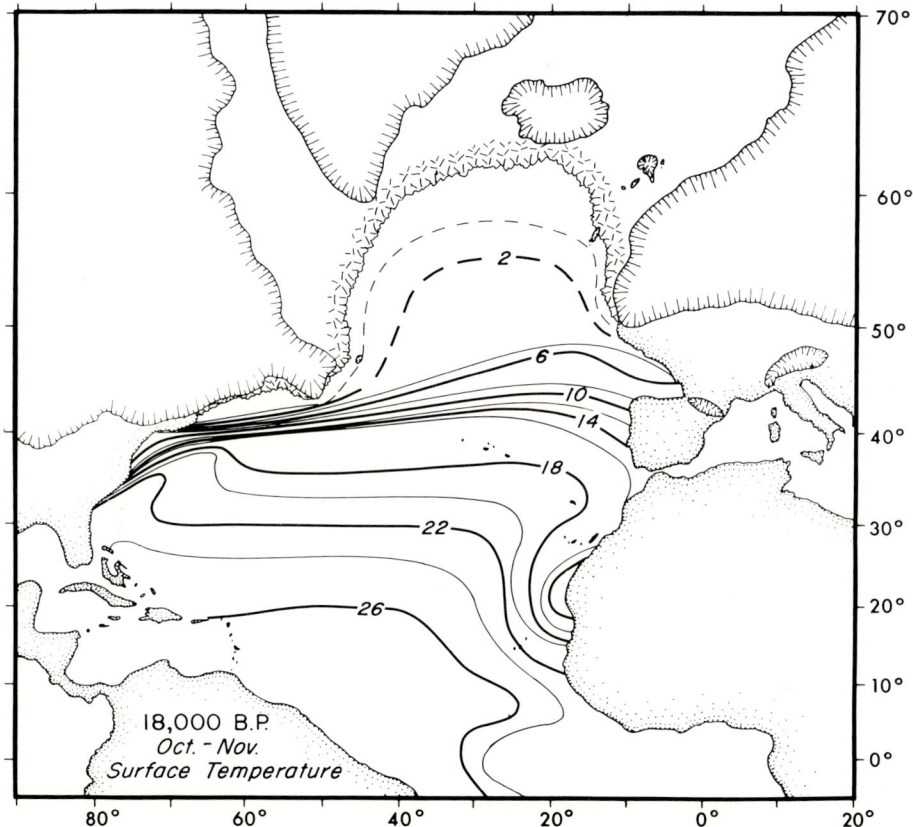

Figure 14. Surface-water isotherm map for October–November 18,000 B.P. See Figure 10 legend for explanation.

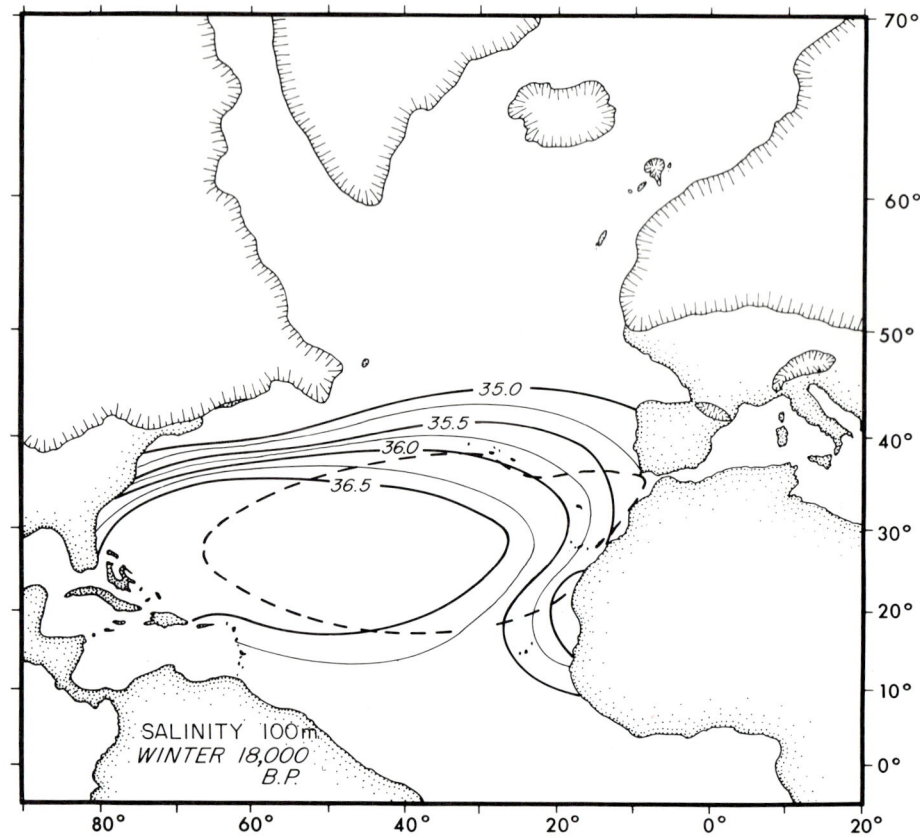

Figure 15. One-hundred-metre isohaline map for February 18,000 B.P. in parts per mil. Contour interval is 0.25‰. Position of the modern 36.5‰ isohaline is denoted by a dashed line for comparison purposes.

gyre, presumably with attendant changes in the surface currents, but the pattern remained basically unaltered from today. North of lat 42°N there was a complete change. This is shown by comparing the surface isotherm maps of today with those of 18,000 B.P.

Surface Currents

The microfossil assemblages and thermal patterns derived from the surface isotherm maps give us an indirect first approximation of 18,000-B.P. surface currents. There are, however, some direct indicators of paleocurrent direction. North of lat 42°N, in the small basin that was bordered by continental glaciers and solid pack ice, a counterclockwise current existed. This was determined by tracing the transport of ash carried by sea ice (Ruddiman and Glover, 1972a, 1972b; Ruddiman and McIntyre, 1973). While this evidence comes from an ash layer dated at 9300 B.P., examination of lithologies throughout the glacial interval carries with it the same message. Glacial sea-ice transport was controlled by the counterclockwise current of the 18,000-B.P. polar sea. The sediment beneath the southern limb of the gyre where it touched warmer waters shows increasing terrigenous input from west to east, presumably due to melting (Ruddiman and McIntyre, this volume).

This counterclockwise current was not different in sense from that of the present. In our modern interglacial mode, water from the North Atlantic Drift moves northeastward into the Norwegian Sea. This surface water both mixes with Arctic water and is modified by atmospheric interchange and is then carried south along the east coast of Greenland as part of the East Greenland Current. In addition, a seasonally variable gyre separates from the North Atlantic Drift and moves westward to the south of Iceland as the Irminger Drift. These two currents combine off the tip of Greenland, turn north into the Labrador Sea, move through the basin, and are then deflected eastward by the Newfoundland Grand Banks to mix with the North Atlantic Drift. At 18,000 B.P., this irregular counterclockwise gyre was simplified by being latitudinally compressed and longitudinally expanded into a circular current system. South of lat 42°N at 18,000 B.P., the Gulf Stream–North Atlantic Drift flowed from Cape Hatteras eastward across the ocean to impinge on Europe at approximately lat 42°N, then bent southward to form a glacial Canaries Current. The strong southward deflection of the isotherms could indicate either intensified transport of cold glacial surface waters or a marked increase in upwelling off Africa (Gardner and Hays, this volume), which would tend to extend the isotherms toward the Equator. Finally, the water turned west to complete the central gyre of the glacial North Atlantic. The most interesting feature south of lat 42°N is the increase in cold surface water due to both increased equatorial divergence and transport from the Southern Ocean into the North Atlantic by the glacial Benguela Current. This incursion of Southern Hemisphere winter waters into Northern Hemisphere summer waters displaced the thermal equator north to at least lat 12°N. While this condition is similar to that of today, the isotherms extended farther westward, suggesting greater mass transport during the glacial period.

We have little indication of oceanic conditions in nearshore waters, particularly in the area south of New England and Newfoundland. This appears to have been an area of dynamic seasonal change. Autumn and winter were dominated by sea ice, spring and summer by low-salinity surface waters that were displaced in the late summer by warmer, higher salinity waters. Evidence for this can be seen in the high detrital input and the odd biota, which contains subtropical as well as cool-water Foraminifera, while Coccolithophorida are rare to absent. Because most Coccolithophorida are intolerant of low salinities, we ascribe their near absence to low-salinity surface waters during the spring, the time of phytoplankton bloom. (Of the oceanic Coccolithophorida, only *Emiliania huxleyi* shows tolerance for lower salinities, and even this species is rare in these sediments.)

Surface Water Masses

Today, the North Atlantic is characterized by the dominance of the mixed water body known as the transitional surface water mass. This is a triangle-shaped unit with its apex at the Grand Banks and its base along the coast of northern Europe. It is formed by the mixing of warm North Atlantic Drift waters with cold subpolar waters from the counterclockwise gyre. In addition, there is today a well-developed subpolar water mass between the transitional and Arctic water masses. At 18,000 B.P., both subpolar and transitional water masses were compressed into a narrow band parallel to and centered on lat 40°N. North of this, the surface waters were polar in character. There was a limited, yet definite, drift of warmer, more saline waters northward off the European coast into the polar system during the summer season. Examination of both raw species data and assemblages indicates that significant amounts of subtropical to subpolar faunas did not extend north of lat 50°N. The North Atlantic gyre occupied roughly the same position as it does today.

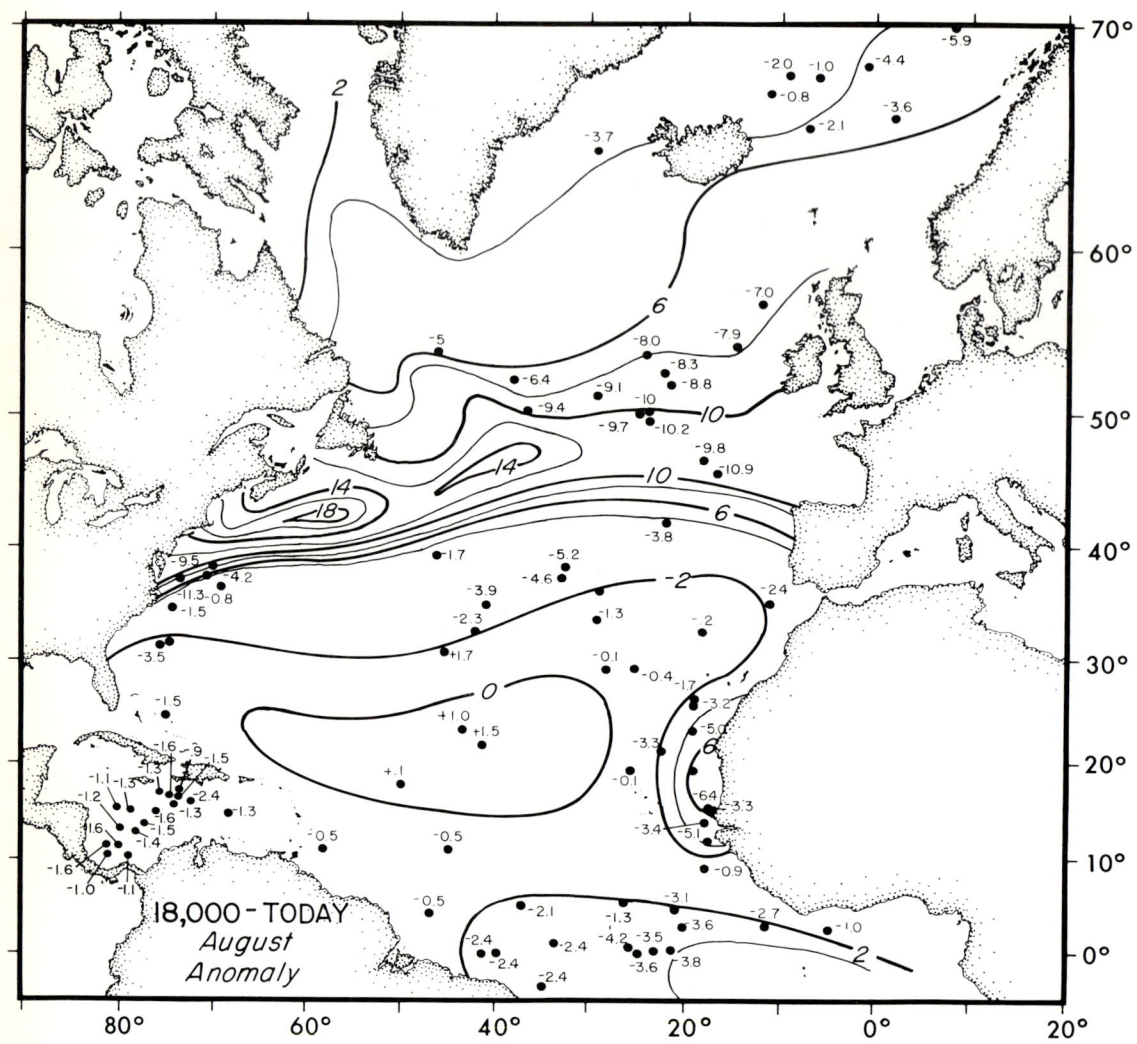

Figure 16. Temperature-anomaly map for August in degrees Celsius. Values were derived by subtracting average modern surface-water temperatures from 18,000 B.P. values (see text for details). Contour interval is 2°C.

Its northern boundary was a few degrees south of its present position, with the center translated westward as shown by the salinity plots. Fossil evidence in the cores indicates that the temperature and salinity characteristics of the central water mass (Sargasso Sea) were also found in the Caribbean 18,000 yr ago (Prell and Hays, this volume). In the tropical region, increased upwelling off the African coast and intensification of the Benguela Current, combined with the drop in world temperature, produced cooler and thinner tropical water masses than found today.

**Temperature Differences:
Glacial Versus Interglacial Periods**

CLIMAP has prepared temperature-anomaly maps for maximum winter (February) and summer (August) temperatures: 18,000 B.P. minus today's temperature (Figs.

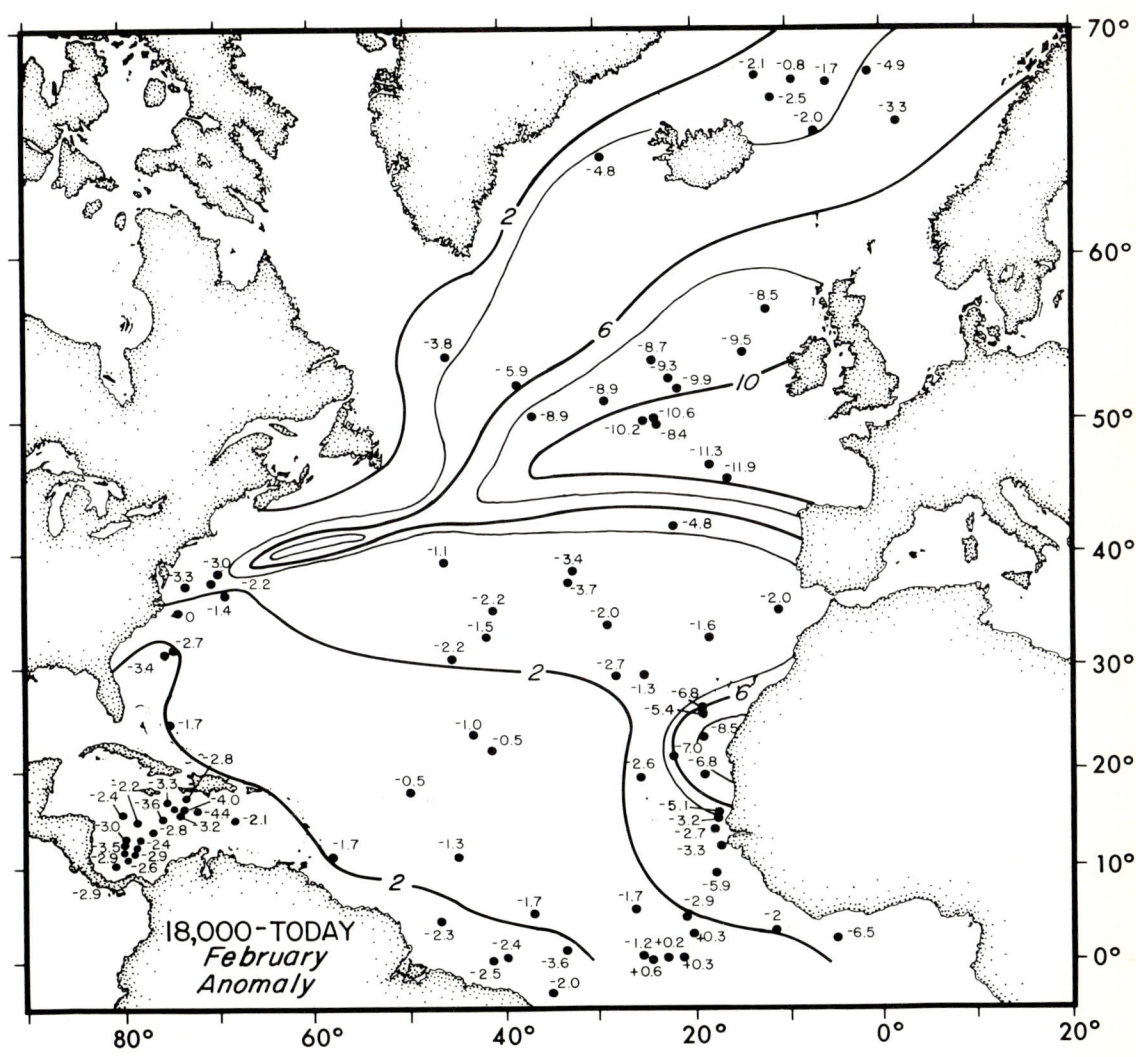

Figure 17. Temperature-anomaly map for February in degrees Celsius. See Figure 16 legend for explanation.

16, 17). These were made by overlaying the 18,000 B.P. isotherm map with core data points on a modern average-surface-temperature map for the appropriate month. The points used to draw the anomaly maps were obtained from both the 18,000 B.P. data points and intersections of 18,000 B.P. and modern isotherms. This latter technique was necessary to continue the pattern across areas lacking core coverage. This allows us to answer the question, How different from today was the last glaciation in degrees Celsius at any point in the Atlantic?

August Anomaly

The greatest variations in temperature from today are found during the Northern Hemisphere summer. The eastern basin shows the widest areal change, while the western basin has the highest absolute difference. A belt of maximum anomaly runs from approximately lat 40°N on the west to lat 50°N on the east, and it abuts the highly populated temperate regions of North America and Europe. The greatest difference from present time, 18°C, is found southeast of Nova Scotia. However, because this is an area without core coverage, it may be an artifact of our technique outlined above. Off the Bay of Biscay and southern England, the difference is 10°C. In the southern part of the subtropical gyre, there is a small positive anomaly; however, the reality of this anomaly is uncertain, since it is within the precision of the transfer functions (± 1.5°C).

The contours have been drawn to best delineate anomalies; thus, the few values that diverged from the general pattern were ignored for this map. Off the coast of Africa, a 6°C anomaly centered on lat 19°N depicts an increased rate of upwelling at 18,000 B.P. South of this, in the area from the Equator to lat 6°N, the increased equatorial divergence and Southern Hemisphere summer Benguela flow produce a small anomaly that reaches to long 43°W.

February Anomaly

The Northern Hemisphere winter temperature anomaly (Fig. 17) has a different pattern from that of summer. The greatest difference occurs in that part of the eastern basin that is presently covered by transitional waters, with a maximum difference off southern England of 11°C. In the western basin, the maximum change is again found southeast of Nova Scotia, but the anomaly covers a smaller area and the maximum is only 8°C. There are no positive anomalies. The upwelling off Africa is more intense, yielding a drop of 8°C. The effect of the equatorial divergence and Benguela flow is minimal in the Northern Hemisphere winter, as it is today, and the anomaly noted on the summer map is absent.

Comparing the two maps, it is clear that there is a marked change in these opposing seasons. The greatest variation in summer is in the western basin, whereas for the winter it is in the eastern basin. The summer band of maximum anomaly is equidimensional across the North Atlantic, while the winter anomaly forms a triangle with its apex in the pod off Nova Scotia and its base in the east, reaching from northern Norway to southern Portugal. The variation in the eastern basin from lat 40°N to 60°N remains essentially the same for both seasons, approximately 10°C, whereas off the North American coast, there is a difference between winter and summer anomalies of 10°C. In short, the glacial summers off North America were much colder than today, but the winter temperatures were almost the same. At Cape Cod, for example, the winter temperature was only 2°C colder, and the summer was 16°C colder than today.

Comparison with Continental Vegetation

The maps constructed by the CLIMAP researchers should be capable of being validated by other parameters, both organic and inorganic. The data for diatoms, radiolarians, and silicoflagellates in oceanic sediments will be used, when completed, for comparison with as well as augmentation of the paleotemperature and salinity data. Distribution of ice-rafted materials has already aided in the delineation of surface water masses and currents (Ruddiman and Glover, 1972a; Ruddiman and McIntyre, 1973, this volume). Other mineralogic tracers will soon be available.

An alternate source of evidence is present in the vegetational patterns described by palynologists from continental areas bordering the North Atlantic. A comparison of these patterns from the coastal regions of the United States (late Wisconsin) and Europe (Weichsel-Wurm) shows a striking correlation with our 18,000 B.P.

maps (Fig. 18). In North America, the boundary between temperate and boreal vegetation (Flint, 1971, Figs. 19 to 24) occurred at the same latitude as the North Atlantic Polar Front. Along the European coast the polar front abutted a thin band of boreal forest in northern Spain (Flint, 1971, Figs. 23 to 28). North of this point, tundra extended to the edge of the glaciers.

Today in Europe, tundra abuts polar waters, boreal meets subpolar waters, and mixed boreal-temperate to temperate border transitional waters (Flint, 1971, Figs. 23 to 27). Today, the dominant vegetational forms from southern Scandinavia to northern Spain are mixed boreal and temperate. At 18,000 B.P. these zones were compressed into two narrow bands in Portugal and Spain. This is directly analogous to the squeezing of the subpolar and transitional water masses between the polar gyre north of lat 45°N and the subtropical gyre to the south. Thus, a climatic continental analogue to our oceanic geometries not only exists but corroborates the pattern we have described.

Finally, one of the characteristics of land adjacent to glaciated areas is the formation of frost cracks in soils. These features are the result of permafrost. In Europe, this structure characterizes the area from southern France into central England and Ireland (Flint, 1971, Figs. 10 to 15). Pewe (1966) showed that the mean annual temperature of areas in which these frost cracks form is −6° to −8°C. Flint (1971, p. 280), using this criterion, noted that "glacial-age mean temperatures were lower than today's means by as much as 15 to 18° in western Europe"; this is slightly higher than our maximum anomalies of 10°C off the European coast.

Temperature Estimates

The 18,000 B.P. isotherm and isohaline maps are reconstructions of a thin layer of water, sitting on a much larger mass. In fact, we have described only a very

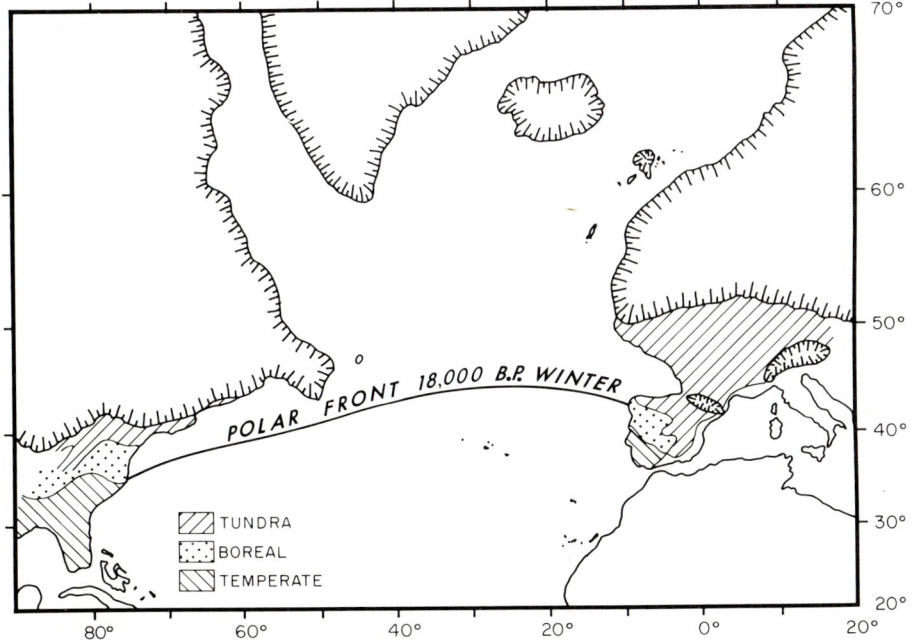

Figure 18. Continental glacial vegetational zones compared with 18,000 B.P. polar front.

small part of the entire ocean-atmosphere system. The results of our research do produce a cohesive picture of the distribution of surface water masses at a finite time in the past. Although the actual temperature at any one particular spot has an uncertainty of 1.5°C, this does no real injury to the reconstruction. In our calibration of the microfossil assemblages, we have used the mean surface-water temperature map for the particular month that we wished to portray. In fact, the temperatures chosen have a natural range over a period of years that averages out to about 4.4°C, or roughly ±2.2°C. For example, the actual ranges along long 20°W are 4.4°C at lat 60°N, 4.4°C at lat 45°N, 4.2°C at lat 30°N, and 4.5°C at lat 15°N (U.S. Naval Oceanographic Office, 1967). Kipp (this volume) has determined the accuracy of temperature estimates from the transfer-function analysis, which is described as a standard error of estimate of ±1.5° to ±1.9°C. This is actually the 80% confidence interval. This error is very close to the observed range of variation in surface-water temperature today. Thus, we can say that our estimates are as accurate for 18,000 B.P. as the observed values and mean-temperature maps are for today.

Subsurface Water Masses

The maps are a basis for further research, either for atmospheric studies or for reconstruction of the three-dimensional oceanography of the North Atlantic Ocean. The former is already being done. CLIMAP has joined with L. Gates of the Rand Corporation to reconstruct the atmospheric structure at 18,000 B.P. using the Minz-Arakawa two-level model (Gates, 1976).

The latter goal of producing a snapshot in time of the entire structure of an ocean basin will take longer, because at the time of writing, there was no statistical ocean model ready to handle our data. Nor have we fully developed techniques to obtain temperatures and salinities for the deep water masses, although Streeter (1973) has pointed the way. Nevertheless, we can extrapolate and make some comments about the possible deep-water structure and its genesis at 18,000 B.P. This following discussion is concerned more with future lines of research and ideas than proven structure.

The North Atlantic is the only Northern Hemisphere ocean that contributes significant deep water to the world ocean. Today, most of the North Atlantic Deep Water is produced in the Norwegian Basin (Worthington and Wright, 1970) by the northward flow of warm saline waters, which are cooled, slightly diluted, and then sink to form the densest water mass in the Atlantic (Reid, 1971). Some of the water passes over the Greenland-Scotland Ridge and in the process mixes with intermediate waters of lower density; it then flows south along the North American continental margin. The production of large volumes of deep water requires a continuous input of saline surface waters as occurs today around parts of Antarctica (the Weddell Sea) and in the Norwegian Sea. If this condition does not exist, then autumn and winter sea-ice growth will occur rapidly without attendant thermohaline flow. The present sources of North Atlantic Deep Water (Norwegian and Labrador Seas) were covered at 18,000 B.P. by permanent pack ice. The polar North Atlantic from lat 60°N to 45°N was surrounded on three sides by glaciers and pack ice. Circulation was dominantly internal, which precludes large surface-water exchange with warmer, more saline water south of the polar front. Under these conditions, low surface salinities would prevail and, during the autumn-winter interval, rapid pack-ice formation would occur without the formation of large volumes of deep water. Examination of all microfossil data available does not show significant incursion northward of surface waters into the polar gyre, except for the area between lat 42°N and 45°N off the European coast.

There was no analogue to today's massive northward transport of surface waters in the North Atlantic Drift and Irminger Current that is necessary for the formation of North Atlantic Deep Water. Yet, there is strong evidence that the North Atlantic has been an important source of subsurface water throughout the climatic cycle. There was no turning off during the glacial maximum of deep-water production; however, the type of deep water produced was different. Streeter (1973), in a study of North Atlantic benthic foraminifers, has shown that during the maximum glaciation, Atlantic Deep Water was approximately 2°C warmer. This is in agreement with the views of Weyl (1968) and Newell (1974), who, in their models of the glacial ocean, hypothesized less vigorous thermohaline circulation and warmer deep waters. There are two possible mechanisms for North Atlantic Deep Water formation at 18,000 B.P.: (1) the entire polar gyre could be the source area, with surface waters receiving some input by advection from the south during summer, followed by rapid autumn-winter cooling, which produced open pack and sea ice and yielded a diffuse but areally wide scale production of deep water slightly warmer than today's; (2) discrete areas within the southern part of the polar seas could have been the source; in these areas, large amounts of warmer, more saline southern waters could replace cooler, more dense units injected into thermohaline flow by autumn-winter cooling. The two areas that best fit this latter category were the nearshore waters from Spain to Rockall Bank along Europe and the area from Cape Cod to the Grand Banks. In a sense, these were North Atlantic equivalents of the present Weddell Sea; however, they are some 20° of latitude closer to the Equator.

North Atlantic Arctic Intermediate Water

Modern production of intermediate water masses in the North Atlantic is restricted by basinal geometry and existing current patterns. Only small amounts are formed in the northwestern Atlantic by atmospheric cooling of subantarctic waters. Sverdrup and others (1942) described these waters as having a temperature of 3° to 5°C and salinity of 34.7 to 34.9‰. In fact, it is a volumetrically insignificant and poorly defined unit in today's oceans. Reconstructions of North Alantic surface water masses and isotherms for 18,000 B.P. produce a polar frontal system perhaps similar to that of the Southern Ocean today. Thus, the North Atlantic, in a band from Cape Hatteras to northern Spain, could have been an important zone of Arctic Intermediate Water formation during the maximum glacial period. Although we cannot give an absolute measure of the increased production of Arctic Intermediate Water from this area at 18,000 B.P. until more parameters are measured, we can say on the basis of a purely areal comparison that intermediate-water production could have increased by at least 16%.

It appears from the reconstructions by Lozano and Hays (this volume) that Antarctic Intermediate Water formation remained the same or increased somewhat, at 18,000 B.P. The glacial world ocean could thus have been enriched in intermediate water over the nonglacial ocean of today.

Northern Border of the Subtropical Gyre

Along the path of the Gulf Stream and North Atlantic Drift, two processes active today have produced the wide mixing zone evident in the surface-sediment record. First, the Gulf Stream's position shifts seasonally (Stommel, 1965). From April to November, when North Atlantic winds are weak and the atmospheric high-pressure cell centered over the North Atlantic gyre is at its most southwestward position, the Gulf Stream's mean path is displaced northward. From December to May, when North Atlantic winds are strongest and the atmospheric high has moved to a more northeastward position, the Gulf Stream returns to its more southern path. Second, exchange of water masses across the northwestern margin of the subtropical gyre occurs when meanders of considerable size detach from the Gulf Stream to form eddies in the region from long 68°W to 58°W. Both cold-water cyclonic gyres and warm-water anticyclonic gyres are produced at the rate of 5 to 8 of each type per year (Fuglister, 1972). At 18,000 B.P., since (1) gradational boundaries were nearly absent between assemblage areas and (2) the structure of the North Atlantic was materially altered by the appearance of a strong polar frontal system running nearly parallel to latitude, we presume that this type of wide-scale mixing through eddy formation was absent or severely suppressed.

There is no disagreement with the comment by Stommel (1965, p. 201) "that the interior of the subtropical and equatorial Atlantic Ocean is essentially wind-driven, from a dynamic point of view." The subtropical gyre is, in a sense, driven by two wind systems: its northern part by the westerlies blowing off North America and its southern part by the trade winds blowing from Africa.

During the maximum glacial period, eastern North America north of lat 40°N and Europe north of lat 50°N were covered by thick ice caps. These produced their own indigenous weather patterns—a vast southward extension of the polar wind systems. The presumption, shared by most meteorologists, is that during the maximum glaciation there was intensification of the westerlies due to the compression of the climatic belt. Lamb (1971, Fig. 5) has produced a reconstruction of the atmosphere over the Northern Hemisphere for the last maximum glaciation in which the structure of atmospheric pressure indicates the possibility of a channeled and increased flow of westerlies from North America across the North Atlantic to Europe. In addition, he pointed out that there was a greater predominance of northerly surface-wind flow east of Greenland over all the higher latitudes of the eastern Atlantic. If this latter idea is correct, then there was a wind vector pushing polar surface waters farther to the south than at present. Two important points to come out of Lamb's paper that bear upon the question of our reconstruction are as follows: (1) the movement southward of the surface polar anticyclone, which would squeeze the westerly belt to the south, and (2) less seasonal change in vigor of mean circulation than occurs at present. The former would intensify the energy put into the northern boundary of the subtropical gyre, and the latter would lessen the effect that we see today of the migration of the atmospheric high that drives the subtropical gyre. Thus, we would be left with a subtropical gyre that was far more stable in its position and in which mass transport was intensified by increasing wind stress. Iselin (1940) noted that increased transport of the Gulf Stream and the attendant subtropical gyre would lower the thermocline in the Sargasso Sea and radically shrink the entire system, which would pull the present Gulf Stream–North Atlantic Drift away from its northeastward course. Stommel (1965) also pointed out that increasing the energy input into the subtropical gyre would cause it to tighten up, with an attendant decrease in the formation of eddies along its northwestern boundary. There is evidence from work done by CLIMAP members that this picture of increased westerly effect is true. Ruddiman and Glover (1975) stress this point in their discussion of subpolar currents as recorded in North Atlantic sediments of glacial and early postglacial age.

CONCLUSIONS: WATER-MASS CHANGE

The North Atlantic Ocean appears to be the most variable of all the world oceans as far as changes in water-mass structure through time are concerned. This comment is made in light of the data prepared by CLIMAP (Ruddiman and McIntyre, this volume; Lozano and Hays, this volume; Sachs, 1973; Moore, 1973; McIntyre, 1974). The North Atlantic surface-water geometric patterns change markedly between glacial and interglacial modes, with the Atlantic divided roughly into two parts—a dynamic area to the north of approximately lat 42°N and a relatively stable area from lat 42°N to the Equator. The northern part undergoes temperature changes between glacial and interglacial modes of as much as 18°C in some areas, whereas in the southern part the variation seldom exceeds 3°C except in upwelling areas. Not only do these changes affect the surface waters, the life in them, and the climate of the surrounding continents, but they materially alter the deep-water structure of the Atlantic (Streeter, 1973) and, perhaps, the world ocean as well.

In a nonglacial mode, the North Atlantic produces a significant part of the world-ocean deep water by the production of North Atlantic Deep Water in the Norwegian and Greenland Seas and, perhaps to a lesser extent, in the ocean southeast of Greenland. Intermediate-water formation in the North Atlantic is minimal, being limited to a part of the northwestern basin around the Labrador Sea. This condition is markedly changed during the glacial mode. The Norwegian and Greenland areas where North Atlantic Deep Water is produced have disappeared, and the type of deep water produced is presumably less in quantity and warmer in character (Streeter, 1973). Although glacial production of North Atlantic Deep Water is suppressed, the production of Arctic Intermediate Water may have been markedly increased over present values during this period.

ACKNOWLEDGMENTS

We thank S. Bowen, R. M. Cline, J. E. Damuth, G. Denton, M. Doncourt, L. K. Glover, R. Heath, J. Imbrie, J. Kostecki, R. Matthews, B. Molfino, T. C. Moore, M. Perry, D. G. Roberts, M. Roche, and J. Thiede for their scientific and editorial aid in the preparation of this paper. This research was funded by National Science Foundation International Decade of Ocean Exploration Grant IDO71-04204. Part of the study was funded by National Science Foundation Grant DES 73-00467. Core curating and repository services are funded by Office of Naval Research Grant N00014-67-A-0108-0004 and National Science Foundation Grant DES 72-01568.

REFERENCES CITED

Bé, A.W.H., Damuth, J., Lott, L., and Free, R., 1976, Late Quaternary climatic record in western equatorial Atlantic sediments, *in* Cline, R. M., and Hays, J. D., eds, Investigation of late Quaternary paleoceanography and paleoclimatology: Geol. Soc. America Mem. 145 (this volume).

Belderson, R. H., Kenyon, N. H., and Wilson, J. B., 1972, Iceberg plough marks in the northeast Atlantic: Palaeogeography, Palaeoclimatology, Palaeoecology, v. 13, p. 215-224.

Bloom, A. L., 1971, Glacial-eustatic and isostatic controls at sea level since the last glaciation, *in* Turekian, K. K., ed., The late Cenozoic glacial ages: New Haven, Conn., Yale Univ. Press, p. 355-379.

Bloom, A. L., Broecker, W. S., Chappell, J.M.A., Matthews, R. K., and Mesolella, K. J., 1974, Quaternary sea level fluctuations on a tectonic coast: New 230Th/234U dates from the Huon Peninsula, New Guinea: Quaternary Research, v. 4, p. 185-205.

Broecker, W. S., and van Donk, J., 1970, Insolation changes, ice volumes and the O^{18} record in deep-sea cores: Rev. Geophysics and Space Physics, v. 8, p. 169-198.

Dreimanis, A. and Karrow, P. F., 1972, Glacial history of the Great Lakes-St. Lawrence region, the classification of the Wisconsin(an) Stage, and its correlatives: Internat. Geol. Cong., 24th, Montreal 1972, sec. 12, p. 5-15.

Emery, K. O., 1967, The Atlantic continental margin of the United States during the past 70 million years: Geol. Assoc. Canada Spec. Paper 4, p. 53-70.

Flint, R. F., 1971, Glacial and quaternary geology: New York, John Wiley & Sons, Inc., 892 p.

Fuglister, F. C., 1972, Cyclonic rings formed by the Gulf Stream in 1965-1966, *in* Gorden, A., ed., Studies in physical oceanography, Vol. 1: New York, Gordon and Breech, p. 137-168.

Gardner, J. V., and Hays, J. D., 1976, Responses of sea-surface temperature and circulation to global climatic change during the past 200,000 years in the eastern equatorial Atlantic Ocean, *in* Cline, R. M., and Hays, J. D., eds., Investigation of late Quaternary paleoceanography and paleoclimatology: Geol. Soc. America Mem. 145 (this volume).

Gates, L., 1975, The January global climate simulated by the two-level Mintz-Arakawa model: A comparison with observation: Jour. Atmos. Sci., v. 32, p. 449-477.

——1976, Modeling the ice-age climate: Science, v. 191, p. 1138-1144.

Imbrie, J., and Kipp, N., 1971, A new micropaleontological method for quantitative paleoclimatology: Application to a late Pleistocene Caribbean core, *in* Turekian, K. K., ed., Late Cenozoic glacial ages: New Haven, Conn., Yale Univ. Press, p. 71-181.

Iselin, C. O'D., 1940, Preliminary report on Long-period variations in the transport of the Gulf Stream system: Papers Phys. Oceanography and Meteorology, v. 8, p. 40.

Kellogg, T. B., 1975, Late Quaternary climatic changes in the Norwegian and Greenland Seas, *in* Weller, G., ed., Climate of the Arctic: 24th Alaskan Sci. Conf., 1973, Fairbanks, Alaska, Proc., p. 3-36.

King, L. H., 1967, On the sediments and stratigraphy of the Scotian shelf: Geol. Assoc. Canada Spec. Paper 4, p. 71-92.

Kipp, N. G., 1976, New transfer function for estimating past sea-surface conditions from sea-bed distribution of planktonic foraminiferal assemblages in the North Atlantic, *in* Cline, R. M., and Hays, J. D., eds., Investigation of late Quaternary paleoceanography and paleoclimatology: Geol. Soc. America Mem. 145 (this volume).

Klovan, J. E., and Imbrie, J., 1971, An algorithm and FORTRAN-IV program for large-scale Q-mode factor analysis and calculation of factor scores: Internat. Assoc. Math. Geology Jour., v. 3, p. 61-77.

Kögler, F. C., 1963, Das Kastenlot: Meyniana, v. 13, p. 1-7.

Ku, T. L., Bischoff, J. L., and Boersma, A., 1972, Age studies of Mid-Atlantic Ridge sediments near 42°N. and 20°N.: Deep-Sea Research, v. 19, p. 233-247.

Lamb, H. H., 1971, Climates and circulation regimes developed over the Northern Hemisphere during and since the last ice age: Palaeogeography, Palaeoclimatology, Palaeocology, v. 10, p. 125-162.

Lozano, J., and Hays, J. D., 1976, Relationship of radiolarian assemblages to sediment types and physical oceanography in the Atlantic and western Indian Ocean sectors of the Antarctic Ocean, *in* Cline, R. M., and Hays, J. D., eds., Investigation of late Quaternary paleoceanography and paleoclimatology: Geol. Soc. America Mem. 145 (this volume).

Manley, G., 1951, The range of variation of the British climate: Geog. Jour., v. 8, p. 43-68.

McIntyre, A., 1967, Coccoliths as paleoclimatic indicators of Pleistocene glaciation: Science, v. 158, p. 1314-1317.

———1974, World-ocean isotherm map during an ice-age, 18,000 years ago [abs.]: EOS (Am. Geophys. Union Trans.), v. 55, p. 259.

McIntyre, A., Ruddiman, W. F., and Jantzen, R., 1972, Southward penetration of the North Atlantic polar front: Faunal and floral evidence of large-scale surface water mass movements over the last 225,000 years: Deep-Sea Research, v. 19, p. 61-77.

Moore, T. C., Jr., 1973, Late Pleistocene/Holocene oceanographic changes in the northeastern Pacific: Quaternary Research, v. 3, p. 99-109.

Newell, R. E., 1974, Changes in the poleward energy flux by the atmosphere and ocean as a possible cause of ice ages: Quaternary Research, v. 4, p. 117-127.

Pewe, T. L., 1966, Paleoclimatic significance of fossil ice wedges: Biul. Peryglacjalny, no. 15, p. 65-73.

Prell, W. L., and Hays, J. D., 1976, Late Pleistocene faunal and temperature patterns of the Columbia Basin, Caribbean Sea, *in* Cline, R. M., and Hays, J. D., eds., Investigation of late Quaternary paleoceanography and paleoclimatology: Geol. Soc. America Mem. 145 (this volume).

Reid, J. L., 1971, General circulation patterns in the world ocean, *in* Matthews, W. H., and others, eds., Man's impact on terrestrial and oceanic ecosystems: Cambridge, Mass., M.I.T. Press, p. 448-459.

Roche, M. B., McIntyre, A., and Imbrie, J., 1975, Quantitative paleo-oceanography of the late Pleistocene-Holocene North Atlantic: Coccolith evidence, *in* Saito, T., and Burckle, L. H., eds., Late Neogene Epoch boundaries: Am. Mus. Nat. History Micropaleontology Spec. Pub. 1, p. 191-225.

Ruddiman, W. F., and Glover, L. K., 1972a, Ice-rafted volcanic ash: A tracer of North Atlantic paleo-circulation [abs.]: EOS (Am. Geophys. Union Trans., v. 53, p. 423.

———1972b, Vertical mixing of ice-rafted volcanic ash in North Atlantic sediments: Geol. Soc. America Bull., v. 83, p. 2817-2836.

———1975, North Atlantic subpolar circulation 9,300 yrs. B.P.: Quaternary Research, v. 5, p. 361-389.

Ruddiman, W. F., and McIntyre, A., 1973, Time-transgressive deglacial retreat of polar water from the North Atlantic: Quaternary Research, v. 3, p. 117-130.

———1976, Northeast Atlantic paleoclimatic changes over the past 600,000 years, *in* Cline, R. M., and Hays, J. D., eds., Investigation of late Quaternary paleoceanography and paleoclimatology: Geol. Soc. America Mem. 145 (this volume).

Sachs, H. M., 1973, Late Pleistocene history of the North Pacific: Evidence from a quantitative study of Radiolaria in core V21-173: Quaternary Research, v. 3, p. 89-98.

Shackleton, N. J., and Opdyke, N., 1973, Oxygen isotopic and paleomagnetic stratigraphy of equatorial Pacific core V28-238: Oxygen isotope temperatures and ice volume on a 10^5 yr and 10^6 yr scale: Quaternary Research, v. 3, p. 39-55.

Stommel, H., 1965, The Gulf Stream, 2nd ed.: Los Angeles, Calif., California Univ. Press, 201 p.

Streeter, S. S., 1973, Bottom water and benthonic foraminifera in the North Atlantic-glacial-interglacial contrasts: Quaternary Research, v. 3, p. 131-141.

Sverdrup, H. U., Johnson, M. W., and Fleming, R. H., 1942, The oceans, their physics, chemistry and general biology: Englewood Cliffs, N.J., Prentice-Hall, Inc., 1,059 p.

U.S. Naval Oceanographic Office, 1967, Oceanographic atlas of the North Atlantic Ocean, Physical oceanography, Sec. 2, Publ. 700: Washington, D.C., U.S. Govt. Printing Office.

van der Hammen, Th., Wijmstra, T., and Zagwijn, Th., 1971, The floral record of the late Cenozoic of Europe, *in* Turekian, K. K., ed., The late Cenozoic glacial ages: New Haven, Conn., Yale Univ. Press, p. 391-424.

Weyl, P. K., 1968, The role of the oceans in climatic change: A theory of the ice ages: Meteorol. Mons., v. 8, p. 37-62.

Worthington, L. V., and Wright, W. R., 1970, North Atlantic Ocean atlas of potential temperature, salinity and oxygen profiles from the *Erika Dau* cruise of 1962: Woods Hole Oceanographic Inst. Atlas Ser. II, v. 24, p. 58.

Manuscript Received by the Society January 27, 1975
Revised Manuscript Received July 14, 1975
Manuscript Accepted July 23, 1975
Lamont-Doherty Geological Observatory Contribution No. 2268

Printed in U.S.A.

Late Quaternary Climatic Changes: Evidence from Deep-Sea Cores of Norwegian and Greenland Seas

THOMAS B. KELLOGG*
Lamont-Doherty Geological Observatory
Columbia University
Palisades, New York 10964

ABSTRACT

The present temperature regime of the Norwegian and Greenland Seas results largely from the warm Norwegian Current. This current is partially responsible for the maritime climates of northern Europe and Scandinavia, and it controls the distribution of planktonic Foraminifera and the extent of sea-ice cover in the Norwegian and Greenland Seas.

Analyses of 6 piston cores show that Norwegian Sea temperatures during most of the past 150,000 yr have been much lower than they are now. Only between 127,000 and 110,000 B.P. did temperatures approach or surpass present-day temperatures. For the remaining time, foraminiferal faunas were similar to or even less diverse than those of today in the northern Greenland Sea, where ice cover is present in winter. This suggests that during most of the last 150,000 yr, ice covered all of the Norwegian and Greenland Seas, probably on a year-around basis. As a result, northern Europe and Scandinavia did not receive air warmed by the Norwegian Current as they do now. Additionally, the presence of total sea-ice cover prevented the formation of Norwegian Sea overflow water, thus altering the deep circulation of the Atlantic.

INTRODUCTION

Among the least studied deep-sea sediments are those from the Norwegian and Greenland Seas (referred to collectively in this paper as "Norwegian Sea"). This region is bounded on the east by Norway, Spitsbergen, and the Barents Sea; on the south by the Iceland-Faeroe Ridge, Iceland, and the Denmark Strait; on the

---
*Present address: Institute for Quaternary Studies, University of Maine at Orono, Orono, Maine 04473.

west by Greenland; and on the north by the Arctic Ocean. Both Greenland and the Arctic have a glacial climate.

Knowledge of the climatic history of the Norwegian Sea is necessary to understand the reasons for world climatic changes during the Quaternary Period. (1) The Norwegian Current, a warm surface current, is a branch of the Gulf Stream-North Atlantic Current system which today transports heat from subtropic regions making possible the temperate climate of Scandinavia at the same latitudes where Greenland supports an ice cap. There is evidence that during the last ice age the Norwegian Current was either much weaker or may have been completely absent from the region; this may partially explain the formation of the Scandinavian ice cap. (2) The deep circulation of the world oceans presently depends on dense water-mass formation in two distinct regions. Antarctic Bottom Water (AABW) forms around the Antarctic Continent as freezing surface water yields salts to the underlying water, thus increasing its density (Jacobs and others, 1970; Gordon, 1971). In the Norwegian Sea, saline Norwegian Current water cools by evaporation. This extremely dense water spills over the Denmark Strait and Iceland-Faeroe Ridge to form lower North Atlantic deep water (Dietrich, 1956a, 1956b) or Norwegian Sea overflow water (Worthington, 1970). Evidence is presented that the Norwegian Sea could not have been a major source area for Norwegian Sea overflow-water formation during the last glacial age. (3) Within the northern and western Norwegian Sea, sediments are being deposited today under glacial conditions. These sediments grade into nonglacial deposits to the south and east. For these reasons, recent Norwegian Sea sediments provide both glacial and nonglacial analogs for comparison with older Norwegian Sea and adjacent North Atlantic sediments. (4) Much climatic work has been done in the regions surrounding the Norwegian Sea. Arctic Ocean sediments have been studied by Bé (1960), Clark (1969, 1970), Darby (1971), Ericson and others (1964a), Mullen and others (1972), Steuerwald and others (1968), Steuerwald and Clark (1972), and others. The deep-ice core from Camp Century, Greenland, reveals a climatic record spanning what may be more than 100,000 yr (Dansgaard and others, 1971). European continental sediments have been studied by numerous workers for over 100 yr. In the North Atlantic, many climatic studies of deep-sea sediments have been made, including those of Bramlette and Bradley (1941), Ericson and others (1964b), McIntyre (1967), Ruddiman and others (1970), Sancetta and others (1972), and McIntyre and others (1972). These studies have left a large gap: the Norwegian Sea.

Prior to 1973, most studies of Norwegian Sea sediments were concerned with either a fossil-assemblage description retrieved from one to several cores (Saito and others, 1967; Stadum and Ling, 1969; Bjørklund and Kellogg, 1972) or sedimentological aspects of a limited number of cores covering a small portion of the region (Böggild, 1907; Holtedahl, 1959; Ericson and others, 1964a; Schreiber, 1967). The most comprehensive of these studies is Ericson and others (1964a). In 26 cores they noted the predominance of "glacial-marine sediment" and suggested that the last ice age ended 11,000 yr ago based on frequencies of foraminiferal species in 6 cores. Kellogg (1975) analyzed carbonate and coarse terrigenous material in 169 surface-sediment samples, carbonate in 29 cores, and both coarse and fine noncarbonate material in 6 cores. In addition, quantitative paleotemperature estimates were made using the Imbrie and Kipp (1971) technique on foraminiferal counts in the same 6 cores discussed in this paper. Data presented by Kellogg (1975) confirm the conclusions presented below.

This paper will show the pattern of climatic change during the past 150,000 yr through study of Norwegian Sea sediments. The paper has four sections. (1) Surface-sediment distributions of various sedimentary and faunal parameters are presented as a base representing modern nonglacial conditions for comparison with older sediments. (2) Stratigraphic analyses of sedimentary and faunal parameters in 6 Norwegian Sea cores spanning at least the last 150,000 yr are compared with surface-sediment distributions. (3) Time control is established for climatic changes observed in the 6 cores. (4) Observed climatic phenomena and their meanings are discussed with reference to late Quaternary climatic change in the larger region surrounding the Norwegian Sea.

SURFACE SEDIMENTS

Cores used for this study were taken by the Lamont-Doherty Geological Observatory research vessel *Vema* on cruises to the Norwegian Sea in 1966, 1969, 1970, and 1972. Trigger-weight cores (TW), obtained from the tripping mechanisms of the longer piston cores, were sampled to study surface sediments. These TW tops represent the last 2,000 to 4,000 yr of sedimentation (Imbrie and Kipp, 1971). Thirty-four TW tops were chosen to provide good geographic coverage (Fig. 1, Table 1).

Techniques

A sample was taken from the top centimeter of each TW. About 0.5 g was set aside for carbonate analysis. The remainder was dried, weighed, and then soaked in water containing a small amount of sodium hexametaphosphate (Calgon) to disaggregate the material. The sample was agitated sonically and washed on

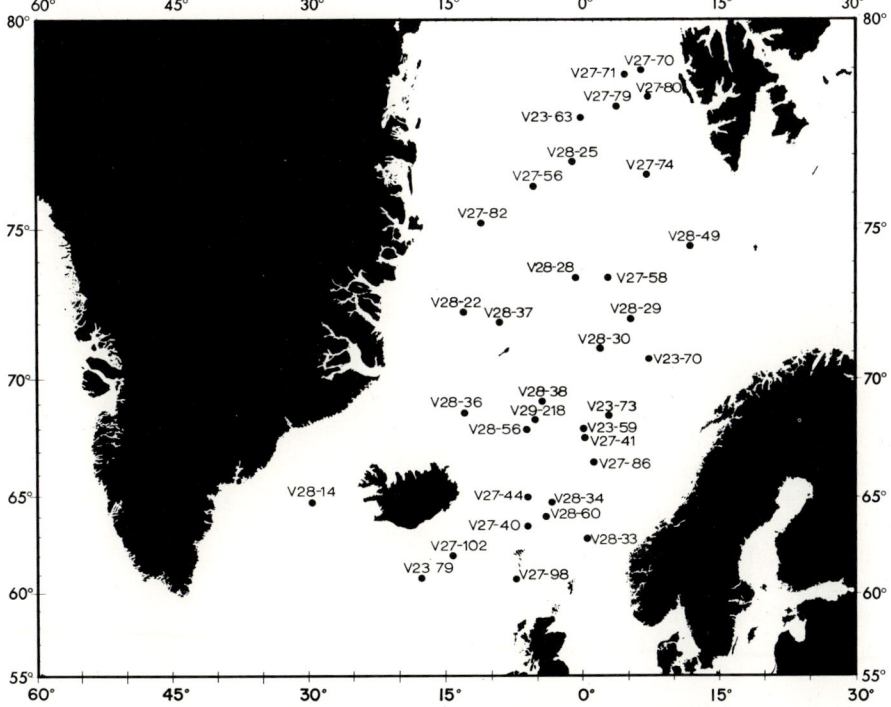

Figure 1. Locations of 34 trigger-weight cores studied for faunal and sedimentary content.

TABLE 1. POSITION AND WATER DEPTH FOR CORES DISCUSSED IN THIS PAPER

Locations of Trigger Cores			
Core no.	Lat (N)	Long	Depth (m)
V23 59	68°02'00"	0°01'00" E	3,083
V23 63	77°57'24"	0°12'12" E	3,050
V23 70	70°59'06"	6°41'24" E	3,047
V23 73	68°32'48"	2°43'00" E	2,970
V23 79	60°56'00"	17°34'00" W	2,478
V27 40	63°26'54"	6°07'48" W	1,652
V27 41	62°29'30"	3°15'48" W	635
V27 44	65°01'18"	6°13'12" W	2,772
V27 56	76°09'30"	5°23'54" W	2,736
V27 58	73°32'24"	2°39'48" W	2,754
V27 70	78°54'06"	7°03'12" E	1,300
V27 71	78°49'54"	4°37'54" E	2,439
V27 74	76°27'54"	6°54'30" E	2,778
V27 79	78°07'06"	3°37'24" E	2,308
V27 80	78°23'30"	7°25'24" E	2,915
V27 82	74°59'36"	10°48'24" W	2,895
V27 86	66°36'24"	1°07'06" E	2,900
V27 98	60°39'12"	7°25'30" W	980
V27 102	62°07'42"	14°12'18" W	1,641
V28 14	64°47'00"	29°34'00" W	1,855
V28 22	72°26'00"	13°39'00" W	1,284
V28 25	76°49'00"	1°20'00" W	3,136
V28 28	73°29'00"	0°50'00" W	2,288
V28 29	72°11'00"	5°16'00" E	2,547
V28 30	71°10'00"	1°37'00" E	2,915
V28 33	62°54'00"	0°35'00" E	1,170
V28 34	64°50'00"	3°35'00" W	3,217
V28 36	68°43'00"	12°43'00" W	1,816
V28 37	72°04'00"	9°04'00" W	2,395
V28 38	69°23'00"	4°24'00" W	3,411
V28 49	74°29'00"	11°40'00" E	2,327
V28 56	68°02'00"	6°07'00" W	2,941
V28 60	64°05'00"	4°02'00" W	3,244
V29 219K	68°23'06"	5°27'36" W	3,325
Locations of Piston Cores			
V27 47	68°27'42"	13°32'30" W	1,717
V27 60	72°11'00"	8°34'48" E	2,525
V27 86	66°36'24"	1°07'06" E	2,900
V28 14	64°47'00"	29°34'00" W	1,855
V28 25	76°49'00"	1°20'00" W	3,136
V28 56	68°02'00"	6°07'00" W	2,941

a 62 μm sieve. Both coarse and fine fractions were saved; the coarse fractions were dried and weighed.

Carbonate analyses were made by the Hülsemann (1966) gasometric technique. Accuracy is ±3% for samples containing more than 50% $CaCO_3$; for samples with less than 50% $CaCO_3$, uncertainty is probably ±5% because of the small volume of CO_2 generated (Siesser and Rogers, 1971).

All foraminifera were picked from an aliquot of the fraction of sample greater than 149 μm. The aliquot was obtained by splitting the sample repeatedly until 300 to 500 planktonic foraminifera remained. All specimens were identified (taxonomy of Bé, 1967) and mounted by species on a cardboard microslide. Each species was counted and the number recorded along with the number of times the fraction

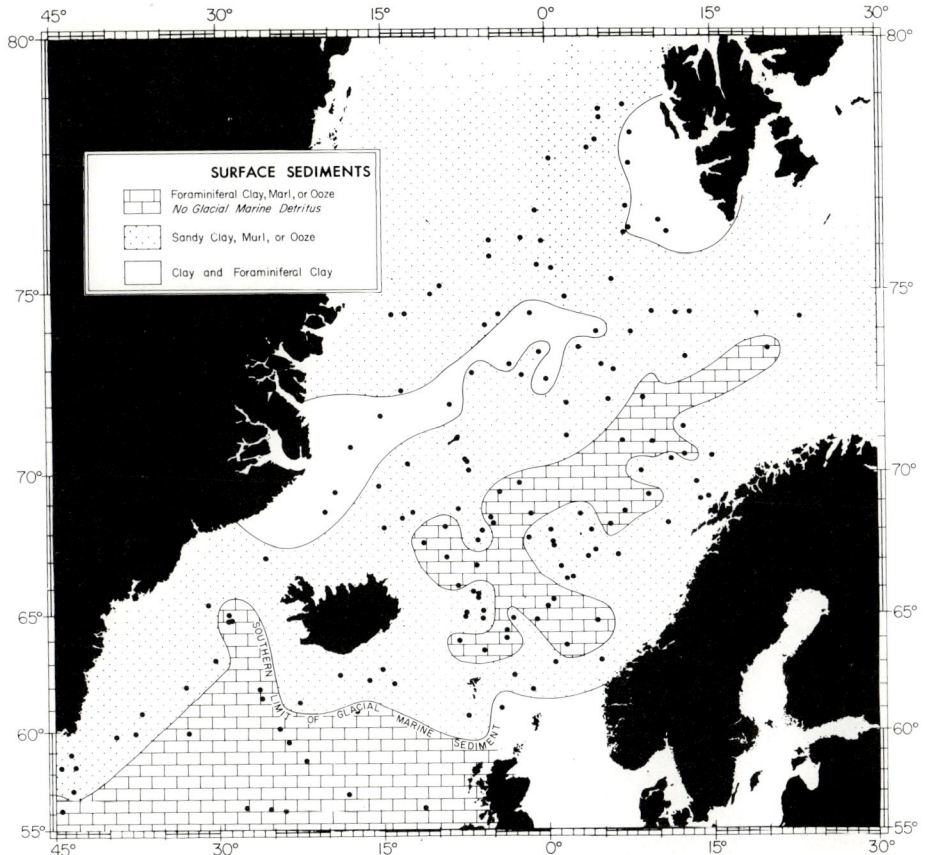

Figure 2. Distribution of sediment types in Norwegian Sea surface sediments (after Kellogg, 1975, Fig. 3). Core locations are indicated by dots.

greater than 149 μm was split. Benthonic foraminifera were separated and the total number determined; species were not identified. (All raw data will be found in Appendixes I–IV.)[1]

Surface Sediment Types

As noted by Ericson and others (1964a), Norwegian Sea sediments usually contain considerable amounts of "glacial-marine" detritus (Philippi, 1912). The sediments also contain variable amounts of biogenic material, mostly planktonic foraminiferal shells composed of $CaCO_3$, and clay-size particles derived from erosion on land. The other important sedimentary material is volcanic ash. Excluding ash, Norwegian Sea sediments result from a three-component sedimentary system in which the prevailing climatic regime is reflected in the relative proportions of the components.

The distribution of sediment types is shown in Figure 2 (after Kellogg, 1975). Sediments are classified as foraminiferal oozes >60% $CaCO_3$, foraminiferal marls 30 to 60% $CaCO_3$, foraminiferal clays 10% to 30% $CaCO_3$, and clay <10% $CaCO_3$. A brief microscopic examination was made of each TW top to estimate the proportion

---

[1] On microfiche in pocket inside back cover.

of coarse noncarbonate material (mostly ice-rafted detritus). Samples containing 10% to 40% terrigenous sand were termed "sandy" clay, marl, or ooze, depending on the carbonate content. No evidence of turbidity current deposition was found in any of the TW tops.

Most of the Norwegian Sea floor is covered by sandy clays, marls, and oozes (Fig. 2). The sandy material is largely attributable to ice rafting. There is little or no terrigenous sand in samples from the southern portion of the region. Foraminiferal clays, marls, and oozes lie directly under the Norwegian Current (Figs. 3, 4). Two areas, one to the west of Spitsbergen and the other in a band trending northeast from Greenland about halfway toward Spitsbergen, have surface sediments with low carbonate content and little or no ice-rafted material. These areas have winter ice cover at present. Having sampled an ice floe carrying rafted debris in the region off Greenland during the summer of 1970, I am surprised to find so little evidence of ice rafting in the underlying sediment.

The terrigenous sands off Norway and on the Iceland-Faeroe Ridge are probably not the result of ice rafting. Rather, they are relict Pleistocene deposits similar to those found along the continental margin off eastern North America (Stetson, 1939) caused by bottom scour. Elsewhere, sandy material is attributed to ice rafting.

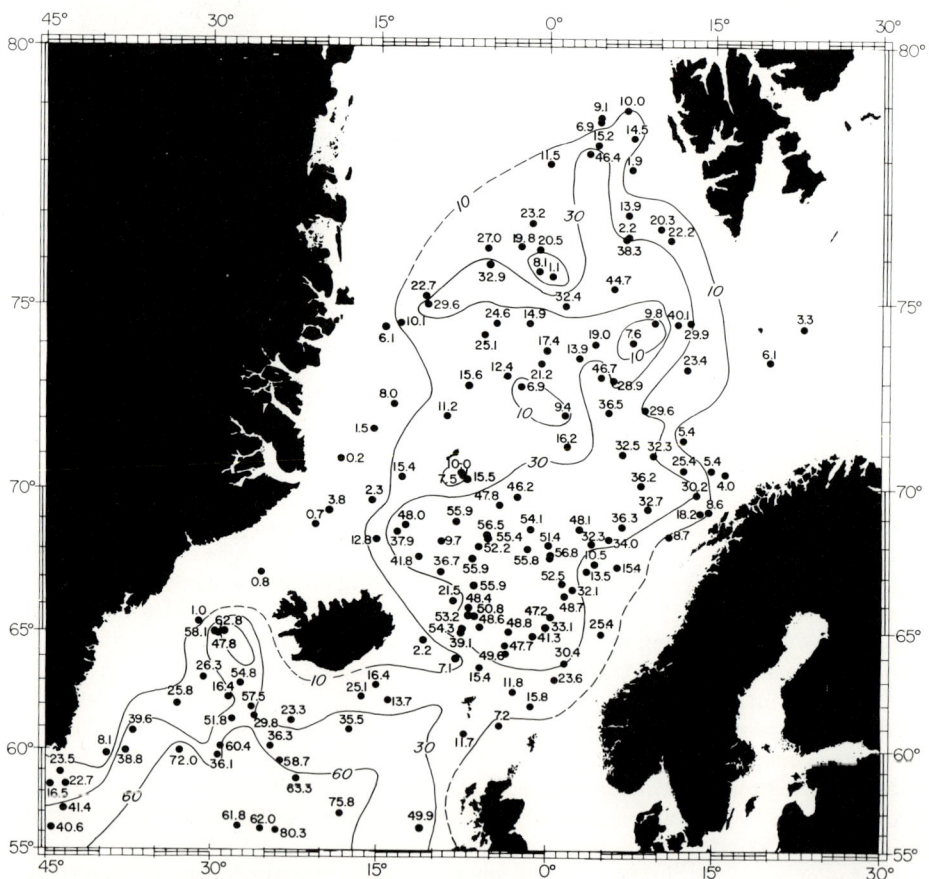

Figure 3. Distribution of calcium carbonate (weight percent) in Norwegian Sea surface sediments (after Kellogg, 1975, Fig. 7).

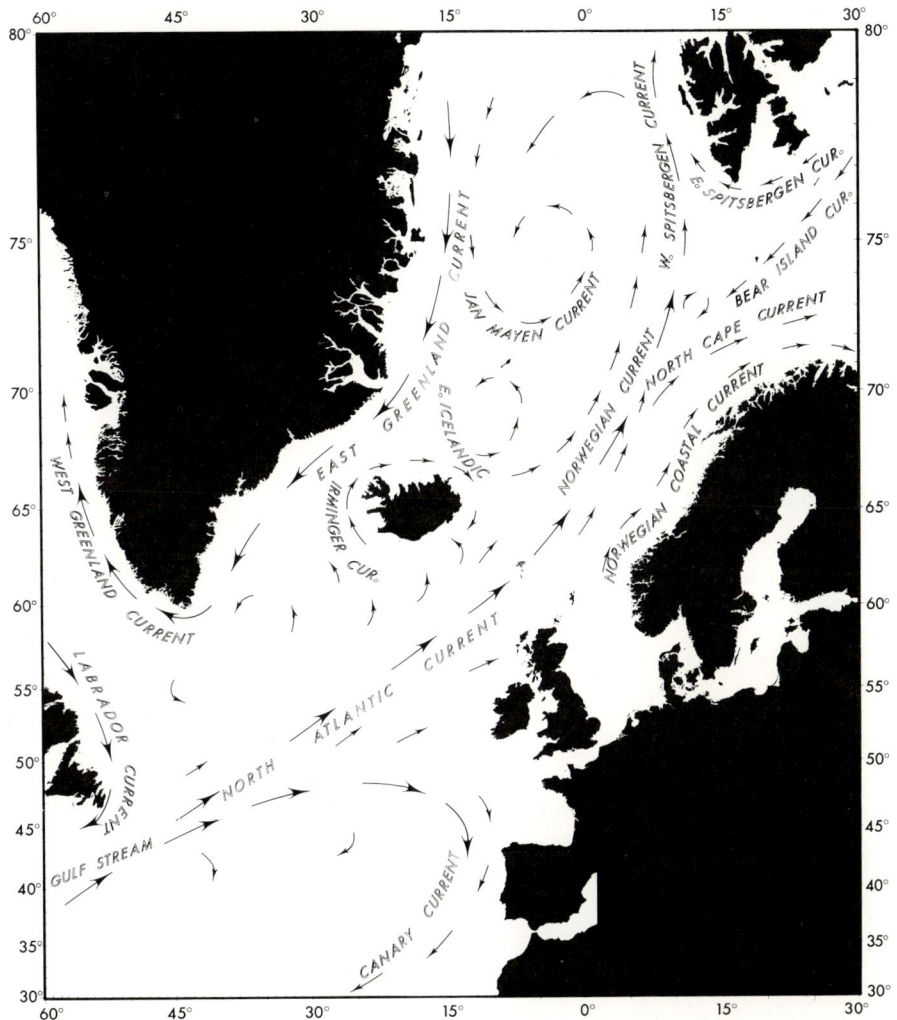

Figure 4. Generalized surface circulation of the North Atlantic and Norwegian Seas (from U.S. Naval Oceanographic Office, 1965; U.S. Navy Hydrographic Office, 1958).

The distribution of ice-rafted detritus in the Norwegian Sea fits closely with our knowledge of ice cover (Figs. 5, 6, 9).

Distribution of $CaCO_3$ in Surface Sediments

Because foraminiferal shells are the most abundant biogenic component in Norwegian Sea sediments, analyses of calcium carbonate content should measure both total biogenic input and foraminiferal productivity in surface waters. Other sources of $CaCO_3$ are usually not significant. Benthonic foraminifera are present at levels usually about two orders of magnitude less than planktonic species. At present coccoliths are absent in surface waters north of the 2.0°C (35.6°F) isotherm (McIntyre and Bé, 1967). Most of the Norwegian Sea is north of this isotherm for at least part of the year, so coccoliths should not be an important source

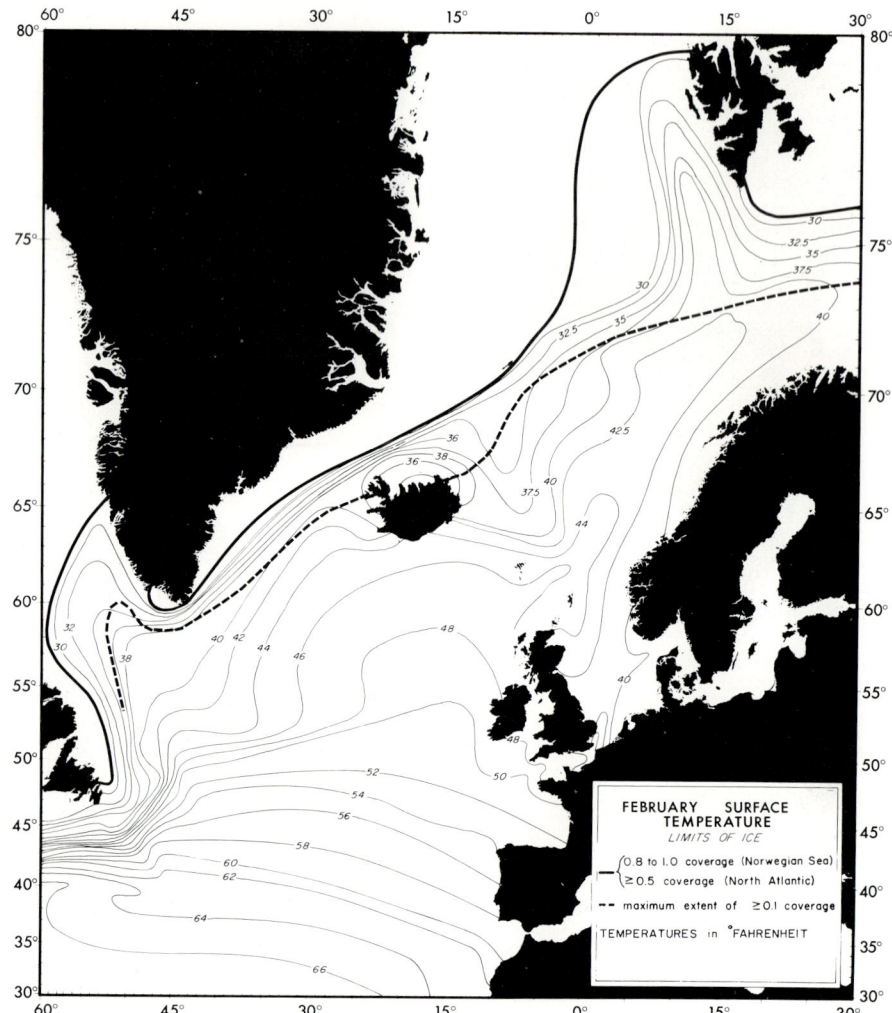

Figure 5. February sea surface temperature and ice limits (from U.S. Naval Oceanographic Office, 1967; U.S. Navy Hydrographic Office, 1958).

of $CaCO_3$ for underlying sediments. The other possible source of $CaCO_3$ is detrital material from the surrounding lands. Considerable amounts of dolomite and calcite are present in Paleozoic rocks in Greenland (Raasch, 1961). Glacial scour might transport this material to the sea. A number of x-ray diffraction measurements were made on surface-sediment samples and on samples taken at depth in several cores, but only trace amounts of magnesium calcium carbonate were found (Table 2). The predominant carbonate occurring in Norwegian Sea sediments is planktonic foraminiferal shells.

Surface-sediment $CaCO_3$ distribution in the Norwegian Sea is shown in Figure 3 (after Kellogg, 1975) based on 169 points, including the 34 TW tops discussed in this paper. This distribution reflects the surface-water circulation (Fig. 4). High carbonate values tend to underlie warm surface currents, especially the Norwegian Current. High carbonate values also tend to cut across topographic features under

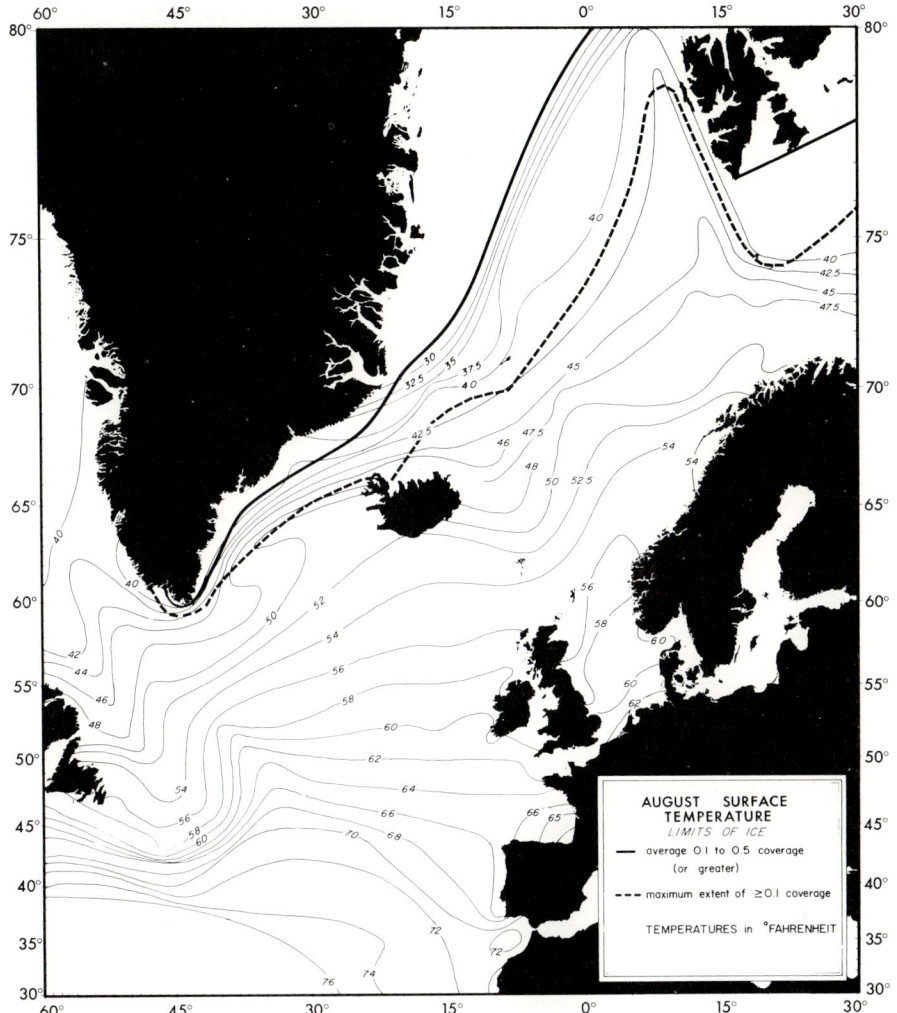

Figure 6. August sea surface temperature and ice limits. Sources as for Figure 5.

warm currents. This suggests that carbonate dissolution is not a major factor in the region (dissolution is discussed in more detail below). Rather, the carbonate content of the sediments reflects productivity in warm surface waters.

Planktonic Foraminifera

In most Norwegian Sea sediment samples, planktonic foraminiferal faunas are dominated by left-coiling *Globigerina pachyderma*. Other species present include right-coiling *G. pachyderma*, *G. bulloides*, *G. quinqueloba*, *Globorotalia inflata*, and *Globigerinita glutinata*.

Two faunal indices derived from species counts provide useful information. These indices are (1) percent right-coiling *G. pachyderma* (Ericson, 1959, showed that *G. pachyderma* coils dextrally in warm waters and sinistrally in cold) and (2)

TABLE 2. NORWEGIAN SEA SAMPLES ANALYZED FOR DOLOMITE

Core	Sample depth (cm)	Lat (N)	Long	Water depth (m)	Dolomite	CaCO$_3$ (%)
V23–62TW	0	74°54'	1°36.5'E	3,713	. .	32.4
V23–63TW	0	77°57.4'	0°12.2'E	3,050	. .	11.5
V23–64TW	0	77°51.7'	7°16.2'E	2,529	tr	1.9
V27–40TW	0	63°26.9'	6°07.8'W	1,652	. .	15.4
V27–47TW	0	68°27.7'	13°32.5'W	1,717	. .	37.9
V27–60TW	0	72°11'	8°34.8'E	2,525	. .	29.6
V27–69TW	0	76°41.9'	10°00.7'E	2,237	tr	20.3
V27–69	160	76°41.9'	10°00.7'E	2,237	?	5.2
V27–69	180	76°41.9'	10°00.7'E	2,237	?	10.1
V27–69	270	76°41.9'	10°00.7'E	2,237	. .	15.2
V27–69	280	76°41.9'	10°00.7'E	2,237	?	4.4
V27–69	400	76°41.9'	10°00.7'E	2,237	present	4.9
V27–69	890	76°41.9'	10°00.7'E	2,237	tr	8.5
V27–74TW	0	76°27.9'	6°54.5'E	2,778	. .	38.3
V27–76TW	0	76°25.4'	10°48'E	2,113	. .	22.2
V27–83TW	0	74°11.6'	5°42.9'W	3,416	?	25.1
V27–86TW	0	66°36.4'	1°07.1'E	2,900	?	52.5
V27–86	70	66°36.4'	1°07.1'E	2,900	present	14.4
V27–86	90	66°36.4'	1°07.1'E	2,900	tr	9.9
V27–86	190	66°36.4'	1°07.1'E	2,900	. .	12.2
V27–86	360	66°36.4'	1°07.1'E	2,900	. .	1.2
V28–14TW	0	64°47'	29°34'W	1,855	. .	47.8
V28–14	160	64°47'	29°34'W	1,855	tr	8.2
V28–14	230	64°47'	29°34'W	1,855	. .	8.5

percent "warm-water" species (all other species except *G. pachyderma*). These two indices are mapped in Figures 7 and 8. Both increase under the Norwegian Current. Right-coiling *G. pachyderma* is most abundant under the axis of the current; the warm species appear to be most abundant under its western boundary.

Left-coiling *G. pachyderma* comprises more than 95% of the total foraminiferal fauna in all samples taken away from the influence of the Norwegian Current. A large part of this region is covered by winter ice (Fig. 5). Historical records suggest that formerly the area covered by sea ice was much more extensive. Figure 9 (after Lamb, 1963) shows ice maxima for the last 200 yr. Comparison of Figures 7, 8, and 9 suggests that left-coiling *G. pachyderma* percentages of 95 or greater are found in cores within the Norwegian Sea with historical records of ice cover. Thus, left-coiling percentages of 95 or greater appear characteristic of sample localities that are near or under winter pack ice. *G. pachyderma* should therefore be a useful index of ice cover for older samples.

Planktonic foraminiferal abundance in Norwegian Sea surface sediments is also controlled by surface circulation (Fig. 10). Abundances were calculated by multiplying the number of specimens in each aliquot by the number of times the greater than 149-μm fraction was split to get 300 to 500 specimens. This number was divided by the original sample weight (before washing) to get the approximate number of specimens larger than 149 μm/g of sediment. W. F. Ruddiman (1973, oral commun.) suggested that abundances calculated in this manner are unreliable if samples were split more than four times. I feel that the calculated abundances provide useful order-of-magnitude estimates, because the resulting geographic pattern (Fig. 10) closely follows known oceanographic parameters.

Benthonic foraminiferal abundances were calculated in the same manner and show a similar pattern (Fig. 11). This similarity is attributed to surface circulation, because sediments underlying highly productive regions receive large amounts of organic matter falling from the surface which may be used as food by the benthonic fauna.

Dissolution

One would not expect dissolution to be very active in destroying $CaCO_3$ because of the relatively shallow depths in the Norwegian Sea. According to Li and others (1969), the carbonate compensation depth in the Atlantic averages about 4,500 m. The deepest core used in this study was 3,400 m, and most cores were less than 3,000 m. Planktonic foraminifera show no indication of dissolution, and fragments are rare. *G. pachyderma* is one of the species most resistant to dissolution (Berger, 1971). In some samples, benthonic foraminifera are broken and (or) have a chalky texture (S. S. Streeter, 1974, oral commun.).

Percent benthonic foraminifera is a faunal indicator of dissolution (Oba, 1969). Normally abyssal sediments have percentages of less than 5.0. Comparison of Figures 10 and 11 shows that percentages in Norwegian Sea surface sediments vary from 0.46 to 53.3, with about half the values exceeding 5.0, suggesting that dissolution may have been active in a number of cores. Most of the samples

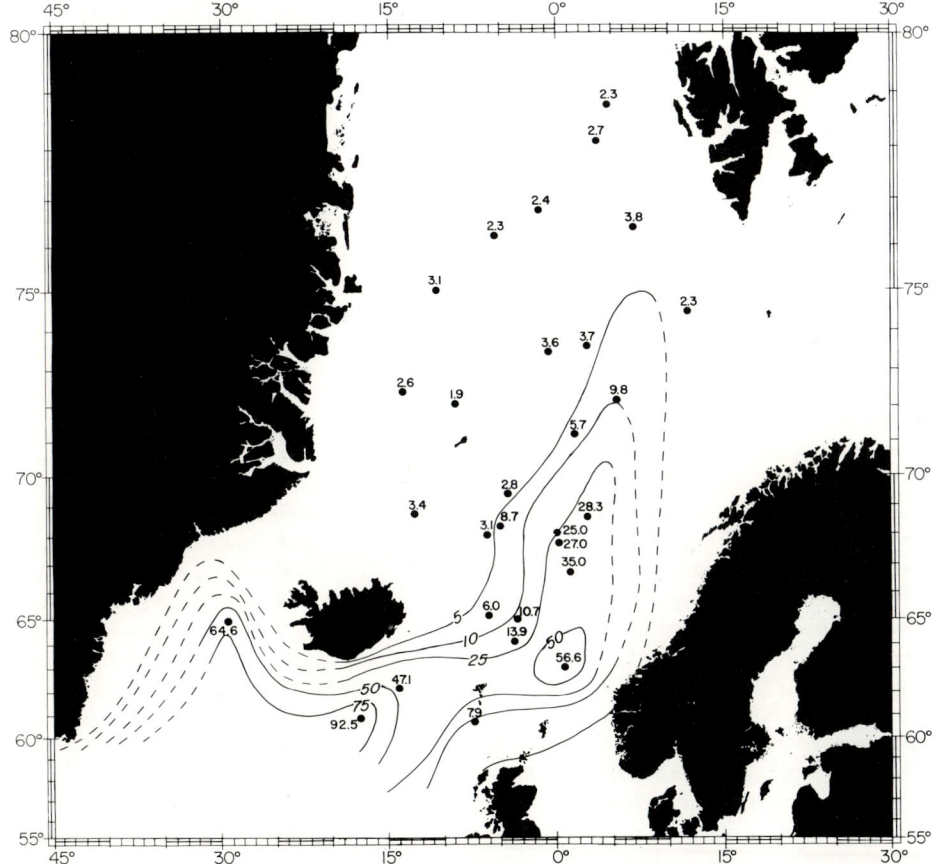

Figure 7. Distribution of right-coiling *Globigerina pachyderma* in surface sediments of the Norwegian Sea. Numbers plotted are percent of the total *G. pachyderma*.

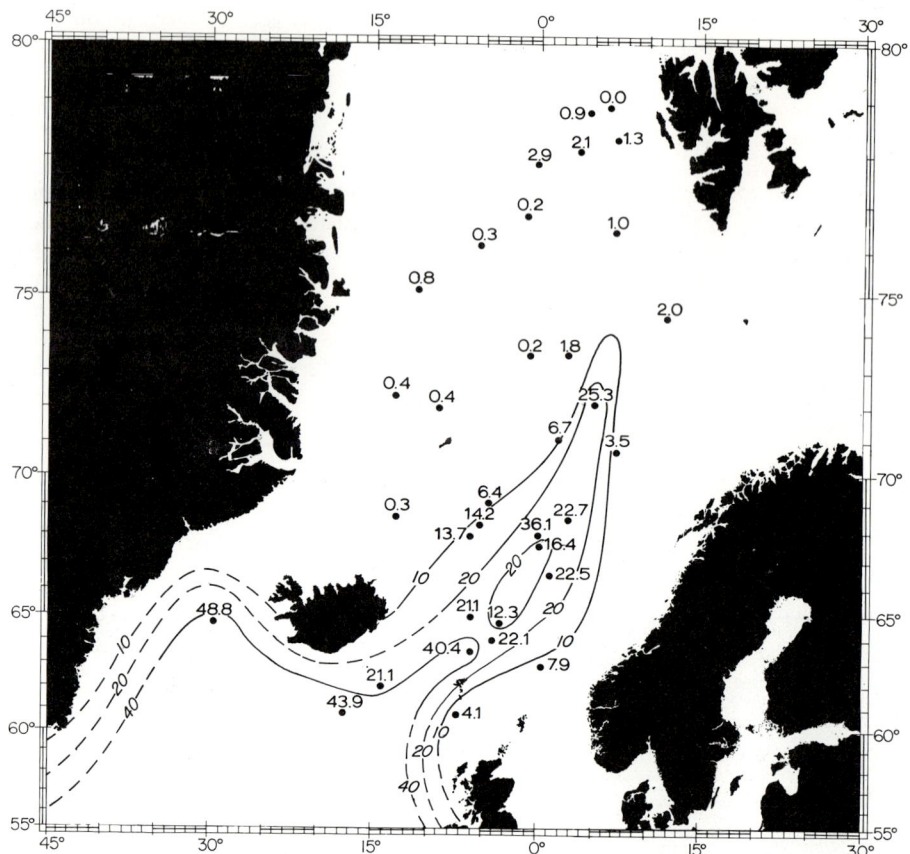

Figure 8. Percentage warm planktonic foraminifera in Norwegian Sea surface sediment samples as percent of the total planktonic foraminiferal fauna.

with large percent benthonic foraminifera were taken from shallow depths. Because benthonic abundance is known to be inversely related to water depth (Upshaw and Stehli, 1962), large percentages in shallow Norwegian Sea surface sediments are probably caused by this factor. The few remaining samples with large percent benthonic foraminifera were taken from the Greenland Sea. These are the samples with fresh planktonic material and broken or chalky benthonic specimens noted by Streeter. S. S. Streeter suggested (1974, oral commun.) that if sedimentation rates in this region were very slow, benthonic specimens may have been subject to slight or moderate dissolution while resting at the sediment-water interface for long periods of time, and they may have been covered with a very thin veneer of Holocene planktonic material. Sedimentation rates of 1.50 cm/10^3 yr in the Holocene and 2.66 cm/10^3 yr in the last glacial stage were calculated for V28-25 from the Greenland Basin (Table 4). These rates would appear to disprove Streeter's hypothesis, but no better explanation is available.

In conclusion, dissolution appears to be insignificant in the Norwegian Basin. In the slightly deeper Greenland Basin, some samples show evidence of dissolution, as recorded by the percentage and condition of benthonic specimens. Dissolution may be only locally important in this region.

Summary

The single most important factor influencing sediment and plankton distributions in the Norwegian Sea is the warm Norwegian Current. This current limits the present extent of sea-ice cover and thus controls the distribution of ice-rafted detritus. Carbonate in the sediments is highest under and adjacent to the Norwegian Current. Carbonate dissolution is a minor factor today because of shallow water depths. Finally, the warm surface water controls the distributions of right-coiling *G. pachyderma* and the "warm" species. Away from the influence of the Norwegian Current, left-coiling *G. pachyderma* dominate the faunas.

STRATIGRAPHIC STUDY OF SIX NORWEGIAN SEA CORES

Six cores were chosen to form a north-south traverse of the Norwegian Sea (Fig. 12, Table 1). The sediments at the core tops reflect all the diverse sedimentary

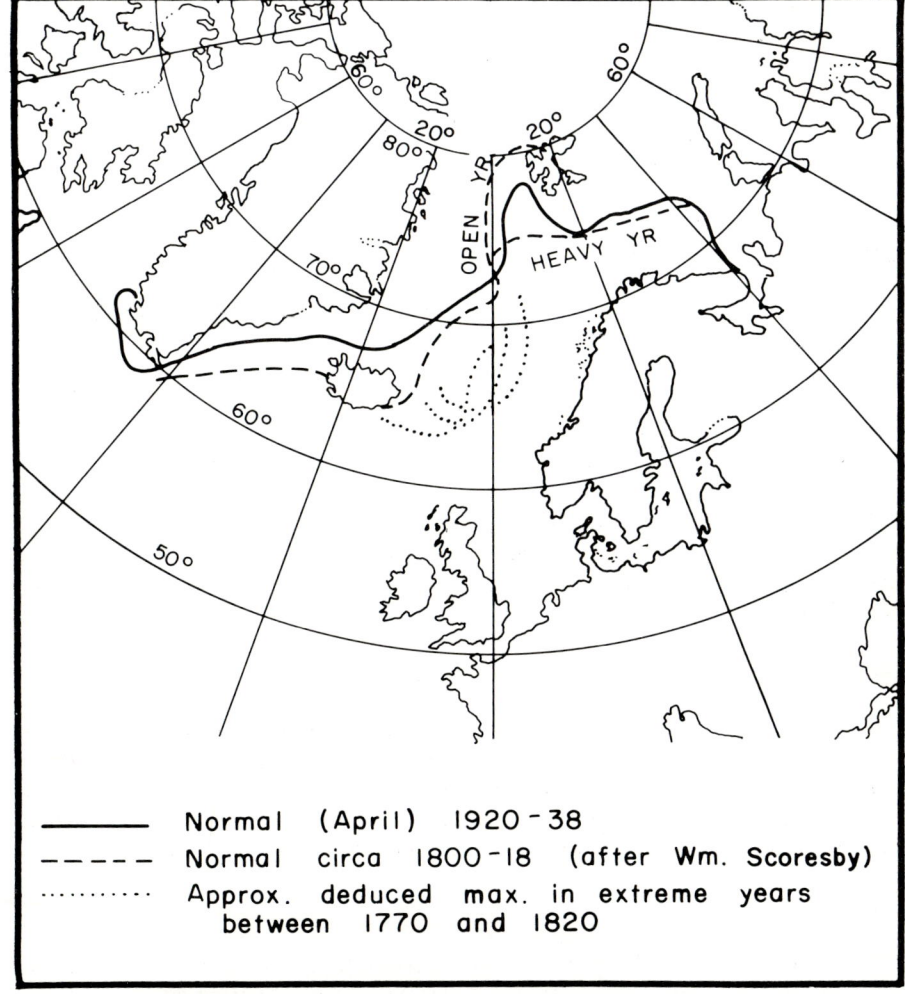

Figure 9. Maximum extent of spring pack-ice 18th to 20th centuries (after Lamb, 1968).

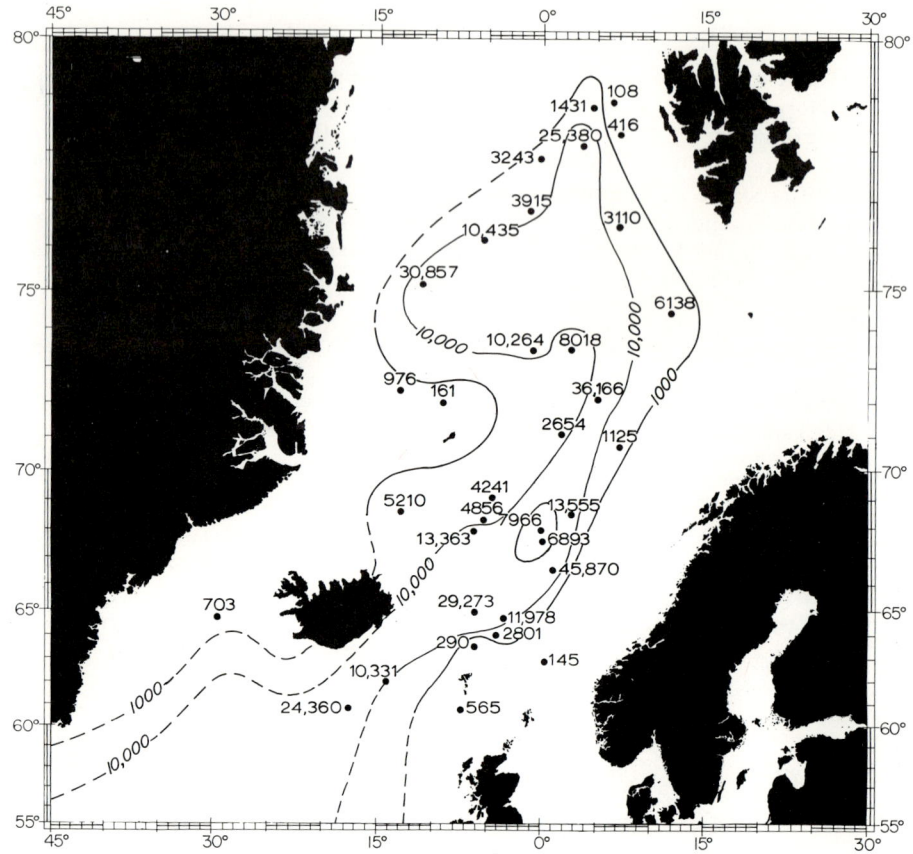

Figure 10. Planktonic foraminiferal abundance of surface sediments. Numbers plotted are number of specimens per gram of sediment.

conditions present in the Norwegian Sea. These conditions form a complete gradation from sediments with polar faunas being deposited under essentially glacial conditions in the northern Greenland Sea (V28-25) to sediments with subpolar faunas being deposited under nonglacial conditions at present in the southern part of the region (V28-14 and V27-86).

Techniques

Each core was sampled at 10-cm intervals from the top to bottom, with samples averaging about 4 to 5 g. Carbonate, coarse-fraction, and faunal analyses were made as described for the TW tops. Percent right-coiling *G. pachyderma*, percent warm fauna, planktonic and benthonic foraminiferal abundance, and carbonate and coarse-fraction content curves for each core are shown in Figures 14 to 19.

Sediments

Sediments range from foraminiferal ooze through foraminiferal marl and foraminiferal clay to clay, with varying amounts of sandy, glacial-marine detritus. Several cores have volcanic ash layers, and V27-86 contains a layer rich in manganese

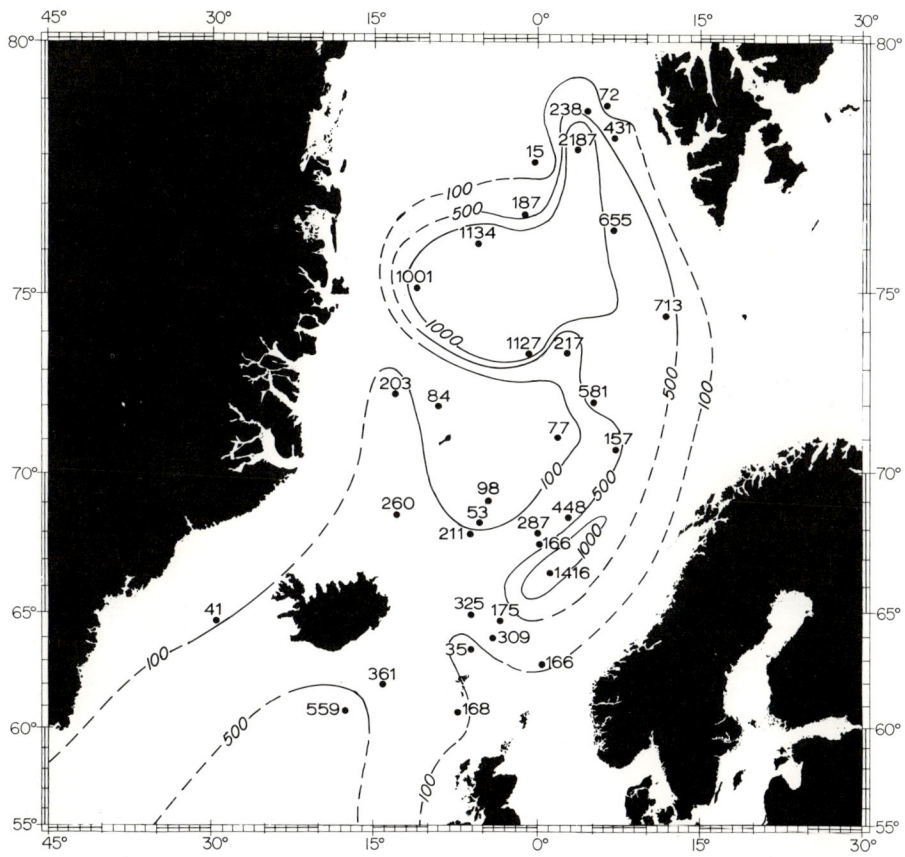

Figure 11. Benthonic foraminiferal abundance of surface sediments. Numbers plotted are number of specimens per gram of sediment.

nodules. Stratigraphic sections for each core are shown in Figure 13. Foraminiferal and carbonate content is highest at the top. Below the high carbonate zone is a zone with considerable glacial-marine detritus and lower foraminiferal and carbonate content, which is, in turn, preceded by a zone in which sedimentary conditions similar to those in the tops of each core are encountered. Below this last zone, sediments in all cores studied, except V28-56, have low foraminiferal and carbonate content and large amounts of glacial-marine detritus. V28-56 has several additional zones similar to the top of the core.

Percent coarse-fraction content (larger than 62 μm) was calculated for each sample. This was done to draw curves showing input of glacial-marine detritus. These curves are not a useful indicator of ice rafting, because foraminiferal shells are found in the same size fraction. Rather, they combine glacial-marine and foraminiferal inputs to produce curves that often appear almost random (Figs. 14–19). Ideally, each sample should be calculated on a carbonate-free basis, or a different size fraction should be analyzed. The latter technique was employed by Kent and others (1971) who analyzed the >250-μm fraction in North Pacific cores in order to avoid radiolarian tests. Kellogg (1975) calculated noncarbonate coarse-fraction curves for Norwegian Sea cores but found that an even better indicator of ice-rafted detritus was the noncarbonate fine-fraction (less than 62-μm) curves.

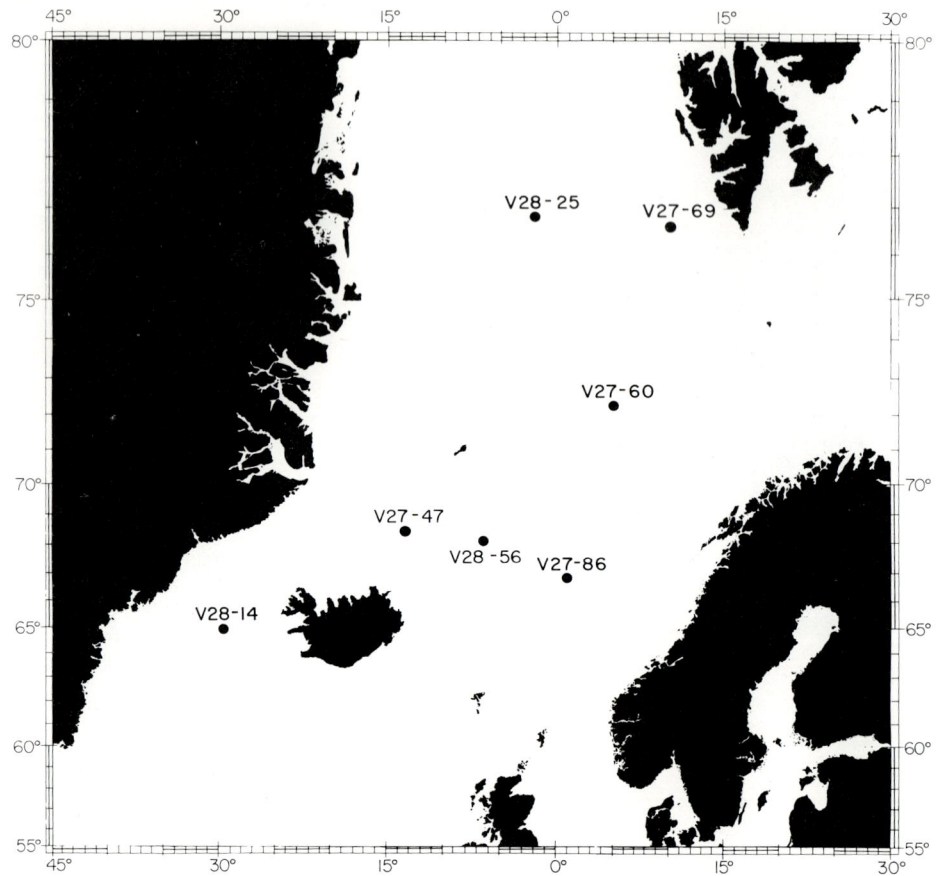

Figure 12. Locations of piston cores discussed in this paper.

Carbonate Curves

Calcium carbonate analyses have been made on deep-sea sediments by many workers, including Arrhenius (1952), Hays and others (1969), Olausson (1967), McIntyre and others (1972), Oba (1969), and Broecker and others (1968). Atlantic cores generally have a cyclic pattern of highs and lows, with high values at the top. Equatorial Pacific cores show similar cyclic patterns, with low values at the top. Interpretation of carbonate curves is usually on a climatic basis, although carbonate deposition is a complicated process with many controlling factors (Broecker, 1971).

In Atlantic cores, high carbonate values are correlated with warm climate. Schott (1935) attributed the correspondence of high carbonate values with warm periods to higher input of terrigenous material during cold periods, thus masking the carbonate. Broecker and others (1958) showed in one North Atlantic core that net carbonate input increased during the glacial stage but was masked or diluted by an even greater increase in terrigenous lutite. In a number of North Atlantic cores, Ruddiman and McIntyre (this volume) find higher absolute rates of biogenous input during interglacial climates but greater delivery of terrigenous sediments during

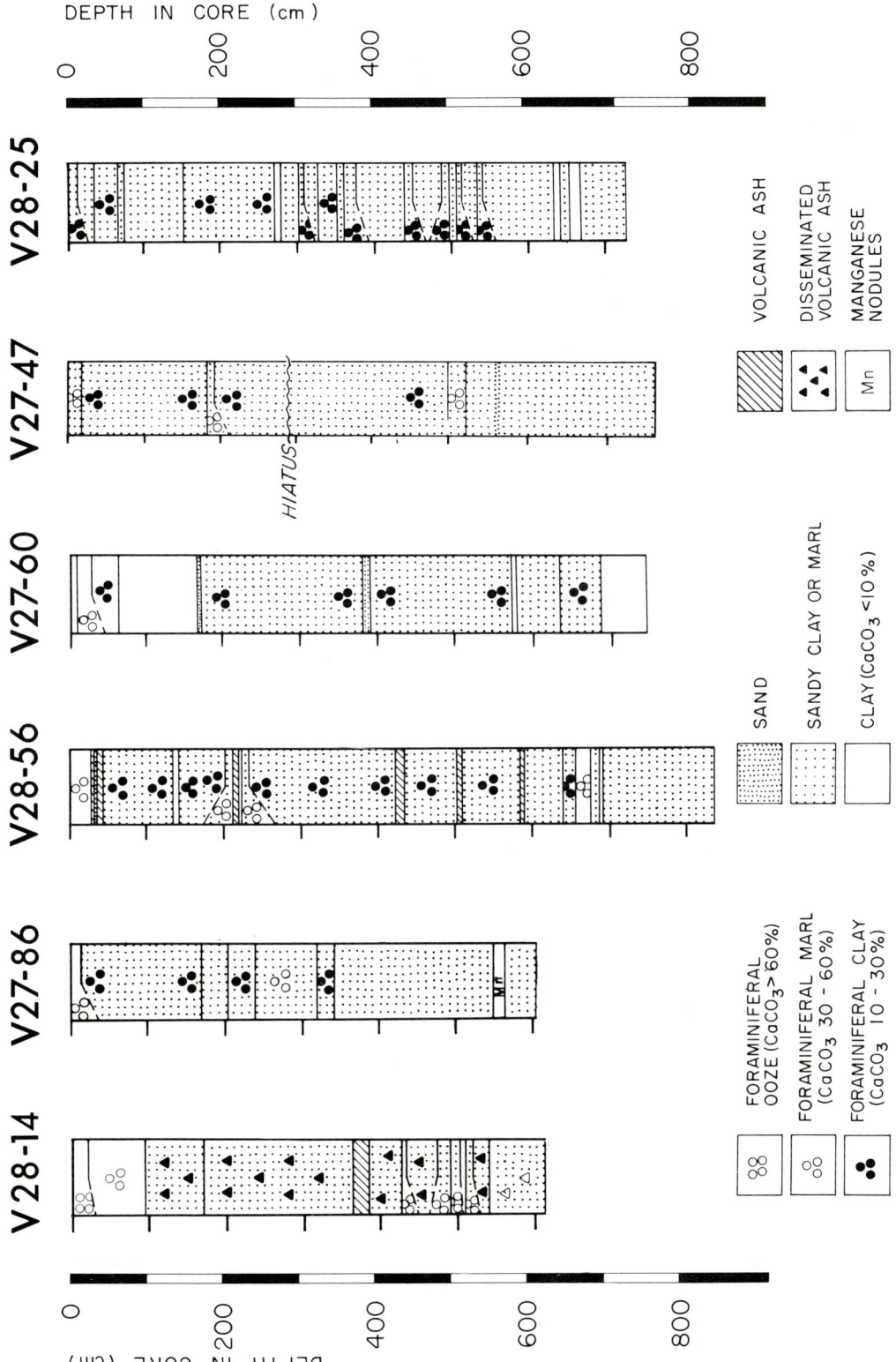

Figure 13. Sedimentary sections for six Norwegian Sea piston cores.

V28 - 14

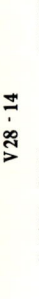

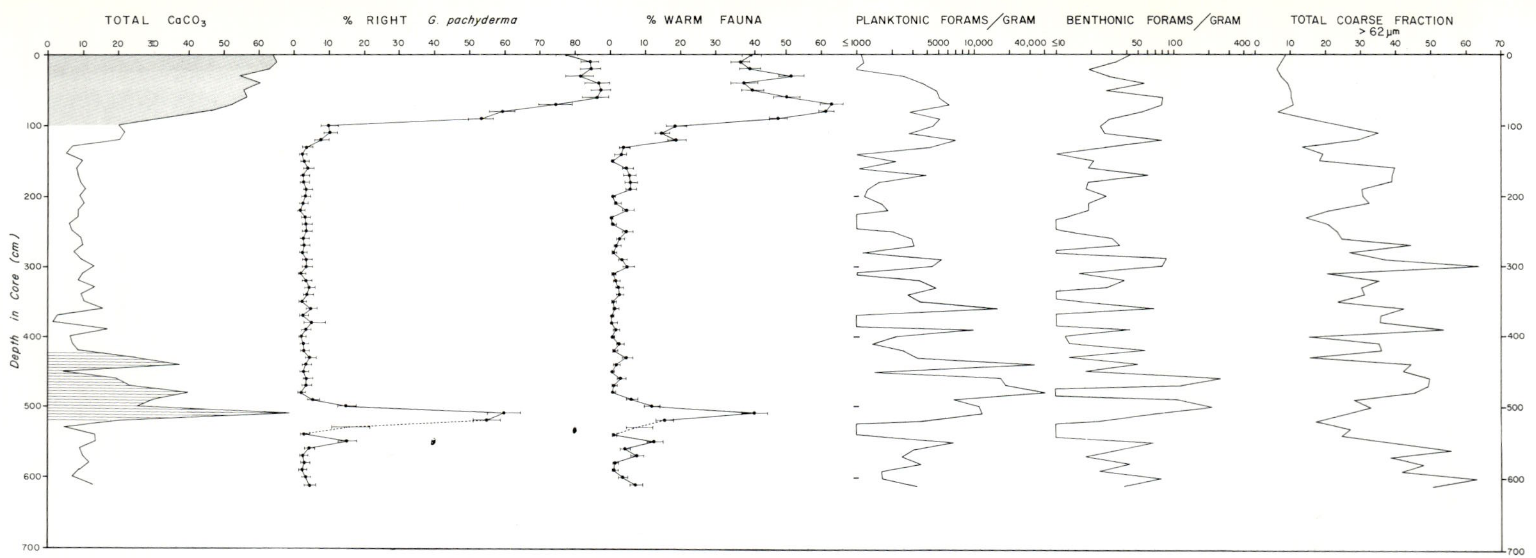

Figure 14. Total carbonate, faunal indices, planktonic and benthonic foraminiferal abundance, and total coarse fraction (larger than 62-μm) for V28-14. The Recent high-carbonate zone is indicated by shading; the PCH by horizontal lines.

V27 - 86

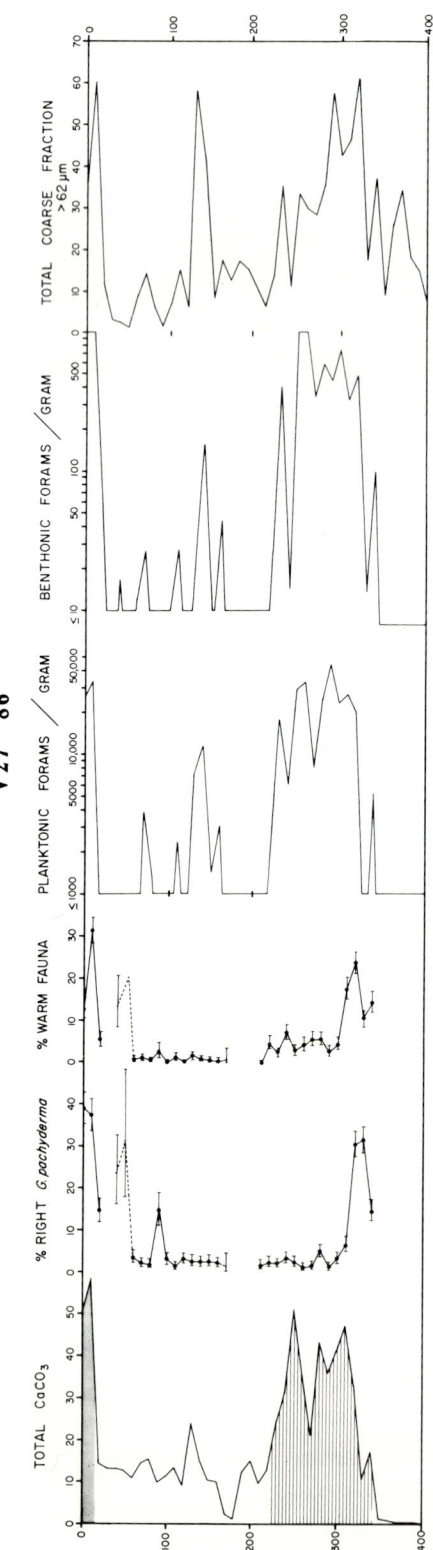

Figure 15. Carbonate, coarse fraction and faunal curves for V27–86. Symbols as in Figure 14.

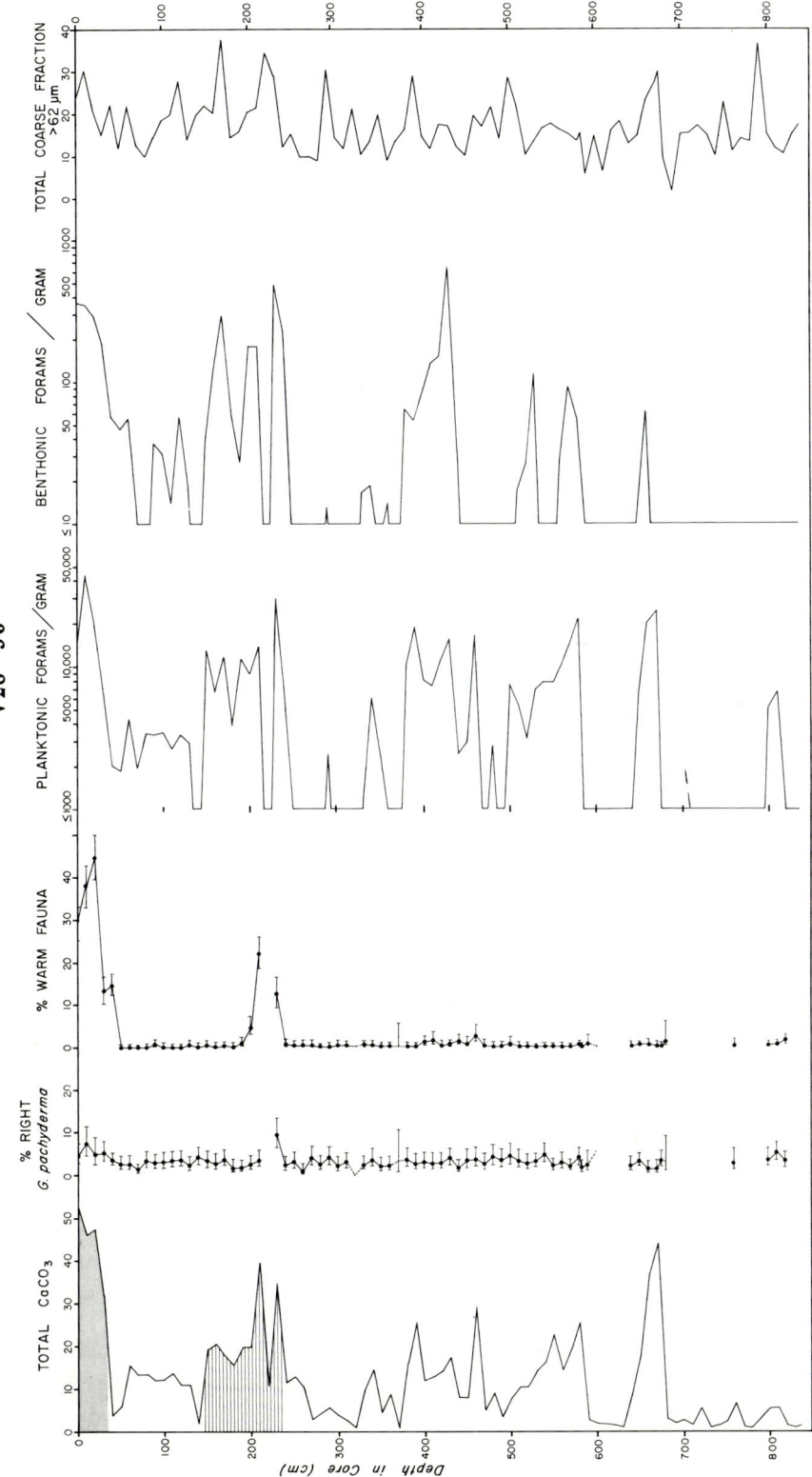

Figure 16. Carbonate, coarse fraction and faunal curves for V28-56. Symbols as in Figure 14.

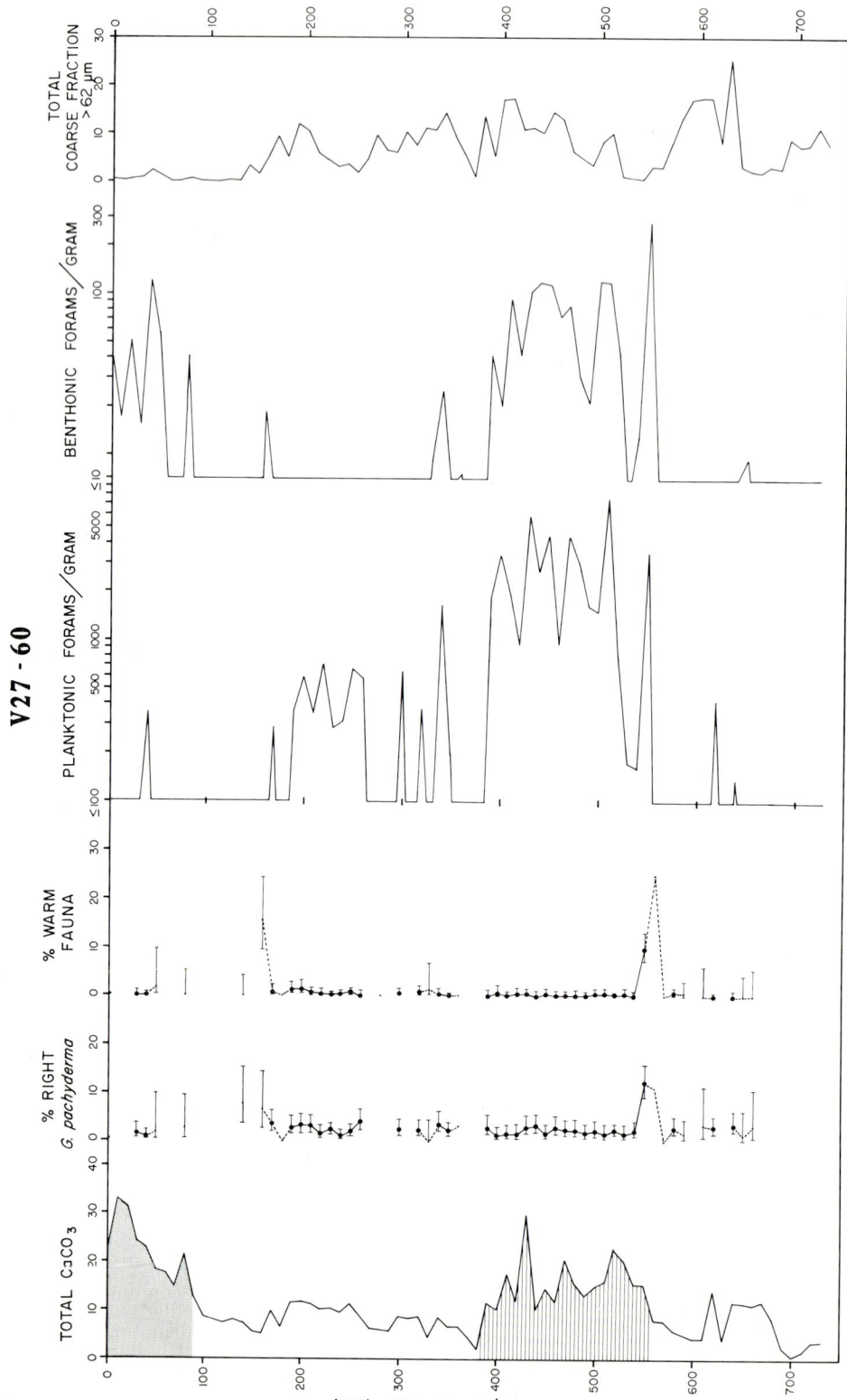

Figure 17. Carbonate, coarse fraction and faunal curves for V27–60. Symbols as in Figure 14.

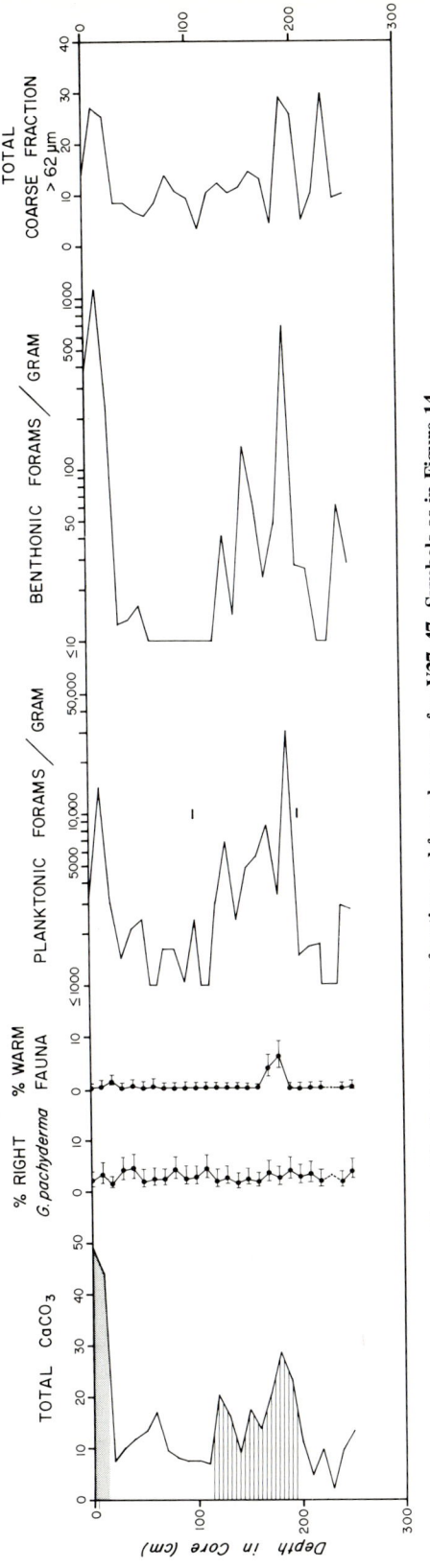

Figure 18. Carbonate, coarse fraction and faunal curves for V27-47. Symbols as in Figure 14.

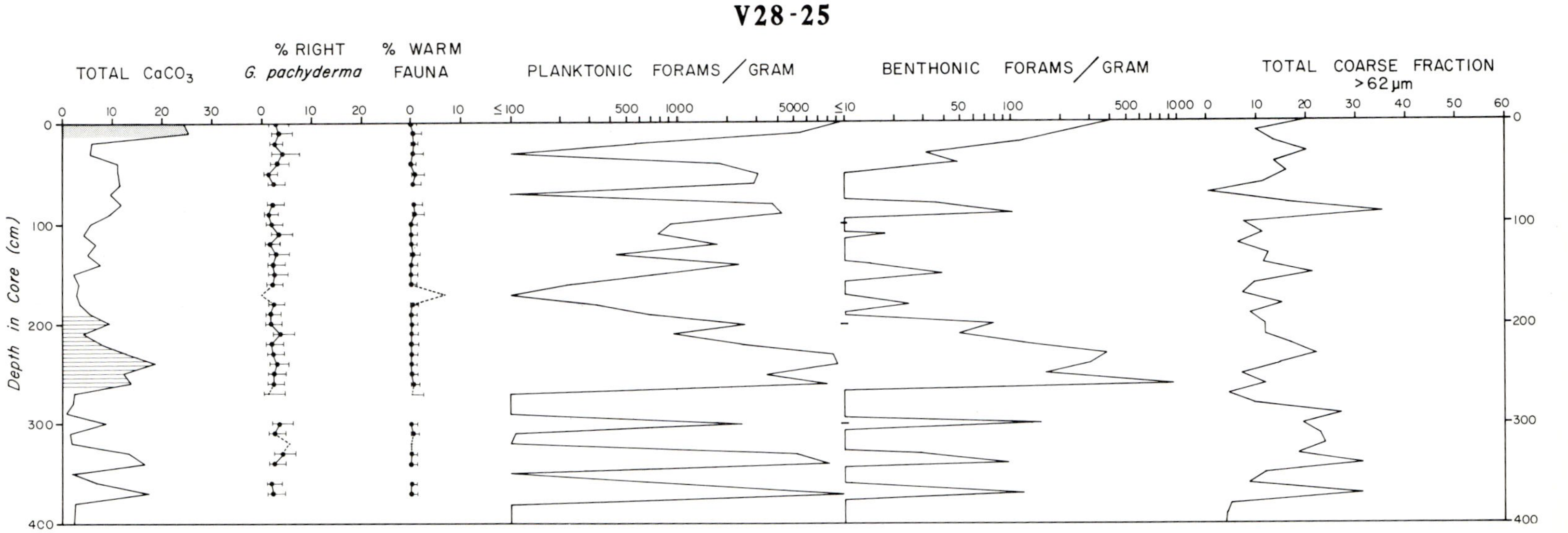

Figure 19. Carbonate, coarse fraction and faunal curves for V28-25. Symbols as in Figure 14.

glacial climates. They explain the constancy of sedimentation rate between glacial and interglacial climates by a coincidental balancing of oceanically produced biogenous and continentally derived terrigenous input.

All 6 Norwegian Sea cores have carbonate curves with typical cyclic Atlantic oscillation patterns (Kellogg, 1975). These are interpreted as follows: The high-carbonate zone at the top termed the "Recent high-carbonate zone," varies in thickness from 10 to 140 cm and is preceded by a low-carbonate zone called "last glacial stage," varying in length from 100 to 335 cm. The last glacial is preceded by the "Penultimate high-carbonate zone" (PCH) varying in thickness from 75 to 180 cm. Several earlier cycles are present in the carbonate curve from V28–56 but have not been interpreted. All the remaining cores end in glacial sediments deposited prior to the PCH.

The interpretation presented above is based on comparison with the carbonate distribution in surface sediments. Carbonate values in the 6 piston-core tops are almost identical with values in corresponding TW tops. High carbonate values are considered indicative of relatively warm surface water and high productivity, because high values in the TW tops are directly related to these parameters. In the last glacial stage, low carbonate values in all 6 cores are lower than the highest values in V28–25, the northernmost core. Thus, during this cold period, conditions throughout the Norwegian Sea were at least as severe as those in the northern Greenland Sea today.

Faunal Indices

Percent right-coiling *G. pachyderma* and percent warm-species curves for the 6 cores are shown in Figures 14 to 19. These indices were calculated as described for surface-sediment samples. Error bars were calculated for each sample based on the normal approximation to the binomial distribution (Meyer, 1965). This statistical technique gives upper and lower bound confidence limits on the percent recorded. Calculations were made at the 95% confidence level. Counts of 300 or more specimens were made for most samples. Samples with less than 25 specimens were considered barren and were not plotted. Samples with between 25 and 200 specimens were plotted with small dots and connected by dashed lines on Figures 14 to 19 because of statistical uncertainty of small number counts.

Values of faunal indices in the 6 cores are generally lower than 5.0% indicating very cold conditions. The only exceptions occur in the Recent high-carbonate zone and in the basal portion of PCH. In the Recent high zone, V28–25, V27–47, and V27–60 taken closest to the present limit of sea ice show no indication of warming. In V28–14, V27–86, and V28–56, high faunal values are found in the Recent high-carbonate zone, indicating relatively warm conditions throughout the zone.

In the basal portion of the PCH, all cores except V28–25 have relatively high values of faunal indices. The values are highest in the southernmost cores and decrease toward the present limit of ice cover. These high values of faunal indices occur as short, sharp pulses. The duration of the pulses are considerably shorter than the PCH. The duration is longest in the southernmost cores and decreases to the north toward the present ice limit.

Foraminiferal Abundance

Planktonic and benthonic foraminiferal abundances were calculated for each sample in the 6 cores in the same manner as surface-sediment samples (Figs. 14 to 19).

Planktonic abundance curves follow the carbonate curves, with two exceptions—V28-14 and V27-60. High abundances coincide with high carbonate values, further strengthening the assertion that high carbonate values are caused by high productivity in warm surface waters. In V28-14 and V27-60, planktonic abundance follows carbonate content except in the Recent high-carbonate zone. Both these cores have anomalous Recent high-carbonate zones in a number of respects. (1) The Recent high zone is considerably longer (sedimentary length) than in any of the other cores studied. (2) Much of the carbonate in the Recent high zone is fine material less than 62 μm, and foraminiferal shells are relatively rare. (3) The biogenic assemblage present contains large numbers of sponge spicules, diatoms, and radiolaria. It is not surprising that the planktonic foraminiferal abundances do not fit the pattern established for the other cores in light of these anomalous features.

Benthonic foraminiferal abundance curves also parallel carbonate content (Figs. 14–19). The one exception occurs in the Recent high-carbonate zone of V28-14 and is probably caused by the same factor that caused low values of planktonic abundance in the same samples.

Summary and Discussion

The curves presented in Figures 14 to 19 make possible the documentation of four climatic regimes in the Norwegian Sea. (1) The present regime is characterized by high carbonate, foraminiferal abundance and faunal index values in the southern cores. This regime is considered as a full interglacial stage, and only at the base of the PCH are these conditions duplicated. Thus, the basal portion of the PCH is considered the last interglacial stage. (2) Intermediate climatic conditions, between full glacial and full interglacial stages, are recognized by the coincidence of high carbonate values with low values of the faunal indices. Such intermediate conditions exist today at the locations of V28-25 and V27-47. Both these cores have polar faunas in their tops, but both have high carbonate values and high foraminiferal abundances in the same samples. V28-25 is located in the northern Greenland Sea where winter sea ice is present; V27-47 is close to the extreme limit of winter sea ice. These conditions, introduced in this paper as "glacial," are, rather, the intermediate climatic conditions typically found in Norwegian Sea cores for most of the time represented by the PCH. (3) Full glacial conditions are characterized by the coincidence of low carbonate values with low values of the faunal indices and foraminiferal abundance. (4) Barren zones, when present, coincide with ultralow carbonate values and are usually found in the cores, relative to the carbonate curves, at the top and bottom of the last glacial sequence and just below the base of the PCH. The first two barren zones may be correlative with the early and late Wisconsin ice advances. The last zone probably corresponds to the pre-Sangamon or pre-Eemian ice maximum. These correlations are strengthened by the correlations between coccolith barren zones in V23-82 and foraminiferal barren zones in V27-86.

The cause of barren-zone formation is not certain. It seems unlikely that surface-water productivity would stop completely during a glacial maximum, even if the Norwegian Sea had been completely ice covered throughout the year. Today the Arctic Ocean supports a foraminiferal fauna under essentially complete ice cover (Bé, 1960). Sedimentation rates are almost twice as high in the last glacial interval as in the high-carbonate zones (Table 4); thus, masking of low carbonate values and low foraminiferal abundances by increased ice rafting certainly contributes to the formation of the barren zones.

TIME CONTROL

The time framework used for the six Norwegian Sea cores is based on dating of core V28-14 and correlation of the carbonate curve with dated carbonate curves and ash horizons from northern North Atlantic cores.

A radiocarbon age of 4018 B.P. ± 120 yr was obtained for the interval between 40 and 47 cm in V28-14 (Fig. 20). Using 44 cm as the level dated gives a sedimentation rate of 9.1 cm/10^3 yr for the top of the core. Ruddiman and Glover (1972) described their ash layer 1 in a number of North Atlantic cores, giving an age of 9300 B.P. based on numerous radiocarbon dates. Ruddiman and McIntyre (1973), on the basis of shard counts in V28-14, considered a disseminated ash layer at 100 cm to be ash layer 1. When the radiocarbon-determined sedimentation rate of 9.1 cm/10^3 yr for the top of V28-14 is extrapolated to ash layer 1, an age of 9100 B.P. is obtained, confirming the work of Ruddiman and McIntyre (1973). The difference of 200 yr is not significant but, if real, might be attributed to compaction in the lower part of the Recent high-carbonate zone. Thus, a convincing time framework is established for the top meter of V28-14.

The top meter of V28-14 appears to have been stretched, probably during the coring operation, relative to the remainder of the core. For this reason, sedimentation rates determined for the top of the core were not extrapolated. Sedimentation rates for the last glacial zone and the PCH were obtained using the ionium method (Ku, 1966; Goldberg and Koide, 1962). Excess Th^{230} is plotted versus depth in V28-14 in Figure 20 (Table 3) for the five samples analyzed. One sample taken from the PCH has a much higher carbonate content than the others. Because

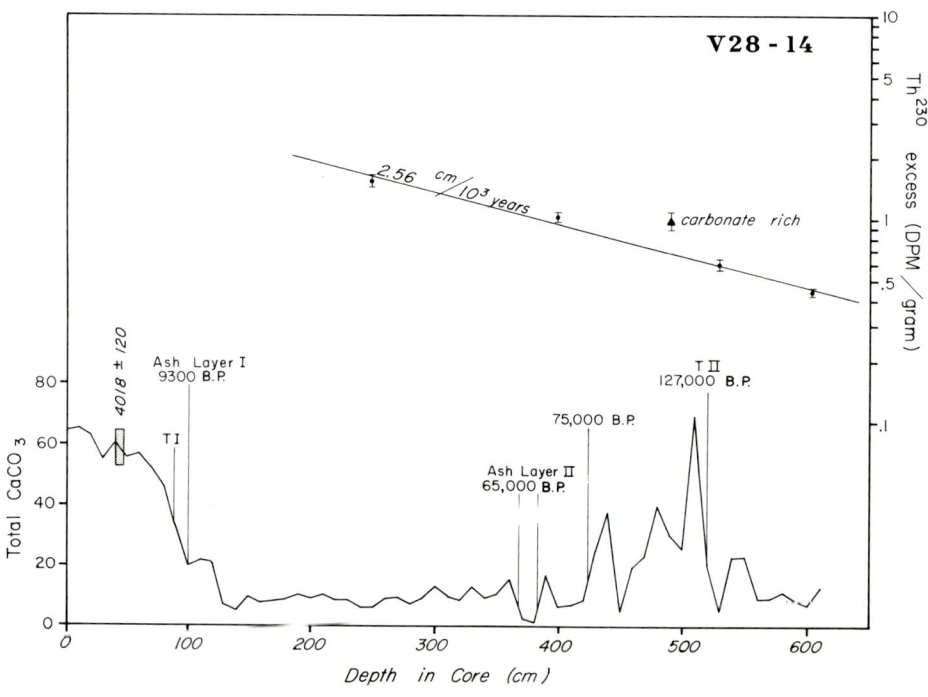

Figure 20. Total carbonate curve for V28-14 showing locations of carbonate terminations I and II, ash layers 1 and 2, and the sample dated by the radiocarbon method at 40 to 47 cm. Excess Th^{230} values are plotted versus depth (right-hand scale) above the carbonate curve to show where samples were taken for ionium measurement.

TABLE 3. V28-14 IONIUM METHOD DATA

Sample depth (cm)	U (ppm)	Th (ppm)	U^{234}/U^{238}	Th^{230}/U^{234}	Th^{230}/Th^{232}	Th^{230} (excess) (DPM/g)
250	1.73±0.05	9.52±0.48	0.89±0.03	2.43±0.15	1.11±0.05	1.52±0.08
398	1.85±0.06	10.4±0.6	1.20±0.03	1.95±0.12	0.82±0.05	1.02±0.06
490*	0.66±0.04	2.45±0.17	1.01±0.06	3.09±0.31	2.39±0.18	0.98±0.10
529	1.23±0.04	4.86±0.29	0.92±0.03	1.77±0.11	1.18±0.06	0.60±0.04
603	1.37±0.04	5.04±0.25	1.12±0.03	1.44±0.04	1.25±0.05	0.44±0.02

*Total carbonate = 30.0%.

no corrections were made for carbonate content, this sample was not considered in the least-squares regression to determine the best-fit line. The line, based on the four low-carbonate samples, has a correlation coefficient of 0.995, suggesting an excellent fit. From the slope of the line and the half life of Th^{230} of 75,000 yr, a sedimentation rate of 2.56 cm/10^3 yr was determined for the segment of V28-14 between 250 and 603 cm. This is about 25% of the rate determined for the top of the core, suggesting that there may be major changes in sedimentation rate between high- and low-carbonate intervals.

The sedimentation rate determined by the ionium method was used to calculate the age at the base of the PCH. The rate was extrapolated from a conspicuous volcanic ash layer occurring between 364 and 382 cm in V28-14. W. F. Ruddiman (1973, oral commun.) believes this ash is ash layer 2 dated in a number of North Atlantic cores at 65,000 B.P. (Ruddiman and Glover, 1972). The extrapolated age of 119,000 B.P. for the base of the PCH is close to the 127,000 B.P. age of termination II (Broecker and van Donk, 1970).

The term "termination" as originally defined by Broecker and van Donk (1970) referred to the sudden change in oxygen isotope values found at the end of glacial sequences. Broecker (1971) has extended usage to include similar sharp changes in carbonate values at the same times. Terminations I and II are dated at about 11,000 and 127,000 B.P., respectively (Broecker and van Donk, 1970). The base of the PCH is a sharp change at the end of a glacial sequence. The extrapolated age of this termination (119,000 B.P.) is close to the age given by Broecker and van Donk for termination II; therefore, their time scale has been applied to conspicuous features of the carbonate curve from V28-14 (Fig. 21).

McIntyre and others (1972) and Ruddiman and McIntyre (this volume) present carbonate curves for a number of North Atlantic cores. Their time frame, based on extrapolation of numerous radiometric analyses made by Ku and others (1972), agrees closely with the time scale of Broecker and van Donk (1970). The same basic pattern of carbonate fluctuations is present in both Norwegian Sea and northern North Atlantic cores (north of 43°N). Carbonate curves from V28-14 and V27-86 are compared with the fine-fraction carbonate curve from V23-82 in Figure 21. The similarity of the patterns is obvious, and correlation of major features of the curves is justified. Therefore, the time scale of Broecker and van Donk (1970) is adopted for all six Norwegian Sea cores (Fig. 22).

Sedimentation Rates in Norwegian Sea Cores

Sedimentation rates were calculated for each core for the Recent high-carbonate zone, the last glacial zone (low-carbonate zone), and the PCH (Table 4). These zones are bounded by carbonate terminations I and II and the 75,000 B.P. level. In all cases, the top of the core is considered to represent the present. Approximate

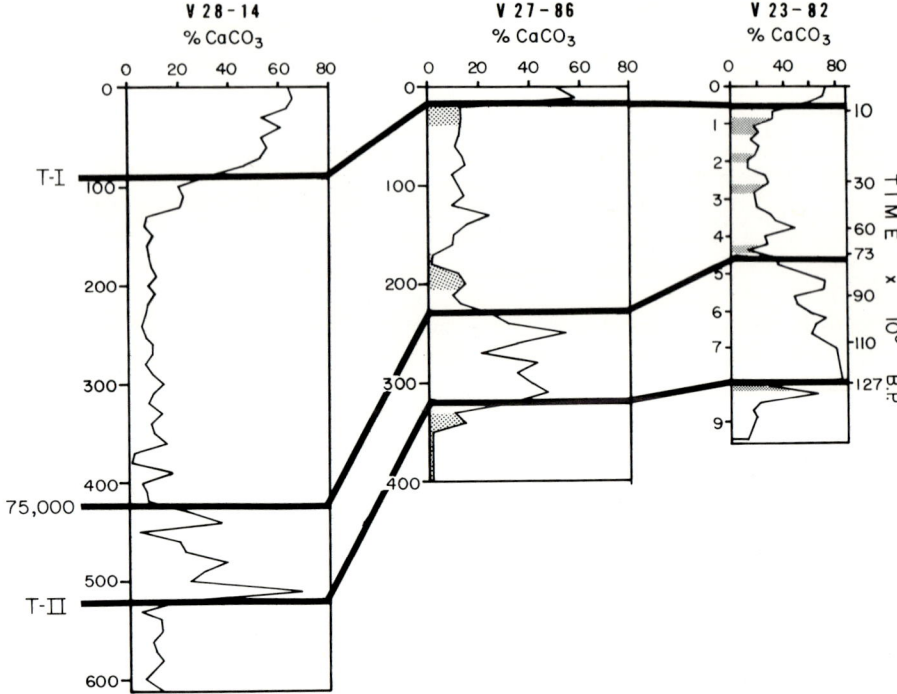

Figure 21. Total carbonate curves from V28-14 and V27-86 and fine-fraction carbonate curve from V23-82 (after McIntyre and others, 1972) correlated at carbonate terminations I and II and at the 75,000-yr level. Stippled areas in V23-82 are barren of coccoliths. Stippled areas in V27-86 are barren of foraminifera.

overall rates were calculated from the top of each core to termination II and were extrapolated to get rough ages at the base of each core (Table 4).

Ruddiman and McIntyre (1973) show that the age of termination I is time transgressive in North Atlantic cores. Broecker and van Donk (1970) give the age of termination I as 11,000 B.P. but in V28-14 the age is clearly 8200 B.P. as shown by Ruddiman and McIntyre (1973). Thus, the age of termination I in the remaining five Norwegian Sea cores is problematic. In order to calculate sedimentation rates for the Recent high-carbonate and last low-carbonate zones, the age of this termination must be established. I have taken Figure 4 from Ruddiman and McIntyre (1973) and extended their 11,000- and 9,000-yr isochrons into the Norwegian Sea (see Kellogg, 1973; Fig. 5-12). The 9,000-yr isochron crosses the Norwegian Sea from the northeast tip of Iceland to northern Norway; the 11,000-yr isochron passes through the Faeroe Islands and intersects the Norwegian coast at about 65°N. On the basis of this first approximation, the age of termination I in V27-86 should be about 11,000 yr. For the remaining cores, the termination occurred about 8,000 yr ago or less.

Calculated sedimentation rates vary by about 1.0 cm/10^3 yr between the last glacial zone and the PCH. Rates calculated for the Recent high-carbonate zone are 2.0 cm/10^3 yr higher than in the last glacial zone. This figure is considered inaccurate because it is biased by the long Recent high-carbonate zones in V28-14 and V27-60, both of which are anomalous, as discussed above. When the average rate for this zone is calculated, omitting these cores, an average rate of 2.30 cm/10^3

yr is obtained. This is almost identical to the rate calculated for the PCH and is thus considered a more reliable figure. In summary, sedimentation rates in the last glacial stage (low-carbonate zone) are almost twice as high as rates in the Recent and PCH.

Dating and Duration of Faunal Pulses

The durations of faunal pulses (percent right-coiling *G. pachyderma* and percent warm fauna) are easily calculated, because the sedimentary intervals over which they occur as well as the sedimentation rates in the zones are now known (Table 5).

The Holocene faunal pulse has lasted longest and indicates warmest temperatures in the southernmost cores (V28-14 and V27-86) and has not yet occurred in V28-25 and V27-47. The pulse is obscured in the barren zones of V27-60.

The warm faunal pulse associated with the base of the PCH follows the same basic trends as in the Holocene with longest durations in the south decreasing toward the north. No pulse is observed in V28-25. Evidently the pulse in the PCH penetrated slightly further into the Norwegian Sea and lasted longer than the Holocene pulse. These relationships are summarized in Figure 23.

In the southernmost cores studied, the warm faunal pulse associated with the base of the PCH lasted about 13,000 to 17,000 yr during the period of time between 110,000 and 127,000 B.P. This time period is correlated by Sancetta and others (1972) with the Pangaion interglacial of eastern Macedonia (van der Hammen and others, 1971, Fig. 2). Van der Hammen and others (1971), in turn, correlate the Pangaion with the Eemian interglacial of western Europe on the basis of pollen stratigraphy. Thus, the warm faunal pulse observed in Norwegian Sea cores at the base of the PCH is probably the Eemian interglacial and may be further correlated with the Barbados III high-sea stand (Mesolella and others, 1969).

The Eemian is the last interglacial period in northwestern Europe. It is immediately followed by a short, intensely cold period (Matthews, 1972; Sancetta and others, 1972; Kukla and Koči, 1972) and, in turn, by a return to warmer conditions (but not as warm as during the Eemian) termed "early glacial" (van der Hammen and others, 1971, Fig. 2). In Norwegian Sea cores, this "early glacial" is represented

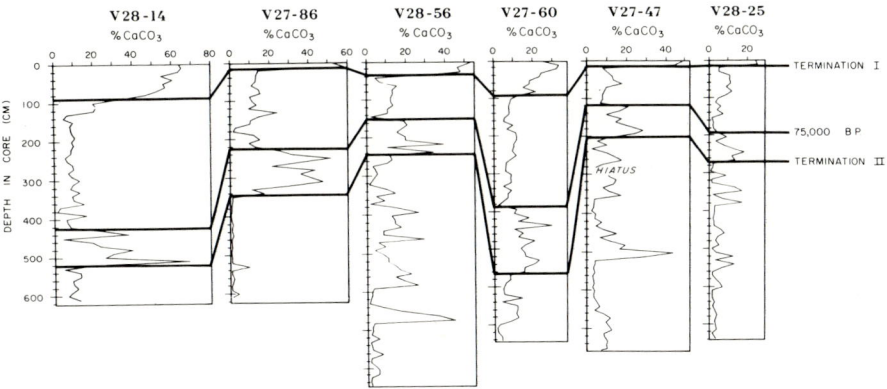

Figure 22. Total carbonate curves for the six Norwegian Sea cores correlated at carbonate terminations I and II and at the 75,000-yr level. Cores are arranged in order of increasing latitude from left to right except that V27-47, which shows a very cold fauna because of its position north of Iceland, is placed between V27-60 and V28-25 instead of between V28-56 and V27-60.

TABLE 4. DEPTHS OF TERMINATIONS I AND II AND THE 75,000-YR LEVEL, AND SEDIMENTATION RATES CALCULATED BETWEEN THESE ISOCHRONS IN SIX NORWEGIAN SEA CORES

Core	Depth in core (cm)			Sedimentation rates (cm/10 yr)				Age at base of core (yr)
	TI	75,000	TII	0 - TI	TI - 75,000	75,000 - TII	0 - TII	
V28-14	90	425	520	10.1	5.00	1.82	4.09	149,000
V27-86	16	225	340	1.45	3.27	2.21	2.67	227,000
V28-56	35	145	235	4.38	1.64	1.73	1.85	452,000
V27-60	90	385	565	11.22	4.40	3.46	4.45	164,000
V27-47	15	115	195	1.88	1.49	1.54	1.53	610,000
V28-25	12	190	265	1.50	2.66	1.44	2.08	347,500
Average				5.19	3.08	2.03	2.78	
Average excluding V28-14 and V27-60				2.30				

by the upper part of the PCH. This is a time of intermediate temperatures in which faunal indices have low values, but productivity, measured by foraminiferal abundance and carbonate content, remains high.

SUMMARY AND DISCUSSION

The basic results of this study are summarized in Figure 23. Planktonic foraminifera found in six Norwegian Sea deep-sea cores form two major assemblages. The first is a polar assemblage dominated by left-coiling G. pachyderma, in which other species comprise less than 5% of the total fauna. The polar fauna has been dominant in the Norwegian Sea for most of the past 127,000 yr and possibly for the past 450,000 yr. Only during the Holocene and about 120,000 B.P. (Eemian) was the polar fauna replaced or displaced by a subpolar fauna.

The subpolar fauna is characterized by the presence of more than 5% of the right-coiling morphotype of G. pachyderma and by the presence of the warm species: G. bulloides, G. quinqueloba, G. inflata, G. glutinata, and others. This assemblage has occurred in the Norwegian Sea only twice in the last 150,000 yr. The occurrences appear as pulses (Fig. 23) in the Holocene and about 120,000 B.P. The pulses

TABLE 5. SEDIMENTARY THICKNESS AND DURATION OF WARM FAUNAL PULSES

Core	Recent high carbonate zone		Penultimate high carbonate zone	
	Thickness (cm)	Duration (yr)	Thickness (cm)	Duration (yr)
V28-14	120	11,900	30	16,700
V27-86	20	13,800	30	13,500
V28-56	40	9,140	21§	12,100
V27-60	†		20	5,780
V27-47	*	0	20	7,700
V28-25	*	0	*	0

*Pulse not present.
†Pulse obscured by barren zones.
§Assuming the ash layer at 200 to 216 cm is an instantaneous event.

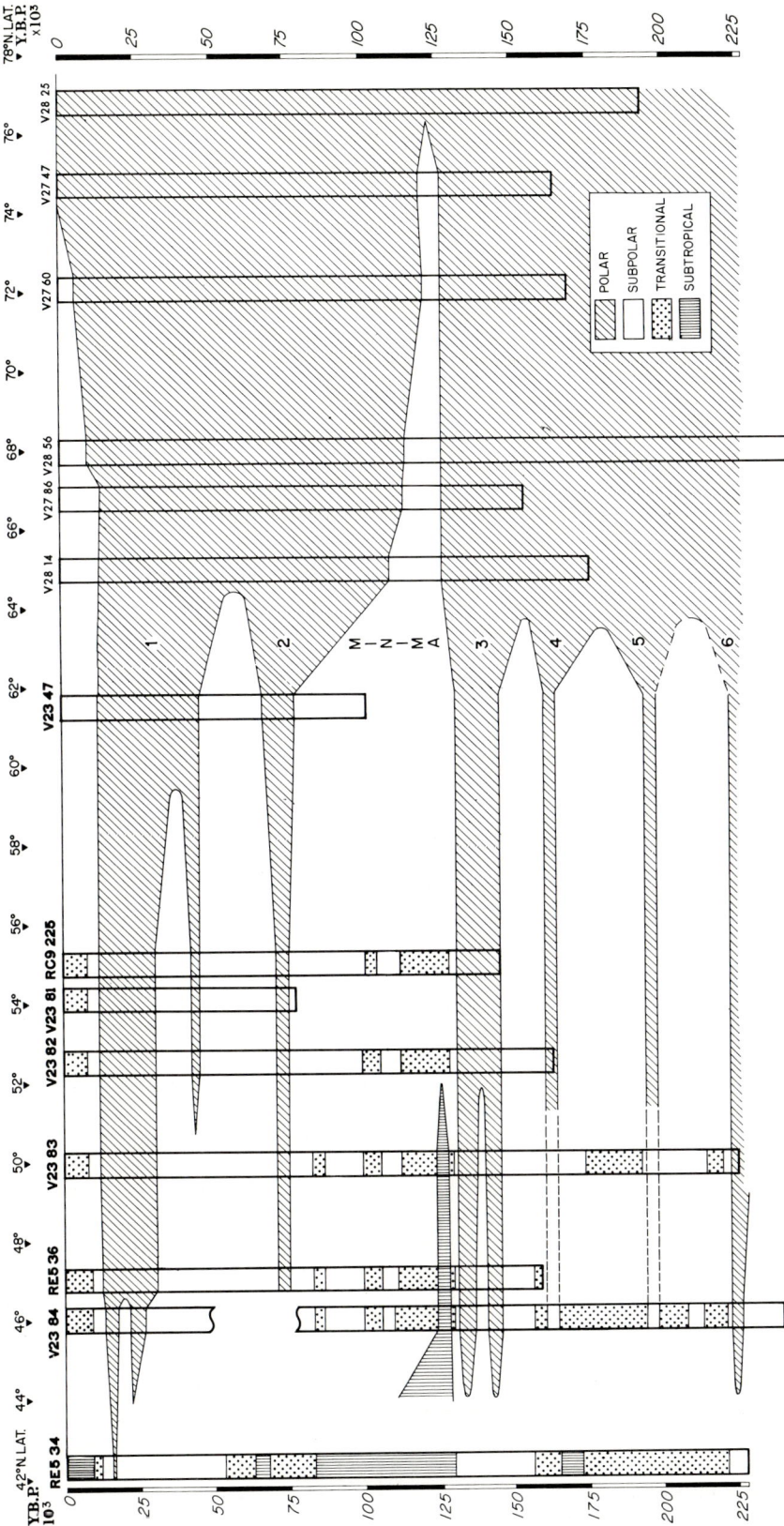

Figure 23. Variations in the faunal and floral composition in the Norwegian Sea and northern North Atlantic. Climatic zones south of 62°N are based on coccoliths (McIntyre and others, 1972). Note that subpolar faunas have been present in the Norwegian Sea only twice in the last 150,000 yr; at present and about 120,000 B.P.

are strongest to the south and disappear toward the north. The Holocene pulse has not yet lasted as long as the 120,000 B.P. pulse.

Because surface-sediment distributions of sedimentary and faunal parameters are known and because the core samples in both the Holocene and 120,000 B.P. pulses are similar in sedimentary and faunal respects, it seems reasonable to equate conditions at these two times. For the remaining 100,000 (or more) of the past 127,000 yr, sedimentary and faunal parameters show conditions similar to or more severe than those of the northern Greenland Sea today. During most of the PCH in each core (except in the 120,000 B.P. pulse), conditions were almost identical to those at present at the location of V28-25. High carbonate and foraminiferal abundance values with low warm-faunal indices characterize these conditions. This regime is the "early glacial" of van der Hammen and others (1971).

In the low-carbonate zones, foraminifera are rare and are dominated by the polar fauna. These conditions are even more severe than those of the northern Greenland Sea today, suggesting that the Norwegian Current, shown to have an extremely important effect on climate of the region today, was either completely absent or at least much weaker. This change in surface-water circulation probably permitted year-around ice cover of the entire Norwegian Sea. Ice certainly covered all of the region during the winters.

These conclusions have a number of important ramifications concerning climate in the larger regions of the North Atlantic, Europe, and Scandinavia, and for the deep circulation of the oceans. The Gulf Stream and its surface extension, the Norwegian Current, at present form an important mechanism for the transfer of heat from the tropics toward the poles. Heat is transferred to northern Europe and Scandinavia by winds blowing across the warm water. This effect is responsible for the temperate climate of Scandinavia at the same latitude where Greenland today supports an ice cap. Data presented above strongly suggest that the Norwegian Current did not exist or was much weaker for most of the past 127,000 yr or more. Thus, a major source of heat for the surrounding land masses was not present. This is not to suggest that the Würm or Wisconsin Glaciation was caused by the Norwegian Current leaving the region. Rather, the changing climate probably caused the exodus of the current. The observations made here probably represent a feedback mechanism that helped to accentuate in Scandinavia and Northern Europe the effects of the worldwide cooling event.

The exodus of the Norwegian Current also permitted the spread of polar waters into presently subpolar regions, thus permitting sea-ice formation throughout the Norwegian Sea. Both faunal and sedimentary evidence show that during the last glacial stage, ice covered the entire region, at least on a seasonal basis. The data do not prove year-around ice cover was present but suggest that this may have been the case. This observation is important for the deep-water circulation of the North Atlantic. Today Norwegian Sea overflow water forms in the Norwegian Sea as saline Norwegian Current water is made denser by evaporation. This dense water sinks to the bottom and flows to the north forming polar bottom water. It also overflows the Iceland-Faeroe Ridge and Denmark Strait to form Norwegian Sea overflow water. If the Norwegian Current, the source of saline water, were absent and ice covered the region effectively prohibiting evaporation, Norwegian Sea overflow water would not have formed in the Norwegian Sea. This appears to have been the case during much of the past 127,000 yr.

Works by Ruddiman and Glover (1972) and Ruddiman and McIntyre (1973) show that ice was present in much of the North Atlantic north of 45°N during the last glacial stage. The extreme southern limit of sea-ice was probably in contact with the northern edge of the clockwise-flowing subtropic gyre. This suggests that the region near the ice margin was probably characterized by melting during most of the year. Deep water might have formed by freezing at the base of the ice north of the ice margin.

Streeter (1973) presents benthonic foraminiferal data from the North Atlantic which suggest that bottom water formed under the ice in the large region north of 45°N (see also Weyl, 1968) during the last glacial stage. This bottom water was characterized by temperatures of about 4° to 5°C. This is considerably warmer than presently forming Norwegian Sea overflow water. Thus, if the bottom water of the North Atlantic formed in the Norwegian Sea during the last glacial stage, it did not do so by the same mechanism as it does today.

ACKNOWLEDGMENTS

I thank Stephen Streeter and Andrew McIntyre for their critical review of the manuscript. Financial support was provided by Office of Naval Research Contract N00014-67-A-0108-0004 and National Science Foundation Grants IDO71-04204, DES-72-01568, GA-29460, and GA-19690. This work is part of a Ph.D. thesis prepared under the helpful supervision of James Hays and submitted to Columbia University.

REFERENCES CITED

Arrhenius, G., 1952, Sediment cores from the east Pacific: Swedish Deep-Sea Exped., 1947-48, Rept., v. 5, pt. 1, 89 p.

Bé, A.W.H., 1960, Some observations on Arctic planktonic foraminifera: Cushman Found. Foram. Research Contr., v. 6, pt. 2, p. 64-68.

——1967, Foraminifera families: Globigerinidae and Globorotalidae, in Fraser, J. H., ed., Fiches d'identification du zooplankton: Charlottenlund Slot, Denmark, Cons. Internat. Explor. Mer, fiche 108.

Berger, W. H., 1971, Sedimentation of planktonic Foraminifera: Marine Geology, v. 11, p. 325-358.

Bjørklund, K. R., and Kellogg, D. E., 1972, Five new Eocene radiolarian species from the Norwegian Sea: Micropaleontology, v. 18, p. 386-396.

Böggild, O. B., 1907, Sediments sous-marins recueillis dans la Mer du Grönland, in Duc d'Orleans, ed., Croisière océanographique accomplie à bord de la Belgica dans la Mer du Grönland (1905): Brussels, Bulens, p. 85-98.

Bramlette, M. N., and Bradley, W. H., 1941, Lithology and geologic interpretations, Pt.I, Geology and biology of North Atlantic deep-sea cores between Newfoundland and Ireland: U.S. Geol. Survey Prof. Paper 196A, p. 1-34.

Broecker, W. S., 1971, Calcite accumulation rates and glacial to interglacial changes in oceanic mixing, in Turekian, K., ed., The late Cenozoic glacial ages: New Haven, Conn., Yale Univ. Press, p. 239-265.

Broecker, W. S., and van Donk, J., 1970, Insolation changes, ice volumes, and the O^{18} record in deep-sea cores: Rev. Geophysics and Space Physics, v. 8, p. 169-198.

Broecker, W. S., Turekian, K. K., and Heezen, B. C., 1958, The relation of deep-sea sedimentation rates to variations in climate: Am. Jour. Sci., v. 256, p. 503-517.

Broecker, W. S., Thurber, D. L., Goddard, J., Ku, T-L., Matthews, R. K., and Mesolella, K. J., 1968, Milankovitch hypothesis supported by precise dating of coral reefs and deep-sea sediments: Science, v. 159, p. 297-300.

Clark, D. L., 1969, Paleoecology and sedimentation in part of the Arctic basin: Arctic, v. 22, p. 233-245.

——1970, Magnetic reversals and sedimentation rates in the Arctic Ocean: Geol. Soc. America Bull., v. 81, p. 3129-3134.

Dansgaard, W., Johnsen, S. J., Clausen, H. B., and Langway, C. C., 1971, Climatic record revealed by the Camp Century ice core, in Turekian, K., ed., The late Cenozoic glacial ages: New Haven, Conn., Yale Univ. Press, p. 37-56.

Darby, D. A., 1971, Carbonate cycles and clay mineralogy of Arctic Ocean sediment cores [Ph.D. thesis]: Madison, Univ. Wisconsin, 117 p.

Dietrich, G., 1956a, Überströmung des Island-Färoër-Rückens in Bodennähe nach Beobachtungen mit dem Forschungsschiff "Anton Dohrn," 1955/56: Deutsch., Hydrographische Zeitschr., v. 9, no. 2, p. 78-89.

——1956b, Schichtung und Zirkulation der Irminger See im Juni 1955: Deutsch. Wissenschaftliche Komm. Meeresforschung Ber., v. 15, no. 4, p. 255-312.

Ericson, D. B., 1959, Coiling direction of *Globigerina pachyderma* as a climatic index: Science, v. 130, p. 219-220.

Ericson, D. B., Ewing, M., and Wollin, G., 1964a, Sediment cores from the Arctic and subarctic seas: Science, v. 144, p. 1183-1192.

——1964b, The Pleistocene epoch in deep-sea sediments: Science, v. 146, p. 723-732.

Goldberg, E. D., and Koide, M., 1962, Geochronological studies of deep-sea sediments by the ionium/thorium method: Geochim. et Cosmochim. Acta, v. 26, p. 417.

Gordon, A. L., 1971, Recent physical oceanographic studies of Antarctic waters, in Research in the Antarctic: Am. Assoc. Adv. Sci., p. 609-629.

Hays, J. D., Saito, T., Opdyke, N. D., and Burckle, L. H., 1969, Pliocene-Pleistocene sediments of the equatorial Pacific: Their paleomagnetic, biostratigraphic, and climatic record: Geol. Soc. America Bull., v. 80, p. 1481-1514.

Holtedahl, H., 1959, Geology and paleontology of Norwegian Sea bottom cores: Jour. Sed. Petrology, v. 29, p. 16-29.

Hülsemann, J., 1966, On the routine analysis of carbonates in unconsolidated sediments: Jour. Sed. Petrology, v. 36, p. 622-625.

Imbrie, J., and Kipp, N. G., 1971, A new micropaleontologic method for quantitative paleoclimatology: Application to a late Pleistocene Caribbean core, in Turekian, K. K., ed., Late Cenozoic glacial ages: New Haven, Conn., Yale Univ. Press, p. 71-191.

Jacobs, S. S., Amos, A. F., and Bruchhausen, P. M., 1970, Ross Sea oceanography and Antarctic bottom water formation: Deep-Sea Research, v. 17, p. 935-962.

Kellogg, T. B., 1973, Late Pleistocene climatic record in Norwegian and Greenland Sea deep-sea cores [Ph.D. thesis]: New York, Columbia Univ. 544 p.

——1975, Late Quaternary climatic changes in the Norwegian and Greenland Seas, in Weller, G., and Bowling, S. A., eds., Climate of the Arctic: Fairbanks, Univ. Alaska Press, p. 3-36.

Kent, D., Opdyke, N. D., and Ewing, M., 1971, Climatic change in the North Pacific using ice-rafted detritus as a climatic indicator: Geol. Soc. America Bull., v. 82, p. 2741-2754.

Ku, Teh-Lung, 1966, Uranium series disequilibrium in deep-sea sediments [Ph.D. thesis]: New York, Columbia Univ.

Ku, T-L., Bischoff, J. L., and Boersma, A., 1972, Age studies of Mid-Atlantic Ridge sediments near 42°N and 20°N: Deep-Sea Research, v. 19, p. 233-247.

Kukla, G. J., and Kočí, A., 1972, End of the last interglacial in the loess record: Quaternary Research, v. 2, p. 347-383.

Lamb, H. H., 1968, Mapping methods applied to the study of climatic variations and vicissitudes, in The changing climate, selected papers by H. H. Lamb: London, Methuen & Co. Ltd., 1968, chap. 4.

Li, Yuan-Hui, Takahashi, T., and Broecker, W. S., 1969, Degree of saturation of $CaCO_3$ in the oceans: Jour. Geophys. Research, v. 74, p. 5507-5525.

Matthews, R. K., 1972, Dynamics of ocean-cryosphere system: Barbados data: Quaternary Research, v. 2, p. 368-373.

McIntyre, A., 1967, Coccoliths as paleoclimatic indicators of Pleistocene glaciation: Science, v. 158, p. 1314-1317.

McIntyre, A., and Bé, A.W.H., 1967, Modern Coccolithophoridae of the Atlantic Ocean—I; Placoliths and cyrtoliths: Deep-Sea Research, v. 14, p. 561-597.

McIntyre, A., Ruddiman, W. F., and Jantzen, R., 1972, Southward penetrations of the North Atlantic polar front: Faunal and floral evidence of large-scale surface water mass movements over the last 225,000 years: Deep-Sea Research, v. 19, p. 61-77.
Mesolella, K. J., Matthews, R. K., Broecker, W. S., and Thurber, D. L., 1969, The astronomical theory of climatic changes: Barbados data: Jour. Geology, v. 77, p. 250-274.
Meyer, P. L., 1965, Introductory probability and statistical applications: Reading, Mass., Addison-Wesley Pub. Co., 339 p.
Mullen, R. E., Darby, D., and Clark, D. L., 1972, Significance of atmospheric dust and ice rafting for Arctic Ocean sediment: Geol. Soc. America Bull., v. 83, p. 205-212.
Oba, T., 1969, Biostratigraphy and isotopic paleotemperature of some deep-sea cores from the Indian Ocean: Tohoku Univ. Sci. Repts., 2d ser., v. 41, no. 2, p. 129-195.
Olausson, E., 1967, Climatological, geochemical and paleo-oceanographical aspects of carbonate deposition: Progress in Oceanography, v. 5, p. 245-265.
Philippi, E., 1912, Die Grundproben der deutschen Südpolar Expedition, *in* von Drygalski, E., ed., Deutsche Südpolar Expedition (1901-1903), v. 2, p. 431-434.
Raasch, G. O. (editor), 1961, Geology of the Arctic: Toronto, Canada, Univ. Toronto Press, 1196 p.
Ruddiman, W. F., and Glover, L. K., 1972, Vertical mixing of ice-rafted volcanic ash in North Atlantic sediments: Geol. Soc. America Bull., v. 83, p. 2817-2836.
Ruddiman, W. F., and McIntyre, A., 1973, Time-transgressive deglacial retreat of polar waters from the North Atlantic: Quaternary Research, v. 3, p. 117-130.
———1976, Northeast Atlantic paleoclimatic changes over the past 600,000 years, *in* Cline, R. M., Hays, J. D., eds., Investigation of late Quaternary paleoceanography and paleoclimatology: Geol. Soc. America Mem. 145 (this volume).
Ruddiman, W. F., Tolderlund, D. S., and Bé, A.W.H., 1970, Foraminiferal evidence of a modern warming of the North Atlantic Ocean: Deep-Sea Research, v. 17, p. 141-155.
Saito, R., Burckle, L. H., and Horn, D. R., 1967, Paleocene core from the Norwegian basin: Nature, v. 216, p. 357-359.
Sancetta, C., Imbrie, J., Kipp, N. G., McIntyre, A., and Ruddiman, W. F., 1972, Climatic record in North Atlantic core V23-82: Comparison of the last and present interglacials based on quantitative time series: Quaternary Research, v. 2, p. 363-367.
Schott, W., 1935, Die Foraminiferen in dem äquatorialen Teil des Atlantischen Ozeans, *in*, Wissenschaftliche Ergebnisse der deutschen Atlantic Expedition Meteor (1925-1927): Berlin, W. de Gruyter & Co., v. 3, p. 43-134.
Schreiber, B. C., 1967, Area SF, volume 8, core, sound velocimeter, hydrographic, and bottom photographic stations—cores: Marine Geophysical Survey Program 65-67 Western North Atlantic and Eastern and Central North Pacific Oceans, Alpine Geophysical Associates, Inc. for U.S. Naval Oceanographic Office, 20 p.
Siesser, W. G., and Rogers, J., 1971, An investigation of the suitability of four methods used in routine carbonate analysis of marine sediments: Deep-Sea Research, v. 18, p. 135-139.
Stadum, C. J., and Ling, H-Y., 1969, Tripylean Radiolaria in deep-sea sediments of the Norwegian Sea: Micropaleontology, v. 15, p. 481-489.
Stetson, H. C., 1939, Summary of sedimentary conditions on the continental shelf off the east coast of the United States, *in* Trask, P., ed., Recent marine sediments: Am. Assoc. Petroleum Geologists, p. 230-244.
Steuerwald, B. A., and Clark, D. L., 1972, *Globigerina pachyderma* in Pleistocene and Recent Arctic Ocean sediments; Jour. Paleontology, v. 46, p. 573-580.
Steuerwald, B. A., Clark, D. L., and Andrew, J. A., 1968, Magnetic stratigraphy and faunal patterns in Arctic Ocean sediments: Earth and Planetary Sci. Letters, v. 5, p. 79-85.
Streeter, S. S., 1973, Bottom water and benthonic Foraminifera in the North Atlantic—glacial-interglacial contrasts: Quaternary Research, v. 3, p. 131-141.
Upshaw, D. F., and Stehli, F. G., 1962, Quantitative biofacies mapping: Am. Assoc. Petroleum Geologists Bull., v. 46, p. 694-699.
U.S. Naval Oceanographic Office, 1965, Oceanographic atlas of the North Atlantic Ocean,

Sec. 1, Tides and currents: Washington, D.C., pub. no. 700.

——1967, Oceanographic atlas of the North Atlantic Ocean, Sec. 2, Physical properties: Washington, D.C., pub. no. 700.

U.S. Navy Hydrographic Office, 1958, Oceanographic atlas of the polar seas, Pt. II, Arctic: Washington, D.C., pub. no. 705.

van der Hammen, T., Wijmstra, T. A., and Zagwijn, W. H., 1971, The floral record of the late Cenozoic of Europe, *in* Turekian, K. K., ed., The late Cenozoic glacial ages: New Haven, Conn., Yale Univ. Press, p. 391-424.

Weyl, P. K., 1968, The role of the oceans in climatic change: A theory of the ice ages: Meteorol. Mons., v. 8, (30), p. 37-62.

Worthington, L. V., 1970, The Norwegian Sea as a Mediterranean basin: Deep-Sea Research, v. 17, p. 77-84.

MANUSCRIPT RECEIVED BY THE SOCIETY DECEMBER 20, 1973
REVISED MANUSCRIPT RECEIVED MAY 28, 1974
MANUSCRIPT ACCEPTED JUNE 7, 1974
LAMONT-DOHERTY GEOLOGICAL OBSERVATORY CONTRIBUTION NO. 2278

Northeast Atlantic Paleoclimatic Changes over the Past 600,000 Years

W. F. Ruddiman
U.S. Naval Oceanographic Office
Chesapeake Beach, Maryland 20732

AND

A. McIntyre
Lamont-Doherty Geological Observatory
Columbia University
Palisades, New York 10964
and
Department of Earth and Environmental Sciences
Queens College
and
University Institute of Oceanography
City University of New York
New York, New York 10021

ABSTRACT

In the subpolar Atlantic Ocean during the Quaternary Period, water-mass environments have migrated across more than 20° of latitude, which is equivalent to temperature oscillations of the ocean surface of at least 12°C. The migrations have occurred along a northwest-trending axis at mean rates of approximately 100 m/yr sustained over intervals of several centuries. During peak glaciations, polar water moved south to lat 42°N, where an abrupt frontal system separated the cyclonic subpolar gyre from the anticyclonic subtropical gyre.

Seven complete climatic cycles have occurred in the past 600,000 yr, within which at least 11 separate major southward advances of polar water have occurred. Both in number and shape, these cycles are correlative to oxygen isotopic cycles in the western equatorial Pacific Ocean and to palynologic cycles determined from a core from Macedonia. The northeast Atlantic cycle geometries are not so uniformly saw-toothed in form as isotopic curves from the equatorial Atlantic Ocean and Caribbean Sea because of interruptions by short but severe cold climatic pulses lasting for intervals as short as a few thousand years. One such pulse, which lasted only 7,000 yr, retained at least 90% of its original peak intensity despite vertical mixing.

Quantitative determination of the absolute input rates of the major sediment fractions over the glacial and interglacial portions of the last major climatic cycle shows that coccoliths and foraminifera were deposited two to three times more rapidly during interglaciations than glaciations; in converse proportions, coarse and fine terrigenous detritus was preferentially rafted into the northeast Atlantic Ocean during glaciations. The absence of coccoliths in polar water accounts for the existence of glacial coccolith-barren zones.

At the scale of local sediment redistribution (related to siting factors), fine coccolith carbonate is most easily redistributed. The absolute abundance of all coarse and fine components increases at higher net sedimentation rates, but fine carbonate increases most rapidly.

INTRODUCTION

Man's historical adjustment to the Earth's climate has been during the relative warmth of the past 6,000 yr. Although the global climate during this time has been cooler than that prevailing over most of geologic time, it has been atypically warmer than most of the last million years of the Pleistocene Epoch. Man's foothold on northern temperate latitudes (particularly his circum–North Atlantic civilizations) is temporary and can be expected to be periodically dislodged by future advances of glacial ice, assuming no technologic intervention. Because the sequence of ice-sheet movements on land is fragmentary and difficult to date, we have examined in deep-sea sediment cores the more continuous record of accompanying polar-water movements across the North Atlantic Ocean. In this paper we summarize both new and previously published evidence that delineates the geography, intensity, and periodicity of these water-mass movements, and we attempt to relate the oceanic data to continental ice-sheet fluctuations.

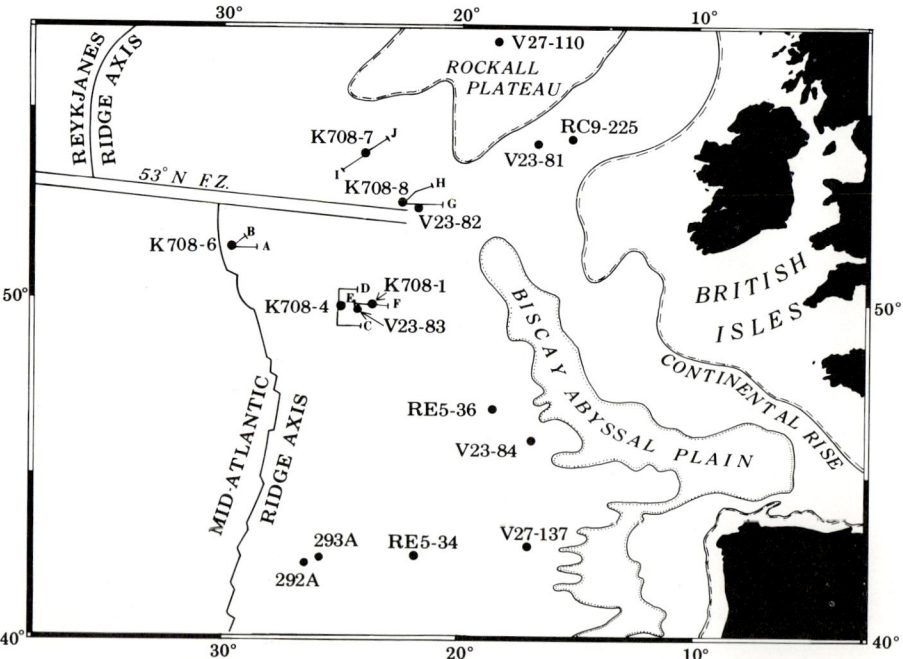

Figure 1. Locations of northeast Atlantic cores and selected seismic profiles.

CORE LOCATIONS AND PHYSIOGRAPHIC PROVINCES

The northeast Atlantic Ocean is ideal for Pleistocene paleoclimatology (Bramlette and Bradley, 1941; McIntyre and others, 1972a). Its shallow depths contain sequences of virtually undissolved calcareous ooze, which provide both stratigraphic control and biotic information free of significant problems caused by selective species removal. Because of productive surface waters and abundant nearby sources of ice-rafted detritus, the high average sedimentation rate permits paleoclimatic resolution at intervals as brief as a few thousand years. Finally, the subpolar North Atlantic Ocean is flanked on three sides by expanses of land that were ice covered during glaciations; its waters are thus an optimally situated paleothermometer of ice fluctuations in the two major temperate-latitude ice sheets of the Pleistocene Epoch (Laurentide and Scandinavian). The interlayering in North Atlantic cores of detritus eroded from the continents by ice sheets, and calcareous ooze reflecting changing sea-surface temperatures, permits simultaneous monitoring of oceanic and continental responses to global climatic change.

The cores selected for this study (Fig. 1; Table 1) were collected aboard the U.S. Naval Oceanographic Office vessel U.S.N.S. *Kane* in 1970 and on many cruises of Lamont-Doherty Geological Observatory vessels in the past two decades. The cores were taken in the following four physiographic provinces, which are distinguished primarily by basement structure and secondarily by the distributional pattern and thickness of sedimentary cover (Ruddiman, 1972):

1. Core K708-6, retrieved from the sparse sediment cover of the Mid-Atlantic Ridge crest, contains several turbidites (Figs. 2, 3). In this youthful (0 to 10 m.y. old) province, sediments are commonly dislodged from steep slopes and deposited in small ponds between basement outcrops.

2. Cores K708-1, K708-4, V23-83, V23-84, V27-137, RE5-34, and RE5-36 were taken from the Mid-Atlantic Ridge flank (10 to 60-m.y.-old crust). The ridge flank is characterized by large-scale basement macrorelief (400 to 800 m) and by sediment thicknesses averaging almost 1 sec of two-way acoustic travel time. Most of the sediment is moderately transparent and conformably draped over the crustal relief. Two of the seven cores from this province (K708-1 and K708-4) contain thin turbidites

TABLE 1. CORE LOCATIONS AND DEPTHS

Core no.	Lat (N)	Long (W)	Depth (m)
V27-110	56°53'	18°23'	1,264
RC9-225	54°18'	15°23'	2,334
V23-81	54°15'	16°50'	2,393
K708-7	53°56'	24°05'	3,502
K708-8	52°44'	22°33'	4,009
V23-82	52°35'	21°56'	3,974
K708-6	51°34'	29°34'	2,469
K708-1	50°00'	23°45'	4,053
K708-4	49°59'	25°01'	3,346
V23-83	49°52'	24°15'	3,871
RE5-36	46°55'	18°35'	4,500
V23-84	46°00'	16°55'	4,513
V27-137	42°42'	17°04'	4,883
RE5-34	42°23'	21°58'	3,750
293A*	42°23'	26°02'	3,485
292A*	42°16'	26°36'	3,216

*From Phleger and others (1953).

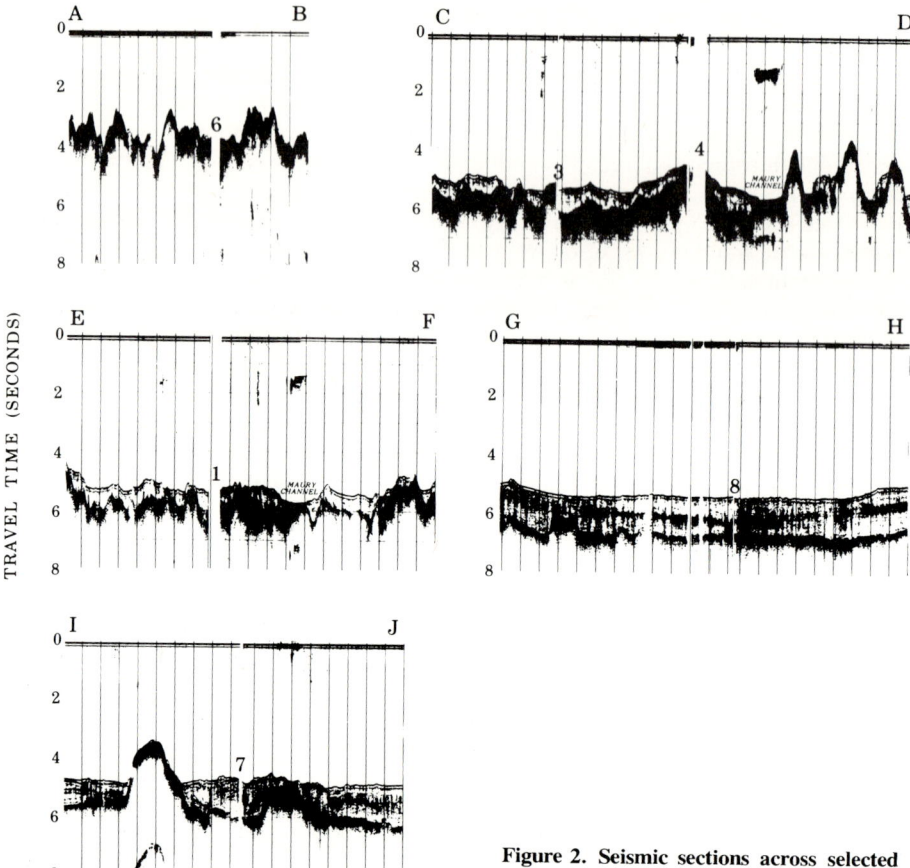

Figure 2. Seismic sections across selected core sites located in Figure 1.

defined by sharp basal contacts, but pelagic contacts are predominant and are burrowed over 5- to 20-cm intervals.

3. Cores K708-7, K708-8, V23-81, V23-82, and RC9-225 were taken from the sedimentary rise flanking Rockall plateau on the south and east. Basement relief is extremely subdued (Fig. 2) except for local seamounts, and the age of the crust exceeds 63 m.y. (Pitman and Talwani, 1972). Across this province, the sediment cover averages more than 1 sec, reaching thicknesses greater than 3 sec beneath the Feni drift, a sediment wedge formed by bottom currents flowing clockwise around the southeastern margin of Rockall plateau (Johnson and Schneider, 1969; Jones and others, 1970).

4. Core V27-110 was taken from a thick, transparent sedimentary sequence lying in a shallow structural saddle between Rockall and Hatton Banks (Roberts and others, 1970).

STRATIGRAPHIC CONTROL

Historical treatment of climatic change necessitates the development of a time-stratigraphic framework. In part this has been accomplished by radiochemical dating of these cores and in part by correlations with control levels from studies in other areas. The data are multifaceted, with input from floral, faunal, and tephrachronologic sources.

The gross stratigraphic control was established from visual matching of fine-fraction carbonate, foraminiferal, and coccolith curve geometries with climatic cycles in previous studies (McIntyre and others, 1972a). The primary pattern is a series of major climatic cycles, labeled A through H after Kukla (1970). We placed these cycles within a time-stratigraphic frame by emphasizing two very abrupt cold-to-warm climatic shifts called terminations I and II by Broecker and van Donk (1970), which were dated at about 11,000 and 127,000 B.P., respectively, by Broecker and Ku (1969) and Ku and others (1972). (Emiliani [1955] used the term "anathermal" to describe intervals of rising temperature; "termination" is preferable, because it specifies the rapidity of the warming trends relative to most periods of cooling.) Terminations I and II occur at the end of major climatic cycles B and C, respectively.

Using a radiocarbon-dated ash zone discovered by Bramlette and Bradley (1941) and studied quantitatively by Ruddiman and Glover (1972b), we have shown that termination I is time-transgressive in the Atlantic Ocean above lat 40°N, occurring at about 13,500 B.P. in the northeast Atlantic, but at only 7000 B.P. in the northwest Atlantic near Greenland (Ruddiman and McIntyre, 1973). Coarse-fraction lithologic data (Fig. 4) and faunal data (Fig. 9) were thus plotted against time by alignment at three reference levels: 0 B.P. for the core top, 13,500 B.P. for termination I, and 127,000 B.P. for termination II. Between these approximate reference levels, time plots were set by linear interpolation. (The dates for terminations I and II cited in this paper are approximate; the imprecisions involved are 5 to 10%. As

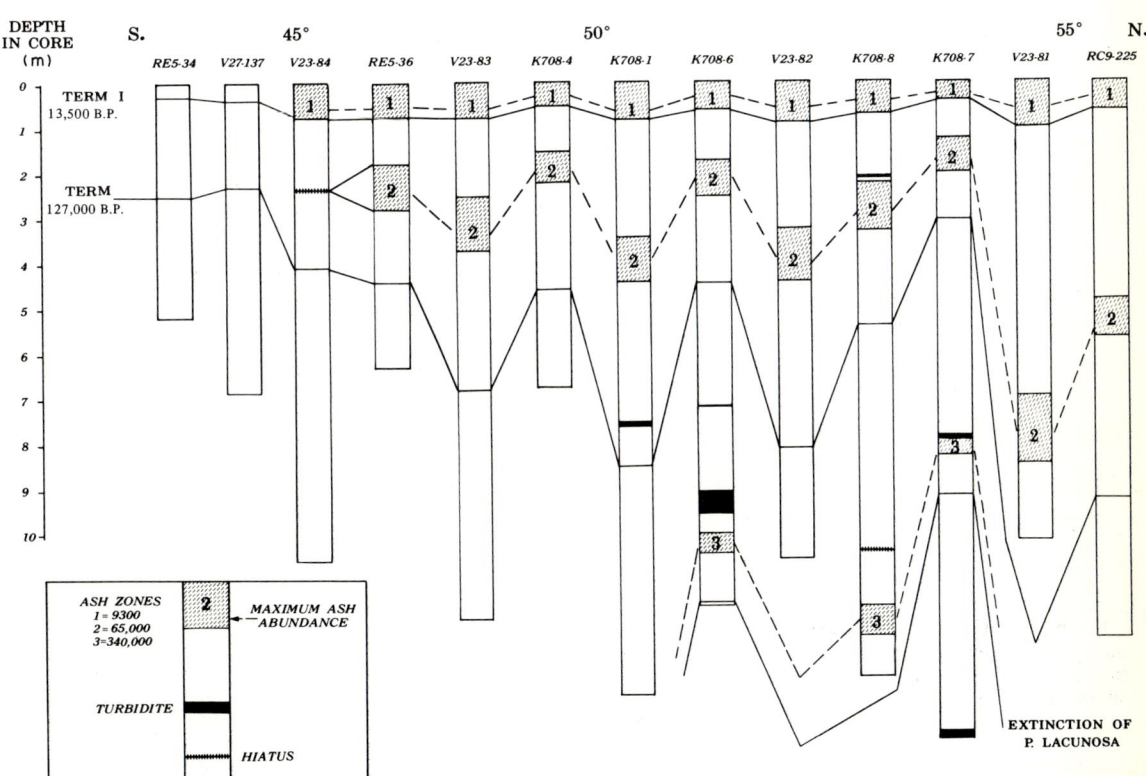

Figure 3. **Stratigraphic control on cores analyzed in detail in this study.**

a result, the uncertainties due to extrapolation and interpolation from these critical levels are added to the initial imprecisions.)

In addition to these paleoclimatic transitions, McIntyre and others (1972a) placed the level of presumed evolutionary shift in coccolithophorid dominance from *Emiliania huxleyi* to *Gephyrocapsa caribbeanica* (Fig. 3) at approximately 73,000 B.P. in the northeast Atlantic Ocean. This additional control level was used to

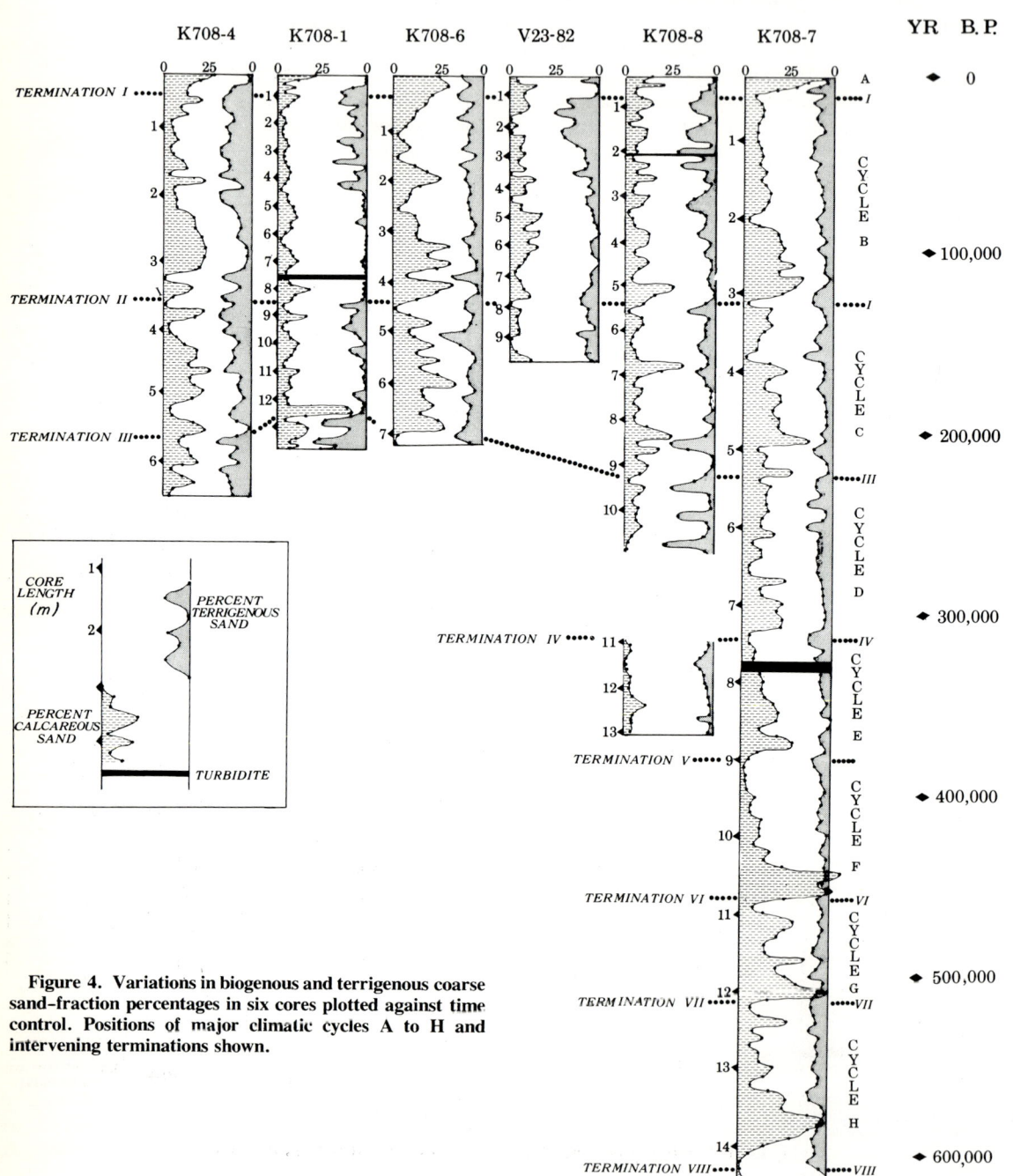

Figure 4. Variations in biogenous and terrigenous coarse sand-fraction percentages in six cores plotted against time control. Positions of major climatic cycles A to H and intervening terminations shown.

plot fine-fraction lithologic changes between terminations I and II in Figure 5. The time scale below termination II at roughly 127,000 B.P. was set in each core except K708-6 and K708-8 by extrapolating the mean sedimentation rate between terminations I and II. Despite minor core-to-core variations, this linear plotting scheme results in a north-to-south sequence of strikingly similar faunal and lithologic trends for the past two major climatic cycles (Fig. 4, 5, 9).

In three cores extending well beyond 200,000 B.P., these obvious correlations break down. Core K708-7 appears to contain the most linear sedimentation rate, based on the extrapolated placement of the *Pseudoemiliania lacunosa* extinction level (coincident with the Ericson U-V boundary) at approximately 405,000 B.P. in core K708-7, as compared to an extrapolated Pa^{231}/Th^{230} date in the Caribbean Sea of approximately 392,000 B.P. (Broecker and Ku, 1969) and an interpolated paleomagnetic age in the equatorial Atlantic Ocean of 390,000 to 440,000 B.P. (Glass and others, 1967; Ericson and Wollin, 1968). The lower sequence in core K708-7 has thus been used as the standard against which more complex variations in cores K708-6 and K708-8 are aligned.

The position of ash zone III in core K708-8 (Ruddiman and Glover, 1972b) indicates that part of one cycle is missing. On the basis of correlations of the fine-carbonate curves (Fig. 5) and the faunal trends (Fig. 9), we have detached the lower 340-cm section of core K708-8, aligned its peak ash abundance (1,200

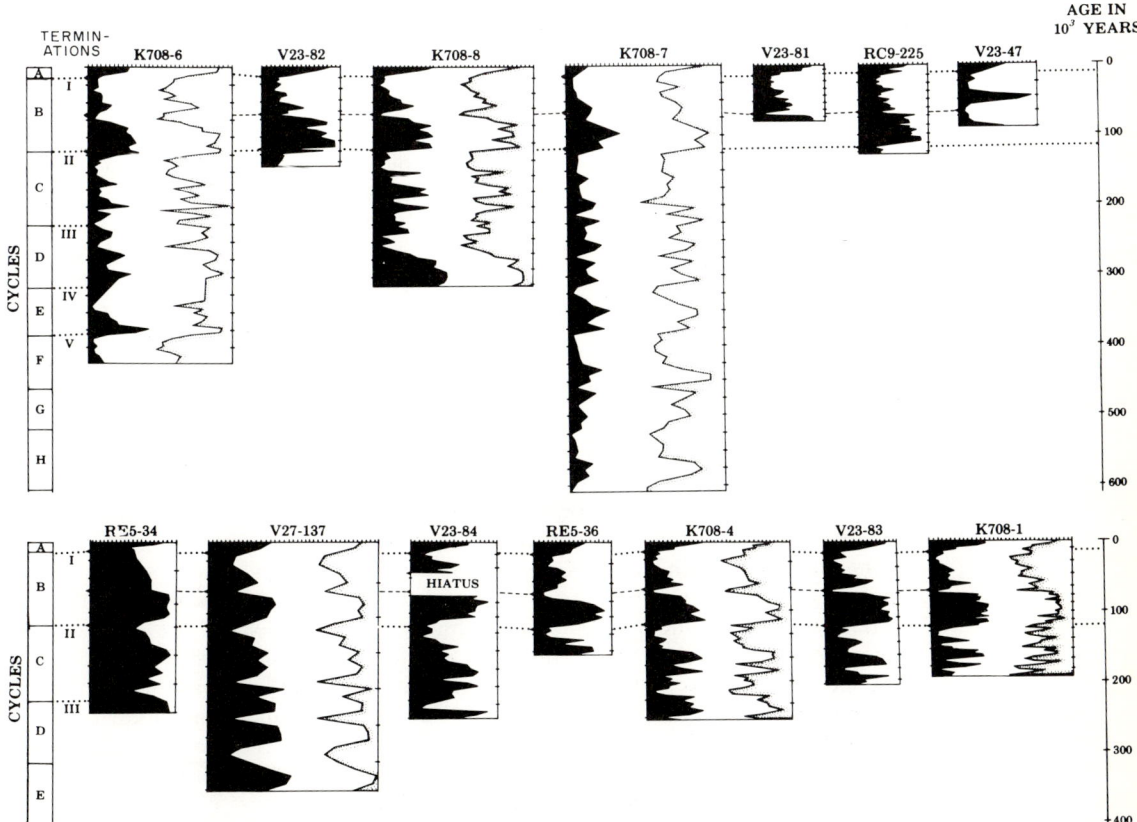

Figure 5. Variations in biogenous and terrigenous lutite fraction percentages plotted against time control. Positions of major climatic cycles A to H, terminations I to VIII, and carbonate minimums of McIntyre and others (1972a) shown. Less than 74-μm carbonate is shown in black; fine-fraction noncarbonate is stippled.

cm) in zone III against that in core K708-7 (812 cm), and plotted the lower part of core K708-8 at the sedimentation rate found above the hiatus. The hiatus in core K708-8 spans the upper (glacial) part of climatic cycle E and the lower (interglacial) part of cycle D.

Core K708-6 contains two turbidite units, a thin 1-cm layer at 720 cm, and a thick unit from 960 to 910 cm. The latter section has been removed from the time plot because its appreciable thickness would have a significant effect on the chronologic scale. Because the sedimentation rate in core K708-6 is slower in the section beneath the thin upper turbidite at approximately 200,000 B.P., we plotted this lower part at the reduced rate, which permitted alignment of the zone-III ash with that in core K708-7. This single adjustment produced a convincing correlation of cores K708-6 and K708-7.

This stratigraphic framework is the basis for the chronologic plots of downcore trends in Figures 4, 5, and 9.[1] Within this framework, the sequence of major climatic cycles (A to H) ending in abrupt terminations (I to VIII) provides the simplest descriptive reference scheme.

MASS-INPUT RATE OF SEDIMENT FRACTIONS

Broecker has shown (Broecker and others, 1958; Broecker, 1971) that variations in the sediment fractions can be meaningfully interpreted only by determining absolute input rates; percentage variations alone are not diagnostic. Three factors control relative biogenous and terrigenous abundances in North Atlantic sediment (where siliceous components are usually negligible): (1) biogenous input (a function of surface-water productivity), (2) biogenous removal (a function of bottom-water corrosiveness and exposure time of calcareous tests on the sea floor), and (3) terrigenous detrital input (turbidity currents, ice-rafting, bottom currents, or wind).

To isolate these factors, we have separately processed four sediment fractions, two of biogenous origin and two of terrigenous derivation (Figs. 4, 5): coarse carbonate (foraminifers), fine carbonate (coccoliths and fragmented foraminifers), coarse terrigenous detritus (glacial marine sand), and fine terrigenous detritus (noncalcareous silt and clay). Cores V23-83, RE5-34, RE5-36, and RC9-225 were available only for fine-fraction (<74 μm) analysis, but six other cores (V23-82, K708-1, K708-4, K708-6, K708-7, and K708-8) were analyzed for all four sediment fractions. The sand-lutite size boundary initially chosen was 74 μm; however, the coarse fractions from the five *Kane* cores were later processed at 62 μm.

Fine-carbonate concentrations were determined using the gasometric analysis of Hülsemann (1966). Percentages of coarse-fraction carbonate were analyzed by the insoluble-residue method. Aside from the normal inaccuracies inherent in each technique (Siesser and Rogers, 1971), an additional complication is that part of the terrigenous detritus from the continents found in these cores is detrital limestone (Bramlette and Bradley, 1941). Thus, although we will discuss the "biogenous" carbonate fractions, a small portion is in fact terrigenous detritus.

Following Broecker (1971), we examined variations in the four components over climatic cycle B between termination II (127,000 B.P.) and termination I (13,500 B.P.). With cycle B split into an interglacial half-cycle (127,000 to 75,000 B.P.) and a glacial half-cycle (75,000 to 13,500 B.P.), we compared influxes of biogenous and terrigenous sediment under these two climatic regimes. (The "glacial" and

---

[1]All carbonate, faunal, and floral data in these cores are included in Appendixes I-III on microfiche in pocket inside back cover.

"interglacial" intervals in this study do not necessarily equate with the continental usage, as explained below. They do, however, include different mean climatic regimes and thus are indicative of the direction of differences during extreme glacial or interglacial conditions.) Also following Broecker (1971), we expressed the sediment-fraction input rates in cm/1,000 yr, because bulk densities for these cores are unavailable. Broecker's technique permits a close approximation of absolute input rates for the separate sediment fractions.

By choosing relatively shallow cores with little apparent carbonate solution, we minimized the influence of bottom-water solution and could ignore it in the analysis of these cores; only sediment-input factors need to be evaluated. Comparative glacial and interglacial input rates of the biogenous and terrigenous fractions are shown in Figure 6. Every point plotted shows higher absolute input rates of biogenous

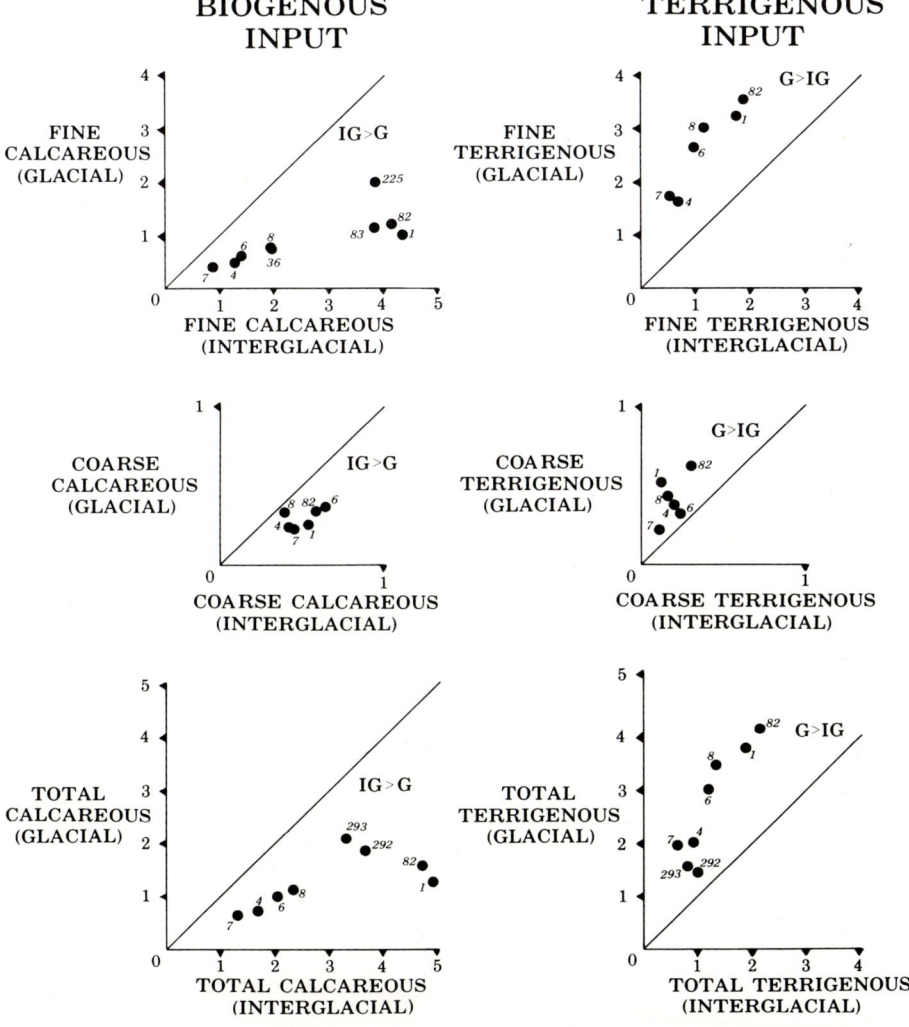

Figure 6. Absolute input rates in cm/1,000 yr of coarse and fine biogenous and terrigenous sediment fractions for interglacial (IG) half-cycle B (127,000 to 75,000 B.P.) relative to glacial (G) half-cycle B (75,000 to 13,500 B.P.).

TABLE 2. TERRIGENOUS DETRITAL SEDIMENT INPUT INTO NORTHEAST ATLANTIC CORES

Core no.	Glacial (G)			Interglacial (IG)		
	Sand	Lutite	Total	Sand	Lutite	Total
K708-7	0.23	1.82	2.05	0.11	0.51	0.62
K708-8	0.45	3.15	3.60	0.16	1.15	1.31
V23-82	0.65	3.69	4.34	0.30	1.84	2.14
K708-6	0.36	2.78	3.14	0.24	0.96	1.20
K708-1	0.54	3.40	3.94	0.12	1.73	1.85
K708-4	0.38	1.71	2.09	0.20	0.70	0.90
Average	0.44	2.76	3.20	0.19	1.15	1.34

Note: Measured in cm/1,000 yr. Mean G/IG terrigenous sand input ratio: 2.32; mean G/IG terrigenous lutite input ratio: 2.40.

sediment in the northeast Atlantic Ocean during interglacial climates but greater delivery of terrigenous sediment during glacial climates. The almost constant sedimentation rates in these cores must then be due to a coincidental balancing of the oceanically produced biogenous input and the continentally derived terrigenous input: as one decreases, the other increases proportionally. Most of this balancing occurs in the fine fraction, which accounts for 70 to 80% of most samples.

The mean glacial/interglacial (G/IG) ratio of absolute input rates for both coarse and fine terrigenous detritus is about 2.5/1, with the analogous fine carbonate ratio nearly the inverse (Tables 2, 3). By comparison, the IG/G input ratio of coarse biogenous material varies by a factor of only 1.5/1, indicating a less marked climatic variation in rates of foraminiferal input.

Biogenous Input

The single climatic cycle we analyzed may not be precisely representative of all such cycles in these cores. For this reason, we emphasize not the exact numbers and ratios derived, but the direction or sense of change. Our use of cm/1,000 yr as a measure instead of g/cm^2/1,000 yr also introduces some error into the

TABLE 3. BIOGENOUS SEDIMENT INPUT INTO NORTHEAST ATLANTIC CORES

Core no.	Glacial (G)			Interglacial (IG)		
	Sand	Lutite	Total	Sand	Lutite	Total
RC9-225	. .	(2.11)	. .	. .	(3.85)	. .
K708-7	0.24	0.45	0.69	0.45	0.87	1.32
K708-8	0.34	0.84	1.18	0.39	1.93	2.32
V23-82	0.36	1.30	1.66	0.58	4.14	4.72
K708-6	0.42	0.61	1.03	0.64	1.39	2.03
K708-1	0.27	1.07	1.34	0.54	4.36	4.90
K708-4	0.26	0.51	0.77	0.42	1.26	1.68
V23-83	. .	(1.21)	. .	. .	(3.81)	. .
RE5-36	. .	(0.81)	. .	. .	(1.94)	. .
293A	. .	. .	(2.17)	. .	. .	(3.32)
292A	. .	. .	(1.90)	. .	. .	(3.65)
Average	0.32	0.80	1.12	0.50	2.33	2.83

Note: Measurements in cm/1,000 yr. Mean IG/G biogenous sand input ratio: 1.59; mean IG/G biogenous lutite input ratio: 2.91. Values in parentheses not included in average.

relative input rates. Hamilton (1969) indicated a bulk density for (interglacial) calcareous ooze (1.53 g/cm³) that is 8% lower than the density of the sand-silt-clay mixture that most nearly approximates glacial marine sediment (1.65 g/cm³). Even with this correction, no data point would contradict the conclusion that biogenous input is greatest during interglacial climates and that terrigenous input is greatest during glacial climates between lat 40°N and 60°N in the northeast Atlantic Ocean.

At high northern latitudes, carbonate productivity variations are dominated by southward movements of polar water to latitudes as low as 42°N to 45°N during glacial periods (McIntyre and others, 1972a). At such times, Coccolithophoridae (and, to a lesser extent, Foraminifera) are prohibited from flourishing, and productivity of the carbonate-secreting biota declines. The total abundance of Foraminifera (Fig. 4) is less markedly reduced during the glacial fine-carbonate minimums (Fig. 5), because the one polar species that persists through the glaciations (*Globigerina pachyderma*, s.) apparently becomes sufficiently abundant to counterbalance partially the loss of subpolar and temperate foraminifers. No species of Coccolithophoridae can inhabit polar water, and at severe glacial maximums the coccoliths disappear completely, leaving coccolith-barren zones within the carbonate minimums.

Terrigenous Input

Both coarse and fine terrigenous detritus are delivered to these cores preferentially during glacial periods (Figs. 4 to 6). The fine terrigenous detrital fraction (Fig. 5) mirrors in detail the climatic interpretation developed independently by coccolith and foraminiferal species assemblages. Coarse terrigenous detritus is a less predictable climatic indicator (Fig. 4), although it is most abundantly supplied during glacial maximums.

Turbidity currents, bottom currents, wind, and ice are potential transporters of terrigenous detritus. Because most of the detritus has a continental source (with Iceland an important exception), it must be horizontally moved several thousand kilometres to reach the middle of the North Atlantic Ocean. This transport occurs near either the upper or lower boundaries of the water column.

Turbidity currents and bottom currents are the predominant near-bottom sediment transporters. Locally derived turbidites are infrequent and clearly marked in these cores and have been eliminated from this analysis. Several provinces in the study area (Fig. 7A) are dominated by long-range downslope sedimentation both from the continents and from Iceland (Ruddiman, 1972). Although Maury Channel passes very near many of the cores in this study (Fig. 7A), it follows deep topography several hundred metres below the cores and has not affected their records.

Bottom currents are active in the northern and eastern parts of the study area (Fig. 7A). Thick sediment drifts and prominent scour effects have been mapped around the southeastern margin of Rockall plateau and on the east flank of the Reykjanes Ridge (Johnson and Schneider, 1969; Jones and others, 1970; Ruddiman, 1972). Two cores near the axis of Feni drift (RC9-225 and V23-81) have very high sedimentation rates but almost no coarse terrigenous detritus. Despite the lack of coarse ice-rafted debris, the glacial parts of these cores are thicker by at least 20% than those in cores from other provinces, owing to the marked increase in terrigenous fine fraction apparently delivered by bottom currents.

Ice and wind are the potential transporting agents of detritus near the air-water interface at these latitudes. Clearly, wind cannot have carried sand and coarse silt to mid-ocean areas; however, it may have contributed minor amounts of fine terrigenous detritus to sediment of the northeast Atlantic Ocean above lat 45°N. There are two possible eolian sources. The extensive European loess fields are east-southeast of the area of this study; detritus from this source would require an unlikely southeasterly wind during glaciations (Lamb and Woodruffe, 1970). The North American loess fields are too distant (5,000 km or more) to have contributed much detritus from the westerlies. Even in low latitudes, where wind-blown sediment may form a large part of the fine detrital input (Folger, 1970), the eolian-lutite input rate 6,000 km downwind from known sources is only about 0.05 cm/1,000 yr (Delaney and others, 1967). Eolian depositional rates in modern high-altitude snowfields also average only 0.02 cm/1,000 yr (Windom, 1969). These modern eolian input rates must be compared with the mean interglacial and glacial lutite depositional rates of 1.15 and 2.76 cm/1,000 yr in the mid-ocean northeast Atlantic cores (Table 2). The two-order-of-magnitude discrepancy suggests that, in percentage terms, wind transport has been insignificant in high-latitude northeast Atlantic sedimentation.

Ice is the dominant transporting agent of terrigenous detritus in this study. Icebergs

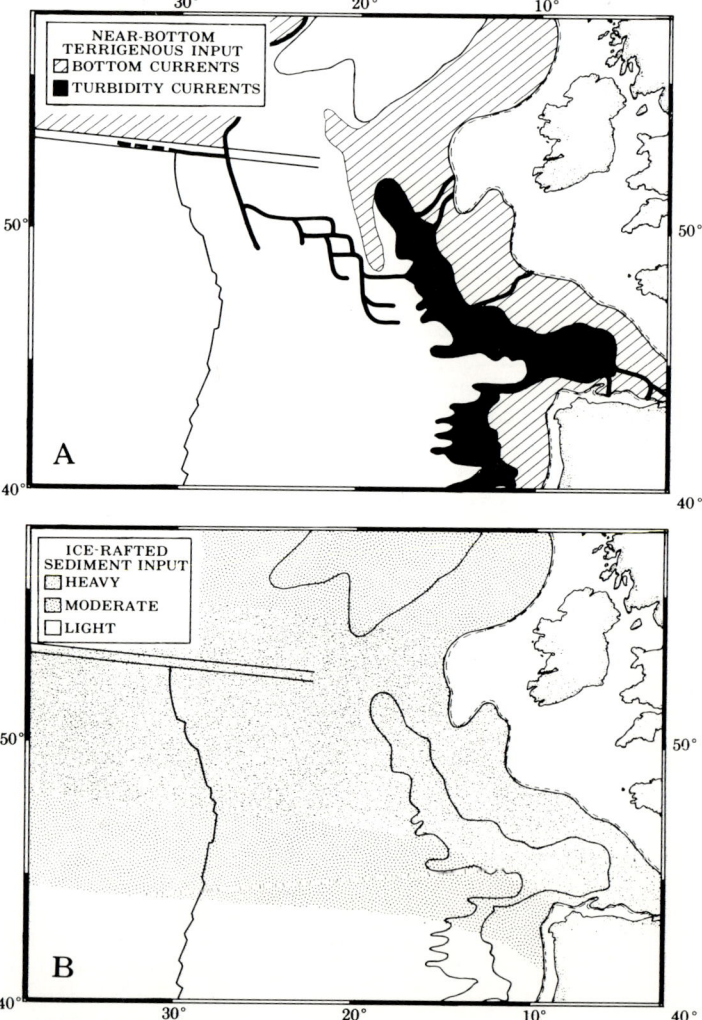

Figure 7. Schematic regional pattern of terrigenous sediment input by A, bottom and turbidity currents, and B, ice rafting.

carry the full range of grain sizes found in glaciers and are the primary rafting agent. Sea ice also provides several means of transport. When grounded, sea ice may pick up detritus near river mouths (Bramlette and Bradley, 1941) or may receive large volumes of detritus in spring thaws along continental margins where streams wash debris onto fast ice. Eolian detritus may also be blown from continental sources across marginal waters covered with pack ice and transported on ice far out to sea. In summary, terrigenous detritus was supplied to the areas in this study primarily by ice; the increased input during glaciations balanced the loss of fine carbonate from coccoliths.

The geographic depositional pattern of ice-rafted detritus can be tentatively inferred from this and previous studies. By tracing abundance patterns of volcanic ash apparently dropped from ice in two posteruptive episodes, Ruddiman and Glover (1972a, 1972b) defined a counterclockwise flow of cold water southward through the Denmark Strait west of Iceland and then eastward toward Europe along lat 48°N to 54°N. The present circulation in the Atlantic Ocean north of lat 45°N is also best described as a weak cyclonic gyre (Warren, 1967). If the two ash-traced patterns are indicative of ice movement during the Pleistocene Epoch, ice-rafted detritus would have been carried in a counterclockwise gyre. The relative scarcity during glacial sequences of ice-rafted sand in cores both to the south (RE5-34) and to the north (V27-110, RC9-225, V23-81) of lat 45°N to 55°N suggests that icebergs dropped terrigenous detritus primarily along an east-west axis at about 50°N (Fig. 7B). If the transport were generally west to east, this latitudinal axis could mean a dominance of Greenlandic and North American detritus in sediment in much of the eastern North Atlantic Ocean, despite the greater proximity of the smaller Scandinavian ice sheet. Further regional quantitative data are needed to test this hypothesis.

Local Sedimentologic Variability

The numerous cores in this study permit an estimate of the variability between cores, in addition to regional average input rates of the sediment fractions. Core-to-core variability must be largely a function of topographic setting. Aside from gross and easily detected downslope movements such as turbidity currents and slumps, more subtle downslope displacement of sediment into depressions results from bottom currents moving sediment across the sea floor and from bottom infauna and epifauna disturbing sediment. As a result, slopes and local highs gradually lose sediment to depressions (Bramlette and Bradley, 1941; Moore and Heath, 1967).

For the glacial and interglacial half-cycles, we have plotted in Figure 8 the mean percentage of each component (coarse and fine, biogenous and terrigenous sediment) against the total sedimentation rate for the chosen half-cycle. Linear regression lines of the component percentages onto the sedimentation rate, as well as the correlation coefficients of the two parameters, are shown. Within the obvious constraint of only six data points, there is a decided increase in the percentage of fine carbonate for both the interglacial and glacial half-cycles for the higher sedimentation-rate cores. The other components either hold at constant percentages or decrease slightly with increasing influxes of sediment.

This increase in the percentage of fine carbonate suggests that the agents that redistribute sediment are able to move coccoliths and other fine carbonate more easily than either terrigenous lutite or the two coarse-fraction components. The absolute amount of all components increases as the sedimentation rate increases (Tables 2, 3), but fine carbonate increases more quickly and thus becomes more abundant in terms of percentage. Because of the limited competence of locally effective bottom currents and the restrictions of gut diameter in some burrowing organisms (such as polychaetes), it is not surprising that fine detritus is more easily redistributed than coarse debris. The implied lesser mobility of terrigenous lutite relative to fine carbonate may indicate that electrostatic charges on the terrigenous clays give them greater cohesiveness than the uncharged coccolith plates.

FORAMINIFERAL FAUNAS

The species of fossil foraminifera in these cores have been investigated in great detail in North Atlantic water masses and surface sediments (Bé, 1959; Bé and Hamlin, 1967; Ruddiman, 1969a; Imbrie and Kipp, 1971). The assemblages from the counted samples in each core are, by analogy with modern distributions, indicative of a set of ecologic water-mass conditions. ("Ecologic" water masses—as distinguished from normal oceanographic use of the term—refer to surface-water environments covering a broad range of temperature and salinity values but having a specific environmental significance.) From the combined pattern of almost 2,000 samples in 13 cores, in which each sample characterizes the ecologic water mass present at the given coring location for one brief interval of geologic time, the movement of water-mass environments across the North Atlantic Ocean through late Pleistocene time can be reconstructed.

Foraminiferal Assemblages

Counting techniques and methods of sample preparation for Foraminifera follow those described in McIntyre and others (1972a), but designation of the species to ecologic water masses has been somewhat modified. Species termed subpolar

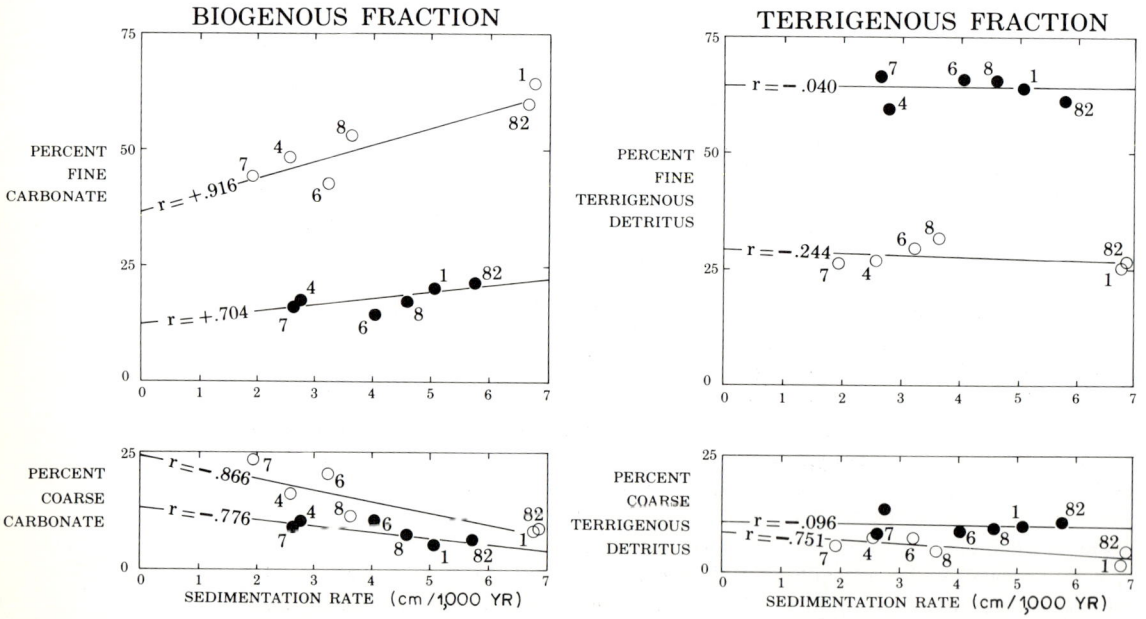

Figure 8. Change in integrated mean percentage of sediment fractions across intervals of glacial (blackened circle) and interglacial (open circle) half-cycles plotted against sedimentation rate over chosen cycle (B).

TABLE 4. ECOLOGICAL WATER MASS DESIGNATIONS OF FORAMINIFERA AND COCCOLITHS

Water mass	Foraminiferal species*	Coccolith species
Polar	Globigerina pachyderma, s.	Barren
Subpolar	Globigerina bulloides Globigerina pachyderma, d. Globigerina quinqueloba	Coccolithus pelagicus
Transitional	Globorotalia inflata Globoquadrina dutertrei	Cyclococcolithus lepoporus type C

Subtropical
  Northern
 	
  Southern
	Globorotalia hirsuta
	Globorotalia truncatulinoides
	Globorotalia scitula
	Globigerinoides ruber
	Globigerinoides conglobatus
	Globigerina rubescens
	Globorotalia crassaformis
		Calciasolenia sinuosa
		Umbilicosphaera mirabilis
		Umbellosphaera tenuis
		Rhabdosphaera clavigera
		Syracosphaera pulchra
		Umbellosphaera irregularis

Cosmopolitan	Orbulina universa Globigerinita glutinata	Emiliania huxleyi Gephyrocapsa caribbeanica

*Species listed constitute more than 0.4% of any sample.

in that paper have been subdivided in this study (Table 4) into subpolar and transitional faunas that reflect modern distributions (Bé and Hamlin, 1967). A distinction has also been made between cold subtropical and warm subtropical foraminifers, following a prominent ecologic water-mass boundary at lat 30°N noted by Ruddiman (1969b) and Bé and others (1971). The cold subtropical foraminifers are *Globorotalia truncatulinoides, G. hirsuta,* and *G. scitula* (the latter was termed subpolar in McIntyre and others, 1972a); these are found in surface sediment beneath northern central water at latitudes south of the Gulf Stream and north of lat 30°N. The warm subtropical foraminifers currently occupy southern central water at latitudes south of Bermuda but north of the shallow thermocline of equatorial water. *Globigerina pachyderma,* s. remains the sole polar species.

These modifications bring to five the number of foraminiferal groups distinguished in these North Atlantic cores (Table 4). Assemblages of these five species groups were assigned to ecologic water masses according to the following percentage limits: polar water—polar species, ≥75%; subpolar water—polar species, <75%, and transitional species, <35%; transitional water:—transitional species, ≥35%, and cold subtropical species, <15%; and cold subtropical water:—cold subtropical species, ≥15%. The warm subtropical species never approach the dominance values (75%) necessary to designate any sample as a warm subtropical water-mass indicator.

Other species recognized but occurring as insignificant parts of the total are *Globorotalia menardii* and *G. tumida, Globigerinoides sacculifer, Globigerina tenellus, Globigerinella aequilateralis, Pulleniatina obliquiloculata, Sphaeroidinella dehiscens, Globigerinita humilis,* and *Hastigerina pelagica.*

Two common species (*Orbulina universa* and *Globigerinita glutinata*) have such a cosmopolitan geographic distribution in modern oceans that they are of minor value as water-mass indicators. We counted these but excluded them from the total in computing the faunal percentage composition of each sample.

Among the taxonomically complex intergradations, we have lumped the form often designated as *Globigerina calida* with *Globigerinella aequilateralis* and the form *Globigerina falconensis* with *G. bulloides*. Both *G. calida* and *G. falconensis* were rare at most levels in these cores. Intergrades between *Globoquadrina dutertrei* and *Globigerina pachyderma*, d. were assigned to the latter species unless a prominent tooth or a very open, high-arched aperture was observed. The taxonomy used by Phleger and others (1953) for cores 292A and 293A differs from ours primarily in the designations of warm-water (equatorial and warm subtropical) species that are extremely rare in the northeast Atlantic Ocean and do not affect the assemblage abundances in Figure 9 (in pocket). Among the cold-water species, the major difference in the studies is that no distinction is made based on the coiling direction of *G. pachyderma*. Based on the coiling percentages in nearby core RE5-34, we made the assumption that one third of the relatively minor *G. pachyderma* population in cores 292A and 293A was left-coiled at all levels.

We initially selected samples for foraminiferal analysis at 15- to 25-cm intervals, representing time intervals of about 5,000 yr. Later samples taken midway between the initial counts were counted if, from brief examination, they suggested intervening climatic shifts. The final sample intervals shown in Figure 9 average less than 5,000 yr, and the scanned but uncounted samples further tighten the resolution of faunal shifts.

The foraminiferal variations in Figure 9 show eight major cycles (seven of which are complete) over the past 600,000 yr, corresponding to the carbonate cycles in Figure 5. These oscillations range between the glacial extreme of total dominance of polar fauna and the interglacial extreme in which transitional and cold subtropical species may exceed 50% of the total.

The extremes of cold and warmth through all eight cycles recorded by the assemblages in Figure 9 are similar, except for a slight tendency toward smaller percentages of cold subtropical species during peak interglacial warmth prior to 225,000 B.P. In general, however, the regular oscillation between similar end-member faunal assemblages suggests little paleoenvironmental adaptation among high-latitude North Atlantic foraminifera over the past 600,000 yr.

Viewed in a lateral (geographic) sense, the most striking feature of the faunal variations in Figure 9 is the separation of the 13 cores into two distinct groups. The nine cores taken north of lat 45°N contain very similar faunal records along synchronous horizons, with only slightly colder assemblages at higher latitudes. The three cores taken south of lat 45°N (including two from Phleger and others, 1953) reveal similar but less extreme faunal oscillations and no dominance of polar fauna. Core V23-84 is gradational between the two groups and lies near the southernmost latitude of polar-front penetration during Pleistocene time (McIntyre and others, 1972a).

Mixing of Faunal Trends

Ruddiman and Glover (1972b) examined vertically disseminated ash zones in five K708 cores and determined that analogous paleoclimatic events of less than 100-yr duration could leave greatly weakened but possibly still-visible spikes expressed in terms of foraminiferal percentages in the sedimentary record. They predicted that episodes exceeding 1,000 yr should be left largely intact in percentage terms. The relatively brief cold faunal pulse at about 195,000 B.P. (carbonate minimum 5 of McIntyre and others, 1972a) can be used to infer the effect of mixing on episodes lasting several thousand years.

McIntyre and others (1967) and Berger and Heath (1968) first showed that an initially abrupt scarp of changing species concentration along an appearance or extinction boundary would be blurred by mixing into a more gradual (exponential) curve of changing concentrations. For evolutionary or climatic episodes with both an upper and lower boundary (Fig. 10A), episode material is mixed across the two margins (A_1 and B_1) into overlying and underlying sediments (A_2 and B_2). For moderately short episodes, the two margins begin to merge (Fig. 10A); for very short episodes and thin instantaneously deposited volcanic ash layers, the peak concentration of event material is substantially diluted from its original value by mixing with overlying and underlying layers (Ruddiman and Glover, 1972b).

The "episode material" in this instance is the percentage of polar fauna in and around the brief carbonate minimum at about 195,000 B.P. in cores K708-6

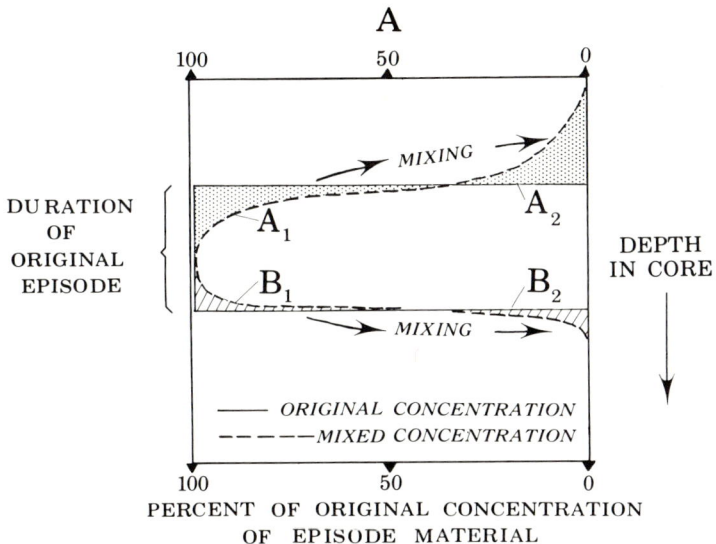

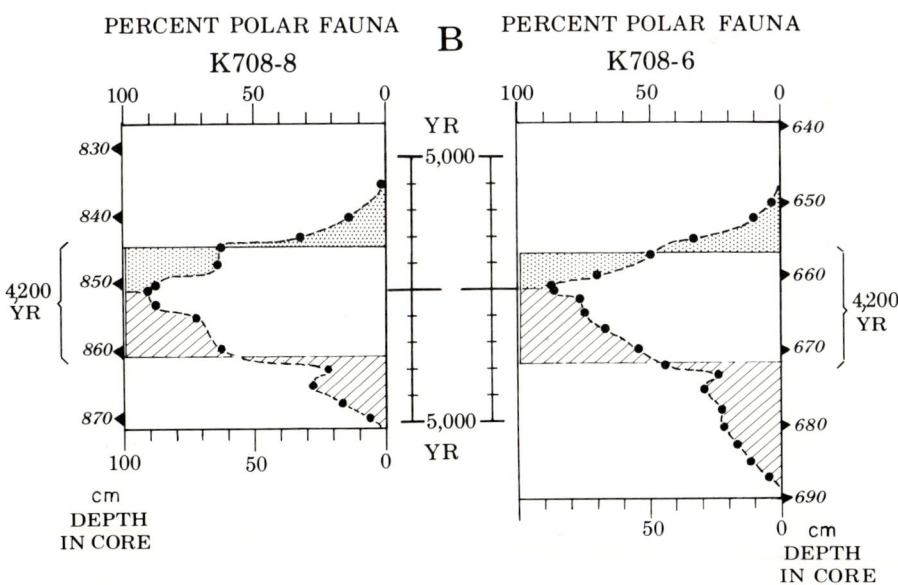

Figure 10. A, Schematic vertical mixing of short climatic episode with instantaneous beginning and end; B, actual form of abrupt cooling (carbonate minimum 5 in major cycle C) in two northeast Atlantic cores.

and K708-8. Because the percentage of polar fauna changes from almost zero at the margins of the episode to about 90% at its peak development (Fig. 10B), this faunal pulse has retained most of its original peak concentration despite mixing. An assumption critical to this analysis is that the sedimentation rate remained constant over the analyzed interval. Accordingly, we have plotted this faunal episode in cores K708-6 and K708-8 against both depth in core and a time scale set by the sedimentation rate between terminations I and II in each core (Fig. 10B).

The shape of the original faunal episode is not known, but we can postulate the two extreme configurations. The present faunal record of the episode (Fig. 10B) represents the unlikely extreme assumption that there has been no mixing since deposition. In this case, the percentage of polar fauna grew gradually to a peak and then slowly decreased, the entire episode lasting 9,000 to 10,000 yr, with the original peak concentration still intact.

For the other extreme, we assume that there was originally an instantaneous increase from an absence of polar fauna above and below the episode margins to 100% polar fauna within the episode (heavy solid lines in Fig. 10). Because we are assuming constant sedimentation rates, these hypothetical original episode margins can be placed roughly at that level in the core where areas $A_1 = A_2$ and $B_1 = B_2$. This gives a minimum possible duration for the original (unmixed) episode of 4,200 yr in both cores K708-8 and K708-6. The present peak concentrations of polar fauna are 91% in core K708-8 and 88% in core K708-6, indicating a maximum decrease of about 10% from this assumed original 100% maximum under this extreme assumption.

Neither extreme assumption is likely to be correct. The episode probably developed and dissipated somewhat more gradually over 6,000 to 7,000 yr, with the original peak concentration having been diminished by a value between 0 and 10%. In summary, we find that paleoclimatic events of about 6,000- to 7,000-yr duration are recorded with essentially undiminished peak intensities in northeast Atlantic cores by foraminiferal variations.

COCCOLITH FLORAS

Live Coccolithophorida are limited to the euphotic zone, about which sufficient ecologic studies of modern species have been made (McIntyre and Bé, 1967; McIntyre and others, 1970; Gaarder, 1971) to determine the biogeography of most species found in the northeast Atlantic cores. Less than 25% of all Coccolithophorida species, however, are preserved in oceanic sediment, which primarily contains robust forms. Most species produce fragile coccoliths prone to mechanical and chemical destruction in the water column and on the bottom. Thus, studies dealing with modern surface-water distributions have not been sufficiently applicable to studies of fossil coccoliths. To obviate this problem, the distribution of modern species of paleontologic value in surface sediment of the North Atlantic Ocean has been investigated (Roche and others, 1975). Eighty-three trigger-weight core tops, which were factor analyzed by the technique of Imbrie and Kipp (1971), produced four assemblages related to surface water masses. Three of these assemblages are used in this paper to describe late Pleistocene paleoceanography.

Characteristics of Coccolithophorida

Our interpretation of paleoceanographic and paleoclimatic history is constructed from the succession and dominance of Foraminifera and Coccolithophorida and from lithologic variations. Foraminifera are the key paleoclimatic indicators, and Coccolithophorida are used to refine the divisions. Comparison between these two groups is limited by the following factors:

1. Coccolithophorida are not viable in polar waters; thus, we lack a floral analogue to *Globigerina pachyderma*, s., the sole faunal end member in polar waters.

2. The ecology and biogeography of some coccolith species important in our cores, notably *Gephyrocapsa ericsonii*, is still poorly documented.

3. Coccolithophorida assemblages are dominated by one or two species, with all other species averaging less than 25% of the total.

4. Evolution has occurred in a number of coccolith species within the time interval under study. A rapidly evolving species is not ecologically stable and will expand its biogeographic range during its early existence, making climatic interpretation spurious. Three of the most common species in the time period covered by our cores have apparently gone through such a process. The most ubiquitous species today, *Emiliania huxleyi*, evolved only a few hundred thousand years ago and appears to have supplanted another species in its rise to dominance.

5. Coccolithophorida (phytoplankton) live only in the upper few hundred metres of the ocean, whereas Foraminifera occupy a much greater depth and temperature range. Thus, vertical layering within the water column is likely, with Coccolithophorida and foraminifers with warmer temperature preferences above and foraminifers adapted to colder temperatures below. Sedimentation may result in a colder (deeper) signature for the foraminifers than for the coccolithophores.

6. Many open-ocean coccolithophores may experience an alternation of generations, with a motile feeding and reproductive stage during seasonal ecologic optimums and a nonmotile spore stage during ecologically inhospitable periods throughout the rest of the year. If the species used in this paper experience alternation of generations, our paleotemperature interpretations may be muted.

Floral Assemblages

Today, the area of study is characterized by Coccolithophorida floral input from three water masses—subpolar, transitional, and subtropical. Tropical species are brought in by surface-current flow in the Gulf Stream-North Atlantic Drift as a summer pulse but are insignificant in comparison with other assemblages. Three assemblages characterize these water masses (Table 4): subpolar water is delineated by *Coccolithus pelagicus*, the transitional assemblage is dominated by *Cyclococcolithus leptoporus* var. C, and the subtropical assemblage is marked by *Umbilicosphaera mirabilis*, *Rhabdosphaera clavigera*, *Calciosolenia sinuosa*, *Syracosphaera pulchra*, *Umbellosphaera irregularis*, and *Umbellosphaera tenuis*.

Samples were counted for 350 individuals in the transmission electron microscope. Factor analysis of the resulting counts produced the three assemblages plotted against depth in Figure 11. This display is not analogous to that of the foraminiferal samples (Fig. 9), because the floral cold end member is subpolar instead of polar. The two most common species, *Emiliania huxleyi* and *Gephyrocapsa caribbeanica*, are not included. In part, this is an artifact of the transfer-function technique, which does not segregate the eurythermal species as a cold subpolar end member, despite their dominance (95%) in the coldest subpolar waters. As a result, we are dependent on a group of species that seldom constitutes more than 30% of the flora. These are the stenothermal forms, whose narrow ecologic limits better define discrete surface water masses than do eurythermal types that exist successfully from arctic to antarctic convergences. Future study of these eurythermal species will probably result in water-mass delineation by means of ecomorphic variations

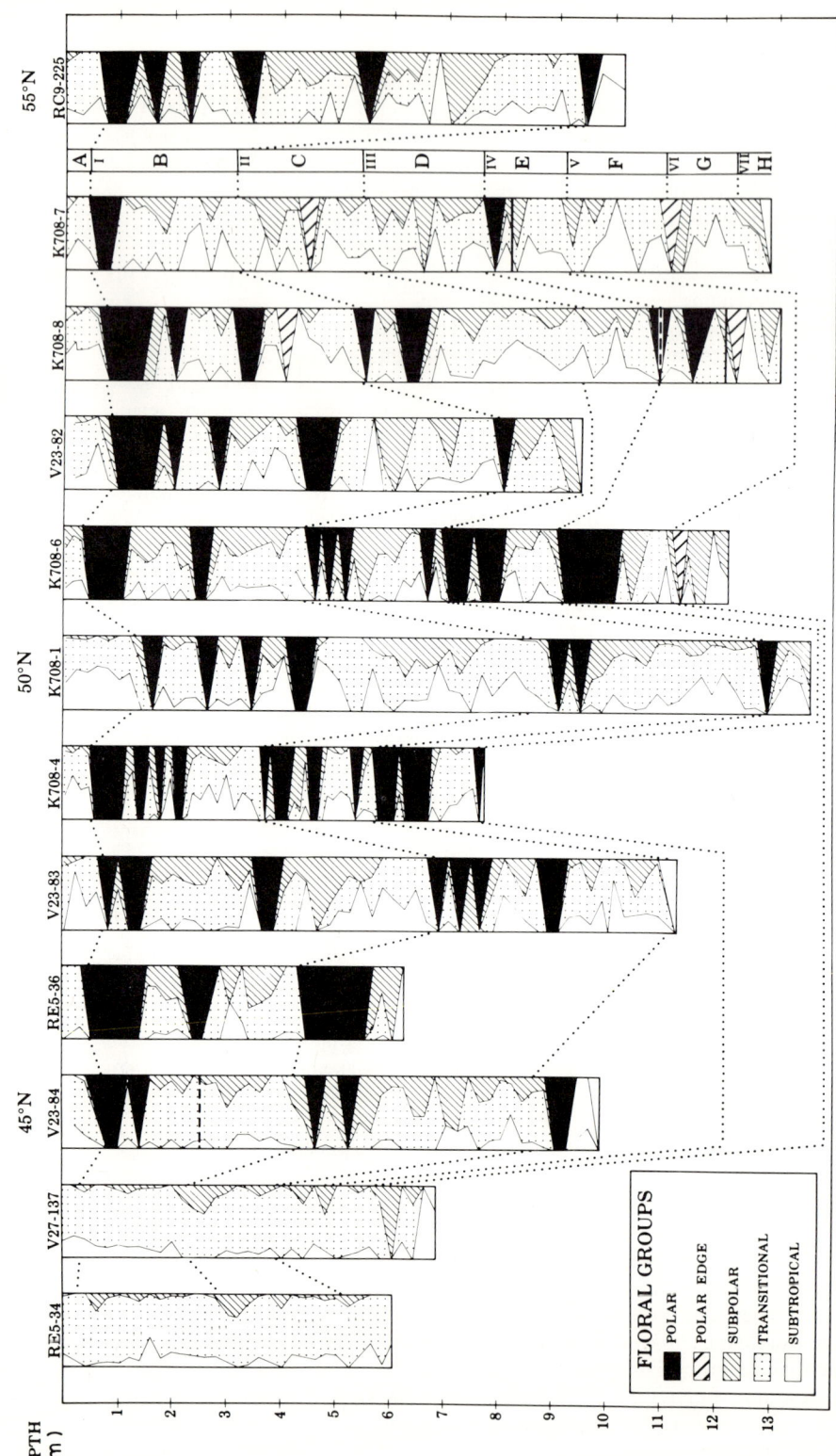

Figure 11. Variation in coccolith flora in the northeast Atlantic; major cycles A to H. Cores plotted to depth. Floral groups, subpolar through subtropical, are transfer-function factors from Roche and others (1975). Polar group is diatom-rich and coccolith-barren. The polar edge contains only *Emiliania huxleyi* and (or) *Gephyrocapsa caribbeanica*.

as shown for *Emiliania huxleyi* (McIntyre and others, 1970).

All cores north of lat 45°N contain coccolith-barren zones during maximum glaciations (Fig. 11), but the number and thickness varies. Because coccoliths are redistributed by bottom currents and vertical mixing, the barren zones may be obliterated in cores with lower sedimentation rates or in cores from provinces with strong bottom currents. Core K708-7, with an average sedimentation rate of 2.5 cm/1,000 yr, contains only two barren zones above termination IV, whereas nearby core K708-8, with a rate of 4.2 cm/1,000 yr, contains six. Only the longest and most intense glacial intervals are represented in core K708-7. These coccolith-barren zones are correlative with the maximum development of the *G. pachyderma* fauna and represent intervals when polar surface waters overlay the cores.

At times, the coccolith flora within less intense carbonate minimums contained only *Emiliania huxleyi* and *Gephyrocapsa caribbeanica*. These represent periods when the coldest (most northern) zone of the subpolar water mass overlay the core. Although the transfer-function treatment does not include these species, we have delineated these intervals (Fig. 11).

Aside from the dominant eurythermal species, the next most abundant floral assemblage in these cores is transitional. During warmer parts of the glacial periods when coccoliths are present, the subpolar assemblage is well developed; during interglaciations, the subtropical assemblages substitute as the subdominant group. The subpolar assemblage (*C. pelagicus*) varies in phase with the carbonate curve, the higher values correlating with the less intense carbonate minimums within major climatic cycles C, D, and E.

In cores south of lat 45°N, where no barren zones occur, maximums in the subpolar assemblage are directly correlative with glacial intervals. Another difference between cores north and south of lat 45°N is the magnitude of fluctuation in the assemblages; as with the Foraminifera, the two southern cores RE5-34 and V27-137 have a more subdued response to climatic variation than the northern cores.

The history of the last major climatic cycle (B) is reflected in the response of the three coccolith assemblages (Fig. 11). The hypsithermal interval (4000 to 6000 B.P.) occurs in cores RE5-36, V23-82, and V23-83 and *Kane* cores 708-1, 708-4, 708-6, and 708-7 as a peak in the subtropical assemblage. The last glaciation is well defined in the north by barren zones and in the south, where coccolith floras are present, by dominance of transitional and subpolar species. In most cores this glacial interval is bracketed by coccolith-barren zones at the peaks of carbonate minimums 1 and 2 of McIntyre and others (1972a).

In 7 of the 14 cores, a short cold floral pulse occurs near the top of the last interglaciation (Fig. 11). Its absence in the other cores is apparently an artifact of the wide sample interval. The rapid migration southward of subpolar species, followed by a similar northward movement of subtropical assemblages, can be dated by interpolation within our time frame at approximately 89,000 B.P. This is correlative in time and intensity with the pre–St. Pierre (North America) and Brörup (Northern Europe) stades. Dansgaard and others (1972) noted an abrupt 10‰ drop in the O^{18}/O^{16} ratio, which they felt was indicative of rapid cooling in the Camp Century ice core (Greenland) at 89,500 B.P. The cooling indicated by coccolith variations appears to be not so abrupt as Dansgaard noted. We do not consider this event the interglacial end, because the subsequent warm interval was not appreciably different from the two preceding warm intervals. The real change to glacial regimen in cycle B does not occur until carbonate minimum 2 at 73,000 B.P. (Figs. 5, 9, 11).

Problem of *Gephyrocapsa ericsonii*

Gephyrocapsa ericsonii is not included in this presentation (Fig. 11) despite its apparent value as an indicator of warmer climate (McIntyre and others, 1972a). In a transfer-function analysis of North Atlantic trigger-weight cores (Roche and others, 1975), this species was not accepted by the equation owing to lack of control from the trigger-weight samples. Further research now indicates that this species prefers near-shore waters, being abundant along continental shelves and around islands in subtropical waters. During interglacial intervals in these cores, *G. ericsonii* is more abundant, particularly during rapid fluctuations in cycles C and D; this suggests that its nutrient and habitat requirements are probably best satisfied by increased land runoff or marine transgression onto the continents.

G. ericsonii decreased in abundance during the very warm interglacial climate following termination II to about 25% of its previous interglacial percentages. At present, it constitutes no more than 15% of the flora in the core-top samples examined. This diminution in abundance since cycle C could be due to ecologic pressure, perhaps leading to extinction.

Coccolith-Barren Zones: Solution, Dilution, and Productivity

The characteristic absence of coccoliths and the virtually monospecific foraminiferal fauna (*G. pachyderma*, s.) in cores north of lat 45°N have been interpreted as indicating polar surface water with reduced productivity (McIntyre and others, 1972a). There is, however, a possible alternative explanation for the genesis of these zones: theoretically, solution by corrosive bottom water could produce a similar thanatocoenose.

Berger (1968, 1970) and Ruddiman and Heezen (1967) have documented the effects of solution on foraminiferal assemblages in deep-sea sediment. The most resistant species in higher latitudes is *G. pachyderma*, s., the dominant form in the barren zones. Although knowledge of the effect of solution on coccoliths is not as complete, Schneidermann (1971) and McIntyre and McIntyre (1971) have listed the following as relatively solution resistant among coccoliths abundant in the North Atlantic Ocean, in order of increasing resistance: *Emiliania huxleyi, Cyclococcolithus leptoporus, Gephyrocapsa caribbeanica,* and *Coccolithus pelagicus*. Also widely applied as a solution index is the ratio of benthic to planktonic foraminifera (Oba, 1969), the benthic forms being the more resistant to solution. We have used both floral and faunal evidence to evaluate the effect of solution on North Atlantic barren zones.

If the barren zones were the result of increasing severity of solution, the coccoliths

TABLE 5. PLANKTONIC/BENTHIC FORAMINIFERAL RATIO IN CORE V27-110

Depth (cm)		Ratio
65		390/1
75		460/1
85	Termination I	340/1
90		523/1
95	Coccolith barren	26/1
105		420/1
110		324/1
115		276/1
125		974/1
130		281/1

should show a predictable sequence of species loss near the margins of the barren intervals. Initially, there should be a decrease in *E. huxleyi* and an increase in *C. pelagicus*. At the edge of the barren zone, *C. pelagicus* should dominate or even be the only species present. In addition, the ratio of benthic to planktonic species should increase to a maximum within the barren zones.

The coccolith floras in McIntyre and others (1972a) did not show such trends; however, the sample spacing was relatively large (20 cm). The first barren zone (100 to 160 cm) and the overlying postglacial sediment in core V23-82 have been examined at 5-cm intervals (65 to 130 cm) for coccolith species and benthic/planktonic foraminiferal ratios (Table 5). The increased detail confirms that no enrichment in resistant coccolith forms occurred and reveals that the benthic/planktonic ratios remained low within the barren zone. The mean value and range of variation obtained within the barren zone are comparable to that in nonbarren postglacial sediment (Table 5), which suggests that no significant solution occurred in barren zones even in this core from 3,974 m.

Core V27-110 from Hatton-Rockall basin on Rockall plateau provides the converse proof of the genesis of northeast Atlantic barren zones. Sediments in this core, which was taken from a depth of 1,264 m, were deposited far above the present compensation depth. A total carbonate curve for the section of core through termination II is shown in Figure 12. Even at this shallow depth and reduced sedimentation rate, a coccolith-barren zone occurs between 44 and 49 cm, with thin, almost-barren zones at 53, 55, and 59 cm. In addition, the coccolith counts in core V27-110 show no significant increase of solution-resistant species near the barren zones or within the almost-barren zones (Fig. 12). Both *Coccolithus pelagicus* and *Cyclococcolithus leptoporus* occur in lower percentages in the last glaciation than in the Holocene Epoch or the last interglaciation. In the North Atlantic Ocean today, where solution is known to occur on the bottom beneath subpolar surface water in the Labrador Basin, the core tops contain between 50

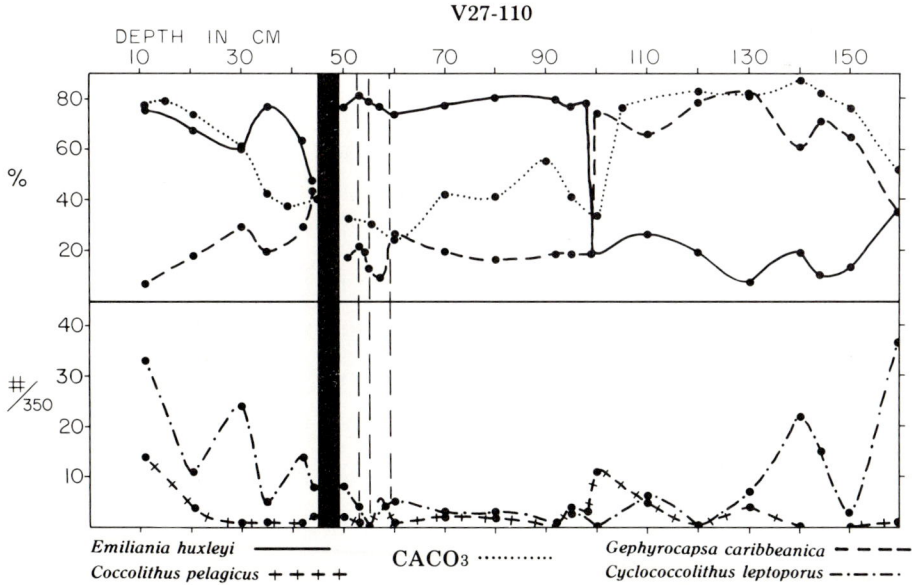

Figure 12. Comparison of selected coccolith species abundance as solution indicators against less than 62-μm carbonate in core V27-110. Vertical black band and dashed lines are coccolith-barren zones.

and 100% of the resistant *C. pelagicus*. The highest percentage in the glacial part of core V27-110 is only 1.7%.

We conclude that barren zones exist in northeast Atlantic depths too shallow for major carbonate solution, and conversely, that faunal and floral indices in and near barren zones even at 4,000 m indicate very little solution. Core intervals that are coccolith barren must represent periods during the Pleistocene glacial maximums when the overlying surface waters were devoid of coccoliths. We thus reiterate that productivity variations are the primary cause of coccolith-barren zones in the Atlantic Ocean north of lat 40°N.

HISTORY OF NORTH ATLANTIC CLIMATIC CHANGES

Frequency of Climatic Oscillations

Long-term climatic cycles varying in length from 56,000 to 113,000 yr are evident in the faunal, floral, and lithologic data (Figs. 4, 5, 9, 11), with the closest sampling intervals and, hence, most diagnostic detail in the foraminiferal variations (Fig. 9). Ignoring climatic pulses with durations of less than 20,000 yr, we note seven complete major cycles in the past 605,000 yr (Table 6) averaging 84,000 yr in length. The same approximate number of major cycles has been noted in previous deep-sea studies of comparable intervals (Emiliani, 1966; Hays and others, 1969; Ruddiman, 1971; Kent and others, 1971; Shackleton and Opdyke, 1973). Because core K708-7 contains all seven cycles, it is the basis for estimating, by linear interpolation, the ages of terminations separating the cycles (Table 6).

We can also detect additional short but severe cold climatic pulses lasting less than 20,000 yr. Sequential numbering of the resulting carbonate minimums in the

TABLE 6. DEPTH AND EXTRAPOLATED AGES OF MAJOR CLIMATIC CYCLES AND THEIR TERMINATIONS IN CORE K708-7 FROM THE NORTHEAST ATLANTIC OCEAN

Major climatic cycles	Termination depths (m)	Termination numbers and extrapolated ages (B.P.)	Alternate termination ages*
A			
	45	(I) 13,500	11,000
B (1-2)†			
	310	(II) 127,000	127,000
C (3-5)†			
	560	(III) 225,000	225,000
D (6-7)†			
	740	(IV) 311,000	300,000
E			
	920	(V) 384,000	380,000
F			
	1,090	(VI) 457,000	. .
G			
	1,220	(VII) 513,000	. .
H			
	1,430	(VIII) 603,000	. .

*From Broecker and van Donk (1971).
†Numbers in parentheses are the equivalent carbonate minimums from McIntyre and others (1972).

sense of McIntyre and others (1972a) is listed in Table 6 and shown in Figure 5. Most prominent among these are the brief cold pulses at roughly 195,000 B.P. (carbonate minimum 5) in the lower interglacial part of major cycle C and at 72,000 B.P. (carbonate minimum 2) in the middle of major cycle B. We also have found analogous short warm climatic episodes that appear to represent significant deglaciations, particularly in the glacial parts of major cycles C and D.

If the numbering of oceanic glacial maximums is extended to encompass these separate shorter episodes, there were 11 major polar-water advances in the past 600,000 yr, any of which could correspond to ice advances on continents. Because there are significant amounts of coarse ice-rafted terrigenous debris at each of these maximums, we infer that continental glaciers may have reached almost full size despite the brevity of these climatic pulses. Within the full 1.2 m.y. of major Pleistocene ice-rafting in mid-latitudes of the Northern Hemisphere (Kent and others, 1971), we estimate that there may have been 20 or more significant ice-sheet advances.

Beyond these, there are several even shorter episodes (such as the cold pulse at 40,000 B.P. in McIntyre and others, 1972a) that may represent discrete glacial advances during generally moderate glaciations. Their number is hard to estimate, because they may be very brief in duration and thus largely blurred by mixing in deep-sea sediment.

We noted several particularly prominent warming periods among the seven complete major cycles of the past 600,000 yr. The warmest periods occur in the lowest parts of major cycles B and D, at 120,000 and 380,000 B.P., respectively. The former has been correlated with the well-known high sea level dated from coral reefs in many regions (McIntyre and others, 1972a).

Cycle Geometry

Broecker and van Donk (1970) described a primary "saw-toothed" cycle characteristic of Caribbean and equatorial Atlantic oxygen-isotopic variations and inferred that the fundamental glacial cycles (the form of the global ice-volume changes) must also be saw-toothed in character, with gradual glacial build-up followed by rapid deglaciation or termination. None of the cycles generated in the northeast Atlantic Ocean by variations in foraminifera, coccoliths, or lithology are so unidirectionally saw-toothed as the low-latitude isotopic curves, nor are all deglacial terminations so abrupt.

The most striking departure from the saw-toothed asymmetry is found in the nearly symmetrical geometry of the interglacial part of cycle C. Its inception is "termination III," an erratic interval of a gradual warming trend building to a full interglaciation, rather than the sudden warming characteristic of terminations. This interglaciation was bisected by the intense and very abrupt cold episode,—carbonate minimum 5,—as shown in Figure 10B, and it ended with a moderately abrupt cooling trend in carbonate minimum 4. Cycle G also fails to fit the idealized geometry; it never reached fully glacial conditions and was only 56,000 yr in length. Its rank as a major cycle can even be questioned. Cycle B contains two subcycles, one from 127,000 to 75,000 B.P., biased toward interglacial warmth, and one from 75,000 to 13,500 B.P., biased toward glacial cold. The latter does not begin in interglacial conditions and thus should not be considered separately as a major cycle.

In general, the major cycles show an erratic saw-toothed geometry. The form of the terminations that separate the major climatic cycles ranges from the striking abruptness of II and IV to the gradual and erratic warming trends across III, V, and VI.

Comparison with Other Oceanic Data

Combining data from many cores, we constructed a single curve representative of northeast Atlantic climatic variations over the past 600,000 yr. This has been compared with other deep-sea paleoclimatic curves spanning substantial portions of the Brunhes normal paleomagnetic epoch (Fig. 13).

Four of the other curves are isotopic studies (Emiliani, 1966, 1972; Shackleton and Opdyke, 1973). One is based on foraminiferal variations in the equatorial Atlantic Ocean (Ruddiman, 1971); this was generalized from two cores with records spanning the entire Brunhes epoch. The curve from Shackleton and Opdyke (1973) also reached the Matuyama epoch. All other cores bottomed within the middle to lower Brunhes epoch, including the composite northeast Atlantic climatic curve from this paper.

For the two cores with paleomagnetic control, the Brunhes-Matuyama boundary at 690,000 B.P. was used as the lower chronologic control point for linear time interpolations to the core top at 0 B.P. The four other cores were plotted against time (Fig. 13) by alignment of terminations I and II to the adjusted time scale of Broecker and Ku (1969) and linear stretching as in Figures 4, 5, and 9. As a result, the dates exceed by 25% those derived by Emiliani (1966) for the Caribbean cores. One additional datum level used for verifying correlations but not as control in the chronologic plotting is the extinction level of *Pseudoemiliania lacunosa*, roughly coincident with the U-V boundary of Ericson and Wollin (1968).

The most impressive correlation among the climatic curves shown in Figure 13 is between the northeast Atlantic record (this paper) and the west-central Pacific record (Shackleton and Opdyke, 1973). Even the very short and visually diagnostic cooling in cycle C (carbonate minimum 5) and the weak cooling of cycle G appear in the Pacific isotopic curve. Also quite similar is the equatorial Atlantic curve (Ruddiman, 1971), which has major peaks that correlate well with the northeast Atlantic curve. The Caribbean isotopic records from Emiliani (1966, 1972) contain the same number of major climatic cycles as the three longer records through at least the *P. lacunosa* (U-V) data.

Significant differences between the three long-record curves exist in the lower Brunhes epoch. Our northeast Atlantic dates for the middle and lower Brunhes cycles derived from the extrapolated time scale of Broecker and Ku (1969) (Fig. 13) are intermediate between the shorter scheme of Rona and Emiliani (1969) and the older time scale that Shackleton and Opdyke (1973) derived for the South Pacific isotopic curve. Apparently, the lengthy chronologic interpolation from 690,000 B.P. to the present allows long-term changes in regional sedimentation rates to create offsets in timing. The offsets in the northeast Atlantic and equatorial Pacific curves are such that there appears to be a difference of one cycle in the bottom part of the Brunhes record (Fig. 13). The last interglacial peak in core K708-7 (northeast Atlantic) looks as if it should correlate with the interglaciation that spans the Brunhes-Matuyama boundary in core V28-238; however, we found no reversal in that interglaciation in core K708-7. Recent evidence from core V28-239 (N. Schackleton, 1974, oral commun.) shows that the correlation between the northeast Atlantic foraminiferal curve (K708-7) and the southwest Pacific isotopic curve (V28-238) shown in Figure 13 is correct in the lower Brunhes section. The weak interglaciation at approximately 620,000 B.P. in core V28-238 is much stronger in core V28-239, just as in core K708-7.

In summary, we conclude tentatively that there are eight full climatic cycles in the Brunhes epoch (B-I) and the beginning of a ninth (A). Additional cores are needed to resolve remaining differences between these basically very similar climatic records.

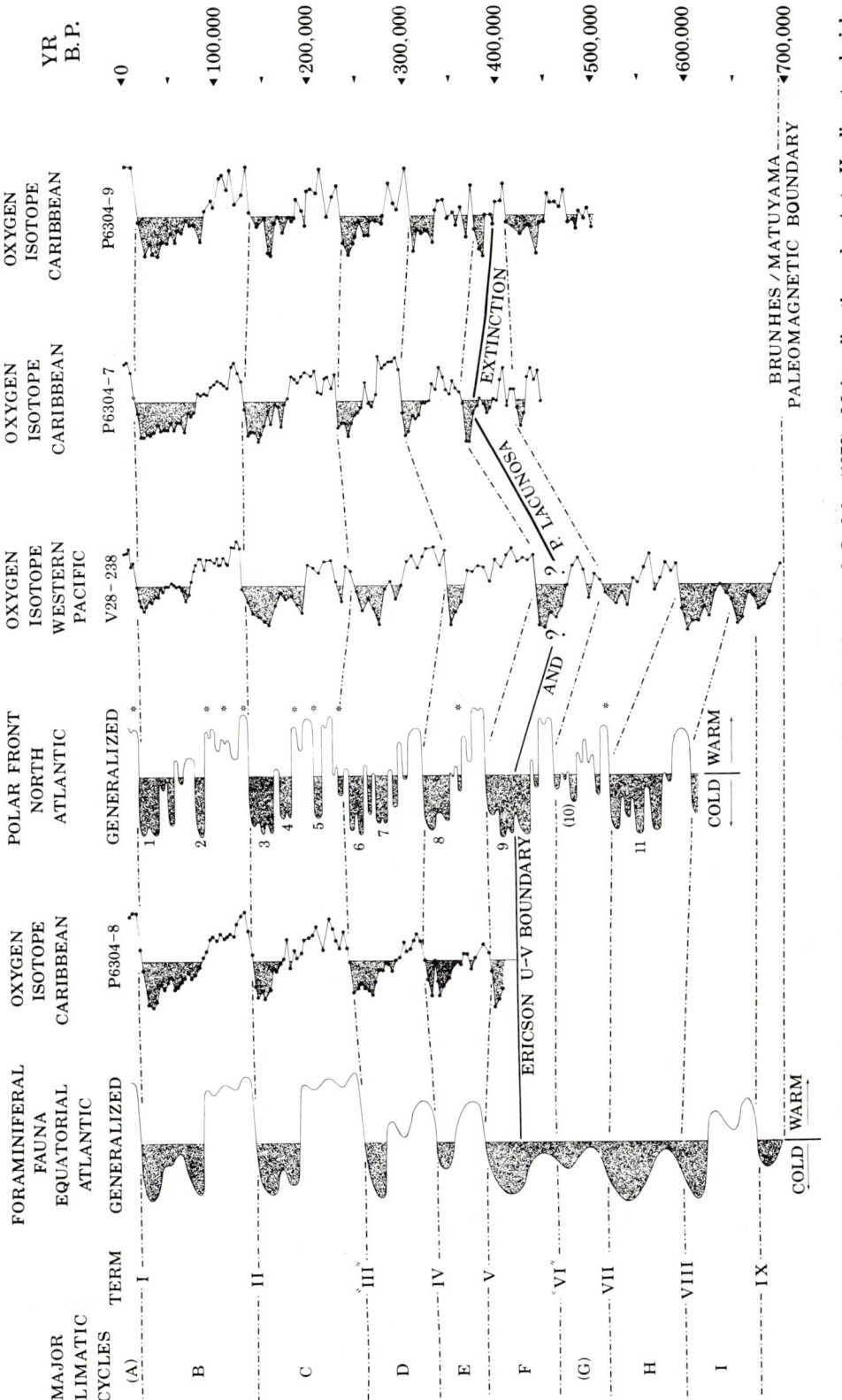

Figure 13. Comparison of cyclic trends in deep-sea cores. Northeast Atlantic curve from this paper; equatorial Atlantic faunal curve from Ruddiman (1971); Caribbean isotopic curves from Emiliani (1966, 1972); Pacific isotopic curve from Shackleton and Opdyke (1973). Major climatic cycles A to H, discrete glacial maximums 1 to 11, terminations I to IX, and asterisks are high sea levels from Mesolella and others (1969).

Land Comparisons

Climatic change modifies adjacent regions of land and sea; therefore, our 600,000-yr record of climatic change should appear in climatically controlled land indicators. Sequential continental climatic data is preserved in palynologic vegetative histories. Lacking an oceanic core in close proximity to land in which both oceanic and palynologic indicators could be compared, we have no jointly occurring time-stratigraphic indicators. Beyond the range of carbon-14 dating, only subjective correlations (curve matching) can be used. Unfortunately, the pollen sections in higher latitudes close to the northeast Atlantic Ocean have recorded either short discontinuous time intervals or palynologic barren zones during maximum ice cover (the "polar desert" of van der Hammen and others, 1971). In southeastern Europe, however, a fortunate exception is found.

Wijmstra (1969) and van der Hammen and others (1971) described an apparently continuous pollen record from Phillippi in eastern Macedonia that seems directly comparable with our long oceanic record in core K708-7 (Fig. 14). We have simplified Figure 7 from van der Hammen and others for display. All herbs have been combined and considered characteristic of open dry to desert environments. Both sections contain climatic cycles A through H. The short, intense oceanic coolings (equivalent to high herb concentrations in the pollen record, indicating dryer climate) within

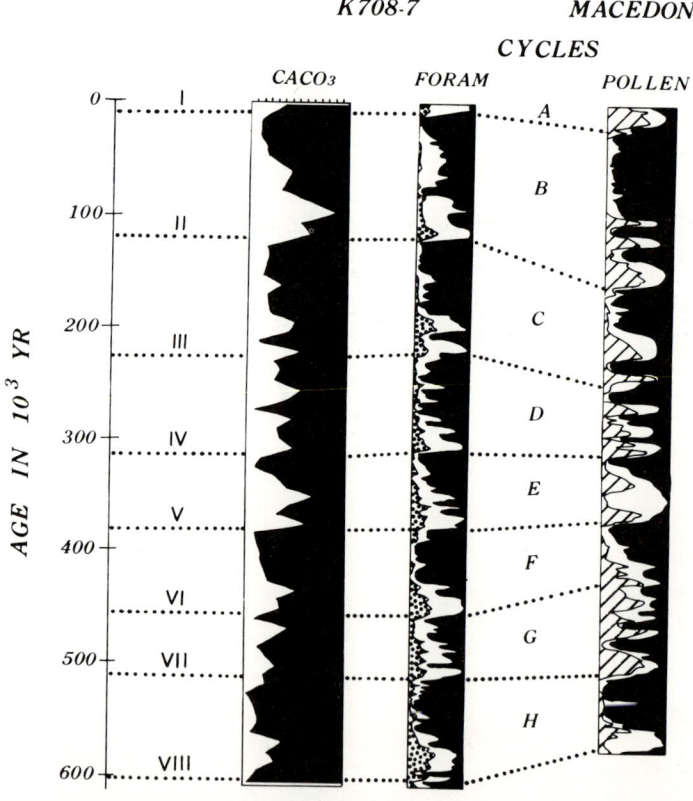

Figure 14. Comparison and proposed chronostratigraphic correlation of oceanic climate record (as delineated by fine-fraction carbonate and foraminifers in core K708-7) with the pollen record from Macedonia (van der Hammen and others, 1971).

the interglaciations of cycles B, C, D, and G are clearly depicted in both sections. The longer, more intense glaciations of cycles B, C, F, and H are recorded in both sequences. Cycle E in Atlantic and Pacific cores contains a short glaciation and a long, well-developed interglaciation, the latter marked by unusually high carbonate percentages (Fig. 5), O^{18}/O^{16} values (Shackleton and Opdyke, 1973), and subtropical coccolith abundances (Fig. 11). This very warm interglaciation just postdates the *Pseudoemiliania lacunosa* and *Stylatractus universus* extinctions within the glacial part of cycle F. In the Macedonian record, cycle E is similar in that its interglacial period is the longest.

There are, however, basic discrepancies between what continental geologists and oceanographers consider an interglaciation, the latter combining short-term (10^4 yr) warm-cold oscillations into a single interglaciation that continental workers may treat as discrete units. The interglaciation of cycle B in cores K708-7 (310 to 210 cm) and K708-8 (535 to 340 cm) is equivalent in Europe to the combined Eemian, Amersfoort, and Brörup (earliest Würm) interglaciations and in North America to the combined Sangamon and St. Pierre interglaciations. This type of discrepancy presumably recurs in older interglaciations.

The complex (gradual) character of terminations "III" and "VI" noted in core K708-7 is repeated in the Phillippi record but results in the only difficult crosscorrelation. Our placement of termination "VI" in the Phillippi section is open to question. We can correlate our carbonate minimums 1 through 9 and 11 with maximum development of dryer (herb) flora. Carbonate minimum 10, which precedes termination "VI," cannot be positively aligned with any of the higher herb-concentration intervals within the Boz-Dagh complex. The last appearance of a Tertiary floral relic (*Zelkova*) occurred no later than cycle F (G. Kukla, 1973, oral commun.). In the Phillippi section, this last appearance is near the top of the Boz-Dagh complex. Using this as a criterion, we correlate termination "VI" with the climatic warming trend just predating the last appearance of *Zelkova*.

There is a major discrepancy between the time frames of core K708-7 and the Phillippi record. The Macedonian section has been carbon-14 dated at 50,000 B.P. Extrapolation from this level gives an age of 400,000 B.P. at 120 m. Taking compaction of lake sediment into account, van der Hammen and others (1971) estimated a more realistic age of 500,000 B.P. for the 120-level. Our time scheme for the oceanic record gives an age of approximately 600,000 B.P. for termination VIII at the base of the Phillippi section. Because the Brunhes-Matuyama boundary at 690,000 B.P. appears to equate with the onset of cycle I, we feel that our longer time scheme should be applied to the Macedonian record.

Insolation Comparison

Attention has been drawn to the possible correlation between curves of deep-sea climate and of summer insolation in high northern latitudes (Emiliani, 1955). Because of the imprecise correlations between these parameters, Broecker (1966) first assumed a two-mode (glacial-interglacial) system with mode changes dependent on extreme peaks of high or low insolation. Subsequently, Broecker and van Donk (1970) hypothesized that the equilibrium climatic state toward which the Northern Hemisphere has tended to drift during Pleistocene time has been one of moderately cold (but not fully glacial) conditions and that extreme summer insolation maximums trigger very rapid short-term shifts (terminations) to warm interglacial conditions.

We have plotted in Figure 15 the insolation curve for lat 55°N from Broecker and van Donk (1970) against the North Atlantic paleoclimatic curve from Figure 13. All insolation maximums since 225,000 B.P. are phased slightly ahead of

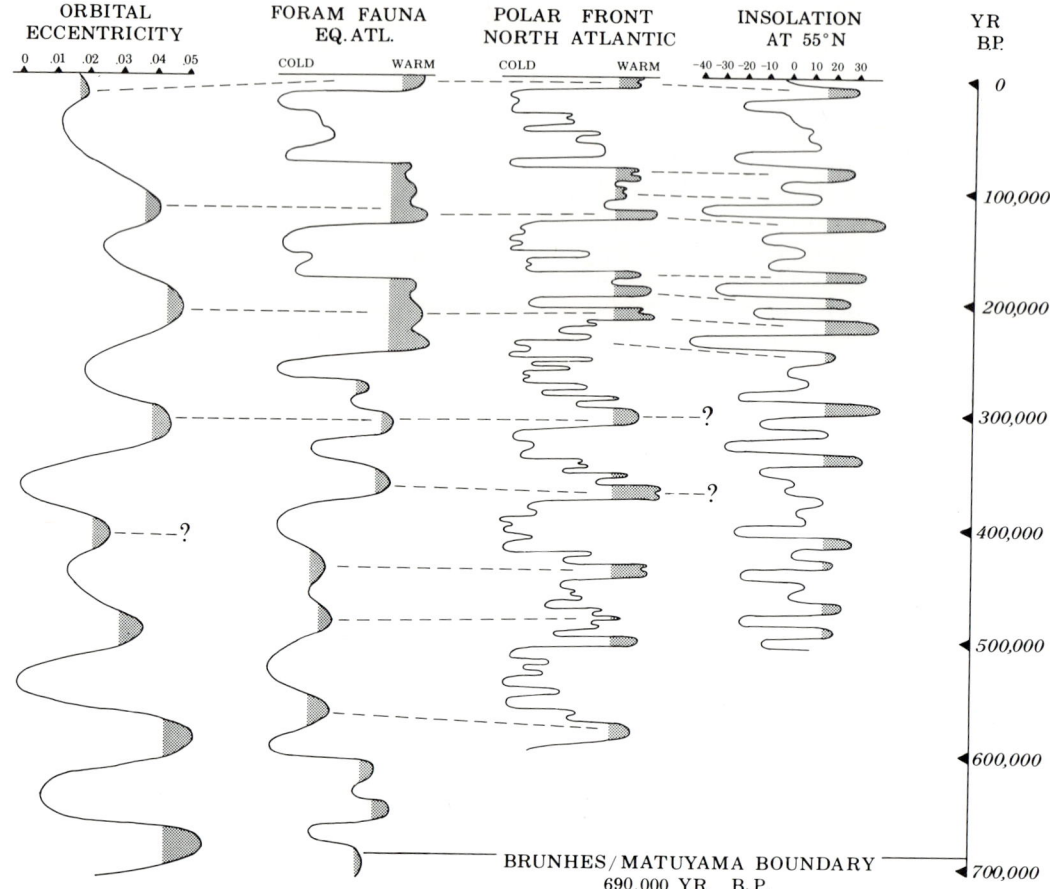

Figure 15. Comparison of deep-sea climatic records against orbital eccentricity of Earth and calculated insolation curve at lat 55°N.

paleoclimatic warming trends. This agreement since 225,000 B.P. is suggestive of a causal relationship. From 225,000 to 600,000 B.P., however, there is no clear correlation. This mismatch may reflect actual divergence between insolation and paleoclimate; however, before 225,000 B.P., it is also possible that the critical tilt and precession components are improperly phased and that the computed insolation curves are thus not dependable below that level (Broecker and van Donk, 1970). A third possibility is that imprecisions in dating and extrapolations of the deep-sea climatic curves have offset the two parameters. Still, no simple linear stretching or compression of the oceanic curves will match the terminations with the insolation maximums in the time range 600,000 to 225,000 B.P. (Fig. 15).

Insolation minimums since 225,000 B.P. are closely time-correlative with cold climatic pulses, but the intensities do not match. One example is the lack of a severe oceanic cooling trend to compare with the extreme insolation minimum at 115,000 B.P. (Fig. 15). The detailed North Atlantic curve shows a moderate cooling trend (McIntyre and Ruddiman, 1972; Sancetta and others, 1972) but not the severe cooling predicted from the intensity of the insolation low. The assumption that this is because of a slow oceanic response would therefore fail to explain the rapid oceanic cooling periods achieved near other less intense insolation minimums. There seems to be no simple way of weighting the orbital parameters so as to make the insolation curve predict the intensities of the North Atlantic oceanic response.

Terminations, according to Broecker and van Donk (1970), only occur in response to extreme solar insolation maximums during Northern Hemisphere summers. Such maximums, in turn, only occur during orbital eccentricity maximums, when seasonal contrasts are largest. Over the past 300,000 yr, the terminations shown in Figure 15 clearly fall within eccentricity maximums; below this level, there is a systematic mismatch between eccentricity maximums and terminations. The Shackleton and Opdyke (1973) curve provides a much better correlation in this respect.

In summary, the correlative timing of oceanic warming trends and Northern Hemisphere summer insolation maximums over the past 225,000 yr suggests a causal relationship, but earlier Brunhes variations weaken this conclusion.

ECOLOGIC WATER–MASS MOVEMENTS AND PALEOTEMPERATURE VARIATIONS

Latitudinal Extent of Water-Mass Movements

North-south ecologic water-mass movements across the eastern North Atlantic Ocean, reconstructed from foraminiferal assemblages in Figure 9 and coccolith assemblages in Figure 11, are shown in Figure 16 (in pocket). The most significant aspect of the migrations is their large horizontal extent across 10° to 20° of latitude, as inferred by McIntyre (1967). Polar water in the eastern North Atlantic Ocean retreated to lat 65°N of Iceland during interglaciations (McIntyre and others, 1972a) and penetrated southward to beyond lat 45°N during glacial maximums. This defines displacements of cold faunal and floral assemblages of over 20° of latitude (2,000 km), with warm fauna and flora moving more than 10° to 15° of latitude (transitional assemblages from below lat 40°N to above lat 55°N; cold subtropical groups from below lat 40°N to just above lat 50°N).

The rate of water-mass shifts can be extremely rapid. At latitudes between 45°N and 50°N, a succession of four distinct ecologic water masses has rapidly moved across the area at numerous times during the Pleistocene Epoch. At the level of termination II, the four water masses moved successively in a cold-to-warm south-to-north geographic shift of 15° to 20° of latitude in less than 10,000 yr. This implies rates of water-mass migration averaging in excess of 150 to 200 m/yr and continuing uninterrupted for thousands of years. Measurements of continental glaciers on land, by comparison, indicate advance and retreat rates on the order of 10 to 500 m/yr (Goldthwait, 1958; Willman and Frye, 1970; Kempton and Gross, 1971). Sustained oceanic rates exceeding 1 km/yr are likely in the North Atlantic Ocean, on the basis of the abrupt climatic changes in optimal records of cores with very high sedimentation rates (McIntyre and Ruddiman, 1972).

In contrast, the lesser faunal and floral variations in the southernmost cores indicate a greater stability of the North Atlantic subtropical gyre throughout Pleistocene time and less significant migrations.

Areal Distribution of Water Masses

North Atlantic water-mass migrations cannot be analyzed only as north-south displacements. In the modern interglacial warmth, the strongest thermal gradient lies in a northwest-southeast direction. Cold water flowing southward along the southeastern coast of Greenland is just as proximal to our study area in the northeast Atlantic Ocean as that directly to the north near Iceland. This configuration and evidence shown by Ruddiman and McIntyre (1973) suggest that west-to-east (or northwest-to-southeast) migrations of cold water should be as important as north-to-south displacements.

We have delineated in Figure 17 (in pocket) the regional distribution of water masses at four different chronologic levels in the Quaternary Period. For reference, we have first shown the present distribution of ecologic water masses, along with mean modern circulation patterns (Fig. 17A).

During conditions of full glaciation, ice-bearing polar water completely enveloped the North Atlantic Ocean above lat 45°N in a cold counterclockwise gyre (Fig. 17B). (Sea-ice limits have not been inferred but have been proposed by McIntyre and others [1972b].) Weyl (1968) essentially inferred such a glacial configuration. Unlike the weak modern oceanographic gradient oriented obliquely to lines of latitude in the eastern North Atlantic Ocean, there developed during full glaciation a latitudinally oriented frontal system along lat 45°N with extreme gradients analogous to that of the modern southern (circumantarctic) ocean. The subpolar and transitional waters were essentially eliminated by being compressed into a narrow band between the clockwise subtropical gyre and the counterclockwise polar gyre. This pattern signifies minimal northward heat transport in the near-surface flow.

For a time-synchronous view of intermediate climates afforded by our core coverage, we have included a pattern developed from samples at the level of peak abundance of ash zone I at 9300 B.P. (Ruddiman and Glover, 1972a, 1972b). At that time in the Pleistocene Epoch, ice-bearing polar waters extended farther to the south and east from Newfoundland across the North Atlantic Ocean than during interglaciations, with relatively warm waters still flowing north along the eastern Atlantic margin, but to the south and east of modern flow (Fig. 17C); combined, these flows formed a net counterclockwise gyre (Ruddiman and Glover, 1972a).

One of the two warmest climatic episodes in the past 600,000 yr is the interglacial peak at 120,000 B.P. following termination II. At this level, which has been correlated with the Barbados III and Pacific Ocean high sea levels (Mesolella and others, 1969), warm water masses (transitional and cold subtropical) dominated virtually the entire map area (Fig. 17D). Presumably, waters in the northwest Atlantic Ocean maintained cooler temperatures, and we have inferred polar and subpolar water masses in the coastal currents along the coast of Greenland. Cold subtropical water penetrated northward as far as lat 55°N in response to this unusual warmth.

Considered together, these maps confirm that advances and retreats of faunally and florally defined ecologic water masses have occurred along a northwest-southeast axis similar to that traced by Ruddiman and McIntyre (1973) in the last deglacial retreat of polar water from the North Atlantic Ocean.

North Atlantic Paleotemperature Variations

The geographic movement of the polar front can be used to infer the intensity of paleotemperature changes of North Atlantic surface waters. At the present, the polar front is marked by the 0°C winter isotherm and the 6°C summer isotherm. Because the 45°N polar-front position during full glaciations (Fig. 17B) overlies the modern 12.5°C winter and 19°C summer isotherms, we surmise that glacial surface-water temperatures fell by 12.5°C in the winter and 13°C in the summer. Using transfer-function techniques on faunas in maximum Wisconsin samples,

McIntyre and others (1972b) suggested a 12°C winter cooling trend in this area.

During peak interglacial warmth at 120,000 B.P. (Fig. 17D), the polar front moved north at least to the present positions of the 2°C summer and −1°C winter isotherms. Such an offset implies that the Barbados III high sea-level stand at 120,000 B.P. occurred during conditions warmer than those at present by 3°C in the winter and 4°C in the summer. On a scale ranging between these two Pleistocene climatic extremes, the present surface-sediment conditions (actually a mix of the past several thousand years) lie 75 to 80% toward the extreme warm end of Pleistocene glacial-interglacial variations.

This amplitude of climatic variation, combined with the extent of water-mass shifts that we have shown in the North Atlantic Ocean, clearly has global significance in the heat budget of the earth. The area of the North Atlantic Ocean, bounded on the south and north by lat 45°N and lat 65°N and by the continents on the east and west, is approximately 5×10^6 km^2, or 5% of the Earth's surface. With surface-water temperatures varying as much as 12°C and with the entire area periodically subject to abundant drifting icebergs and sea ice, the back-reacting consequences to the planetary albedo of climatically induced variations in this ocean area are as significant as the radically changing albedo patterns on land in areas of primary continental glaciation. In addition, the glacial diminutions and interglacial enhancements of poleward heat transport in the oceanic surface waters must be important in the total Equator-to-pole heat transport of the Earth's fluid system.

ACKNOWLEDGMENTS

We thank B. Grosvenor, C. Fruik, and M. Perry for illustrations, and L. Glover, M. Roche, and J. Kostecki for laboratory assistance and helpful discussions. The research was jointly supported by the U.S. Naval Oceanographic Office and by National Science Foundation Grants GA-14177 and GA-40053X. We were also aided by Office of Naval Research Contract N00014-67-A-0108-0004 and National Science Foundation Grant DES-72-01568 to Lamont-Doherty Geological Observatory.

REFERENCES CITED

Bé, A.W.H., 1959, Ecology of Recent planktonic foraminifera. Pt. I. Areal distribution in the western North Atlantic: Micropaleontology, v. 5, p. 77–100.

Bé, A.W.H., and Hamlin, W. H., 1967, Ecology of Recent planktonic Foraminifera. Pt. III. Distribution in the North Atlantic during the summer of 1962: Micropaleontology, v. 13, p. 87–106.

Bé, A.W.H., Vilks, G., and Lott, L., 1971, Winter distribution of planktonic foraminifera between the Grand Banks and the Caribbean: Micropaleontology, v. 17, p. 31–42.

Berger, W. H., 1968, Planktonic Foraminifera: Selective solution and paleoclimatic interpretation: Deep-Sea Research, v. 15, p. 31–43.

——1970, Planktonic foraminifera: Selective solution and the lysocline: Marine Geology, v. 8, p. 111–138.

Berger, W. H., and Heath, G. R., 1968, Vertical mixing in pelagic sediments: Jour. Marine Research, v. 26, p. 135–143.

Bramlette, M. N., and Bradley, W. H., 1941, Geology and biology of North Atlantic deep-sea cores between Newfoundland and Ireland. Pt. 1. Lithology and geologic interpretations: U.S. Geol. Survey Prof. Paper 196-A, p. 1–34.

Broecker, W. S., 1966, Absolute dating and the astronomical theory of glaciation: Science, v. 151, p. 299-304.

——1971, Calcite accumulation rates and glacial to interglacial changes in oceanic mixing, *in* Turekian, K. K., ed., The late Cenozoic glacial ages: New Haven, Conn., Yale Univ. Press, p. 239-265.

Broecker, W. S., and Ku, T. L., 1969, Caribbean cores P6304-8 and P6304-9: New analysis of absolute chronology: Science, v. 166, p. 404-406.

Broecker, W. S., and van Donk, J., 1970, Insolation changes, ice volumes, and the O^{18} record in deep-sea cores: Rev. Geophysics and Space Physics, v. 8, p. 169-198.

Broecker, W. S., Turekian, K. K., and Heezen, B. C., 1958, The relation of deep-sea sedimentation rates to variations in climate: Am. Jour. Sci., v. 256, p. 503-517.

Dansgaard, W., Johnsen, S. J., Clausen, H. B., and Langway, C. C., 1972, Speculation about the next glaciation: Quaternary Research, v. 2, p. 396-398.

Delaney, A. C., Parkin, D. W., Griffin, J. J., Goldberg, E. D., and Reimann, B.E.F., 1967, Airborne dust collected at Barbados: Geochim. et Cosmochim. Acta, v. 31, p. 885-909.

Emiliani, C., 1955, Pleistocene temperatures: Jour. Geology, v. 63, p. 538-578.

——1966, Paleotemperature analysis of the Caribbean cores P6304-8 and P6304-9 and a generalized temperature curve for the last 425,000 years: Jour. Geology, v. 74, p. 109-126.

——1972, Quaternary paleotemperatures and the duration of the high-temperature intervals: Science, v. 178, p. 398-401.

Ericson, D. B., and Wollin, G., 1968, Pleistocene climates and chronology in deep-sea sediments: Science, v. 162, p. 1227-1234.

Folger, D. W., 1970, Wind transport of land-derived mineral, biogenic, and industrial matter over the North Atlantic: Deep-Sea Research, v. 17, p. 337-352.

Gaarder, K. R., 1971, Comments on the distribution of Coccolithophorids in the oceans, *in* Funnell, B. M., and Riedel, W. R., eds., The micropaleontology of oceans: Cambridge, England, Cambridge Univ. Press, p. 97-103.

Glass, B. P., Ericson, D. B., Heezen, B. C., Opdyke, N. D., and Glass, J. A., 1967, Geomagnetic reversals and Pleistocene chronology: Nature, v. 216, p. 437-442.

Goldthwait, R. P., 1958, Wisconsin age forests in western Ohio. I. Age and glacial events: Ohio Jour. Sci., v. 58, p. 24-27.

Hamilton, E. L., 1969, Sound velocity, elasticity, and related properties of marine sediments, North Pacific III: North Pacific Naval Undersea Research and Development Center Tech. Pub. no. 145, 79 p.

Hammen, T. van der, see van der Hammen.

Hays, J. D., Saito, T., Opdyke, N. D., and Burckle, L. H., 1969, Pliocene-Pleistocene sediments of the equatorial Pacific: Their paleomagnetic, biostratigraphic, and climatic record: Geol. Soc. America Bull., v. 80, p. 1481-1514.

Hülsemann, J., 1966, Notes: On the routine analysis of carbonate in unconsolidated sediments: Jour. Sed. Petrology, v. 36, p. 622-625.

Imbrie, J., and Kipp, N., 1971, A new micropaleontological method for quantitative paleoclimatology: Application to a late Pleistocene Caribbean core, *in* Turekian, K. K., ed., The late Cenozoic glacial ages: New Haven, Conn., Yale Univ. Press, p. 71-181.

Johnson, G. L., and Schneider, E. D., 1969, Depositional ridges in the North Atlantic: Earth and Planetary Sci. Letters, v. 6, p. 416-422.

Jones, E.J.W., Ewing, M., Ewing, J. I., and Eittreim, S. L., 1970, Influences of Norwegian Sea overflow water on sedimentation in the northern North Atlantic and Labrador Sea: Jour. Geophys. Research, v. 75, p. 1655-1680.

Kempton, J. P., and Gross, D. L., 1971, Rate of advance of the Woodfordian (late Wisconsinan) glacial margin in Illinois: Stratigraphic and radiocarbon evidence: Geol. Soc. America Bull., v. 82, p. 3245-3250.

Kent, D., Opdyke, N. D., and Ewing, M., 1971, Climatic change in the North Pacific using ice-rafted detritus as a climatic indicator: Geol. Soc. America Bull., v. 82, p. 2741-2754.

Ku, T. L., Bischoff, J. L., and Boersma, A., 1972, Age studies of Mid-Atlantic Ridge sediments near 42°N and 20°N: Deep-Sea Research, v. 19, p. 233-247.

Kukla, J., 1970, Correlation between loesses and deep-sea sediments: Geol. Fören. Stockholm Förh., v. 92, p. 148-180.

Lamb, H. H., and Woodruffe, A., 1970, Atmospheric circulation during the last ice age: Quaternary Research, v. 1, p. 29-58.

McIntyre, A., 1967, Coccoliths as paleoclimatic indicators of Pleistocene glaciation: Science, v. 158, p. 1314-1317.

McIntyre, A., and Bé, A.W.H., 1967, Modern Coccolithophoridae of the Atlantic Ocean—1. Placoliths and cyrtoliths: Deep-Sea Research, v. 14, p. 561-597.

McIntyre, A., and McIntyre, R., 1971, Coccolith concentrations and differential solution in oceanic sediments, *in* Funnel, B. M., and Reidel, W. R., eds., The micropaleontology of oceans: Cambridge, England, Cambridge Univ. Press, p. 253-261.

McIntyre, A., and Ruddiman, W. F., 1972, Northeast Atlantic post-Eemian paleooceanography: A predictive analog of the future: Quaternary Research, v. 2, p. 350-354.

McIntyre, A., Bé, A.W.H., and Preikstas, R., 1967, Coccoliths and the Pliocene-Pleistocene boundary, Progress in oceanography, Vol. 4, The Quaternary history of the ocean basins: London and New York, Pergamon Press, p. 3-25.

McIntyre, A., Bé, A.W.H., and Roche, M. B., 1970, Modern Pacific Coccolithophorida—A paleontologic thermometer: New York Acad. Sci. Trans., v. 32, p. 720-731.

McIntyre, A., Ruddiman, W. F., and Jantzen, R., 1972a, Southward penetrations of the North Atlantic polar front: Faunal and floral evidence of large-scale surface water mass movements over the last 225,000 years: Deep-Sea Research, v. 19, p. 61-77.

McIntyre, A., Bé, A.W.H., Biscaye, P., Burckle, L., Gardner, J., Geitzenauer, K., Goll, R., Kellog, T., Prell, W., Roche, M., Imbrie, J., Kipp, N., Ruddiman, W., Moore, T., and Heath, R., 1972b, The glacial North Atlantic 17,000 years ago: Paleoisotherm and oceanographic maps derived from floral-faunal parameters by CLIMAP: Geol. Soc. America Abs. with Programs, v. 4, no. 7, p. 590-591.

Mesolella, K. J., Matthews, R. K., Broecker, W. S., and Thurber, D. L., 1969, The astronomical theory of climatic change—Barbados data: Jour. Geology, v. 77, p. 250-274.

Moore, T. C., and Heath, G. R., 1967, Abyssal hills in the central equatorial Pacific: Detailed structure of the sea floor and subbottom reflectors: Marine Geology, v. 5, p. 161-179.

Oba, T., 1969, Biostratigraphy and isotopic paleotemperature of some deep-sea cores from the Indian Ocean: Tohoku Univ. Sci. Repts., v. 41, p. 129-195.

Phleger, F. B., Parker, F. L., and Peirson, J. F., 1953, North Atlantic Foraminifera: Swedish Deep-Sea Expedition Repts., v. 7, no. 1, 122 p.

Pitman, W. C., and Talwani, M., 1972, Sea-floor spreading in the North Atlantic: Geol. Soc. America Bull., v. 83, p. 619-646.

Roberts, D. G., Bishop, D. G., Laughton, A. S., Ziolkowski, A. M., Scrutton, R. A., and Matthews, D. A., 1970, New sedimentary basin on Rockall plateau: Nature, v. 225, p. 170-172.

Roche, M. B., McIntyre, A., and Imbrie, J., 1975, Quantitative paleo-oceanography of the Late Pleistocene-Holocene North Atlantic: Coccolith evidence, *in* Saito, T., and Burckle, L., eds., Late Neogene Epoch boundaries: Micropaleontology Spec. Pub. 1, p. 199-224.

Rona, E., and Emiliani, C., 1969, Absolute dating of Caribbean cores P6304-8 and P6304-9: Science, v. 163, p. 66-68.

Ruddiman, W. F., 1969a, Planktonic foraminifera of the subtropical North Atlantic gyre [Ph.D thesis]: New York, Columbia Univ., 291 p.

——1969b, Recent planktonic foraminifera: Dominance and diversity in North Atlantic surface sediments: Science, v. 164, p. 1164-1167.

——1971, Pleistocene sedimentation in the equatorial Atlantic: Stratigraphy and faunal paleoclimatology: Geol. Soc. America Bull., v. 82, p. 283-302.

——1972, Sediment redistribution of the eastern Reykjanes Ridge flanks: Seismic evidence: Geol. Soc. America Bull., v. 83, p. 2039-2062.

Ruddiman, W. F., and Glover, L. K., 1972a, Ice-rafted volcanic ash: A tracer of North Atlantic paleocirculation [abs.]: Am. Geophys. Union Trans., v. 53, p. 423.

——1972b, Vertical mixing of ice-rafted volcanic ash in North Atlantic sediments: Geol. Soc. America Bull., v. 83, p. 2817-2836.

Ruddiman, W. F., and Heezen, B. C., 1967, Differential solution of planktonic foraminifera: Deep-Sea Research, v. 14, p. 801-808.

Ruddiman, W. F., and McIntyre, A., 1973, Time-transgressive deglacial retreat of polar water from the North Atlantic: Quaternary Research, v. 3, p. 117–130.

Sancetta, C., Imbrie, J., Kipp, N., McIntyre, A., and Ruddiman, W. F., 1972, Climatic record in North Atlantic deep-sea core V23-82: Comparison of the last and present interglacials based on quantitative time series: Quaternary Research, v. 2, p. 363–367.

Schneidermann, N., 1971, Selective dissolution of recent coccoliths in the Atlantic Ocean: Geol. Soc. America Abs. with Programs, v. 3, no. 7, p. 695.

Shackleton, N. J., and Opdyke, N. D., 1973, Oxygen isotope and paleomagnetic stratigraphy of equatorial Pacific core V28-238: Oxygen isotope temperatures and ice volumes on a 10^5 year and 10^6 year scale: Quaternary Research, v. 3, p. 39–55.

Siesser, W. G., and Rogers, J., 1971, An investigation of the suitability of four methods used in routine carbonate analysis of marine sediments: Deep-Sea Research, v. 18, p. 135–139.

van der Hammen, T., Wijmstra, T. A., and Zagwijn, W. H., 1971, The floral record of the late Cenozoic of Europe, *in* Turekian, K. K., ed., The late Cenozoic glacial ages: New Haven, Conn., Yale Univ. Press, p. 391–424.

Warren, B., 1967, Oceanic circulation, *in* Fairbridge, R. W., ed., The encyclopedia of oceanography: New York, Reinhold Pub. Corp., p. 590–597.

Weyl, P. K., 1968, The role of the oceans in climatic change: A theory of the ice ages: Meteorol. Mons., v. 8, p. 37–62.

Wijmstra, T. A., 1969, Palynology of the first 30 meters of a 120 m deep section in northern Greece: Acta Botanica Neederland, v. 18, p. 511–527.

Willman, H. B., and Frye, J. L., 1970, Pleistocene stratigraphy of Illinois: Illinois Geol. Survey Bull., v. 94, p. 204.

Windom, H. L., 1969, Atmospheric dust records in permanent snowfields: Implications to marine sedimentation: Geol. Soc. America Bull., v. 80, p. 761–782.

MANUSCRIPT RECEIVED BY THE SOCIETY FEBRUARY 4, 1974
REVISED MANUSCRIPT RECEIVED OCTOBER 3, 1974
MANUSCRIPT ACCEPTED OCTOBER 25, 1974
LAMONT-DOHERTY GEOLOGICAL OBSERVATORY CONTRIBUTION NO. 2269

Printed in U.S.A.

Geological Society of America
Memoir 145
© 1976

O^{18} Record of the Atlantic Ocean for the Entire Pleistocene Epoch

JAN VAN DONK*
*Lamont-Doherty Geological Observatory and
Department of Geological Sciences
Columbia University,
Palisades, New York 10964*

ABSTRACT

Twenty-one isotopically determined "interglacial" and an equal number of isotopically determined "glacial" or near-"glacial" stages are recognized in the isotope record of *Globigerinoides sacculifer*, a planktonic foraminiferal species, from a well-dated equatorial Atlantic core representing the past 2.3 m.y. Many of the glacial stages (especially before 1 m.y. B.P.) are less pronounced than the most recent glacial maximum.

The observed maximum change in oxygen-isotope values is 1.1‰. At least 90% of the changes in the isotopic composition is attributable to variation in the isotopic composition of ocean water, which is due to the waxing and waning of large continental glaciers. Thus, the isotope record could more appropriately be called an ice-volume record rather than a temperature record, although the two records are closely related. A long period of relatively decreased ice volume occurred from 1.2 to 1.0 m.y. B.P. Low sea-level stands (that is, periods of maximum glaciation) comparable to the Wisconsin maximum occurred at 145,000, 240,000, 530,000, and 750,000 B.P.

Correlation between this record and isotope records from the Caribbean Sea and equatorial Pacific Ocean is good. Because the oxygen-isotope record primarily represents an ice-volume record, it is an extremely useful correlative tool. Ice-volume changes should be recorded in the oceans almost synchronously, because complete oceanic mixing occurs in less than 1,500 yr.

INTRODUCTION

Since the introduction of the oxygen-isotope method (Urey, 1947) as a tool in paleoclimatic studies, it has been widely used in reconstructions of late Pleistocene

---

*Present address: University of Botswana, Lesotho and Swaziland, Private bag 22, Gaborone, Botswana, Africa.

climate based on deep-sea cores (Emiliani, 1955, 1966; Broecker and van Donk, 1970; Shackleton and Opdyke, 1973; Imbrie and others, 1973). However, no study has presented a complete record of oxygen-isotope fluctuations during Pleistocene time of calcareous shells from deep-sea sediment. Most of the cores studied have been from the Caribbean or equatorial Atlantic area and only cover the last 500,000 yr. The one Pacific core to which the oxygen-isotope method has been applied is thought to represent about 850,000 yr (Shackleton and Opdyke, 1973).

In this paper, I present the first oxygen-isotope record for the entire Pleistocene Period. The record is of the equatorial Atlantic Ocean and covers the past 2.3 m.y. Correlation with the Pacific Ocean record is good, although discrepancies occur in placing some events in the time reference frame. It will be shown that the time scale of events used by Shackleton and Opdyke (1973) is probably less accurate than the scale adopted here. The reason for this discrepancy is the assumption by Shackleton and Opdyke (1973) of a constant sedimentation rate to the first magnetic boundary. This assumption does not appear to be valid.

Temperature fluctuations have been determined by Briskin and Berggren (1975) for this core on the basis of the quantitative faunal method first proposed by Imbrie and Kipp (1971). Because this technique is independent of the isotope method, an evaluation could be made of the ice-volume contribution to the oxygen-isotope fluctuations.

This particular core was chosen because it represents a continuous and complete record of the Pleistocene Epoch, and it has a good paleomagnetic chronology. Detailed faunal studies that have been made of this core show that dissolution effects are minimal (Briskin and Berggren, 1975).

ANALYTICAL TECHNIQUE

Core V16-205, with a total length of 1,232 cm, was raised in the tropical Atlantic Ocean (lat 15°24′N, long 43°24′W) from a depth of 4,045 m. A single species of planktonic foraminifera, *Globigerinoides sacculifer*, was extracted from samples at 10-cm intervals and in several sections at 3-cm intervals. A total of 136 samples were analyzed. The samples were lightly crushed and cleaned ultrasonically. They were then acidified with 100% H_3PO_4 at 25°C and converted into CO_2 gas according to the procedure established by McCrea (1950), but without having been roasted. The gas was analyzed in a commercial Nuclide dual-collector mass spectrometer. The overall analytical precision of the isotopic values is 0.12‰ (1 σ) for a single analysis.

RESULTS

The results are given in Table 1 and Figure 1. The observed maximum change in the oxygen-isotope values is 1.1‰. In this paper I have adopted the numbering system for climatic stages that was first introduced by Emiliani (1955) and extended by Shackleton and Opdyke (1973). Below stage 21, the numbering has been extended by assigning odd numbers to significant isotopically determined interglacial peaks or zones. The numbering for core V12-122 in this paper is different below stage 7 from that of Ryan (1972), because he suspected a hiatus between 480 to 500 cm based on correlation of climatic peaks and magnetic events in other Caribbean and Mediterranean cores. Imbrie and others (1973) did not find evidence for such

TABLE 1. OXYGEN ISOTOPE RESULTS ON EQUATORIAL ATLANTIC CORE V16-205

Depth in core (cm)	δO^{18} vs. PDB	Depth in core (cm)	δO^{18} vs. PDB	Depth in core (cm)	δO^{18} vs. PDB
0	-0.78 ± 0.09 (6)*	340	-0.18	750	-0.85
10	-0.39 ± 0.07 (3)	350	-0.06	760	-0.84
20	$+0.30 \pm 0.03$ (2)	360	-0.32 ± 0.08 (3)	770	-0.79
30	-0.09 ± 0.10 (4)	370	-0.41	780	-0.38
40	$+0.04$	380	-0.55	790	-0.38
50	-0.65	390	-0.33	800	-0.48
60	-0.80	395	-0.34 ± 0.11 (4)	810	-0.85
70	-0.73	400	$+0.02$	820	-0.36
75	-0.47 ± 0.06 (2)	405	-0.36 ± 0.09 (2)	830	-0.20
80	$+0.01 \pm 0.15$ (5)	410	-0.08	840	-0.90
90	$+0.05$	420	-0.56	850	-0.81
100	-0.47	430	-0.23	860	-0.68
110	-0.67	440	-0.19	870	-0.88
120	-0.44	450	-0.62	880	-0.59
130	-0.16	460	-0.59	885	-0.66 ± 0.03 (3)
140	-0.01 ± 0.14 (5)	470	-0.54	890	-0.55
150	-0.13	480	-0.28	900	-0.42
160	-0.79	485	-0.57 ± 0.09 (3)	910	-0.53
170	-1.06 ± 0.05 (2)	490	-0.59 ± 0.10 (3)	918	-0.60 ± 0.16 (2)
180	-0.64	495	-0.72 ± 0.02 (2)	920	-0.66
190	-0.44	500	-0.58	930	-0.56
200	-0.78 ± 0.26 (2)	510	-0.59	932	-0.76
210	-0.90	520	-0.58	940	-0.91
215	-0.82 ± 0.09 (2)	530	-0.60	950	-0.82
220	-0.63 ± 0.09 (2)	540	-0.66	958	-0.51 ± 0.10 (2)
225	-0.49 ± 0.19 (2)	550	-0.53 ± 0.11 (2)	960	-0.53
230	-0.40 ± 0.11 (2)	560	-0.65	970	-0.42
235	-0.83	570	-0.62	980	-0.43
240	-0.74	580	-0.69 ± 0.09 (4)	990	-0.56
245	-0.43	590	-0.65	1,000	-0.43
250	-0.26	600	-0.68	1,005	-0.55 ± 0.05 (2)
260	-0.31	610	-0.45	1,010	-0.17
270	-0.19 ± 0.03 (2)	620	-0.50	1,018	-0.08
275	-0.75 ± 0.10 (3)	630	-0.95	1,020	-0.47
283	-0.81 ± 0.09 (2)	640	-0.80	1,030	-0.49
290	-0.81	650	-0.90	1,032	-0.39
293	-0.09	660	-0.68	1,040	-0.65
297	$+0.10$	670	-0.47	1,050	-0.67
300	-0.30	680	-0.86	1,063	-0.72
303	-0.30	690	-0.45	1,070	-0.63
307	-0.19 ± 0.17 (3)	700	-0.71	1,077	-0.62 ± 0.04 (2)
310	$+0.17 \pm 0.06$ (5)	710	-1.12	1,095	-0.53
313	-0.20 ± 0.12 (3)	720	-0.46	1,133	-0.45
317	-0.01	730	-0.58	1,187	-0.70 ± 0.06 (2)
320	-0.18	740	-0.86	1,212	-0.53 ± 0.04 (2)
330	-0.41				

*Numbers in parentheses indicate number of replicate analyses.

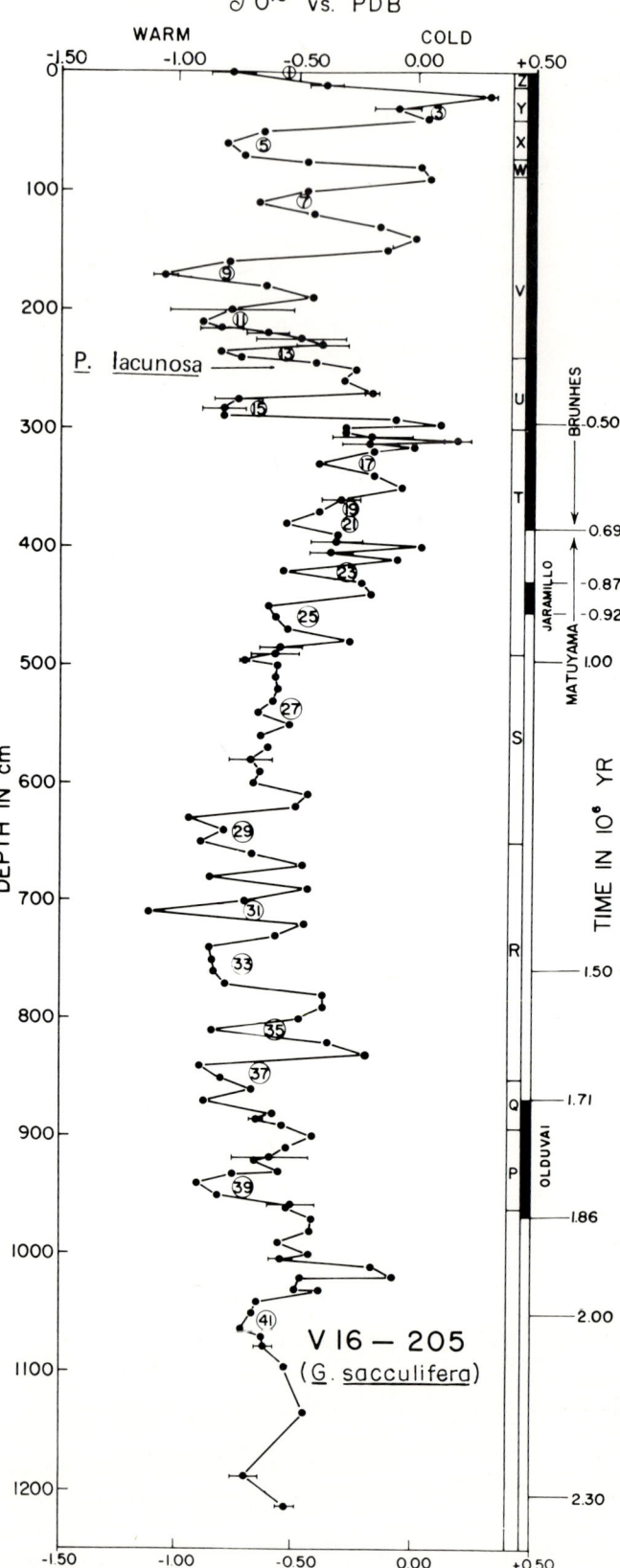

Figure 1. Oxygen-isotope record of *G. sacculifer* in core V16-205 as a function of depth and time. Small circled numbers represent stage numbers; first vertical column at right represents faunal zones as defined by Ericson and Wollin (1968). Black areas in second column represent time periods of normal magnetic field polarity; white areas represent periods of reversed magnetic field polarity.

a hiatus at this level, but they did suspect a hiatus of 1 m at about a depth of 800 cm.

Twenty-one isotopically determined interglacial stages are defined for the past 2.1 m.y., with an equal number of isotopically determined glacial or near-glacial stages, although many of the latter are much less pronounced than the most recent glacial maximum. The most striking feature of the isotope record is the long, near-interglacial stage that is numbered 27 (Fig. 1). This stage lasted from about 1.2 to 1.0 m.y. B.P. and has also been recognized in the isotope record of a South Pacific core (van Donk, 1970). The average yearly temperature, as computed from summer and winter temperatures based on the quantitative faunal method (Briskin and Berggren, 1975), remained nearly constant at an intermediate temperature level over this period (Figs. 5, 6). Below the Brunhes-Matuyama boundary, isotope fluctuations are smaller than those within the Brunhes normal epoch; overall, the isotope values tend to cluster around lighter weights. The sedimentation rate in core V16-205 is rather low (0.55 ± 0.05 cm/1,000 yr); therefore, some extreme conditions may have been missed in the record owing to mixing or insufficient close sampling. Closer sampling might reveal such extremes unless reworking of bottom sediment by burrowing organisms has taken place. In slow-sedimentation-rate cores, one can expect the record of any measured parameter to be somewhat subdued as compared to the actual fluctuations in the particular parameter. The significant conclusion that may be drawn from this oceanic record is that glacial-interglacial cycles have occurred throughout Pleistocene time, although many are of lesser magnitude than the most recent ones.

ABSOLUTE CHRONOLOGY

Good magnetic stratigraphic control is available for core V16-205 (Glass and others, 1967). The Brunhes-Matuyama boundary and the Jaramillo and Olduvai normal events within the Matuyama reversed epoch are clearly recorded (Table 2). The best dates available for these magnetic events are 0.69 m.y. for the Brunhes-Matuyama boundary, 0.87 to 0.92 m.y. for the Jaramillo event, and 1.71 to 1.86 m.y. for the Olduvai event (Opdyke, 1972). There is some controversy as to the true age of the Olduvai event (called "Gilsa event" by Cox, 1969).

TABLE 2. DEPTHS OF FAUNAL AND PALEOMAGNETIC BOUNDARIES IN CORE V16-205

Faunal boundaries*	Depth† (cm)	Paleomagnetic boundaries	Depth§ (cm)
Y–Z	12	Brunhes-Matuyama	385
X–Y	39		
W–X	72	Jaramillo	430–455
V–W	87		
U–V	240	Olduvai	870–970
T–U	300		
S–T	490		
R–S	650		
Q–R	850		
P–Q	890		

*Defined by the presence or absence of *Globorotalia menardii* complex (Ericson and Wollin, 1968).
†Briskin and Berggren (1975).
§Glass and others (1967).

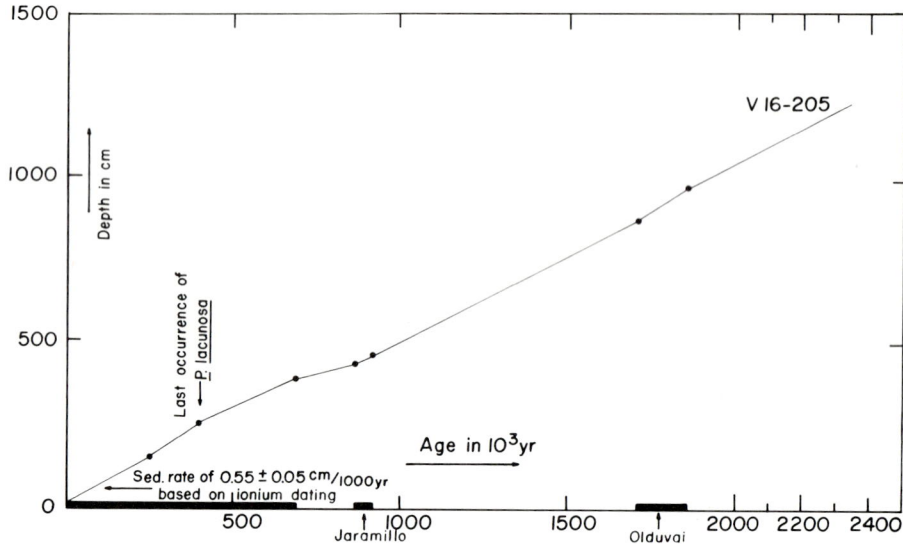

Figure 2. Depth-age relationship in core V16-205, based on paleomagnetic boundaries, ionium dating, and a faunal boundary.

In this paper I have adopted Opdyke's (1972) time scale, but for alternate opinions regarding this event the reader is referred to Cox (1969) and Watkins (1972). Within the Brunhes epoch (upper 385 cm) the sedimentation rate is nearly uniform, as determined from three independent lines of evidence (Fig. 2). Using the ionium dating technique, T.-L. Ku (1973, oral commun.) found the sedimentation rate for the top 150 cm to be uniform at 0.55 ± 0.05 cm/1,000 yr. The sedimentation rate for the upper 385 cm, based on the first magnetic reversal boundary, is 0.56 cm/1,000 yr. A faunal boundary—the extinction of the coccolithophoridae species *Pseudoemiliania lacunosa*—occurs between 240 and 245 cm in this core (M. Roche and A. McIntyre, 1973, oral commun.). The age of this extinction has been determined in the North Atlantic Ocean at 400,000 B.P. by A. McIntyre (1973, oral commun.) and in the equatorial Pacific Ocean at about 380,000 yr by Geitzenauer and others (this volume); this yields an average sedimentation rate of 0.63 ± 0.02 cm/1,000 yr. The ages for this extinction level were based on interpolated magnetic ages from 7 cores in the North Atlantic Ocean and 5 cores in the equatorial Pacific Ocean. The extinction level of *P. lacunosa* occurs within stage 13 of core V16-205 and has also been reported within stage 13 of the isotope record of equatorial Pacific core V28-238 (Shackleton and Opdyke, 1973) and Caribbean core P6304-9 (Gartner, 1972). It is not present in core P6304-8 as erroneously reported by Shackleton and Opdyke (1973), because this core does not reach stage 13.

Gartner (1972) quoted an age of 275,000 yr for the extinction level of *P. lacunosa* based on the time scale adopted by Emiliani and other workers (Emiliani, 1955, 1958; Rosholt and others, 1961; Rona and Emiliani, 1969). As the independent dating techniques used for core V16-205 (as well as the time scale derived by Shackleton and Opdyke, 1973) show, such an age is at least 100,000 yr too young. Figure 2 shows that the sedimentation rate is nearly uniform throughout the core at 0.55 ± 0.05 cm/1,000 yr, with the exception of the section representing the period between the Brunhes-Matuyama boundary and the end of the Jaramillo event, when the sedimentation rate was 0.25 cm/1,000 yr. Additionally, the sedimentation rate within the Olduvai was slightly higher at 0.66 cm/1,000 yr.

Faunal evidence and magnetic stratigraphy indicate that core V16-205 represents a continuous section of Pleistocene age (Ericson and Wollin, 1968; Briskin and Berggren, 1975). Briskin and Berggren (1975) found that solution-sensitive foraminifera are abundant and that the percentage of benthic foraminifera is less than 1%, which indicates that this core has undergone little or no alteration due to solution effects.

CORRELATION OF THE V16-205 ISOTOPE RECORD WITH OTHER ISOTOPE RECORDS

In Figures 3 and 4, the V16-205 oxygen-isotope record is compared with late Pleistocene records from other reports (V28-238, Shackleton and Opdyke, 1973; V12-122, Broecker and van Donk, 1970; Imbrie and others, 1973; P6304-9, Emiliani, 1966). In Figure 4 the time scale of core V28-238 has been adjusted so that the extinction level of $P.$ $lacunosa$ occurs at 400,000 B.P.; it also takes into account the ages of two other dated faunal levels at 160 cm (75,000 ± 5,000 yr) and 320 cm (200,000 ± 20,000 yr) (Geitzenauer and others, this volume).

It is reassuring that in cores V16-205 (equatorial Atlantic), V28-238 (equatorial Pacific), and P6304-9 (Caribbean), the extinction of $P.$ $lacunosa$ occurred at or near the peak of the isotopically determined interglacial stage 13. This supports the conclusion that isotope records are very useful correlative tools, as pointed out by Shackleton and Opdyke (1973).

The advantage of using O^{18} records rather than paleontologic criteria as a correlative tool is that O^{18} fluctuations in fossil foraminiferal shells primarily represent changes in the O^{18} composition of ocean water (as discussed below). Because oceanic mixing is relatively fast, with residence time of deep-sea water less than 1,500 yr (Broecker and others, 1960; Bien and others, 1963), O^{18} changes in ocean water occur nearly synchronously throughout the oceans. On the other hand, paleontologic criteria for correlation often are not synchronous because of the transgressive or regressive nature of appearance and extinction of the various organisms. This is especially true when comparing faunal boundaries in cores from very different latitudes or different ocean basins. However, the extinction of $P.$ $lacunosa$ seems to have occurred synchronously in the Caribbean, equatorial North Atlantic, and equatorial Pacific areas, on the basis of independent dating as cited above.

In core V12-122 (Caribbean) the extinction of $P.$ $lacunosa$ occurs just below interglacial stage 11, but at a level where the isotope value is rather heavy. This could possibly be due to a 1-m section missing in the sedimentary record below isotopically determined warm stage 11, as pointed out earlier by Imbrie and others (1973). This section would have contained isotopically determined interglacial stage 13 with the extinction level of $P.$ $lacunosa$ within it. It is clear that equivalent isotope minimums and maximums can be recognized in all cores from the three oceanic regions. The validity of this correlation is evident because the same number of interglacial stages are recognized from the top of the core to the extinction level of $P.$ $lacunosa$ in cores from the three different oceanic regions. Similarly, a comparison with the level of the Brunhes-Matuyama magnetic reversal boundary in cores V28-238 and V16-205 reveals an equal number of isotopically determined interglacial and glacial stages. However, some differences do exist. Interglacial stage 7 appears to have been interrupted briefly by a near-glacial event. This is evident in both Caribbean and equatorial Pacific cores but not in the equatorial Atlantic core. At the level of the Brunhes-Matuyama boundary, core V28-238 reveals

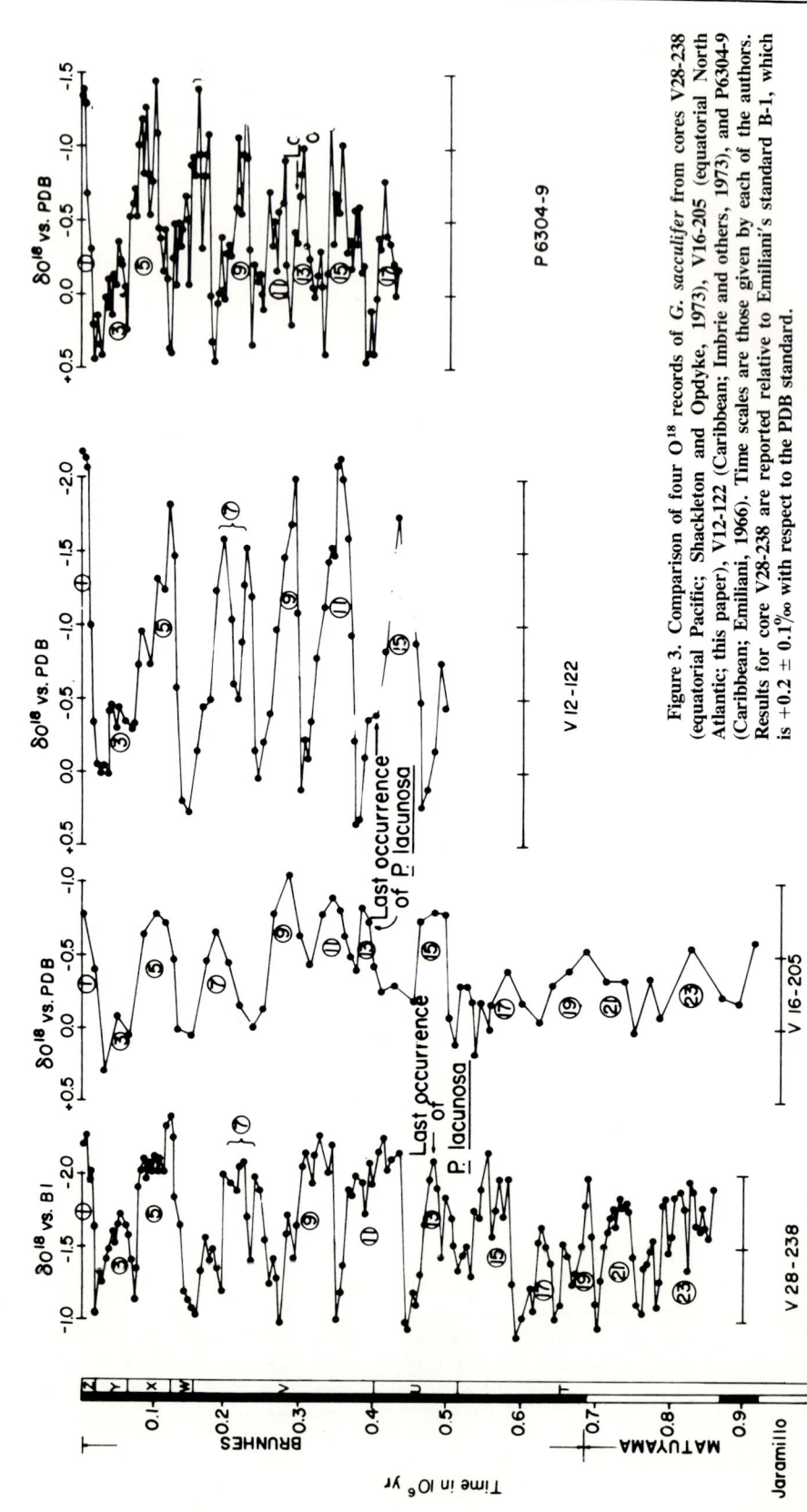

Figure 3. Comparison of four O¹⁸ records of *G. sacculifer* from cores V28-238 (equatorial Pacific; Shackleton and Opdyke, 1973), V16-205 (equatorial North Atlantic; this paper), V12-122 (Caribbean; Imbrie and others, 1973), and P6304-9 (Caribbean; Emiliani, 1966). Time scales are those given by each of the authors. Results for core V28-238 are reported relative to Emiliani's standard B-1, which is +0.2 ± 0.1‰ with respect to the PDB standard.

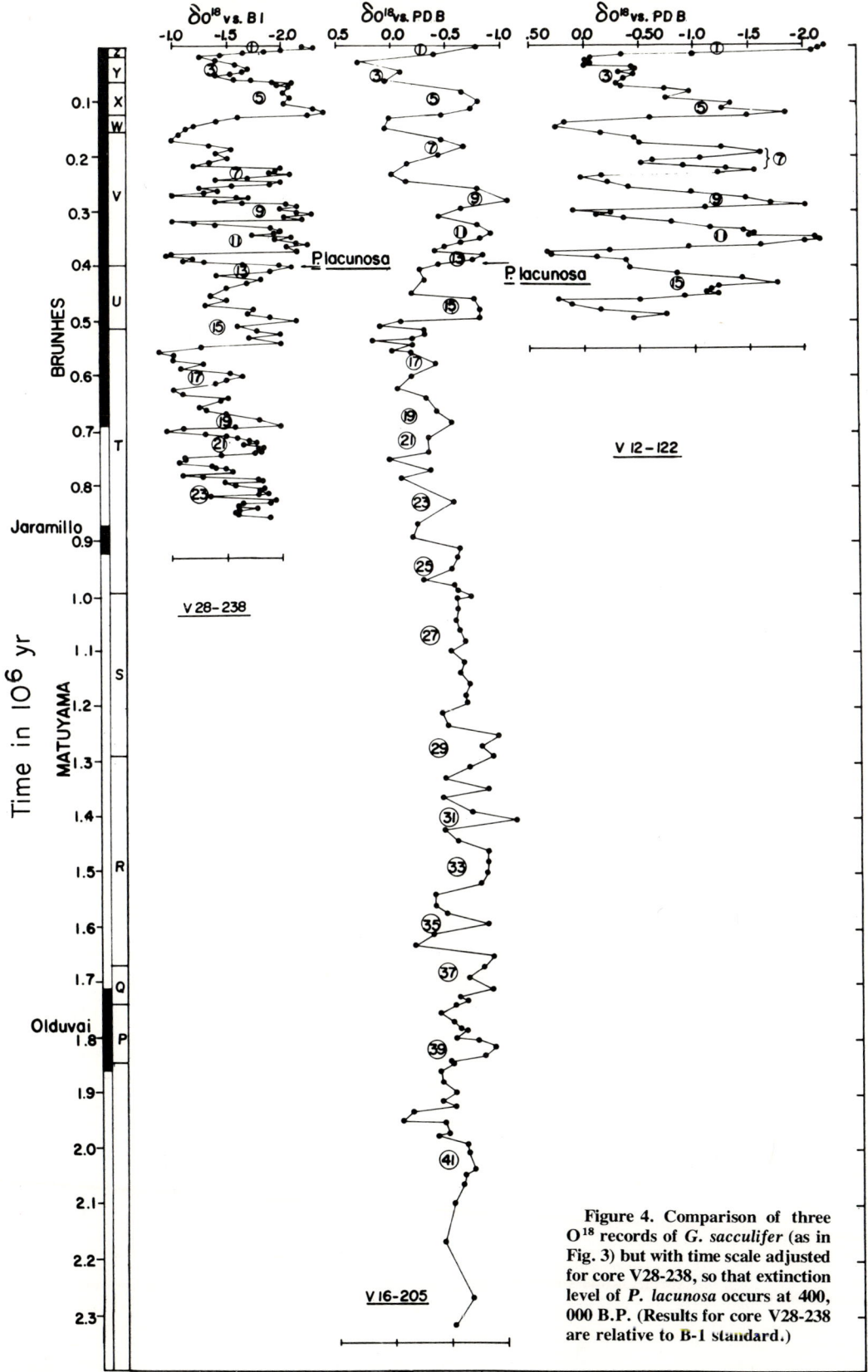

Figure 4. Comparison of three O^{18} records of *G. sacculifer* (as in Fig. 3) but with time scale adjusted for core V28-238, so that extinction level of *P. lacunosa* occurs at 400,000 B.P. (Results for core V28-238 are relative to B-1 standard.)

two distinct interglacial stages (19 and 21; Fig. 4). This appears as one long interglacial stage in core V16-205. Some of the glacial stages do not appear to be as extreme in the V16-205 record as in the Caribbean and Pacific records (especially between interglacial stages 9, 11, and 13; Fig. 4). The slow rate of sedimentation in core V16-205 accounts for this difference. Most samples are 10 cm apart and have a sedimentation rate of 0.5 to 0.6 cm/1,000 yr; this represents a sampling interval of about 20,000 yr. Thus, it is very likely that some of the more extreme parts in the record may have been missed.

Figure 3 clearly shows that there is excellent agreement between cores V16-205 and V12-122 in the timing of the isotopically determined interglacial and glacial

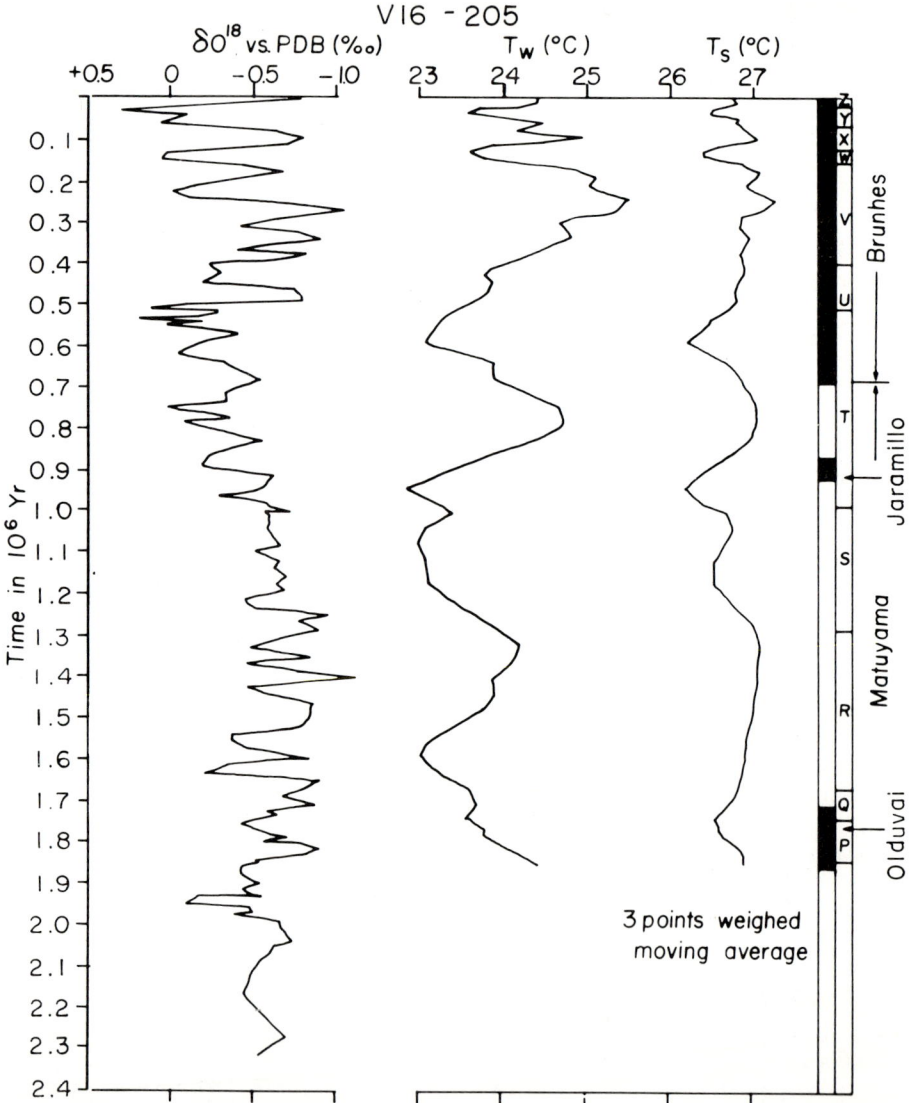

Figure 5. Comparison of O^{18} record of core V16-205 with surface-water winter temperature (T_w) and surface-water summer temperature (T_s) records, as deduced from quantitative faunal method (Briskin and Berggren, 1975).

stages. The chronology of core V12-122 has been well documented by several independent lines of evidence, as discussed extensively by Broecker and van Donk (1970), whereas the chronology of core V16-205 (discussed above) is based on ionium dating of the top 150 cm, five paleomagnetic boundaries, and one well-dated faunal extinction boundary. The timing of events in core V28-238, given by Shackleton and Opdyke (1973), does not always agree with core V16-205. Their interglacial stages 1, 3, 5, 19, 21, and 23 occur at nearly the same times as in core V16-205, but the other stages seem to occur earlier in core V28-238 than in V16-205. For example, the estimated age in core V28-328 for the extinction level of *P. lacunosa* in stage 13 is 485,000 B.P. (Shackleton and Opdyke, 1973), which is 85,000 yr older than the age assigned to it by A. McIntyre (1973, oral commun.) in the Atlantic Ocean and by Geitzenauer and others (this volume) in the Pacific Ocean. However, the peak of stage 19 does occur at the Brunhes-Matuyama boundary in both cores V28-238 and V16-205. The evidence presented above suggests that the assumption of a constant sedimentation rate between the top of the core sample and the first magnetic boundary in core V28-238 (Shackleton and Opdyke, 1973) is not valid. Whereas the ages given in Shackleton and Opdyke appear too old, the ages for the various isotope peaks in core P6304-9 given in Emiliani (1966) appear much too young (Broecker and Ku, 1969; Broecker and van Donk, 1970; Shackleton and Opdyke, 1973).

O^{18} FLUCTUATIONS OF PLEISTOCENE OCEAN WATER

The interpretation of the O^{18} record of foraminiferal shells in terms of a paleotemperature record on one hand or a paleo–ice-volume record on the other has been a much-debated issue. Emiliani (1955, 1966, 1971) adhered to the view that the fluctuation in the isotopic composition of ocean water during changes from glacial to interglacial extremes is only a minor contribution to the O^{18} fluctuation seen in the isotope record of foraminiferal shells (about 0.5‰). His interpretations were primarily in terms of temperature. On the other hand, Olausson (1965), Craig (1965), Shackleton (1967), Dansgaard and Tauber (1969), Broecker and van Donk (1970), Shackleton and Opdyke (1973), and Imbrie and others (1973) have maintained that the fluctuation in the isotopic composition of the oceans has been much larger than Emiliani's estimate of 0.5‰. Their estimates have varied from 1.2 to 1.7‰, and they have interpreted the isotope record of foraminiferal shells primarily in terms of ice-volume changes during transitions from glacial to interglacial conditions.

Imbrie and Kipp's (1971) quantitative paleoenvironmental method, based on statistical treatment of faunal data, has been widely applied in paleoclimatic investigations (Imbrie and others, 1973; Sachs, 1973). Briskin and Berggren (1975) applied this method to the abundance data of planktonic foraminifera in core V16-205 and generated three curves with the following variables as a function of time: (1) summer temperature, (2) winter temperature, and (3) salinity. Figure 5 reproduces the summer and winter temperature curves given by Briskin and Berggren (1975) as well as the O^{18} record. The time scale adopted in Figure 5 is slightly different from that of the above authors in that the Olduvai event in this paper is from 1.71 to 1.86 m.y. B.P., whereas Briskin and Berggren (1975) accepted 1.69 to 1.76 m.y. B.P. There is good agreement between the O^{18} lows and the summer and winter temperature highs for the past 350,000 yr. The cool summer and winter temperatures at 600,000 B.P. are reflected in a high O^{18}, but the warm summer and winter temperatures at 800,000 B.P. are not reflected in the O^{18} record. Summer and winter temperatures do not parallel each other well before 1 m.y. B.P.

Because it is not known in which season foraminifera build their shells, I have averaged summer and winter temperatures as the best estimate for temperature variations during Pleistocene time at the location of core V16-205. Assuming that the temperature record as generated by the quantitative faunal method reflects the true temperature variations at this site during Pleistocene time, we can use the temperature equation of Epstein and others (1953), as modified by Craig (1965),

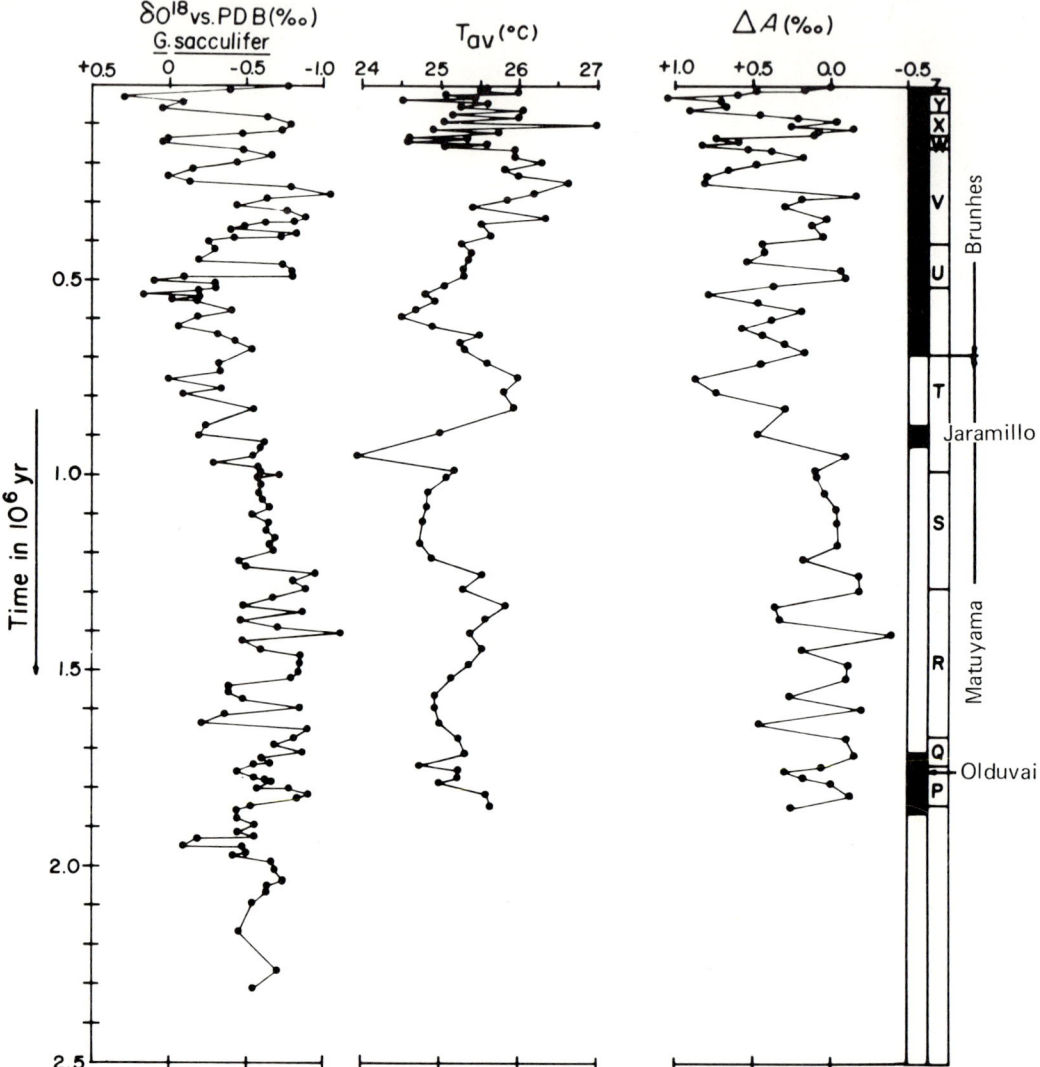

Figure 6. Comparison of O^{18} record of *G. sacculifer* with surface-water average temperature (T_{av}) and the O^{18} record of surface water given as ΔA at site V16-205. $T_{av} = (T_s + T_w)/2$, where T_s and T_w are based on the quantitative faunal method (Briskin and Berggren, 1975). ΔA is calculated for each depth level from the following relationship: $\Delta A = \delta O^{18}$ carb.$_D - \delta O^{18}$ carb.$_O + 0.2(T_{av_D} - T_{av_O})$, where D and O denote values of parameters in question at depth and at top of core, respectively. The number 0.2 is the value of temperature effect on isotope fractionation between carbonate and water (0.20‰/°C; Epstein and others, 1953). ΔA is change in O^{18}/O^{16} ratio of surface water in past relative to present O^{18}/O^{16} ratio of such water at site V16-205; it primarily represents change in volume of glacial ice.

to generate a record for the O^{18} variations of the ocean water at this site. The results, given as ΔA (‰), are shown in Figure 6. The maximum variation in the isotopic composition of the sea water is 1‰. Thus, 80 to 90% of the variation in the oxygen-isotope composition of the foraminiferal shells is due to a change in the isotopic composition of sea water. This, in turn, results from the waxing and waning of large continental glaciers. Therefore, the O^{18} curves based on isotope analyses of foraminiferal shells represent ice-volume curves. This conclusion has been reached by many other workers (Olausson, 1965; Shackleton, 1967; Dansgaard and Tauber, 1969; Imbrie and others, 1973; Shackleton and Opdyke, 1973). The ΔA curve in Figure 6 indicates that sea-level stands almost as high or higher than today occurred 12 times in the past 1.9 m.y. A very prominent one occurred at about 1.4 m.y. B.P., whereas more recently, prominent high stands occurred at 120,000, 280,000, and 490,000 B.P. Times of maximum glaciation comparable to the Wisconsin maximum occurred only 4 times, all within the past 750,000 yr. Prior to that time the amount of glacial ice appears to have been considerably less. These 4 glacial maximums occurred at 145,000, 240,000, 530,000, and 750,000 B.P. The maximum glaciation in Yellowstone Park (Bull Lake) has been found to be approximately 145,000 B.P. (K. Pierce, J. Obradovich, and I. Friedman, 1973, oral commun.). Sharp (1968) reported on one of the most prominent glaciations in the Sierra Nevada, as evidenced by thick till deposits, and placed it at approximately 750,000 B.P. (Sherwin glaciation).

Another interesting feature in Figure 6 is the warm peak in the average temperature curve, which does not always coincide with the low peak in ΔA. For example, the warm peak at 250,000 B.P. coincides with a high peak in the ΔA curve, and the cold peak at 950,000 B.P. coincides with a low peak in ΔA. The explanation for these seemingly contradictory conditions is not clear, but they could be related to an influx of warm, relatively saline water at the surface in the former case and an influx of cold, relatively less saline water in the latter. It should be noted that relatively small temperature fluctuations (less than 2°C) are involved. Briskin and Berggren (1975) did not determine a high-salinity ocean water at 250,000 B.P., but they did find evidence for low-salinity ocean water at 950,000 B.P. Imbrie and others (1973) were the first workers to generate a ΔA curve as described above for Caribbean core V12-122. However, they only plotted this parameter for two short sections in that core. In Figure 7, I have generated the complete ΔA curve for this Caribbean core on the basis of the data of Imbrie and others (1973). The most notable difference between the ΔA curves is the larger amplitude at the Caribbean site (about 1.8 to 2‰), which also shows a greater amplitude in the average temperature curve (about 4 to 5°C). The reason for the greater amplitude in ΔA may be a local effect of an increase in net evaporation relative to precipitation during glacial periods in the Caribbean Sea superimposed on the world-wide increase in δO^{18} of ocean water (at least 1.2‰; Imbrie and others, 1973) during glacial maximums. Bonatti and Gartner (1973) found evidence for increased aridity of the Caribbean region during glacial periods. The local additional increase in the δO^{18} value of the Caribbean water of about 0.6‰ over the world-wide increase in δO^{18} of the ocean water of 1.2‰ would mean that the Caribbean water would be about 1‰ more saline than expected solely from the decrease in eustatic sea level at times of maximum glaciation (assuming the δO^{18}-salinity relationship in North Atlantic surface waters as given by Craig and Gordon, 1965).

A comparison between the two ΔA curves (Caribbean Sea and equatorial North Atlantic Ocean) is made in Figure 8. There is a good correlation between the changes in the O^{18}/O^{16} ratios of the ocean water in these two areas, as well as in the timing of the changes, especially in the past 275,000 yr. As mentioned

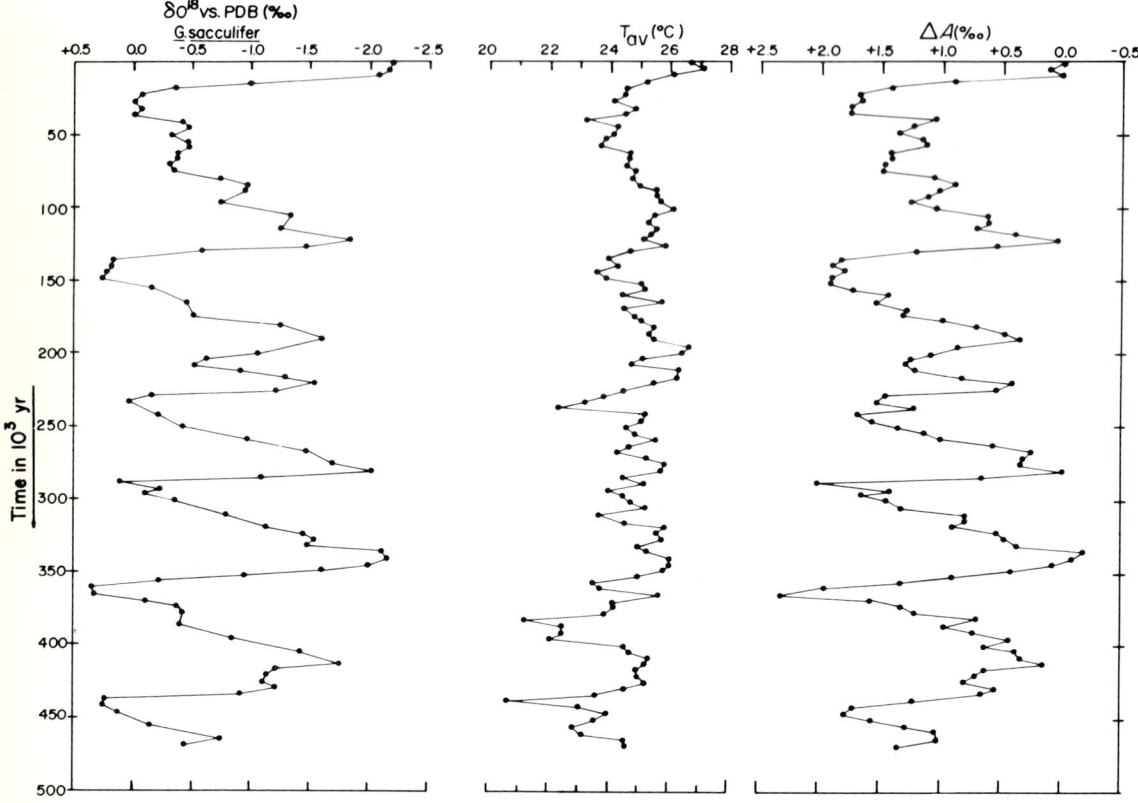

Figure 7. Comparison of O^{18} record of *G. sacculifer* with surface-water average temperature (T_{av}) and O^{18} record of surface water (ΔA) at site of Caribbean core V12-122. Meaning of symbols is same as in Figure 6. Data from Imbrie and others (1973).

above, the double peak 7 in core V12-122 is not seen in V16-205, and the low sea-level stand (high ΔA value) between peaks 9 and 11 in core V12-122 is not well defined in core V16-205. The missing peak 13 in core V12-122 is the cause for the apparent offset in timing of the low and high sea stands at stage 15. It needs to be emphasized once more that the changes seen in any parameter in core V16-205 will always be greatly subdued and less detailed because of the low sedimentation rate. On the basis of core V16-205 data presented here, it is clear that the change in the O^{18}/O^{16} ratio of ocean water is at least 1‰ between a period of maximum glaciation and a period of extreme interglacial conditions. Because δO^{18} variations in foraminiferal carbonate shells from deep-sea cores primarily reflect changes in the δO^{18} of ocean water, the timing of the changes in sea-level stand based on the O^{18} records correlates well from the Caribbean Sea to the equatorial Atlantic and the equatorial Pacific Oceans.

ACKNOWLEDGMENTS

I am indebted to T.-L. Ku for doing the ionium dating, made possible by National Science Foundation, Oceanographic Section, Grant AG-35407.

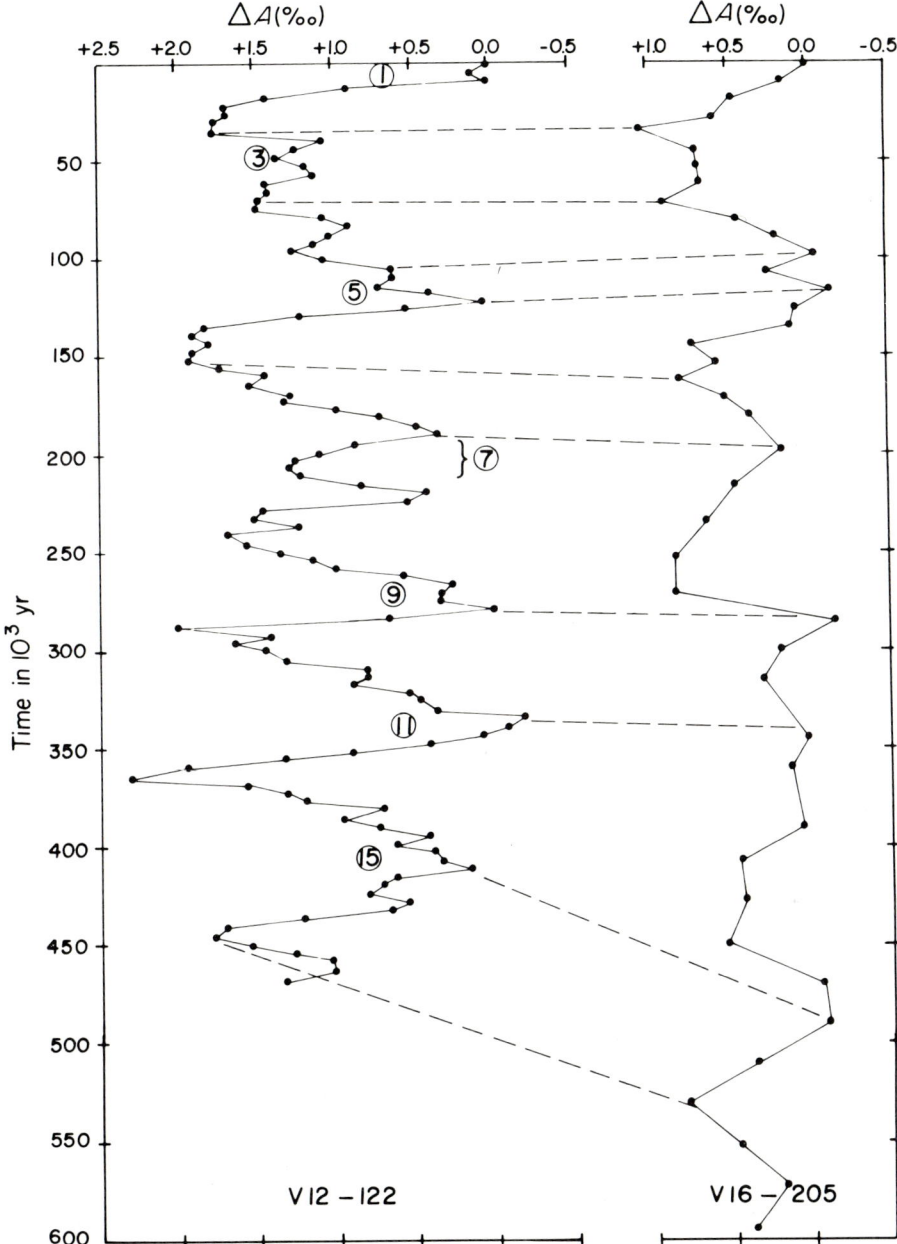

Figure 8. Correlation of ΔA records at sites V12-122 and V16-205.

The valuable assistance of M. Roche in establishing the *P. lacunosa* extinction level in this core and of A. McIntyre and K. Geitzenauer in establishing the absolute age for this extinction is gratefully acknowledged. I thank M. Briskin and W. A. Breggren for making their temperature results based on the quantitative faunal method available to me. Permission by K. Pierce, J. Obradovich, and I. Friedman to cite their as yet unpublished finding of a maximum glaciation in Yellowstone

Park at around 145,000 B.P. is appreciated. J. Imbrie has been most helpful in providing me with *G. sacculifer* samples. J. Lawrence critically read the manuscript.

Support for this research was provided by National Science Foundation Grants IDO71-04204, GA-28507, and GA-35381. The core collection and curating at Lamont-Doherty Geological Observatory were supported by Office of Naval Research Contract N00014-67-A-0108-0004 and National Science Foundation Grant DES-72-01568.

REFERENCES CITED

Bien, G. S., Rakestraw, N. W., and Suess, H. E., 1963, Radiocarbon dating of deep water of the Pacific and Indian Oceans, *in* Radioactive dating: Internat. Atomic Energy Agency, Athens 1962, Proc., p. 159–173.

Bonatti, E., and Gartner, S., Jr., 1973, Caribbean climate during Pleistocene ice ages: Nature, v. 244, p. 563–565.

Briskin, M., and Berggren, W. A., 1975, Pleistocene stratigraphy and quantitative paleoceanography of tropical North Atlantic core V16-205, *in* Saito, T., and Burckle, L. H., eds., Late Neogene Epoch boundaries: Am. Mus. Nat. History. Micropal. Press Spec. Pub., no. 1, p. 167–198.

Broecker, W. S., and Ku, T.-L., 1969, Caribbean cores P6304-8 and P6304-9: New analysis of absolute chronology: Science, v. 166, p. 404–406.

Broecker, W. S., and van Donk, J., 1970, Insolation changes, ice volumes, and the O^{18} record in deep-sea cores: Rev. Geophysics and Space Physics, v. 8, p. 169–198.

Broecker, W. S., Gerard, R., Ewing, M., and Heezen, B. C., 1960, Natural radiocarbon in the Atlantic Ocean: Jour. Geophys. Research, v. 65, p. 2903–2931.

Cox, A., 1969, Geomagnetic reversals: Science, v. 163, p. 237–245.

Craig, H., 1965, The measurement of oxygen isotope paleotemperatures, *in* Stable isotopes in oceanographic studies and paleotemperatures: Pisa, Consiglio Nazionale delle Ricerche Laboratorio di Geologia Nucleare, 23 p.

Craig, H., and Gordon, L. I., 1965, Deuterium and O^{18} variations in the ocean and the marine atmosphere, *in* Symposium on marine geochemistry: Narragansett Marine Lab. Occasional Pub., no. 3, p. 277–374.

Dansgaard, W., and Tauber, H., 1969, Glacier O^{18} content and Pleistocene ocean temperatures: Science, v. 166, p. 499.

Emiliani, C., 1955, Pleistocene temperatures: Jour. Geology, v. 63, p. 538–578.

——1958, Paleotemperature analysis of core 280 and Pleistocene correlations: Jour. Geology, v. 66, p. 264–275.

——1966, Paleotemperature analysis of Caribbean cores P6304-8 and P6304-9 and a generalized temperature curve for the past 425,000 years: Jour. Geology, v. 74, p. 109–124.

——1971, The amplitude of Pleistocene climatic cycles at low latitudes and the isotopic composition of glacial ice, *in* Turekian, K., ed., The late Cenozoic glacial ages: New Haven, Conn., Yale Univ. Press, 606 p.

Epstein, S., Buchsbaum, R., Lowenstam, H. A., and Urey, H. C., 1953, Revised carbonate-water isotopic temperature scale: Geol. Soc. America Bull., v. 64, p. 1315–1326.

Ericson, D. B., and Wollin, G., 1968, Pleistocene climate and chronology in deep-sea sediments: Science, v. 162, p. 1277–1284.

Gartner, S., 1972, Late Pleistocene calcareous nannofossils in the Caribbean and their interoceanic correlation: Palaeogeography, Palaeoclimatology, Palaeoecology, v. 12, p. 169–191.

Geitzenauer, K. R., Roche, M. B., and McIntyre, A., 1976, Modern Pacific coccolith assemblages: Derivation and application to late Pleistocene paleotemperature analysis, *in* Cline, R. M., and Hays, J. D., eds., Investigation of late Quaternary paleoceanography and paleoclimatology: Geol. Soc. America Mem. 145 (this volume).

Glass, B., Ericson, D. B., Heezen, B. C., Opdyke, N. D., and Glass, J. A., 1967, Geomagnetic reversals and Pleistocene chronology: Nature, v. 216, p. 437–441.

Imbrie, J., and Kipp, N. G., 1971, A new micropaleontological method for quantitative paleoclimatology: Application to a late Pleistocene Caribbean core, *in* Turekian, K., ed., The late Cenozoic glacial ages: New Haven, Conn., Yale Univ. Press, 606 p.

Imbrie, J., van Donk, J., and Kipp, N. G., 1973, Paleoclimatic investigation of a late Pleistocene Caribbean core: Comparison of isotopic and faunal methods: Quaternary Research, v. 3, p. 10-38.

McCrea, J. M., 1950, On the isotopic chemistry of carbonates and a paleotemperature scale: Jour. Chem. Physics, v. 18, p. 849-857.

Olausson, E., 1965, Evidence of climatic changes in North Atlantic deep-sea cores, with remarks on isotopic paleotemperature analysis: Prog. Oceanog., v. 3, p. 221-252.

Opdyke, N. D., 1972, Paleomagnetism of deep-sea cores: Rev. Geophysics and Space Physics, v. 10, p. 213-249.

Rona, E., and Emiliani, C., 1969, Absolute dating of Caribbean cores P6304-8 and P6304-9: Science, v. 163, p. 66-68.

Rosholt, J. N., Emiliani, C., Geiss, J., Koczy, F. F., and Wangersky, J., 1961, Absolute dating of deep-sea cores by Pa^{231}/Th^{230} method: Jour. Geology, v. 69, p. 162-185.

Ryan, W.B.F., 1972, Stratigraphy of late Quaternary sediments in the eastern Mediterranean, *in* Stanley, D. J., ed. The Mediterranean Sea: Stroudsburg, Pa., Dowden, Hutchinson & Ross, Inc., p. 149-169.

Sachs, H. M., 1973, Late Pleistocene history of the North Pacific: Evidence from a quantitative study of radiolaria in core V21-173: Quaternary Research, v. 3, p. 89-98.

Shackleton, N., 1967, Oxygen isotope analyses and Pleistocene temperatures reassessed: Nature, v. 215, p. 15-17.

Shackleton, N. J., and Opdyke, N. D., 1973, Oxygen isotope and paleomagnetic stratigraphy of equatorial Pacific core V28-238: Oxygen isotope temperatures and ice volumes on a 10^5 year and 10^6 year scale: Quaternary Research, v. 3, p. 39-55.

Sharp, R. P., 1968, Sherwin till-Bishop tuff geological relationships, Sierra Nevada, California: Geol. Soc. America Bull., v. 79, p. 351-364.

Urey, H. C., 1947, The thermodynamic properties of isotopic substances: Jour. Chem. Soc., p. 562.

van Donk, J., 1970, The oxygen isotope record in deep-sea sediments, [Ph.D. thesis]: New York, Columbia Univ., 228 p.

Watkins, N. D., 1972, Review of the development of the geomagnetic polarity time scale and discussion of prospects for its finer definition: Geol. Soc. America Bull., v. 83, p. 551-574.

MANUSCRIPT RECEIVED BY THE SOCIETY MARCH 11, 1974
REVISED MANUSCRIPT RECEIVED SEPTEMBER 23, 1974
MANUSCRIPT ACCEPTED OCTOBER 7, 1974
LAMONT-DOHERTY GEOLOGICAL OBSERVATORY CONTRIBUTION NO. 2270

Printed in U.S.A.

Geological Society of America
Memoir 145
© 1976

Late Quaternary Climatic Record in Western Equatorial Atlantic Sediment

ALLAN W. H. BÉ
JOHN E. DAMUTH
LEROY LOTT
AND
ROSEMARY FREE
Lamont-Doherty Geological Observatory
Columbia University
Palisades, New York 10964

ABSTRACT

Ten cores contain successions of planktonic foraminiferal assemblages and sedimentary components that record late Quaternary climatic changes in the western equatorial Atlantic Ocean. The climatic cooling that began at the end of the last interglaciation (X zone) and continued throughout the last glaciation (Y zone) led to the sequential disappearance of tropical *Globorotalia menardii flexuosa*, *Globoquadrina hexagona*, and *Pulleniatina obliquiloculata* and the incursion of cool-water *Globoquadrina dutertrei*, *Globorotalia truncatulinoides*, *G. inflata*, and *Globigerina bulloides*. Despite a twofold increase in cool equatorial species, a tropical climate prevailed throughout the last glaciation. Paleotemperature estimates derived by factor analysis and regression techniques indicate only a small (0.1° to 3.6°C) difference between glacial and postglacial winter temperatures. The coldest sea-surface temperatures occurred at about 73,000 B.P. Concurrently, calcareous remains underwent extensive dissolution, which is reflected in the cores by $CaCO_3$ and coarse-fraction minimums, excessive fragmentation of planktonic foraminiferal shells, absence of pteropods, and an increase in the ratio of benthic to planktonic foraminifera. Faster water-mass circulation during the early part of the last glaciation as compared to the Holocene Epoch is postulated to account for the increased dissolution.

Sea-level lowering (approximately 100 m) during the last glaciation and most of the last interglaciation allowed South American rivers to discharge large quantities of terrigenous sediment, which was continuously transported to the continental rise and abyssal plains by gravity-controlled sediment flows. Sea-level rise during the Holocene Epoch shut off the terrigenous-sediment supply to the deep sea, and the continental rise and abyssal plains became a regime of pelagic sedimentation. Prior to the Holocene Epoch, the supply of terrigenous sediment was shut off only for a brief period at the beginning of the last interglaciation. This indicates that a warm (interglacial) period similar to the Holocene Epoch occurred at the beginning of the last glacial–interglacial cycle.

Comparison of the timing of calcium-carbonate fluctuations of the cores with the timing of stadial-interstadial periods inferred from the continental stratigraphy of the eastern Great Lakes region reveals an excellent correlation. Interstadial (warm) periods correlate with carbonate maximums, whereas stadial (cold) periods correlate with carbonate minimums.

INTRODUCTION

The present study is part of a broader CLIMAP investigation of the paleoclimatology and paleoceanography of the North and South Atlantic Ocean during the past 125,000 yr. In particular, it serves as a geographic link between similar, concurrent studies of the Caribbean Sea by Prell (1974) and of the eastern equatorial Atlantic Ocean by Gardner (1973). Together, these studies form a cohesive unit, in which fluctuations in climate during latest Pleistocene and Holocene time can be analyzed. This paleoclimatic comparison has been made for an east-west traverse along the equatorial belt by Prell and others (this volume).

The objectives of the present study are (1) to determine the temporal successions of planktonic foraminiferal assemblages for the past 80,000 yr and to compare the observed faunal and sedimentary records with the timing of known global climatic changes and (2) to decipher the climatic history of the western equatorial Atlantic by means of quantitative estimates of summer and winter paleotemperatures obtained by the Imbrie and Kipp (1971) technique of paleoecological analysis.

REGIONAL SETTING

The area of the present study lies between lat 20°N and 5°S in the western equatorial Atlantic Ocean. The major physiographic provinces and features of the western equatorial Atlantic are shown in Figure 1. The following description of the region is summarized from Damuth (1973).

The continental shelf reaches a maximum width of 350 km off the mouth of the Amazon River. The shelf break is at a depth of 100 to 115 m between the Orinoco and Amazon Rivers, but south of the Amazon River the shelf break becomes progressively shallower and averages 50 to 75 m in depth east of long. 43°W. The continental slope has gradients averaging 1:4 to 1:13 and is incised by numerous, closely spaced submarine canyons east of long. 45°W.

The continental rise extends seaward from the continental slope for 200 to 600 km. Regional gradients average 1:600 to 1:900 north and 1:200 to 1:400 south of the Amazon Cone. The part of the continental rise to the south of the Amazon Cone is crossed by numerous deep-sea channels (many with natural levees), which are as much as 100 m deep and several kilometres wide.

The continuity of the continental rise is interrupted off the Amazon River by the steeper gradients (1:150 to 1:200) of the Amazon Cone—a vast accumulation of sediment that extends northward from the continental shelf for 650 to 700 km (Damuth and Kumar, 1975). Huge quantities of terrigenous sediment discharged by the Amazon River have been spread across the cone and adjacent abyssal plains via a complex network of distributary channels. The thickness of the sedimentary deposits within the upper cone may be as great as 13 km (Edgar and Ewing, 1968).

The abyssal floor of the western equatorial Atlantic Ocean is dominated by two northwest-trending abyssal plains. The Demerara Abyssal Plain slopes northwestward from the base of the Amazon Cone. A gap between the Barracuda Ridge and the Mid-Atlantic Ridge allows the Demerara Abyssal Plain to join the Barracuda Abyssal Plain, which trends westward to the Puerto Rico Trench. The sedimentary deposits under the abyssal plains are as thick as 2 km and are composed largely of terrigenous sediment derived from the Amazon River. The Ceara Abyssal Plain trends southeastward from the Amazon Cone and Ceara Rise and has sediment derived from both the Amazon River and the minor rivers to the south.

The abyssal plains are bounded on the east by the precipitous morphology of the Mid-Atlantic Ridge. The ridge is offset by several fracture zones, the most prominent of which are the Vema, Doldrums, and St. Paul's. Isolated aseismic rises and topographic highs ascend above the continental rise and abyssal plains. The Ceara Rise separates the Demerara and Ceara Abyssal Plains. The seaward continuity of the continental rise is broken between long. 46°W and 35°W by the rugged seamounts of the North Brazilian Ridge. Many small isolated seamounts and regions of rugged basement morphology occur throughout the region and are shown in black in Figure 1.

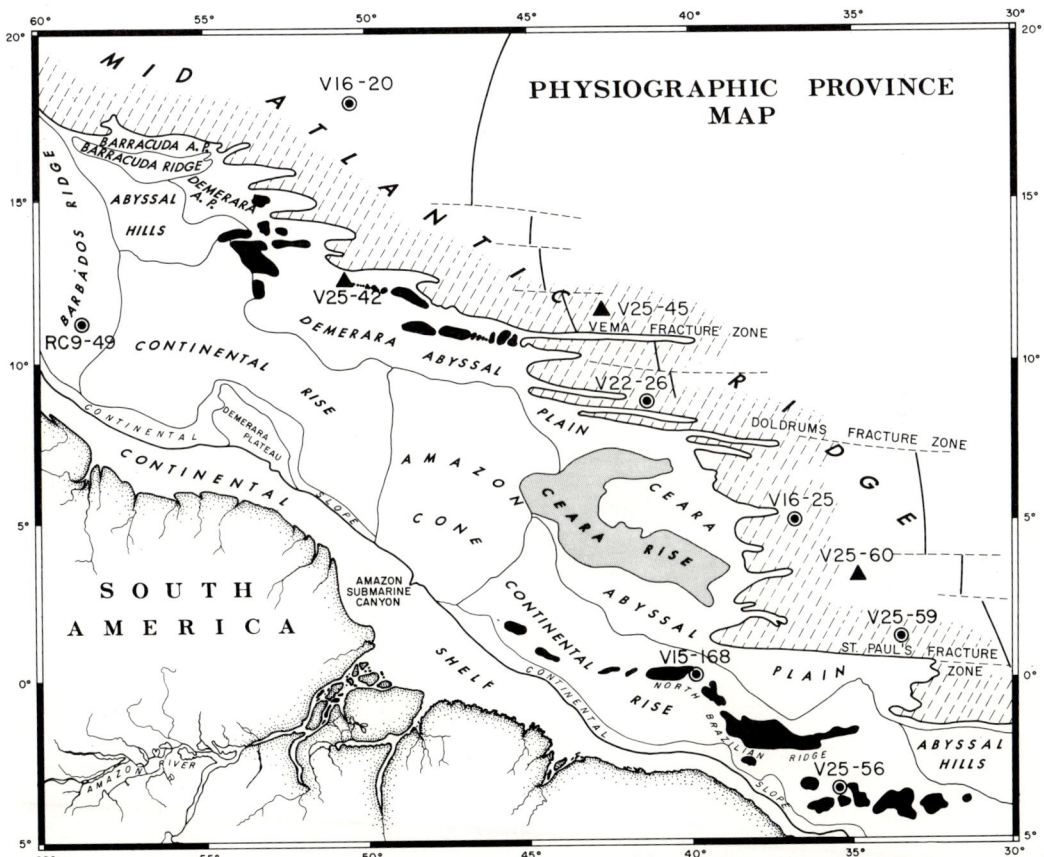

Figure 1. Map showing the physiographic provinces and features of the western equatorial Atlantic Ocean (modified from Damuth, 1973). Locations of cores used in the present study are shown. Circles denote the 7 cores selected for paleoenvironmental analysis. Triangles denote additional cores described in text.

The water masses of the cold-water sphere within the western equatorial Atlantic Ocean are the Antarctic Bottom Water, the North Atlantic Deep Water, and the Antarctic Intermediate Water. The physicochemical properties of the Antarctic Bottom Water are responsible for the dissolution of planktonic foraminifera and nannoplankton in abyssal sediment of this region. The axis of the Antarctic Bottom Water lies along the continental rise of Argentina and southern Brazil, but north of the Equator it shifts eastward to the Mid-Atlantic Ridge flank because of the Coriolis force.

CORE SELECTION AND LABORATORY PROCEDURE

Two constraints were placed on the selection of cores: (1) the cores should contain rich and well-preserved assemblages of planktonic foraminifera, and (2) they should have a continuous upper Quaternary fossil record without major hiatuses. Although all available cores (approximately 250) from the western equatorial Atlantic Ocean were examined, only a few met these two requirements. Most cores from the continental margin and abyssal plains are composed of sediment in which foraminiferal remains are too sparse for meaningful total faunal studies; this is because of dilution by terrigenous material and (or) dissolution by reactions between $CaCO_3$ and organic detritus. Cores from deep (>4,500 m) abyssal regions have foraminiferal assemblages severely altered or destroyed by dissolution. Cores from the Mid-Atlantic Ridge and isolated topographic highs (depths <4,000 m) provide well-preserved faunal assemblages, but the majority of these cores are unsuitable because of hiatuses caused by slumping.

Seven cores were found to contain continuous upper Quaternary sections that could be utilized for quantitative analysis of the planktonic foraminiferal assemblages (Table 1, Fig. 1). The four cores (V16-20, V16-25, V22-26, and V25-59) from the western flank of the Mid-Atlantic Ridge are composed of light-brown foraminiferal marl and ooze; however, cores V16-20 and V22-26 are outside the influence of the productive equatorial currents and thus have such low accumulation rates that they are of marginal value for resolving climatic fluctuations. They are, nevertheless, included for comparison with the other cores.

Two cores from the continental rise (V15-168 and V25-56) and one from the Barbados Ridge (RC9-49) contain gray hemipelagic sediment. $CaCO_3$ fluctuations and biostratigraphic zonations were determined for three additional cores (V25-42, V25-45, and V25-60) but were not included in the paleoecological analysis.

The piston cores were sampled at 10-cm intervals for $CaCO_3$, coarse-fraction, and fáunal analyses. The percentage of coarse fraction (>63 μm) for each sample was computed, and values of coarse-fraction content for each core are graphically presented in Figure 2.

For total faunal analysis, the >149-μm fraction of each sample was repeatedly divided in a microsplitter to obtain an aliquot of at least 300 specimens of planktonic foraminifera. The maximum, average, and minimum numbers counted were 733, 432, and 56 specimens, respectively, from a total of 248 samples examined. A total of 35 taxonomic categories, including 30 species, 1 subspecies, and 4 coiling varieties, were enumerated.[1] The species nomenclature and taxonomic key of Bé (1967) was adopted for the CLIMAP program.

Fluctuations in the abundance of the *Globorotalia menardii* complex were

---

[1] Appendix I is a chart of species expressed as the percent of total planktonic foraminifera. See microfiche in pocket inside back cover.

TABLE 1. LOCATIONS, WATER DEPTHS, AND CORE LENGTHS OF TEN CORES FROM THE WESTERN EQUATORIAL ATLANTIC

Core	Location (lat, long)		Depth (m)	Length (cm)	Area
RC9-49	11°11′N	58°35′W	1,851	663	Barbados Ridge
V15-168	00°12′N	39°54′W	4,219	1,101	Continental rise, north of North Brazilian Ridge
V16-20	17°56′N	50°21′W	4,540	871	West flank of Mid-Atlantic Ridge
V16-25	05°04′N	36°48′W	4,255	1,166	West flank of Mid-Atlantic Ridge
V22-26	08°43′N	41°15′W	3,270	356	West flank of Mid-Atlantic Ridge, near crest
V25-56	03°33′S	35°14′W	3,512	735	Continental rise, near base of a seamount
V25-59	01°22′N	33°29′W	3,824	841	West flank of Mid-Atlantic Ridge
V25-42	12°33′N	50°39′W	4,707	1,035	Small hill that protrudes above Demerara Abyssal Plain
V25-45	11°33′N	42°37′W	3,455	1,136	Crest of Mid-Atlantic Ridge
V25-60	3°17′N	34°50′W	3,749	877	West flank of Mid-Atlantic Ridge

Note: The first seven cores were selected for paleoenvironmental analysis.

determined by using the frequency-to-weight ratio method of Ericson and Wollin (1956b, 1968). The frequency of abundance of the *G. menardii* complex down each core as determined by this method is shown in Figure 2.

The percentage of $CaCO_3$ in each sample was measured by the rapid gasometric technique of Hülsemann (1966). The experimental error using this technique is less than 3%. Measurements were made on reagent grade $CaCO_3$ (100%) standard samples prior to the first and after every few determinations to ensure proper functioning of the apparatus. No correction was made for salt content of the dry cores. The total $CaCO_3$ content down each core is shown in Figure 2.

BIOSTRATIGRAPHIC ZONATION AND CORRELATION OF UPPER QUATERNARY SEDIMENT

Deciphering synchronous events in the climatic and sedimentary history of the western equatorial Atlantic Ocean requires a reliable time-stratigraphic framework. The following criteria were used for stratigraphic zonation and correlation of the cores: (1) abundance fluctuations of individual species of planktonic foraminifera (most useful for biostratigraphic zonation are fluctuations in abundance of the *Globorotalia menardii* complex); (2) fluctuations of total $CaCO_3$ and coarse-fraction (>63 μm) contents within the sediment; and (3) carbon-14 dates (for cores V25-59, V15-168, and RC9-49).

The following list contains names considered to represent equivalent periods of time and will be used interchangeably: (1) "Z zone" or "postglaciation" or "Holocene" for the period 11,000 B.P. to the present; (2) "Y zone" or "last glaciation" or "Wisconsin" for the period 75,000 to 11,000 B.P.; and (3) "X zone" or "last interglaciation" for the period 127,000 to 75,000 B.P. Dates for these climatic zones are those given by Broecker and van Donk (1970).

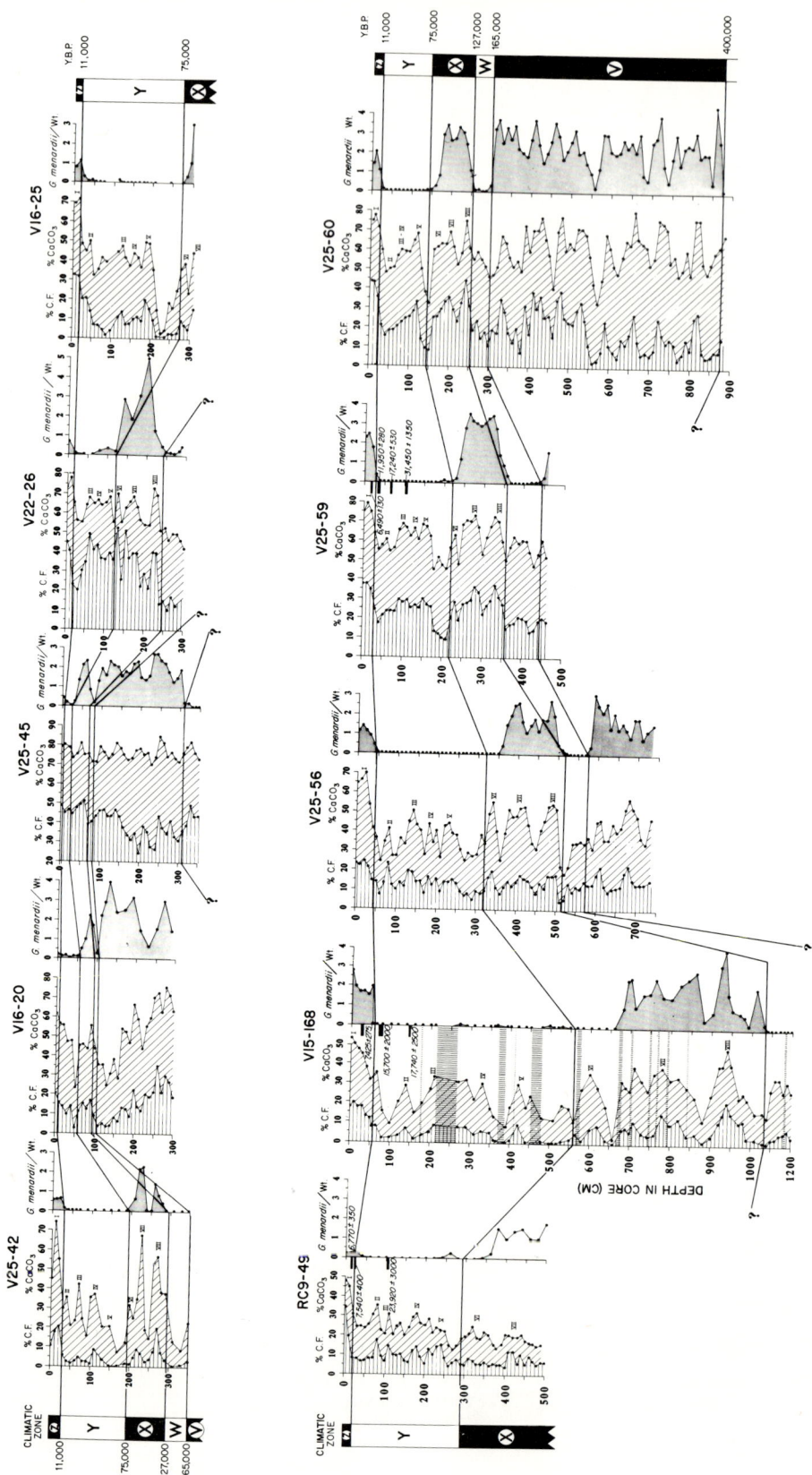

Figure 2. CaCO₃, coarse-fraction, and *Globorotalia menardii* complex variations for the 10 cores of this study. Solid lines correlate Ericson and Wollin's climatic-zone boundaries from core to core. Dates for climatic-zone boundaries are from Broecker and van Donk (1970). Radiocarbon-dated intervals for cores V25-59, V15-168, and RC9-49 are indicated. Correlative CaCO₃ peaks are denoted by Roman numerals for each core.

Foraminiferal Biostratigraphy

Studies of equatorial Atlantic sediment (Schott, 1935; Cushman and Henbest, 1941; Phleger and others, 1953; Phleger, 1942; Ovey, 1950; Wiseman, 1954; Ericson and Wollin, 1956a, 1956b; Ericson and others, 1961) have revealed that geographic shifts in the distribution of planktonic foraminiferal species have occurred at periodic intervals during late Quaternary time and that the shifts are a consequence of changes in climatic conditions. Ericson and Wollin (1956b) were able to construct characteristic, reproducible climatic curves for the entire Quaternary Period based on changes in abundance of the temperature-sensitive *Globorotalia menardii* complex in sediment of the equatorial Atlantic Ocean, Caribbean Sea, and Gulf of Mexico (Ericson and Wollin, 1956a; Ewing and others, 1958; Ericson and others, 1961, 1964). Ericson (1961) identified a series of warm and cold climatic zones and designated them by the letters Z to Q in order of increasing age. The validity and usefulness of the frequency oscillations of *G. menardii* as a broad, but geographically restricted, biostratigraphic and paleoclimatic tool have been confirmed by subsequent workers (for example, Imbrie and Kipp, 1971; Ruddiman, 1971; Kennett and Huddlestun, 1972a, 1972b), although the oscillations are less useful in recording local temperature changes.

Ericson and Wollin's (1968) frequency-to-weight-ratio method was used to delineate biostratigraphic zones and to correlate our piston cores (Fig. 2). The climatic zones (Z, Y, X, and so forth) revealed by the *Globorotalia menardii* curves are correlated from core to core. The Z zone (0 to 11,000 B.P.) contains abundant *G. menardii* and *G. tumida* but no *G. menardii flexuosa* and represents the Holocene Epoch. The Y zone (11,000 to 75,000 B.P.) contains no *G. menardii* complex and represents the last glaciation (Wisconsin). The X zone (75,000 to 127,000 B.P.) contains abundant tests of the *G. menardii* complex, especially the subspecies *G. menardii flexuosa*, and represents the last interglaciation.

Other foraminiferal species are also reliable stratigraphic markers. *Pulleniatina obliquiloculata* is abundant throughout the X zone and the lower one-half of the Y zone in equatorial Atlantic and Caribbean cores, but it disappears consistently in the middle of the Y zone at approximately 40,000 B.P. (Damuth, 1973). *P. obliquiloculata* reappears in the uppermost part of the Y zone (approximately 15,000 B.P.) and is abundant throughout the Holocene Epoch. A prominent pulse of left-coiling *Globorotalia truncatulinoides* occurs slightly above the X-Y boundary at approximately 73,000 B.P. *Globoquadrina hexagona* is present throughout the X zone and lowermost Y zone. In the overlying sediment, this species, as well as *Globorotalia menardii flexuosa*, is absent, indicating that both were presumably driven away by the unfavorable marine environment of the last glaciation in the North Atlantic Ocean.

$CaCO_3$ and Coarse-Fraction Fluctuations

In addition to faunal oscillations, temporal fluctuations of $CaCO_3$ and coarse-fraction (>63 μm) contents were used for stratigraphic correlation. Close-interval carbonate and coarse-fraction analyses down piston cores from the equatorial Atlantic Ocean (Damuth, 1973, 1975; Gardner, 1973, 1975) and Caribbean Sea (Prell, 1974) have revealed characteristic peaks and depressions that are synchronous from core to core and region to region.

$CaCO_3$ and coarse-fraction variations are shown in Figure 2. Eight prominent carbonate–coarse-fraction peaks (identified by Roman numerals beside the $CaCO_3$ curve of each core) are recognized in both the pelagic sediment of the Mid-Atlantic

Ridge and abyssal hills (cores V25-59, V25-60, V25-42, V22-26, and V16-25) and hemipelagic sediment of the continental margin (cores V15-168, V25-56, and RC9-49). Two cores (V25-56 and V16-20) have sedimentation rates so low (<1 cm/10^3 yr) that individual peaks cannot be resolved because of the sampling interval and (or) bioturbation and sediment reworking.

Core V25-59 best reflects the $CaCO_3$ and coarse-fraction fluctuations in the western equatorial Atlantic, because it contains relatively undissolved pelagic foraminiferal marl and ooze deposited at a moderate (approximately 3 cm/10^3 yr) rate. The other pelagic cores generally show less regular curves because of selective dissolution or reworking. Nevertheless, most peaks are recognizable in each core. The three hemipelagic cores show sections that are expanded in comparison to sections of the pelagic cores because of terrigenous dilution. Although $CaCO_3$ and coarse-fraction curves of these cores show many secondary oscillations, the eight major peaks are recognizable in each core (Fig. 2).

The Z zone (Holocene) of each core has high $CaCO_3$ and coarse-fraction content. Peak I generally contains the maximum $CaCO_3$ and coarse-fraction values observed within each core and occurs at about mid-Holocene. The upper one-third of the Y zone is marked by a broad $CaCO_3$–coarse-fraction depression. This depression is broken in most cores by a small but prominent peak (II). The middle one-third of the Y zone has relatively higher $CaCO_3$ and coarse-fraction content than the upper and lower thirds of the Y zone and shows three well-defined peaks (III, IV, and V) separated by minor depressions. The lower one-third of the Y zone is characterized by very sharp $CaCO_3$ and coarse-fraction depressions, which often contain the lowest values observed within the Z, Y, and X zones of each core.

The X zone (last interglaciation) of each core is characterized by relatively high content of $CaCO_3$ and coarse fraction; these values often approach those observed for the Z zone. Three prominent peaks (VI, VII, and VIII) separated by two very sharp depressions are recognized in the X zone.

Radiocarbon Dates

Radiocarbon dates for intervals of sediment of the uppermost Y zone and lower Z zone were determined for cores RC9-49, V15-168, and V25-59. The sample intervals and their ages (based on total $CaCO_3$) are given in Table 2 and plotted in Figure 2.

Lithostratigraphic Relationships

Many cores from the western equatorial Atlantic Ocean contain lithostratigraphic relationships within their upper 2 m that can be easily recognized and correlated from core to core. These cores are composed of terrigenous sediment and were raised from the continental rise, Amazon Cone, or abyssal plains (see Damuth, 1973, or McGeary and Damuth, 1973, for core locations). The upper 30 to 80 cm of each core consists of light-brown to orange-tan pelagic foraminiferal marl or ooze and is Holocene in age. The remainder of each core consists of upper Pleistocene gray terrigenous hemipelagic clay with interbedded turbidite composed of silt- to gravel-sized particles. A very prominent rust-colored, iron-rich crust as thick as 10 cm separates the light-brown pelagic sediment from underlying gray terrigenous sediment in nearly all cores. Recent studies (Damuth, 1973; McGeary and Damuth, 1973) have revealed that the rust-colored crust and the associated lithologic change occurred at approximately the Pleistocene-Holocene boundary. These lithostratigraphic relationships provide an easy means of identifying the

TABLE 2. RADIOCARBON DATES FOR SEVERAL INTERVALS IN THREE CORES

Core	Depth in core (cm)	C^{14} age (yr)
RC-49	13.5–16.5	6,770 ± 350
	17–22	7,540 ± 400
	101–109	23,920 ± 3,000
V15-168	23–29	7,425 ± 275
	67–78	15,700 ± 2,000
	142.5–147.5	17,740 ± 2,500
V25-59	15–18	6,490 ± 130
	33–38	11,950 ± 280
	64–68	17,240 ± 530
	101–108	31,450 ± 1,350

Pleistocene-Holocene boundary in sediment from the continental rise, Amazon Cone, and abyssal plains.

PALEOCLIMATIC INTERPRETATIONS

Fluctuations in Faunal Assemblages

Previous studies of the distribution and abundance of planktonic foraminifera from plankton tows as well as surface sediment samples in the equatorial and North Atlantic Ocean have established the latitudinal zonations and ranges of ecological tolerances for each species (for example, Schott, 1935, 1966; Phleger and others, 1953; Boltovskoy, 1964; Jones, 1967; Ruddiman, 1969; Cifelli and Smith, 1970; Barash, 1971; Bé and Tolderlund, 1971; Tolderlund and Bé, 1971). Despite the complexities in the ecological and post-mortem behavior of each species in terms of relative abundance, seasonal productivity and mortality, geographic and depth habitat preferences, and selective dissolution effects of the skeletons on the sea floor, micropaleontologists have been able to identify species groups with similar ecological responses. One subdivision that groups 27 species of living planktonic foraminifera into 5 faunal assemblages (tropical, subtropical, transitional, subpolar, and polar) was proposed by Bé and Tolderlund (1971) for the Atlantic and Indian Oceans. A faunal grouping of 6 sea-bed assemblages, derived by factor analysis, has been specifically applied to estimating past sea-surface conditions (Kipp, this volume).

Because the mid- and high-latitude species constitute a minor element of all the assemblages in the western equatorial region, we prefer to follow Ruddiman's (1971) simpler approach of grouping the species into "warm-water" and "cool-water" groups according to their known ecological tolerances and inverse relationships in abundance. Table 3 lists the species composition of these two categories.

In general, the cool-water group is composed of 9 species whose closest proximity today to the western equatorial Atlantic Ocean is in the eastern tropical Atlantic, in upwelling regions off West Africa and Venezuela, in the northern Sargasso Sea, or in deep water (Schott, 1935, 1966; Ruddiman, 1969; Bé and Tolderlund, 1971; Miro, 1971). The warm-water group consists of 21 tropical and subtropical species, of which the surface-dwelling *Globigerinoides ruber* and *G. sacculifer* are by far the dominant species.

Our present observations largely confirm the climatic oscillations and magnitude

TABLE 3. PLANKTONIC FORAMINIFERAL SPECIES COMPOSING COOL-WATER
AND WARM-WATER ASSEMBLAGES

Cool-water species	
Globigerina pachyderma (Ehrenberg) (1)	*Globorotalia truncatulinoides* (d'Orbigny) (7)
Globigerina bulloides d'Orbigny (2)	*Globorotalia scitula* (Brady) (8)
Globigerina falconensis Blow (3)	*Globorotalia crassaformis* (Galloway & Wissler) (9)
Globigerina quinqueloba Natland (4)	
Globoquadrina dutertrei (d'Orbigny) (5)	
Globorotalia inflata (d'Orbigny) (6)	

Warm-water species	
Globigerinoides ruber (d'Orbigny) White variety (10)	*Globigerina rubescens* Hofker (19)
Globigerinoides ruber (d'Orbigny) Pink variety (11)	*Globigerinoides tenellus* Parker (20)
	Globigerinita glutinata (Egger) (21)
Globigerinoides sacculifer (Brady) (12)	*Orbulina universa* d'Orbigny (22)
Globigerinoides conglobatus (Brady) (13)	*Globigerinita humilis* (Brady) (23)
Globoquadrina hexagona (Natland) (14)	*Globigerina digitata* (Rhumbler) (24)
Globorotalia menardii (d'Orbigny) (15, ruled)	Other species (25):
	Globorotalia hirsuta (d'Orbigny)
Globorotalia menardii flexuosa (Koch) (15, black)	*Globorotalia crassula* Cushman & Stewart
Globorotalia tumida (Brady) (15, white)	*Sphaeroidinella dehiscens* (Parker & Jones)
Pulleniatina obliquiloculata (Parker & Jones) (16)	*Hastigerina pelagica* (d'Orbigny)
Globigerinella aequilateralis (Brady) (17)	*Hastigerinella digitata* (Rhumbler)
Globigerinella calida Parker (18)	*Candeina nitida* (d'Orbigny)

Note: Frequency patterns for these species are shown in middle diagrams of Figures 4 through 10. Numbers in brackets are for individual species shown in bottom diagrams of Figures 4 through 10.

of the temperature shifts in the tropical and subtropical Atlantic Ocean and adjacent areas that have been reported previously by Ericson and Wollin (1956a, 1956b), Emiliani (1966), Imbrie and Kipp (1971), Ruddiman (1971), Ku and others (1972), Gardner (1973), Kennett and Huddlestun (1972a, 1972b), Lynts and others (1973), Prell (1974), Briskin and Berggren (1975), and others.

Broad oscillations in the proportions of warm- and cool-water species groups during latest Quaternary time are apparent in the seven cores studied (Figs. 3 through 9). Although the faunal assemblages reflect a predominantly tropical-subtropical climate for the past 80,000 yr, a cooler aspect is evidenced by a twofold greater abundance of the cool-water species during the last glaciation than during the last interglaciation or the Holocene Epoch. During the last glaciation, the cool-water species generally constituted about 40% of the total planktonic foraminifera in the equatorial region, diminishing northward to about 10% in core V16-20 located near lat 18°N.

Ruddiman's (1971) observation that during cold intervals the nonequatorial cool fauna may outnumber warm species may be true in his central equatorial cores, but it is not substantiated in the western equatorial region. The logical explanation is that as the cool-water species are transported westward from the eastern equatorial Atlantic Ocean and Benguela Current region by the North and South Equatorial Currents, their influence gradually diminishes. Evidence for this is presented by Prell and others (this volume).

Occasional pulses in which cool-water species outnumber the warm-water group in abnormally high proportions may in part be artifacts caused by intense dissolution activity, which favors the preservation of the solution-resistant shells of the cool-water group. Such an interval is observed in cores V25-56, V25-59, and V16-25 in the lowermost Y zone and coincides with a distinct $CaCO_3$ minimum as well as Emiliani's (1966) stage 4 (70,000 to 75,000 B.P.).

The maximum relative abundance of the cool-water species is observed in the middle to upper Y zone (approximately 17,000 to 50,000 B.P.); during this interval the warm-water species were at their lowest ebb, and *Globorotalia menardii*, *Globoquadrina hexagona*, and *Pulleniatina obliquiloculata* were conspicuously absent. This is in close agreement with Kennett and Huddlestun's (1972a, 1972b) observations in the Gulf of Mexico.

Cool-Water Species. The most prominent species that lives in cool-water conditions (see Figs. 3 through 9) is *Globoquadrina dutertrei*, which at present constitutes less than 5% of the total planktonic foraminifera in the western equatorial Atlantic Ocean and more than 10% in the eastern Atlantic surface sediment (Schott, 1966; Ruddiman, 1969). Ruddiman (1971) also found *G. dutertrei* to be the most common cool-water species in the glacial and interglacial sedimentary deposits of the central equatorial Atlantic Ocean. In our area, it is about two to five times more abundant in the Y zone than in the X or Z zones. Its proliferation during glacial periods was apparently caused by faster oceanic circulation, which resulted in a more intense flow of the South and North Equatorial Currents and permitted increased numbers of *G. dutertrei* (and other cool-water species) to reach the western equatorial Atlantic.

Globorotalia inflata, *G. truncatulinoides*, and *G. scitula* are also indicators of

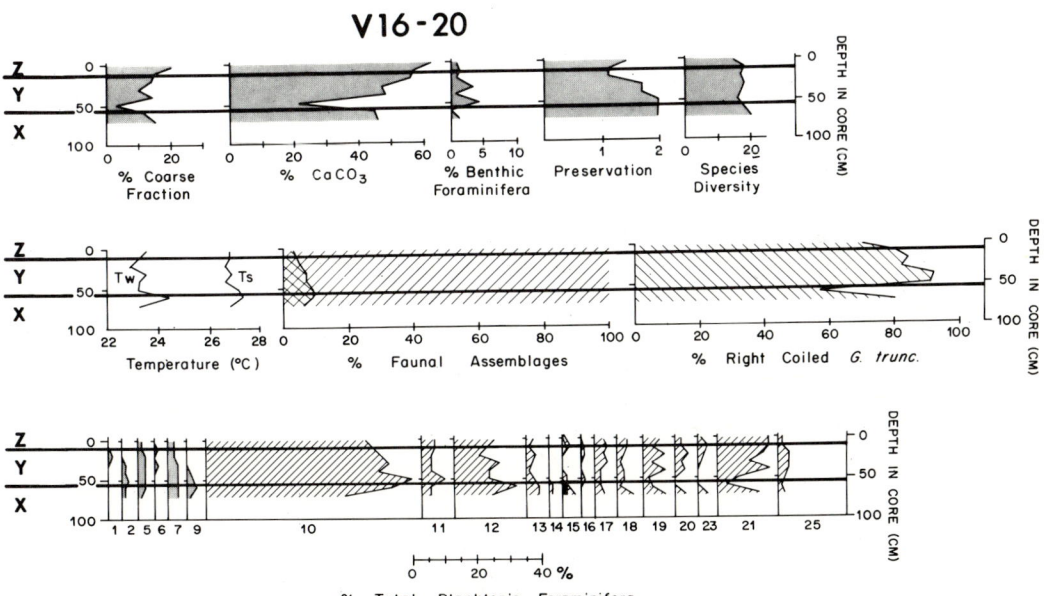

Figure 3. Coarse-fraction, $CaCO_3$, faunal, and paleoclimatic analyses of core V16-20. Percentage benthic foraminifera: total benthic and planktonic foraminiferal shells. Preservation: condition ranges from scale 1 (good) to scale 4 (poor). Species diversity: total number of species of planktonic foraminifera. Temperature: Winter (T_w) and summer (T_s) sea-surface temperatures determined by factor analysis. Percentage faunal assemblages: "cool" and "warm" assemblages are made up of cumulative percentages of cool-water and warm-water species, respectively. Percentage right-coiled *Globorotalia truncatulinoides* (ruled area). Frequency patterns of individual species: Numbers in bottom row refer to species listed in Table 2. Stippled areas represent cool-water species. Ruled areas represent warm-water species. Species 15 is *Globorotalia menardii* complex and includes *G. menardii flexuosa* (black), *G. tumida* (white), and *G. menardii* (ruled).

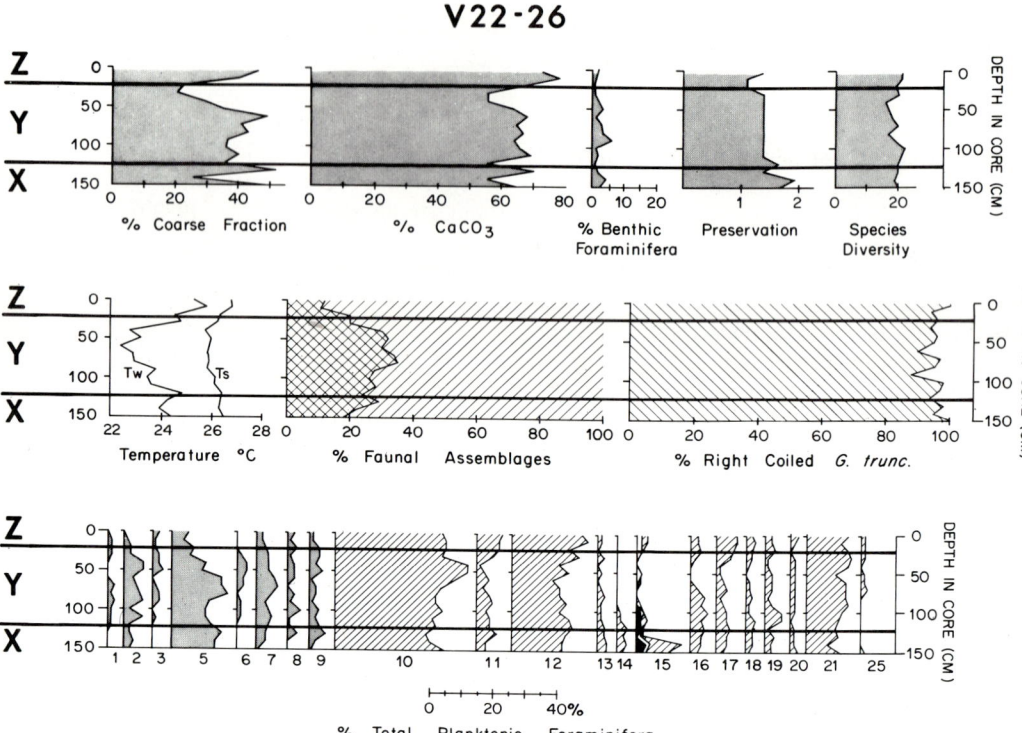

Figure 4. Coarse-fraction, CaCO$_3$, faunal, and paleoclimatic analyses of core V22-26 (see Fig. 3 legend for explanations).

cool water and are morphologically distinct and easily recognizable. *G. inflata* now proliferates in the upwelling region off West Africa (Schott, 1966; Ruddiman, 1969) as well as in the transitional zone in the middle latitudes (Bé and Tolderlund, 1971). Its presence in the western equatorial Atlantic Ocean would require transport over several thousand kilometres from either an eastern or northern source. Similarly, a predominance of left-coiling *G. truncatulinoides* in the equatorial area would indicate transportation from source areas either north or south that are at least 1,900 km away. Ericson and others (1954) delineated a left-coiling province of *G. truncatulinoides* in the central North Atlantic Ocean north of about lat 20°N and between northwest Africa and North America.

Brief dominance of the left-coiling over the right-coiling variety of *G. truncatulinoides* occurs slightly above the X-Y boundary (approximately 73,000 B.P.) and coincides with minimum CaCO$_3$-coarse-fraction values and high proportions of cool-water species (Figs. 3, 5 through 8). The most pronounced left-coiling pulses are observed in equatorial cores V15-168 and V25-59 (65% and 82% left-coiling, respectively.). They are less marked in cores from the northern sites RC9-49 and V22-22, which border the present-day left-coiling province. The brief intervals in which left-coiling *G. truncatulinoides* was prevalent in the tropical Atlantic Ocean apparently indicate cooler subtropical conditions similar to those found today at about lat 30°N or 30°S. Whether the migration of the left-coiling population was primarily from the North Atlantic, South Atlantic, or both regions is not known.

Two important cool-water species that have less distinctive morphologies are *Globigerina pachyderma* and *G. bulloides*. The former is believed to belong to

V16-25

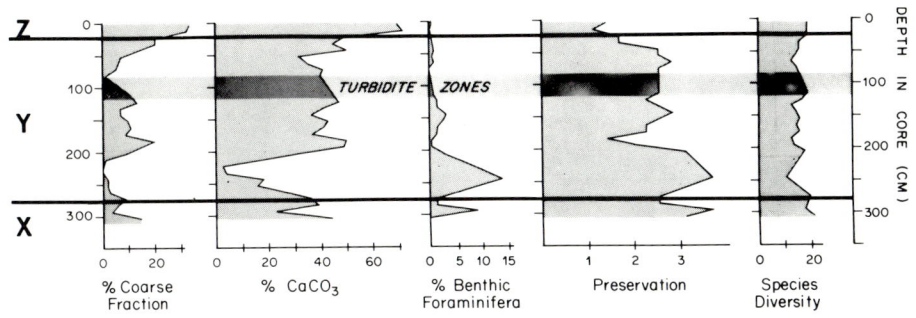

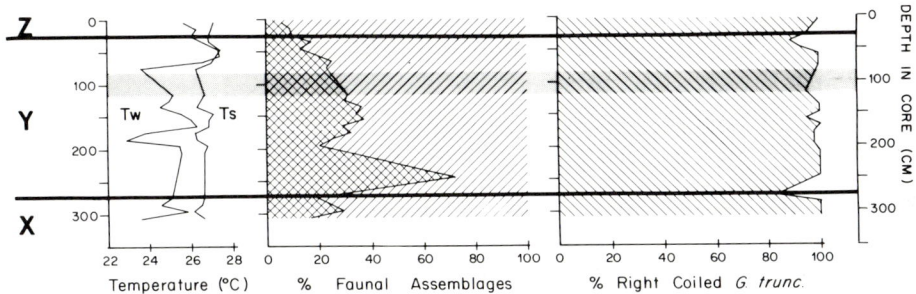

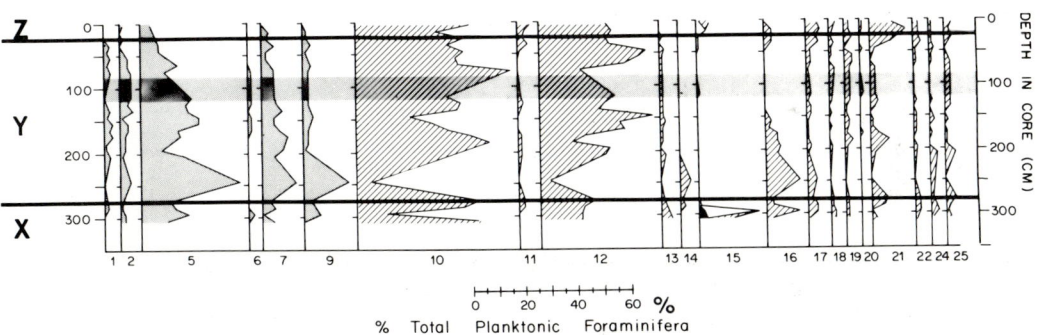

Figure 5. Coarse-fraction, CaCO₃, faunal, and paleoclimatic analyses of core V16-25 (see Fig. 3 legend for explanations).

a cline that grades morphologically from left-coiling *G. pachyderma* in polar water to the right-coiling variety in subpolar areas and to the right-coiling *Globoquadrina dutertrei* in subtropical and tropical waters. The morphologic intergradation between right-coiling *Globigerina pachyderma* and *Globoquadrina dutertrei* is most pronounced in the transitional zone, as noted by Bé and Tolderlund (1971).

Globigerina bulloides occurs throughout the Z and Y zones in western equatorial Atlantic sediment. Although generally considered to be a subpolar species, it has great morphologic plasticity. A closely related species, *Globigerina falconensis*, is commonly encountered in subtropical water of the Sargasso Sea. Recently, Miro (1971) reported the occurrence of *G. bulloides*, *G.* cf. *quinqueloba*, and

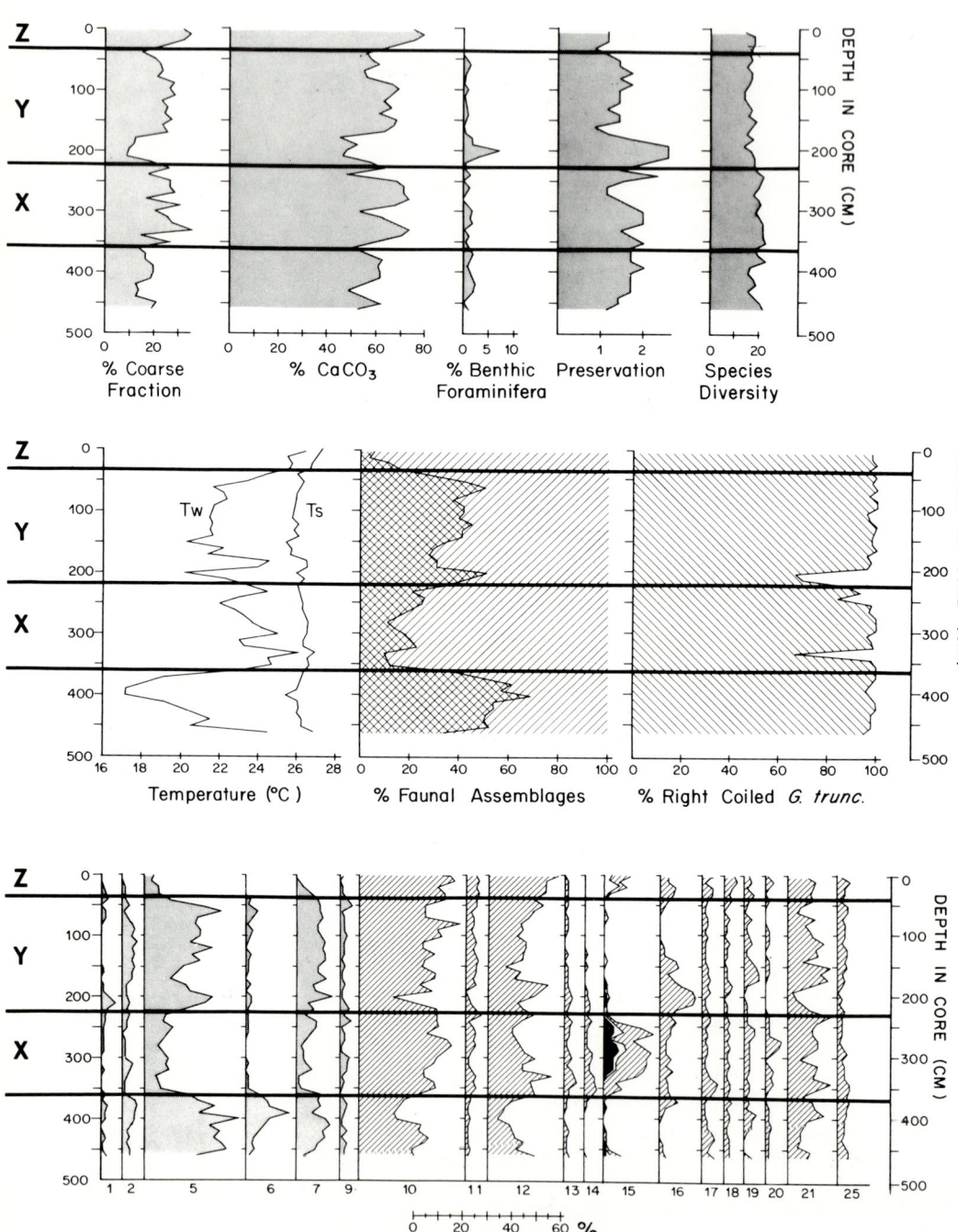

Figure 6. Coarse-fraction, CaCO$_3$, faunal, and paleoclimatic analyses of core V25-59 (see Fig. 3 legend for explanations).

V25-56

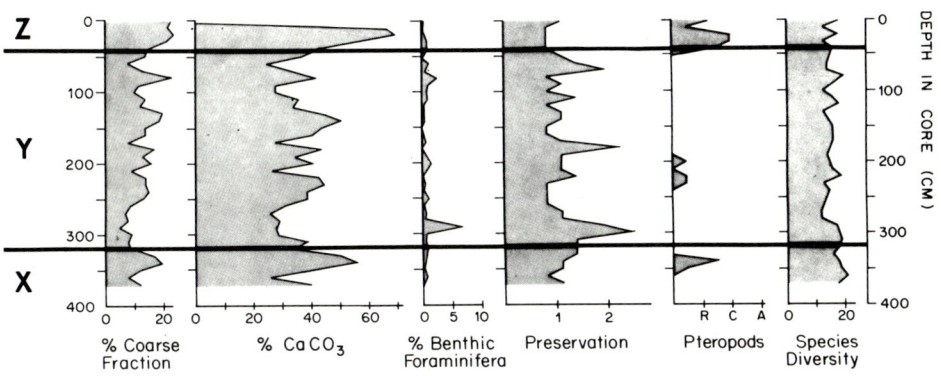

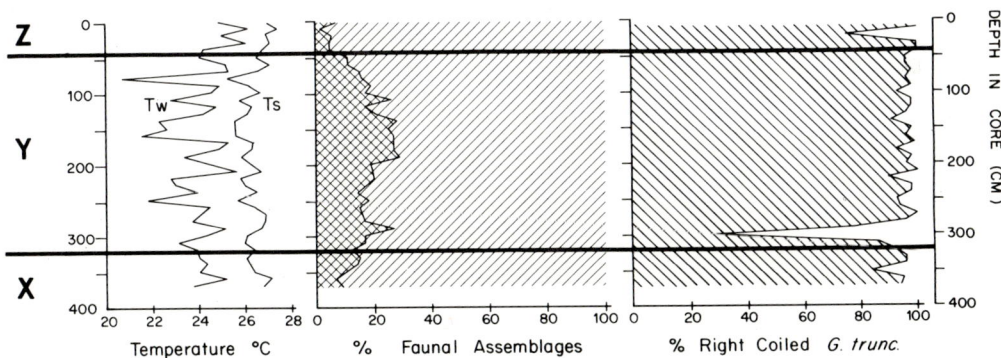

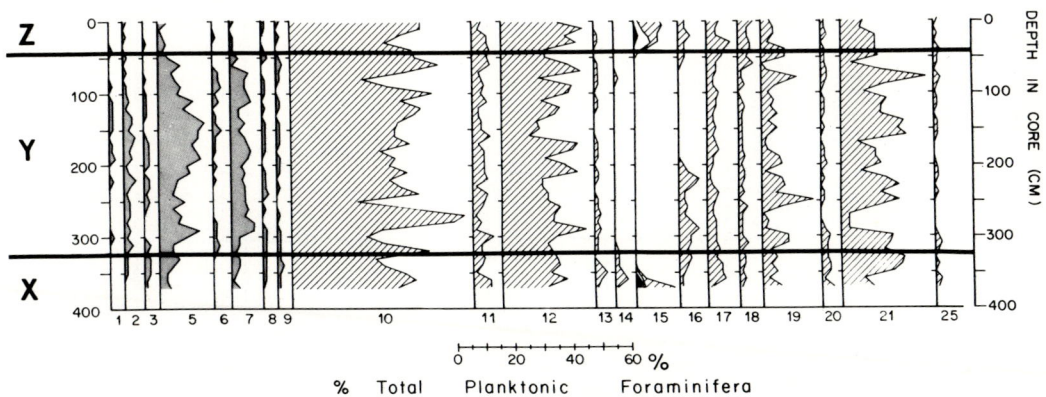

Figure 7. Coarse-traction, CaCO$_3$, faunal, and paleoclimatic analyses of core V25-56. Pteropods: R = rare; C = common; A = abundant (see Fig. 3 legend for other explanations).

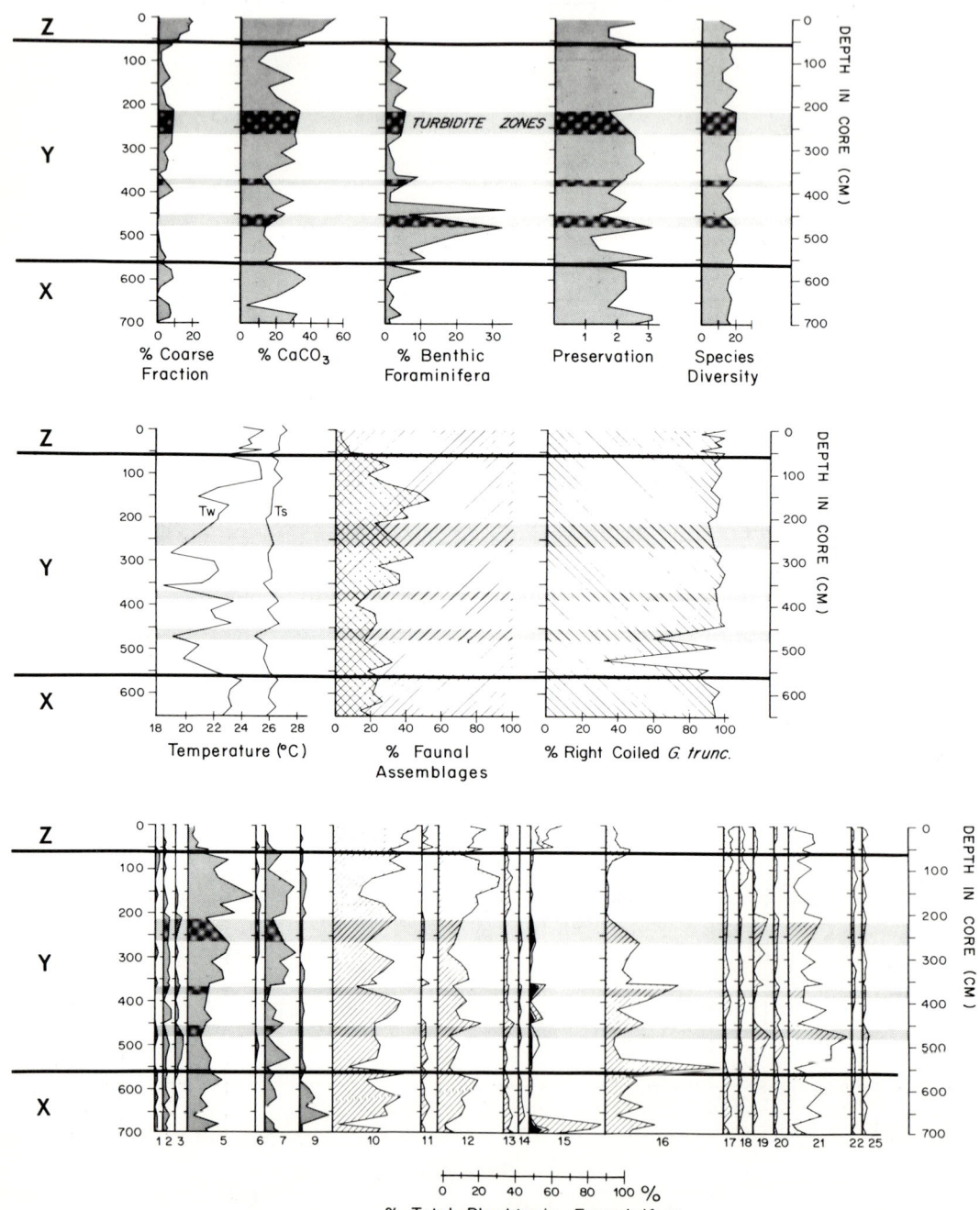

Figure 8. Coarse-fraction, CaCO$_3$, faunal, and paleoclimatic analyses of core V15-168 (see Fig. 3 legend for explanations).

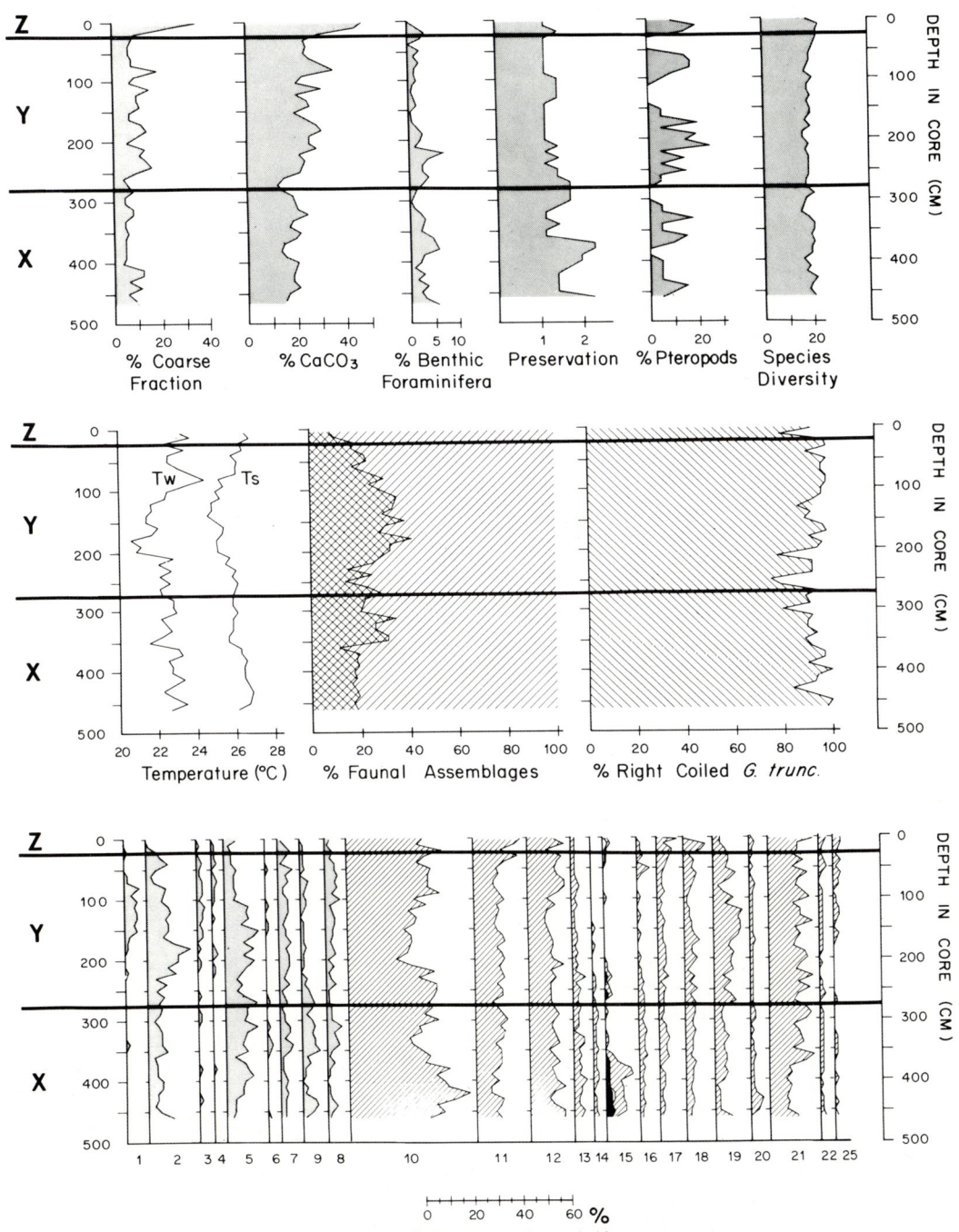

Figure 9. Coarse-fraction, CaCO$_3$, faunal, and paleoclimatic analyses of core RC9-49 (see Figs. 3 and 7 legends for explanations).

G. crassaformis in the upwelling area and the 200- to 700-m depth range in the southern Caribbean Sea (Cariaco trench). *G. bulloides* is also known to comprise more than 25% of the planktonic foraminiferal assemblage in the surface sediment off northwest Africa (Ruddiman, 1969), although it is not common there in the overlying water (Bé and Tolderlund, 1971). It is conceivable that *G. bulloides* in the western equatorial Atlantic is either transported by the North Equatorial Current from the eastern Atlantic or is associated with tropical upwelling zones off South America. In either case it is not necessarily representative of subpolar or high-latitude conditions.

Warm-Water Species. The three dominant warm-water species (see Figs. 3 through 9) are *Globigerinoides ruber, G. sacculifer,* and *Globigerinita glutinata.* Variants of *G. ruber* with white tests comprise 30 to 40% of the total planktonic foraminifera in both interglacial and glacial sediments, whereas variants with pink tests are considerably less abundant.

Globigerinoides sacculifer, a widespread and abundant species in present-day equatorial waters, occurs in frequencies of about 15% in both the Z and Y zones. The common occurrence of *G. glutinata* (approximately 10%) is difficult to explain in terms of ecological significance, because this ubiquitous species is present in subpolar as well as tropical waters of the Atlantic Ocean.

The persistent abundance of *G. ruber* and *G. sacculifer* in the Y zone indicates relatively slight environmental contrasts between the glacial and interglacial epochs in the western equatorial Atlantic. Our paleotemperature estimates (Figs. 3 through 9) reveal that glacial and interglacial differences in winter surface temperature were small and ranged from 0.1° to 3.6°C.

Globorotalia menardii, Pulleniatina obliquiloculata, and *Globoquadrina hexagona* are excellent indicators of tropical marine environments (Bé and Tolderlund, 1971). Their great abundance in sedimentary deposits of the last interglaciation (X zone) is generally attributed to warm climatic conditions. The sequential disappearance of these three species in the lower Y zone and their absence in the middle and upper Y zone may reflect progressive climatic cooling.

The other 15 species of planktonic foraminifera contribute relatively small percentages to the total faunal assemblages. Although they are generally considered to have tropical and subtropical affinities, their frequency patterns do not show any significant fluctuations between glacial and interglacial periods.

Quantitative Paleoenvironmental Estimates

The validity of paleoclimatic interpretations of the Pleistocene and Holocene Epochs depends on the assumption that ecological tolerances and responses of shell-bearing species to physicochemical conditions in the oceans are the same today as they have been in the past. On this assumption, Imbrie and Kipp (1971) introduced a technique of quantitatively estimating physical parameters of surface water from planktonic foraminiferal assemblages in North Atlantic surface sediment. Gardner (1973) applied Q-mode factor analysis (Imbrie and van Andel, 1964; Klovan and Imbrie, 1971) to the relative abundances of planktonic foraminifera in 199 core-top samples from the North Atlantic and resolved these foraminifera into four factors or varimax assemblages that accounted for 90.6% of the variability of the data (see Table 3 in Gardner and Hays, this volume). Each factor is a statistically independent faunal assemblage and a unique combination of 24 species categories defined by the relative positive or negative contribution of each species. The factor loadings of the surface-sediment samples were mapped by Gardner (1973) and delineated well-defined biogeographic zones (Gardner and Hays, this volume, Fig. 1).

Factor 1 (tropical varimax assemblage) showed high factor loadings (>0.9000) in the region between the Equator and approximately lat 30°N. The tropical assemblage is composed primarily of *Globigerinoides ruber*, *G. sacculifer*, *Globigerinita glutinata*, and *Globigerinella aequilateralis*.

Factor 2 (subpolar-polar varimax assemblage) has high factor loadings in high-latitude areas north of approximately lat 40° to 50°N. The reason for the subpolar-polar grouping is that left-coiling *Globigerina pachyderma* (a polar variety) and right-coiling *G. pachyderma* (a subpolar variety) were considered a single species. Because species diversity in high-latitude regions is very low, the very significant contributions of both varieties of *G. pachyderma* resulted in a composite subpolar-polar factor. In addition to *G. pachyderma*, this assemblage is composed of *Globigerina bulloides*, *Globorotalia inflata*, *Globigerinita glutinata*, and others.

Factor 3 (gyre-margin varimax assemblage) occurs mainly in the Gulf Stream region and, less distinctly, in the equatorial zone. It is predominantly represented by *Globorotalia menardii*, *G. tumida*, *Globoquadrina dutertrei*, *Globigerinoides sacculifer*, and *Pulleniatina obliquiloculata*.

Factor 4 (subtropical varimax assemblage) has its highest factor loadings in the northern Sargasso Sea. *Globorotalia inflata*, *Globigerina falconensis*, *Globorotalia truncatulinoides*, and *Globigerina bulloides* are dominant. The relatively low factor loadings of this assemblage indicate a significant mixing of tropical and subpolar-polar assemblages, and hence, it may also be considered a transitional fauna (Bé and Tolderlund, 1971).

The foraminiferal species in our seven cores were analyzed in terms of these four varimax assemblages by using an F' matrix derived from factor analysis of the surface-sediment samples. Figure 10 clearly indicates that the tropical assemblage is dominant and the subpolar-polar, gyre-margin, and subtropical assemblages are relatively insignificant elements in all cores. A reciprocal relationship is evident in the fluctuations between the tropical assemblage and subpolar-polar–gyre-margin assemblages. A climatic cooling is indicated when the tropical assemblage decreases, whereas the other two assemblages increase concurrently.

By applying the transfer functions (regression equations) developed by Imbrie and Kipp (1971) and Imbrie and others (1973), winter (February) and summer (August) sea-surface temperatures were calculated from the species-abundance data of planktonic foraminifera in our cores. The following interpretations were made from the temperature curves in Figures 3 through 9:

1. Summer sea-surface temperature (T_s) fluctuations have lower amplitudes than winter sea-surface temperatures (T_w) at all seven core locations.

2. The T_w curves probably reflect more closely the magnitude of climatic variations than do the T_s curves, because T_w displays greater regional contrast and less seasonal stability than T_s throughout the latest part of the Quaternary.

3. The climatic records for cores V25-59, V22-26, V25-56, and RC9-49 are more distinct than those for cores V16-20, V16-25, and V15-168. The slow sedimentation rate in core V16-20 does not permit resolution of small-scale climatic variations. Cores V16-25 and V15-168 contain intervals in which the foraminiferal assemblages have been altered by severe dissolution; as a result, meaningful paleoclimatic results cannot be obtained from these intervals.

4. Core V25-59 apparently contains a climatic record that most closely parallels temperature fluctuations in the higher latitudes of the North Atlantic Ocean. At this equatorial site, temperatures as warm as during mid-Holocene time occurred only one other time during the past 127,000 yr—at the beginning of the last interglaciation, approximately 125,000 B.P. (Fig. 6). A climate colder than that during the Holocene Epoch prevailed during most of the last interglaciation. Very

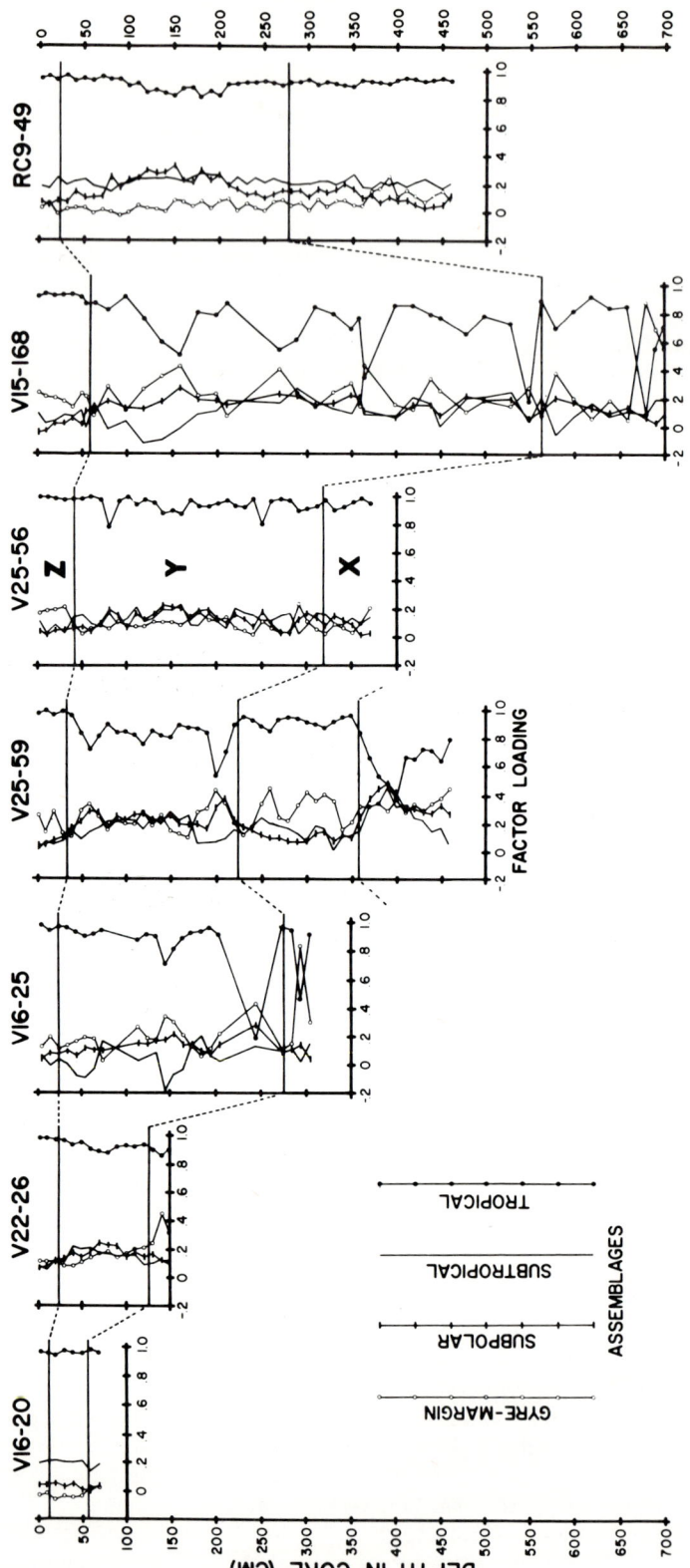

Figure 10. Factor loadings for the 4 varimax assemblages in 7 cores. Each factor is derived from the surface faunal data, whose present-day distribution is shown in Figure 11.

sharp T_w gradients mark the Pleistocene-Holocene boundary (termination I, approximately 11,000 B.P.) and the W-X boundary (termination II, approximately 127,000 B.P.).

5. During the last glaciation (Y zone) the coldest winter temperatures occurred at different times for different core locations. The observed differences may depend on geographic location or circulation patterns. At the two equatorial sites (V25-59 and V15-168), the coldest winter temperatures occurred during the early and middle parts of the last glaciation. At site RC9-49, near Barbados, the coldest winter temperatures were observed during the middle part of the last glaciation. At the other three sites (V22-26, V16-25, and V25-56) the coldest temperatures prevailed from the middle to the end of the last glaciation.

The means of the winter and summer sea-surface temperatures for the Holocene Epoch, last glaciation, and last interglaciation, as well as the differences between these mean values, were calculated. The following observations are based on the results shown in Table 4:

1. Mean winter temperatures during the Holocene Epoch ranged from 23.3° to 26.0°C. Mean summer temperatures during the Holocene ranged from 25.2° to 27.0°C. Mean winter temperatures during the last glaciation ranged from 22.2° to 25.5°C. Mean summer temperatures during the last glaciation ranged from 25.6° to 26.8°C.

TABLE 4. COMPARISON OF MEANS AND DIFFERENCES OF ESTIMATED WINTER AND SUMMER SEA-SURFACE TEMPERATURES FOR THE HOLOCENE EPOCH AND WISCONSIN GLACIATION

Core	Seasonal temperature	Holocene Epoch (Z zone)	Wisconsin glaciation (Y zone)	$T_Z - T_Y$*
V16-20	T_w†	23.33	23.21	0.12
	T_s†	26.73	26.75	−0.02
	$T_s - T_w$	3.4	3.54	
V16-25	T_w	25.99	25.45	0.54
	T_s	27.03	26.71	0.32
	$T_s - T_w$	1.04	1.26	
V25-56	T_w	25.28	23.82	1.46
	T_s	27.05	26.22	0.83
	$T_s - T_w$	1.77	2.4	
RC9-49	T_w	23.33	22.32	1.01
	T_s	26.56	25.55	1.01
	$T_s - T_w$	3.23	3.23	
V22-26	T_w	25.25	23.58	1.67
	T_s	26.68	26.08	0.60
	$T_s - T_w$	1.43	2.5	
V15-168	T_w	24.95	22.16	2.79
	T_s	26.88	26.18	0.70
	$T_s - T_w$	1.93	4.02	
V25-59	T_w	25.96	22.40	3.56
	T_s	27.02	26.09	0.92
	$T_s - T_w$	1.06	3.69	

Note: Differences in mean temperatures are calculated within and between Z and Y zones.
*T_Z = temperature during Holocene Epoch; T_Y = temperature during Wisconsin glaciation.
†T_w = estimated winter temperature; T_s = estimated summer temperature.

2. The difference in mean winter temperatures between the Holocene Epoch and the last glaciation ranged from 0.1° to 3.6°C. The greatest mean winter-temperature increase (>3.6°C.) between these intervals occurred in the equatorial core V25-59, and the smallest increase (0.1°C) occurred in the northernmost core V16-20 near the southern Sargasso Sea.

3. The difference in mean summer temperatures between the Holocene Epoch and the last glaciation was very small. At site V16-20, there was a very slight increase (−0.02°C) from the last glacial mean T_s to the Holocene mean T_s. Prell (1974) noted a similar climatic trend in the Caribbean Sea. At the other locations, the increase in T_s ranged from 0.3° to 1.7°C.

4. The increase in mean winter temperatures from the last glaciation to the Holocene Epoch amounts to 1.6°C., whereas the increase in mean summer temperatures is 0.6°C. Thus, the overall temperature contrast between the glacial and interglacial western equatorial Atlantic surface waters is small.

Preservation and Dissolution of Planktonic Foraminiferal Assemblages

The factors controlling the dissolution of planktonic foraminifera have been studied by Berger (1968, 1970, 1971). The degree of dissolution and preservation of planktonic foraminifera in our sediment samples was evaluated from the following characteristics: $CaCO_3$ content; relative abundance of shell fragments of planktonic foraminifera; relative abundance of solution-resistant and solution-prone species of planktonic foraminifera; relative abundance of benthic foraminifera; and relative abundance of pteropods.

Each factor alone does not reveal the total extent of dissolution; however, a combination of these interrelated factors provides a qualitative method for estimating the degree of dissolution that a planktonic foraminiferal assemblage has undergone.

Calcium Carbonate Content. It is not possible to determine quantitatively the amount of carbonate dissolution within a given interval of sediment merely by determining the total $CaCO_3$ content of the sediment, because two additional factors—productivity of calcareous organisms and dilution by noncalcareous sediment—also affect the ultimate $CaCO_3$ composition of pelagic sediment. However, sediment deposited in deep abyssal regions is attacked by cold bottom water and undergoes extensive dissolution. In the western equatorial Atlantic, pelagic sediment deposited in depths shallower than approximately 4,200 m shows little or no carbonate dissolution and consists of foraminiferal marl and ooze ($CaCO_3$ >25%). Sediment deposited at depths between 4,200 m and 5,000 m shows intervals of moderate to high dissolution and is composed of alternating beds of marl ($CaCO_3$ >25%) and brown clay ($CaCO_3$ >25%). Planktonic foraminiferal tests are sparse in brown clay, and if present at all, they are generally only fragments of very resistant species, which implies extensive dissolution. Brown clay predominates in cores V25-42 (4,704 m) and V16-15 (4,254 m). The low $CaCO_3$-coarse-fraction content of the Y zone seems to indicate greater carbonate dissolution during the last glaciation. Maximum dissolution occurred during the earliest part of the last glaciation just above the X-Y boundary; it is recorded in these two cores by an interval of brown clay with very low $CaCO_3$ (<10%) and coarse-fraction (0 to 2%) content (Fig. 2). Pelagic sediment deposited in depths greater than 5,000 m exhibits complete dissolution and consists entirely of homogeneous brown clay.

Relative Abundance of Shell Fragments of Planktonic Foraminifera. Because the number and size frequency of shell fragments correspond in a general manner to the degree of carbonate dissolution (Phleger and others, 1953), we have attempted to make a visual estimation of the percentage of fragmented shells in a total assemblage of (fragmented and unfragmented) planktonic foraminiferal shells. Thus, a sample containing 0 to 24% test fragments is designated as preservation scale 1 (little or no dissolution); 25 to 49% test fragments is scale 2 (moderate dissolution); 50 to 74% test fragments is scale 3 (severe dissolution); and 75 to 100% test fragments is scale 4 (very severe to complete dissolution). The visual estimation method is subjective and might underestimate the smaller fragments; however, actual counting of the smaller fragments is too time consuming, if not counterproductive. We recommend the method only for a gross approximation of the preservation state of planktonic foraminifera.

Figure 11 shows that two of the relatively shallow hemipelagic cores, RC9-49 (1,851 m) and V25-56 (3,512 m), have the least fragmentation. Progressively worse preservation occurs in the following cores in order of increasing fragmentation: V22-26 (3,270 m), V25-59 (3,812 m), V16-20 (4,540 m), V16-25 (4,254 m), and V15-168 (4,219 m). The relatively high organic content in the hemipelagic core V15-168 may be responsible for the high degree of dissolution in this core.

Relative Abundance of Solution-Resistant and Solution-Prone Species of Planktonic Foraminifera. From field and experimental observation, Ruddiman and Heezen (1967) and Berger (1967) have shown that the dissolution process attacking planktonic foraminifera is selective with respect to species, shell size, shell thickness, and shell porosity. Berger (1968, 1970) ranked the foraminiferal species according to their relative solubility; he showed, in general, that the nonspinose *Globorotalia*, *Globoquadrina*, and *Pulleniatina* species are more solution-resistant than the spinose *Globigerina*, *Globigerinella*, and *Globigerinoides* species.

In our samples, a low degree of dissolution is indicated by high percentages of *Globigerinoides ruber*, *G. sacculifer*, *Globigerinella aequilateralis*, and *Orbulina universa* and low percentages of *Globorotalia menardii*, *G. truncatulinoides*, *Globoquadrina dutertrei*, and *Pulleniatina obliquiloculata*. Examples of such assemblages are found at 185 cm and 195 cm in core V16-25 (Fig. 5). By contrast, inverse proportions of these two groups are indicative of a high degree of dissolution; such assemblages are found at 245 cm in core V16-25.

Berger (1968) noted that dissolution tends to reduce the percentage of more fragile warm-water species (for example, *G. ruber*) and to increase the percentage of more resistant cool-water or deep-water species (for example, *G. dutertrei*, *G. menardii*, and *G. truncatulinoides*). This selective process would give a "colder aspect" to a faunal assemblage in warm-water regions and is an important factor to be considered for the high-dissolution zones in the lower last glacial sediment of cores V15-168, V25-59, and V16-25, when winter paleotemperatures were anomalously depressed.

Relative Abundance of Benthic Foraminifera. In general, deep-water benthic foraminifera are more solution resistant than planktonic foraminiferal shells (Oba, 1969). Some factors that may account for this are the higher magnesium content, smaller pores, and perhaps more robust ultrastructure of benthic foraminifera. Low percentages of benthic species indicate a masking effect by planktonic foraminifera, whereas high percentages of benthic species reflect the removal of planktonic shells by dissolution. Thus, the percentage of benthic foraminifera in a sample of more than 300 specimens of planktonic and benthic foraminifera can be used as a relative index of $CaCO_3$ dissolution. In pelagic sediment, the presence of more than 2% benthic foraminifera indicates moderate to high dissolution (Figs. 3 through 9, 11).

Relative Abundance of Pteropods. Because aragonitic shells of pteropods are more susceptible to dissolution than the calcitic shells of planktonic foraminifera, a large

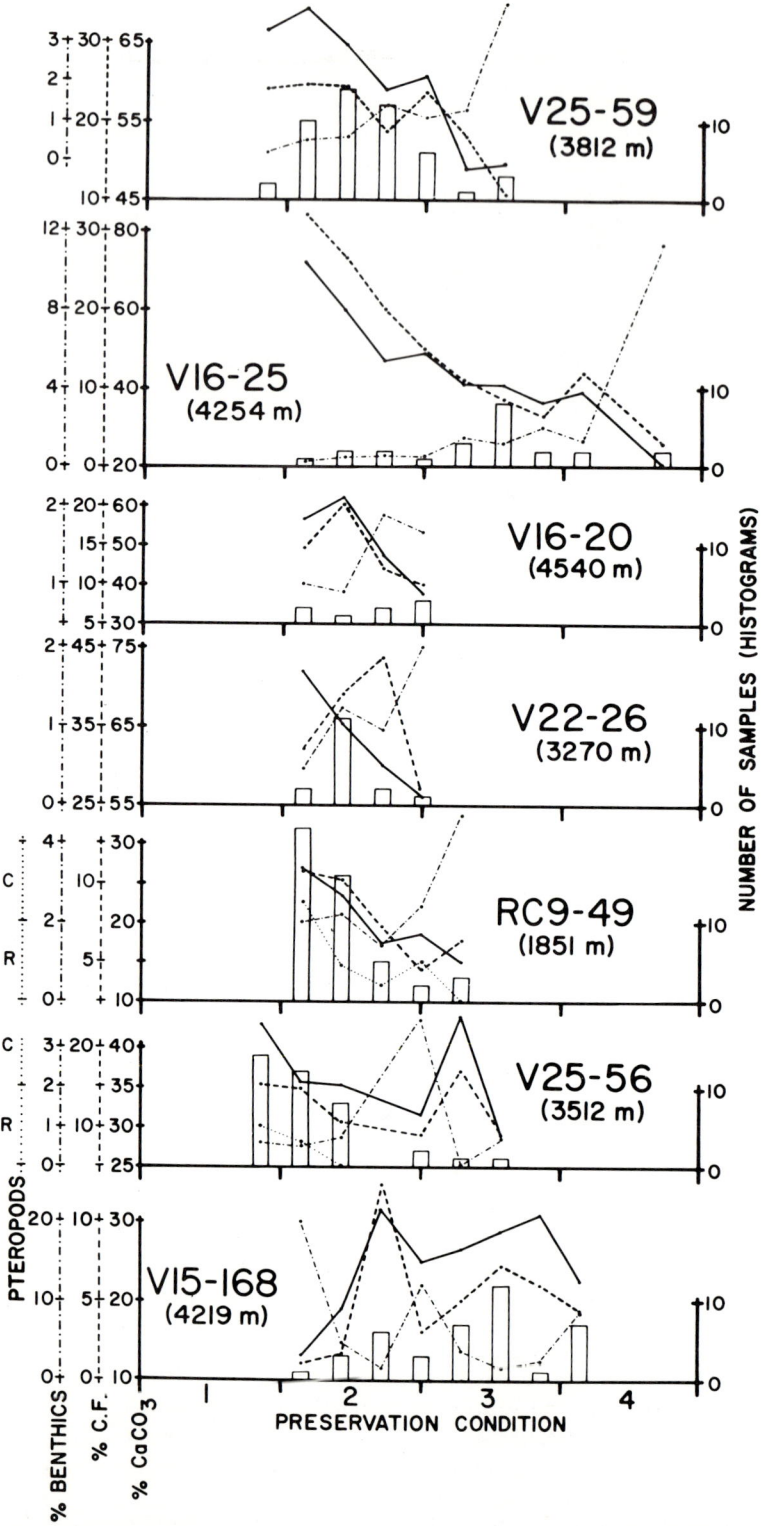

Figure 11. Comparison of mean values of $CaCO_3$, coarse fraction, benthic foraminifera, and pteropods for samples ranked in order of their preservation condition.

diversity of species and high percentages of unbroken shells of pteropods are associated with excellent preservation conditions. In the tropical and subtropical North Atlantic Ocean, pteropod ooze occurs in depths shallower than the aragonite compensation depth of about 2,200 m (Chen, 1964, 1971).

Pteropod shells are absent in five cores (V22-26, V25-59, V16-20, V16-25, and V15-168). Pteropods are present in cores V25-56 and RC9-49, and their abundance fluctuations are shown in Figures 7 and 9. Core RC9-49 (1,851 m) was raised from a depth shallower than the aragonite compensation depth. Core 25-56 (3,512 m) has pteropod shell remains only in the Z and X zones and a short interval in the middle of the Y zone.

**Correlation between Preservation Condition,
$CaCO_3$-Coarse-Fraction Values,
Benthic Foraminifera, and Pteropoda**

The interrelationships between preservation condition, $CaCO_3$ and coarse-fraction values, and the relative abundances of benthic foraminifera and pteropods are shown in Figure 11. By grouping the samples of each core according to their preservation condition and using the four-scale method described above, mean values of the latter four parameters were calculated for all samples having similar preservation states *regardless* of age. The percentage of fragmented shells of planktonic foraminifera was determined twice by one of us (Bé) and checked independently by another (Free), and the combined results allowed each sample to be plotted along smaller subdivisions within the four-scale range. The histograms indicate the total number of samples in each preservation category ranked in order of preservation condition.

In five of the seven cores, $CaCO_3$, coarse-fraction, and pteropod abundances decrease, whereas relative abundance of benthic foraminifera increases as the preservation condition deteriorates. These relationships are particularly distinct in cores V25-59, V16-25, V16-20, and RC9-49 but are less distinct in core V22-26. In the two hemipelagic cores, V15-168 and V25-56, these trends are erratic because of terrigenous influx and displaced benthic foraminifera from the continental shelf.

The fact that $CaCO_3$, coarse-fraction, and pteropod abundances are inversely related to benthic foraminifera abundance and preservation condition (regardless of sample age) strongly suggests that dissolution activity is an important factor in controlling the $CaCO_3$ content of the sediment.

Maximum dissolution activity seems to have occurred during the early part of the last glaciation. In six of the seven cores, minimum $CaCO_3$ and coarse-fraction values correspond with peak frequencies of benthic foraminifera and shell fragmentation in the lowermost Y zone (about 70,000 to 73,000 B.P.; Figs. 3 through 9). This period of intense dissolution activity at the beginning of the last glaciation coincides with the abrupt climatic change reflected by the transition from a predominantly subtropical foraminiferal assemblage to one showing a considerable influx of cool-water species. In addition, cores V25-42 and V16-25 contain brown clay ($CaCO_3$ <10%), and planktonic foraminifera are absent or very rare in this interval (Fig. 2). The cores also show a similar but less severe episode of dissolution activity during the last part of the last glaciation (upper Y zone, approximately 20,000 to 14,000 B.P.)

Intervals of least dissolution activity occurred during the Holocene Epoch and the middle part of the last glaciation and correspond with $CaCO_3$ maximums. In cores V25-56 and RC9-49 these periods also coincide with peak pteropod abundances.

The above relationships suggest that dissolution activity controls the degree of preservation of planktonic foraminiferal shells and strongly influences the $CaCO_3$ variations. Increased dissolution activity was probably caused by relative increases in water-mass circulation (especially of the Antarctic Bottom Water) during glacial maximums.

Other evidence of episodes of increased dissolution activity have been reported for various regions of the North Atlantic. Mörner (1972b, Fig. 3, p. 345) noted that core 280A from the central North Atlantic contains red clay at core levels assigned to Emiliani's (1955) stages 2 and 4. Kennett and Huddlestun (1972a, 1972b) presented faunal evidence of an abrupt climatic change at 90,000 B.P. that coincided with lowering of the lysocline in the Gulf of Mexico. Gardner (1973, 1975) reported that carbonate dissolution was more intense during the glacial than the interglacial epochs in the eastern equatorial Atlantic. Damuth (1973, 1975) presented evidence of increased circulation and water-mass extent during glacial periods for the western equatorial Atlantic.

Relationship of Terrigenous Sedimentation to Late Quaternary Climate Change

Examination and biostratigraphic zonation of approximately 250 piston cores raised from the continental slope and rise, the Amazon Cone, and abyssal plains have revealed that terrigenous sedimentation was modulated by climatic fluctuations during late Quaternary time (Damuth and Fairbridge, 1970; McGeary and Damuth, 1973; Damuth, 1973; Damuth and Kumar, 1975). Two types of terrigenous sediment are recognized in the cores on the basis of the process of deposition: hemipelagic and redeposited. Hemipelagic sediment is composed of gray, terrigenous clay (often rich in silt) that contains abundant disseminated organic detritus. This sediment contains a pelagic biogenic component, which indicates that the sediment was deposited by a slow, continuous process (5 to 25 cm/10^3 yr) rather than episodically. Redeposited sediment, which consists largely of terrigenous silt, sand, and gravel, is commonly interbedded with the hemipelagic sediment. Redeposited beds are as much as several metres thick. The composition and structure of most beds indicate that transportation and deposition were by turbidity currents.

Biostratigraphic zonation of the hemipelagic sediment using planktonic foraminifera (*Globorotalia menardii* complex and *Pulleniatina obliquiloculata*) has revealed the relationship between terrigenous sedimentation and climatic fluctuations. It has permitted the calculation of sedimentation rates for the continental rise and abyssal plains.

Holocene Sedimentation. The recession of continental glaciers at the end of the Holocene Epoch (approximately 11,000 B.P.) caused sea level to rise approximately 100 m (Flint, 1971). The sea-level rise displaced the locus of river sedimentation from the shelf break landward as much as 350 km. Thus, coarse sediment (silt, sand, and gravel) is now trapped on the inner continental shelf, because the low gradient prevents seaward movement of the sediment beyond the 20 to 40-m contours. In addition, strong northwesterly longshore currents transport nearly all suspended clay and silt (which are not trapped in river estuaries) along the South American coast. The sediment is confined to the inner shelf and is gradually deposited along the coast as far northwest as the Orinoco River (Reyne, 1961; Diephuis, 1966; Allersma, 1971). The wide continental shelf and the strong northwesterly currents have thus prevented nearly all terrigenous sediment discharged by the Amazon and smaller rivers from reaching the deep ocean throughout the Holocene Epoch. This is reflected by the Holocene sections (Z zone) of piston cores, which consist entirely of light-brown, high-carbonate pelagic foraminiferal marl and ooze, raised from the continental rise, Amazon Cone, and abyssal plains.

Holocene (Z zone) sedimentation rates are shown in Figure 12. Highest rates (7 to 14 cm/10^3 yr) are observed on the continental slope and uppermost continental rise and apparently reflect relatively high organic production. Rates on the middle to lower continental margin generally range from 4 to 7 cm/10^3 yr, whereas rates on the abyssal provinces (<4,500 m) and the Mid-Atlantic Ridge generally range from 2 to 4 cm/10^3 yr. Rates of less than 2 cm/10^3 yr are observed in deep (>5,000 m) abyssal regions.

Sedimentation during the Last Glaciation. The expansion of continental glaciers during the last glaciation lowered sea level at least 100 m (Flint, 1971). During this time the continental shelf along northeast South America was partially or entirely emergent, and rivers discharged their sediment directly into the heads of submarine canyons, where the sediment could easily be transported to abyssal depths by gravity-controlled sediment flows. Thus, huge volumes of terrigenous sediment derived primarily from the Amazon River were continuously deposited on the continental rise, Amazon Cone, and abyssal plains throughout the last glaciation. As a result, gray hemipelagic sediment comprises the Y zones of all cores from these regions and is interrupted only by redeposited silt-sand beds.

Sedimentation rates for the Y zone are shown in Figure 13. Cores containing

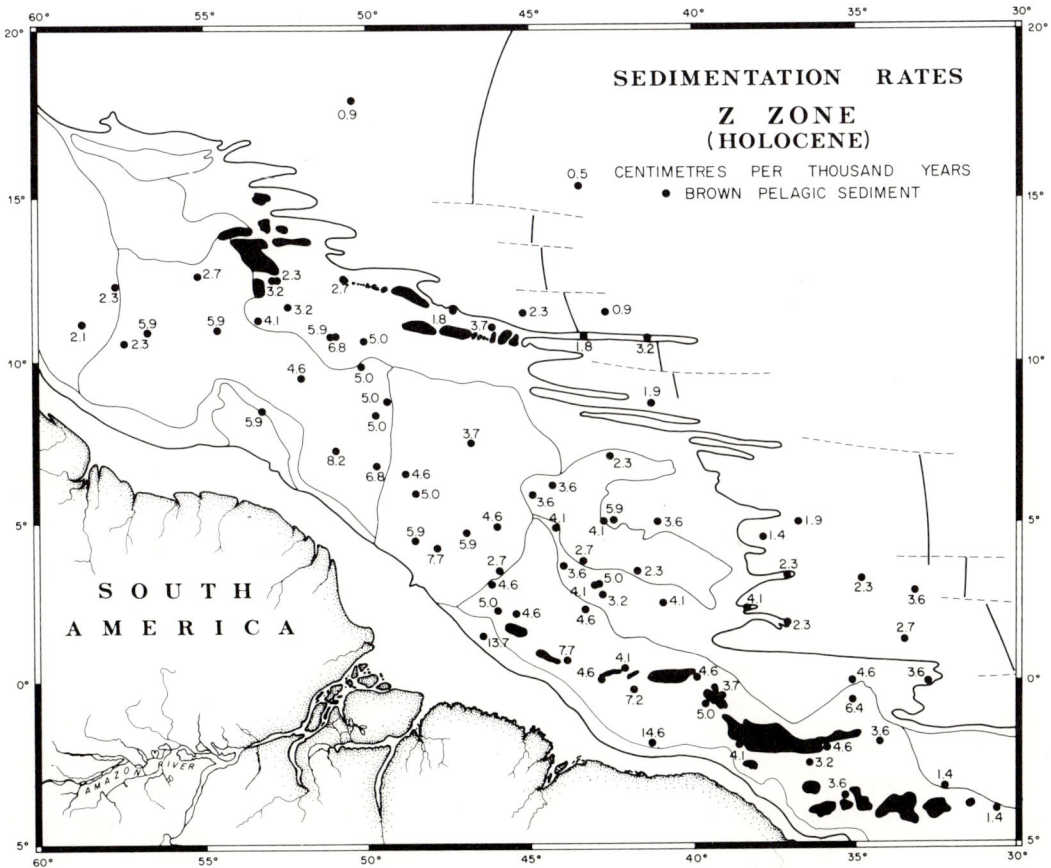

Figure 12. Sedimentation rates during the Holocene Epoch (Z zone) (modified from Damuth, 1973).

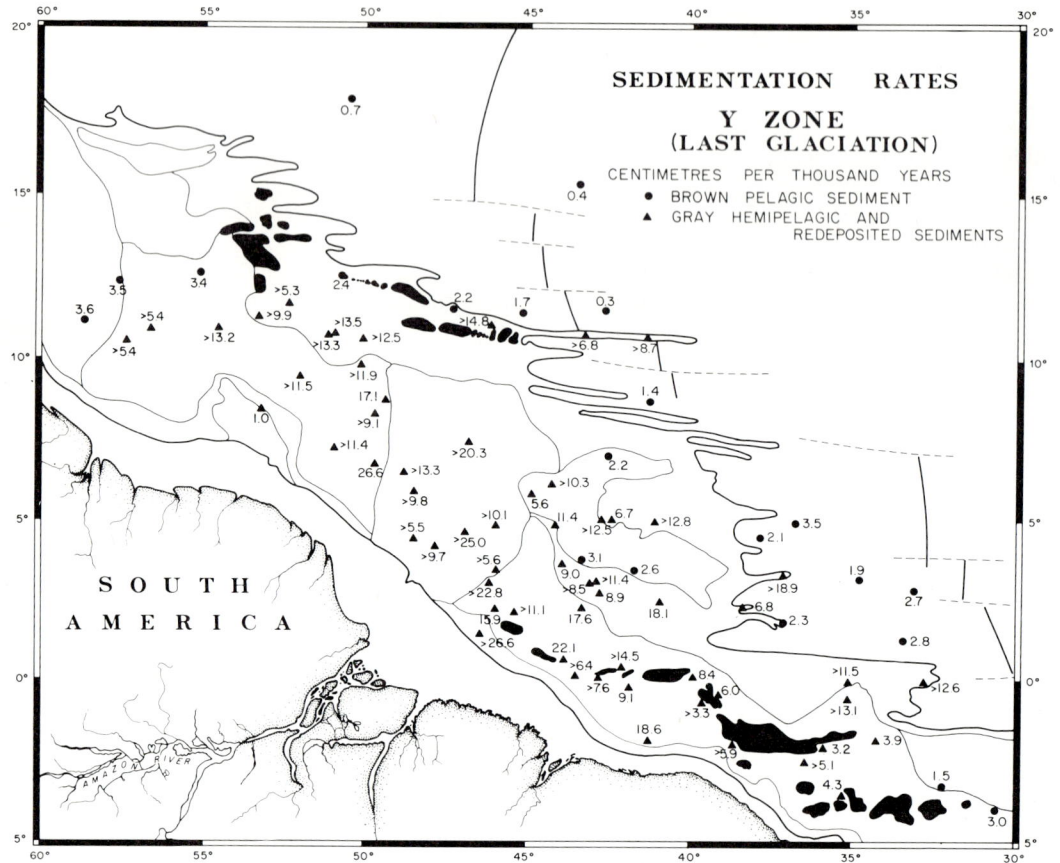

Figure 13. Sedimentation rates during the last glaciation (Y zone) (modified from Damuth, 1973).

Y zones composed of gray terrigenous sediment (continental margin provinces and abyssal plains) are distinguished from cores containing brown pelagic sediment. Many cores from the continental margin and abyssal plains do not penetrate the entire Y zone. For these cores only minimal rates could be calculated (Damuth, 1973). Sedimentation rates on the continental rise, Amazon Cone, and abyssal plains range from approximately 4 to more than 25 cm/10^3 yr. Comparison of these rates with Z zone rates for the same cores (Figs. 12, 13) clearly reveals that increased sedimentation rates were prevalent during the Wisconsin glaciation because of the great influx of terrigenous sediment. Sedimentation rates for brown pelagic sediment in the Y zone of the Mid-Atlantic Ridge, abyssal provinces, and topographic highs normally ranged from 1 to 4 cm/10^3 yr (Fig. 13); these are quite comparable to pelagic rates observed for these provinces in Holocene sediment (Fig. 12).

Data from the piston cores demonstrate that terrigenous sediment accumulated continuously on the continental margin and abyssal plains throughout the last glaciation; however, the postglacial sea-level rise (approximately 11,000 B.P.) abruptly shut off the supply of terrigenous sediment. This cessation was marked by the formation of a thin (<10 cm) rust-colored, iron-rich crust on most of the continental margin and abyssal plains (McGeary and Damuth, 1973).

Sedimentation during the Last Interglaciation (X Zone). Unfortunately, most piston cores from the continental margin and abyssal plains only penetrate sediment of the Z and Y zones because of the high sedimentation rates. A few cores from the continental rise (including V15-168 and V25-56) in the region of the North Brazilian Ridge do penetrate the entire X zone (see Damuth, 1973, for a more complete discussion of these cores). These cores reveal that terrigenous hemipelagic clay and redeposited silt, sand, and gravel accumulated throughout most of the last interglaciation (X zone), just as they did during the last glaciation. The only break in terrigenous sedimentation took place at the beginning of the last glaciation (approximately 125,000 B.P.) and is marked in the cores by the occurrence in the lowermost X zone of a thin bed (~30 cm) of light-brown pelagic sediment. In some instances this pelagic bed contains thin rust-colored laminae at the base, which may be genetically similar to the rust-colored crusts that mark the Y-Z boundary. These pelagic beds at the base of the X zone apparently represent a period similar to the Holocene Epoch when a glacio-eustatic sea-level rise briefly halted terrigenous sedimentation. Paleotemperature estimates presented above show that sea-surface temperatures were as warm as present-day temperatures *only* during this short interval at 125,000 B.P.

As previous studies (for example, Kukla, 1972; Mörner, 1972a; Broecker and others, 1968) as well as $CaCO_3$ fluctuations in western equatorial Atlantic sediment (Damuth, 1973, 1975; and this paper) have demonstrated, the last interglaciation was not one continuous warm period, but rather a series of interstades and stades. Recent studies of ancient Barbados coral terraces suggest that during the last interglaciation sea level was elevated as high or higher than the present level only at the very beginning of the interglaciation (approximately 125,000 B.P.; Steinen and others, 1973) and that during the rest of the last interglaciation, sea level reached only to within approximately 15 m of the present level during two brief periods (approximately 82,000 and 103,000 B.P.). Most of the last interglaciation was thus characterized by a sea level that was moderately to considerably lower than at present, and therefore, deposition of terrigenous sediment during this interval is to be expected.

If the glacial mode predominated throughout the major part of any glacial-interglacial cycle as hypothesized by Broecker and van Donk (1970), then high stands of sea level similar to the present would be limited to relatively short intervals (10,000 to 20,000 yr) at the beginning of each glacial-interglacial cycle. Thus, throughout the major part of the last interglaciation (X zone) and last glaciation (Y zone), sea level would have been considerably lower than it is now, and most of the continental shelf would have been emergent. If, during this period of low sea level, rivers could not flow all the way to the shelf break, they could have easily extended their deltas to the shelf break after a short period of time and thus provided an uninterrupted supply of terrigenous sediment to the deep ocean.

Data from piston cores raised from the continental margin and abyssal plains suggest that the glacial mode did predominate throughout most of the last glacial-interglacial cycle of the Pleistocene Epoch. Although both the deep-sea and continental sedimentary records reveal that several relatively warm periods or interstades occurred during this period, the magnitude and duration of each associated sea-level rise was apparently not sufficient to stop the supply of terrigenous sediment to the deep ocean, as did the sea-level rises at the beginning of the last interglaciation (termination II of Broecker and van Donk, 1970) and during the Holocene Epoch (termination I).

The terrigenous sedimentary record of the western equatorial Atlantic thus suggests that during a complete glacial-interglacial cycle, the glacial mode predominates throughout most of the cycle, with interglacial conditions such as at present restricted only to a brief interval (10,000 to 20,000 yr) at the beginning (or end) of each cycle. The terrigenous sedimentary record is in agreement with fluctuations of climate inferred from the sea-surface-temperature estimates for the past 125,000 yr (this study) as well as sea-level fluctuations inferred by other investigators.

Relationship of $CaCO_3$ Fluctuations to Quaternary Climatic Events

Because many investigators (Schott, 1935; Correns, 1937; Kullenberg, 1953; Wiseman, 1954, 1965; Broecker and others, 1958; Turekian, 1965; Olausson, 1967; Needham and others, 1969; Broecker, 1971; Ruddiman, 1971; Hays and Peruzza, 1972) have demonstrated that a relationship exists between apparent rates of $CaCO_3$ accumulation in Atlantic sediment and Quaternary climatic fluctuations, it is of interest to determine to what extent the $CaCO_3$ fluctuations correlate with warm and cold periods of the past 130,000 yr as inferred from upper Quaternary continental stratigraphy of eastern North America.

Figure 14 shows the correlation between the $CaCO_3$ fluctuations of core V25-59 and recognized stadial-interstadial periods of North America. A fairly accurate time scale for core V25-59 was needed to make this correlation and was developed in the following manner. Dates for the Z-Y (11,000 B.P.), Y-X (75,000 B.P.), and X-W (127,000 B.P.) boundaries are from Broecker and van Donk (1970). A constant sedimentation rate was assumed for the entire X zone, and dates were assigned by interpolation. Radiocarbon dates were obtained from three levels within the upper Y zone and are listed in Table 2. The radiocarbon dates reveal that the apparent sedimentation rate within the Y zone was not constant, but was greater during the 40,000- to 11,000-B.P. interval than prior to 40,000 B.P. This apparent increase in sedimentation rate in the upper 1 m of the core probably is a result of the sediment not being completely dewatered yet or not having yet been compacted the same amount as the sediment in the rest of the core. Constant sedimentation rates had to be assumed for each interval between the Z-Y boundary, radiocarbon-dated levels, and the Y-X boundary. Dates were then assigned by interpolation.

Broecker and others (1968) and Mesolella and others (1969) have radiometrically dated raised coral-reef terraces on Barbados at approximately 82,000, 105,000, and 125,000 B.P. and thus have demonstrated that three high sea-level stands occurred during the last interglaciation. These high sea-level stands are plotted in Figure 14 and correlate in age with three prominent carbonate peaks (VI, VII, and VIII) that characterize the X zone. Steinen and others (1973) recently presented evidence from the subsurface of Barbados for a low sea-level stand (that is, cold period) between 105,000 and 125,000 B.P. This low sea-level stand is represented by a sharp depression in the carbonate curve between peaks VII and VIII. In addition, James and others (1971) reported the existence of a 60,000-yr-old reef terrace, which implies a high sea-level stand during the middle Wisconsin glaciation (last glaciation) and approximately correlates with peak V of the carbonate curve (Fig. 14).

Climatic fluctuations during the Wisconsin glaciation have been inferred from continental deposits in the eastern Great Lakes-St. Lawrence region of North America. These deposits record ice-margin fluctuations of the Laurentide Ice Sheet, which expanded or shrank in response to climate oscillations (Goldthwaite and others, 1965; Dreimanis, 1969; Dreimanis and Karrow, 1972). Mörner (1972a) proposed a chronology for the past 130,000 yr for the cold and warm periods that is inferred from these deposits and is based on astronomic dates for the solar insolation fluctuations presented by Kukla (1969). Mörner's proposed stadial-interstadial periods and subdivisions for the Wisconsin glaciation and last interglaciation are plotted in Figure 14 along the time scale for core V25-59. This plot reveals an excellent correlation between the stades and interstades (ice-margin fluctuations) recognized in continental deposits and the $CaCO_3$ fluctuations of western equatorial Atlantic sediment. $CaCO_3$ highs correlate peak for peak with interstades (warm periods), whereas carbonate depressions correlate with stades (cold periods).

The large $CaCO_3$ depression of the upper one-third of the Y zone correlates with the late Wisconsin glacial maximum. Peak II within this depression probably represents the Erie Interstade (approximately 15,500 to 16,500 B.P.). The high-carbonate interval of the middle one-third of the Y zone correlates peak for peak with the series of middle Wisconsin interstades. Peak III is equivalent to the Plum Point Interstade, peak IV records the Port Talbot II Interstade, and peak V correlates with the Port Talbot I Interstade. Depressions between peaks III to V correlate

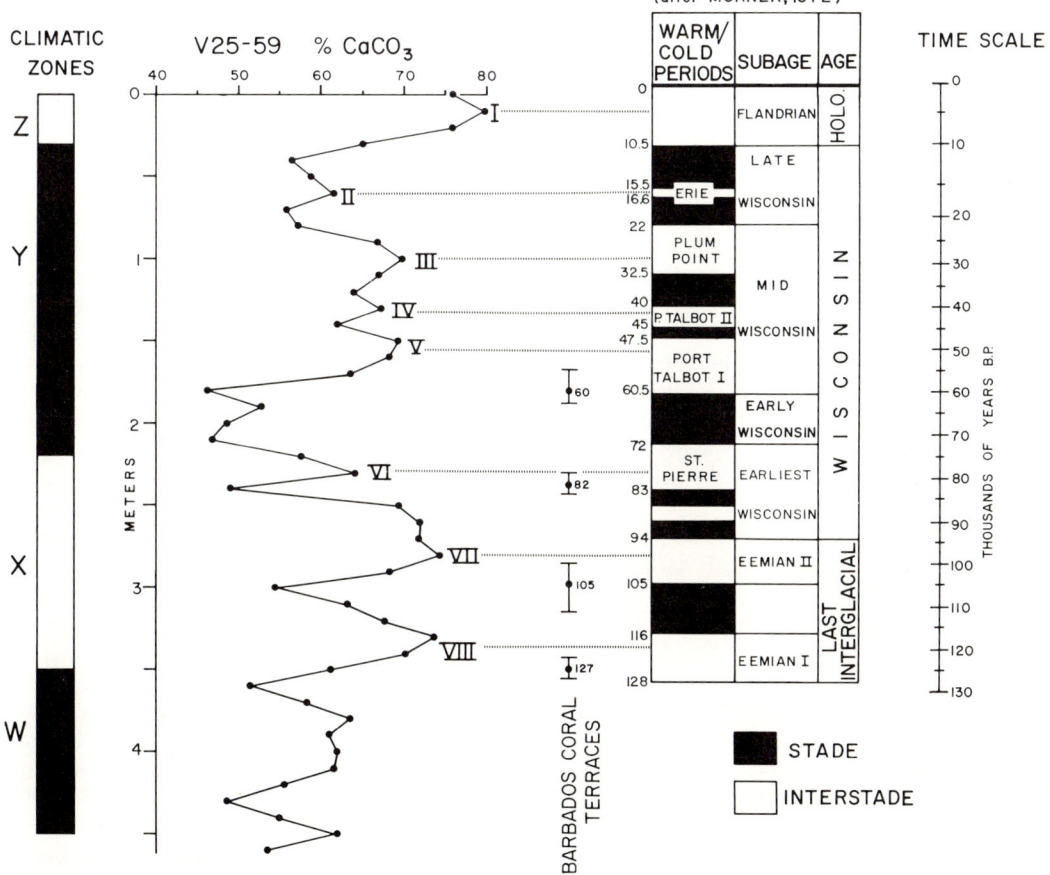

Figure 14. Correlation of $CaCO_3$ fluctuations in western equatorial Atlantic sediment, as represented by core V25-59, during the last 130,000 yr with radiometrically dated high sea-level stands reported for Barbados by Broecker and others (1968), Mesolella and others (1969), and James and others (1971) and with stadial-interstadial periods (ice-margin fluctuations) proposed by Mörner (1972a) for the eastern Great Lakes region of North America. *G. menardii* climatic zones for core V25-59 are shown on the left.

with short stades. The prominent carbonate depression of the lower Y zone correlates with the early Wisconsin glacial maximum. Peak VI correlates with the Barbados I (82,000 B.P.) high sea-level stand as well as the St. Pierre Interstade of the earliest part of the Wisconsin glaciation. Peaks VII and VIII correlate with the Eemian II and Eemian I warm periods, respectively, of the last interglaciation. Depressions between these peaks correlate with stades (Fig. 14).

The consistent reproducibility of the $CaCO_3$ fluctuations in western equatorial Atlantic sediment and their excellent correlation with warm and cold periods inferred from continental stratigraphy and radiometrically dated high sea-level stands suggest that these $CaCO_3$ fluctuations accurately reflect the magnitude and timing of stadial-interstadial periods and Northern Hemisphere ice-margin fluctuations during the past 130,000 yr. The continental stratigraphy is somewhat uncertain because of gaps in the sedimentary record and limitations in radiocarbon dating methods. The fact that western equatorial Atlantic sediment provides a continuous sedimentary record and that $CaCO_3$ fluctuations in this sediment apparently correlate with the continental record offers encouragement that carbonate determinations in deep-sea sediment may provide a continuous, accurate record of climatic fluctuations for at least the late Quaternary time.

CONCLUSIONS

The planktonic foraminiferal record of seven deep-sea cores from the western equatorial Atlantic Ocean indicates that the environmental contrast between the Holocene Epoch, the last glaciation, and the last interglaciation is relatively slight and that the climate retained a decidedly tropical aspect throughout the past 80,000 yr. Some cooling during the last glaciation is evident, but the difference between glacial and postglacial winter sea-surface temperatures amounted to less than 1.7°C at five core sites. A somewhat greater temperature contrast was noted at the other two core locations (V25-56 and V15-168). These two sites are closest to the Equator. During the last glaciation the faunal assemblages had a higher proportion of cool-water species than during the last interglaciation or the Holocene Epoch. The cool-water species are believed to have been transported from the eastern equatorial Atlantic Ocean by an intensified Equatorial Current.

At the Equator (core V25-59), conditions as warm as during the Holocene Epoch occurred only once in the past 127,000 yr, at the beginning of the last interglaciation (approximately 125,000 B.P.).

Piston cores from the continental rise, Amazon Cone, and abyssal plains consist largely of gray hemipelagic and redeposited sediment that is rich in terrigenous detritus. Deposition of terrigenous sediment was continuous throughout the last glaciation and most of the last interglaciation. However, sea-level rise at the beginning of the Holocene Epoch interrupted the supply of terrigenous sediment to these provinces, and as a result, light-brown foraminiferal ooze marks the Holocene sections of the cores. The only other time that deposition of terrigenous sediment was interrupted during the past 175,000 yr was briefly at the beginning of the last interglaciation (approximately 125,000 B.P.). This interval coincides with the only time during the past 175,000 yr when sea-surface temperatures were as great as those of the Holocene Epoch; it indicates that a brief period of warm climate similar to the Holocene occurred at the beginning of the last glacial–interglacial cycle.

Calcium carbonate determinations for the piston cores demonstrate that carbonate fluctuations during the past 130,000 yr are time stratigraphic throughout the western equatorial Atlantic Ocean. Comparison of the timing of the carbonate fluctuations with the timing of stadial-interstadial periods reported for the Laurentide Ice Sheet in the eastern Great Lakes region of North America reveals an excellent correlation. Interstadial (warm) periods correlate with carbonate maximums, whereas stadial (cold) periods correlate with carbonate minimums.

ACKNOWLEDGMENTS

We are grateful to William Hutson, John Imbrie, Brian Tucholke, Michael Roche, and Andrew McIntyre for helpful discussions, and to Mary Mathews for drafting assistance. This study was supported by National Science Foundation Grants IDO-71-04204 and GA-38276X. During this research J. E. Damuth was aided by Office of Naval Research Contract 0014-67-A-0108-0004 and National Science Foundation Grant GA-27281. The sediment cores are from the Lamont-Doherty Geological Observatory Core Library, which is maintained by Office of Naval Research Contract 0014-67-A-0108-0004 and National Science Foundation Grant DES-72-01568.

REFERENCES CITED

Allersma, E., 1971, Mud on the oceanic shelf off Guiana, in Symposium on investigations and resources of the Caribbean Sea and adjacent regions: New York, UNESCO, p. 193–203.
Barash, M. S., 1971, The vertical and horizontal distribution of planktonic foraminifera in Quaternary sediments of the Atlantic Ocean, in Funnell, B. M., and Riedel, W. R., eds., The micropaleontology of oceans: London, Cambridge Univ. Press, p. 433–441.
Bé, A.W.H., 1967, Foraminifera, families: Globigerinidae and Globorotaliidae, in Fraser, J. H., ed., Fiches d'identification du zooplancton: Charlottenlund, Denmark, Conseil Permanent Internat. pour L'Exploration de la Mer, no. 108, 8 p.
Bé, A.W.H., and Tolderlund, D. S., 1971, Distribution and ecology of living planktonic foraminifera in surface waters of the Atlantic and Indian Oceans, in Funnell, B. M., and Riedel, W. R., eds., The micropaleontology of oceans: London, Cambridge Univ. Press, p. 105–149.
Berger, W. H., 1967, Foraminiferal ooze: Solution at depths: Science, v. 156, p. 383.
——1968, Planktonic foraminifera: Selective solution and paleoclimatic interpretation: Deep-Sea Research, v. 15, p. 31–43.
——1970, Planktonic foraminifera: Selective solution and the lysocline: Marine Geology, v. 8, p. 111–138.
——1971, Sedimentation of planktonic foraminifera: Marine Geology, v. 11, p. 325–358.
Boltovskoy, E., 1964, Distribución de los foraminiferos planctonicos vivos en el Atlántico Equatorial, parte oeste (Expedición Equalant): Argentina Servicio Hidrografía Naval Publ. H639, p. 1–54.
Briskin, M., and Berggren, W. A., 1975, Pleistocene stratigraphy and quantitative paleooceanography of tropical North Atlantic core V16-205; in Saito, T., and Burckle, L. H., eds., Late Neogene Epoch boundaries: Am. Mus. Nat. History, Micropaleontology Press Spec. Pub. 1, p. 167–198.
Broecker, W. S., 1971, Calcite accumulation rates and glacial to interglacial changes in ocean mixing, in Turekian, K. K., ed., The late Cenozoic glacial ages: New Haven, Yale Univ. Press, p. 239–265.
Broecker, W. S., and van Donk, J., 1970, Insolation changes, ice volumes and the O^{18} record in deep-sea cores: Rev. Geophysics and Space Physics, v. 8, p. 169–198.
Broecker, W. S., Turekian, K. K., and Heezen, B. C., 1958, The relation of deep sea sedimentation rates to variation in climate: Am. Jour. Sci., v. 256, p. 503–517.
Broecker, W. S., Thurber, D. L., Goddard, J., Ku, T.-L., Matthews, R. K., and Mesolella,

K. J., 1968, Milankovitch hypothesis supported by precise dating of coral reefs and deep-sea sediments: Science, v. 159, p. 297–300.

Chen, C., 1964, Pteropod ooze from the Bermuda pedestal: Science, v. 144, p. 60–62.

——1971, Occurrence of pteropods in pelagic sediments, in Funnell, B. M., and Riedel, W. R., eds., The micropaleontology of oceans: London, Cambridge Univ. Press, p. 351.

Cifelli, R., and Smith, R. K., 1970, Distribution of planktonic foraminifera in the vicinity of the North Atlantic Current: Smithsonian Contr. Paleobiology, v. 4, p. 1–52.

Correns, C. W., 1937, Die Sedimente des Aquatorialen Atlantischen Ozeans: Wissenschaft. Ergebnisse der Deutschen Atlantischen Expedition auf dem Forschungs- und Vermessungsschiff "Meteor," 1925-1927, v. 3, pt. 3, p. 1–298.

Cushman, J. A., and Henbest, L. G., 1941, Geology and biology of North Atlantic deep-sea cores between Newfoundland and Ireland. Pt. 2, Foraminifera: U.S. Geol. Survey Prof. Paper 196-A, p. 35–56.

Damuth, J. E., 1973, The western equatorial Atlantic: Morphology, Quaternary sediments and climatic cycles [Ph.D. thesis]: New York, Columbia Univ., 602 p.

——1975, Quaternary climate change as revealed by calcium-carbonate fluctuations in western equatorial Atlantic sediments: Deep-Sea Research, v. 22, p. 725–743.

Damuth, J. E., and Fairbridge, R. W., 1970, Equatorial Atlantic deep-sea arkosic sands and ice-age aridity in tropical South America: Geol. Soc. America Bull., v. 81, p. 189–206.

Damuth, J. E., and Kumar, N., 1975, Amazon Cone: Morphology, sediments, age, and growth pattern: Geol. Soc. America Bull., v. 86, p. 863–878.

Diephuis, J.G.H.R., 1966, The Guiana coast: Ned. Aardrijksk. Genoot. Tijdschr., v. 83, p. 145–152.

Dreimanis, A., 1969, Late Pleistocene lakes in the Ontario and the Erie Basins: 12th Conf. Great Lakes Research, Proc., p. 170–180.

Dreimanis, A., and Karrow, P. F., 1972, Glacial history of the Great Lakes–St. Lawrence region, the classification of the Wisconsin Stage, and its correlatives: Internat. Geol. Cong., 24th, Montreal 1972, sec. 12, p. 5–15.

Edgar, T., and Ewing, J., 1968, Seismic refraction measurements on the continental margin of northeastern South America [abs.]: Am. Geophys. Union Trans., v. 49, p. 197–198.

Emiliani, C., 1955, Pleistocene temperatures: Jour. Geology, v. 63, p. 538–578.

——1966, Isotopic paleotemperatures: Science, v. 154, p. 851–857.

Ericson, D. B., 1961, Pleistocene climatic record in some deep-sea sediment cores: New York Acad. Sci. Annals, v. 95, art. 1, p. 537–541.

Ericson, D. B., and Wollin, G., 1956a, Correlation of six cores from the equatorial Atlantic and the Caribbean: Deep-Sea Research, v. 3, p. 104–125.

——1956b, Micropaleontological and isotopic determinations of Pleistocene climates: Micropaleontology, v. 2, p. 257–270.

——1968, Pleistocene climates and chronology in deep-sea sediments: Science, v. 162, p. 1227–1234.

Ericson, D. B., Wollin, G., and Wollin, J., 1954, Coiling direction of *Globorotalia truncatulinoides* in deep-sea cores: Deep-Sea Research, v. 2, p. 152–158.

Ericson, D. B., Ewing, M., Wollin, G., and Heezen, B. C., 1961, Atlantic deep-sea sediment cores: Geol. Soc. America Bull., v. 72, p. 193–286.

Ericson, D. B., Ewing, M., and Wollin, G., 1964, The Pleistocene Epoch in deep-sea sediments: Science, v. 146, p. 723–732.

Ewing, M., Ericson, D. B., and Heezen, B. C., 1958, Sediments and topography of the Gulf of Mexico, in Weeks, L., ed., Habitat of oil: Am. Assoc. Petroleum Geologists, p. 995–1053.

Flint, R. F., 1971, Glacial and Quaternary geology: New York, John Wiley & Sons, 892 p.

Gardner, J. V., 1973, The eastern equatorial Atlantic: Sedimentation, faunal, and sea-surface temperature responses to global climatic changes during the past 200,000 years [Ph.D. thesis]: New York, Columbia Univ., 301 p.

——1975, Late Pleistocene carbonate dissolution cycles in the eastern equatorial Atlantic, in Sliter, W. V., Bé, A.W.H., and Berger, W. H., eds., Dissolution of deep-sea carbonates: Cushman Found. Foram. Research Spec. Pub. 13, p. 129–141.

Gardner, J. V., and Hays, J. D., 1976, Responses of Sea-surface temperature and circulation to global climatic change during the past 200,000 years in the eastern equatorial Atlantic Ocean, *in* Cline, R. M., and Hays, J. D., eds. Investigation of late Quaternary paleoceanography and paleoclimatology: Geol. Soc. America Mem. 145 (this volume).

Goldthwaite, R. P., Dreimanis, A., Forsyth, J. L., Karrow, P. F., and White, G. W., 1965, Pleistocene deposits of the Erie Lobe, *in* Wright, H. E., and Frey, D. G., eds., The Quaternary of the United States: Princeton, N.J., Princeton Univ. Press, p. 85-97.

Hays, J. D., and Peruzza, A., 1972, Late Pleistocene climates inferred from calcium carbonate content of equatorial Atlantic deep-sea cores: Geol. Soc. America Abs. with Programs, v. 3, p. 595.

Hülsemann, J., 1966, On the routine analysis of carbonates in unconsolidated sediments: Jour. Sed. Petrology, v. 36, p. 622-625.

Imbrie, J., and Kipp, N. G., 1971, A new method for quantitative paleoclimatology: Application to a late Pleistocene Caribbean core, *in* Turekian, K. K., ed., The late Cenozoic glacial ages: New Haven, Yale Univ. Press, p. 71-181.

Imbrie, J., and van Andel, Tj. H., 1964, Vector analysis of heavy-mineral data: Geol. Soc. America Bull., v. 75, p. 1131-1156.

Imbrie, J., van Donk, J., and Kipp, N. G., 1973, Paleoclimatic investigation of a late Pleistocene Caribbean deep-sea core: Comparison of isotopic and faunal methods: Quaternary Research, v. 3, p. 10-38.

James, N. P., Mountjoy, E. W., and Omura, A., 1971, An early Wisconsin reef terrace at Barbados, West Indies, and its climatic implications: Geol. Soc. America Bull., v. 82, p. 2011-2018.

Jones, J. I., 1967, Significance of distribution of planktonic foraminifera in the equatorial Atlantic undercurrent: Micropaleontology, v. 13, p. 489-501.

Kennett, J. P., and Huddlestun, P., 1972a, Late Pleistocene paleoclimatology, foraminiferal biostratigraphy and tephrochronology, western Gulf of Mexico: Quaternary Research, v. 2, p. 38-69.

——1972b, Abrupt climatic change at 90,000 yr B.P.: Faunal evidence from Gulf of Mexico cores: Quaternary Research, v. 2, p. 384-395.

Kipp, N. G., 1976, New transfer function for estimating past sea-surface conditions from sea-bed distribution of planktonic foraminiferal assemblages in the North Atlantic, *in* Cline, R. M., and Hays, J. D., eds., Investigation of late Quaternary paleoceanography and paleoclimatology: Geol. Soc. America Mem. 145 (this volume).

Klovan, J. E., and Imbrie, J., 1971, A logarithm and Fortran IV program for large scale Q-mode factor analysis: Internat. Assoc. Math. Geol. Jour., v. 3, p. 61-77.

Ku, T.-L., Bischoff, J. L., and Boersma, A., 1972, Age studies of Mid-Atlantic Ridge sediments near 42°N and 20°N: Deep-Sea Research, v. 19, p. 233-247.

Kukla, G. J., 1969, The cause of Holocene climatic changes: Geol. Mijnbouw, v. 48, p. 307-334.

——1972, Insolation and glacials: Boreas, v. 1, p. 63-96.

Kullenberg, B., 1953, Absolute chronology of deep-sea sediments and deposition of clay on the ocean floor: Tellus, v. 5, p. 302-305.

Lynts, G. W., Judd, J. B., and Stehman, C. F., 1973, Late Pleistocene history of Tongue of the Ocean, Bahamas: Geol. Soc. America Bull., v. 84, p. 2665-2684.

McGeary, D.F.R., and Damuth, J. E., 1973, Post-glacial iron-rich crusts in hemipelagic deep-sea sediment: Geol. Soc. America Bull., v. 84, p. 1201-1212.

Mesolella, K. J., Matthews, R. K., Broecker, W. S., and Thurber, D. L., 1969, The astronomical theory of climate change, Barbados data: Jour. Geology, v. 77, p. 250-274.

Miro, M. D. de, 1971, Los foraminíferos planctónicos vivos y sedimentados del margen continental de Venezuela [abs.]: Acta Geol. Hispánica, v. 6, p. 102-108.

Mörner, N. A., 1972a, World climate during the last 130,000 years: Internat. Geol. Cong., 24th, Montreal 1972, sec. 12, p. 72-79.

——1972b, When will the present interglacial end?: Quaternary Research, v. 2, p. 341-349.

Needham, H. D., Conolly, J. R., Ruddiman, W. F., Bowles, F. A., and Heezen, B. C., 1969, Continental sediment in equatorial Atlantic ooze: A climatic record of the Pleistocene: Geol. Soc. America Abs. with Programs, v. 1, no. 7, p. 158.

Oba, T., 1969, Biostratigraphy and isotopic paleotemperature of some deep-sea cores from the Indian Ocean: Tohoku Univ. Sci. Repts., ser. 2, v. 41, p. 129-195.

Olausson, E., 1967, Climatological, geoeconomical, and paleo-oceanographic aspects of carbonate deposition: Prog. Oceanography, v. 4, p. 245-265.

Ovey, C. D., 1950, On the interpretation of climatic variations as revealed by a study of samples from an equatorial Atlantic deep-sea core, *in* Centenary symposium on climatic change: Royal Meteorol. Soc. Centenary Proc., p. 211-215.

Phleger, F. B., 1942, Foraminifera of submarine cores from the continental slope, Pt. 2: Geol. Soc. America Bull., v. 53, p. 1073-1098.

Phleger, F. B., Parker, F. L., and Peirson, J. F., 1953, North Atlantic Foraminifera: Swedish Deep-Sea Exped. Repts., 1947-1948, v. 7, p. 1-122.

Prell, W. L., 1974, Late Pleistocene faunal, sedimentary and temperature history of the Colombia Basin, Caribbean Sea [Ph.D. thesis]: New York, Columbia Univ., 518 p.

Prell, W. L., Gardner, J. V., Bé, A.W.H., and Hays, J. D., 1976, Equatorial Atlantic and Caribbean foraminiferal assemblages, temperatures, and circulation: Interglacial and glacial comparisons; *in* Cline, R. M., and Hays, J. D., eds., Investigation of late Quaternary paleoceanography and paleoclimatology: Geol. Soc. America Mem. 145 (this volume).

Reyne, A., 1961, On the contribution of the Amazon River to accretion of the coast of the Guianas: Geologie en Mijnbouw, v. 40, p. 219-226.

Ruddiman, W. F., 1969, Foraminifera of the subtropical North Atlantic gyre [Ph.D. thesis]: New York, Columbia Univ. 282 p.

——1971, Pleistocene sedimentation in the equatorial Atlantic: Stratigraphy and faunal paleoclimatology: Geol. Soc. America Bull., v. 82, p. 283-302.

Ruddiman, W. F., and Heezen, B. C., 1967, Differential solution of planktonic foraminifera: Deep-Sea Research, v. 14, p. 801-808.

Schott, W., 1935, Die Foraminiferen in dem aquatorialen Teil des Atlantischen Ozeans: Wissenschaft. Ergebnisse der Deutschen Atlantischen Expedition auf dem Forschungs- und Vermessungsschiff "Meteor," 1925-1927, v. 3, p. 43-134.

——1966, Foraminiferenfauna und Stratigraphie der Tiefseesedimente im Nordatlantischen Ozean, *in* Sediment cores from the North Atlantic Ocean: Swedish Deep-Sea Exped. Repts., v. 7, no. 8, p. 357-469.

Steinen, R. P., Harrison, R. S., and Matthews, R. K., 1973, Eustatic low stand of sea level between 125,000 and 105,000 B.P.: Evidence from the subsurface of Barbados, West Indies: Geol. Soc. America Bull., v. 84, p. 63-70.

Tolderlund, D. S., and Bé, A.W.H., 1971, Seasonal distribution of planktonic foraminifera in the surface waters of five stations in the western North Atlantic: Micropaleontology, v. 17, p. 297-329.

Turekian, K. K., 1965, Some aspects of the geochemistry of marine sediments, *in* Riley, J. P., and Skirrow, O., eds., Chemical oceanography, Vol. 2: New York, Academic Press, Inc., p. 81-126.

Wiseman, J.D.H., 1954, The determination and significance of past temperature changes in the upper layer of the equatorial Atlantic Ocean: Royal Soc. Proc., ser. A, v. 222, p. 296-323.

——1965, The changing rate of calcium carbonate sedimentation on the equatorial Atlantic floor and its relation to continental late Quaternary stratigraphy, *in* Sediment cores from the North Atlantic Ocean: Swedish Deep-Sea Exped. Repts., v. 7, p. 288-354.

MANUSCRIPT RECEIVED BY THE SOCIETY JUNE 17, 1974
REVISED MANUSCRIPT RECEIVED JANUARY 27, 1975
MANUSCRIPT ACCEPTED FEBRUARY 20, 1975
LAMONT-DOHERTY GEOLOGICAL OBSERVATORY CONTRIBUTION NO. 2271

Late Pleistocene Faunal and Temperature Patterns of the Colombia Basin, Caribbean Sea

WARREN L. PRELL*
AND
JAMES D. HAYS†
*Lamont-Doherty Geological Observatory
Columbia University
Palisades, New York 10964*

ABSTRACT

High-salinity, low-productivity Sargasso Sea type water developed in situ within the Colombia Basin during glacial extremes. Evidence for "ice-age aridity" in northern South America supports this hypothesis. Q-mode factor analysis of total planktonic foraminiferal faunas from three Colombia Basin sediment cores defines equatorial-tropical and southern Sargasso faunal assemblages, which correspond to interglacial and glacial conditions in the Colombia Basin. The oceanographic conditions represented by these assemblages were deduced from the present faunal distributions in the North Atlantic Ocean. The interglacial assemblage has maximum abundances in the equatorial zone, which is characterized by moderate salinities, strong currents, and high nutrients and productivity. The glacial assemblage occurs today in the distinctive southern Sargasso Sea water of high salinity, low nutrients, and low productivity. The alternation between interglacial-equatorial and glacial-Sargasso Sea assemblages in the Colombia Basin suggests that during glacial periods, Sargasso type water was formed in situ within the Colombia Basin.

Paleotemperature estimates derived by the Imbrie and Kipp technique (1971) reveal only 2° to 3°C average temperature changes between interglacial and glacial stages. During glacial stages, however, seasonal contrast ($Ts-Tw$) was increased; the winters had much larger variations (4° to 5°C) than did the summers (1° to 2°C). Because the evaporation/precipitation ratio (E/P) is greatest today during the Caribbean winter, increased seasonal contrast during glacial stages may have increased the E/P, resulting in high salinities similar to present Sargasso water.

---

*Present address: Department of Geological Sciences, Brown University, Providence, Rhode Island 02912.
†Also Department of Geological Sciences, Columbia University.

The increased E/P suggests that the Intertropical Convergence Zone was on the average located over South America more during the glacial stages than at present.

INTRODUCTION

Much of the early knowledge about late Pleistocene oceans and climates came from Caribbean Sea sediments. Interpretations of these sediments have been based on analysis of planktonic foraminiferal faunas (Ericson and Wollin, 1956; Lidz, 1966; Imbrie and Kipp, 1971) and on measurements of the oxygen-isotope content of planktonic foraminifera (Emiliani, 1955, 1966, 1972; Broecker and van Donk, 1970). Most work has been directed toward understanding the changes observed in one section rather than toward understanding the regional changes. This study attempts to reconstruct both individual stratigraphic and regional faunal and temperature patterns within the Colombia Basin during late Pleistocene time.

This study first defines the faunas living within the Colombia Basin during late Pleistocene time and then maps those faunas in the present-day Atlantic Ocean to establish their oceanographic associations. Using this approach, late Pleistocene faunal events may be interpreted in terms of present-day oceanographic conditions. The late Pleistocene faunal data are also used to make quantitative estimates of paleotemperature and paleosalinity. This study attempts to understand the regional faunal and oceanographic responses to climatic change as recorded by planktonic foraminifera of the Colombia Basin.

Physical Setting

The Colombia Basin, Caribbean Sea, lies north of Colombia and Panama, south of Jamaica and Hispaniola, and east of Costa Rica and Nicaragua. The Beata Ridge is its eastern boundary and separates the Colombia and Venezuela Basins, except at the southern boundary where Aruba Gap connects the basins (Fig. 1).

The bathymetric framework of the Colombia Basin contains two types of depositional sites. One type accumulates terrigenous sediment and includes the deep abyssal plains and some continental margins where downslope sediment transport is active. The second is pelagic and includes the ridges and rises above the zone of downslope terrigeneous transport. The Nicaragua Rise, the Beata Ridge, and the small elevations in the western Colombia Basin are examples of pelagic-type accumulation sites. The western Colombia Basin sites are hemipelagic, because they accumulate much more clay than carbonate.

Oceanography

The Caribbean Current and its branches dominate the surface circulation of the Colombia Basin. This current exhibits a maximum seasonal surface-termperature range of only 2° to 3°C and is coolest in February and warmest in August or September. These seasonal variations decrease with depth and disappear below 100 m (Colon, 1963). Seasonal surface-temperature maps of the Caribbean Sea (Wust, 1964; Perlroth, 1971) show low-temperature zones along the coasts of Colombia and Venezuela. These zones represent upwelled water that is cooler and more saline. The upwelling adjacent to the Colombia coast is more intense in January than in August.

Maps of winter-spring and summer-autumn average surface salinity (Wust, 1964) also show seasonal variations. The winter season has higher average salinities

(35.75 to 36.75‰) but contains patches of lower salinity Amazon River water. The summer season has different conditions with lower salinities (average 35.17‰) and has a continuous tongue of low-salinity Amazon-Orinoco water that extends across the Caribbean. Interpretation of these salinity variations requires an understanding of regional and seasonal conditions of evaporation and precipitation. Wust (1964) and Gordon (1967) showed that the winter ratio of evaporation/precipitation (E/P) is two and one-half times greater than the summer ratio, and that E/P ratio is the most important factor determining seasonal salinity fluctuations in the Caribbean. Caribbean winter salinity averages 0.72‰ higher than the summer average, in contrast to the open Atlantic where the seasonal range is 0.20‰.

The dry Caribbean winter season is characterized by a more subtropical, semiarid climate with abundant sunshine, high evaporation, low precipitation, and low runoff, which result in higher surface salinities. By contrast, the rainy Caribbean summer season is characterized by a more tropical semihumid climate with increased cloudiness, low evaporation, high precipitation, and high runoff, which result in lower surface salinities. This seasonal pattern is related to the position of the Intertropical Convergence Zone, which is over the Caribbean Sea during the summer and moves south over South America during the winter.

Selection and Correlation of Cores

More than 100 cores from the Colombia Basin were examined for stratigraphically continuous sections. Cores were judged continuous by comparing the stratigraphic position of faunal datums to the carbonate and coarse-fraction (>62μm) curves. Only ten cores were judged continuous enough for detailed faunal analysis. Data

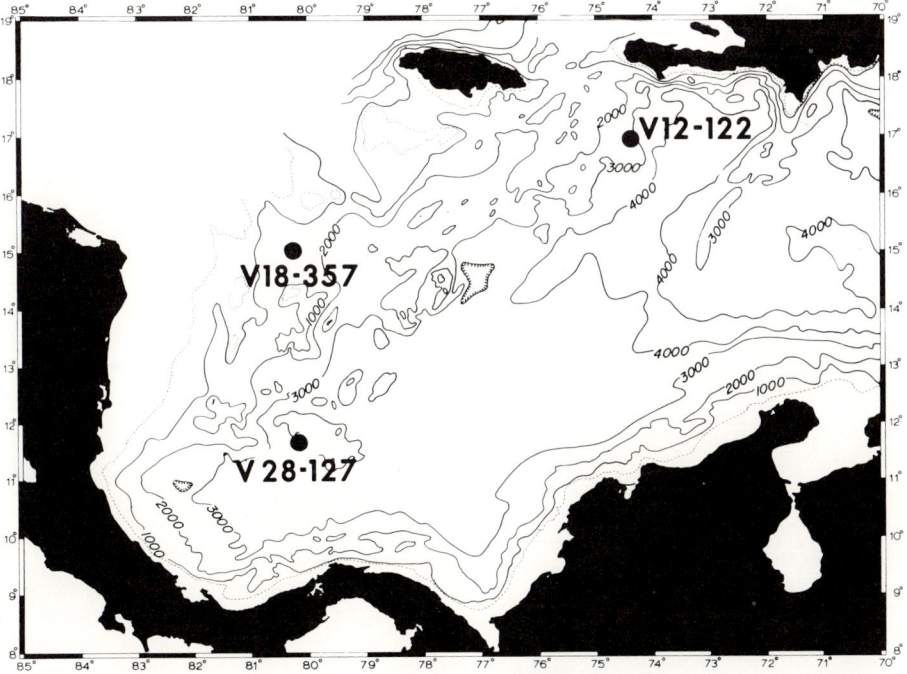

Figure 1. Bathymetric map of Colombia Basin showing locations of cores V12-122, V18-357, and V28-127.

from two of these cores (V18-357 and V28-127) plus the data for V12-122 (Imbrie and Kipp, 1971) are used for the stratigraphic aspects of this study.

Cores V18-357 and V28-127 come from different depositional environments within the Colombia Basin. From the shallow (1,818 m) carbonate-rich Nicaragua Rise, V18-357 represents pelagic accumulation at approximately 2 cm/1,000 yr. V12-122, also from the Nicaragua Rise, is deeper (2,800 m) but very similar to V18-357. V28-127 is located on a small rise (3,227 m) in the deeper western Colombia basin and represents hemipelagic sedimentation with about one-half the carbonate content and twice the accumulation rate of V18-357.

The stratigraphic nomenclature of late Pleistocene equatorial Atlantic sediments is a mixture of faunal zones (Ericson and others, 1961) and oxygen-isotope cycles (Emiliani, 1955, 1966; Broecker and van Donk, 1970). Figure 2 shows the major zonations in use and compares them to the oxygen-isotope, carbonate, and coarse-fraction curves for core V12-122. The Ericson zonation is based on the presence of the *G. menardii* complex. The Emiliani zonation (1964) uses the oxygen-isotope curve and assigns odd numbers to low O^{18} events and even numbers to high O^{18} events. Broecker and van Donk define terminations as the sharp decrease in O^{18} content that marks the boundary between major cycles, which they number with Roman numerals. Terminations I and II mark Emiliani's Stage 1-2 and 5-6 boundaries. McIntyre and others (1972) used carbonate curves in their North Atlantic zonation to identify three major carbonate minima (minima 1, 2, and 3), which correlate roughly to Emiliani's Stages 2, 4, and 6.

Comparison of these zonations with O^{18}, carbonate, and coarse-fraction curves of core V12-122 (Fig. 2) shows that the isotopic stages correlate with the carbonate and coarse-fraction (a reflection of foraminiferal abundance) curves and faunal boundaries, with minor exceptions. One should note that the X-Y and Stage 4-5 boundaries are not synchronous. This study uses a combination of Emiliani's Stage numbers to describe oxygen-isotope, carbonate, and coarse-fraction curves and uses Ericson's faunal zones to correlate the curves.

Correlation of sediment cores is based on Ericson and others' (1961) *G. menardii* complex zonation and the disappearance of *P. obliquiloculata* within the Y zone. Secondary use is made of carbonate and coarse-fraction curves. The usefulness of *G. menardii* zones in the tropical and subtropical Atlantic Ocean is well established (Ericson and others, 1961; Kennett and Huddlestun, 1972). The presence of *P. obliquiloculata* in the lower Y zone was originally recognized by Ericson and Wollin (1956) and was later confirmed to occur in Stages 3 and 4 (Emiliani, 1964). Kennett

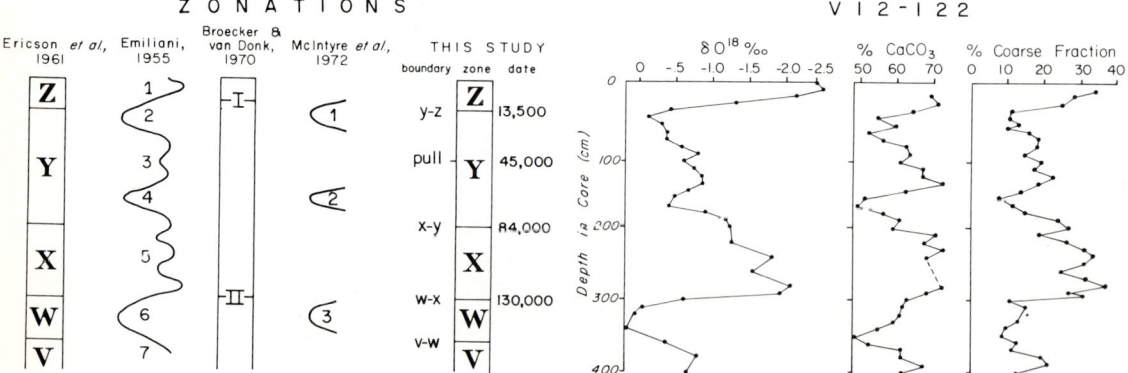

Figure 2. Summary of late Pleistocene deep-sea zonations compared to O^{18}, $CaCO_3$, coarse-fraction curves of core V12-122.

and Huddlestun (1972) observed that in the Gulf of Mexico, *P. obliquiloculata* disappeared within the lower part of their subzone Y6. In the Colombia Basin, however, *P. obliquiloculata* consistently disappears in the middle of Stage 3, between two characteristic carbonate maxima approximately correlating with Kennett and Huddlestun's substage Y4. The presence of *P. obliquiloculata* in the lower Y zone indicates that the Colombia Basin remained warm while the Gulf of Mexico became cooler during that period.

We used the C^{14} and thorium-ionium dates of Broecker and van Donk (1970) to interpolate the absolute ages of all faunal datums and stage boundaries. A summary of the time scale for the faunal datums and carbonate stages used in this study is given in Figure 3.

Sample Preparation

Sediment samples of approximately 5g were disaggregated. The resulting sediment was agitated and cleaned ultrasonically for no longer than 10 sec, which does not fragment fragile species. Wet sieving at 62μm separated the foraminifera from the clay and silt fractions, which were retained for further study. A Calgon wash was used to remove most clay particles from the tests. The foraminifera were oven dried at low temperature and weighed to obtain the amount of coarse fraction greater than 62 μm. All samples were dry sieved at 149 μm to eliminate juvenile forms. The resulting fraction was split repeatedly with a Soiltest microsplitter to obtain 300 to 400 specimens. All specimens were identified, counted, and mounted on standard paleontological slides. The percent abundance of each species was calculated relative to total planktonic foraminifera. The taxonomy follows that of Bé (1967) and Parker (1962) and is given in Table 1 for all species >2% in at least one sample.

LATE PLEISTOCENE FAUNAL VARIATIONS

Quantitative faunal analyses of Caribbean sediments began with Lidz's (1966) study of core P6304-8. He confirmed that the abundance of *G. menardii* shows a strong correlation with the isotopic temperature curve, and he established several species ratios that approximated the oxygen-isotope curve. Imbrie and Kipp (1971) used another quantitative approach in which they correlated sea-surface temperatures and surface-sediment faunal assemblages for the entire Atlantic Ocean. They then used these correlations to estimate paleotemperatures for faunal assemblages in Colombia Basin core V12-122. They defined a tropical assemblage dominated by *G. menardii*, and *P. obliquiloculata* and noted that the assemblage dominated all samples in V12-122. From these data, Imbrie and Kipp (1971) suggested that the winter temperature of the Caribbean Sea surface has not fallen below 20°C during the past 400,000 yr. Hecht (1973) estimated glacial/interglacial surface-temperature variations of 5°C for the Caribbean Sea, using a quantitative model based on direct comparison of Pleistocene foraminiferal assemblages with Recent assemblages whose geographic distributions have been correlated with modern ocean-surface temperatures.

Faunal Assemblages

This study uses factor analysis to quantitatively define the composition and abundance of the important faunal assemblages living within the Colombia Basin during late Pleistocene time. The resulting assemblages are then used to reconstruct the oceanographic conditions experienced in the Colombia Basin during glacial periods.

Because cores V18-357 and V28-127 represent slightly different depositional and oceanographic environments, their faunal data, representing 5,000-yr intervals (33 samples from V18-357 and 35 samples from V28-127), were analyzed together, which gave an average for Colombia Basin faunal composition over 150,000 yr.[1] Data from V12-122 were not used because they are based on counts of approximately 200 individuals and are not of the same statistical validity as the data used in this study. Using the program CABFAC (Klovan and Imbrie, 1971), a Q-mode factor analysis reduced the 18 species >2% to four independent varimax factor assemblages. These factors were almost identical to those derived from the complete fauna of each core (Prell, 1974). The species composition and importance in each of the four factor assemblages is given in Table 2.

The first factor assemblage has dominant *G. ruber* with *G. sacculifer* and *G. dutertrei*, which accounts for 35.4% of the data. The second factor assemblage is dominated by *G. menardii–G. tumida* with lesser *G. crassaformis* and *G. ruber* and composes only 9.9% of the combined data. The third factor assemblage has dominant *G. sacculifer* with lesser *G. ruber*, totaling 28.2% of the combined data. The fourth factor assemblage contains abundant *G. dutertrei* and *G. glutinata* with lesser *G. ruber* and accounts for 25.3% of the combined data.

Significance of the Faunal Assemblages

Four factor assemblages accurately describe the faunal variation of the Colombia Basin during the past 150,000 yr. Some of these assemblages however, are scarce or absent in present-day Colombia Basin (Prell, 1974). To understand the variation of late Pleistocene faunal assemblages in the Colombia Basin, we sought outside of the Caribbean Sea in the more diverse faunas and environments of the North Atlantic.

Planktonic foraminiferal faunas for 215 core tops (141 North Atlantic samples, N. Kipp, Brown University, 1973; 43 equatorial Atlantic samples, J. V. Gardner, Lamont-Doherty Geological Observatory, 1973; 31 Colombia Basin samples, W. L. Prell, Lamont-Doherty Gelogical Observatory, 1973) have been analyzed in terms of the four late Pleistocene assemblages of the Colombia Basin. This analysis proportions the fauna of each Atlantic sample into the four late Pleistocene assemblages of Colombia Basin and computes a communality that measures the amount of data accounted for by the four assemblages. In terms of the Imbrie and Kipp (1971) notation, the F matrix (species-factor assemblage) derived from the Colombia Basin is multiplied times the U matrix (percent species per sample) from the North Atlantic to give a B′ matrix (sample-factor assemblage) for the North Atlantic samples. This approach defines only the faunal variation (in terms of faunal assemblages) actually experienced by the Colombia Basin and then maps those assemblages within the present-day North Atlantic Ocean. Assemblage distributions and environmental associations within the North Atlantic Ocean can thus be directly correlated to the late Pleistocene samples of Colombia Basin. Interpretation of these factor assemblages is based on their species composition and their distribution within the North Atlantic and then by comparison to the

---

[1] Appendixes I and II containing faunal data for cores V18-357 and V28-127 are on microfiche in pocket inside back cover.

plankton and oceanographic data associated with their distributions.

Numerical data for the species composition of the factor assemblages, the sample-factor loadings, and the communalities for the 215 surface-sediment faunas from the Caribbean and Atlantic are given in Prell (1974); only the faunal distribution maps are presented here (Figs. 4, 5, 6).

Distribution of Faunal Assemblages

Well-preserved samples from the Caribbean Sea, the Gulf of Mexico, and the equatorial Atlantic Ocean have communalities >0.90, which means that 90% of their faunal composition can be explained by the four factor assemblages of the Colombia Basin. North of the 0.90 contour near 30°N, the communalities decrease rapidly as the faunas become dominated by northern subtropical and subpolar species that are not found in the Colombia Basin. The communality contours demonstrate that the factor assemblages of the Colombia Basin represent accurately only the faunas and waters south of 30°N.

The *G. sacculifer* assemblage, with its highest factor loadings in the Caribbean and equatorial Atlantic, is interpreted as an equatorial-tropical assemblage (Fig. 4). The importance of this assemblage in sediments is greatly reduced by carbonate dissolution, which partially accounts for the low values in the southern Colombia Basin and in the eastern equatorial Atlantic. However, because Bé and Tolderlund (1971) have shown that *G. sacculifer* abundance in plankton tows is highest in the central equatorial Atlantic and decreases toward the margins, the relative low abundance of this assemblage along the margins of South America and Africa is probably due to sea-surface conditions. The equatorial nature of this assemblage

TABLE 1. LIST OF SPECIES USED IN FACTOR ANALYSIS

Orbulina universa
Globigerinoides conglobatus
Globigerinoides ruber
Globigerinoides tennellus
Globigerinoides sacculifer
Globigerinella aequilateralis
Globigerina calida
Globigerina bulloides
Globigerina falconensis
Globigerina rubescens
Pulleniatina obliquiloculata
Globorotalia inflata
Globorotalia crassaformis
Globigerinita glutinata
Globoquadrina dutertrei
 + "PD intergrade"*
Globorotalia truncatulinoides
Globorotalia menardii
 + *Globorotalia tumida*
Globigerina pachyderma

Note: Taxonomy follows Bé (1967) and Parker (1962).

*"PD intergrade" describes specimens transitional between *G. pachyderma* and *G. dutertrei*, which are rare in the Caribbean but common in the eastern equatorial Atlantic (Gardner, 1973).

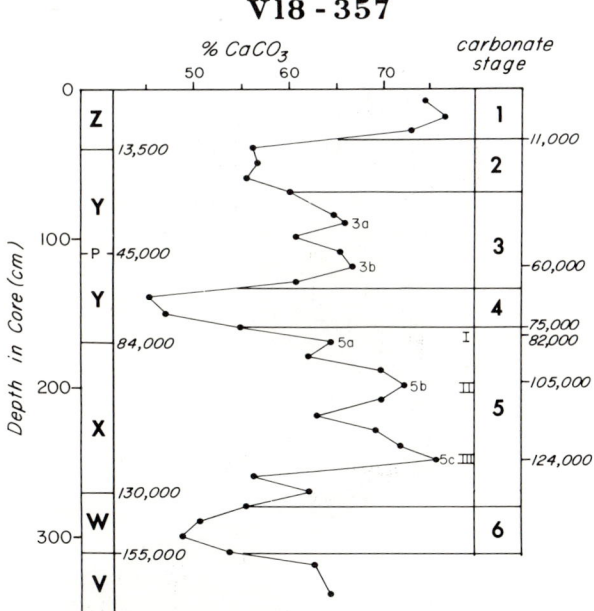

Figure 3. Summary of time scale used for faunal datums and carbonate stages of late Pleistocene. Barbados terraces are indicated as I, II, and III.

TABLE 2. F' MATRIX OF LATE PLEISTOCENE COLOMBIA BASIN FAUNAS

Var.	1	2	3	4
G. univ.	0.044	−0.095	−0.133	0.183
G. cglob.	0.074	0.400	−0.157	−0.168
G. ruber	3.583	1.155	−1.414	1.177
G. ten.	0.104	0.083	−0.170	−0.033
G. sacc.	−1.446	−0.526	−3.889	0.044
G. aequi.	0.222	−0.204	−0.612	0.005
G. calid.	0.005	0.012	−0.090	−0.045
G. bull.	−0.220	0.261	0.050	0.381
G. falc.	0.038	0.132	0.032	0.049
G. rubes.	−0.115	−0.036	0.121	0.559
P. obliq.	−0.184	0.694	0.049	0.001
G. infla.	−0.076	0.358	−0.004	0.553
G. crasf.	−0.188	1.054	0.296	0.151
G. glut.	−0.364	−0.123	0.014	2.411
G. dut. + PD*	−1.262	0.613	0.466	3.055
G. trunc.	0.380	−0.169	−0.040	0.115
G. men. + G. tum.	−0.999	3.734	−0.308	−0.764
G. pac.	−0.006	−0.021	0.050	0.125

Note: Factor 1 is *G. ruber* (southern Sargasso); factor 2 is *G. menardii* (part of equatorial-tropical); factor 3 is *G. sacculifer* (equatorial-tropical); and factor 4 is *G. dutertrei* (transitional).

The importance of a species is indicated by its value under each factor. Although their values are negative, *G. sacculifer* and *G. ruber* are the most important species in factor 3.

*Specimens transitional between *G. pachyderma* and *G. dutertrei* (Gardner, 1973).

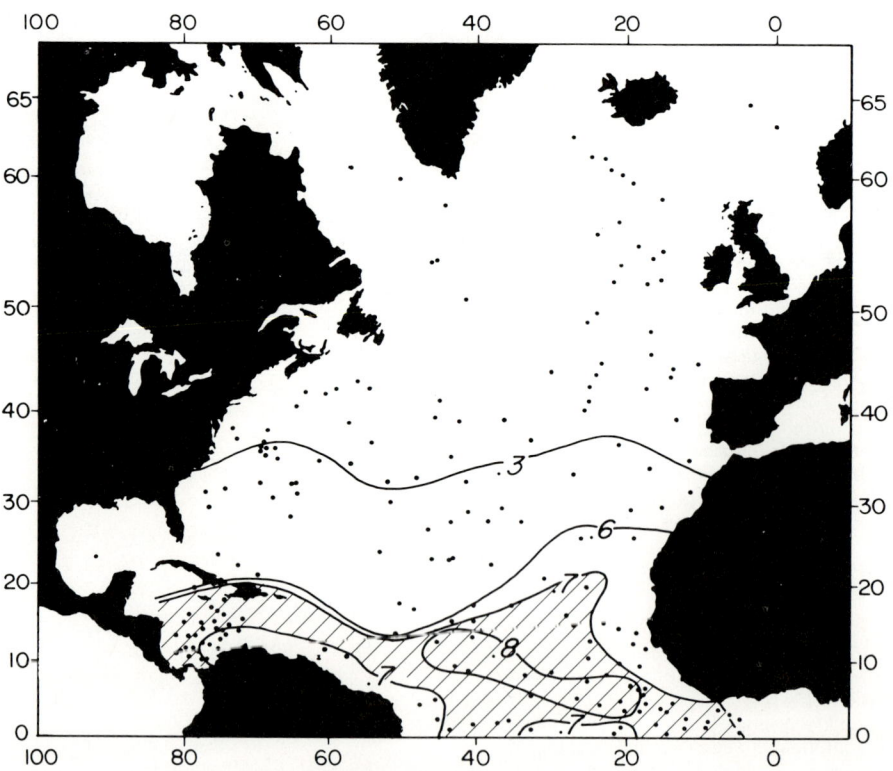

Figure 4. Contour map of *G. sacculifer* (equatorial-tropical) assemblage-factor loadings in Atlantic and Caribbean surface sediments.

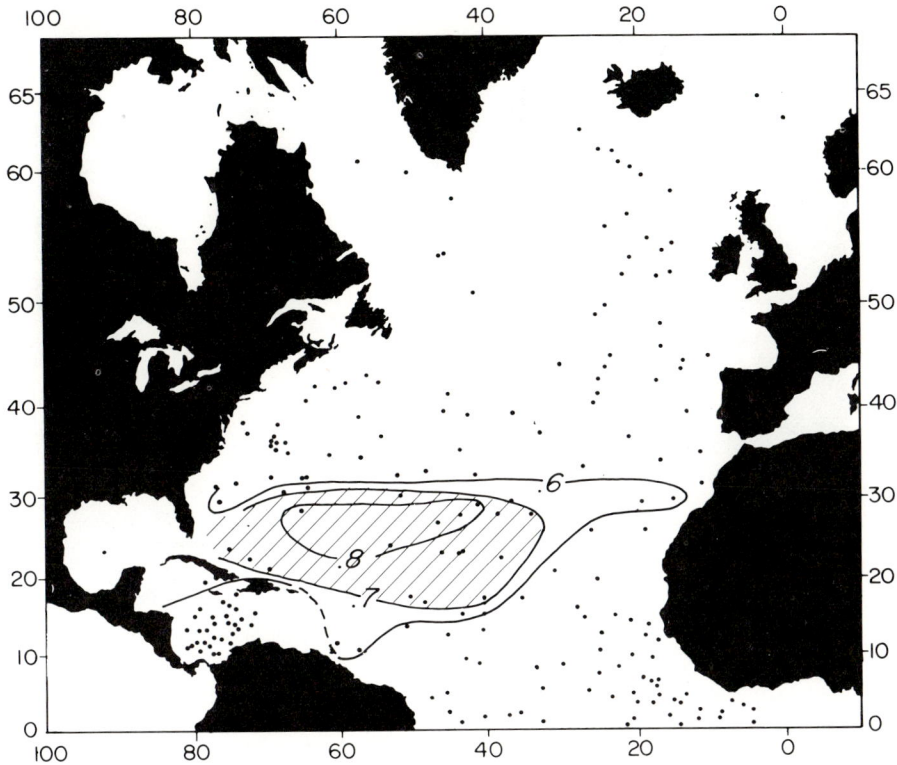

Figure 5. Contour map of *G. ruber* (southern Sargasso) assemblage-factor loadings in Atlantic and Caribbean surface sediments.

is clear from its species composition and its distribution. The zone of maximum abundance is south of 15°N and is characterized by waters warmer than 24°C and salinities <36.00‰. A similar assemblage occurs in the plankton data of this area (Bé and Tolderlund, 1971). Ruddiman (1969) used species gradients to identify a transitional zone between 16° and 20°N, which approximates the northern limit of maximum abundance of *G. sacculifer*. This zone was used by Ruddiman to divide equatorial water from the southern Sargasso central water. The tropical factor of Imbrie and Kipp (1971) extended to 30°N and is more general than the equatorial-tropical assemblage of this study.

The *G. ruber* assemblage, interpreted as a southern Sargasso assemblage, has its highest values in the central and western Atlantic between 17° and 30°N (Fig. 5). Abundances south of 17°N are moderate and variable, although the assemblage decreases rapidly north of 30°N. This assemblage plots adjacent to the equatorial-tropical assemblage and clearly defines the southern Sargasso central water of Ruddiman (1969), which is characterized by strong dominance of *G. ruber*. The northern limit of the abundant *G. ruber* assemblage coincides with the subtropical convergence described by Bé and others (1971) from both oceanographic and plankton data. Bé (1960) and Ruddiman (1969) show that cool-water species are dominant north of the subtropical convergence.

The equatorial-tropical assemblage is characterized by the greater importance of *G. sacculifer* relative to *G. ruber*. Both species prefer warm water, as demonstrated by the latitudinal distribution, but they have differing salinity tolerances. Bé and

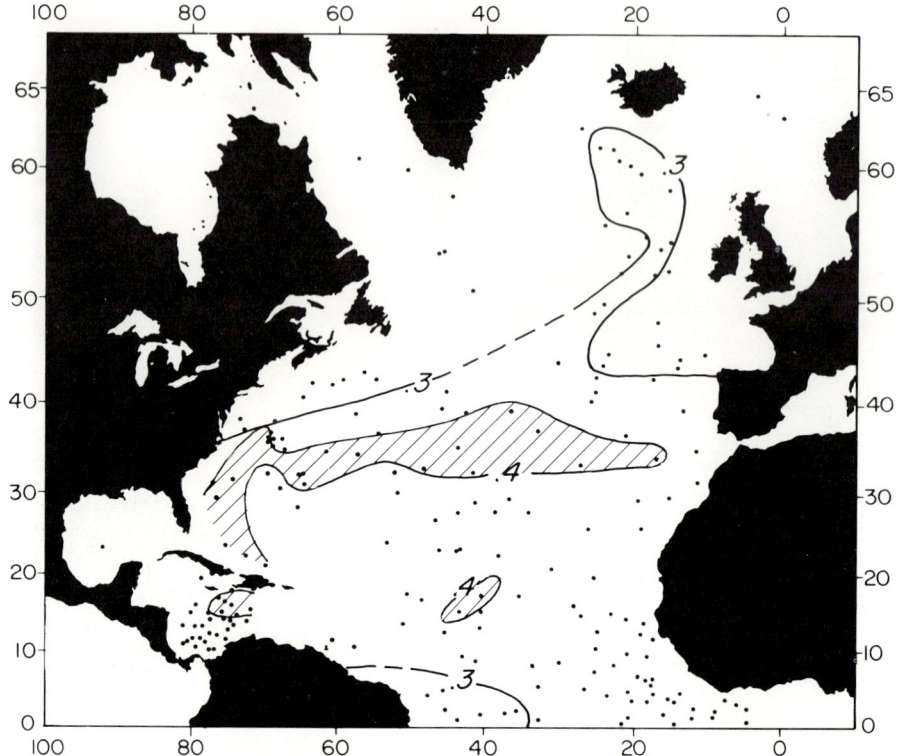

Figure 6. Contour of *G. dutertrei* (transitional) assemblage-factor loadings Atlantic and Caribbean surface sediments.

Tolderlund (1971) show that *G. sacculifer* is stenohaline preferring salinities of 34.5‰ and that *G. ruber* is euryhaline and abundant at both high (>36.0‰) and low (<34.5‰) salinities. Additionally, *G. ruber* appears to tolerate low-nutrient water better than does *G. sacculifer*. The *G. sacculifer* assemblage reflects warm, moderately saline equatorial water that is nutrient rich compared to the Sargasso Sea central water. The *G. ruber* assemblage reflects the warm, more saline nutrient-poor Sargasso Sea water.

The distribution and significance of the *G. dutertrei–G. glutinata* assemblage, interpreted as transitional, is more difficult to interpret because the assemblage is not so abundant in the present Atlantic Ocean as it was in late Pleistocene time in the western Caribbean. The highest values of this assemblage form a moderately well-defined tongue at about 35°N (Fig. 6). Small patches of values >0.40 occur in the central Atlantic at 15°N and in the northern Colombia Basin. Because dissolution tends to increase the importance of this assemblage, other means must be taken to isolate the effects of dissolution.

Although this assemblage is found adjacent to and north of the southern Sargasso assemblage, it is not identical to the northern subtropical fauna of either Ruddiman (1969) or Bé and others (1971). Ruddiman (1969) noted that *G. dutertrei* formed a ring enclosing the central waters, but his *G. dutertrei* abundance zone does not coincide with high values of this assemblage. One reason for this lack of correspondence is the abundance of *G. glutinata*, which also controls the factor distribution. Because this assemblage seems to have no direct analog in the present North Atlantic, two interpretations of its significance are possible. It may represent the central water margins as suggested by Ruddiman on the basis of *G. dutertrei*. Bé and Tolderlund (1971) suggest that *G. glutinata* is also adapted to transitional and adjacent subtropical waters. Thus the assemblage may represent transitional waters, possibly between the equatorial-tropical and southern Sargasso assemblages. The other possibility is that the assemblage represents increased mixing or upwelling. The preference of *G. dutertrei* for strong current systems and regions of upwelling has been shown by Bé and Tolderlund (1971). The stratigraphic maxima of this assemblage (Figs. 7–9) usually occur between the maxima of the equatorial-tropical and southern Sargasso assemblages and may represent a transitional phase between those two faunal and oceanographic conditions.

The *G. menardii* factor assemblage has low and extremely variable values throughout the equatorial and subtropical North Atlantic Ocean. Dissolution strongly enhances this assemblage. Because this assemblage reflects mainly tropical conditions and is most abundant near the equator, it is combined with the *G. sacculifer* equatorial-tropical assemblage in the study of past assemblage variations. The combined assemblage gives a better tropical indicator because the high abundances of *G. menardii* in the X zone tends to depress the *G. sacculifer* equatorial-tropical assemblage in that zone.

In summary, two important faunal assemblages (equatorial-tropical and southern Sargasso) that represent late Pleistocene faunas of the Colombia Basin have been mapped in the North Atlantic Ocean and coincide with recognized water masses and faunal provinces. Therefore, these assemblage-oceanographic associations may be applied to the interpretation of late Pleistocene events in the Colombia Basin.

Late Pleistocene Variations of Faunal Assemblages

The proportions of factor assemblages of the Colombia Basin versus depth in cores V18-357, V12-122, and V28-127 are given in Figures 7, 8, and 9. Examination of these plots shows that the factor-assemblage composition of the Colombia Basin fluctuated sharply during late Pleistocene time. To establish a present-day baseline, each factor assemblage was averaged over 23 well-preserved Colombia Basin trigger-core top samples (Prell, 1974). These averages represent the importance of each assemblage in the surface sediments.

In the surface-sediment samples, the *G. sacculifer–G. menardii* equatorial-tropical

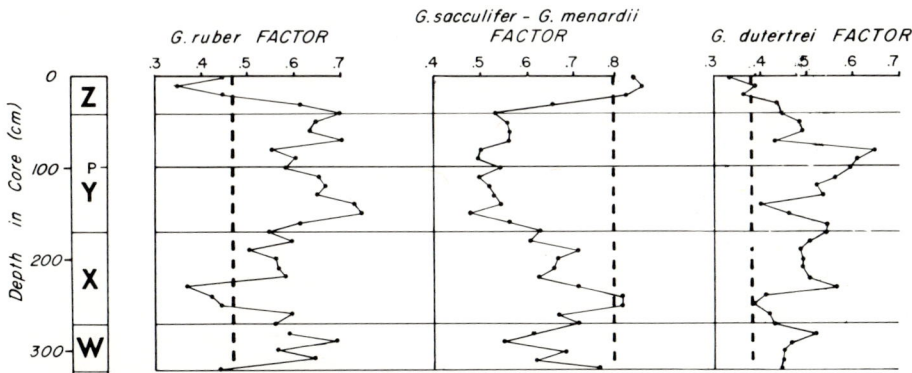

Figure 7. Colombia Basin assemblages versus depth in core V18-357. *G. ruber* factor = southern Sargasso assemblage; *G. sacculifer–G. menardii* factor = equatorial-tropical assemblage; and *G. dutertrei* factor = transitional assemblage. Dashed lines indicate average present-day factor loadings of each assemblage.

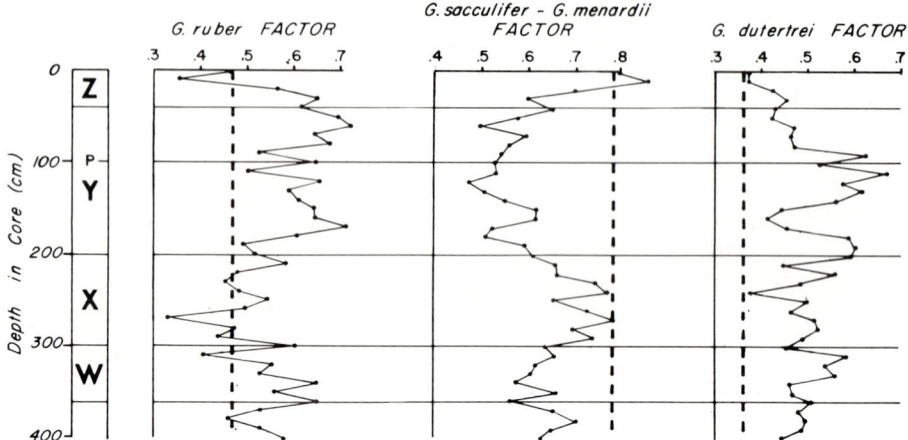

Figure 8. Colombia Basin assemblages versus depth in core V12-122. For dashed lines, see Figure 7.

factor is dominant with an average factor loading of 0.783 representing 61% of the present-day fauna. The southern Sargasso *G. ruber* factor has an average loading of 0.464, representing about 22%, and the *G. dutertrei* transitional factor averages 0.380, representing 14% of the present-day fauna.

Figures 7, 8, and 9 show that the Colombia Basin experienced a consistent alternation or succession of faunal assemblages during the past 150,000 yr. The fauna has changed from the presently dominant equatorial-tropical assemblage through a minor maximum of the transitional assemblage to a dominant southern Sargasso assemblage during the glacial stage at approximately 18,000 B.P. The assemblage curves also show the equatorial-assemblage was more abundant within the Z zone (Stage 1), and the curves support the existence of a climatic optimum during Holocene time. The glacial Y zone is characterized by a change from southern Sargasso (Stage 2) to transitional (Stage 3) to southern Sargosso (Stage 4) assemblage with minor fluctuations superimposed upon this pattern. The transitional *G. dutertrei*

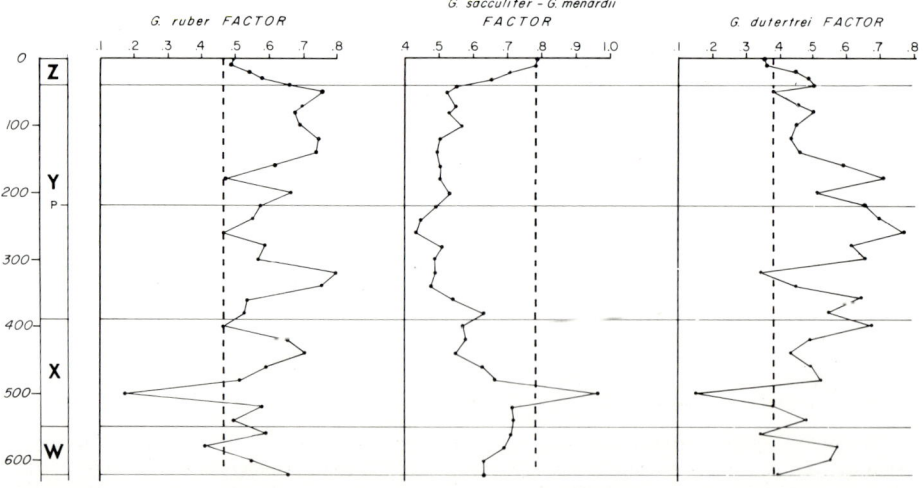

Figure 9. Colombia Basin assemblages versus depth in core V28-127. For dashed lines, see Figure 7.

assemblage attains its maximum abundance in carbonate Stage 3, which is midway in intensity between full glacial and full interglacial. The assemblage succession of the last interglacial (Stage 5 or the X zone) was more complex but generally exhibited three equatorial-tropical maxima flanked by transitional assemblages with minor southern Sargasso maxima between the equatorial-tropical maxima. Of the three equatorial-tropical maxima, only the oldest one at approximately 124,000 B.P. reaches abundances similar to the present-day sediments.

Because the oceanographic associations of these assemblages have been determined, the faunal patterns of Figures 7, 8, and 9 can be interpreted as being representative of changing surface-water masses within the Colombia Basin.

Relation of Faunal Assemblages to Carbonate Curves

The faunal-assemblage variations reflect changing oceanographic conditions and have a direct relationship to the carbonate and coarse-fraction curves of the Colombia Basin. This relationship suggests that the carbonate and coarse-fraction curves represent carbonate productivity, specifically foraminiferal productivity, of the surface waters.

Comparison of faunal assemblage proportions with carbonate and coarse-fraction curves for the same cores shows a pattern of high-carbonate and coarse-fraction content when the equatorial-tropical assemblage is abundant; and low-carbonate and coarse-fraction content when the southern Sargasso assemblage is abundant. To illustrate this relationship, the factor loadings for the southern Sargasso assemblage were plotted adjacent to the precent-carbonate content for the same samples in core V18-357 (Fig. 10). Zones of abundant southern Sargasso assemblage and low-carbonate content are easily correlated. Carbonate glacial Stages 2, 4, and 6 correlate almost exactly with maxima of the southern Sargasso assemblage. The shallow core V18-357 also shows correlation of three secondary carbonate

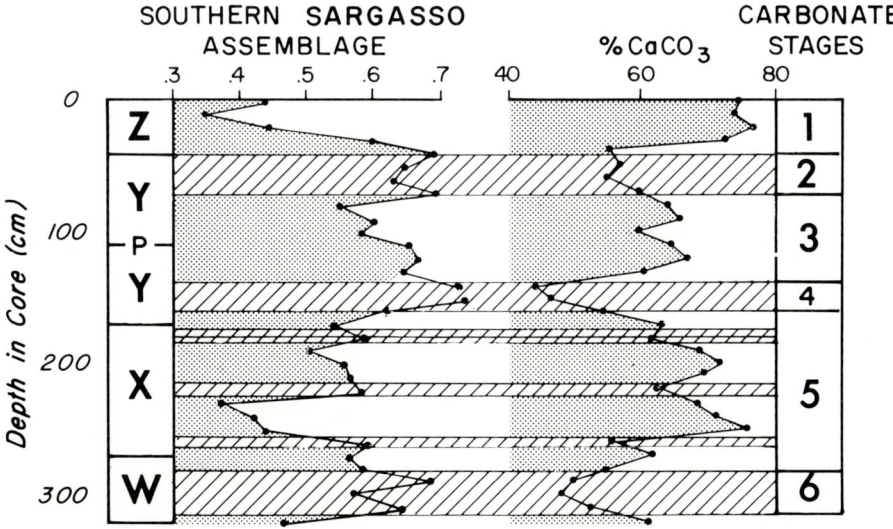

Figure 10. Plot of southern Sargasso assemblage-factor loadings and $CaCO_3$ content for same samples versus depth in core V18-357. Faunal datums are indicated in left column, carbonate stages in right column. Zones of abundant southern Sargasso assemblage and low $CaCO_3$ are diagonally shaded.

lows with Sargasso maxima in carbonate Stage 5. The relationship of the low coarse-fraction content and the abundant southern Sargasso assemblage is also consistent. The same relationships occur in V12-122 and V28-127.

Interpretation of Faunal-Assemblage Variations

The present-day distribution of the southern Sargasso and equatorial-tropical assemblages suggests a hypothesis for interpretation of both the faunal variations and the carbonate curves of the Colombia Basin. In the North Atlantic, the southern Sargasso assemblage occupies water with relatively high salinity and low productivity characterized by low diversity. Bé and others (1971) have shown that the minimum Atlantic plankton abundances occur in the southwestern sector of the Sargasso Sea. This low productivity results from low nutrient supply due to the permanent thermocline and high salinity that blocks upward transport of nutrients into the euphotic zone. These data suggest that, during glacial extremes, conditions similar to present southern Sargasso Sea occurred in the western Caribbean, resulting in high concentrations of *G. ruber*, low productivity, and, consequently, decreased foraminiferal carbonate accumulation.

If this hypothesis is correct, what climate changes could have produced Sargasso type water in the Colombia Basin? Southern Sargasso Sea water must be shifted at least 10°S from its present position for the trade winds to transport it into the Caribbean (Fig. 5). This shift agrees with McIntyre's (1967) conclusion that the northern boundary of the subtropical gyre was displaced, at the height of the last glaciation from its present position, at 40°N, off North America, to approximately 30°N. If this southward shift was combined with a decrease in the strength of the South Equatorial Current or its southward diversion, as was suggested by Imbrie and Kipp (1971), the Colombia Basin may not have been dominated by the Caribbean Current during glacial stages. The faunal data presented here generally support this hypothesis and we conclude that the Colombia Basin was characterized by saline low-productivity waters during the glacial maxima. The scarcity of the equatorial-tropical assemblage during these periods also suggests that the water temperature was <24°C.

The saline, low-productivity, Sargasso type water might also have been produced in situ by increased evaporation relative to precipitation during the glacial stages. This hypothesis has an analog in the present Caribbean climatic pattern, in which the winter is characterized by a more semiarid climate with high evaporation and low precipitation resulting in higher surface salinities in winter. Newell (1973) has shown that the latitude of the Intertropical Convergence Zone (ITCZ) is controlled by the temperature gradient across the midlatitudes and that, during January, the ITCZ mean position of rising motion is approximately 6°S. He concluded that the ITCZ was farther south at 20,000 B.P. Sanchez and Kutzbach (1974) noted that a southward shift of the ITCZ or descending and ascending branches of the Hadley circulation correspond to a southward shift of dry and wet zones, respectively. Therefore, a southward shift of the ITCZ during the glacial stages, and especially during glacial winters, may position the descending Hadley circulation over the Caribbean Sea. Production of the more saline Sargasso type water suggests an intensification of this pattern so that the ITCZ was, on the average, located over South America for a longer time during glacial periods than it is today, which resulted in a longer dry season for the Caribbean. Both increased evaporation/precipitation rate and decreased current flow are probably necessary to maintain the Sargasso type water in the Colombia Basin. For without a decreased Equatorial Current flow into the Caribbean, equatorial and Amazon water of lower salinity would be transported into the Caribbean as it is today during the summer. These changes would be local and additive to a world-wide salinity increase resulting from the storage of sea water in glacial ice.

The faunal data document the alternation of interglacial equatorial-tropical and glacial southern Sargasso faunas and surface oceanographic conditions in the Colombia Basin. The presence of the Sargasso type water in the Colombia Basin suggests greater evaporation/precipitation rates caused by a southward shift of the ITCZ and decreased Caribbean Current inflow during the glacial stages, along with a possible southern shift of the present southern Sargasso Sea.

PALEOCEANOGRAPHIC ESTIMATES

Paleoceanographic estimates of winter (Tw) and summer (Ts) sea-surface temperature and summer sea-surface salinity (Ss) were made using the late Pleistocene faunas of cores V18-357, V12-122, and V28-127. The estimates were derived using Imbrie and Kipp's technique (1971), which quantitatively correlates parameters and surface-sediment faunal assemblages by regression equations and then uses the equations to estimate the oceanographic parameters in Pleistocene samples (see Imbrie and Kipp, 1971, for complete explanation of this technique; and Prell, 1974, for explanation of application). The estimates in this study are based on 27 foraminiferal species in 155 well-preserved Atlantic and Caribbean surface-sediment samples. Oceanographic data were obtained for each sample location from the U.S. Naval Oceanographic Office (1967).

Estimates of sea-surface winter temperature (Tw), summer temperature (Ts), and summer salinity (Ss) over the past 150,000 yr for cores V18-357, V12-122, and V28-127 are presented in Figures 11, 12, and 13. Numerical values for these estimates are given in Prell (1974).

The core V18-357 estimates (Fig. 11) show a Tw range of 4.2°C varying from 27.1°C in Stage 1 to 22.9°C in Stage 4. The Ts range is only 2.6°C ranging from 28.6°C to 26.0°C. Because the single sample at 230 cm is 0.8°C warmer than any other sample, the real summer-temperature range may be only about 2.0°C. The

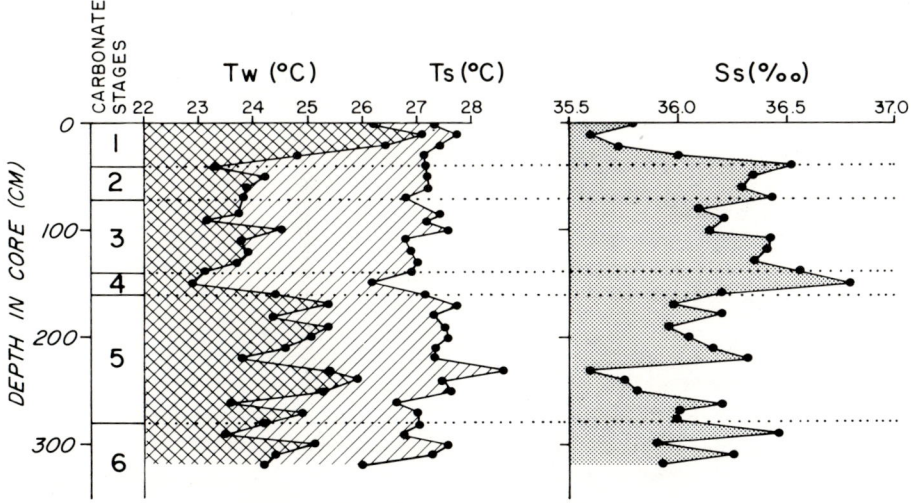

Figure 11. Paleoceanographic estimates of sea-surface winter and summer temperature (Tw and Ts) and summer surface salinity (Ss) versus depth in core V18-357.

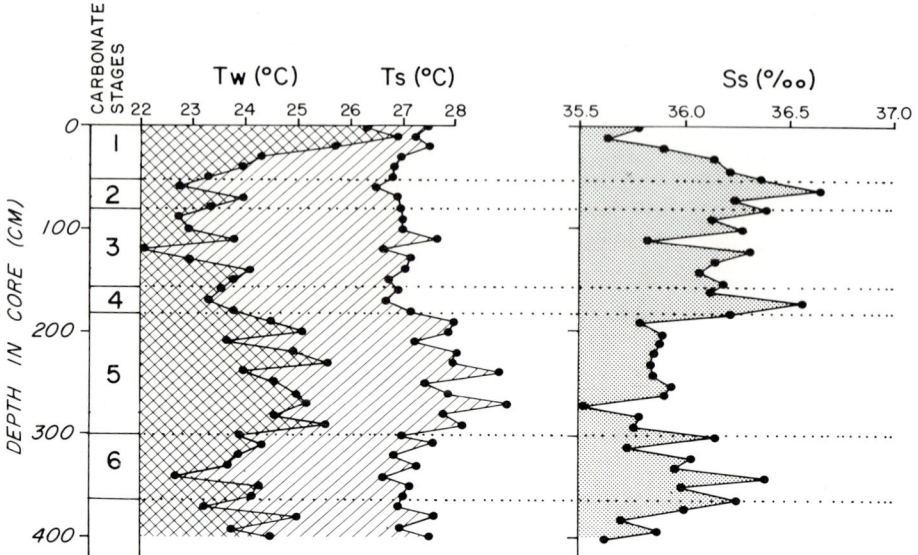

Figure 12. Paleoceanographic estimates of Tw, Ts, and Ss versus depth in core V12-122.

Ss variation is 1.2‰ ranging from 36.8‰ to 35.6‰. The pattern of Ss variation is inverse to the Tw pattern and is most saline in the coldest samples.

Core V12-122 (Fig. 12) has a total Tw range of 5.1°C varying from 26.9° to 21.8°C. The Ts variation is 2.6°C with extremes of 29.1° and 26.5°C. Ss varies from 35.5‰ to 36.6‰ giving a range of 1.1‰.

Because core V28-127 has twice the accumulation rate of V18-357 and V12-122, it is plotted at half scale to ease comparison with the other cores. V28-127 has glacial temperature estimates approximately 0.5°C cooler than V18-357 and V12-122 and has no temperature maxima observed in Stage 1 (Fig. 13). The lack of Stage 1 climatic optimum results in lower temperature ranges and especially lower Tw. The Tw varies from 25.9°C at the core top to 22.1°C, giving a range of 3.8°C.

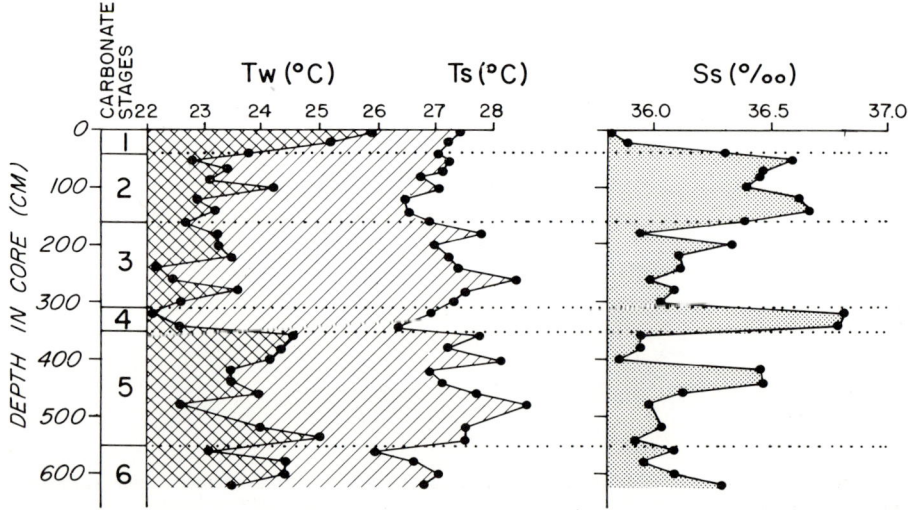

Figure 13. Paleoceanographic estimates of Tw, Ts, and Ss versus depth in core V28-127.

The Ts has a range of 2.6°C varying from 28.6° to 26.0°C. The trigger-core top should represent recent conditions; however, the piston-core sample at 10 cm may be older than the climatic optimum. The difference between Ts and Tw during the glacial stage is about three times greater than present-day and may have been more. Summer salinity varies from 35.8‰ at the core top to 36.8‰, giving a range of 1.0‰.

Several observations may be made concerning the amount and pattern of variation shown by the above parameters: (1) Winter temperatures are higher during Stage 1 than at any time in Stage 5 but summer temperatures of Stage 1, 3, and 5 are similar, with Stage 5 being the warmest. (2) Seasonal contrast ($Ts-Tw$) is much greater during the glacial stages because Tw varies more from its present range than does Ts. Compared to the present seasonal contrast in these cores, the seasonal contrast during the glacial stages is almost three times as great. (3) In all cores and especially, in core V18-357, the Tw curve and the inverse of the Ss curve closely reflect the carbonate and oxygen-isotope curves. These curves match peak for peak in Stage 5 of V18-357 to define all three Barbados terrace correlations (Broecker and van Donk, 1970). (4) Stage 2 and especially Stage 4 represent major fluctuations of all three parameters, particularly Ss. The Ss curve of V28-127 exhibits a short but intense Stage 4 and a longer, less-intense Stage 2. This relationship is also observed to a lesser extent in the other parameters and cores. The well-defined Stage 4 represents a significant departure from the "sawtooth pattern" of Broecker and van Donk (1970). The estimates of paleotemperatures and paleosalinities are reasonable in all cores and closely correlate with known patterns of oxygen-isotope and carbonate content.

SUMMARY AND CONCLUSIONS

Detailed faunal analysis of three sediment cores from the Colombia Basin has defined the variation of planktonic foraminiferal assemblages during late Pleistocene time. Estimates of paleotemperatures and paleosalinities have also been determined for the same faunal data. A combination of these data permits reconstruction of the oceanographic conditions in the Colombia Basin over the past 150,000 yr.

During the maximum glacial phases represented by Stage 2 and 4, the Colombia Basin was dominated by the southern Sargasso assemblage. This assemblage is characterized by a dominance of *G. ruber* and represents the high-salinity and low-productivity conditions of the present-day southern Sargasso Sea. The estimated winter sea-surface temperatures for glacial Stages 2 and 4 are 22° to 23°C, and the summer temperatures are 26° to 27°C. The resulting seasonal contrast ($Ts-Tw$) of 4°C is approximately three times that estimated for the core tops (Fig. 11-13). The estimated summer salinities for glacial Stages 2 and 4 are 36.6‰ to 36.8‰. The stratigraphic record reveals that these southern Sargasso conditions were the most extreme experienced by the Colombia Basin during the past 150,000 yr.

Comparison of the reconstructed glacial conditions with the present-day Caribbean climatic pattern of higher evaporation/precipitation ratios during the winter, when the ITCZ is over South America, implies that the Caribbean experienced less rainfall and thus evaporation/precipitation was increased during the glacial stages. This pattern probably results because the ITCZ was located, on the average, over South America longer than it is during the present-day Caribbean winter. Southward shifts of the ITCZ have been suggested by Newell (1973) and Sanchez and Kutzbach (1974) on the basis of meteorological data. The conclusion of an arid Caribbean Basin is supported by the vegetation patterns of South America, which also responded to glacial climatic changes. Van der Hammen (1961) and van der Hammen and others (1971) noted that the mangrove forests of coastal British Guiana were replaced by grass savannas during periods of lowered sea level. Likewise, the case for increased savannas due to ice-age aridity in tropical South America has been summarized by Damuth and Fairbridge (1970).

Based on cement types and mineralogy of Pleistocene and Holocene eolianities, Ward (1973) inferred an arid or semiarid climate with sparce rainfall and intense evapotranspiration along the northeastern coast of the Yucatan Peninsula during Pleistocene time. Using variations of quartz/kaolinite ratios in Caribbean sediments, Bonatti and Gartner (1973) suggested that high aridity prevailed in the Caribbean Basin and that arid climatic belts expanded toward both the Equator and the middle latitudes during Pleistocene cold periods. This study used the oceanographic associations of the southern Sargasso assemblage, the estimate of greatly increased seasonal contrast during the glacial stages, and the present Caribbean climatic pattern to suggest that the Colombia Basin climate was more arid during the glacial stages.

The stratigraphic record of the Colombia Basin over the past 150,000 yr reveals close correspondence of the faunal-assemblage and temperature-estimate variations with the variations of carbonate and coarse fraction in the sediments. All parameters show that present-day climatic and oceanographic conditions represent a small fraction of the past 150,000 yr. Both the abundance of the equatorial-tropical assemblage and the paleotemperature estimates document that only a small part of the last interglacial (equivalent to the Barbados III terrace at 124,000 B.P.) has faunas and temperatures comparable to the present-day Colombia Basin. Comparison of the Stage 1 climatic optimum to the Barbados III maxima however, shows that the winter temperature estimates are higher and that the equatorial-tropical assemblage is consistently more abundant in Stage 1. Thus, real differences do exist between these two climatic optimums, and predictions based on the 124,000-yr temperature maxima may not be applicable to the present interglacial stage.

ACKNOWLEDGMENTS

We thank Andrew McIntyre, Ted C. Moore, Jr., and Steve Streeter for reviewing this paper, and J. V. Gardner, B. Molifino, and M. Roche for their support and helpful discussions. John Imbrie and Nilva Kipp provided assistance and North Atlantic faunal data.

This work has been supported by National Science Foundation Grant IDO71-04204 under the International Decade for Ocean Exploration Program. Sediment cores for this study were collected under Office of Naval Research Contract N00014-67-A-0108-0004 and National Science Foundation Grants GA-35454 and GA-19690.

REFERENCES CITED

Bé, A.W.H., 1960, Ecology of Recent planktonic foraminifera. Pt. 2—Bathymetric and seasonal distributions in the Sargasso Sea off Bermuda: Micropaleontology, v. 6, p. 373–392.
——1967, Foraminifera families: Globigerinidae and Globorotalidae, *in* Fraser, J. H., ed., Fiches d'Identification du zooplankton: Charlottenlund, Denmark. Cons. Internat. Explor. Mer, fiche 108.
Bé, A.W.H., and Tolderlund, D. S., 1971, Distribution and ecology of living planktonic

foraminifera in surface waters of the Atlantic and Indian Oceans, *in* Funnell, B., and Riedel, W. R., eds., Micropaleontology of oceans: London, Cambridge Univ. Press, p. 105–149.

Bé, A.W.H., Vilks, G., and Lott, L., 1971, Winter distribution of planktonic foraminifera between the Grand Banks and the Caribbean: Micropaleontology, v. 17, p. 31–42.

Bonatti, E., and Gartner, S., 1973, Caribbean climate during Pleistocene ice ages: EOS (Am. Geophys. Union, Trans.), v. 54, no. 4, p. 327–328.

Broecker, W. S., and van Donk, J., 1970, Insolation changes, ice volumes and the O^{18} record in deep-sea cores: Rev. Geophysics, v. 8, p. 169–198.

Colon, F. A., 1963, Seasonal variations in heat flux from the sea surface to the atmosphere over the Caribbean Sea: Jour. Geophys. Research, v. 68 (5), p. 1421–1430.

Damuth, J. E., and Fairbridge, R. W., 1970, Equatorial Atlantic deep-sea arkosic sands and ice-age aridity in tropical South America: Geol. Soc. America Bull, v. 81, p. 189–206.

Emiliani, C., 1955, Pleistocene temperatures: Jour. Geology, v. 63, p. 538–578.

——1964, Paleotemperature analysis of the Caribbean cores A254-B-R-C and CP-28: Geol. Soc. America Bull., v. 75, p. 129–144.

——1966, Paleotemperature analysis of Caribbean cores P6304–8 and P6304–9 and a generalized temperature curve for the past 425,000 years: Jour. Geology, v. 74, p. 109–126.

——1972, Quaternary paleotemperatures and the duration of the high-temperature intervals: Science, v. 178, p. 398–401.

Ericson, D. B., and Wollin, G., 1956, Correlation of six cores from the equatorial Atlantic and the Caribbean: Deep-Sea Research, v. 3, 104–125.

Ericson, D. B., Ewing, M., Wollin, G., and Heezen, B. C., 1961, Atlantic deep-sea sediment cores: Geol. Soc. America Bull., v. 72, p. 193–276.

Gardner, J. V., 1973, The eastern equatorial Atlantic: Sedimentation, faunal, and sea-surface temperature responses to global climatic changes during the past 200,000 years [Ph.D. dissert.]: New York, Columbia Univ., 387 p.

Gordon, A. L., 1967, Circulation of the Caribbean Sea: Jour. Geophys. Research, v. 72, p. 6207–6223.

Hecht, A. D., 1973, Faunal and oxygen isotopic paleotemperatures and the amplitude of glacial/interglacial temperature changes in the equatorial Atlantic, Caribbean Sea, and Gulf of Mexico: Quaternary Research, v. 3, no. 4, p. 671–690.

Imbrie, J. E., and Kipp, N., 1971, A new micropaleontological method for quantitative paleoclimatology: Application to a late Pleistocene Caribbean core, *in* Turekian, K. K., ed., The late Cenozoic glacial ages: New Haven, Yale Univ. Press, p. 71–179.

Imbrie, J. E., van Donk, J., and Kipp, N., 1973, Paleoclimatic investigation of late Pleistocene Caribbean deep-sea cores: Comparison of isotopic and faunal methods: Quatenary Research, v. 3, p. 10–38.

Kennett, J. P., and Huddlestun, P., 1972, Late Pleistocene paleoclimatology, foraminiferal biostratigraphy and tephrochronology, western Gulf of Mexico: Quaternary Research, v. 2, p. 38–69.

Klovan, J., and Imbrie, J. E., 1971, An algorithm and Fortran IV program for large scale Q-mode factor analysis: Internat. Assoc. Mathematical Geology Jour., v. 3, p. 61–77.

Lidz, L., 1966, Deep-sea Pleistocene biostratigraphy: Science, v. 154, p. 1448–1452.

McIntyre, A., 1967, Coccoliths as paleoclimatic indicators of Pleistocene glaciation: Science, v. 158, p. 1314–1317.

McIntyre, A., Ruddiman, W. F., and Jantzen, R., 1972, Southward penetration of the North Atlantic polar front: Faunal and floral evidence of large-scale surface water mass movements over the last 225,000 years: Deep-Sea Research, v. 19, p. 61–77.

Newell, R. E., 1973, Climate and the Galapagos Islands: Nature, v. 245, p. 91–92.

Parker, F. L., 1962, Planktonic foraminiferal species in Pacific sediments: Micropaleontology, v. 8, p. 219–254.

Perlroth, I., 1971, Distribution of mass in the near surface waters of the Caribbean: Symposium on investigations and resources of the Caribbean Sea and adjacent regions: Paris, UNESCO, 545 p.

Prell, W. L., 1974, Late Pleistocene faunal, sedimentary, and temperature history of the Colombia Basin, Caribbean Sea [Ph.D. dissert.]: New York, Colombia Univ., 400 p.

Ruddiman, W. F., 1969, Foraminifera of the subtropical North Atlantic gyre [Ph.D. thesis]: New York, Columbia Univ., 282 p.

Sanchez, W. A., and Kutzbach, J. E., 1974, Climate of the American tropics and subtropics in the 1960s and possible comparisons with climatic variations of the last millenium: Quaternary Research, v. 4, no. 2, p. 128–135.

U.S. Naval Oceanographic Office, 1967, Oceanographic atlas of the North Atlantic Ocean, Sec. II, Physical properties: U.S. Naval Oceanographic Office Pub. no. 700, 300 p.

van der Hammen, Th., 1961, The Quaternary climatic changes on northern South America: New York Acad. Sci. Annals v. 95, p. 676, 683.

van der Hammen, Th., Wijmstra, T. A., and Zagwijn, W. H., 1971, The floral record of the late Cenozoic of Europe, *in* Turekian, K. K., ed., The late Cenozoic glacial ages: New Haven, Yale Univ. Press, p. 391–424.

Ward, W. C., 1973, Influence of climate on the early diagenesis of carbonate eolianities: Geology, v. 1, no. 4, p. 171–174.

Wust, G., 1964, Stratification and circulation in the Antillean-Caribbean Basins: New York, Columbia Univ. Press, 201 p.

Manuscript Received by the Society March 22, 1974
Revised Manuscript Received September 23, 1974
Manuscript Accepted October 7, 1974
Lamont-Doherty Contribution no. 2272

Geological Society of America
Memoir 145
© 1976

Responses of Sea-Surface Temperature and Circulation to Global Climatic Change During the Past 200,000 Years in the Eastern Equatorial Atlantic Ocean

James V. Gardner[*]

and

James D. Hays
Lamont-Doherty Geological Observatory
and
Department of Geological Sciences
Columbia University
New York, New York 10927

ABSTRACT

Analyses of deep-sea cores from the equatorial Atlantic Ocean suggest strong variation in the intensity of atmospheric and oceanic circulation in response to the waxing and waning of ice sheets during the past 200,000 yr.

Comparisons of estimates of paleotemperature, determined by multivariate statistical analysis, between Holocene sediments and sediments from an 18,000 B.P. datum in an equatorial core show only small changes (1° to 2°C) for estimated temperatures for February, but changes of 2° to 10°C for temperature estimates for August. The average "annual" paleotemperatures for this equatorial core agree well with paleotemperatures calculated from oxygen isotope data.

By contrast, the zone of upwelling off northwest Africa shows almost a 10°C decrease in temperatures for February 18,000 B.P. but only a 1° to 4°C difference in August. The region between the equator and the upwelling zone shows very little change in sea-surface temperature for the past 200,000 yr.

Inferences drawn from estimates of paleotemperatures and faunal analyses suggest that during glacial stages the Intertropical Convergence Zone (ITCZ) was essentially in its interglacial position with seasonal migrations comparable to today's. The difference between glacial and interglacial modes is one of intensity. The glacial

---
[*]Present address: Pacific-Arctic Branch of Marine Geology, U.S. Geological Survey, Menlo Park, California 94025.

mode was more intense than the interglacial mode with the southeast and northeast trade winds strengthening seasonally during the winters of the Southern and Northern Hemisphere respectively.

INTRODUCTION

Global climate is continuously changing; during the past 200,000 yr at least two complete glaciations, one complete interglaciation, and the beginning of another interglacial stage have occurred. Our aim in this paper is to document the oceanic response in the equatorial Atlantic to these changes by detailed study of deep-sea sediments. The region lies in the trade winds belt and contains the Benguela and Canary Currents and their westward confluence, the South Equatorial Current system.

Some of the earliest studies of the climatic record of deep-sea sediments were in the equatorial Atlantic, and the area has been one of the most actively studied during the past 40 yr. Schott (1935) suggested that the presence and absence of *Globorotalia menardii*, a planktonic foraminifer, reflects alternation of past warm and cool climates, respectively. Schott's work was extended and amplified by Cushman and Henbest (1940), Phleger (1942), Ovey (1950), Phleger and others (1953), Olausson (1965), Ericson and Wollin (1968), and Ruddiman (1971). These authors demonstrated that warm-water assemblages of planktonic foraminifers alternate stratigraphically with cooler-water assemblages, suggesting changing conditions in the surface and near-surface water. Ericson and others (1961) established for late Quaternary time a biostratigraphic zonation that is based on planktonic foraminifers and allows detailed correlation of cores throughout the equatorial and temperate Atlantic.

Emiliani (1955) brought studies of the equatorial Atlantic one step further by analyzing cores for variations of oxygen isotopes in planktonic foraminifers and by calculating paleotemperatures from these isotope data. Although a controversy developed (Shackleton, 1967; Shackleton and Opdyke, 1973) regarding the amount of isotope change that resulted from extraction of large amounts of water from the ocean during glacial periods, Emiliani's work initiated the era of quantitative Pleistocene paleoclimatology.

The work of Emiliani (1955, 1966), Broecker and others (1958, 1968), Broecker (1966), Ku and Broecker (1966), Broecker and Ku (1969), and Broecker and van Donk (1970) led to a correlation of the biostratigraphic zonation with an absolute time scale.

Our work is an attempt to refine the prediction of paleotemperatures using multivariate statistical analyses of foraminiferal populations and closely spaced samples (2,000 to 3,000-yr intervals) to infer paleo-oceanographic and paleometeorologic circulation during the past 200,000 yr from these estimates.

We selected 11 piston cores that have continuous records at least back to 200,000 B.P. and have accumulation rates greater than $1.5 \text{ cm}/10^3$ yr (Gardner, 1973). Our criteria were calcium carbonate analyses and a planktonic foraminiferal biostratigraphy that departs from that of Ericson and others (1961) by defining the X-Y boundary as the last appearance of *G. hexagona*. This places our X'-Y' boundary somewhat later than that of Ericson and others (1961). We also selected a datum level that represents full glacial conditions, is easily identified, and is synchronous in each of the cores in this study and correlative with cores throughout the tropical and subtropical Atlantic. On the basis of planktonic foraminiferal biostratigraphy and C^{14} dating, we estimate the datum to be 18,000 B.P. (see McIntyre and others, this volume, for detailed discussion). Total counts of the planktonic foraminifers were made on samples from the 18,000 B.P. level in 11 cores and on closely spaced samples from 3 cores that contain a continuous record of the past 200,000 yr. Transfer functions for sea-surface temperatures for August and February were formulated and used to predict August and February sea-surface temperatures for the 18,000 B.P. level and at 3 sites for a record of the past 200,000 yr.

QUANTITATIVE ESTIMATES OF SEA-SURFACE PALEOTEMPERATURE

Quantitative estimates of sea-surface temperatures can be calculated using the technique developed by Imbrie and Kipp (1971). Briefly, their method involves counting all individuals of species constituting more than 2% of the assemblage of a faunal or floral group throughout a wide geographical area. These data are factor analyzed and resolved into assemblages that define independent factors (factor assemblages). A linear or curvilinear least-squares regression is then made between factor assemblages and observed oceanic parameters. This regression represents an equation (or transfer function) for the parameter in question that can be used with matrices of factor assemblages of stratigraphic data to generate estimates of paleoenvironmental parameters. The underlying implicit assumptions are (1) fossil species reacted to the same environmental parameters and with the same degree of response as do their modern representatives, and (2) the species (assemblages) have a linear or near-linear response either to the parameter being estimated or to some other unknown parameter that is in turn linearly or nearly linearly related to it.

We used several modifications of the method of Imbrie and Kipp (1971) in this study. (1) For their data base, Imbrie and Kipp selected only samples having indications of very little dissolution. Our study included samples showing moderate to severe dissolution effects; we rejected samples in which fewer than 200 planktonic foraminifers could be counted from a standard sample size (25 cc) or samples deemed "dissolved" using the criteria discussed below. By including faunas from dissolution facies, a small amount of precision was sacrificed but many more data points could be used to generate transfer functions. This yielded some detail in the numerous intervals that have undergone dissolution. (2) We followed J. Imbrie's suggestion that we eliminate the percent range transformation step in the calculations. (3) Imbrie and Kipp used 61 piston core tops from the whole Atlantic Ocean for a reference base whereas we used 199 trigger-weight core tops from all parts of the North Atlantic Ocean. (See Table 1 for locations of all cores used in this study.) The 199 trigger-weight core-top analyses included 31 of Imbrie and Kipp's original 61 samples. The remaining 168 analyses represent a group effort by workers from Brown University (under the direction of N. Kipp) and Lamont-Doherty Geological Observatory (W. L. Prell, T. Kellogg, and J. V. Gardner) to generate a comprehensive North Atlantic coverage.[1] (4) The final difference between Imbrie and Kipp (1971) and our study is that some species, subspecies, or varieties have been added together and considered as one variable. From the original 30 species and varieties counted (Table 2 and Taxonomic Notes), the clearest results were obtained when the following combinations were used: (1) Species that never reached abundances of 2% were eliminated. These species simply represented noise in

---
[1] Faunal counts made on all samples are included in Appendixes I and II on microfiche in pocket inside back cover.

TABLE 1. LOCATION AND WATER DEPTH OF THE CORES IN THIS STUDY

Core	Lat (N)	Long	Depth (m)	Core	Lat (N)	Long	Depth (m)

North Atlantic

Core	Lat (N)	Long	Depth (m)	Core	Lat (N)	Long	Depth (m)
A152 84	4421	3016 W	2750	RC10 22	2118	6940 W	3832
A153 154	2800	3847 W	4020	RC10 49	1634	7932 W	1238
A157 3	5056	4145 W	4025	RC10 256	1329	8138 W	1146
A164 13	3543	6720 W	4940	RC11 10	0458	4558 W	3834
A164 14	3606	6719 W	4810	RC11 11	0213	4524 W	3759
A164 15	3608	6855 W	4480	RC11 12	0055	4346 W	4145
A164 16	3608	6908 W	4624	RC11 13	0141	4131 W	4378
A164 17	3547	6856 W	4665	RC13 146	1154	7502 W	3387
A164 23	3613	6924 W	4512	RC13 147	1253	7451 W	3941
A167 1	3739	7257 W	2706	RC13 150	1537	7140 W	3788
A167 12	3150	7421 W	4700	RC13 152	1642	7526 W	2027
A167 13	3139	7521 W	2880	RC13 153	1504	7557 W	3334
A167 18	2946	7648 W	914	RC13 154	1453	7745 W	2308
A173 12	4215	6326 W	2103	RC13 156	1002	7657 W	3103
A179 6	1924	7833 W	4731	RC13 157	1115	7809 W	3588
A179 13	2356	7545 W	1847	RC13 158	1311	7950 W	3016
A179 20	3047	6740 W	4993	RC13 191	0111	2056 W	4960
A179 24	3546	6905 W	4663	RC13 194	0431	1341 W	4770
A180 13	3908	4239 W	4880	RC13 196	0342	0743 W	3735
A180 15	3916	3642 W	4610	SP9 3	5352	2106 W	2740
A180 20	3334	2722 W	4040	SP10 5	6329	0004 W	2012
A180 39	2550	1918 W	3470	V2 4	4315	5617 W	4076
A180 58	1140	1735 W	2240	V2 9	4228	5454 W	4702
A180 70	0339	1818 W	4846	V3 128	2346	9229 W	3495
A180 72	0036	2147 W	3841	V4 8	3714	3308 W	1655
A180 73	0010	2300 W	3749	V4 12	3651	2101 W	4510
A181 7	1033	5730 W	3766	V4 32	3503	1137 W	2230
A181 9	2044	5912 W	5215	V5 1	3236	6930 W	5135
ALB230	1012	2636 W	5500	V5 31	3217	6507 W	1554
ALB232	0753	2443 W	4975	V5 40	3213	6443 W	1902
ALB233	0713	2238 W	4125	V7 42	3855	5708 W	5260
ALB234	0545	2143 W	3577	V7 53	3654	5402 W	5258
ALB235	0312	2028 W	4560	V7 67	3440	6128 W	4308
ALB236	0206	2007 W	5065	V7 68	4046	6436 W	4133
ALB240	0059	1837 W	7500	V7 77	3822	7339 W	582
ALB242	0030	2538 W	3620	V8 7	1224	7746 W	3823
ALB243	0027	2745 W	3740	V9 31	0814	3752 W	4204
R9 7	5939	2246 W	2770	V10 80	3123	1137 W	2550
RC7 21	1416	7535 W	4094	V10 88	2258	3812 W	4971
RC8 106	1400	7453 W	3904	V10 89	2302	4348 W	3523
RC9 61	1011	7713 W	3191	V10 98	3126	6411 W	4299
RC9 222	4506	1034 W	4903	V12 4	2417	5304 W	5009
RC9 225	5458	1523 W	2334	V12 80	0145	1603 W	4918
V12 122	1700	7424 W	2800	V22 186	0323	2007 W	4471
V14 4	1529	4031 W	4473	V22 188	0440	2055 W	2600
V14 5	0051	3251 W	3255	V22 193	0955	2058 W	4956
V16 20	1756	5021 W	4410	V22 196	1350	1858 W	3728
V16 21	1717	4825 W	3975	V22 197	1410	1835 W	3167
V16 23	1315	4040 W	4886	V22 202	1424	2109 W	4310
V16 200	0158	3704 W	4092	V22 204	1501	2314 W	1723
V16 205	1524	4324 W	4043	V22 211	2042	3127 W	4402
V16 206	2320	4629 W	3734	V22 219	2755	4338 W	2582
V16 209	3000	5152 W	4673	V22 230	3239	5218 W	5048
V16 227	6003	5050 W	3305	V22 232	3445	5715 W	4685
V17 1	2829	6503 W	5027	V23 13	4131	4508 W	4845
V17 158	1223	1855 W	4358	V23 22	5412	4558 W	3669
V17 162	2458	2856 W	5480	V23 38	6240	2732 W	1426
V17 163	2728	3408 W	5132	V23 81	5415	1650 W	2393
V17 164	2937	3655 W	4433	V23 82	5235	2156 W	3974
V17 165	3245	4154 W	3924	V23 83	4952	2415 W	3871
V17 167	4230	5906 W	4352	V23 84	4600	1655 W	4513
V17 196	6044	5750 W	2818	V23 96	2948	1505 W	3471
V18 16	1344	5107 W	4931	V23 101	1953	2532 W	4482
V18 21	0414	4745 W	2374	V23 105	1712	3550 W	5009
V18 357	1502	8014 W	1818	V23 107	1729	4040 W	5104
V19 19	1314	7822 W	3760	V24 8	2236	7236 W	4887
V19 21	1036	7921 W	3168	V24 28	1519	7757 W	2274
V19 284	0016	0446 E	3937	V24 31	1215	7638 W	3623
V19 291	0200	0515 W	4956	V25 24	2647	4642 W	4563
V19 293	0219	0447 W	4814	V25 46	0919	4300 W	4310

TABLE 1. - Continued

Core	Lat (N)	Long	Depth (m)	Core	Lat (N)	Long	Depth (m)

North Atlantic - Continued

Core	Lat (N)	Long	Depth (m)	Core	Lat (N)	Long	Depth (m)
V19 295	0257	0607 W	4605	V26 46	0934	1811 W	2898
V19 296	0125	0905 W	5017	V26 50	0616	1755 W	4826
V19 297	0237	1200 W	4122	V26 51	0602	1815 W	4572
V19 298	0339	1503 W	4792	V26 52	0550	1805 W	4903
V19 299	0439	1729 W	5000	V26 53	0018	1448 W	4323
V19 300	0653	1928 W	4263	V26 124	1608	7427 W	3005
V19 301	0810	2245 W	4724	V26 125	1704	7614 W	2210
V19 302	1015	2522 W	5583	V27 10	4216	6030 W	3895
V19 308	2901	4124 W	3197	V27 20	5400	4612 W	3510
V19 310	3318	4816 W	4607	V27 28	5811	4437 W	2308
V20 7	1133	6031 W	1018	V27 111	5604	2405 W	2809
V20 233	0200	3536 W	3884	V27 122	4818	1658 W	4696
V20 234	0519	3302 W	3133	V27 137	4241	1704 W	4883
V20 235	0828	3008 W	5242	V27 144	3935	1333 W	4894
V20 242	2322	4339 W	4565	V27 162	3411	1651 W	4281
V20 253	3817	6832 W	4889	V27 164	2929	1937 W	4623
V21 222	1427	7714 W	4060	V27 167	2556	2635 W	5099
V22 24	1245	4538 W	4321	V27 172	1632	2851 W	4938
V22 26	0843	4115 W	3720	V27 173	1457	2713 W	4931
V27 174	1251	2500 W	4912	V29 176	4033	2600 W	3064
V27 176	0711	2522 W	4232	V29 177	4132	2542 W	3391
V27 178	0506	2639 W	4327	V29 179	4401	2432 W	3331
V27 179	0412	2334 W	4592	V29 180	4518	2352 W	3179
V27 245	0144	0904 W	5229	V29 183K	4908	2530 W	3629
V27 249	0114	1222 W	5212	V29 184	4908	2530 W	3645
V27 251	0323	1433 W	4746	V29 189	5222	1732 W	4039
V27 266	3535	4354 W	4804	V29 190	5240	1510 W	1458
V28 34	6450	0335 W	3217	V29 193	5524	1844 W	1326
V28 41	6741	0014 E	1904	V29 194	5700	2119 W	2085
V28 111	1456	7355 W	4010	V29 198	5844	1534 W	1139
V28 115	1356	7222 W	3970	V29 200	5957	1912 W	2650
V28 116	1322	7418 W	4027	V29 202	6023	2858 W	2658
V28 122	1146	7841 W	3623	V29 203	6048	2226 W	2085
V28 126	1059	8109 W	3162	V29 204	6111	2300 W	1849
V28 127	1139	8008 W	3237	V29 205	6133	2507 W	1450
V28 128	1136	8045 W	3264				

Core	Lat (S)	Long	Depth (m)	Core	Lat (S)	Long	Depth (m)

South Atlantic

Core	Lat (S)	Long	Depth (m)	Core	Lat (S)	Long	Depth (m)
A180 76	0046	2602 W	3512	V 14 67	3912	1330 E	3016
A180 78	0130	2701 W	4261	V 14 77	2939	3252 E	1818
RC 11 16	0246	3513 W	3835	V 14 81	2826	4347 E	3634
RC 11 21	1716	3558 W	4021	V 14 90	1623	6109 E	3314
RC 11 26	2835	3004 W	2349	V 15164	0945	3424 W	3588
RC 11 37	3159	3532 W	2578	V 16 33	1520	1943 W	4360
RC 11 59	3805	5407 W	1849	V 16 41	2752	0106 W	4462
RC 11 79	4900	0436 W	3385	V 16 50	3321	1626 E	2376
RC 11 80	4645	0003 W	3656	V 16189	2850	4102 W	3781
RC 11 86	3547	1827 E	2829	V 16190	2757	4227 W	2919
RC 12234	5509	6400 W	2027	V 18 34	3121	3649 W	3252
RC 12241	4328	5740 W	3449	V 18110	5355	4442 W	2610
RC 12291	4235	1748 W	3508	V 19216	2520	3647 E	2206
RC 12294	3716	1006 W	3308	V 19222	3322	3424 E	2005
RC 12297	3609	0643 W	3528	V 19245	2612	0441 E	2725
RC 12298	3457	0445 W	2213	V 20167	2103	7230 E	3634
RC 12304	3313	2911 E	3290	V 20213	2820	1309 E	2175
RC 12305	3018	3402 E	2794	V 20230	0157	3902 W	3294
V 12 18	2842	3430 W	4021	V 22 71	2925	3325 W	3314
V 12 43	4519	5759 W	3880	V 22122	3935	2435 E	3272
V 12 53	4054	2023 W	3797	V 22172	1240	0949 W	4127
V 12 56	3630	0806 E	3222	V 26 54	0528	1156 W	2906
V 12 66	2259	0701 E	2760	V 26 55	1136	1534 W	3457
V 12 79	0131	1147 W	3823	V 26 63	2358	3757 W	3619
V 14 61	5428	0236 W	1835	V 26 68	3021	1548 W	3233

Note: Latitudes and longitudes are in degrees and minutes; water depths are in corrected metres.

TABLE 2. SPECIES USED IN THIS STUDY

Orbulina universa	*Globoquadrina dutertrei*
Globigerinoides conglobatus	*Globoquadrina hexagona*
Globigerinoides ruber	*Globorotalia inflata*
Globigerinoides tenellus	*Globorotalia truncatulinoides*
Globigerinoides sacculifer	*Globorotalia hirsuta*
Globigerina rubescens	*Globorotalia crassaformis*
Globigerina quinqueloba	*Globorotalia scitula*
Globigerina pachyderma	*Globorotalia menardii*
Globigerina bulloides	*Globorotalia tumida*
Globigerina falconensis	*Globorotalia tumida flexuosa*
Globigerina humilis	*Pulleniatina obliquiloculata*
Globigerina calida	*Candeina nitida*
Globigerina digitata	*Sphaeroidinella dehiscens*
Globigerinella aequilateralis	*Hastigerina pelagica*
Globigerinita glutinata	*G. pachyderma–G. dutertrei* intergrade ("PD intergrade")

the final solution. (2) Pink and white varieties of *Globigerinoides ruber* were combined. No distinctive patterns were recognized when they were separated, and the overall importance of *G. ruber* was diminished. (3) *Globigerinoides sacculifer* with and without the sac-like final chamber were added together. (4) *Globoquadrina dutertrei* and a *G. dutertrei–Globigerina pachyderma* intergrade ("PD intergrade"; see Taxonomic Notes) were merged. The abundance of *G. dutertrei* relative to *G. pachyderma* indicates that the "PD intergrade" encountered in this study is probably more closely related to *G. dutertrei*. Again, we combined the two for a stronger signal. (5) *Globorotalia menardii* and *Globorotalia tumida* were combined. These species are most important as dissolution indicators. Maps of plankton abundance (Bé and Tolderlund, 1971) and undissolved fossil assemblages (Ruddiman, 1969) show these two species represent between 5% and 15% of the total fauna. When these two species were combined, a strong dissolution factor appeared and those cores could be eliminated. (6) Right- and left-coiling varieties of *G. pachyderma* were combined. Only one or two individuals of left-coiling *G. pachyderma* were found in the entire area; thus, this grouping did not appreciably change the *G. pachyderma* influence. (7) Right- and left-coiling varieties of *Globorotalia truncatulinoides* were combined. The remarks concerning *G. pachyderma* apply to *G. truncatulinoides* as well. Using the above modifications, we reduced the total number of variables for each sample from 34 to 24. The 199 analyses from the data base were processed by Q-mode factor analysis using the program CABFAC of Klovan and Imbrie (1971). The general theory of factor analysis is detailed in Imbrie and van Andel (1964), Manson and Imbrie (1964), and Krumbein and Graybill (1965). This process reduces the number of variables (species) to a minimum number of factors (composite species assemblages) while maximizing the cumulative variance of the original data. A regression analysis was then performed on the results of the factor analysis against sea-surface temperatures for August (T_{Aug}) and then sea-surface temperatures for February (T_{Feb}) for each of the 199 core locations. The temperatures were obtained from the *Oceanographic Atlas of the North Atlantic* (U.S. Naval Oceanographic Office, 1967). Second-order regression equations were found for T_{Aug} and T_{Feb}. Multiple correlation coefficients for the regressions and standard error of estimates, which are essentially the precision of the regressions, are given in Table 3.

To test the accuracy of these transfer functions, we used 50 faunal analyses of surface samples from the South Atlantic, with a latitudinal range of 0° to 55°S

TABLE 3. STATISTICS ON NORTH ATLANTIC EQUATIONS AND SOUTH ATLANTIC TEST OF THE EQUATIONS

Number of samples	199
Number of factors	4
Multiple correlation coefficient	$T_{Feb} = 0.965$
	$T_{Aug} = 0.937$
Standard error of estimate	$T_{Feb} = \pm 1.96°C$
	$T_{Aug} = \pm 1.91°C$
Standard deviation of	$T_{Feb} = \pm 1.08°C$
South Atlantic residuals	$T_{Aug} = \pm 1.84°C$

(see Table 1). The observed sea-surface temperatures for August and February were compared to the estimated T_{Aug} and T_{Feb} for each core. The standard deviations of the differences between the estimated and observed T_{Aug} and T_{Feb} were calculated and are also shown in Table 3. This test for the accuracy of the equations based on surface samples indicates that the estimated temperatures are highly reliable; however, the test may not be completely valid for the accuracy of the paleotemperatures. This possibility implies that glacial faunas reacted in a different way or to different environmental parameters than do interglacial faunas, which seems unlikely.

The Q-mode factor analysis of the faunal data from the 199 trigger-weight core tops resolved four factors that account for 90.6% of the original data. The identification of the resolved factors (tropical, subpolar-polar, gyre margin, and subtropical) is based on the important species for each factor (Table 4) and their correlated distribution with the plankton data of Bé and Tolderlund (1971) and

TABLE 4. Q-MODE FACTOR ANALYSIS RESULTS FROM 199 SURFACE SAMPLES FROM THE NORTH ATLANTIC SPECIES VERSUS FACTOR MATRIX*

	FACTORS			
Species	Tropical	Subpolar-polar	Gyre margin	Subtropical
O. universa	0.026	0.015	0.057	0.008
G. conglobatus	0.017	0.000	0.028	0.008
G. ruber	0.885	−0.039	−0.106	0.238
G. tenellus	0.019	−0.003	−0.016	0.039
G. sacculifer	0.430	0.029	0.240	−0.395
S. dehiscens	−0.004	−0.006	0.050	0.006
G. aequilateralis	0.083	0.002	0.029	0.022
G. calida	0.010	0.002	−0.002	0.033
G. bulloides	−0.039	0.380	−0.027	0.295
G. falconensis	−0.003	0.020	−0.030	0.376
G. digitata	0.004	0.002	0.011	0.003
G. rubescens	0.019	−0.001	−0.012	0.018
G. hexagona	0.000	0.000	0.000	0.000
P. obliquiloculata	0.000	−0.009	0.187	0.026
G. inflata	−0.064	0.261	0.089	0.553
G. crassaformis	0.009	−0.002	0.042	−0.001
G. hirsuta	−0.010	0.010	0.002	0.166
G. scitula	−0.004	0.028	−0.006	0.064
C. nitida	0.010	0.001	−0.004	−0.008
G. glutinata	0.125	0.225	−0.048	0.074
"PD Intergrade" and G. dutertrei	0.016	0.305	0.439	0.018
G. truncatulinoides (R&L)†	0.015	0.013	−0.015	0.308
G. menardii & G. tumida	−0.009	−0.083	0.820	0.063
G. pachyderma (R&L)	0.024	0.796	−0.097	−0.336

*The higher the value, the more important that species is to the factor.
†(R&L) = Right and left-coiling.

the surface-sediment assemblages of Ruddiman (1969) and Imbrie and Kipp (1971). The geographic distributions of these four factors (Fig. 1A–E) show that the high factor loadings (>0.90) for the tropical factor cover the region from the equator north to lat 30°N and those for the subpolar-polar factor extend south from the Norwegian Sea to lat 40° to 50° N. Using plankton data, Bé and Hamlin (1967) found a boundary, at about 25°N, that separates North Atlantic subtropical from tropical faunal zones. Ruddiman (1969) recognized this boundary in surface-sediment distributions, and Imbrie and Kipp (1971) saw this boundary as separating tropical from subtropical factor assemblages. Our subpolar-subtropical boundary (at lat 40° to 50°N) also correlates roughly with a boundary in the plankton data of Bé and Hamlin (1967).

A narrow belt of subtropical factor loadings occurs between the tropical and subpolar-polar factors. The low values of the subtropical factor loadings probably indicate that this assemblage differs somewhat from that found by Imbrie and Kipp (1971). Our data suggest that the gyre margin assemblage is predominantly a Gulf Stream-type assemblage, matching what Ruddiman (1969) called the Warm Peripheral Extended Water Mass. Although weaker in strength, the same gyre margin assemblage is found in the equatorial zone.

The significant point is that the factor analysis of the 199 samples grouped the data into four assemblages that, when mapped, separated into ecologically distinct, well-known zoogeographic areas.

Cores located south of the thermal equator experience summer conditions in February and winter conditions in August because they are influenced by events taking place in the Southern Hemisphere. In order to avoid confusion, predictions for August and February sea-surface temperatures for cores south of the thermal equator (lat ~8°N) have been reversed to February and August temperatures, thereby giving a synoptic view of equatorial conditions. This reversal is confirmed in the test of the equations using South Atlantic core tops. Therefore, here T_{Feb} will correspond to Northern Hemisphere winters and T_{Aug} to Northern Hemisphere summers.

18,000 B.P. (GLACIAL) SEA-SURFACE TEMPERATURES

Estimates of sea-surface temperatures for T_{Feb} (Fig. 2A) during the glacial maximum show the same broad pattern found today (Fig. 2C). The estimates of T_{Feb} during the glacial maximum indicate extensive upwelling off northwest Africa with sea-surface temperatures below 20°C, which is 5° cooler than temperatures found there today. The estimates of T_{Feb} during the glacial maximum for the remainder of the area are comparable to sea-surface temperatures in that region today.

The distribution of sea-surface temperatures for August during the glacial maximum (Fig. 2B) also shows a general pattern that is consistent with the present configuration (Fig. 2D). The upwelling that occurs in February is reduced, but a tongue of cool surface water invades the equatorial belt from the southeast. The estimates of glacial T_{Aug} indicate an intrusion of surface water, with temperatures less than 20°C and a core of 16°C water. The sea-surface temperatures during the glacial maximum are as much as 8°C cooler than those found there today. The two glacial stage maps also indicate that the location of the thermal equator in August during the glacial maximum was in the approximate location of the present August thermal equator (lat 8°N), but the thermal equator during February was about 4° farther south and was centered around the geographic equator.

The sea-surface temperatures during the glacial maximum are interpreted as

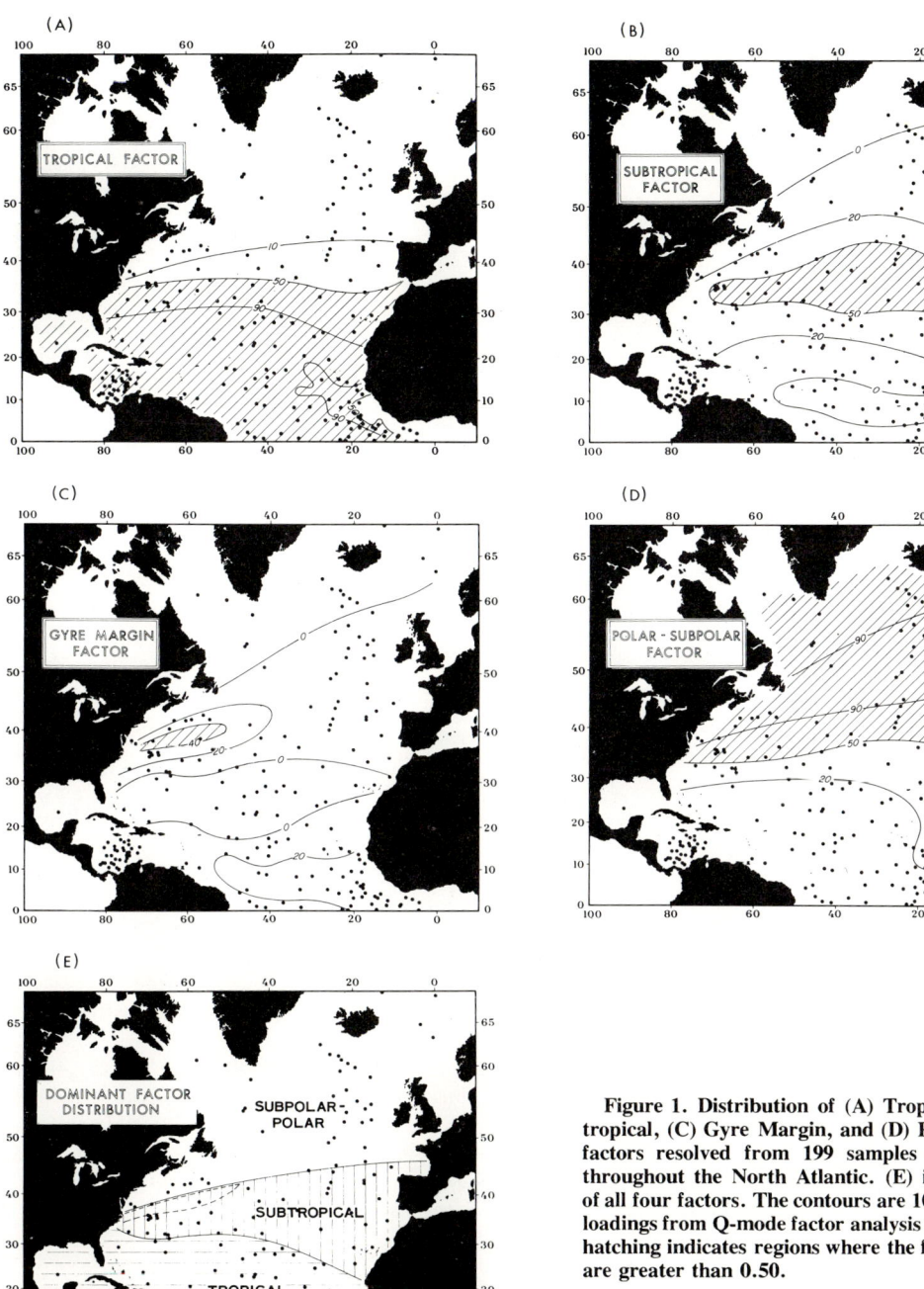

Figure 1. Distribution of (A) Tropical, (B) Subtropical, (C) Gyre Margin, and (D) Polar-Subpolar factors resolved from 199 samples (solid circles) throughout the North Atlantic. (E) is a composite of all four factors. The contours are 100 times factor loadings from Q-mode factor analysis results. Cross-hatching indicates regions where the factor loadings are greater than 0.50.

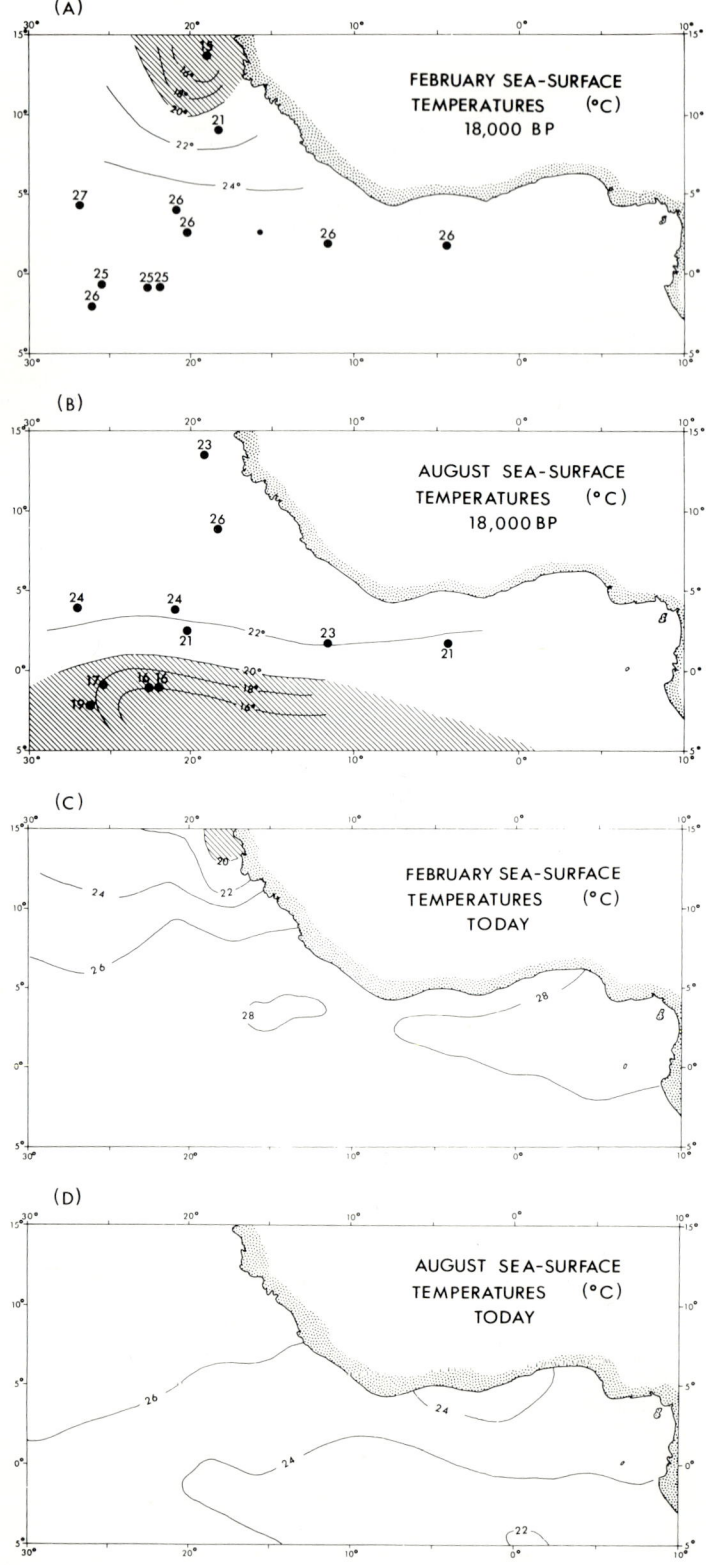

Figure 2. Comparison of (A) February and (B) August estimates of sea-surface temperatures for 18,000 B.P. with present-day (C) February and (D) August observed values. Temperatures are in °C and the contour interval is 2°C. The cross-hatched area represents the region of surface water cooler than 20°C.

responses to seasonal intensification of atmospheric circulation that caused increased upwelling off Africa during February and strengthened the Benguela–South Equatorial Current system during the August season. By contrast, the central region was barely affected by glacial conditions. The present seasonal contrast results from the migration of the Intertropical Convergence Zone (ITCZ), which is controlled by two large African atmospheric high-pressure centers. The similarities of glacial and present seasonality, upwelling off northwest Africa, and intensification of the Benguela–South Equatorial Current system indicate that the ITCZ during glacial stages must not have migrated very far outside its present seasonal boundaries.

The datum used above represents a maximum glacial configuration and relatively stable conditions. Most of the recent studies of late Pleistocene glaciation (Dansgaard and others, 1971; Porter, 1971; Bloom, 1971; Dreimanis and Karrow, 1972, among others) indicate that by 18,000 B.P. glacial conditions had persisted for over 4,000 yr. Therefore, the pattern for 18,000 B.P. (Fig. 2) can be used as the circulation configuration for glacial conditions, and today's circulation pattern can be used for the interglacial mode. The two patterns can then be used to represent extremes between which late Pleistocene conditions fluctuated.

ESTIMATES OF PALEOTEMPERATURE FOR THE PAST 200,000 YEARS

We chose three cores from the eastern equatorial Atlantic (Fig. 3) for close stratigraphic study in order to detail the oceanic response to late Pleistocene climatic changes in three different and critical oceanographic provinces: the South Equatorial Current, the tropical water mass, and the upwelling zone off northwest Africa. Within these three areas, specific core selection was governed by (1) the presence of at least one full climatic cycle, interpreted from carbonate curves and biostratigraphy; (2) work already performed on the cores that could be used for comparison (that is, oxygen-isotope analyses); (3) the abundance of planktonic foraminifers throughout the core so that at least 300 individuals could be obtained per standard

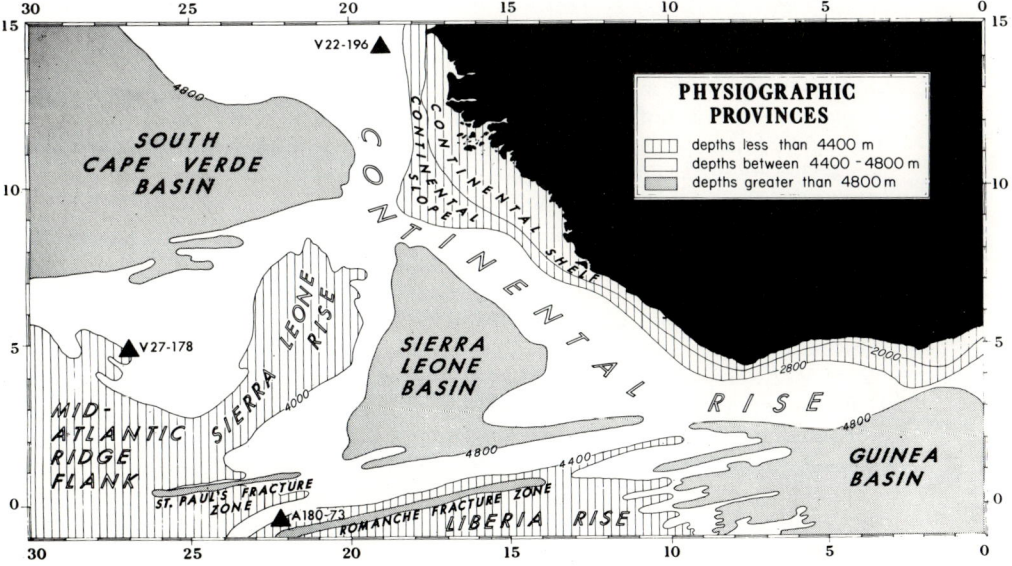

Figure 3. Location of the three cores used in the stratigraphic study.

sample; and (4) the lack of any reworked, older material in the samples. The three cores best meeting the above criteria are A180-73, V22-196, and V27-178.

Core A180-73 is located just north of the Romanche Fracture Zone (Fig. 3) on the gently sloping northeast flank of the Mid-Atlantic Ridge at 3,750 m depth. The core is composed of a uniform gray-tan biogenic ooze with no visual signs of burrowing or deformation. The sediment shows little or no effects of dissolution and has been analyzed for oxygen isotopes (Emiliani, 1955). The strategic position of the core under the South Equatorial Current allows a resolution of the position and strength of the current with time.

The carbonate curve, biostratigraphy, and dissolution indices for Core A180-73 are shown in Figure 4A. Very minor dissolution could well have occurred in the glacial portions of this core. Alternate explanations for the small variations in the number of benthonic foraminifers encountered are differences in the splits made from the bulk sample or minor productivity changes in planktonic and(or) benthonic foraminifers with time. The percentages of species show that although a decrease in the numbers of *G. ruber* and *G. sacculifer* (species sensitive to dissolution) occurs in the sample with a very low planktonic-to-benthonic ratio, a corresponding increase in any of the species resistant to dissolution does not appear. If the increase from a nondissolved count of 1 to 2 benthonic foraminifers per 300 counts to 13 was the result of dissolution, 85 to 92% of the total foraminifers must have been dissolved. This clearly is not the case with the sample at 200 cm. The drop in abundances of *G. ruber* and *G. sacculifer* is more likely the result of one of the alternate explanations.

Core V27-178 was raised from 4,327 m off the western flank of the Sierra Leone Rise (Fig. 3). The core is composed of a uniform pale yellow-brown biogenic ooze throughout with no apparent burrowing or structures. This core should have recorded the changes that took place in the center of the tropical water mass, away from the influences of current systems.

Figure 4B shows the carbonate curve, biostratigraphy, and dissolution indices for V27-178. The core has evidence of severe dissolution, restricted to glacial intervals. As seen on a histogram of frequency of benthonic foraminifera percentages within the core (Fig. 5A), there are either less than 10% benthonic foraminifers or more than 30%. Samples with less than 10% benthonic foraminifers were generally nonfragmented and contain a diverse assemblage with fragile species well represented. All samples with more than 30% benthonic foraminifers are fragmented, and in some cases it is not possible to find even 100 identifiable planktonic species in a standard sample. Samples with percentages of benthonic foraminifers greater than 30% were not used for paleotemperature estimates.

When data points with evidence of severe dissolution are omitted, only one point is left in the W zone; however, the Y' zone is fairly well sampled (Fig. 4B). The samples used for quantitative estimates have very low percentages of benthonic foraminifera, high planktonic-to-benthonic foraminifera ratios, and, in general, show minimal evidence of dissolution. These samples are therefore considered reliable assemblages and are not appreciably biased by dissolution.

Core V22-196 was chosen mainly because of its location in the upwelling area off northwest Africa (Fig. 3). Although some sections show considerable effects of dissolution, this core is the best available from this region. The core was raised from the African continental slope off Senegal in 3,732 m of water. It consists of alternating yellowish-brown and dark brown foraminifera-bearing clays. The light yellowish-brown layers occur at 0 to 40, 390 to 490, 540 to 580, 980 to 1,010, and 1,055 to 1,075 cm. The intervening layers are dark brown. Burrowing is noticeable, although not intense, at the color boundaries. The carbonate curve, biostratigraphy,

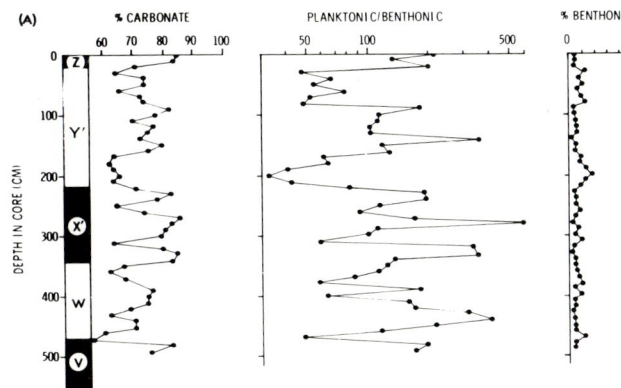
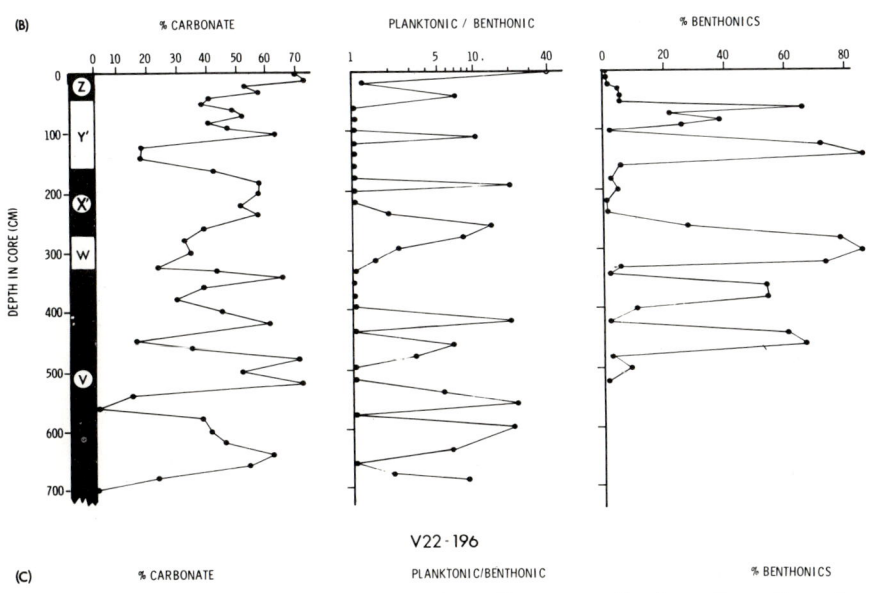

Figure 4. Plots of percent carbonate, planktonic to benthonic foraminifera ratio, and percent benthonic foraminifers for Cores (A) A180-73, (B) V27-178, and (C) V22-196. Percent benthonic foraminifers curves have been plotted only to emphasize the relationship of benthonic to planktonic species.

and dissolution indices are shown in Fig. 4C. Again, the distribution of the percentage of benthonic foraminifers is bimodal with a gap occurring between 15% and 30% (Fig. 5B). Examination of the fragile planktonic species in the samples with less than 15% benthonic foraminiferas reveals a normal distribution of fragile and resistant forms. This suggests a natural division between relatively undissolved (less than 15% benthonic foraminifers) and dissolved assemblages (more than 30% benthonic foraminifers).

Dissolution effects are so severe in the glacial zones of this core, especially between 10 and 240 cm, that some standard-sized samples did not yield a single planktonic foraminifer. These samples are composed of planktonic foraminiferal "hash" with hundreds of benthonic species. However, severe dissolution was not restricted to the glacial zones. Samples at 400, 440, 480, 500, and 520 cm, within the X' interglacial zone, also show severe effects of dissolution. All cores examined from this upwelling zone show poorly preserved sections; however, important information can be gleaned from those sections of the core that are unaffected.

Regrettably, because of dissolution effects, most of the samples from the glacial maximums are unreliable for estimating paleotemperatures. However, a sample at 40 cm, close to the 18,000 B.P. datum, is well preserved and can be used to derive a minimal temperature for full glacial conditions. In general, the interglacial intervals are well preserved except for a time of severe cooling within the lower X' zone.

The estimates of paleotemperatures for each core will be discussed separately. The three cores will then be used to outline a paleo-oceanic and paleometeorologic model of the eastern equatorial Atlantic for the past 200,000 yr. Four nomenclature systems for geologic-climate episodes of the past 200,000 yr are shown in Figure 6.

ESTIMATES OF PALEOTEMPERATURES

Estimates of February and August sea-surface temperatures for Core A180-73 for the past 200,000 yr are shown in Figure 7. The T_{Aug} curve is the more responsive

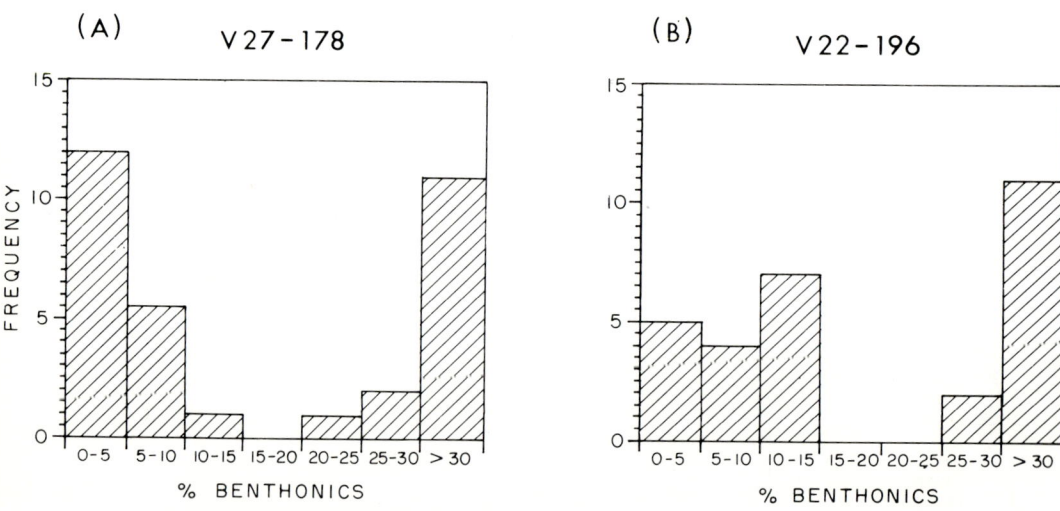

Figure 5. Histograms showing percentage of benthonic foraminifers in Cores (A) V27-178 and (B) V22-196. The bimodal split divides partially dissolved samples (>25%) from nondissolved samples (<15%).

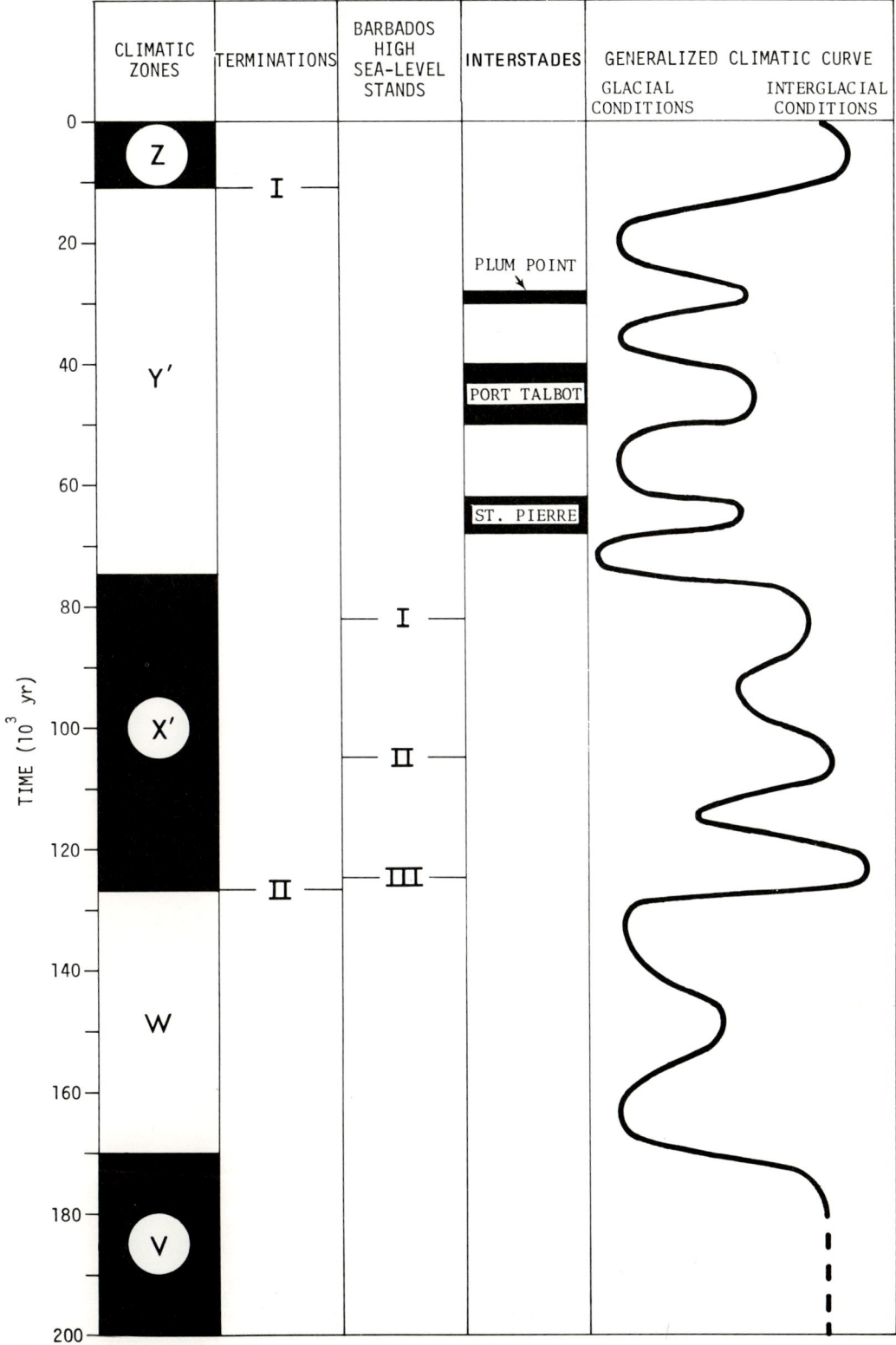

Figure 6. Four nomenclature systems for geologic-climate episodes of the past 200,000 yr.

of the two curves suggesting that the summer season had a more dynamic circulation. Estimated sea-surface temperatures for August for the glacial periods range between 15° and 20°C; estimates for the interglacial stages vary between 22° and 26°C. By contrast, temperature estimates for February range between 23° and 26°C for the glacial stages and 26° to 27°C for the interglacial stages. This indicates that

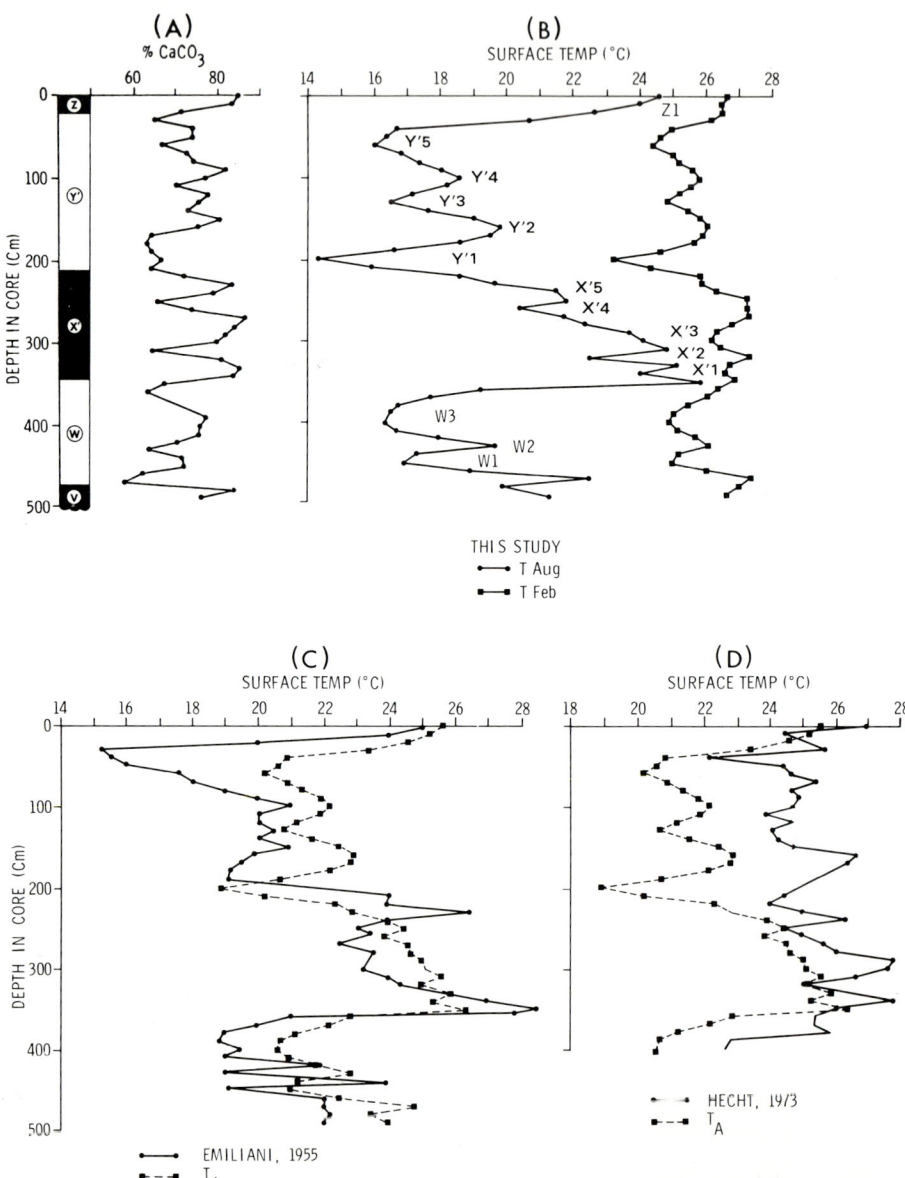

Figure 7. Core A180-73 results: (A) carbonate curve and biostratigraphic section; (B) estimated sea-surface temperatures for August (T_{Aug}, (solid points) and February (T_{Feb}, squares); (C) estimated sea-surface temperatures calculated from oxygen isotopes (Emiliani, 1955) and the average estimated temperatures $[T_A = (T_{Feb} + T_{Aug})/2]$; and (D) estimated sea-surface temperatures from Hecht (1973) and the average estimated temperatures (T_A).

sea-surface temperatures for February only changed a maximum of 4°C, and, in general, the difference between the interglacial and glacial stages was only a degree or two. Sea-surface temperatures for August, however, were greatly affected by glacial conditions. The T_{Aug} changed a maximum of 11°C, and at least a 6°C change is average.

It is interesting that the saw-toothed pattern, discussed by Broecker and van Donk (1970), which appears in oxygen isotope curves, can be seen in the T_{Aug} curve but not in the T_{Feb} curve. The temperature curves show a rough 20,000-yr cycle. Starting from the V-W boundary, the prominent T_{Feb} and T_{Aug} peak in the middle of the W zone (W2 on Fig. 7B) correlates with the inferred insolation high at 170,000 B.P. (Mesolella and others, 1969). The W2 and Y'2 and Y'4 temperature peaks are of nearly identical amplitude (T_{Aug} = 20°C; T_{Feb} = 26°C). This indicates that there was indeed a climatic event at about 170,000 B.P. This insolation high (or interstade) was quickly followed by a return to glacial conditions, represented by W3. An event at about 150,000 B.P., which appears on insolation curves and the carbonate curve, is not evident in the paleotemperature curves. The global warming at Termination II (W-X' boundary 127,000 B.P.) was very abrupt; T_{Aug} increased almost 10°C and T_{Feb} by 2°C. The two estimates are completely in phase with each other. The Barbados III sea-level stand at 125,000 B.P. (Mesolella and others, 1969) is represented on the T_{Aug} curve by two minor peaks on a general high (X'1). Paleotemperature estimates for X'1 represent, for both T_{Feb} and T_{Aug}, conditions within 1°C of conditions found at this site today (Z1). A post-Barbados III low stand of sea level shows up only as a small T_{Aug} decrease (about 3°C). This small temperature response is in sharp contrast to the change in the carbonate curve. Either conditions in this equatorial location did not deteriorate significantly, or the climatic event was too short-lived to allow a full response of the sea-surface temperatures.

Peaks X'3 and X'5 probably correlate with Barbados II and I high sea-level stands, representing 105,000 and 82,000 B.P., respectively. Within the X' zone, T_{Feb} shows a broad temperature plateau with only 1°C variation throughout the 47,000-yr duration, whereas T_{Aug} shows decreasing temperatures with time (~4°C over the 47,000 yr). Then, at about 80,000 B.P., conditions deteriorated almost as abruptly as they did when Termination II occurred. T_{Feb} and T_{Aug} show sharp drops in temperature to a minimum at about 73,000 B.P. (Y'1). In the region of this core, the most drastic conditions of the past 200,000 yr occurred at this time. Sea-surface temperatures during August were less than 15°C and February temperatures were about 23°C, giving a seasonal contrast of 8°C.

Within the Y' zone, two temperature maximums are found, Y'4 and Y'2. This contrasts with three carbonate highs found in the same climatic zone. The estimated temperature highs can be correlated with the 62,000 and 32,000 B.P. insolation maximums found on mid- to low-latitude insolation curves (Vernekar, 1972) and with the Port Talbot (40,000 to 50,000 B.P.) and St. Pierre (62,000 to 68,000 B.P.) interstades (Dreimanis and Karrow, 1972). The youngest interstade, Plum Point, (28,000 to 30,000 B.P.) apparently is not represented in this core (compare Fig. 6 with Fig. 7B). The estimated T_{Aug} indicates that the two interstades never reached interglacial conditions, but rather oscillated within the glacial mode.

The temperature at the 18,000-B.P. level (Y'5) was not the minimum for the Y' zone; however, it was within a degree or two of the maximum severity. Thus, this level again proves to be an adequate choice for a representative glacial datum.

Following the glacial conditions in 18,000 B.P., another abrupt warming occurred, Termination I. The magnitude of the warming is essentially identical to Termination II, in both T_{Feb} and T_{Aug} and in seasonal contrast.

A model to explain these responses of sea-surface temperature must account for the changing seasonal contrasts, the large response of T_{Aug} to global climatic changes, and the lesser response of T_{Feb}. The present location of this core is on the northern edge of the South Equatorial Current system, and it should therefore be very sensitive to the seasonal strengths or small north-south fluctuations in the position of the current. The T_{Feb} and T_{Aug} estimates indicate that although sea-surface temperatures cooled during glacial periods, the decrease was felt most strongly during the Southern Hemisphere winter season. The estimated sea-surface temperatures for glaciations in February (Northern Hemisphere winter season) are not strongly affected. The Benguela–South Equatorial Current system controls the sea-surface temperature in the region from which Core A180-73 was taken. Therefore, the current in the August season during Southern Hemisphere winters 18,000 yr ago was significantly strengthened. During the February season of glacial stages, there was either a change in position of the current system—the current flowed south of Core A180-73—or the circulation slowed. In either case, conditions during the February season of glacial stages were similar to those found during interglacial stages.

A comparison can be made between the estimated temperature from Core A180-73 (Fig. 7B) and the carbonate response (Fig. 7A), as well as the paleotemperatures estimated by Emiliani (1955) (Fig. 7C, solid line) and Hecht (1973) (Fig. 7D). In a gross way, there is a good correlation between the carbonate curve and the paleotemperature estimates. Terminations I and II appear in both data sets, as do peaks X'1, X'3, and X'5 corresponding to Barbados I, II, and III high sea-level stands, and glacial periods at 18,000, 73,000, and 135,000 B.P. (Y'5, Y'1 and W3, respectively). The differences are found in the responses of the carbonate systems and sea surface temperature to minor changes of climate. The carbonate curve shows much more dramatic changes at the post-Barbados III cooling (X'2) and the cooling between Barbados I and II (X'4) than do estimates of sea surface temperature. Major carbonate dissolution can be discounted for this core but minor dilution probably occurred (Gardner, 1973). If this dilution were a response to increased trade-wind intensity, one might expect to find it in those zones with the most severe glacial climates (that is, 18,000, 73,000, and 135,000 B.P.). These are the zones with low carbonate values. The carbonate decrease at the post-Barbados III cooling can be interpreted as increased eolian dilution for a short time interval (perhaps less than 10,000 yr) when climatic conditions cooled, then rapidly warmed.

The techniques used by Emiliani (1955) and Hecht (1973) were attempts to estimate annual sea-surface temperatures; to compare our results with theirs we have averaged the T_{Feb} and T_{Aug} values (Fig. 7B) to compute an annual (T_A) curve. Emiliani's (1955) curve (Fig. 7C) shows a rapid warming at Termination II (W3 to X'1 transition on Fig. 7B) and at the prominent peaks X'5 and X'1 (Fig. 7B), but it lacks an X'3 peak. The only significant difference between our T_A curve and Emiliani's is a cooling at 18,000 B.P. (Y'5 on Fig. 7B). The cooling at 73,000 B.P. (Y'1) from Emiliani's data is much less severe than the cooling at 18,000 B.P. Our curve indicates the opposite occurred. Emiliani's 18,000 B.P. value is similar to our T_{Aug} value (16°C) but differs by 5°C from our T_A value (20°C). We cannot explain this one difference; for the remainder of the record, the two curves deviate by no more than about 2°C. In fact, the similarity of the two curves, in both absolute values and pattern, produced by totally independent methods, is remarkable. Hecht's (1973) curve (Fig. 7D), however, shows significant differences in phase and amplitude from our estimates.

The three techniques (the present study, isotope temperatures, and Hecht's distance-coefficient method) can be compared in both amplitude and range (Table

5). As pointed out, the isotope-derived temperatures closely correlate, in both absolute value and pattern, with T_A values, the averaged estimated paleotemperatures. Hecht's estimates more closely compare with our T_{Feb} values, than with our T_{Aug} value but his pattern does not resemble those derived from our T_A or isotopic temperatures.

Estimated T_{Feb} and T_{Aug} for Core V27-178 are shown in Figure 8A. Several samples, mostly from the coldest glacial periods, were omitted from the regressions because of dissolution. Consequently, the coldest glacial conditions are not represented. The estimated T_{Feb} and T_{Aug} values show that only small changes have occurred, and a small seasonal contrast was typical for the past 200,000 yr. The X' zone does not stand out as a period of higher surface temperatures as it did in Core A180-73. There is a small response of T_{Feb} in the Y' zone at about 18,000 B.P. where the T_{Feb} value is almost 4°C cooler than either today's (sample at 2-cm depth on Fig. 8) or the 125,000 B.P. temperature. This lower T_{Feb} value, coupled with almost no response by T_{Aug}, gives a small increase in seasonal contrast during glacial periods. The estimates of T_{Feb} for glacial periods that we assume to be reliable range from 24°C to 25°C, which is considerably warmer than the warmest glacial T_{Feb} value for A180-73. Estimated T_{Aug} for glacial periods were also constant at about 27°C. The paleotemperatures for interglacial stages show a slight response to the post-Barbados III cooling (a 1°C cooling in T_{Feb} values). The rapid cooling found in Core A180-73 at about 73,000 B.P. can be seen in Core V27-178 as the drop from a T_{Feb} value of about 26°C to a low of 24°C.

Th region of the tropical Atlantic in the vicinity of the Sierra Leone Rise (where Core V27-178 was taken) seems to have been subjected to much smaller glacial-interglacial differences than occurred several hundred kilometers to the south. Although the maximum glacial conditions at 73,000 B.P. cannot be determined, the severe glacial conditions at 18,000 B.P. are represented but show only a minor response. The small temperature changes in the core may reflect the general response of a tropical region to climatic change outside the influence of major current systems and upwelling areas.

The estimated paleotemperatures for Core V22-196 were made on samples showing relatively little effects of dissolution. Therefore, glacial maximums are again poorly represented.

TABLE 5. COMPARISON OF SEA-SURFACE TEMPERATURE ESTIMATES DERIVED FROM CORE A180-73 USING TRANSFER FUNCTIONS, ISOTOPIC PALEOTEMPERATURES, AND DISTANCE COEFFICIENTS

		Transfer functions (this study)	Isotopic temperature (Emiliani, 1955)	Distance coefficients (Hecht, 1973)
Interglacial	T_{Aug}	22–26°C	. .	. .
	T_{Feb}	26–27°C	. .	. .
	T_A	24–26°C	22–27°C	24–28°C
Glacial Stage	T_{Aug}	15–20°C	. .	. .
	T_{Feb}	23–26°C	. .	. .
	T_A	19–23°C	17–21°C	22–26°C
Temperature change between glacial and interglacial stages	T_{Aug}	2–11°C	. .	. .
	T_{Feb}	0–4°C	. .	. .
	T_A	1–5°C	1–10°C	2–6°C

Note: T_{Feb} = February sea-surface temperature; T_{Aug} = August sea-surface temperature; T_A = Annual sea-surface temperature.

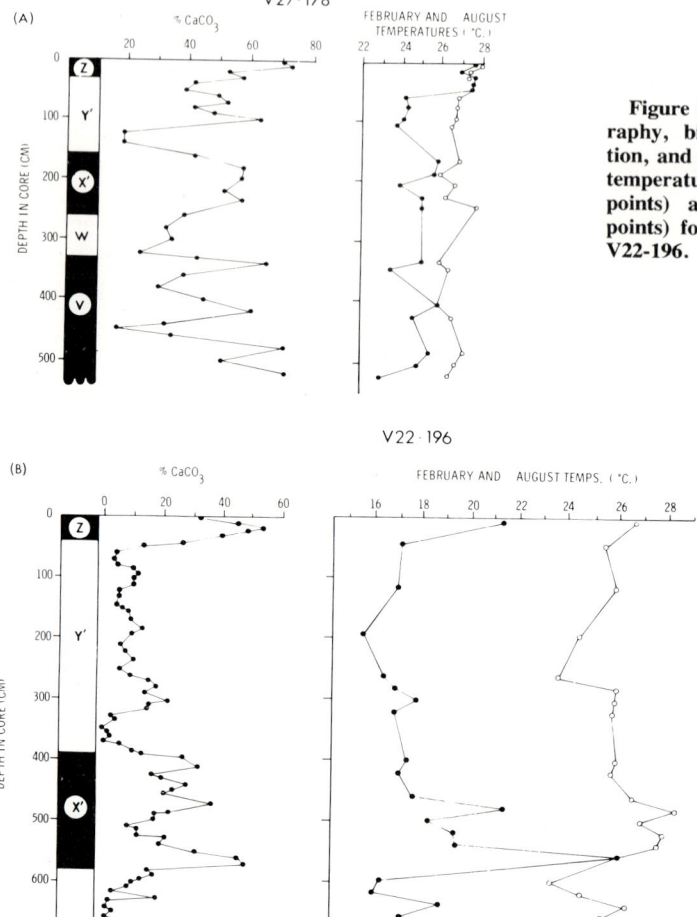

Figure 8. Carbonate stratigraphy, biostratigraphic zonation, and estimated sea-surface temperatures for August (open points) and February (solid points) for Core V27-178 and V22-196.

The calcium carbonate variations (Fig. 8B) in this core show consistent climatically induced fluctuations that have been correlated with sea-level changes recorded on Barbados (three prominent carbonate maximums in the X' zone) and interstades recorded in North America (three maximums in Y' zone) (Hays and Perruzza, 1972). Due to dissolution, samples providing reliable temperature estimates cannot be as closely spaced as the carbonate values. However, the estimated temperatures provide some important information (Fig. 8B). At this core site, August temperatures during the past 200,000 yr fluctuated only a few degrees while estimated temperatures for February show much larger variations (~10°). The maximum variation of February temperatures is between the top of the W zone and the base of the X' zone (Termination II). The maximum February temperature (>26°C) corresponds with the lowest carbonate maximum in the X' zone which probably correlates with the Barbados III high sea-level stand. After this, temperature maximum drop again to relatively low levels with a small maximum that corresponds to the second carbonate maximum (Barbados II). Both the T_{Aug} and T_{Feb} estimates indicate rather constant conditions at this core site during the last 200,000 yr, with an important exception recorded by February temperature estimates at the base of the X' zone, indicating a rapid warming and just as rapid cooling during a short interval in the lower X' zone.

A model that adequately explains the paleoestimates for V22-196 must involve seasonal upwelling throughout most of the past 200,000 yr. The large seasonal contrast implies active upwelling during winter seasons but very little upwelling during summers. We interpret the values of T_{Feb} during glacial periods (16° to 18°C) as a response to vigorous upwelling; cooler temperatures result from either deeper waters being brought to the surface or a longer upwelling season. The T_{Aug} response suggests that a season of little or no upwelling, which allowed warm surface waters to migrate into the area, has been present for at least the past 200,000 yr.

The seasonal upwelling found off northwest Africa today is a direct result of the northeast trade winds that fluctuate with the migration of the Intertropical Convergence Zone (ITCZ) (Sverdrup and others, 1942; Critchfield, 1966). During the Northern Hemisphere winter months, the trade winds blow offshore along the northwest African coast creating upwelling conditions. The northeast Trade Winds become variable and lose much of their intensity during the summers when the ITCZ is located at lat. 16° to 20°N. The estimates of paleotemperature indicate that the northernmost position and the seasonal migration of the ITCZ during most of the past 200,000 yr were comparable to those that occur today.

MODEL DERIVED FROM ESTIMATES OF PALEOTEMPERATURE

A circulation model (Fig. 9) can be constructed that accommodates the results from the three cores as well as the results developed from the 18,000 B.P. datum. During glacial periods, the Benguela-South Equatorial Current system either migrated north-south seasonally or had a seasonal fluctuation of intensity, both of which imply control by the trade winds and the Intertropical Convergence Zone. The intensity of upwelling off northwest Africa also fluctuated with the seasons and was strongest during the season when the equatorial flow was weakest. These conditions are similar to those found in this region today: intense upwelling off Africa occurs during Northern Hemisphere winter months, and the strongest circulation of the South Equatorial Current occurs during the Southern Hemisphere winter.

The paleotemperature data suggests that during glacial periods, the ITCZ migrated approximately within its present seasonal limits. The cooler temperatures during February of glacial stages in the zone of northwest African upwelling reflect seasonal upwelling as well as encroachment of cooler subtropical waters brought into the region by the Canary Current. This situation was probably enhanced by the compression of the North Atlantic gyre circulation during the glacial mode (McIntyre and others, 1972).

The cool conditions along the equator during Southern Hemisphere winters (August) of the glacial periods reflect the strong circulation of the Benguela-South Equatorial Current system. This system transported cool, recently upwelled, mid-latitude waters into the equatorial region. The northward migration of the ITCZ during Northern Hemisphere summers would have shifted the southeast trade winds to the north, thus bringing the Benguela Current into the vicinity of the geographic equator. This position change, together with an increased intensity of the southeast trade winds, could account for the fluctuations.

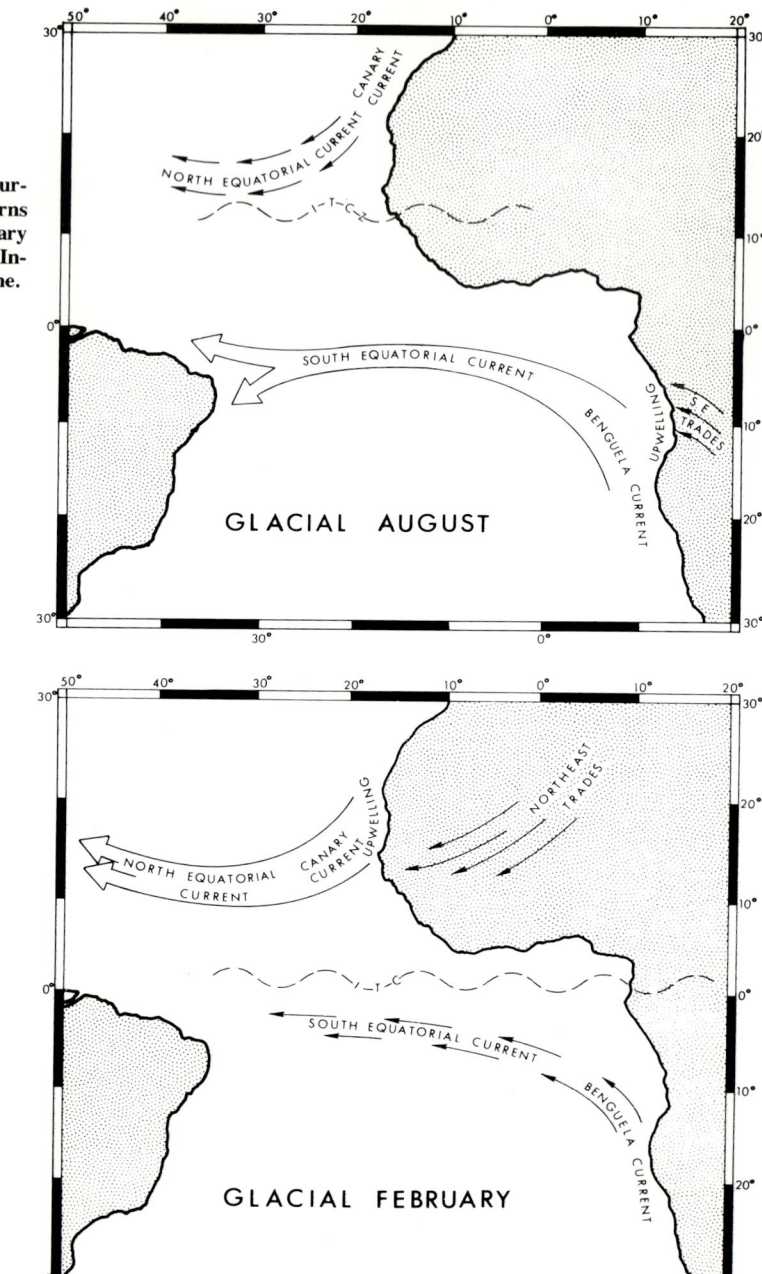

Figure 9. Generalized surface currents and wind patterns for glacial August and February seasons. ITCZ identifies the Intertropical Convergence Zone.

The intensification of both upwelling and surface currents suggests a general strengthening of atmospheric circulation during winters of the glacial stages. Therefore, increased thermal gradients in the winter season of each hemisphere must have increased the vigor of the atmospheric and oceanic circulation. Certainly, the build-up of land ice in the Northern Hemisphere and sea ice (Hays and others, this volume) in the Southern Hemisphere would produce this effect. Although the difference in ice area between today and 18,000 B.P. was significantly greater in the Northern Hemisphere than the Southern Hemisphere, covering a relatively warm ocean (the Antarctic Ocean) with ice could have a very large effect (Fletcher, 1969, unpub. data). The resulting increased thermal gradient between the south polar region and the equator would increase the atmospheric circulation of the Southern Hemisphere, especially during the winter season (Lamb, 1961). This is reflected in sea-surface temperatures for August and interpreted as an intensification of the Benguela–South Equatorial Current system and the strengthening of the southeast trade winds. The strengthened trade winds, in turn, indicate a strengthening of the entire South Atlantic gyre (Lamb, 1961).

SUMMARY AND CONCLUSIONS

During glacial stages, atmospheric circulation intensified although the ITCZ migrated north-south approximately within its present boundaries. The largest changes in oceanic circulation occurred as a response to winter cooling of each hemisphere. The northeast trade winds were intensified during Northern Hemisphere winters, which produced cold upwelling conditions along the northwest African margin. The Benguela–South Equatorial Current system intensified as a response to increased Southern Hemisphere circulation during Southern Hemisphere winters. The tropical region between the North and South Equatorial Currents witnessed very little change.

During glacial stages, the general atmospheric circulation of the winter seasons was intensified because of the increased thermal gradient between the equator and poles. Consequently, the oceanic circulation was strengthened. However, the data seem to indicate that the environmental responses did not occur in a square wave manner that alternated between glacial and interglacial conditions. Rather, the data show that just after a termination (rapid change from glacial to interglacial climates), conditions approximated those found in this region today. These new conditions did not persist, however, and in a short time (a few thousand years), cooling of the environment began anew, leading again to the intense winter circulation mode.

Circulation conditions comparable to today's were very short-lived. During most (90%) of the past 200,000 yr, the winter circulation was significantly stronger than it is now. Past conditions most comparable to today's are found only about 125,000 B.P. At that time, sea-surface temperatures were essentially as they are today for both summer and winter at the three core locations. The environment was more nearly that of a glacial stage at all other times that are represented by these cores.

TAXONOMIC NOTES

A short discussion of the taxonomy is necessary because this study uses total planktonic foraminiferal faunal counts as its primary data base.

Thirty foraminiferal species, subspecies, and varieties were recognized (Table 2) using the taxonomies of Bé (1967) and Parker (1962, 1967). Where conflicts appeared, Bé (1967) was followed. For counting purposes, several species were subdivided. *Globigerinoides ruber* was separated into pink and white varieties; *G. sacculifer* was split into those with a sac-like final chamber (*G. sacculifer*) and those with a spherical final chamber (*G. trilobus*); and *Globorotalia truncatulinoides* and *Globigerina pachyderma* were separated into left- and right-coiling varieties.

A problem of identification appeared in intergradations between *Globoquadrina dutertrei* and *G. pachyderma* (right-coiling). This problem has been encountered by others (Cifelli, 1961; Bé and Hamlin, 1967; Ruddiman, 1969; Bé and Tolderlund, 1971). Criteria for this study were based on the presence or absence of an umbilical tooth and the number of chambers in the last whorl. Specimens with an umbilical tooth were called *G. dutertrei*, regardless of the number of chambers in the final whorl. Specimens without a tooth but with a large, deep, open umbilicus and five or more chambers in the final whorl were also assigned to *G. dutertrei*. Specimens with more than four chambers but with a compact test and lacking an umbilical tooth were called *G. pachyderma–G. dutertrei* intergrade ("PD intergrade"). Compact specimens with four or less chambers and without an umbilical tooth were placed in right-coiling *G. pachyderma*.

Another problem of morphological intergradation (*Globigerina calida* and *Globigerinella aequilateralis*) was resolved by examining the symmetry of the primary aperture relative to the coiling plane. If symmetrical, the specimen was called *G. aequilateralis*; if not symmetrical, *G. calida*.

Inevitably, specimens were encountered that defied classification. They were lumped into an "unknown" category. Counts of at least 300 specimens were made on all samples.

ACKNOWLEDGMENTS

We acknowledge the stimulating discussions and helpful comments of CLIMAP colleagues, especially W. L. Prell, A. McIntyre, M. Roche, B. Molfino, and J. Imbrie, and the careful and constructive reviews by A. G. Kaneps, G. R. Heath, and J. Thiede. This research was supported by the CLIMAP Program, International Decade for Ocean Exploration, National Science Foundation Grant GX 28671; the cores were processed at the Lamont-Doherty Geological Observatory under grants from the Office of Naval Research (N00014-67-A-0108-0004) and the National Science Foundation (DES-72-01568).

REFERENCES CITED

Bé, A.W.H., 1967, Foraminifera, families: *Globigerinidae* and *Globorotaliidae*, in Fraser, J. H., ed., Fiches d'identification du zooplancton: Charlottenlund, Denmark, Conseil Internat. l'Exploration Mer, fiche no. 108.

Bé, A.W.H., and Hamlin, W. H., 1967, Ecology of Recent planktonic foraminifers, pt. 3, Distribution in the North Atlantic during the summer of 1962: Micropaleontology, v. 13, no. 1, p. 87–106.

Bé, A.W.H., and Tolderlund, D. S., 1971, Distribution and ecology of living planktonic foraminifera in surface waters of the Atlantic and Indian Ocean, in Funnell, B. M., and Riedel, W. R., eds., Micropaleontology of the oceans: London, Cambridge Univ. Press, p. 105–149.

Bloom, A. L., 1971, Glacial-eustatic and isostatic controls of sea level since the last glaciation, in Turekian, K. K., ed., The late Cenozoic glacial ages: New Haven, Yale Univ. Press, p. 355–379.

Broecker, W. S., 1966, Absolute dating and the astronomical theory of glaciation: Science, v. 151, p. 299–304.

Broecker, W. S., and Ku, T.-L., 1969, Caribbean cores P6304-8 and P6304-9: New analysis of absolute chronology: Science, v. 166, p. 404–406.

Broecker, W. S., and van Donk, J., 1970, Insolation changes, ice volumes, and the O^{18} record in deep-sea cores: Rev. Geophys. Space Phys., v. 8, no. 1, p. 169–198.

Broecker, W. S., Turekian, K. K., and Heezen, B. C., 1958, The relation of deep-sea sedimentation rate to variations in climate: Am. Jour. Sci., v. 256, p. 503–517.

Broecker, W. S., Thurber, D. L., Goddard, J., Ku, T.-L., Mathews, R. K., and Mesolella,

K. J., 1968, Milankovitch hypothesis supported by precise dating of coral reefs and deep-sea sediments: Science, v. 159, p. 297-300.

Cifelli, R., 1961, *Globigerinia incompta*, a new species of pelagic foraminifera from the North Atlantic: Cushman Found. Foram. Research Contr. 12, no. 3, p. 420.

Critchfield, H. J., 1966, General climatology (2nd ed.): Englewood Cliffs, N. J., Prentice-Hall, Inc., 420 p.

Cushman, J. A., and Henbest, L. G., 1940, Geology and biology of North Atlantic deep-sea cores, Pt. II, Foraminifera: U.S. Geol. Survey Prof. Paper 196-A, p. 35-50.

Dansgaard, W., Johnson, S. J., Clausen, H. B., and Langway, C. C., Jr., 1971, Climatic record revealed by the Camp Century ice core, *in* Turekian, K. K., ed., The late Cenozoic glacial ages: New Haven, Yale Univ. Press, p. 37-56.

Dreimanis, A., and Karrow, P. F., 1972, Glacial history of the Great Lakes-St. Lawrence region, the classification of the Wisconsin (an) Stage, and its correlatives: Internat. Geol. Cong., 24th, Montreal, Sec. 12, p. 5-15.

Emiliani, C., 1955, Pleistocene temperatures: Jour. Geology, v. 63, p. 538-578.

———1966, Paleotemperature analysis of Caribbean cores P6304-8 and P6304-9 and a generalized temperature curve for the past 425,000 years: Jour. Geology, v. 74, p. 109-126.

Ericson, D. B., and Wollin, G., 1968, Pleistocene climates and chronology in deep-sea sediments: Science, v. 162, p. 1227-1234.

Ericson, D. B., Ewing, M., Wollin, G., and Heezen, B. C., 1961, Atlantic deep-sea sediment cores: Geol. Soc. America Bull., v. 72, no. 2, p. 193-286.

Gardner, J. V., 1973, The eastern equatorial Atlantic: Sedimentation, faunal, and sea-surface temperature responses to global climatic changes during the past 200,000 years [Ph.D. thesis]: New York, Columbia Univ., 387 p.

Hays, J. D., and Perruzza, A., 1972, The significance of calcium carbonate oscillations in eastern equatorial Atlantic deep-sea sediments for the end of the Holocene warm interval: Quaternary Research, v. 2, p. 355-362.

Hays, J. D., Lozano, J., Shackleton, N., and Irving, G., 1976, Reconstruction of the Atlantic and Western Indian Ocean sectors of the 18,000-B.P. Antarctic Ocean, *in* Cline, R. M., and Hays, J. D., eds., Investigation of late Quaternary paleoceanography and paleoclimatology: Geol: Soc. America Mem. 145 (this volume).

Hecht, A. D., 1973, Faunal and oxygen isotopic paleotemperatures and the amplitude of glacial/interglacial temperature changes in the equatorial Atlantic, Caribbean Sea, and Gulf of Mexico: Quaternary Research, v. 3, no. 4, p. 671-690.

Imbrie, J., and van Andel, Tj. H., 1964, Vector analysis of heavy-mineral data: Geol. Soc. America Bull., v. 75, p. 1131-1156.

Imbrie, J., and Kipp, N., 1971, A new micropaleontological method for quantitative paleoclimatology: Application to a late Pleistocene Caribbean core, *in* Turekian, K. K., ed., The late Cenozoic glacial ages: New Haven, Yale Univ. Press, p. 71-179.

Klovan, J. E., and Imbrie, J., 1971, An algorithm and Fortran IV program for large scale Q-mode factor analysis: Internat. Assoc. Math. Geology Jour., v. 3, no. 1, p. 61-77.

Krumbein, W. C., and Graybill, F. A., 1965, An introduction to statistical models in geology: New York, McGraw-Hill, 475 p.

Ku, T.-L., and Broecker, W. S., 1966, Atlantic deep-sea stratigraphy: Extension of absolute chronology to 320,000 years: Science, v. 151, p. 448-450.

Lamb, H. H., 1961, Fundamentals of climate, *in* Nairn, A.E.M., ed., Descriptive paleoclimatology: New York, Interscience Pubs., p. 8-44.

Manson, V., and Imbrie, J., 1964, Fortran program for factor and vector analysis of geologic data using IBM 7090 or 7094 computer system: Kansas Geol. Survey Computer Contr., Spec. Pub., no. 13, p. 1-46

McIntyre, A., Bé, A., Biscaye, P., Burckle, L., Gardner, J., Geitzenauer, K., Goll, R., Kellogg, K., Prell, W., Roche, M., Imbrie, J., Kipp, N., Ruddiman, W., and Moore, T., 1972, The glacial North Atlantic 17,000 years ago: Paleoisotherm and oceanographic maps derived from floral-faunal parameters by CLIMAP: Geol. Soc. America, Abs. with Programs, v. 4, no. 7, p. 590-591.

McIntyre, A., Kipp, N., Bé, A.W.H., Crowley, T., Kellogg, T., Gardner, J., Prell, W., and Ruddiman, W. F., 1976, Glacial North Atlantic 18,000 years ago: A CLIMAP reconstruction, *in* Cline, R. M., and Hays, J. D., eds., Investigation of late Quaternary paleoceanography and paleoclimatology: Geol. Soc. America Mem. 145 (this volume).

Mesolella, K. J., Matthews, R. K., Broecker, W. S., and Thurber, D. L., 1969, The astronomical theory of climatic change: Barbados data: Jour. Geology, v. 77, p. 250–274.

Olausson, E., 1965, Evidence of climatic changes in North Atlantic deep-sea cores, with remarks on isotopic paleotemperature analysis, *in* Sears, M., ed., Progress in oceanography (vol. 3): London, Pergamon Press Ltd., p. 221–252.

Ovey, C. D., 1950, On the interpretation of climatic variations as revealed by a study of samples from an equatorial Atlantic deep-sea core: Royal Meteorol. Soc., Centenary Proc., p. 211–228.

Parker, F. L., 1962, Planktonic foraminiferal species in Pacific sediments: Micropaleontology, v. 8, no. 2, p. 219–254.

——1967, Late Tertiary biostratigraphy (planktonic foraminifera) of tropical Indo-Pacific deep-sea cores: Bulls. Am. Paleontology, v. 52, no. 235, p. 115–208.

Phleger, F. B., 1942, Foraminifera of submarine cores from the continental slope: Geol. Soc. America Bull., v. 53, no. 7, p. 1073–1098.

Phleger, F. B., Parker, F. L., and Peirson, J. F., 1953, North Atlantic foraminifera: Swedish Deep-Sea Exped. Repts. (1947–1948), v. 7, no. 1, 122 p.

Porter, S. C., 1971, Fluctuations of late Pleistocene Alpine glaciers in western North America, *in* Turekian, K. K., eds., Late Cenozoic glacial ages: New Haven, Yale Univ. Press, p. 307–329.

Ruddiman, W. F., 1969, Foraminifera of the subtropical North Atlantic gyre [Ph.D. thesis]: New York, Columbia Univ., 282 p.

——1971, Pleistocene sedimentation in the equatorial Atlantic: Stratigraphy and faunal paleoclimatology: Geol. Soc. America Bull., v. 82, no. 2, p. 283–302.

Schott, W., 1935, Die Foraminiferen in dem äquatorialen Teil des Atlantischen Ozeans: Wissenschaft. Ergebn. Dt. Atl. Expedit. Meteor, 1925–1927, v. 3, p. 43–134.

Shackleton, N., 1967, Oxygen isotope analysis and Pleistocene temperatures re-assessed: Nature, v. 215, p. 15–17.

Shackleton, N. J., and Opdyke, N. D., 1973, Oxygen isotope and paleomagnetic stratigraphy of equatorial Pacific core V28-238: Oxygen-isotope temperatures and ice volumes on a 10^5 and 10^6 year scale: Quaternary Research, v. 3, p. 39–55.

Sverdrup, H. U., Johnson, N. W., and Fleming, R. H., 1942, The oceans: Their physics, chemistry, and general biology: Englewood Cliffs, N.J., Prentice-Hall, Inc., 1060 p.

U.S. Naval Oceanographic Office, 1967, Oceanographic atlas of the North Atlantic Ocean, Sec. II, Physical properties: U.S. Naval Oceanographic Office Publ. 700, 300 p.

Vernekar, A. D., 1972, Long-period global variations of incoming solar radiation: Meteorol. Monos., v. 12, 127 p.

MANUSCRIPT RECEIVED BY THE SOCIETY MARCH 28, 1974
REVISED MANUSCRIPT RECEIVED SEPTEMBER 23, 1974
MANUSCRIPT ACCEPTED OCTOBER 7, 1974
LAMONT-DOHERTY GEOLOGICAL OBSERVATORY CONTRIBUTION NO. 2273

Geological Society of America
Memoir 145
© 1976

Equatorial Atlantic and Caribbean Foraminiferal Assemblages, Temperatures, and Circulation: Interglacial and Glacial Comparisons

Warren L. Prell*
James V. Gardner*
Lamont-Doherty Geological Observatory
and
Department of Geological Sciences
Columbia University
Palisades, New York 10964

Allan W. H. Bé
Lamont-Doherty Geological Observatory
Columbia University
Palisades, New York 10964

and

James D. Hays
Lamont-Doherty Geological Observatory
and
Department of Geological Sciences
Columbia University
Palisades, New York 10964

ABSTRACT

Intensification of the North and South Equatorial Current systems and trade winds occurred during glacial periods, according to a comparison of late Holocene (interglacial), 18,000 B.P. (glacial), and late Quaternary (0 to 180,000 B.P.) faunal assemblages and sea-surface temperature estimates from the equatorial Atlantic and Caribbean regions. Faster circulation of the North Equatorial Current system

---

*Present address: (Prell) Department of Geological Sciences, Brown University, Providence, Rhode Island 02912. (Gardner) Pacific-Arctic Branch of Marine Geology, U.S. Geological Survey, Menlo Park, California 94025.

in glacial Northern Hemisphere winters (February) is indicated by increased upwelling of cool (15°C) water off northwest Africa and slightly cooler conditions across the northern tropical Atlantic and Caribbean. Intensification of the South Equatorial Current occurred along the Equator during the Southern Hemisphere winter (August). This interpretation is based on the dominance of a cool-equatorial assemblage, which indicated that waters of 16° to 18°C replaced the tropical assemblage that lives today in 24° to 26°C water in this region. The cool influence of the glacial (August) Benguela–South Equatorial Current decreased rapidly westward along the equatorial belt so that the fauna was dominated by the tropical assemblage in the Caribbean. Sea-surface temperatures increased rapidly from east to west in the equatorial belt, so that at long 35°W, the 16°C water had reached ambient temperatures of 24° to 26°C.

Both faunal assemblages and temperature estimates of eight late Quaternary Atlantic and Caribbean sediment cores show that the equatorial region experienced three maximum incursions of cool Benguela Current water during the past 150,000 yr—at approximately 135,000 B.P., 73,000 B.P., and 18,000 B.P. Differences of glacial to interglacial sea-surface temperatures range from 5° to 10°C in the eastern equatorial Atlantic to 2° to 3°C in the western Atlantic and the Caribbean. During this time, only two periods with similar faunas and surface temperatures occurred—today and 125,000 B.P.

Seasonal temperature contrast (August to February) is three to four times greater in all cores for glacial conditions than for interglacial conditions. The winter temperatures (February to the north of the thermal equator and August south of it) show the greatest changes, and they control the overall temperature pattern. Identical temperature patterns for cores affected by the North and the South Equatorial Currents suggest that the Northern and Southern Hemispheres are generally in phase and that more severe winters control the glacial temperature pattern.

INTRODUCTION

The responses of the Caribbean Sea and the western equatorial and eastern equatorial Atlantic Ocean to Quaternary climatic changes have been examined (Prell and Hays, this volume; Bé and others, this volume; Gardner and Hays, this volume). This study integrates the individual areas and reconstructs the late Quaternary paleocirculation of the equatorial and tropical Atlantic surface currents.

The Atlantic between lat 20°N and 10°S and the Caribbean Sea form a warm-water belt in which relatively stable climatic conditions prevail. No sharp oceanographic boundaries occur, and hence, only small diversity gradients exist for calcareous plankton within the region (Bé and Tolderlund, 1971; Jones, 1967; McIntyre and Bé, 1967). This equatorial-tropical belt also serves as a barrier to extensive exchange of cold-water species between the northern and southern regions of the Atlantic Ocean.

The dominant features of the complex equatorial current system are the westward-flowing South Equatorial and North Equatorial Currents (Defant, 1961). The South Equatorial Current, fed by the Benguela Current, flows across the equatorial Atlantic between lat 20°S and 5°N. The North Equatorial Current flows westward between about lat 5°N and 20°N, continuing along the South American coast north of lat 4° 30′N as the Guiana Current to Trinidad. At Trinidad, part of the flow enters the Caribbean Sea as the Caribbean Current.

The equatorial current system supports distinct planktonic foraminiferal assemblages, which represent different water-mass associations and whose identity can be resolved in the surface sediments by Q-mode factor analysis. The resultant factor assemblages of planktonic foraminifers can be used to interpret water-mass movements in the equatorial and tropical Atlantic.

Samples from trigger-weight core tops are used to represent late Holocene and interglacial conditions, and samples from the 18,000-B.P. level (McIntyre and others, this volume) to represent glacial conditions. The percentage of abundance of species of planktonic foraminifers is the basis for this analysis. The late Holocene (interglacial) and the 18,000-B.P. (glacial) levels were analyzed separately to define their biogeographic patterns. The resultant faunal (factor) assemblages and their distributions were then compared to interpret the difference between glacial and interglacial conditions.

LATE QUATERNARY FAUNAS: INTERGLACIAL AND GLACIAL COMPARISONS

Q-mode factor analysis (see Klovan and Imbrie, 1971) resolved the late Holocene faunal data into four factor assemblages, three of which show strong similarities to present-day plankton assemblages, whereas the fourth indicates carbonate dissolution. Samples with high abundances of resistant species and abnormally low numbers of dissolution-susceptible tests were interpreted as dissolution faunas. Twenty-four of the original 99 samples were eliminated on this basis. The remaining 75 samples were resolved into three independent assemblages that accounted for 97.5% of the original data. Figure 1 gives the location of the late Holocene interglacial samples; Table 1 presents the relative importance of each species in the three factors. The interglacial assemblages are termed tropical, equatorial, and temperate after their composition and general distributions, and they do not necessarily conform to previous oceanographic or faunal classifications in this area. Tropical refers to the Atlantic between approximately lat 5°N and 20°N, whereas equatorial describes the area between lat 5°N and 5°S.

The glacial datum (18,000-B.P.) was selected from 47 cores using the criteria of McIntyre and others (this volume). The planktonic foraminiferal data were independently resolved into four assemblages, which accounted for 97.0% of the data. The location of the 18,000 B.P. (glacial) samples is given in Figure 2, and the species composition of each glacial datum factor is given in Table 2. All assemblages are interpreted as reflecting surface-water conditions, because badly dissolved samples were eliminated prior to the factor analysis. The glacial assemblages are named "glacial tropical," "glacial cool-equatorial," "glacial temperate," and "glacial equatorial" after their species composition and areal distribution.

The tropical factors are dominant in both the interglacial and glacial data. The interglacial tropical assemblage accounts for 61% of the interglacial data and is characterized by abundant *Globigerinoides ruber* and *Globigerinita glutinata*. *G. ruber* is the most prolific foraminifer in subtropical waters and second in abundance only to *Globigerinoides sacculifer* in tropical waters (Bé and Tolderlund, 1971). *G. glutinata* is a cosmopolitan species with highest abundances in the subtropical zone. This assemblage is dominant in the western Caribbean and western Atlantic up to at least lat 20°N, but it is much weaker in the eastern equatorial Atlantic (Fig. 3).

The glacial-tropical factor assemblage comprises 64.3% of the 18,000 B.P. data and is almost identical to its interglacial analogue, the two most important species again being *G. ruber* and *G. glutinata*. The distribution of this glacial factor (Fig.

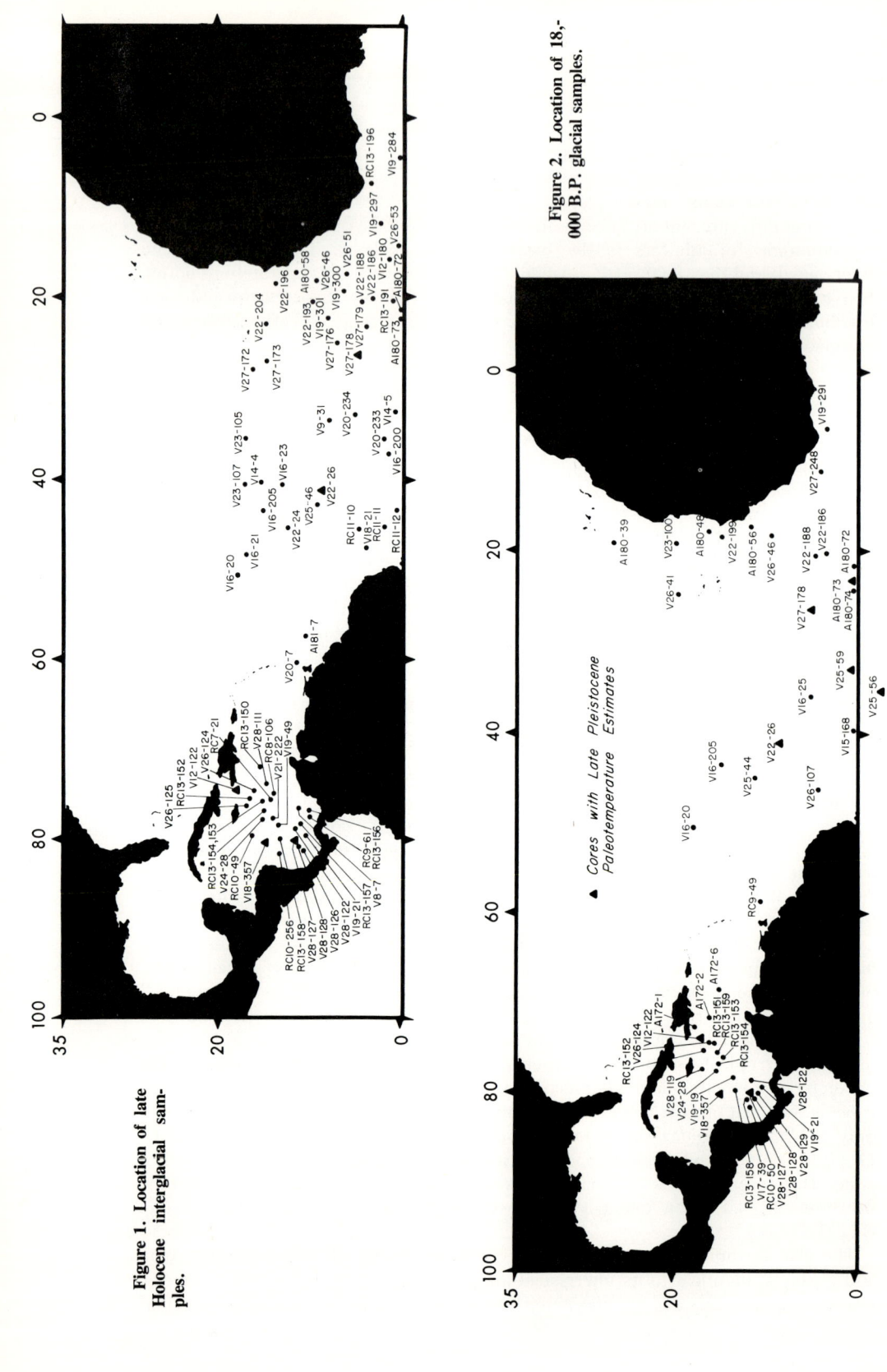

Figure 1. Location of late Holocene interglacial samples.

Figure 2. Location of 18,000 B.P. glacial samples.

4) is similar to its present distribution, but is more abundant in the Caribbean and much less abundant in the eastern equatorial Atlantic. The high Caribbean values have been interpreted by Prell and Hays (this volume) to indicate glacial sea-surface conditions similar to the present-day southern Sargasso Sea.

The equatorial factor assemblages differ sharply from glacial to interglacial data. The interglacial equatorial assemblage accounts for 33.9% of the interglacial data and is dominated by *Globigerinoides sacculifer, Globoquadrina dutertrei,* and *Globorotalia menardii.* The distribution of the assemblage (Fig. 5) shows highest values south of lat 10°N and off Africa in the area where the tropical assemblage is weakest. The inverse correlation between the tropical and equatorial assemblages reflects reversed relative frequencies of *G. ruber* and *G sacculifer.* Bé and Tolderlund (1971) noted this reversal in plankton data and suggested salinity as the controlling mechanism. They observed that *G. ruber* has maximum frequencies at salinities either greater than 36.0‰ or less than 34.5‰ and that *G. sacculifer* has maximum abundances at salinities between 34.5‰ and 36.0‰. Gardner (1973) showed that high *G. sacculifer* abundances in the eastern equatorial Atlantic also correlate with low integrated primary productivity, low phosphate concentrations, and low seasonal

TABLE 1. PLANKTONIC FORAMINIFERAL SPECIES VERSUS FACTOR ASSEMBLAGE MATRIX

Species	Factor 1	Factor 2	Factor 3
Orbulina universa	0.047	0.230	0.703
Globigerinoides conglobatus	0.050	0.100	0.122
Globigerinoides ruber	5.410	0.415	0.483
Globigerinoides tenellus	0.167	−0.119	0.120
Globigerinoides sacculifer	−0.124	4.950	−2.390
Sphaeroidinella dehiscens	−0.022	0.077	0.104
Globigerinella aequilateralis	0.314	0.388	−0.100
Globigerina calida	0.092	−0.037	0.188
Globigerina bulloides	0.043	0.024	0.670
Globigerina falconensis	0.108	−0.045	0.256
Globigerina digitata	−0.018	0.066	0.139
Globigerina rubescens	0.183	−0.076	−0.163
Globigerina humilis	0.004	−0.003	0.003
Globigerina quinqueloba	0.003	0.001	−0.006
Globigerina pachyderma L	−0.004	0.010	0.008
Globigerina pachyderma R	−0.032	0.126	0.662
Globoquadrina dutertrei	−0.599	1.876	3.455
Globoquadrina conglomerata	0.002	−0.001	−0.001
Globoquadrina hexagona	0.006	−0.004	0.011
Pulleniatina obliquiloculata	−0.265	0.701	0.954
Globorotalia inflata	0.014	0.120	1.156
Globorotalia truncatulinoides L	0.061	−0.056	0.157
Globorotalia truncatulinoides R	0.077	−0.002	0.132
Globorotalia crassaformis	−0.057	0.224	0.334
Globigerina pachyderma-Globoquadrina dutertrei intergrade*	−0.069	0.228	0.819
Globorotalia hirsuta	0.016	−0.016	0.038
Globorotalia scitula	0.053	−0.007	0.012
Globorotalia menardii	−0.185	1.225	2.713
Globorotalia tumida	−0.106	0.648	0.735
Candeina nitida	−0.037	0.073	0.752
Globigerinita glutinata	1.013	−0.082	0.014

Note: Data obtained by Q-mode factor analysis of the 75 surface-sediment (late Holocene, interglacial) samples. Factor 1 = tropical; factor 2 = equatorial; factor 3 = temperate.
*See Gardner and Hays (this volume).

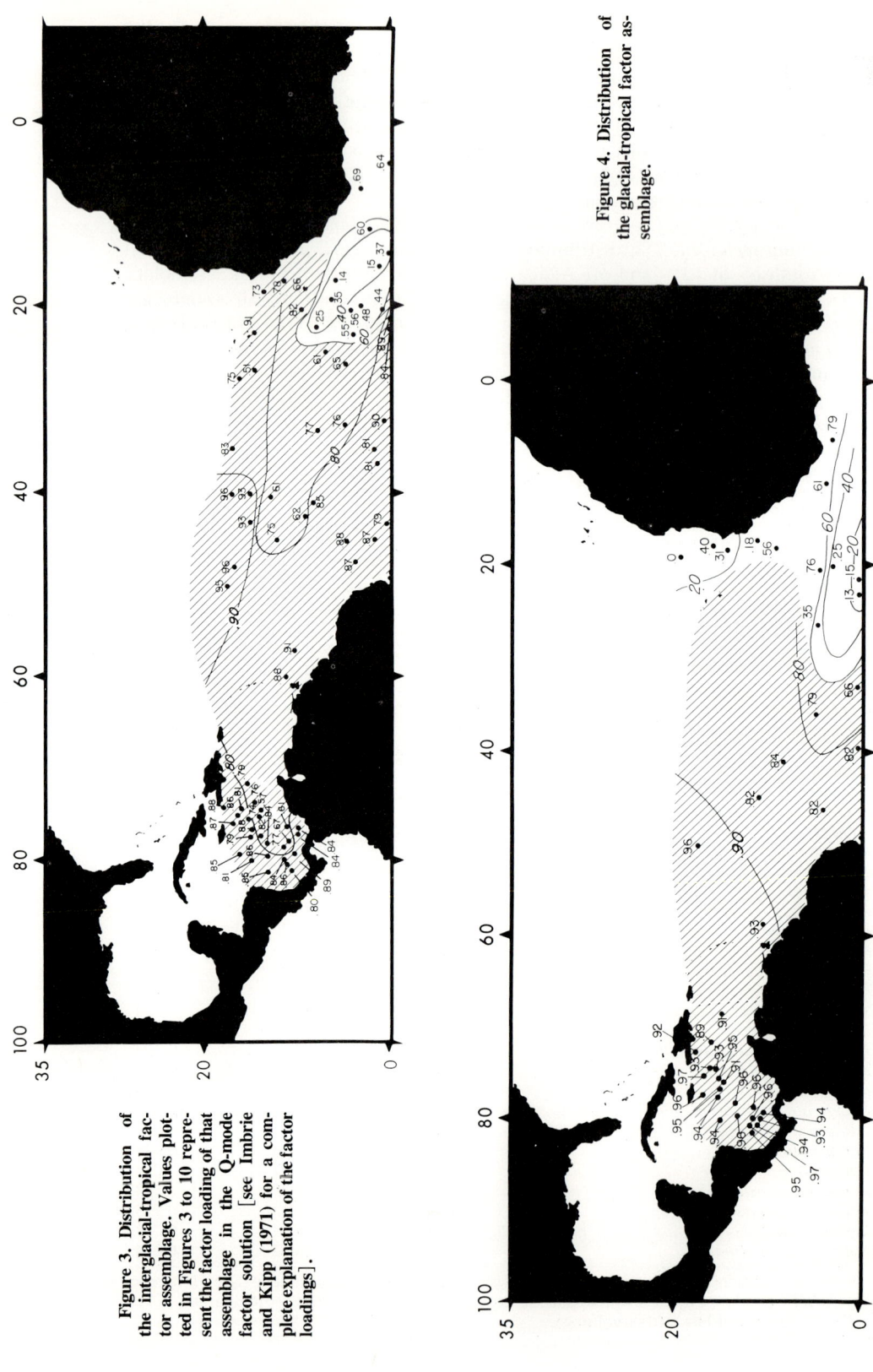

Figure 3. Distribution of the interglacial-tropical factor assemblage. Values plotted in Figures 3 to 10 represent the factor loading of that assemblage in the Q-mode factor solution [see Imbrie and Kipp (1971) for a complete explanation of the factor loadings].

Figure 4. Distribution of the glacial-tropical factor assemblage.

sea-surface-temperature contrast, whereas *G. ruber* showed the opposite affinities.

The glacial-equatorial assemblage is also marked by abundant *G. sacculifer* but accounts for only 4.6% of the 18,000 B.P. data. Compared to its interglacial counterpart, the glacial-equatorial assemblage is less abundant, especially in the eastern equatorial Atlantic and Caribbean, and has its maximum abundance farther to the west (Fig. 6). Although the glacial-equatorial assemblage does not dominate any given area, its similarity to the interglacial-equatorial assemblage implies that similar environmental conditions occurred 1,000 to 1,500 km farther west during glacial phases.

The temperate factor assemblages are less abundant than previous assemblages but are important because they define distinct water masses. The interglacial temperate assemblage is marked by abundant *Globoquadrina dutertrei* and *Globorotalia menardii*, which typically reach maximum abundances in strong current systems in the tropical-subtropical regions. *Globorotalia inflata*, also abundant, is an indicator of transitional zones separating subpolar from subtropical water

TABLE 2. PLANKTONIC FORAMINIFERAL SPECIES
VERSUS FACTOR ASSEMBLAGE MATRIX

Species	Factor 1	Factor 2	Factor 3	Factor 4
Orbulina universa	0.160	0.038	0.115	0.167
Globigerinoides conglobatus	0.169	0.087	−0.072	−0.021
Globigerinoides ruber	5.312	0.482	0.726	0.250
Globigerinoides tenellus	0.235	−0.045	−0.001	0.216
Globigerinoides sacculifer	0.020	1.572	−0.001	−5.134
Sphaeroidinella dehiscens	0.002	−0.000	−0.000	0.004
Globigerinella aequilateralis	0.695	−0.076	−0.070	0.183
Globigerina calida	0.030	−0.060	0.080	−0.370
Globigerina bulloides	−0.643	−0.535	4.929	−0.264
Globigerina falconensis	−0.004	0.199	0.117	0.081
Globigerina digitata	−0.010	0.028	−0.010	−0.150
Globigerina rubescens	0.185	−0.057	0.089	−0.290
Globigerina humilis	0.004	−0.006	0.026	−0.006
Globigerina quinqueloba	−0.012	−0.015	0.087	−0.006
Globigerina pachyderma L	−0.007	0.004	0.050	0.007
Globigerina pachyderma R	−0.426	0.432	1.991	0.181
Globoquadrina dutertrei	−0.532	4.531	−0.160	0.758
Globoquadrina conglomerata	0.000	0.000	0.000	0.000
Globoquadrina hexagona	0.000	0.000	0.000	0.000
Pulleniatina obliquiloculata	0.014	−0.004	0.011	−0.356
Globorotalia inflata	−0.183	1.237	1.347	0.721
Globorotalia truncatulinoides L	0.025	0.028	−0.006	0.027
Globorotalia truncatulinoides R	0.230	0.843	−0.223	0.501
Globorotalia crassaformis	−0.077	0.623	0.070	−0.065
Globigerina pachyderma-Globoquadrina dutertrei intergrade*	−0.289	2.111	0.210	1.519
Globorotalia hirsuta	0.007	0.019	0.000	0.021
Globorotalia scitula	0.016	0.006	0.092	−0.098
Globorotalia menardii	−0.011	0.018	0.100	0.013
Globorotalia tumida	0.000	0.025	0.029	0.016
Candeina nitida	0.007	0.010	−0.005	−0.015
Globigerinita glutinata	1.047	0.385	0.442	−0.572

Note: Data obtained by Q-mode factor analysis of the 47 18,000 B.P. (glacial) samples. Factor 1 = glacial tropical; factor 2 = glacial cool-equatorial; factor 3 = glacial temperate; factor 4 = glacial equatorial.

*See Gardner and Hays (this volume).

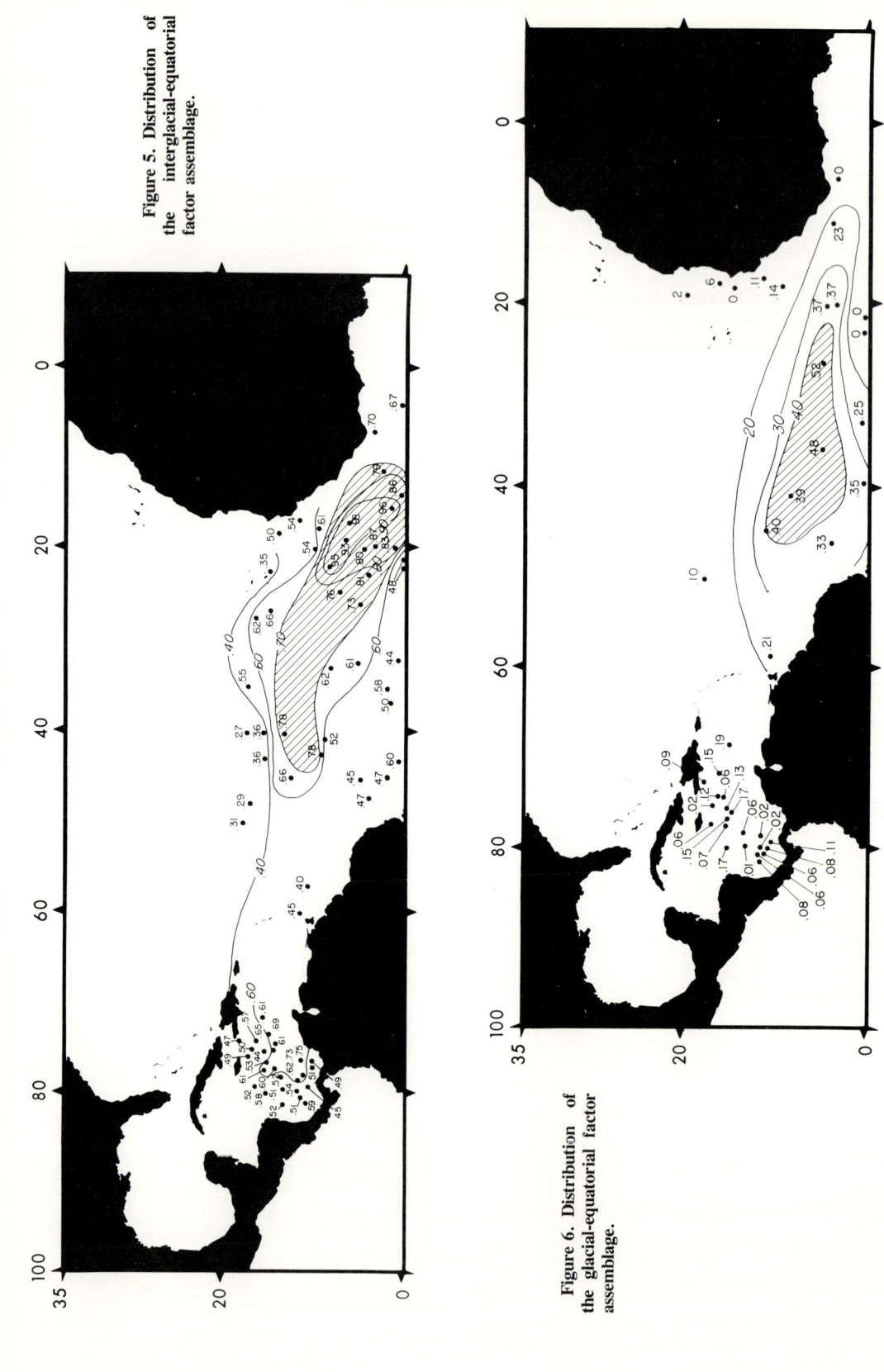

Figure 5. Distribution of the interglacial-equatorial factor assemblage.

Figure 6. Distribution of the glacial-equatorial factor assemblage.

masses (Bé and Tolderlund, 1971). The assemblage accounts for only 2.7% of the interglacial data. This factor occurs today along the South Equatorial Current and, compared to the equatorial and tropical assemblages, reflects more temperate waters (Fig. 7). The temperate assemblage results from two separate phenomena: transport of cooler Benguela Current waters into the equatorial region and coastal upwelling off the northwest African coast.

The glacial-temperate factor (Fig. 8) accounts for 10.2% of the 18,000 B.P. data and represents an assemblage composed predominantly of *Globigerina bulloides*, *Globigerina pachyderma* (right coiling), and *Globorotalia inflata*, all indicative of subpolar to cold-temperate waters (Bé and Tolderlund, 1971). Thus, the glacial-temperate assemblage reflects colder water than the interglacial-temperate assemblage. The increased abundance of these species in upwelling areas off northwest Africa (Gardner, 1973) supports the suggestion of McIntyre and others (1972) that the 18,000 B.P. polar front was much farther south than it is at present. The southward migration of the polar front may have allowed subpolar faunas access to the cooler upwelling waters off northwest Africa and may have extended their range into the tropical zone.

The glacial-cool-equatorial factor was resolved only from the 18,000 B.P. data and has no interglacial analogue. The assemblage accounts for 17.9% of the glacial data and reflects abundant *G. dutertrei*, "PD intergrade" (see Gardner and Hays, this volume, Table 1), *G. inflata*, *G. sacculifer*, and *Globorotalia truncatulinoides* (right coiling). Today, these species (with the exception of *G. sacculifer*) have maximum abundances in current systems along the margins of oceanic gyres or in upwelling zones, and they are especially abundant off southwest Africa in the Benguela Current. The abundance of the glacial-cool-equatorial assemblage (Fig. 9) is inverse to the glacial-tropical assemblage, indicating that an intensified Benguela–South Equatorial Current introduced this assemblage into the equatorial belt (Gardner, 1973).

Comparison of the faunal assemblages shows that the following trends have occurred from 18,000 B.P. (glacial) to the present late Holocene (interglacial): (1) an increase of the tropical factor in the eastern Atlantic, but a slight decrease in the western Caribbean (see Figs. 3, 4); (2) a significant increase and eastward migration of the equatorial factor (see Figs. 5, 6); (3) a decrease and warmer aspect of the temperate factor off northwest Africa (see Figs. 7, 8); (4) the disappearance of the glacial–cool-equatorial factor, which intruded into the equatorial belt from the east-southeast (Fig. 9); (5) larger faunal changes occurred in the eastern equatorial Atlantic, smaller changes in the central and western equatorial Atlantic, and small faunal changes (often the reverse of those in the eastern Atlantic) in the Caribbean.

Using the above glacial and interglacial factor assemblages, faunal data from eight Atlantic and Caribbean sediment cores (Fig. 13) were used to define continuous records of late Quaternary faunal fluctuations. This analysis divides the late Quaternary data into the same assemblages derived from the late Holocene (interglacial) and 18,000-B.P. (glacial) data and allows direct comparison with those assemblage maps. Data for three cores (Fig. 10), A180-73, V25-59, and V18-357, from the eastern equatorial Atlantic, the central equatorial Atlantic, and the Caribbean, respectively, are used to show the pattern of faunal fluctuations across the equatorial and tropical Atlantic during the past 150,000 yr. The interglacial-tropical and glacial-cool-equatorial assemblages are used to represent interglacial and glacial conditions.

Comparison of these assemblages during the past 150,000 yr shows the following relationships: (1) Only two periods occurred when the faunas of all three areas

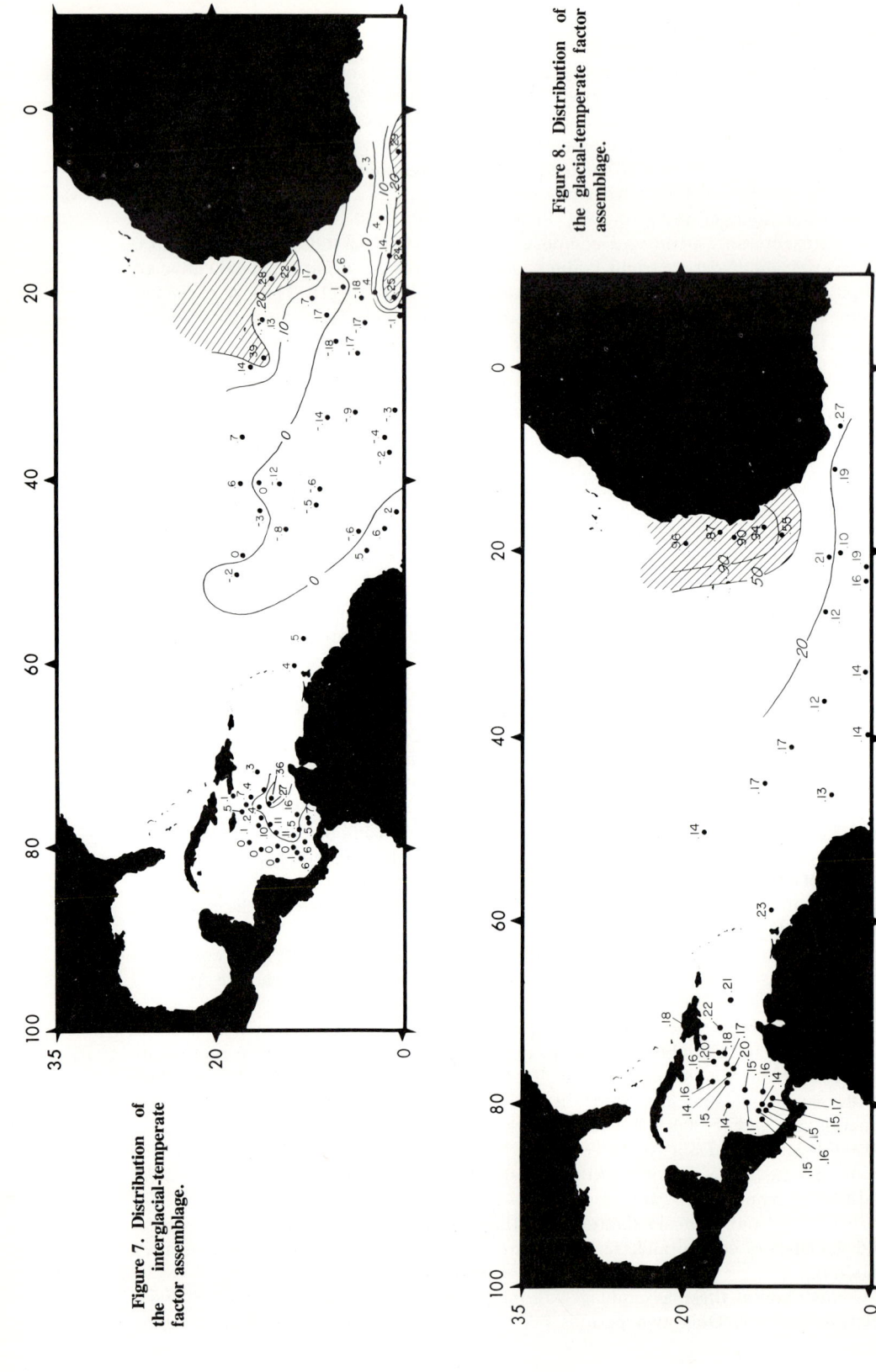

Figure 7. Distribution of the interglacial-temperate factor assemblage.

Figure 8. Distribution of the glacial-temperate factor assemblage.

were almost identical—today and 125,000 B.P. (2) Three glacial episodes occurred in the equatorial Atlantic at approximately 135,000, 73,000, and 18,000 B.P., as shown by the abundance of the cool-equatorial fauna, which indicates the strength of the Benguela Current. (3) The central and western equatorial Atlantic underwent a faunal change similar to that in the eastern equatorial Atlantic, but with a smaller reduction of the tropical fauna during glacial conditions. (4) The tropical fauna increased in the Caribbean during the glacial stages, exhibiting a reversed trend from the equatorial Atlantic. The timing of maximum abundance of the cool-equatorial assemblage also differs. For example, in the Caribbean, the maximum abundance occurred at approximately 45,000 B.P. rather than at 73,000 B.P. or 18,000 B.P. as in the equatorial Atlantic.

ESTIMATES OF SEA-SURFACE TEMPERATURE

Sea-surface temperatures were estimated for conditions typified by August and February using the 18,000 B.P. data (Imbrie and Kipp, 1971). Caribbean paleotemperatures were determined using the equations of Prell (1974), whereas the equatorial Atlantic paleotemperatures were derived from the equations of Gardner (1973). Because these equations are based on Northern Hemisphere data, they assigned the coldest temperatures to February and thus reversed the seasons when applied to Southern Hemisphere samples. To correct this reversal, the seasonality (whether the coldest temperature occurs in February or August) must be determined for samples along the Equator that are influenced by events occurring in the South Atlantic. The average thermal equator (Neumann and Pierson, 1966) has been used to determine the appropriate season for such samples. For example, site A180-73 is in the Northern Hemisphere but has its coldest temperatures in August. Thus, the estimated temperatures along the geographic Equator are coldest in August and warmest in February, thus forming an analogue to today's seasonal temperature distribution.

The distribution of February sea-surface temperatures at 18,000 B.P. (Fig. 11B) is comparable to today's February isotherm pattern (Fig. 11A), except that the upwelling region off northwest Africa is more pronounced and cooler temperatures occur in the western Caribbean. The present 18° to 22°C waters off northwest Africa (Figs. 8, 11B) were 15° to 16°C at 18,000 B.P. These lower temperatures probably resulted from a combination of increased upwelling and a compression of the North Atlantic gyre, as suggested by McIntyre and others (1972), which allowed cool, temperate to subpolar waters to migrate into the region. The February isotherms show only a small gradient between the western equatorial Atlantic and the Caribbean.

The 18,000 B.P. August temperatures (Fig. 12B) are similar to today's (Fig. 12A) except at the equatorial belt. Both maps show cooler waters entering the equatorial region via the Benguela-South Equatorial Current system (Gardner, 1973), but the 18,000 B.P. August temperatures (16°C to 18°C) are 7° to 8°C cooler than either today's August or the 18,000 B.P. February temperatures. However, at long 35°W, this tongue of cool water cannot be differentiated from the western equatorial Atlantic waters (24° to 25°C). Likewise, cores at lat 5°N to 8°N along the equatorial belt show only 2° to 3°C temperature differences during glacial conditions; thus the northern extent of the cool Benguela-South Equatorial Current is limited to south of lat 5°N. The Caribbean and northwestern equatorial Atlantic show temperatures of 27°C, which are similar to today's August temperatures; this suggests that the tropical conditions of the North Equatorial Current were not significantly

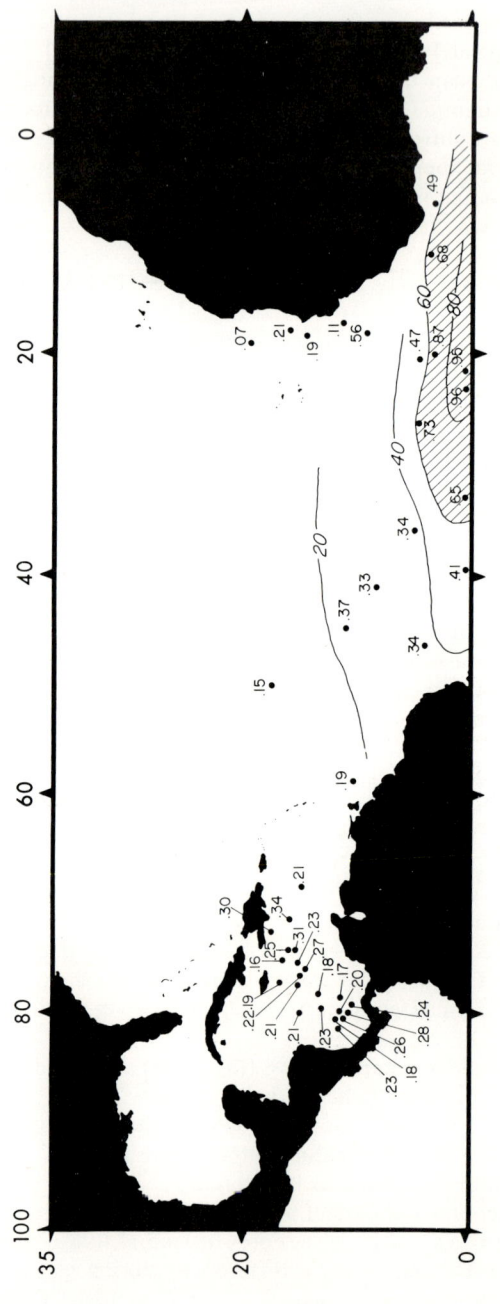

Figure 9. Distribution of the glacial–cool–equatorial factor assemblage.

affected by the incursions of cold water from the Southern Hemisphere.

Paleotemperatures were calculated for eight equatorial Atlantic and Caribbean cores (Fig. 13) to study the continuous variation of sea-surface temperatures over the past 80,000 to 150,000 yr. All eight cores show consistent and similar patterns of temperature change. These patterns are compared in cores V18-357, V25-59, and A180-73 (Fig. 14). The February and August patterns for the past 150,000 yr show the following trends:

1. Large (5° to 10°C) glacial-interglacial temperature differences occurred along the Equator in the eastern area, whereas smaller (2° to 3°C) differences occurred in the northern and western equatorial region and the Caribbean.

2. Large seasonal temperature contrast (August versus February) was characteristic during the glacial stages because of low winter temperatures. The eastern region exhibited the greatest contrast, whereas smaller contrasts occurred in the northern and western equatorial Atlantic and the Caribbean.

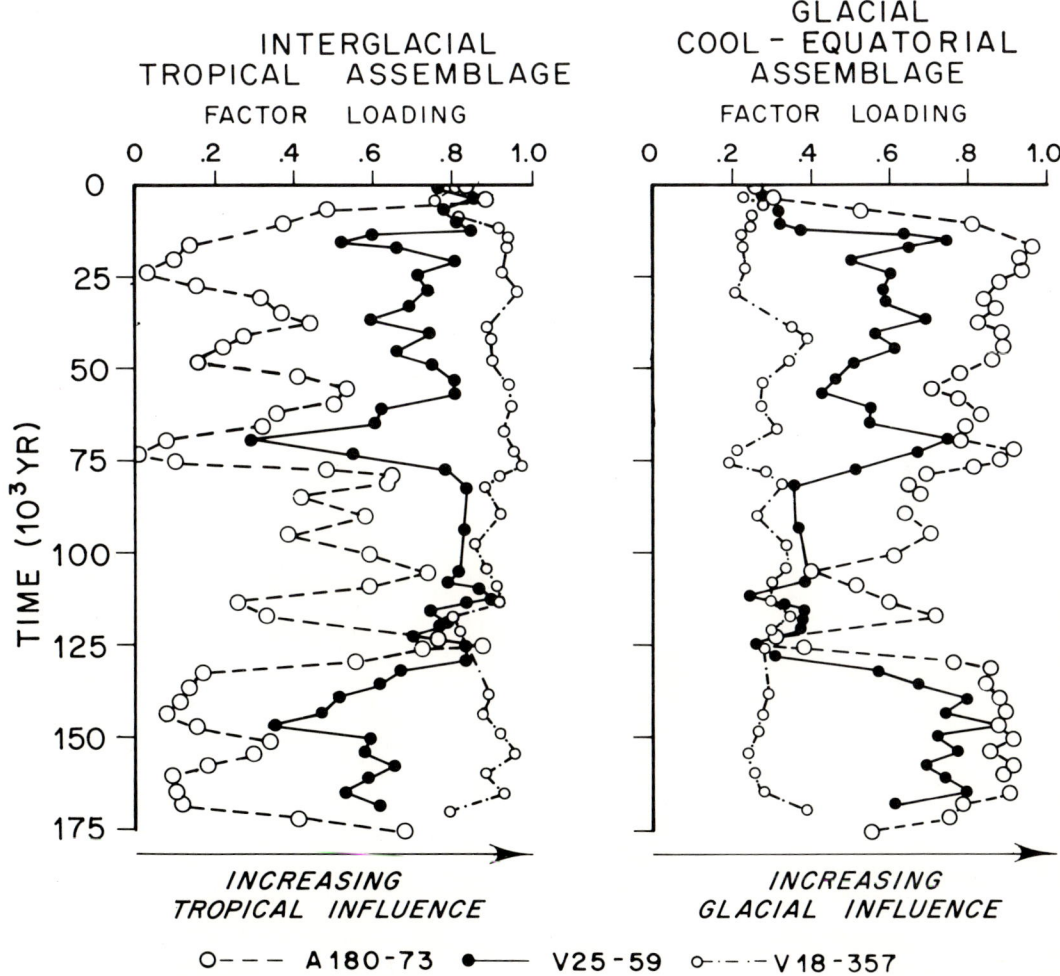

Figure 10. Fluctuation of the interglacial-tropical assemblage and the glacial–cool-equatorial assemblage versus time in cores A180-73, V25-59, and V18-357. Time scale used is that of Broecker and van Donk (1970).

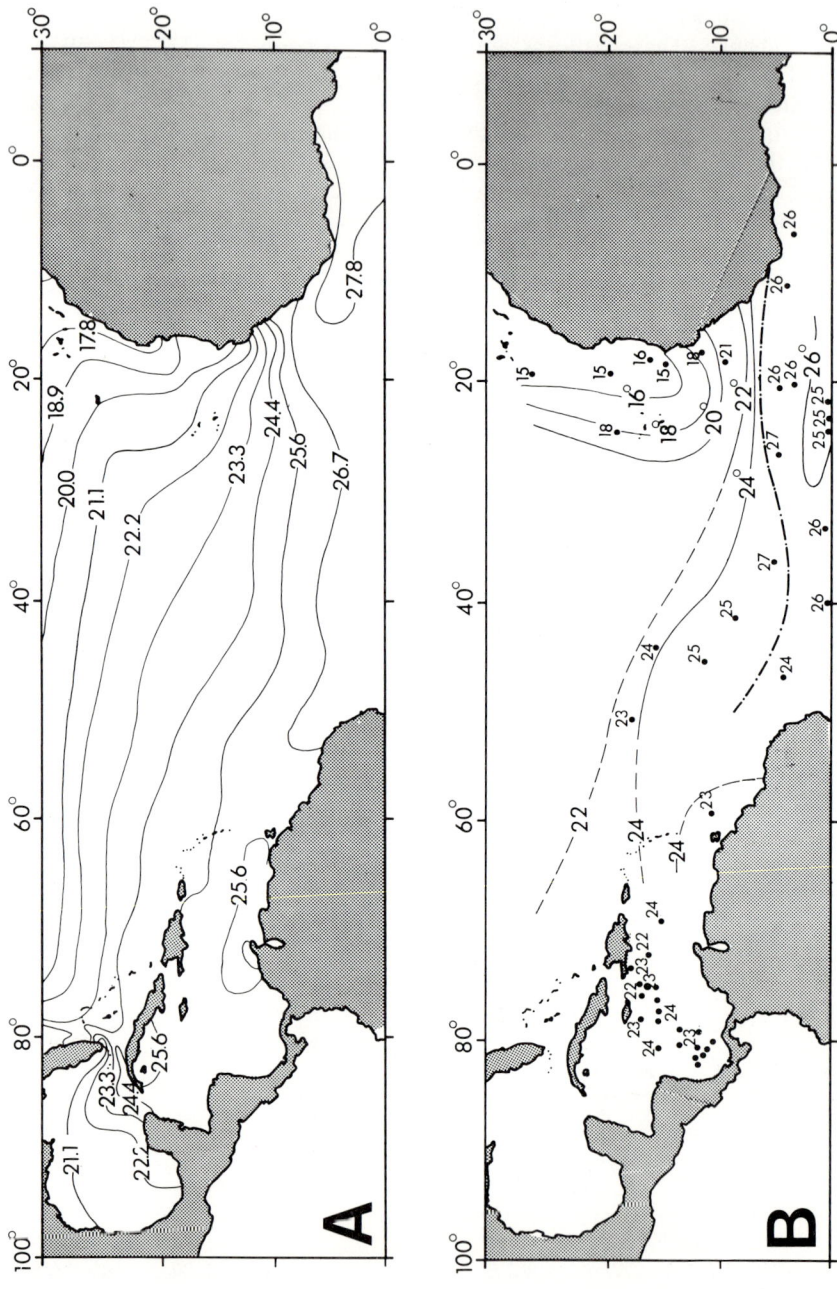

Figure 11. Distribution of February sea-surface temperatures: A, present day (interglacial); B, 18,000 B.P. (glacial).

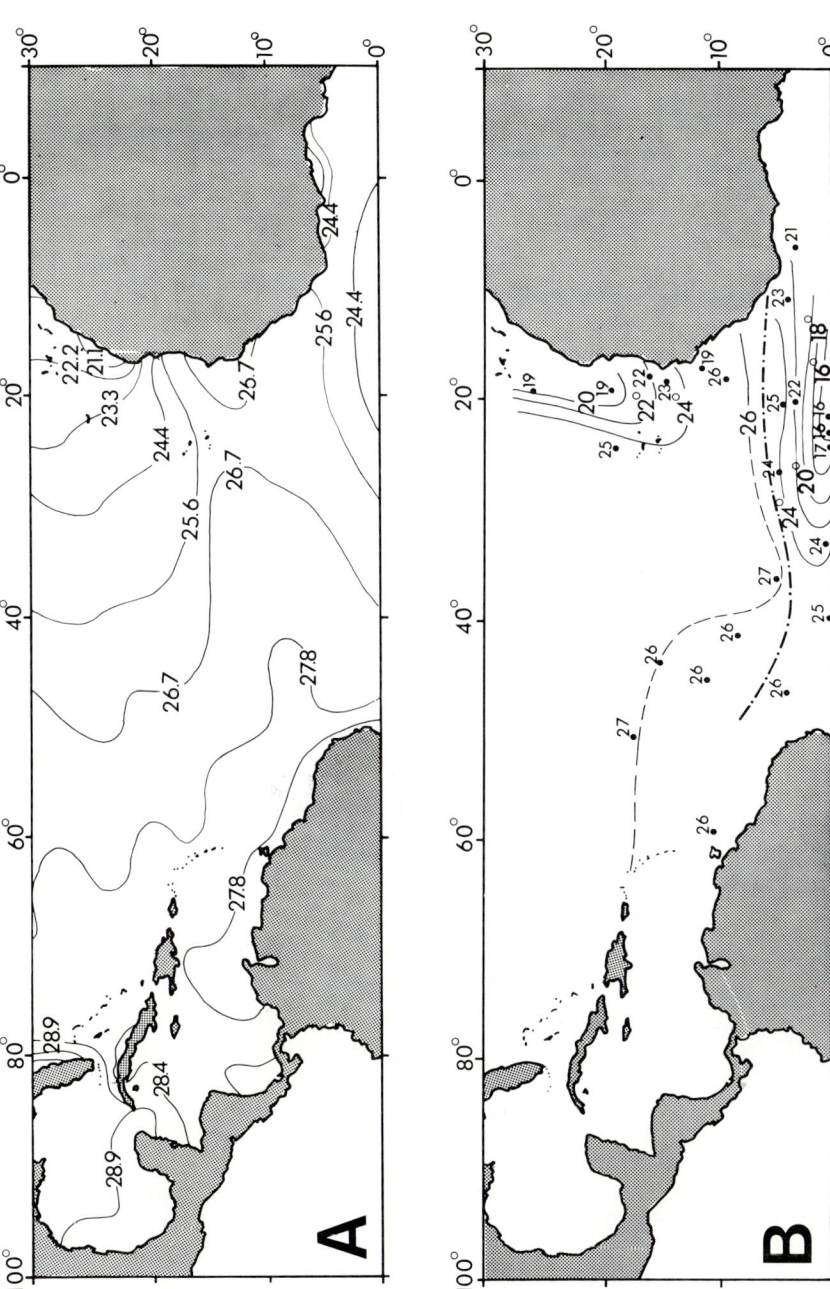

Figure 12. Distribution of August sea-surface temperatures: A, present day (interglacial); B, 18,000 B.P. (glacial).

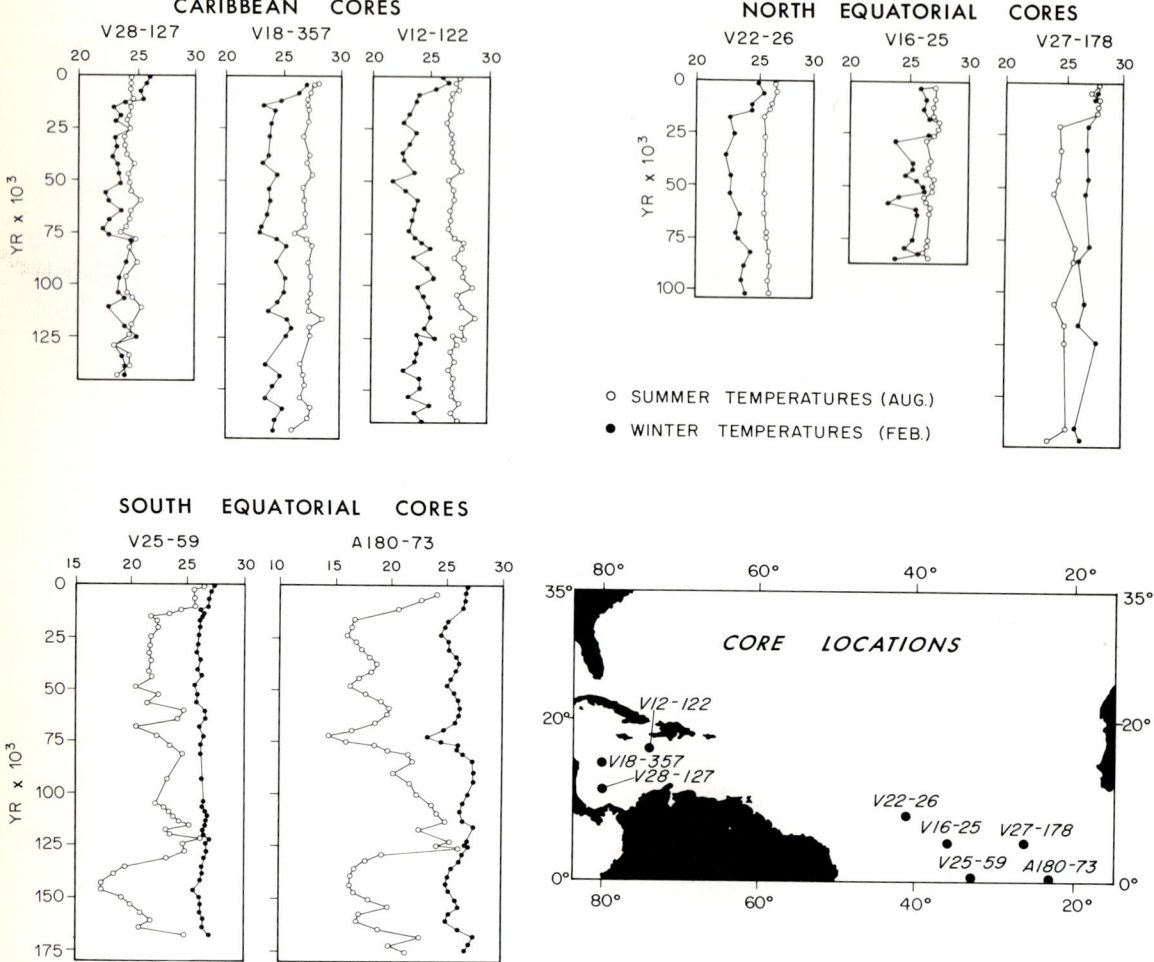

Figure 13. **Fluctuation of estimated February and August sea-surface temperatures (°C) versus time in eight equatorial Atlantic and Caribbean cores.**

3. Along the Equator the coldest temperatures and largest fluctuations occurred in August (cores A180-73 and V25-59). However, in the tropical Atlantic and Caribbean, the coldest temperatures and largest fluctuations occurred in February (cores V22-26 and V18-357). The seasonality (whether the coldest temperatures occurred in August or February) is determined by the position of the core relative to the thermal equator.

4. Sea-surface temperatures 125,000 yr ago were comparable to present equatorial Atlantic conditions. However, glacial winters in the western Caribbean were cooler by 1° to 2°C, as indicated by the consistency of the three Caribbean cores rather than by the precision of the equation.

5. The period centered around 73,000 B.P. exhibited the coolest August and February sea-surface temperatures in both the Caribbean and the equatorial Atlantic. These temperatures were comparable to the previous glacial stage at about 135,000 B.P. and were slightly colder than the most recent glacial maximum at 18,000 B.P.

DISCUSSION

The present isotherm patterns observed for February and August are also found at 18,000 B.P. in the equatorial Atlantic, which suggests that glacial and interglacial circulation patterns are similar. However, glacial temperature and faunal gradients were sharper, which suggests circulation was more intense during glacial conditions. These patterns are consistent for the past 150,000 yr. The similarity in circulation patterns and difference in intensities suggest that the glacial dynamics of equatorial Atlantic circulation were similar to present dynamics and thus were dependent on the northeast and southeast trade winds. Both seasonality and circulation patterns suggest that the Intertropical Convergence Zone (ITCZ) remained within its present zone of seasonal migration in the equatorial Atlantic during glacial phases but may have remained over South America for longer periods. Both seasonal upwelling off northwest Africa and seasonal intensification of the Benguela Current reflect the southern (February) and northern (August) positions of the ITCZ, respectively. The ITCZ must have remained within its present zone of migration in the eastern Atlantic, because these oceanic patterns had the same geographic location at 18,000 B.P. as they do today. Accepting this, the intensification of these systems must have been due to strengthened trade winds at 18,000 B.P. The identification of increased eolian components in glacial-age sediments off northwest Africa by Biscaye

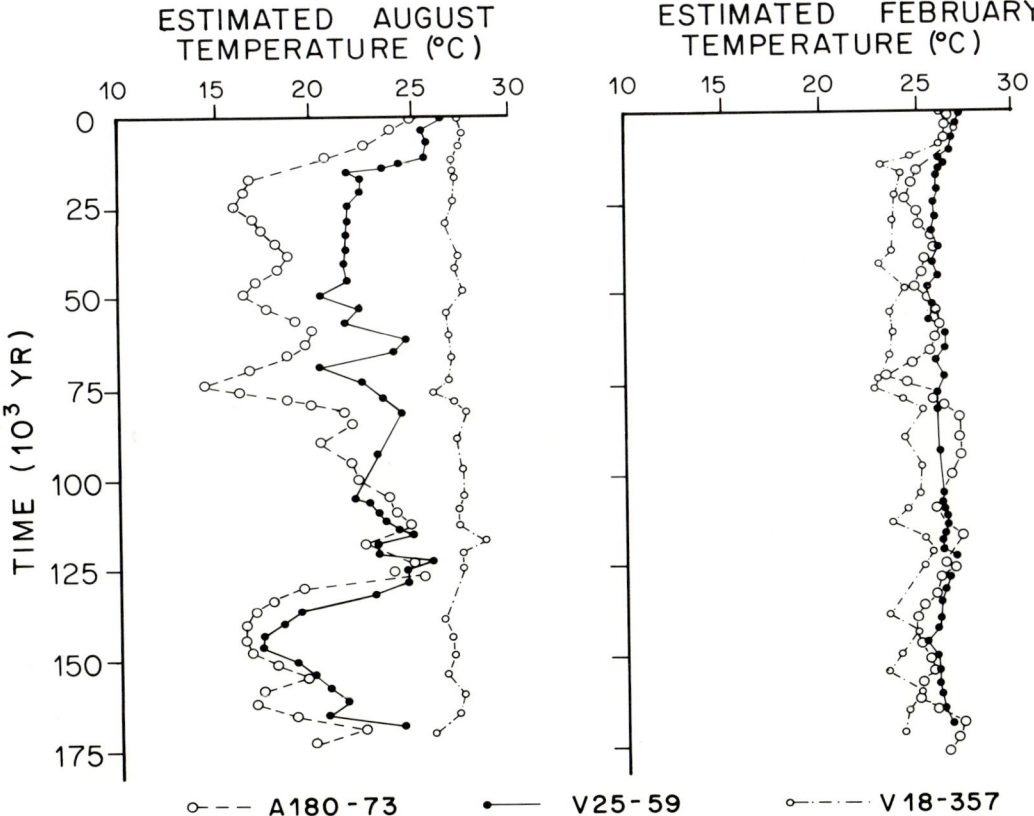

Figure 14. Estimated August and February sea-surface temperatures versus time in cores A180-73, V25-59, and V18-357.

and others (1973) supports the hypothesis of increased glacial trade winds. However, it is possible that compression of the North and South Atlantic gyres could have increased velocities within the equatorial currents.

The sharp gradient in sea-surface temperatures from the eastern to the western equatorial Atlantic and Caribbean, especially in August, is an important feature of equatorial Atlantic glacial circulation. Three explanations for the sharp gradient follow. First, Benguela–South Equatorial Current waters containing the glacial-cool-equatorial assemblage were submerged under the warmer, lighter surface waters containing the interglacial-tropical assemblage. If submergence did occur, two assemblages should be resolved from the underlying sediments. However, the 18,000 B.P. data resolved only the tropical factor in the area of supposed submergence. This indicates that the two assemblages and, thus, the water masses they represent were not superimposed. Second, partial southward deflection of the South Equatorial Current along the South American margin would reduce water transport across the Equator. Although poor core coverage does not allow exclusion of this possibility, the August temperatures predicted from cores RC11-11 and V25-26, located along the South American margin, do not suggest an intensified Brazil Current. Lastly, and most likely, the cool Benguela contribution to the South Equatorial Current may merely have been heated by the intense equatorial sun as it flowed west along the Equator as it is today. The stronger glacial wind stress would produce a thinner mixed layer in the east and a thicker mixed layer in the west, thus allowing rapid heating in the eastern Atlantic which would promote a steep east-west thermal gradient.

In contrast to the equatorial belt, the Atlantic between lat 5°N and 20°N and the Caribbean are not affected by the Benguela–South Equatorial Current and show small temperature fluctuations, with the coldest temperatures occurring February. These areas respond to the flow of the equatorial current, which was also intensified by the glacial trade winds. However, the 18,000 B.P. isotherm maps show that the Canary–North Equatorial Current does not transport the upwelled waters off northwest Africa across the Atlantic as a tongue of cool water. Thus, waters of the Canary–North Equatorial Current reflect ambient tropical temperatures of approximately 24° to 27°C in the western Atlantic and Caribbean. Cores in this area show temperature variations of 2° to 3°C, which we believe represent the general response of the tropical region to glacial-interglacial change.

SUMMARY

Faunal and temperature reconstructions suggest that the surface-water circulation of the equatorial Atlantic responded in phase with Quaternary global climatic events. These responses are considered to be the relative intensities of the North and South Equatorial Currents interpreted from faunas and estimated sea-surface temperatures. This indicates an intensification of global atmospheric circulation and, in this region specifically, the trade winds. The major features of the glacial circulation pattern are increased upwelling off northwest Africa and generally cooler conditions in the Caribbean and North Equatorial Current during February and intensified circulation of the Benguela–South Equatorial Current system during August.

Although the seasons north and south of the thermal equator are reversed, the average temperature patterns are identical throughout the past 150,000 yr, which suggests that the Northern and Southern Hemispheres are in phase and that the temperature patterns are controlled by winter conditions (February in the north and August in the south).

ACKNOWLEDGMENTS

We thank A. McIntyre and S. Streeter for reviewing this paper and B. Molfino and M. Roche for their support and helpful discussions. This work was supported by National Science Foundation Grant IDO71 04204 under the International Decade for Ocean Exploration Program. Sediment cores for this study were collected under Office of Naval Research Contract N00014-67-A-0108-0004 and National Science Foundation Grants DES-72-01568 and GA-19690.

REFERENCES CITED

Bé, A.W.H., and Tolderlund, D. S., 1971, Distribution and ecology of living planktonic foraminifera in surface waters of the Atlantic and Indian Oceans, *in* Funnell, B. M., and Riedel, W. R., eds., Micropaleontology of the oceans: London, Cambridge Univ. Press, p. 105-149.

Bé, A.W.H., Damuth, J. E., Lott, L., and Free, R., 1976, Late Quaternary climatic record in western equatorial Atlantic sediments, *in* Cline, R. M., and Hays, J. D., eds., Investigation of late Quaternary paleoceanography and paleoclimatology: Geol. Soc. America Mem. 145 (this volume).

Biscaye, P., Bé, A.W.H., Ellis, D., Gardner, J., Kellogg, T., McIntyre, A., Prell, W., Roche, M., and Venkatarathnam, K., 1973, Holocene vs. glacial (17,000 Y.B.P.) patterns of sedimentation in the Atlantic Ocean: Geol. Soc. America Abs. with Programs, v. 5, no. 7, p. 551.

Broecker, W. S., and van Donk, J., 1970, Insolation changes, ice volumes and the O^{18} record in deep-sea cores: Rev. Geophysics and Space Physics, v. 8, p. 169-198.

Defant, A., 1961, Physical oceanography, Vol. 1: New York, Pergamon Press, 729 p.

Gardner, J. V., 1973, The eastern equatorial Atlantic: Sedimentation, faunal, and sea-surface temperature responses to global climatic changes during the past 200,000 years [Ph.D. thesis]: New York, Columbia Univ., 387 p.

Gardner, J. V., and Hays, J. D., 1976, Responses of sea-surface temperature and circulation to global climatic change during the past 200,000 years in the eastern equatorial Atlantic Ocean, *in* Cline, R. M., and Hays, J. D., eds., Investigation of late Quaternary paleoceanography and paleoclimatology: Geol. Soc. America Mem. 145 (this volume).

Imbrie, J., and Kipp, N., 1971, A new micropaleontological method for quantitative paleoclimatology: Application to a late Pleistocene Caribbean core, *in* Turekian, K. K., ed., Late Cenozoic glacial ages: New Haven, Yale Univ. Press, p. 71-179.

Jones, J. I., 1967, Significance of distribution of planktonic foraminifera in the equatorial Atlantic undercurrent: Micropaleontology, v. 13, p. 489-501.

Klovan, J. E., and Imbrie, J., 1971, An algorithm and Fortran IV program for large scale Q-mode factor analysis: Internat. Assoc. Math. Geology Jour., v. 3, p. 61-77.

McIntyre, A., and Bé, A.W.H., 1967, Modern coccolithophoridae of the Atlantic Ocean—I, Placoliths and cyrtoliths: Deep-Sea Research, v. 14, p. 561-597.

McIntyre, A., Bé, A.W.H., Biscaye, P., Burckle, L., Gardner, J., Geitzenauer, K., Goll, R., Kellogg, T., Prell, W., Roche, M., Imbrie, J., Kipp, N., Ruddiman, W., and Moore, T., 1972, The glacial North Atlantic 17,000 years ago: Plaeoisotherm and oceanographic maps derived from floral-faunal parameters by CLIMAP: Geol. Soc. America Abs. with Programs, v. 4, no. 7, p. 590–591.

McIntyre, A., Kipp, N., Bé, A.W.H., Crowley, T., Gardner, J., Prell, W., and Ruddiman, W. F., 1976, Glacial North Atlantic 18,000 years ago: A CLIMAP reconstruction, *in* Cline, R. M., and Hays, J. D., eds., Investigation of late Quaternary paleoceanography and paleoclimatology: Geol. Soc. America Mem. 145 (this volume).

Neumann, G., and Pierson, W. J., Jr., 1966, Principles of physical oceanography: Englewood Cliffs, N.J., Prentice-Hall, Inc., 545 p.

Prell, W. L., 1974, Late Pleistocene faunal, sedimentary and temperature history of the Columbia Basin, Caribbean Sea [Ph.D. thesis]: New York, Columbia Univ., 518 p.

Prell, W. L., and Hays, J. D., 1976, Late Pleistocene faunal and temperature patterns of the Columbia Basin, Caribbean Sea, *in* Cline, R. M., and Hays, J. D., eds., Investigation of late Quaternary paleoceanography and paleoclimatology: Geol. Soc. America Mem. 145 (this volume).

MANUSCRIPT RECEIVED BY THE SOCIETY SEPTEMBER 23, 1974
REVISED MANUSCRIPT RECEIVED FEBRUARY 28, 1975
MANUSCRIPT ACCEPTED APRIL 29, 1975
LAMONT-DOHERTY GEOLOGICAL OBSERVATORY CONTRIBUTION NO. 2279

Printed in U.S.A.

Geological Society of America
Memoir 145
© 1976

Corresponding Patterns of Contemporary Pollen and Vegetation in Central North America

T. WEBB III
Department of Geological Sciences
Brown University
Providence, Rhode Island 02912

AND

J. H. MCANDREWS
Department of Geology
Royal Ontario Museum
and
Department of Botany
University of Toronto
Toronto, Ontario, Canada M5S 2C6

ABSTRACT

Use of modern pollen spectra as a basis for interpreting diagrams of fossil spectra requires compilation of the modern spectra in readily accessible form, such as contoured maps of percentage values of individual pollen types. Maps are presented that show the distribution of modern pollen based on 606 samples from central North America (lat 35°N to 70°N, long 75°W to 110°W). Only data published after 1960 are included, and data from 69 sites are presented for the first time. The maps show differences in the pollen percentages among vegetational regions. For example, peak pollen values of Cyperaceae and *Betula* occur in the tundra; high values of *Picea* appear in the northern boreal forest; high values of *Pinus* appear in the southern boreal forest and the adjacent conifer-hardwood forest; high values of *Tsuga, Fagus,* and *Acer* occur eastward in the conifer-hardwood forest; high values of *Quercus, Ambrosia, Fraxinus,* and *Carya* occur in the deciduous forest; and high values of nontree pollen (Gramineae, *Artemisia,* Chenopodiineae, and Compositae) appear in the prairie.

Trend-surface analysis and principal components analysis summarize the regional trends of each pollen type and illustrate the patterns of covarying pollen types within the data. Although these data provide a basis for interpreting the major fossil pollen zones for Holocene time in central North America, additional sampling and more detailed examination of the data are required for description of the fine-scale changes within fossil zones.

INTRODUCTION

Reconstructing past vegetational or climatic patterns from pollen diagrams involves transforming the pollen data into vegetational or climatic terms. A basis for this transformation is the association between the modern distribution of pollen types and the modern distribution of vegetational and climatic regions (Wright, 1967; Davis, 1967; McAndrews, 1966). Assemblages of modern pollen can be correlated with spectra of fossil pollen and thus link the fossil data and the vegetation and climate they reflect. Quantitative methods can aid this interpretative process (see, for examples, Adam, 1970; Davis, 1963; Ogden, 1969; Yarranton and Ritchie, 1972; Webb, 1974b; Webb and Bryson, 1972). For these methods to work well requires ready access to modern pollen spectra and knowledge of the geographic distribution of the various pollen types. The modern data, however, are scattered among many publications and are not generally presented on maps.

To improve this situation, we compiled 606 samples of modern pollen spectra from central North America, presented these data in maps showing the isofrequency contours of each pollen type, and summarized the data using numerical techniques of data analysis. The data set includes pollen spectra from sediment, moss-polster, and soil-surface samples, but not from atmospheric traps.

Although contour mapping of pollen frequencies was introduced by Szafer in 1935 and has since been used by marine palynologists (Muller, 1959; Cross and others, 1966), this method for presenting data has only recently been applied to studies of modern pollen from inland sites (McAndrews and Power, 1973; Webb, 1974a, 1974b; Birks and others, 1975; Davis and Webb, 1975). Previously the modern data were plotted as bar diagrams or pie diagrams.

Bar diagrams and plots are useful for illustrating pollen variation along transects (McAndrews and Wright, 1969; Maher, 1963) and within certain regions (Lichti-Federovich and Ritchie, 1968). Maps with bar diagrams (Curtis, 1959; Cole, 1969) or pie diagrams (King and Kapp, 1963; Kapp, 1965; Lichti-Federovich and Ritchie, 1965) at each sample site illustrate the two-dimensional variation of three to six pollen types. Contoured maps—or isopoll maps—of pollen frequencies, however, show the distributional patterns of individual pollen types. These maps show the magnitude and location of geographic gradients and present the data in a form that can be compared with climatic and vegetational maps. The isopoll maps are similar to contoured maps of foraminiferal abundances in the oceans (Kipp, this volume).

Further summary of the data by trend-surface analysis and by principal components analysis shows the basic structure of the data. Trend-surface analysis (McAndrews and Power, 1973) provides generalized contours and predicts values in regions of sparse data. Our principal components analysis extracts patterns of covarying pollen types and compresses into four maps most of the information spread originally among 21 pollen types. This technique permits examination of the internal structure of pollen data for comparison with the vegetation.

With these three techniques—isopoll mapping, trend-surface analysis, and principal components analysis—we illustrate and summarize the distributional patterns of the main pollen types that compose the spectra of modern pollen collected in central North America. We also list the available samples of modern pollen in central North America. In 1967, M. B. Davis accomplished this task in northeastern

North America, and R. B. Davis and Webb (1975) have updated and expanded her data set for eastern North America.

Our paper is similar to the study by Kipp (this volume) that describes the modern distribution of species and assemblages of foraminifera in the North Atlantic Ocean and that anticipates use of our modern pollen samples for deriving calibration functions which, in a manner comparable to that of McIntyre and others (this volume), can be used to reconstruct maps of past climatic conditions from Holocene pollen records in eastern North America. This anticipated study will complement the studies of marine organisms and climates by providing a detailed synoptic record of Holocene climates over the North American continent.

DATA AND PRELIMINARY STATISTICS

Sources of the Data

The sample sites lie between lat 37°N and 66°N and between long 80°W and 110°W (Fig. 1). Most samples are from surficial sediment in lakes, but some are from *Sphagnum* or other moss polsters, and a few samples in the west are from soil duff or the sediment in cattle-watering tanks.

The data set consists of both previously published and unpublished samples; all were analyzed after 1960 and contain counts of herb pollen. All samples of modern pollen that were collected within the geographic bounds of this study and presented in the following papers are included: Bartley (1967), Bradbury and Waddington (1973), Brubaker (1973), Cole (1969), Davis and others (1971), Fries (1962), Grüger (1973), Kapp (1965), King and Kapp (1963), Lichti-Federovich and Ritchie (1965, 1968), McAndrews (1966, 1968), McAndrews and Wright (1969), Miller (1973), Mott (1969), Ogden (1966, 1969), Skinner (1973), Swain (1973), Watts and Wright (1966), and Webb (1971, 1973a, 1974a, 1974b). Also included are the samples from the inland lakes analyzed by McAndrews and Power (1973); samples 1, 3, 4, and 7 by Mott (1975b); and the samples from the surface sediment under the greatest water depths analyzed by Davis and others (1969) from six Wisconsin lakes.

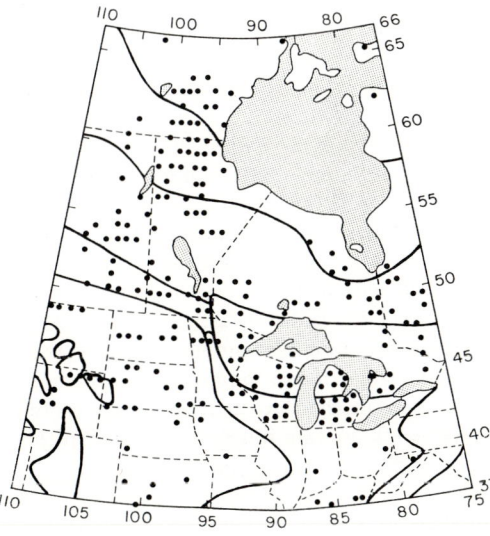

Figure 1. Map showing the general distribution of the 606 samples used. Each dot indicates that, for an area of 1° latitude by 1° longitude at the given location, at least one sample occurs. Areas with particularly high densities of samples occur in lower Michigan, Wisconsin, central Saskatchewan, and northern Manitoba and along transects in eastern Ontario, eastern Minnesota, northern Montana, and northeastern Colorado. Heavy lines are vegetational boundaries (see Fig. 2).

Previously unpublished data are listed in Table 1,[1] which gives the latitude and longitude of each site, the type of material sampled, and the pollen analyst. In each sample analyzed by McAndrews, at least 280 pollen grains were counted.

Most of the pollen numbers are from tables of pollen counts, but some were derived from percentage data in Fries (1962), Grüger (1973), Kapp (1965), King and Kapp (1963), Lichti-Federovich and Ritchie (1965, 1968), Mott (1969), and Watts and Wright (1966).

Pollen Sum and Preliminary Statistics

For the calculation of pollen percentages, the data were sorted into the 34 categories of pollen shown on Table 2, and the total of the counts in these categories is the pollen sum. The 34 categories consist of 31 pollen types (described in McAndrews and others, 1973) plus unknown pollen grains, unidentifiable pollen grains, and pollen grains from miscellaneous types of woody and herbaceous plants. The pollen sum, therefore, is the standard sum used in most pollen studies in central North America and includes no spores or pollen of aquatic taxa. Counts of the 31 pollen types were almost always recorded in published tables and diagrams, but the number of pollen grains in the final three categories were sometimes missing.

Table 2 summarizes the standard statistics for each of the 34 pollen types. *Pinus* is the most abundant and best dispersed pollen type in the data set. It has an average value of 25% and is present in all but one of the samples. *Betula*, *Picea*, *Ambrosia*, *Quercus*, Gramineae, *Alnus*, Chenopodiineae, *Artemisia*, and Cyperaceae are also dominant members of the pollen record. Each appears in more than 70% of the samples, has an average value (over all samples) above 2.5%, and a maximum value of greater than 50%. *Abies*, *Juniperus-Thuja*, *Carya*, *Larix*, *Tilia*, *Platanus*, *Juglans*, *Corylus*, *Myrica*, Ericaceae, *Rumex*, and *Plantago* pollen are only minor members of the pollen record, and the remaining pollen types (*Salix*, *Tsuga*, *Acer*, *Fagus*, *Ulmus*, *Fraxinus*, *Ostrya-Carpinus*, Compositae, and *Populus*) are of intermediate importance.

Vegetation Map

Because the correspondence between the distribution of the various pollen types and the vegetation is discussed, we constructed Figure 2 to show the distribution of the seven main vegetational regions in our study area and then used this map as a base map on which to present the isopolls. The seven regions shown are the tundra, forest-tundra, boreal forest, conifer-hardwood forest, deciduous forest, prairie, and the mountain and basin complex. The composition of the vegetation in these regions is described by Rowe (1972), Weaver and Clements (1938), and Küchler (1964). For each woody-plant pollen type, the range limits of the corresponding tree genus are also shown on the map. These limits (except alder) are from Little (1971).

Table 2 gives the mean percentage value and variance of each pollen type within each vegetational region (Fig. 2). The values on Table 2 are cited within the descriptions of the individual pollen types, and Figure 12 diagrams some of these values. Table 2 also shows that, of the 606 sites, more than a third are in the conifer-hardwood forest and about a sixth are in the deciduous forest, whereas the tundra and mountain and basin complex each contain fewer than 20 samples.

---

[1] Table 1 on microfiche in pocket inside back cover.

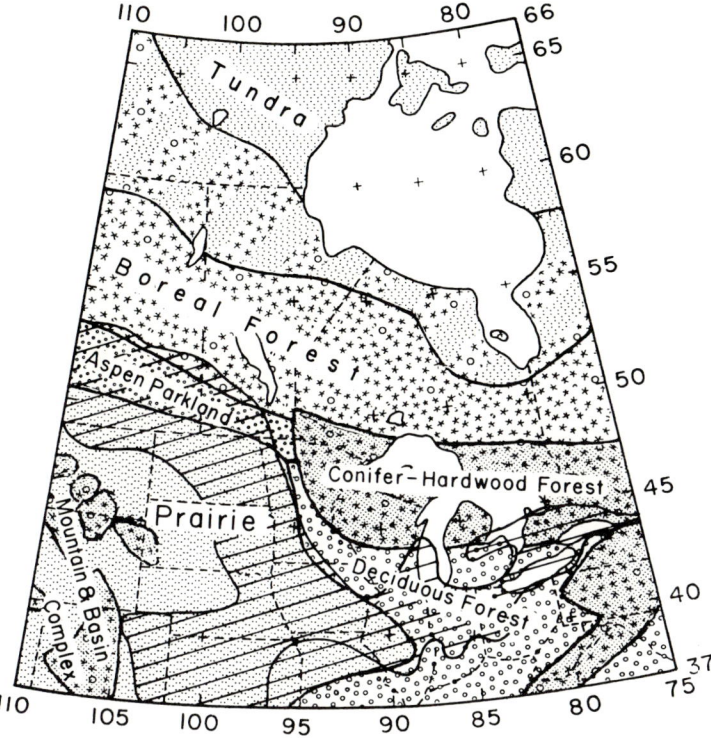

Figure 2. Map of the major vegetational regions in central North America, adapted from Rowe (1972) and Küchler (1964). Cross-hatching indicates areas that are at least 60% cultivated (*Oxford World Atlas*, 1973).

METHODS OF SPATIAL ANALYSIS

Isopoll Maps

Contour maps of the pollen percentages, or isopoll maps (Erdtman, 1943), were made with the SYMAP computer mapping package (Dudnik, 1971) on polar stereographic projections. This program interpolates a grid of pollen percentages from the values at the arbitrarily spaced sampling sites. The estimated pollen percentage at each grid point is converted to a symbol that represents a certain range of percentage values, and the grid of symbols is portrayed on a line-printer. Contour lines drawn between areas of differing symbols on the grid represent the transition from one range of percentages to the next. The number and distance between points for interpolation, the number and sizes of percentage ranges, and the scale of the maps produced are controllable factors that affect the level of detail that can be shown on the maps.

Isopoll maps were produced for all the dominant pollen types and for most of the types of intermediate abundance. Maps were not produced for minor pollen types except for *Abies* and *Carya*, because no regional patterns were apparent.

The zero contour line was added by hand and encompasses all samples within which a pollen type was recorded even if only in trace amounts. Areas circled by zero contour lines beyond the regionally continuous zero contour line indicate the presence of one or two pollen grains at isolated sites (for example, Fig. 3c). On some maps, asterisks indicate the location of particularly high pollen percentages (for example, Fig. 3b).

TABLE 2. AVERAGE POLLEN PERCENTAGES (x) AND STANDARD

	Cyperaceae		Salix		Betula		Alnus	
	x	s.d.	x	s.d.	x	s.d.	x	s.d.
Tundra	19.82	9.71	5.29	11.22	24.03	9.64	9.51	7.15
Forest-Tundra	5.11	6.72	1.27	3.70	13.99	8.78	15.29	10.19
Boreal Forest	1.42	2.68	0.91	1.02	10.29	7.12	7.72	5.48
Conifer-Hardwood Forest	1.99	4.96	1.19	1.66	17.17	11.71	4.56	4.22
Deciduous Forest	1.42	2.03	1.31	1.40	3.63	3.34	1.57	3.14
Aspen Parkland	4.42	3.45	2.82	1.39	8.67	8.07	2.95	1.62
Prairie	4.00	7.83	0.38	0.71	0.72	1.32	0.62	0.81
Mountain and Basin	0.87	1.36	0.13	0.27	0.79	2.74	0.04	0.11
Total x and s.d.	2.77	5.76	1.24	2.59	11.16	10.88	4.88	6.11
Maximum %/no. of sites with nonzero values	53.2	475	40.1	471	83.5	561	55.5	539

	Picea		Pinus		Abies		Tsuga	
	x	s.d.	x	s.d.	x	s.d.	x	s.d.
Tundra	10.38	6.92	19.89	10.47	0.0	0.0	0.01	0.04
Forest-Tundra	32.66	13.18	22.84	10.17	0.55	2.71	0.06	0.28
Boreal Forest	29.27	12.61	42.31	15.48	1.62	3.78	0.04	0.17
Conifer-Hardwood Forest	7.89	10.13	33.08	15.03	0.74	1.01	2.38	3.71
Deciduous Forest	0.42	1.77	7.50	5.69	0.08	0.56	0.29	0.62
Aspen Parkland	3.71	2.41	16.17	9.75	0.04	0.13	0.0	0.0
Prairie	0.52	0.90	10.35	8.79	0.08	0.18	0.01	0.05
Mountain and Basin	0.62	0.85	25.28	18.75	0.21	0.34	0.0	0.0
Total x and s.d.	10.82	14.47	25.78	17.84	0.62	1.90	1.00	2.59
Maximum %/no. of sites with nonzero values	67.1	507	76.4	606	32.7	271	26.2	177

	Fagus		Acer		Ulmus		Ostrya-Carpinus	
	x	s.d.	x	s.d.	x	s.d.	x	s.d.
Tundra	0.0	0.0	0.0	0.0	0.0	0.0	0.0	0.0
Forest-Tundra	0.0	0.0	0.02	0.08	0.12	0.30	0.01	0.04
Boreal Forest	0.01	0.06	0.06	0.20	0.16	0.69	0.08	0.41
Conifer-Hardwood Forest	0.65	1.64	1.50	1.90	1.81	2.17	0.90	1.77
Deciduous Forest	1.26	2.31	1.54	1.82	4.63	3.34	1.45	1.35
Aspen Parkland	0.0	0.0	0.30	0.47	0.57	0.51	0.0	0.0
Prairie	0.0	0.0	0.12	0.30	0.66	1.03	0.10	0.38
Mountain and Basin	0.0	0.0	0.0	0.0	0.0	0.0	0.0	0.0
Total x and s.d.	0.48	1.49	0.89	1.59	1.65	2.53	0.63	1.37
Maximum %/no. of sites with nonzero values	12.7	120	11.7	294	28.4	361	20.7	257

	Quercus		Carya		Ambrosia		Populus	
	x	s.d.	x	s.d.	x	s.d.	x	s.d.
Tundra	0.0	0.0	0.03	0.07	1.07	1.32	0.0	0.0
Forest-Tundra	0.19	0.40	0.03	0.10	1.07	1.29	0.09	0.21
Boreal Forest	0.41	0.66	0.02	0.08	1.08	1.20	0.10	0.17
Conifer-Hardwood Forest	5.70	6.35	0.18	0.32	7.19	7.06	0.68	1.28
Deciduous Forest	25.47	9.89	1.70	1.73	25.15	10.02	0.49	0.92
Aspen Parkland	7.71	9.20	0.0	0.0	3.67	2.40	2.95	6.19
Prairie	2.84	4.88	0.13	0.90	12.17	13.99	0.22	0.56
Mountain and Basin	0.36	0.57	0.0	0.0	1.03	0.91	0.0	0.0
Total x and s.d.	7.42	10.70	0.39	1.01	8.91	11.17	0.53	1.68
Maximum %/no. of sites with nonzero values	49.6	445	8.1	195	56.9	557	25.6	247

	Gramineae		Chenopodiineae		Compositae		Artemisia	
	x	s.d.	x	s.d.	x	s.d.	x	s.d.
Tundra	2.98	4.19	0.36	0.46	0.08	0.12	1.21	1.07
Forest-Tundra	0.95	1.00	0.50	0.66	0.20	0.47	1.05	0.98
Boreal Forest	1.15	1.27	0.60	0.46	0.27	0.48	1.14	0.89
Conifer-Hardwood Forest	3.73	4.40	1.44	1.62	0.68	1.22	1.68	2.50
Deciduous Forest	9.52	8.30	2.36	2.34	1.03	1.31	0.81	0.91
Aspen Parkland	11.95	6.02	11.17	8.28	2.41	1.52	15.51	5.62

DEVIATIONS (s.d.) FOR EACH VEGETATIONAL REGION

	Gramineae		Chenopodiineae		Compositae		Artemisia	
	x	s.d.	x	s.d.	x	s.d.	x	s.d.
Prairie	16.88	9.65	24.06	19.16	4.60	6.64	15.65	16.99
Mountain and Basin	14.42	10.22	15.10	19.20	4.89	5.64	34.69	21.01
Total x and s.d.	6.06	7.71	4.39	10.14	1.20	2.87	4.32	9.91
Maximum %/no. of sites with nonzero values	54.5	562	75.7	526	35.5	433	73.9	501

	Ericaceae		Myrica		Larix		Juniperus-Thuja	
	x	s.d.	x	s.d.	x	s.d.	x	s.d.
Tundra	3.36	4.60	0.20	0.40	0.01	0.05	0.02	0.06
Forest-Tundra	1.75	2.42	0.75	1.54	0.14	0.19	0.02	0.07
Boreal Forest	0.42	0.98	0.04	0.14	0.34	0.78	0.06	0.17
Conifer-Hardwood Forest	0.30	1.89	0.14	0.41	0.23	0.70	0.69	1.40
Deciduous Forest	0.00	0.05	0.03	0.10	0.03	0.15	0.45	0.90
Aspen Parkland	0.0	0.0	0.03	0.13	0.0	0.0	0.04	0.13
Prairie	0.00	0.01	0.00	0.01	0.00	0.01	0.33	0.50
Mountain and Basin	0.0	0.0	0.00	0.0	0.0	0.0	0.0	0.0
Total x and s.d.	0.40	1.73	0.13	0.53	0.16	0.56	0.39	1.01
Maximum %/no. of sites with nonzero values	26.3	126	7.2	106	9.0	142	8.1	213

	Tilia		Juglans		Platanus		Plantago	
	x	s.d.	x	s.d.	x	s.d.	x	s.d.
Tundra	0.0	0.0	0.05	0.13	0.0	0.0	0.0	0.0
Forest-Tundra	0.00	0.03	0.05	0.14	0.0	0.0	0.00	0.03
Boreal Forest	0.0	0.0	0.01	0.04	0.0	0.0	0.01	0.03
Conifer-Hardwood Forest	0.13	0.28	0.12	0.24	0.04	0.14	0.05	0.21
Deciduous Forest	0.40	0.59	0.60	0.61	0.24	0.46	0.25	0.44
Aspen Parkland	0.02	0.05	0.08	0.13	0.0	0.0	0.05	0.09
Prairie	0.02	0.07	0.17	0.42	0.02	0.12	0.23	1.68
Mountain and Basin	0.0	0.0	0.0	0.0	0.0	0.0	0.0	0.0
Total x and s.d.	0.13	0.33	0.18	0.39	0.06	0.23	0.09	0.58
Maximum %/no. of sites with nonzero values	3.1	138	4.0	190	2.4	60	13.0	76

	Rumex		Corylus		Fraxinus		Misc. Herbs	
	x	s.d.	x	s.d.	x	s.d.	x	s.d.
Tundra	0.80	1.65	0.01	0.05	0.0	0.0	0.88	1.61
Forest-Tundra	0.62	1.61	0.12	0.24	0.06	0.19	0.39	0.80
Boreal Forest	0.06	0.14	0.14	0.34	0.04	0.12	0.17	0.52
Conifer-Hardwood Forest	0.27	0.66	0.67	1.53	0.58	0.73	0.58	1.18
Deciduous Forest	0.94	1.45	0.32	0.44	1.90	1.83	1.55	2.83
Aspen Parkland	0.46	2.19	1.16	1.62	0.82	0.98	1.51	0.85
Prairie	0.05	0.19	0.07	0.28	0.50	1.17	1.62	4.30
Mountain and Basin	0.0	0.0	0.09	0.35	0.0	0.0	1.49	2.18
Total x and s.d.	0.37	1.06	0.41	1.09	0.66	1.17	0.85	2.09
Maximum %/no. of sites with nonzero values	11.4	182	11.5	262	10.3	303	31.4	335

	Unidentifiable		Unknown		No. of stations
	x	s.d.	x	s.d.	
Tundra	0.0	0.0	0.0	0.0	16
Forest-Tundra	0.07	0.30	0.0	0.0	45
Boreal Forest	0.02	0.16	0.01	0.10	98
Conifer-Hardwood Forest	0.48	0.89	0.56	1.05	237
Deciduous Forest	0.83	1.07	0.80	1.10	107
Aspen Parkland	0.81	1.79	0.0	0.0	26
Prairie	1.91	2.21	0.95	1.78	59
Mountain and Basin	0.0	0.0	0.0	0.0	18
Total x and s.d.	0.56	1.19	0.45	1.04	606
Maximum %/no. of sites with nonzero values	8.3	184	9.4	215	

Trend-Surface Analysis

Local anomalies in pollen distributions may hinder interpretation of regional distribution patterns. For any geographically distributed set of data, trend-surface analysis seeks to filter out local differences among samples and to emphasize the regional pattern (Davis, 1973). This technique fits a polynomial function of the geographic coordinates (x for latitude and y for longitude) to the percentages of the ith pollen type (P_i) and thus approximates the geographic trend of that pollen distribution by an nth-degree equation of the form

$$P_i = a_1 x^n + a_2 y^n + a_3 x^{n-1} y + \ldots + a_{q-1} y + a_q, \qquad (1)$$

where

$$q = \sum_{k=1}^{n+1} k$$

and the a_j ($j = 1, \ldots, q$) values are regression coefficients. The derived equation has predictive value and can be used to estimate pollen percentages at a station, given its latitude and longitude.

Pollen variation, however, cannot be adequately modeled by a function of latitude and longitude alone, and our choice of the degree of the polynomial is somewhat arbitrary. As the degree of the polynomial is raised, the resulting surface more closely fits the data and smooths out less of the local variation. The multiple correlation coefficient measures goodness of fit between the trend-surface pattern and the observed distribution.

For this study, fifth-degree trend surfaces ($n = 5$) were calculated for 18 pollen types, and contours of these surfaces were plotted. The surfaces are sufficiently general to filter out the effects of individual sites and present a summary of the regional patterns in the data. In each case, the multiple correlation coefficient was 0.5 or higher. Because of a program limitation, only 500 of the 606 samples were used in this analysis. The 106 samples that we deleted were from areas of dense sampling.

Principal Components Analysis

The first two techniques—isopoll mapping and trend-surface analysis—show the distribution of individual pollen types. Because the vegetation map shows regions dominated by groups of trees or herbs, a summary of the data was needed to show the distributions of groups of covarying pollen types. Several numerical techniques exist that provide this type of summary (see Birks and others, 1975), but we used principal components analysis (PCA).

This technique computes eigenvectors of the correlation matrix between all pairs of pollen types. In this study, the eigenvectors are rotated by the varimax criterion (Kaiser, 1958); this procedure results in components that are easier to interpret than the unrotated components (Imbrie and Kipp, 1971). Components having a reasonable interpretation are chosen for plotting (Webb, 1974a). Contour maps are produced by finding the weights or component scores that project each site onto the set of derived components. These maps show the distribution of "pollen assemblages" and are equivalent to the maps of foraminifera assemblages given in the study by Imbrie and Kipp (1971). (For previous use of this technique on spatially distributed pollen data, see Cole [1969], Brubaker [1973], Webb [1973b,

1974a, 1974b], and Birks and others [1975], and for further discussion of this technique, see Adam [1970], Anderson [1958], Birks [1974], Kutzbach [1967], and Morrison [1967].)

RESULTS OF SPATIAL ANALYSIS

The maps of all three forms of spatial analysis are presented and described in this section. The isopoll map for each pollen type is described in detail, and the trend-surface maps are used to summarize the general trends noted on the isopoll maps. The results of principal components analysis are given last.

Maps of the Pollen Types

Figures 3 through 7 present the isopoll maps in geographic order. Maps of pollen types with peak values in the north precede pollen types with peak values in the south, and pollen types with high values in the southeast appear before types with high values in the southwest. Because no samples were taken in an area bounded by northern Missouri, northern Illinois, central Indiana, and western Kentucky, this region lacks contour lines on each of the isopoll maps. (G. Peterson [in prep.] recently prepared a set of pollen data from this area.) When a pollen type within the pollen data is described, the botanical name is used, whereas the common name is used when referring to the plant in the vegetation.

1. Cyperaceae (Fig. 3A). Cyperaceae pollen is present in most samples in the data set except for some samples from soils in the Great Plains. A regional pattern of high percentages appears in the northern tundra and forest-tundra, and the tundra shows the highest average percentages (Table 2). Peaks that occur outside of the north are associated with single sites and reflect local aquatic plants. For example, 29% Cyperaceae was recorded in one of the two samples collected from marshes in lower Michigan (Webb, 1974a), and the high percentages in North Dakota and Minnesota result from bulrush (*Scirpus*) stands in prairie lakes. These local exceptions to the regional distribution have led some palynologists to consider excluding it from the pollen sum (Wright and Patten, 1962). The regional pattern shown on Figure 3A, however, warrants the inclusion of Cyperaceae in late-glacial pollen spectra where high values may reflect tundra vegetation.

2. *Salix* (Fig. 3B). *Salix* is less well represented in the data set (Table 2) than Cyperaceae and, except for the tundra and prairie, is generally 2% or less of the pollen sum. The low values of this pollen type render it only marginally mappable, and caution is required in using this map. The highest values of *Salix* occur in the tundra, but these are isolated samples from surface peat and probably reflect only local occurrences of willows. (No lake sample records a value above 3%.) Because of the sparsity of other sites in this area, these few isolated high percentages have a disproportionate effect on the patterns of contours in the north. Therefore, although the presence-absence pattern of *Salix* has a regional character, the percentage contours largely reflect local occurrences of willows.

3. *Betula* (Fig. 3C). Percentages of more than 20% of *Betula* pollen occur in the far north, where the distribution resembles that of Cyperaceae. Other areas with high values are the northern and southern shores of Lake Superior, east-central Ontario, and southwestern Quebec. In these southern areas, the high values are found where there is a medium to heavy population intensity of birch trees (see Figs. 10 and 11 in Halliday and Brown, 1943).

Because *Betula* is a prominent member of the pollen assemblage of the conifer-

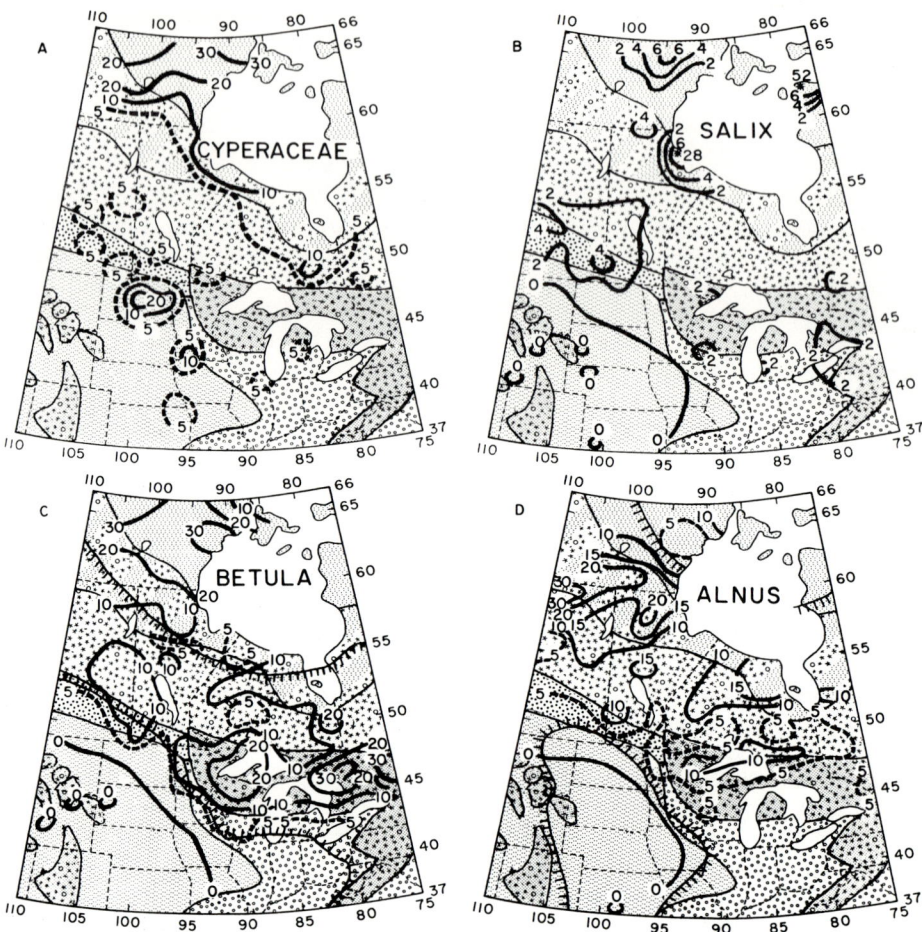

Figure 3. Isopollen maps of the percentage values of (A) sedge, (B) willow, (C) birch, and (D) alder pollen. Hachured lines indicate range limit of the plants producing the pollen. The range limits for birch mark the limits for trees only (as opposed to shrubs). Dashed contours indicate contours spaced at intervals different from the standard interval on the map.

hardwood forest and tundra, high *Betula* values lie both to the north and south of the high values of *Picea* (Fig. 4A). This bimodal distribution of *Betula* reflects the fact that different species produce the pollen for each mode: where distinctions in *Betula* pollen between tree and shrub grains have been made to the west (Ritchie, 1972) and to the east (Davis and Webb, 1975) of our study area, the shrub grains predominate in tundra samples, and the tree grains dominate in forested regions. Little or no *Betula* appears in the prairie or the mountain and basin complex, both of which lie beyond the limit of birch trees. *Betula*, therefore, has a striking geographic pattern.

4. *Alnus* (Fig. 3D). Similar to its distribution in the McKenzie River Delta (Ritchie, 1972), *Alnus* has high percentages (>15%) in the forest-tundra south of peaks of Cyperaceae, *Salix*, and *Betula*. Almost no values of more than 5% occur south of the boreal forest. Little or no *Alnus* occurs in the prairie; this reflects the absence of alder there. An exception to the general regional pattern occurs in western Wisconsin where the samples were collected in man-made lakes (Webb,

1971). The high proportion (20%) of *Alnus* in these samples probably reflects alder growing on the flood plains of rivers that feed these lakes (Curtis, 1959).

5. *Picea* (Fig. 4A). High percentages of *Picea* (>20%) occur in the forest-tundra and the boreal forest (Table 2). The highest values lie between Hudson Bay and the Great Lakes in areas showing high population intensity of spruce trees (see Fig. 2 in Halliday and Brown, 1943; Leopold, 1958). Further comparison of the pollen and tree percentages shows the relative abundance of spruce trees to be somewhat greater than that of *Picea* pollen; this fact indicates the general under-representation of spruce in the pollen record when compared to the vegetation (see also Webb, 1974a).

South of the boreal forest, *Picea* values decrease sharply to less than 5%. Within the tundra, the decrease is less rapid, but the values north of lat 65°N are less than 5%. Almost no *Picea* occurs at sites in the prairie and deciduous forest, and the zero contour line runs east-west at about lat 43°N. Despite the relatively high settling velocity of *Picea* (Gregory, 1973), the zero contour line lies south of the southern limit of spruce trees, especially in the prairie. This occurrence either may reflect the presence of planted spruce, especially in southernmost Canada

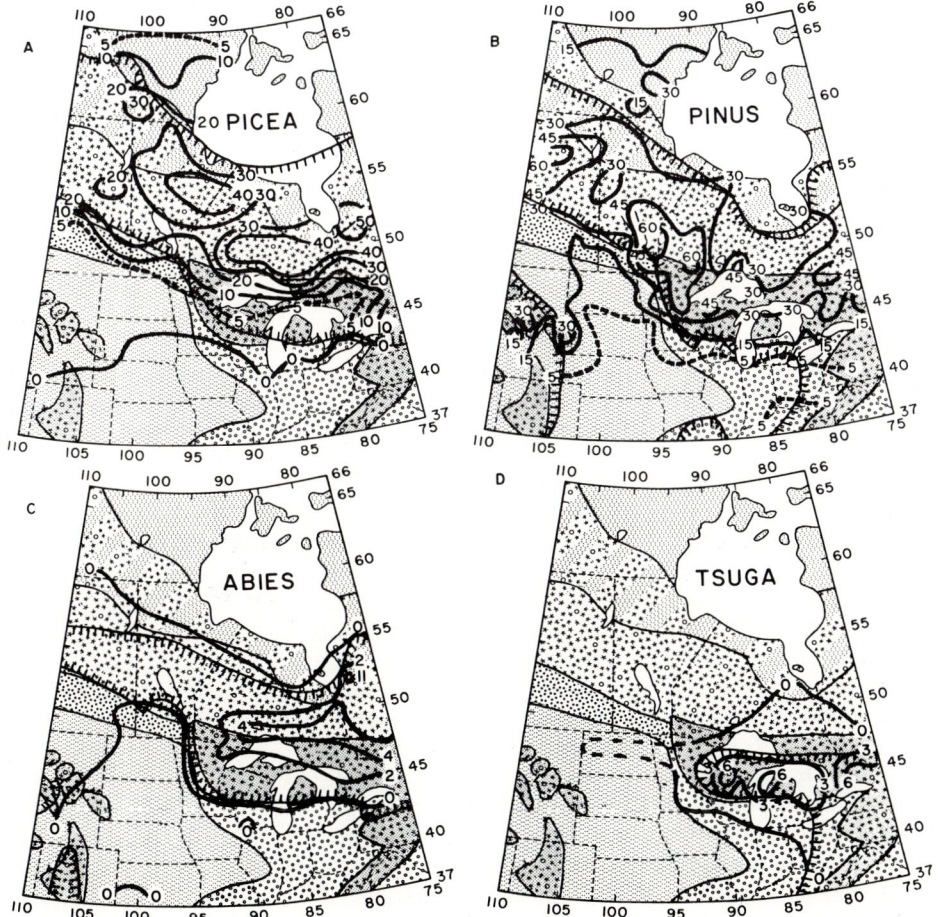

Figure 4. Isopollen maps for (A) spruce, (B) pine, (C) fir, and (D) hemlock pollen. Hachured lines and dashed contours as in Figure 3, except dashed contour in D that indicates uncertainty.

(Ritchie, 1974, written commun.), or may be an artifact of lower total pollen influx within the prairie than within the deciduous forest. Overall, the distribution of *Picea* pollen is closely related to the regional distribution of spruce trees.

6. *Pinus* (Fig. 4B). West of long 95°W, peaks of *Pinus* (>30%) lie within the boreal forest, but east of long 95°W, the high values occur primarily within the conifer-hardwood forest. High values also appear in the few samples within the Black Hills and within the mountain and basin complex. West of Lake Superior, the high values of *Pinus* pollen coincide with the medium to high population intensity for jack pine (see Fig. 4 in Halliday and Brown, 1943). East of Lake Superior, the high pollen values occur where low intensity values of jack pine appear. These high pollen values are related to abundant white and red pine, which were not mapped by Halliday and Brown (1943). Because distinctions between haploxylon and diploxylon *Pinus* types were made in only about one-third of the 606 modern samples, separate maps for these *Pinus* types were not constructed. *Pinus* values fall to 5% or less in samples from the deciduous forest and are small in prairie samples, many of which are distant from the mountains. *Pinus* is present, however, in all but one of the 606 samples and, despite its relatively high settling velocity (Gregory, 1973), is the most widely dispersed of all the pollen types.

7. *Abies* (Fig. 4C). *Abies* pollen occurs mainly in samples from the boreal forest and conifer-hardwood forest, with its highest values (>4%) confined to a narrow zone in the eastern part of the boreal forest, an area with a relatively high population intensity of fir trees (see Fig. 6 in Halliday and Brown, 1943). Unlike the other two vesiculate pollen types, *Picea* and *Pinus*, the zero contour lines for *Abies* do not extend far beyond the range of fir trees. The pollen values underrepresent the tree percentages by about a factor of five.

8. *Tsuga* (Fig. 4D). High values (>3%) of *Tsuga* are within the conifer-hardwood forest. Except for a few trace amounts, *Tsuga* pollen is not found north of lat 50°N or west of long 93°W. These limits coincide in general with the range of hemlock trees (Little, 1971).

9. *Fagus* (Fig. 5A). High percentages (>2%) of *Fagus* are restricted to stations south of lat 46°N and east of long 88°W, and no stations outside of the range of beech trees contain more than trace amounts. Stations in the eastern Upper Peninsula of Michigan and in eastern Wisconsin would probably extend the observations of pollen to the western limit of beech trees.

10. *Acer* (Fig. 5B). Like *Fagus* and *Tsuga*, high percentages (>2%) of *Acer* pollen are restricted to samples within southern Ontario, Michigan, and northern Wisconsin where sugar maple (*A. saccharum*) is abundant. The zero contour line for observations of *Acer* extends outside of this area but lies within the range of box elder (*A. negundo*).

11. *Ulmus* (Fig. 5C). High values (>3%) of *Ulmus* lie within the deciduous forest and extend into the southern part of the conifer-hardwood forest (Table 2). The extension of the 3% contour into northwestern Minnesota reflects the zone of deciduous forest between the prairie and the conifer-hardwood forest. Except for two sites with trace amounts, no *Ulmus* is found north of lat 55°N, and the zero contour line generally parallels the range limit of elm trees.

13. *Ostrya-Carpinus* (Fig. 5D). *Ostrya-Carpinus* pollen occurs almost exclusively at stations within the conifer-hardwood and deciduous forests, the regions where its mean values are largest (1% to 2%; see Table 2). Only trace amounts appear in samples outside these vegetational regions, and the zero contour line parallels the range limit for hornbeam trees. Within its zone of occurrence, however, little pattern is apparent in the location of the isopolls. On this broad scale of analysis, therefore, the presence of this pollen type rather than its relative abundance is

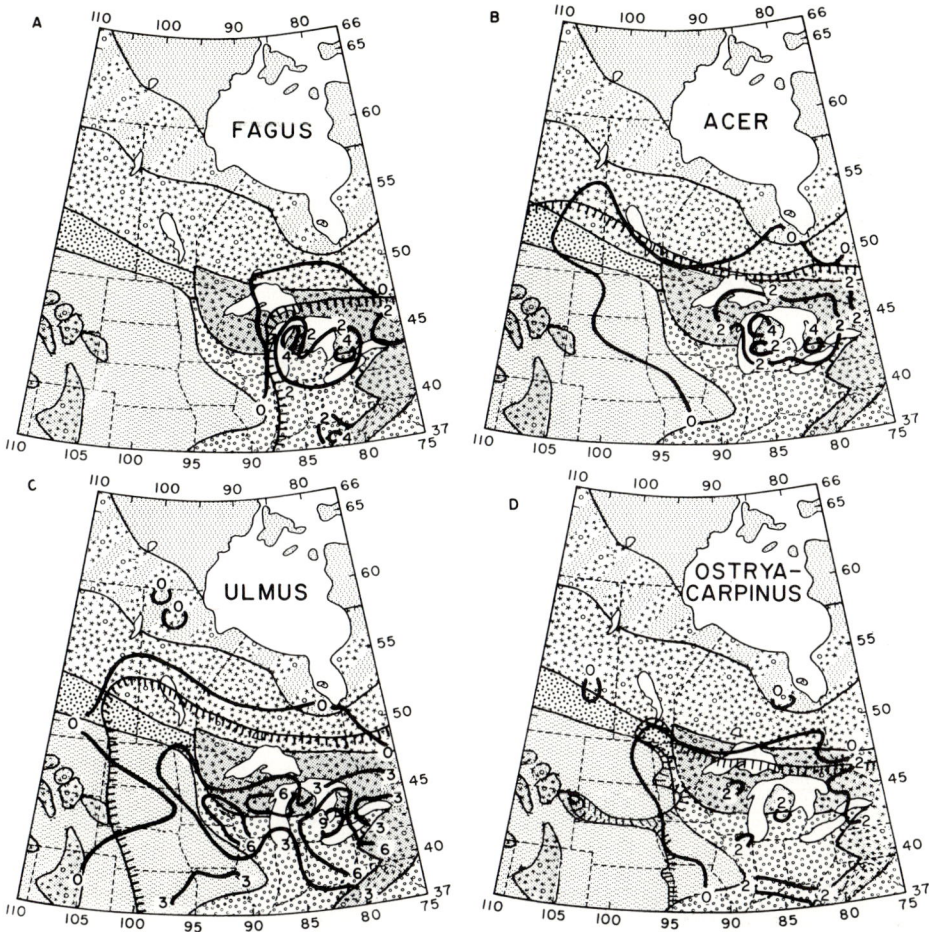

Figure 5. Isopollen maps for (A) beech, (B) maple, (C) elm, and (D) hornbeam pollen. Hachured lines as in Figure 3.

the main conveyor of vegetational information.

14. *Quercus* (Fig. 6A). The peak values (>4%) of *Quercus* lie within the deciduous forest, and the 20% contour line parallels the northern and western borders of this vegetational region. Like *Ulmus*, the pattern of the isopolls in Minnesota and southern Manitoba reflects the northwestward extension of the deciduous forest, although high values of *Quercus* pollen extend well into the prairie region. A greater density of samples in eastern North Dakota and western Minnesota would better define this pattern.

Quercus is the dominant pollen type of the deciduous forest, and here its average value is three times higher than in other vegetational regions (Table 2). Within the boreal forest, *Quercus* values are all less than 5%. *Quercus* is absent north of lat 55°N, and the 5% contour line parallels the range limit of oak trees and lies well within this limit except in southern Kansas.

15. *Carya* (Fig. 6B). Values of *Carya* of more than 1% are mostly confined to the deciduous forest. Although *Carya* is only a minor numerical constituent of the pollen assemblage associated with the deciduous forest, it is a reliable indicator of this forest region (Table 2). The 1% contour line parallels the range limit for

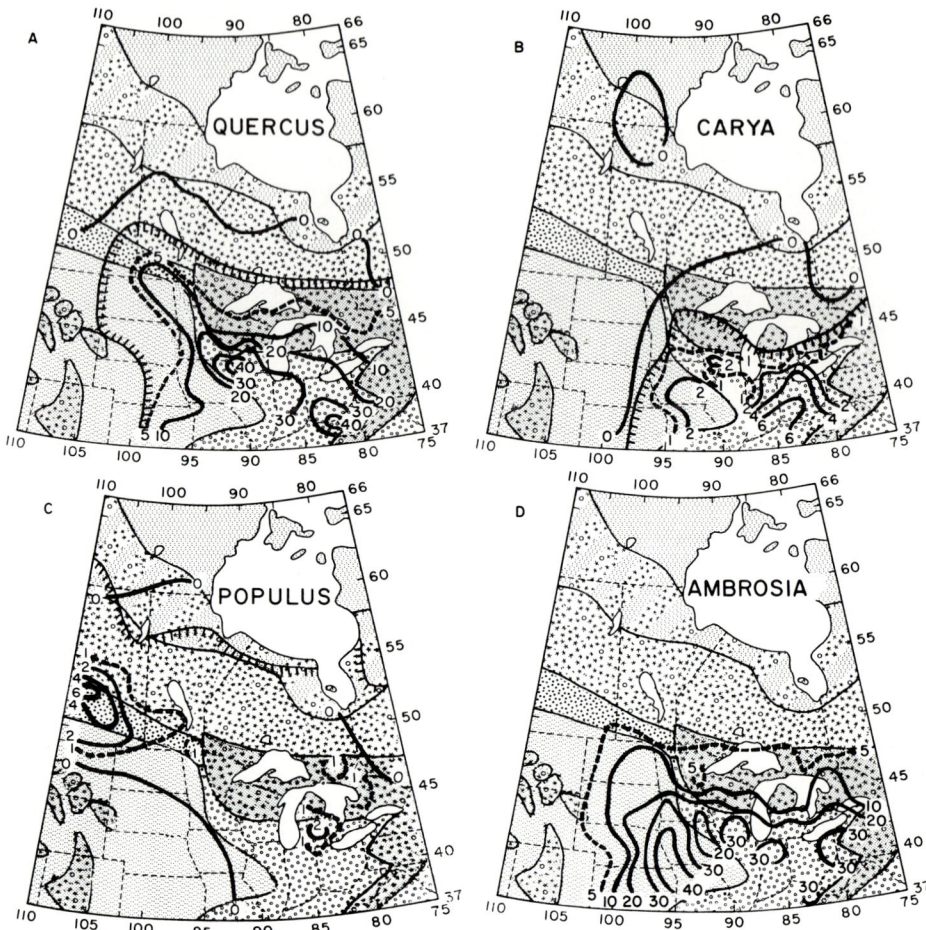

Figure 6. Isopollen maps for (A) oak, (B) hickory, (C) poplar, and (D) ragweed pollen. Hachured lines and dashed contours as in Figure 3.

hickory, and no pollen values occur beyond this boundary except as trace amounts; for example, the zero contour line at lat 60°N encircles sites of several samples in which single grains were counted.

16. *Populus* (Fig. 6C). Peak values (>6%) of *Populus* lie mainly in the area of aspen parkland in Saskatchewan and Manitoba, primarily because of exceptionally high percentages (27% and 23%) at two lake sites. Because most other sites even in this area contain no *Populus* pollen (see Mott, 1969), caution is required in using this map. Values greater than 1% also occur in samples in southern Ontario and lower Michigan. This pattern of pollen values coincides with areas of high population intensity of poplar trees (see Fig. 9 in Halliday and Brown, 1943). This coincidence implies that some vegetational interpretations can be based on the presence of *Populus*. No *Populus* appears north of lat 60°N, and the range of pollen lies generally within the range of poplar trees.

17. *Ambrosia* (Fig. 6D). High values (>20%) of *Ambrosia* pollen occur in the eastern part of the prairie and in the deciduous forest. The latter region has the highest average value (Table 2). The distribution reflects ragweed abundance that is the result of both climatic control and the continued disturbance of the landscape

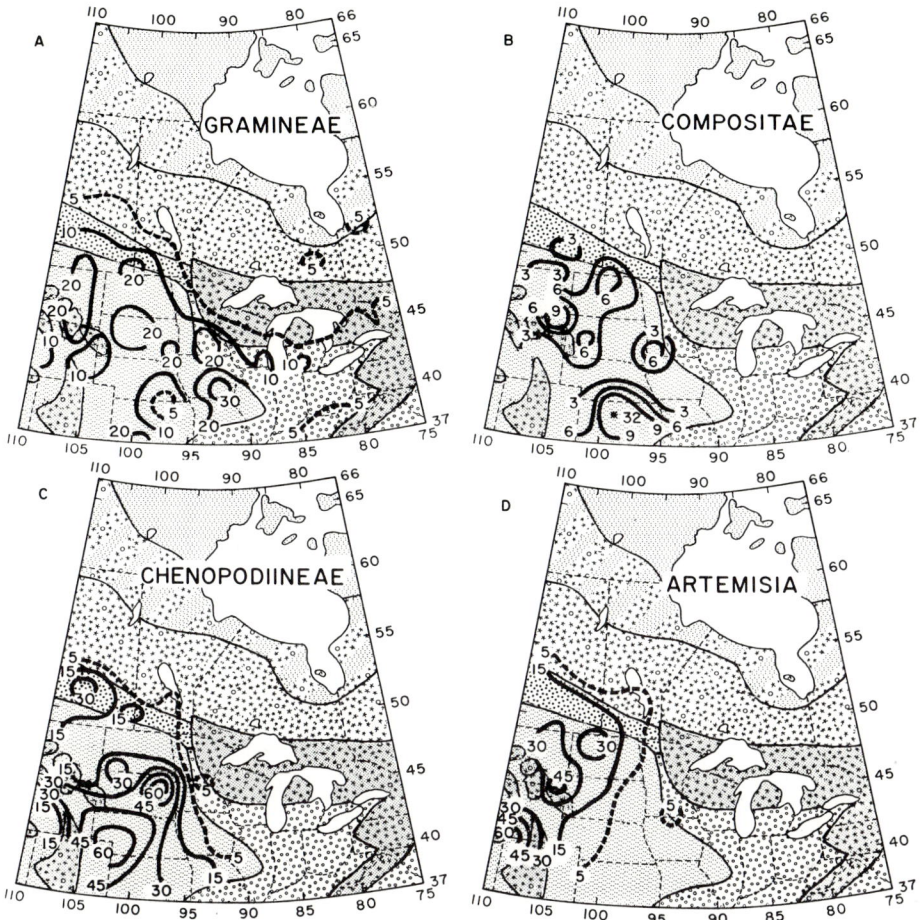

Figure 7. Isopollen maps for (A) grass, (B) composite, (C) pigweed, and (D) sage pollen. Dashed contours as in Figure 3.

by man. In presettlement time (before 1850), the distribution of *Ambrosia* pollen to the west and north, however, seems to have been controlled by climatic factors such as day length, temperature, and precipitation (Lindsay, 1953).

18. Gramineae (Fig. 7A). Values of Gramineae pollen of more than 10% are generally confined to the prairie, and the 10% contour line parallels the boundary between the prairie and deciduous forest. Within this area, however, the stations with values of more than 20% are scattered. Three stations with high values also appear at isolated locations within the deciduous forest. Therefore, although the values of Gramineae pollen have a definite regional distribution, the high pollen values may also reflect local conditions, such as the type of substrate sampled (for example, duff in the prairie), the presence of aquatic grasses (for example, wild rice; McAndrews, 1969), or agricultural weeds.

19. Compositae (Fig. 7B). Compositae pollen includes long-spined pollen of entomophilous species as distinct from short-spined anemophilous *Ambrosia* and *Artemisia*. Most of the high values (>6%) appear within the prairie. Because species of the composite family dominate the flora of the prairie (Curtis, 1959), these high values reflect this vegetational region. Although the high values show no generally consistent pattern, they occur west of the high values of *Ambrosia* pollen.

20. Chenopodiineae (Fig. 7C). Chenopodiineae includes pollen from the families Chenopodiaceae and Amaranthaceae. The modern distribution of this pollen type reflects the prairie region, and like Compositae, the high values (>15%) appear west of the high values of *Ambrosia*. To the north, the 5% contour line lies along the southern border of the boreal forest, and eastward it parallels the boundary between the prairie and deciduous forest. Where Chenopodiineae is present in forested regions, it—like *Ambrosia*—reflects cultivation (Webb, 1973a). Peak values (>60%) in the prairie are derived mostly from weedy species, but Chenopodiineae values are relatively high in presettlement prairie samples (Watts and Bright, 1968; Webb, 1973a). High Chenopodiineae values are therefore an indicator of prairie in Holocene deposits. Within the prairie, the isolated high values partly reflect local conditions about individual sites, for example, large numbers of Chenopodiineae plants around cattle-watering tanks (McAndrews and Wright, 1969).

21. *Artemisia* (Fig. 7D). *Artemisia* pollen dominates most western stations. Peak values are farther west than those of Chenopodiineae or Compositae, but like Chenopodiineae, the 5% contour line lies along the southern border of the boreal forest. Where this contour turns south, however, it is much farther west than that for Chenopodiineae pollen.

Pollen Types Not Mapped

Percentages of Ericaceae, *Myrica, Larix, Juniperus-Thuja, Corylus, Juglans, Platanus, Plantago, Rumex,* and *Fraxinus* were too low to give mappable patterns. Table 2 gives the major distributional information for these pollen types and shows that Ericaceae and *Myrica* pollen have their highest mean values in the tundra and forest-tundra; that *Larix* is largely confined to samples in the forest-tundra, boreal forest, and conifer-hardwood forest; that *Juniperus-Thuja* occurs mainly in the conifer-hardwood forest and deciduous forest; that *Tilia, Juglans, Platanus, Plantago,* and *Rumex* occur mainly in samples from the deciduous forest; and that *Corylus* and *Fraxinus* occur in both of these regions plus the aspen parkland. The presence of these pollen types in a pollen count, therefore, conveys only general information about the vegetation.

The statistics for miscellaneous, unknown, and unidentifiable grains (Table 2) show that the highest mean values for these types occur in the conifer-hardwood forest, deciduous forest, and the prairie. In part, this result reflects agricultural disturbance in these regions, which has increased the diversity of pollen types by addition of cultigens and weeds.

Summary and Trend-Surface Analysis

Figures 8 and 9 show maps of contours derived from trend-surface analysis of the dominant pollen types. Peak values of pollen types in the various vegetational regions are as follows: Cyperaceae in the tundra; *Betula* in the tundra and the northern conifer-hardwood forest; *Alnus* in the forest-tundra; *Picea* in the forest-tundra and boreal forest; *Pinus* in the western boreal forest and the conifer-hardwood forest; *Tsuga* in the eastern conifer-hardwood forest; *Fagus* and *Acer* in the eastern conifer-hardwood forest and eastern deciduous forest; *Quercus, Ulmus,* and *Carya* in the deciduous forest; *Ambrosia* in the eastern prairie and deciduous forest; and Gramineae, Compositae, Chenopodiineae, and *Artemisia* in the prairie. Within the prairie, Gramineae is highest in the east and *Artemisia* is highest in the west.

The multiple correlation coefficient (R), given on the maps in Figures 8 and

9 for each of the above pollen types, indicates how well the trend surface matches the actual distribution of the pollen type. This coefficient is greater than 0.50 for each map shown. In general, taxa with higher percentages have higher correlation values. Among the herbaceous pollen types with high percentages, low correlation values occur in Gramineae and Chenopodiineae, taxa that contain ecologically diverse species ranging from prairie or tundra dominants to agricultural weeds and halophytes depending on regional or local habitat conditions. Among the arboreal pollen types with high percentages that were mapped, *Betula* and *Pinus* have low correlation values that are perhaps related to local pollen abundance within often pure stands. On the other hand, for the size of their percentage values, *Carya* and *Ulmus* have exceptionally high correlation values; this occurrence indicates even pollen dispersal.

As discussed in the section on methods, the degree of the polynomial equation to use depends on the application. A map that gives the pollen percentages at each station but is not contoured shows the measured values but forces the reader to sort out the geographic distribution of the pollen type. Interpolating between samples with contour lines shows the geographic distribution but incorporates several possible scales of variation from local to regional, and the local differences between values at certain sites may obscure the regional distribution. The contour lines from the low-degree trend-surface equation smooth over these local differences and emphasize the regional patterns but less exactly represent the measured data. We attempt to present the distribution of each pollen type and to highlight its regional distribution by presenting both interpolated and trend-surface contours. The trend-surface contours show an excellent fit between the pollen data and the vegetational pattern on the scale of Figure 2.

The next section presents another summary of the data—principal components analysis—that illustrates the patterns of covarying pollen types in the data. It also illustrates the prominence of the vegetational patterns in determining the patterns on the pollen maps.

Principal Components Analysis

Principal components analysis (PCA) with varimax rotation was performed on the matrix of correlations between 21 pollen types. The pollen types used include the 20 pollen types that have average values (over all samples) greater than 0.5% (Table 2) and *Carya* that has an average value of 0.2%.

The first four components, accounting for 47% of the total variance (Table 3), indicate the presence of five groups among the pollen types. This value of "explained variance" is quite typical for PCA based on a correlation matrix of pollen percentage data (Cole, 1969; Webb, 1974a; Birks and others, 1975).

The next three components account for an additional 16% of the variance (Table 3) and are described briefly. No description is given for the remaining 14 components, whose spatial distributions contain no interpretable information. The total variance accounted for by the first seven components is therefore 63%.

The first principal component, which accounts for 23% of the total variance, indicates the strong positive covariance among *Quercus, Ambrosia, Carya, Ulmus, Fraxinus,* and *Ostrya-Carpinus* and their negative covariance with *Pinus* and *Picea* (Table 3). High positive values on Figure 10A appear in the deciduous forest, and values less than −1 appear in the mountain and basin complex. The first principal component reflects the grouping of tree genera in the deciduous forest and the lack of pine and spruce trees in this forest region. This component also indicates the high relative abundance of *Ambrosia* pollen in the area of the deciduous

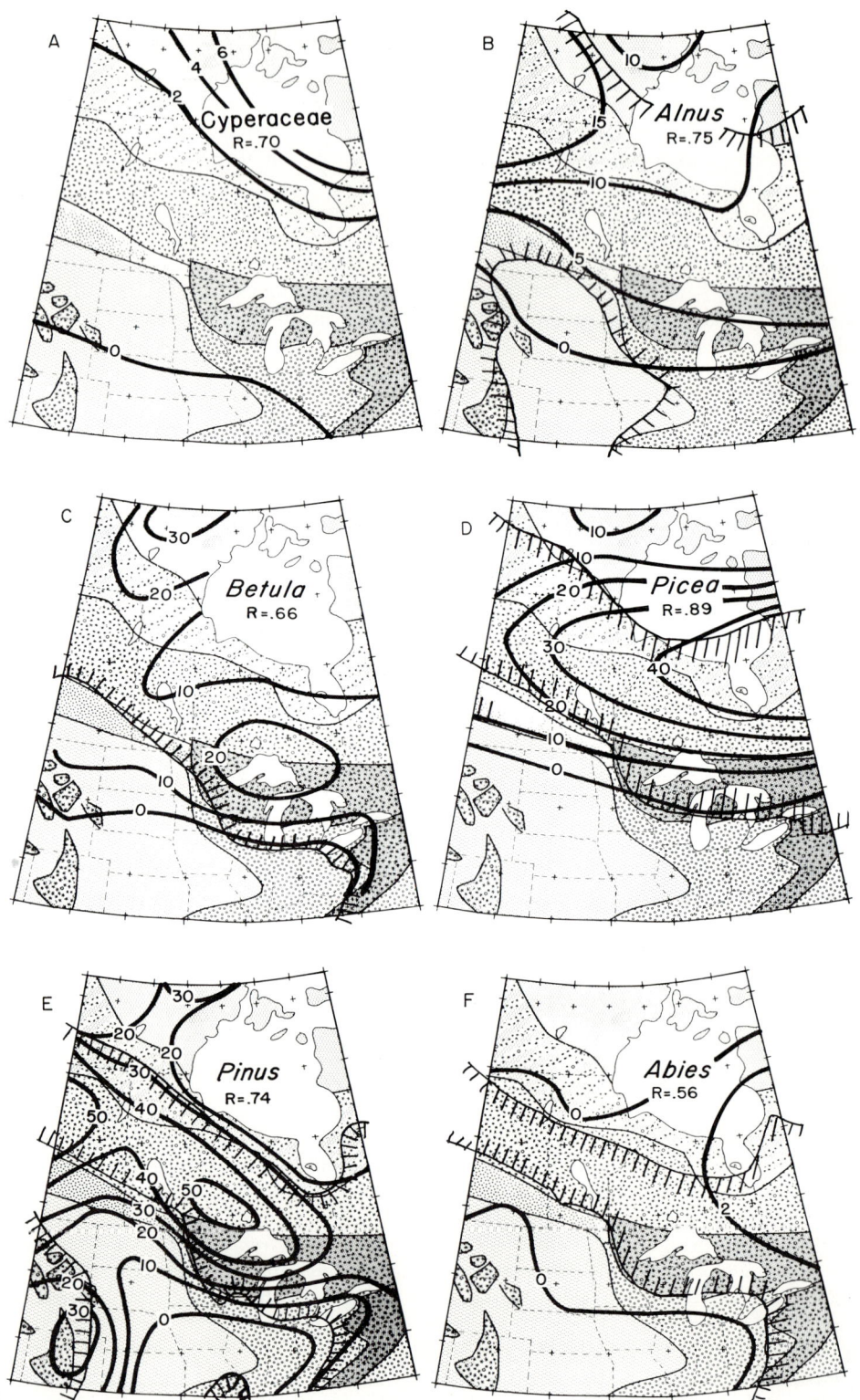

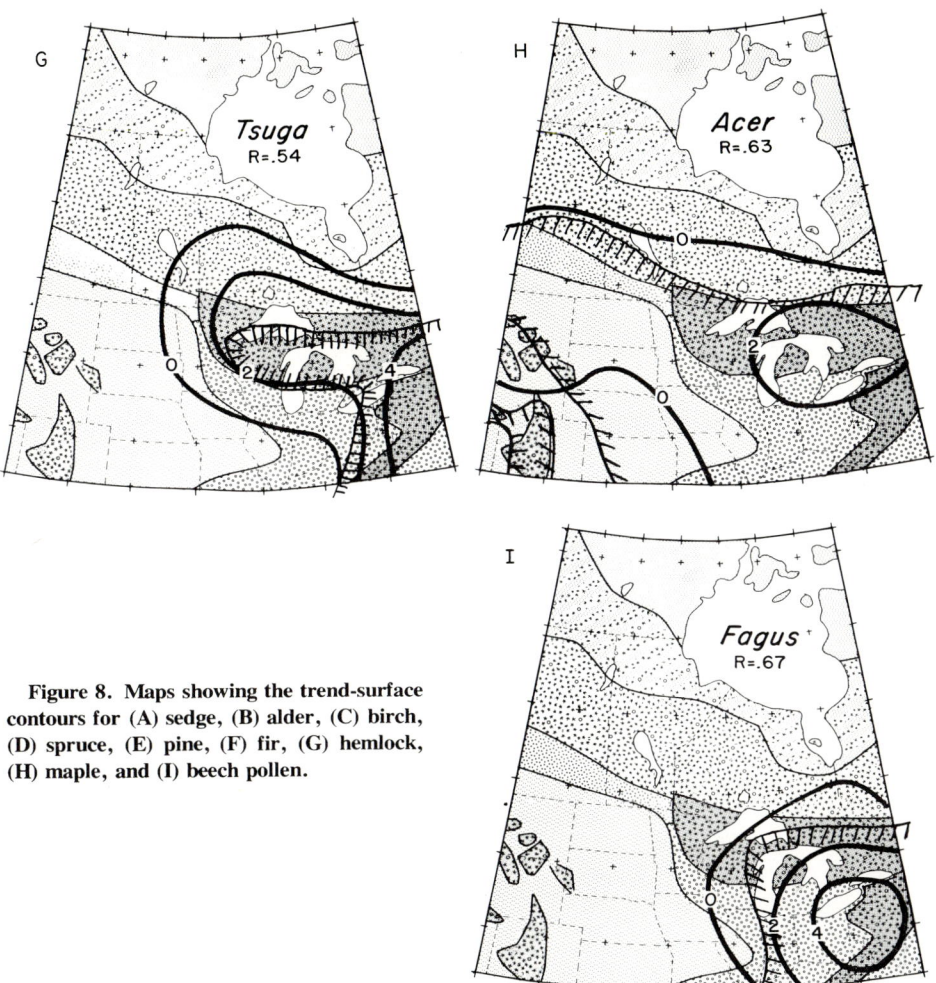

Figure 8. Maps showing the trend-surface contours for (A) sedge, (B) alder, (C) birch, (D) spruce, (E) pine, (F) fir, (G) hemlock, (H) maple, and (I) beech pollen.

forest. As discussed before, this feature reflects human disturbance by land clearance and cultivation that foster ragweed growth on the disturbed soils.

The second principal component, which accounts for 9.5% of the variance, shows a major gradient in the data that lies between the prairie and the regions of forest-tundra and boreal forest. This eigenvector groups together Chenopodiineae, *Artemisia*, Gramineae, and Compositae (all elements of the prairie) in opposition to *Picea*, *Alnus*, *Pinus*, and *Betula* (elements of the northern forests; see Table 3). The map of the scores for this component (Fig. 10B) shows high positive values in the prairie and negative values in the forest-tundra and northern boreal forest. A steep gradient in these values parallels the boundary between the prairie and forest.

The third principal component accounts for 7.7% of the variance and mainly reveals the positive covariance of *Tsuga*, *Acer*, *Fagus*, and *Ulmus* (Table 3), which are all elements of the conifer-hardwood forest. *Picea* and *Alnus* covary negatively with this group of pollen types. The positive component scores in Figure 10C lie primarily within the region of the conifer-hardwood forest, and values less than −1 appear in the northern boreal forest and forest-tundra.

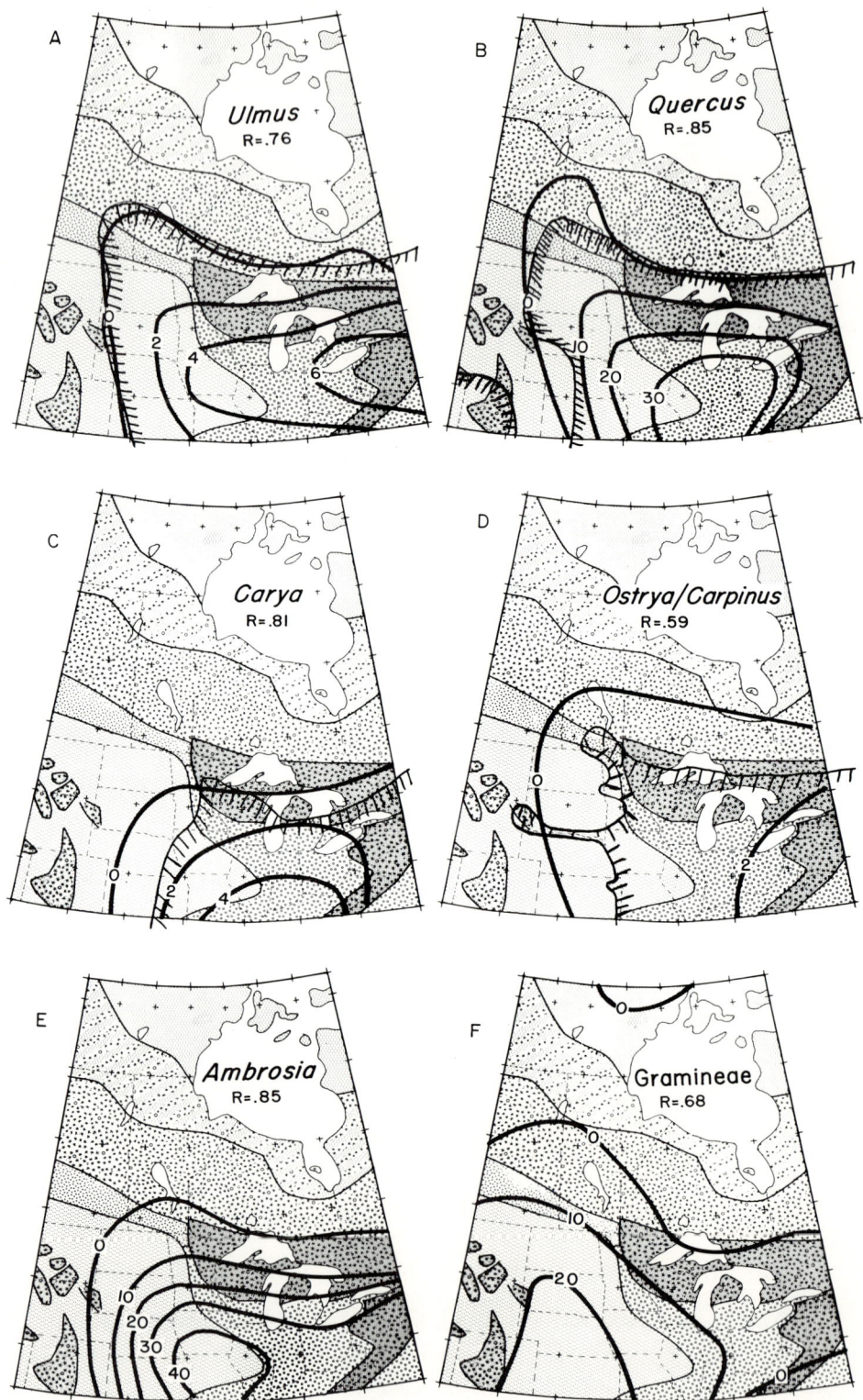

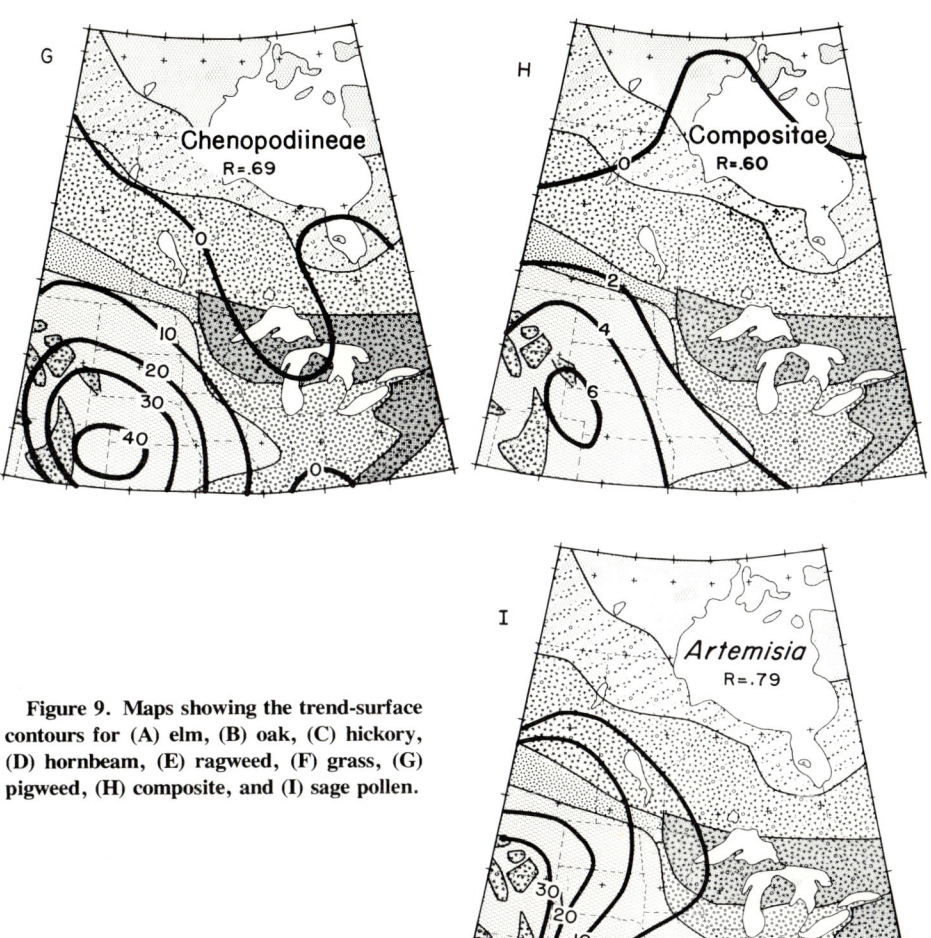

Figure 9. Maps showing the trend-surface contours for (A) elm, (B) oak, (C) hickory, (D) hornbeam, (E) ragweed, (F) grass, (G) pigweed, (H) composite, and (I) sage pollen.

The high positive values for the fourth principal component lie mainly within the tundra (Fig. 10D). This component accounts for 7.2% of the variance and shows the positive correlation between Cyperaceae and *Salix* (Table 3). This component reflects the high values of these two pollen types in the tundra and is especially influenced by the few high values of *Salix* in peat samples in this region. Some large, positive component scores also appear in North Dakota where aquatic sedge-family plants (for example, *Scirpus*) are abundant (Fig. 3A).

Although the next three components account for only 16% of the total variance (Table 3), some interpretation is possible. All three components are dominated by single pollen types, with the coefficients greater than 0.9 for *Populus* and *Abies* on the fifth and seventh components, respectively. The separation of *Populus* and *Abies* into separate components indicates that these pollen types are distributed with spatial patterns unlike any of the other pollen types in the data set. Their highest correlation with any other pollen type is less than 0.2, whereas all other pollen types show a correlation of at least 0.4 or better with some other pollen type. (*Quercus* and *Ambrosia* possess the highest correlation of 0.67.) Presence

TABLE 3. ROTATED PRINCIPAL COMPONENTS

	1	2	3	4	5	6	7
Cyperaceae	−0.10	−0.02	−0.07	0.78	−0.20	0.08	−0.15
Salix	−0.00	−0.04	−0.04	0.76	0.22	−0.09	0.10
Betula	−0.27	−0.38	0.24	−0.00	0.04	0.71	0.04
Alnus	−0.30	−0.46	−0.34	0.02	0.13	0.30	−0.33
Picea	−0.44	−0.49	−0.43	−0.03	−0.01	−0.13	0.14
Pinus	−0.60	−0.42	0.02	−0.24	−0.13	−0.28	0.03
Abies	−0.12	−0.14	−0.02	−0.04	0.01	0.06	0.92
Tsuga	−0.10	−0.10	0.81	−0.11	−0.09	0.15	−0.02
Acer	0.23	−0.05	0.76	−0.01	0.06	0.12	0.07
Fagus	0.23	−0.14	0.66	0.01	0.15	−0.41	−0.07
Ulmus	0.65	−0.05	0.49	−0.03	0.04	−0.02	−0.06
Fraxinus	0.64	−0.07	0.29	0.03	0.24	−0.27	−0.10
Ostrya	0.39	−0.04	0.36	−0.11	0.08	0.32	0.09
Carya	0.71	−0.11	0.01	−0.06	0.02	−0.24	−0.04
Quercus	0.87	−0.03	0.09	−0.07	−0.05	−0.09	−0.05
Ambrosia	0.83	0.18	0.04	−0.03	−0.14	0.07	0.04
Gramineae	0.22	0.70	−0.07	0.17	−0.03	−0.08	−0.04
Chenopodiineae	−0.06	0.71	−0.10	−0.07	−0.07	−0.03	−0.03
Artemisia	−0.27	0.66	−0.08	−0.08	0.20	−0.14	−0.13
Compositae	−0.01	0.61	−0.04	−0.06	0.02	−0.00	0.01
Populus	0.01	0.04	0.04	0.03	0.90	0.03	0.00
Variance (%)	23.0	9.5	7.7	7.2	6.3	5.5	4.2

of *Populus* in a separate component may result from the differential preservation of the pollen at different sites (Webb, 1974a). Because this factor of differential preservation operates in a unique fashion on *Populus*, its spatial distribution is distinct and independent. *Betula* dominates the sixth component, whose distribution reflects the bimodal character of this pollen type. No interpretation seems evident for the other 14 components.

Principal components analysis thus summarizes the data by representing in four maps (Fig. 10) most of the information shown in 20 maps (Figs. 3 through 7) and by indicating the independent nature of the distribution of two of the pollen types and the bimodal distribution of *Betula* pollen. Figure 11 shows the separation of pollen types into four groups (A, B, C, and D) corresponding to groups of taxa in the various vegetational regions. Figure 11 also illustrates the values of the 21 pollen types in Table 3 within a coordinate system with the first two principal components as axes. Although Cyperaceae, *Salix*, *Populus*, and *Abies* are all located within group D, they can be considered separately from this group because their real location among the principal components is along the axes of the fourth, fifth, and seventh principal components. The four groups of pollen types now each contain genera from similar vegetational regions. Group A contains the pollen types that dominate in the samples from the prairie and aspen parkland, group B contains the pollen types from the deciduous forest, group C contains the pollen types from the boreal forest and forest-tundra, and group D—which lies appropriately between groups A and C—contains the pollen types of the conifer hardwood forest. Group D can be further separated from groups A to C because it is located along the third principal component.

Table 3 and Figures 10 and 11 thus show that the first four principal components place the pollen types into five groups, each characteristic of a major vegetational region. The tundra is represented by Cyperaceae and *Salix*; the forest-tundra and boreal forest by *Picea*, *Pinus*, *Betula* and *Alnus*; the conifer-hardwood forest by *Tsuga*, *Fagus*, and *Acer*; the deciduous forest by *Quercus*, *Carya*, *Ulmus*, *Ostrya-Carpinus*, and *Ambrosia*; and the prairie by Chenopodiineae, Compositae, Gramineae, and *Artemisia*. This numerical analysis of the pollen data, therefore, illustrates in a concise form an internal structure of these data that reflects the modern vegetation of central North America (Fig. 2).

DISCUSSION AND CONCLUSIONS

The results of our study summarize the broad-scale features of modern pollen data of central North America and illustrate the main patterns with maps. These maps suggest three main topics for discussion. First, the maps show that, although the pollen data are subject to many sources of variation, the variation in the vegetation dominates over other factors in determining the distribution of most pollen types over the broad scale of our study. Second, the maps show what similarities and differences exist between the pollen data in central North America and the data in other regions, such as eastern North America and the Soviet Union. Finally, the maps can be used both to aid interpretation of fossil data and to identify regions requiring additional sampling.

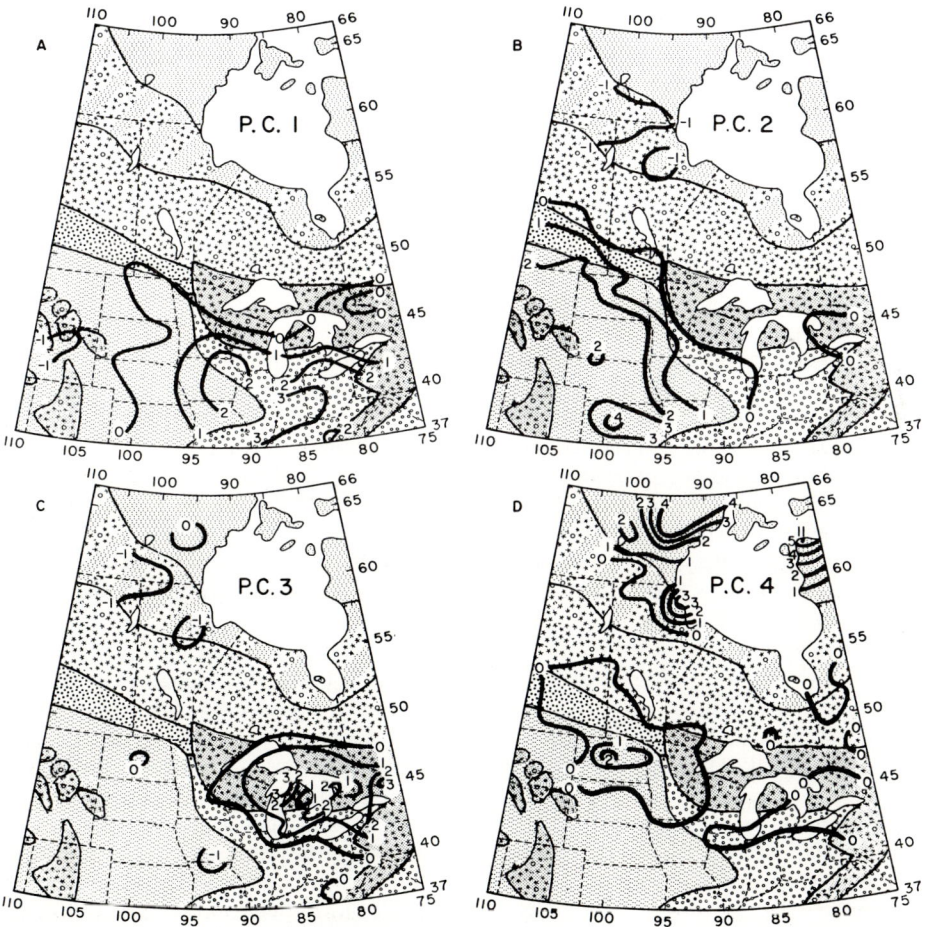

Figure 10. Maps of the component scores for (A) the first, (B) the second, (C) the third, and (D) the fourth principal components.

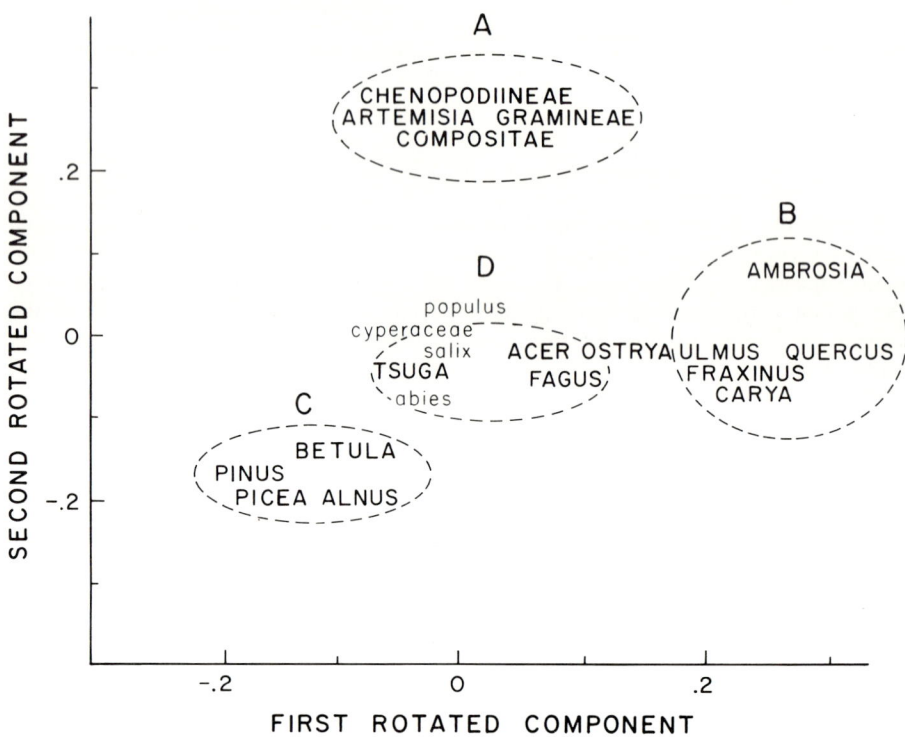

Figure 11. Plot of the pollen types along the first two rotated principal components.

Sources of Variation: An Evaluation

The correspondence between modern pollen data and the contemporary regional vegetation has long been recognized in most palynological work (see von Post, 1967; Aario, 1940; Lichti-Federovich and Ritchie, 1968; Davis, 1967). Few satisfactory illustrations of the relationships between pollen and vegetation, however, appear in textbook chapters or review articles introducing pollen analysis (Cain, 1944; Faegri and Iversen, 1964; Butzer, 1971; Colinvaux, 1973). Because these general introductions to pollen analysis often emphasize the many sources of variation in pollen data, this lack of good illustrations of the pollen-vegetation relationships can leave the impression that pollen analysis is a complex subject based on a shaky foundation.

The maps presented in our study oppose this impression and make the pollen-vegetation relationships clearly visible. We can draw this conclusion despite recognizing that many sources of variation do exist in our data. Several of these sources exist because our data come from a large geographic area and were produced by several palynologists. The sources of variation include differences in the pollen sum among samples, inconsistent identification of pollen grains by various analysts, differences in sampling technique, differences in sediment types, differences in basin size among lakes, and errors in assembling the data set. What we find, however, is that the variation in the pollen data attributable to these sources is often small compared to the variation attributable to the vegetation. The following discussion introduces a format for illustrating this point and expands upon West's (1973) description of his Table 1.

The geographic variation of the ith pollen type (P_i, measured in pollen grains of type i that are analyzed) reflects the variations in the following factors: the geographic distribution of the plants producing the pollen (V_i, measured in number of plants of type i), the rate of production of pollen by each plant (R_i, measured in grains produced per plant), the aerobiological properties of the pollen and the physical factors operating to disperse the pollen (T_i, measured in grains transported to site per grains produced), the deposition of the pollen (D_i, measured in grains deposited at site per grains transported to site), the preservation of the pollen (Q_i, measured in grains preserved per grains deposited at site), and the manipulation of the pollen by palynologists during sampling (M_i, measured in grains sampled per grains preserved) and during preparation, identification, and counting (A_i, measured in grains analyzed per grains sampled). Equation 2 (modified from Fagerlind, 1952) provides a simple way of accounting for the effect of each of these factors.

$$P_i = V_i \times R_i \times T_i \times D_i \times Q_i \times M_i \times A_i, \tag{2}$$

where each term varies with latitude (x) and longitude (y); that is, $P_i = P_i(x,y)$, $V_i = V_i(x,y)$, and so forth. Equation 2 shows that the value of a given pollen type results from its value in the vegetation, which is modified by a relative weighting for pollen production, transport, deposition, preservation, sampling techniques, and analysis. Equation 2, which is basic to the approach of finding r-values (Davis, 1963) for the pollen types, has a major fault because it does not account for the numeric interactions among pollen types due to the use of percentage data. Despite this deficiency, the form of equation 2 is useful for examining the geographic variation of each pollen type in relation to the vegetation. We can note here that if each of the variables R_i, T_i, ..., A_i was randomly distributed about some constant value, then for any area with a dense array of sites, we can average out or smooth over the random variation and find the constant (or r-value) that links P_i and V_i (see Andersen, 1970).

Because of the relative weighting for the various factors at an individual point, the pollen value seldom equals the vegetation value, an observation long recognized by palynologists (von Post, 1967). The information in pollen data, however, lies not so much in their values at individual points but in their variation in time and space; that is, concern with $P_i \neq V_i$ is much less important than concern with $\Delta P_i \neq \Delta V_i$, where ΔP_i is the change in P_i in space (x,y). If we then look at equation 2 in terms of the "variation" of $P_i (\Delta P_i)$ we have

$$\Delta P_i = P_i(\Delta V_i/V_i + \Delta R_i/R_i + ... + \Delta A_i/A_i). \tag{3}$$

Within the context of equation 3, our results show that $\Delta P_i \cong P_i (\Delta V_i/V_i)$ and imply that $\Delta V_i/V_i >> \Delta R_i/R_i + ... + \Delta A_i/A_i$. For a few pollen types, such as *Populus* and *Juniperus-Thuja*, however, $\Delta P_i \neq P_i(\Delta V_i/V_i)$. For these pollen types, the primary sources of variation are probably differential preservation in different sediment types (ΔD_i and ΔQ_i), inconsistent identification (ΔA_i), and high counting errors (ΔA_i). These factors combine to swamp the variation in V_i and result in little resemblance between the mapped distribution of the pollen type and the distribution of the plants producing it.

The results of our study show in general that the variation in V_i is much greater than the variation in most other factors. The scale of our study area and the density of samples within it contribute directly to this result. Firstly, the variation in V_i is maximized in our data set because the samples extend from regions where taxa are absent to regions where they may be dominant. Even for species with fairly low abundances in the pollen record (for example, *Acer*), this vegetational contrast is great enough to outweigh noise due to counting errors. Secondly, the large number of samples allows contouring to smooth the data and average out the local features and thus emphasizes the broad-scale patterns.

Generally, as the size of the region sampled is decreased, the magnitude of the vegetational variation will decrease and the random variation of other factors becomes relatively more important in determining the values of the pollen types. Studies by McAndrews (1966), Lichti-Federovich and Ritchie (1968), and Webb (1974a, 1974b) show that vegetational variation over areas one-fourth to one-hundredth the size of our study area is great enough for pollen samples to record vegetational change when the data are dense enough; but, within each of these regions, anomalies appear when data are introduced that were collected in different sediment types or that were counted by different analysts. For instance, Cain (1944) reported values of 70% oak from Soden Lake in Michigan where Davis and others (1971) and Webb (1974a) found only 50% oak, and Cole (1969) sampled *Sphagnum* near Churchill, Manitoba, and found higher values of Cyperaceae than Lichti-Federovich and Ritchie (1968) extracted from their samples 23 and 24 of lake sediments. Controlling ΔA, ΔQ, and ΔD is therefore more important for finding a good correspondence in the variation of the pollen with the variation in the vegetation for studies on these smaller scales than for our current study of our large data set.

If one assumes that only vegetational variations follow broad-scale patterns and that all other factors in equation 2 do not, then trend-surface analysis can provide the necessary broad-scale filtering to remove the variations due to these local-scale factors. Our results (Figs. 8 and 9) show that this assumption is reasonable. Principal components analysis also acts as a filter on the pollen data by concentrating the broad-scale patterns into the first few components. These components account for most of the variance in the data. The local differences among samples that result from such factors as substrate differences (ΔD and ΔQ) are found among the higher numbered components that individually account for little of the total variance. A major advantage of both of these techniques is that they present subsets of the original data in which the variation in the vegetation (ΔV_i) dominates the other factors even more than in the raw data.

One aspect of the close correspondence between the variation in the pollen and the variation in the vegetation is that, for most pollen types, a positive change in pollen numbers ($\Delta P_i > 0$) from one sample to another implies a positive change in vegetational numbers ($\Delta V_i > 0$). For this result ($\Delta P_i > 0$ implies $\Delta V_i > 0$) to hold through time at a given station, Mosimann and Greenstreet (1971) developed a model that requires highly restrictive conditions for the representation coefficients (a_i), which are the ratio of pollen influx values (x_i) of type i to the number of plants (t_i) on the landscape (see p. 26 in Mosimann and Greenstreet, 1971). Mosimann and Greenstreet's model requires that the representation coefficients be equal for all pollen types included in the pollen sum for a given sample. For different samples in space, however, the empirical evidence from studies of modern pollen—including our study, Andersen (1970), and Webb (1974a)—indicates that similar changes in the pollen and vegetation are the rule with few exceptions, although the data in Davis and others (1973) show that the representation coefficients are not all equal within a given sample. Percentage data, therefore, appear better behaved empirically than Mosimann and Greenstreet (1971) reasoned mathematically. Further study is needed to show how robust percentage data are against different values of the representation coefficients.

Mosimann and Greenstreet (1971) also showed that the conditions on the representation coefficients are less restrictive when influx data are used, but do not mention that changes in influx values can result from factors other than changes in the population of a taxon (see Pennington, 1973). Influx values nevertheless can answer certain ambiguities arising from studies of percentage data (Davis, 1969), and they can even suggest radically different interpretations of the data (Davis, 1963; Ritchie, 1972; Mott, 1975a; Davis and others, 1975). The compilation of a set of modern pollen samples in influx terms comparable to the available set of modern samples in percentage terms is an important future task (Davis and Webb, 1975), but because of high levels of within-lake and between-lake variance (Davis and others, 1973; Pennington, 1973), much work will be required.

Comparison with Other Regions

The broad-scale vegetational patterns are evident from studies of pollen samples from other subcontinental areas (Davis, 1967; Davis and Webb, 1975; Neustadt, 1957). Within North America, the latitudinal changes in pollen from tundra to deciduous forest are basically similar whether the transect is located east or west of Hudson Bay. Comparison of our maps with those of Davis and Webb (1975), however, shows certain differences in the pollen rain between eastern and central North America. This point is also discussed by Davis (1967). For instance, in eastern North America, *Alnus* has higher values in the tundra than it has in central North America, *Abies* percentages are much higher in the boreal forest, and *Nyssa*, Aquifoliaceae, and *Castanea* appear in a number of samples within the deciduous forest.

Modern data from the Soviet Union also show broad-scale changes that reflect the vegetational changes from tundra to semidesert. To compare the basic broad-scale patterns in the pollen data from Eurasia and North America, we plotted the average pollen spectra from comparable vegetational regions in each continent (Fig. 12). Data from V. P. Grichuk's (1941) study of the modern pollen distribution were used to illustrate the regional trends in the Soviet Union. Subsequent studies of modern pollen from this area now exist (Grichuk, 1967, 1970; Nikolaeva-Prokhorova

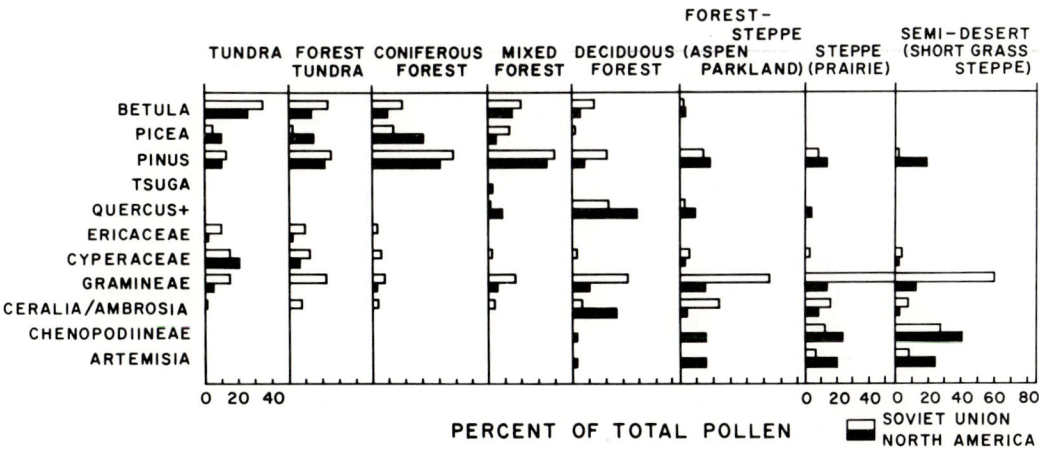

Figure 12. Average pollen spectra from the major vegetational regions of central North America and the European sector of the Soviet Union (from Grichuk, 1941). Pollen percentages calculated from a sum of total pollen excluding aquatic types.

and Shalandina, 1973), but Grichuk's (1941) data were easy to summarize and are quite similar in geographic range to our data.

Although differences exist in the exact composition of the vegetation and in the substrates sampled for pollen (Grichuk [1941] used soil samples), there is a good correspondence between the pollen changes in these two data sets (Lichti-Federovich and Ritchie, 1968). Pollen from woody plants dominates from the forest-tundra into the deciduous forest in both areas. Conifer pollen dominates from the forest-tundra into the mixed forest, and pollen from deciduous trees dominates in the deciduous forest. Herb pollen, including Gramineae, Chenopodiineae, and *Artemisia*, dominates in the prairie and semisteppe regions, and a mixture of herb (Cyperaceae, Ericaceae, and Gramineae) and arboreal (*Betula* and *Salix*) pollen characterizes the tundra region.

Certain differences are also apparent between the continents. In comparison with North America, *Pinus* pollen has higher percentages in the forest-tundra and boreal forest of Eurasia, and the values of *Picea* are lower. Gramineae dominates almost alone in the grasslands of Eurasia, whereas in North America, it shares dominance with Chenopodiineae and *Artemisia*. The highest values of cereal pollen, an indicator of cultivated land, occurs in the forest-steppe of the Soviet Union, whereas *Ambrosia*, the main indicator of cultivation in North America, has its highest values in the deciduous forest. These differences reflect differences in vegetation and in land use between the continents. (For further discussion of the difference in the pollen rain in arctic regions of North America and Eurasia, see Birks, 1973.)

Uses for the Maps

Besides illustrating the correspondence between modern vegetation and modern pollen within central North America, the maps facilitate interpretation of stratigraphic changes in fossil pollen. The maps aid researchers in locating regions in which pollen spectra exist that are similar to fossil spectra. Tables or diagrams of the spectra from these areas can then be consulted for detailed comparisons.

The maps also present the pollen data in a form easily comparable to mapped edaphic and climatic data, as well as to vegetational patterns. These comparisons aid in assessing the relationships among the various sets of data and in selecting the variables to use in calibration functions (see discussion below). In studies of other microfossils, mapping of foraminifera from ocean core-tops has facilitated interpretation of past changes in these organisms during Pleistocene time (Kipp, this volume).

The maps themselves act as direct analogues for maps showing the past distribution of certain pollen types (J. C. Bernabo and T. Webb III, in prep.). These latter maps also present isopolls on a broad scale. Comparison of these previous pollen distributions with the modern maps shows how these earlier distributions differed from those of today. Difference maps can be constructed to show the magnitude and pattern of the differences (compare with Webb, 1973b) and to indicate the scale and probable type of environmental changes that have occurred.

The modern maps, however, are not ideal for these comparisons because they contain the effects of recent human disturbance (see discussion of *Ambrosia* pollen). Especially for agricultural regions, maps showing the distribution of the pollen types in presettlement times will provide a better baseline for evaluating environmental changes during the Holocene Epoch. Such a set of maps is in preparation; they will add to the local and regional studies of Davis and others (1971), McAndrews (1966, 1968), and Webb (1973a, 1973b) in which modern and presettlement pollen distributions are compared.

The maps further provide a guide to future research by showing unsampled or insufficiently sampled areas. For the current data set, the sampling of the broad-scale geographic variation of the pollen data will be complete when more samples exist from the Great Plains, the James Bay lowlands, the tundra north of lat 60°N, and Illinois and the surrounding states. Additional sampling on a finer scale than generally exists for our study is required on several maps in order to position the isopolls more precisely, for example, in central and western Minnesota. This scale of sampling is especially appropriate in the vicinity of ecotones along which steep gradients appear in the isopolls.

Where quantitative vegetational data now exist and are mapped (Halliday and Brown, 1943; Webb, 1974a), a good correspondence exists between the maps of the pollen data and the maps of the vegetational data (see the description of *Picea*, *Abies*, *Pinus*, *Populus*, and *Betula* pollen in this paper and in Webb, 1974a). Canonical correlation analysis, principal components analysis, and multiple regression analysis provide methods for expressing this correspondence as mathematical equations that can estimate vegetational values from linear combinations of two or more pollen types. Recent work with these techniques illustrates their ability to transform microfossil data into quantitative estimates of past climates (Imbrie and Kipp, 1971; Webb and Bryson, 1972; Bryson and Kutzbach, 1974).

Both the even sampling of the modern pollen and the presence of a quantitative measure of the vegetation at each modern sample are required for these quantitative methods to work. The gathering of necessary additional pollen samples and of the quantitative vegetational data are tasks of top priority in the current effort to map the Holocene development of the contemporary ecosystems.

In conclusion, this study shows the importance of mapping pollen data and illustrates the broad-scale vegetational patterns reflected by the pollen data. These maps provide a geographic context for future pollen studies in this area. In sparsely sampled regions, the isopolls derived either from interpolation or trend-surface analysis offer a prediction for the distribution of pollen values in those areas. The data from new samples can then check the accuracy of these initial contours and indicate possible changes in their positions.

ACKNOWLEDGMENTS

National Science Foundation Grants GX28672 (CLIMAP) to Brown University and GA10651X to the University of Wisconsin–Madison and U.S. Department of Defense Applied Research Projects Agency Grants F4462073C0021 to Brown University and Air Force Office of Scientific Research Grant 72-2407A to the University of Wisconsin–Madison supported this work. We wish to thank M. Anderson, D. R. Clark, F. Huhn, J. Pollock, and B. Santamaria for technical assistance and H.J.B. Birks, D. R. Clark, E. B. Leopold, D. A. Livingstone, J. C. Ritchie, and A. M. Swain for critically reading drafts of this article. We also wish to thank R. E. Bailey, R. G. Baker, B. S. Hanson, S. Lichti-Federovich, and A. M. Swain for use of their unpublished data.

REFERENCES CITED

Aario, L., 1940, Waldgrenzen und subrezenten Pollen spektren in Petsamo, Lappland: Acad. Sci. Fennicae Annales, ser. A, v. 54, p. 1-120.

Adam, D. P., 1970, Some palynological applications of multivariate statistics [Ph.D. thesis]: Tucson, Arizona Univ., 132 p.

Andersen, S. Th., 1970, The relative pollen productivity and pollen representation of north European trees, and correction factors for tree pollen spectra determined by surface pollen analyses from forests: Danmarks Geol. Undersögelse [skr.], ser. II, v. 96, p. 1-99.

Anderson, T. W., 1958, An introduction to multivariate statistical analysis: New York, John Wiley & Sons, Inc., p. 374.

Bartley, D. D., 1967, Pollen analysis of surface samples of vegetation from arctic Quebec: Pollen et Spores, v. 9, p. 101-105.

Bernabo, J. C., and Webb, T., III, 1976, Changing patterns in the Holocene pollen record of northeastern North America: A mapped summary: Quaternary Research (in press).

Birks, H.J.B., 1973, Modern pollen rain studies in some arctic and alpine environments, in Birks, H.J.B., and West, R. G., eds., Quaternary plant ecology: Oxford, Blackwell Scientific Pubs., p. 143-168.

———1974, Numerical zonation of Flandrian pollen data: New Phytologist, v. 73, p. 351-358.

Birks, H.J.B., Webb, T., III, and Berti, A. A., 1975, Numerical analysis of pollen samples from central Canada: A comparison of methods: Rev. Palaeobotany and Palynology, v. 20, p. 133-169.

Bradbury, J. P., and Waddington, J.C.B., 1973, The impact of European settlement on Shagawa Lake, northeastern Minnesota, U.S.A., in Birks, H.J.B., and West, R. G., eds., Quaternary plant ecology: Oxford, Blackwell Scientific Pubs., p. 289-307.

Brubaker, L. B., 1973, Ancient and modern forest patterns associated with glacial till and outwash in north-central upper Michigan [Ph.D. thesis]: Ann Arbor, Michigan Univ., 139 p.

Bryson, R. A., and Kutzbach, J. E., 1974, On the analysis of pollen-climate canonical transfer functions: quaternary Research, v. 4, p. 162-174.

Butzer, K. W., 1971, Environment and archeology: An ecological approach to prehistory [2nd ed.]: Chicago, Aldine-Atherton, 703 p.

Cain, S. A., 1944, Foundations of plant geography: New York, Harper and Brothers, 300 p.

Cole, H., 1969, Objective reconstruction of the paleoclimatic record through application of eigenvectors of present-day pollen spectra and climate to the late Quaternary pollen stratigraphy [Ph.D. thesis]: Madison, Wisconsin Univ. 110 p.

Colinvaux, P. A., 1973, Introduction to ecology: New York, John Wiley & Sons, Inc., 621 p.

Cross, A. T., Thompson, G. G., and Zaitzeff, J. B., 1966, Source and distribution of palynomorphs in bottom sediments, southern part of Gulf of California: Marine Geology, v. 4, p. 467-524.

Curtis, J. T., 1959, The vegetation of Wisconsin: Madison, Univ. Wisconsin Press, 657 p.

Davis, J. C., 1973, Statistics and data analysis in geology: New York, John Wiley & Sons, Inc., 550 p.

Davis, M. B., 1963, On the theory of pollen analysis: Am. Jour. Sci., v. 261, p. 897-912.

———1967, Late-glacial climate in northern United States: A comparison of New England and the Great Lakes region, in Cushing, E. S., and Wright, H. E., Jr., eds., Quaternary paleoecology: New Haven, Conn., Yale Univ. Press, p. 11-43.

———1969, Palynology and environmental history during the Quaternary Period: Am. Scientist, v. 57, p. 317-332.

Davis, M. B., Brubaker, L. B., and Beiswenger, J. M., 1971, Pollen grains in lake sediments: Pollen percentages in surface sediments from southern Michigan: Quaternary Research, v. 1, p. 450-467.

Davis, M. B., Brubaker, L. B., and Webb, T., III, 1973, Calibration of absolute pollen influx, in Birks, H.J.B., and West, R. G., eds., Quaternary plant ecology: Oxford, Blackwell Scientific Pubs., p. 9-25.

Davis, R. B., and Webb, T., III, 1975, The contemporary distribution of pollen in eastern North America: A comparison with the vegetation: Quaternary Research, v. 5, p. 395-434.

Davis, R. B., Brewster, L. A., and Sutherland, J., 1969, Variation in pollen spectra within lakes: Pollen et Spores, v. 11, p. 557-571.

Davis, R. B., Bradstreet, T. E., Stuckenrath, R., Jr., and Borns, H. W., Jr., 1975, Vegetation and associated environments during the past 14,000 years near Moulton Pond, Maine: Quaternary Research, v. 5, p. 435-465.

Dudnik, E. E., 1971, SYMAP user's manual for synagraphic computer mapping: Chicago, Univ. Illinois at Chicago Circle, 51 p.

Erdtman, G., 1943, An introduction to pollen analysis: Waltham, Mass., Chronica Botanica Co., 335 p.

Faegri, K., and Iversen, J., 1964, Textbook of pollen analysis: New York, Hafner Pubs. Co., 237 p.

Fagerlind, F., 1952, The real signification of pollen diagrams: Bot. Notiser, v. 1952, p. 185-224.

Fries, M., 1962, Pollen profiles of late Pleistocene and recent sediments at Weber Lake, northeastern Minnesota: Ecology, v. 43, p. 295-308.

Gregory, P. H., 1973, The microbiology of the atmosphere [2nd ed.]: New York, John Wiley & Sons, 377 p.

Grichuk, M. P., 1967, The study of pollen spectra from recent and ancient alluvium: Rev. Palaeobotany and Palynology, v. 4, p. 107-112.

——1970, Zakonomernosti formirovaniia sovremennykh sporovo-pyl'tsevykh spektrov, kak osnova dlia interpretatsii iskopaemykh sporovo-pyl'tsevykh spektrov [The regularities of formation of contemporary spore-pollen spectra as a basis for interpreting fossil spore-pollen spectra]: Akad. Nauk SSSR Sibirsk. Otdeleniye Inst. Geologii i Geofiziki Trudy, v. 92, p. 12-19 (in Russian).

Grichuk, V. P., 1941, Opyt kharakteristiki sostava pyl'tsy v sovremennykh otlozheniiakh razlichnykh rastitel'nykh zon evropeiskoi chasti SSSR [An evaluation of the characteristic pollen composition of contemporary deposits of the different vegetation zones of the European part of the USSR]: Akad. Nauk SSSR Izv. Ser. Geog. Problemy fizicheskoi geografii, v. 11, p. 101-130 (in Russian).

Grüger, J., 1973, Studies on the late Quaternary vegetation history of northeastern Kansas: Geol. Soc. America Bull., v. 84, p. 239-250.

Halliday, W.E.D., and Brown, A.W.A., 1943, The distribution of some important forest trees in Canada: Ecology, v. 24, p. 353-373.

Imbrie, J., and Kipp, N. G., 1971, A new micropaleontological method for quantitative paleoclimatology: Application to a late Pleistocene Caribbean core, in Turekian, K., ed., The late Cenozoic glacial ages: New Haven, Conn., Yale Univ. Press, p. 71-181.

Kaiser, H. F., 1958, The varimax criterion for analytic rotation in factor analysis: Psychometrika, v. 23, p. 187-200.

Kapp, R. O., 1965, Illinoian and Sangamon vegetation in south-western Kansas and adjacent Oklahoma: Michigan Univ. Contr. Mus. Paleontology, v. 19, p. 167-225.

King, J. E., and Kapp, R. O., 1963, Modern pollen rain studies in eastern Ontario: Canadian Jour. Botany, v. 41, p. 243-252.

Kipp, N. G., 1976, New transfer function for estimating past sea-surface conditions from sea-bed distribution of planktonic foraminiferal assemblages in the North Atlantic, in Cline, R. M., and Hays, J. D., eds., Investigation of late Quaternary paleoceanography and paleoclimatology: Geol. Soc. America Mem. 145 (this volume).

Küchler, A. W., 1964, Potential natural vegetation of the conterminous United States: Am. Geog. Soc. Spec. Pub. 36, 38 p., 116 pls., map.

Kutzbach, J. E., 1967, Empirical eigenvectors of sea-level pressure, surface temperature, and precipitation complexes over North America: Jour. Appl. Meteorology, v. 6, p. 791-802.

Leopold, E. B., 1958, Some aspects of late glacial climate in eastern United States: Veroeff. Geobot. Inst. Rubel in Zurich, v. 34, p. 80-85.

Lichti-Federovich, S., and Ritchie, J. C., 1965, Contemporary pollen spectra in central Canada. II. The forest-grassland transition in Manitoba: Pollen et Spores, v. 7, p. 63-87.

——1968, Recent pollen assemblages from western interior of Canada: Rev. Palaeobotany and Palynology, v. 7, p. 297-344.

Lindsay, D. R., 1953, Climate as a factor influencing the mass ranges of weeds: Ecology v. 34, p. 308-321.

Little, E. L., Jr., 1971, Atlas of United States trees. Vol. 1: U.S. Dept. Agriculture Misc. Pub. 1146, 9 p., 200 maps.

Maher, L. J., 1963, Pollen analyses of surface materials from the southern San Juan Mountains,

Colorado; Geol. Soc. America Bull., v. 74, p. 1485-1504.

McAndrews, J. H., 1966, Postglacial history of prairie, savanna, and forest in northwestern Minnesota: Torrey Bot. Club Mem., v. 22, p. 1-72.

——1968, Pollen evidence for the prehistoric development of the "Big Woods" in Minnesota (U.S.A.): Rev. Palaeobotany and Palynology, v. 7, p. 201-211.

——1969, Palaeobotany of a wild rice lake in Minnesota: Canadian Jour. Botany, v. 47, p. 1671-1679.

McAndrews, J. H., and Power, D. N., 1973, Palynology of the Great Lakes: The surface sediments of Lake Ontario: Canadian Jour. Earth Sci., v. 10, p. 777-792.

McAndrews, J. H., and Wright, H. E., Jr., 1969, Modern pollen rain across Wyoming basins and the northern Great Plains (U.S.A.): Rev. Palaeobotany and Palynology, v. 9, p. 17-43.

McAndrews, J. H., Berti, A. A., Norris, G., 1973, Key to the Quaternary pollen and spores of the Great Lake region: Royal Ontario Mus. Life Sci. Misc. Pub., 61 p.

McIntyre, A., Kipp, N., Bé, A., Crowley, T., Kellogg, T., Gardner, J. V., Prell, W., and Ruddiman, W. F., 1976, Glacial North Atlantic 18,000 years ago; A CLIMAP reconstruction, in Cline, R. M., and Hays, J. D., eds., Investigation of late Quaternary paleoceanography and paleoclimatology: Geol. Soc. America Mem. 145 (this volume).

Miller, N. G., 1973, Late-glacial and postglacial vegetation change in southwestern New York state: New York State Mus. and Sci. Service Bull., 420 p.

Morrison, D. F., 1967, Multivariate statistical methods: New York, McGraw-Hill Book Co., Inc., 338 p.

Mosimann, J. E., and Greenstreet, R. L., 1971, Representation—Insensitive method in palaeoecological pollen studies: in Patil, G. P., Pielou, E. C., and Waters, W. E., eds., Statistical ecology, Vol. 1, Spatial patterns and statistical distribution: University Park, Pennsylvania State Univ. Press, p. 23-58.

Mott, R. J., 1969, Palynological studies in central Saskatchewan: Contemporary pollen spectra from surface samples: Canada Geol. Survey Paper 69-32, 13 p.

——1975a, Palynological studies of lake sediment profiles from southwestern New Brunswick: Canadian Jour. Earth Sci., v. 12, p. 273-288.

——1975b, Modern pollen spectra from northwestern Ontario: Canada Geol. Survey Paper 75-1, pt. B, p. 147-150.

Muller, J., 1959, Palynology of Recent Orinoco delta and shelf sediments: Reports of the Orinoco shelf expedition. Vol. 5: Micropaleontology, v. 5, p. 1-32.

Neustadt, M. I., 1957, The history of forests and Holocene paleogeography in the USSR: Moscow Akad. Nauk SSSR, 404 p. (in Russian).

Nikolaeva-Prokhorova, K. V., and Shalandina, V. T., 1973, Opyt sopostavleniia sostava sovremennykh khovinykh i shirokolist-vennyhk lesov Tatarskoi ASSR a subfossil'nymi sporovo-pyl' tsevymi spektrami [Comparison of contemporary composition of coniferous and broadleafed forests in the Tatar ASSR with subfossil spore-pollen spectra]: Bot. Zh., v. 58, p. 1619-1627 (in Russian).

Ogden, J. G., III, 1966, Forest history of Ohio. I. Radiocarbon dates and pollen stratigraphy of Silver Lake, Logan County, Ohio: Ohio Jour. Sci., v. 66, p. 387-400.

——1969, Correlation of contemporary and late Pleistocene pollen records in the reconstruction of postglacial environments in northeastern North America: Mitt Internat. Ver. Limnology, v. 17, p. 64-77.

Oxford World Atlas, 1973 (Cohen, S. B., Geographic ed.): New York, Oxford Univ. Press.

Pennington, W., 1973, Absolute pollen frequencies in the sediments of lakes of different morphometry, in Birks, H.J.B., and West, R. G. eds., Quaternary plant ecology: Oxford, Blackwell Scientific Pubs., p. 79-104.

Ritchie, J. C., 1972, Pollen analysis of late Quaternary sediments from the Arctic treeline of the MacKensie River delta region, Northwest Territories, Canada, in Vasari, Y., Hyvarinen, H., and Hicks, S., eds., Climate change in arctic areas during the last ten-thousand years: Oulu, Finland, Acta Univ. Oulueusis, ser. A, v. 3, Geology no. 1, p. 253-271.

Rowe, J. S., 1972, Forest regions of Canada: Canada Dept. Environment Pub. 1300, 172 p.

Skinner, R. G., 1973, Quaternary stratigraphy of the Moose River basin, Ontario: Canada

Geol. Survey Bull. 225, 77 p.

Swain, A. M., 1973, A history of fire and vegetation in northeastern Minnesota as recorded in lake sediments: Quaternary Research, v. 3, p. 383-396.

Szafer, W., 1935, The significance of isopollen lines for the investigation of the geographical distribution of trees in the postglacial period: Acad. Polonaise Sci. Bull., Ser. B, Sci. Nat., v. I, p. 235-239.

von Post, L., 1967, Forest tree pollen in south Swedish peat bog deposits: Pollen et Spores, v. 9, p. 375-401.

Watts, W. A., and Bright, R. C., 1968, Pollen, seed, and mollusk analysis of a sediment core from Pickerel Lake, northeastern South Dakota: Geol. Soc. America Bull., v. 79, p. 855-876.

Watts, W. A., and Wright, H. E., Jr., 1966, Late Wisconsin pollen and seed analysis from the Nebraska Sandhills: Ecology, v. 47, p. 202-210.

Weaver, J. E., and Clements, F. E., 1938, Plant ecology [2nd ed.]: New York, McGraw-Hill Book Co., Inc., 520 p.

Webb, T., III, 1971, The late- and postglacial sequence of climatic events in Wisconsin and east-central Minnesota: Quantitative estimates derived from fossil pollen spectra by multivariate statistical analysis [Ph.D. thesis]: Madison, Wisconsin Univ., 161 p.

——1973a, Pre- and postsettlement pollen from a short core, Blackhawk Lake, west-central Iowa: Iowa Acad. Sci. Proc., v. 80, p. 41-44.

——1973b, A comparison of modern and presettlement pollen from southern Michigan (U.S.A.): Rev. Palaeobotany and Palynology, v. 16, p. 137-156.

——1974a, Corresponding patterns of pollen and vegetation in lower Michigan: A comparison of quantitative data: Ecology, v. 55, p. 17-28.

——1974b, A vegetational history of northern Wisconsin: Evidence from modern and fossil pollen: Am. Midland Naturalist, v. 92, p. 12-34.

Webb, T., III, and Bryson, R. A., 1972, Late- and postglacial climatic change in the northern Midwest, U.S.A.: Quantitative estimates derived from fossil pollen spectra by multivariate statistical analysis: Quaternary Research, v. 2, p. 70-115.

West, R. G., 1973, Introduction, in Birks, H.J.B., and West, R. G., eds., Quaternary plant ecology: Oxford, Blackwell Scientific Pubs., p. 1-3.

Wright, H. E., 1967, The use of surface samples in Quaternary pollen analysis: Rev. Palaeobotany and Palynology, v. 2, p. 321-330.

Wright, H. E., and Patten, H. L., 1962, The pollen sum: Pollen et Spores, v. 5, p. 445-450.

Yarranton, G. A., and Ritchie, J. C., 1972, Sequential correlations as an aid in placing zone boundaries: Pollen et Spores, v. 14, p. 213-223.

MANUSCRIPT RECEIVED BY THE SOCIETY JANUARY 20, 1975
REVISED MANUSCRIPT RECEIVED JUNE 23, 1975
MANUSCRIPT ACCEPTED JULY 14, 1975

Printed in U.S.A.

ANTARCTIC

Geological Society of America
Memoir 145
© 1976

Relationship of Radiolarian Assemblages to Sediment Types and Physical Oceanography in the Atlantic and Western Indian Ocean Sectors of the Antarctic Ocean

JOSE A. LOZANO*
AND
JAMES D. HAYS
Lamont-Doherty Geological Observatory
and
Department of Geological Sciences
Columbia University
Palisades, New York 10964

ABSTRACT

The relative abundances of 18 selected taxonomic groups of Radiolaria were determined for 145 trigger-weight core-top samples collected between long 80°E and 55°W and lat 35°S and 60°S. Seventy-two samples were considered to represent nonreworked recent sediments. A factor analysis of these 72 samples resolved Antarctic, subantarctic, and subtropical assemblages, with distributions closely corresponding to the main surface water masses in the area.

Using the technique of Imbrie and Kipp, paleoecologic equations have been developed to estimate surface-water temperatures for August (winter) and February (summer). Estimated temperatures range from near 20°C in the northern parts of the area to 0°C at the location of the southernmost cores. The standard error of estimate of the equations is less than 1.5°C. The $CaCO_3$ content of the 145 surface-sediment samples shows that variations in the calcite compensation depth (CCD) for locations north of the Antarctic Polar Front depends mainly on varying dissolution. The CCD is about 4,800 m in the western basin of the Atlantic Ocean but is deeper than 5,200 m, at least locally, in the Indian Ocean. South of the Antarctic Polar Front the CCD is shallower than 3,700 m.

Radiolaria are common in most samples, with the maximum number per gram of sediment being found in samples collected near or beneath the Antarctic Polar Front.

---

*Present address: Departamento de Geologia, Universidad Nacional, Bogota D.E., Colombia, South America.

INTRODUCTION

Our major objective was to develop means by which past marine conditions can be reconstructed for the western Indian and Atlantic sectors of the Antarctic Ocean.

In order to generate paleoclimatic information, it is necessary to establish present-day base lines for climatically sensitive parameters. In this paper, we provide base-line information for the construction and interpretation of an 18,000 B.P. map, presented in Hays and others (this volume), by relating various properties of the sediments to present-day oceanographic conditions. We make an effort to go beyond the use of assemblages and sedimentary properties for qualitative interpretations to a more quantitative approach by estimating temperatures using the techniques developed by Imbrie and Kipp (1971) for foraminiferal assemblages and first used for Radiolaria by Sachs (1973a, 1973b).

PREVIOUS WORK

The first sea-floor sediment description of the Southern Ocean was based on observations made on the H.M.S. *Challenger* expedition. The map of Murray and Renard (1891) shows terrigenous deposits around Antarctica as far north as lat 60°S; northward, a belt of diatom ooze of varying width extends more or less to lat 50°S, followed to the north by *Globigerina* ooze at shallow depths and red clay in the deep basins.

Philippi (1912) recognized the influence of the glacial period in the superposition of *Globigerina* ooze on red clay or on diatom ooze or of diatom ooze on glacial marine sediments (his glacial marine Ablagerungen). He explained the lithologic changes as being produced by ice advances and retreats. Schott (1939) expanded this interpretation to explain lithologic changes by latitudinal displacements of the polar waters induced by fluctuations in ice cover.

Riedel (1958) made some references to the areal and depth distribution of Antarctic species. Hays (1965) distinguished an Antarctic fauna living south of the polar front and a warm-water fauna that showed a gradual increase in species abundance from south to north of the polar front. Hays used the relative changes between these two faunas to estimate past climatic changes and correlated them to the radiolarian stratigraphic zonation he established. For a review of previous stratigraphic work, see Hays and others (this volume).

Petrushevskaya (1967) described 68 species and subspecies of Radiolaria she found in plankton and surface bottom-sediment samples of the Indian and Pacific sectors of the Southern Ocean. She studied their abundance distribution and also drew the northern boundary for an "Antarctic complex" and the southern boundary of the "tropical complex." Nigrini (1967) studied the Radiolaria in Holocene Indian Ocean sediments and distinguished a low-latitude assemblage characterized by twelve species and a middle-latitude assemblage made up of seven species; the boundary between the two is about lat 30°S to 35°S. She also did a preliminary survey of Atlantic Ocean sediments and concluded that the species she described as characteristic of the low- and middle-latitude assemblages in the Indian Ocean appeared to be similarly distributed in the Atlantic Ocean.

Recently, Goll and Bjørklund (1974) studied the abundance of Radiolaria in South Atlantic bottom sediments and compared them with surface nutrients, dissolved silicate throughout the water column, and lithology. They concluded that a clear relationship between nutrients and radiolarian abundance is not evident. Using six species as representative of a general distribution pattern in the South Atlantic, they divided the faunal distribution into polar, subpolar, central, and equatorial provinces.

DESCRIPTION OF AREA

The area of study covers the southern parts of the western Indian Ocean and the Atlantic Ocean from lat 35°S to 60°S and long 80°E and 55°W (Fig. 1, Table 1). The morphology of the area has been presented by Heezen and Tharp (1972). Figure 2 shows the distribution of samples superimposed on a generalized topographic map.

PHYSICAL OCEANOGRAPHY

Detailed oceanographic studies can be found in Deacon (1933, 1937, 1963), Gordon (1967, 1971), and Gordon and Goldberg (1970). Horizontal and vertical circulation patterns of the area are shown in Figures 3 and 4.

The ocean currents are predominantly directed to the east and are driven primarily by prevailing westerly winds. The easterly winds drive a narrow band of westerly flowing water adjacent to Antarctica. The Antarctic divergence resulting from the Ekman drift associated with the wind field occurs at the contact between the two directions of flow. The surface circulation pattern extends to the sea floor with little attenuation due to the lower degree of baroclinicity of the Antarctic waters. The West Wind Drift or Antarctic Circumpolar Current is strongly influenced by changes in bottom topography. On approaching a ridge, the flow is deflected to the north and returns to the south in the lee of a ridge (Kort, 1962; Gordon and Goldberg, 1970).

Figure 4 shows the circulation and water-mass distribution in a vertical section. The Circumpolar Deep Water is composed of relatively warm saline water of low oxygen content; the deep water upwells as it moves east and south. According to Gordon and Goldberg (1970), as this water reaches the near-surface layers it is altered into two different water masses by sea-air exchange of heat and water, which is associated in the southern part of the area with the seasonal freezing and melting of sea ice. The two new water masses are colder, have a high dissolved-oxygen content, and are lower in salinity than the Circumpolar Deep Water. One water mass remains at the surface as the Antarctic Surface Water (formed primarily by melting sea ice and winter cooling); the second, through the process of sea ice formation and (or) by interaction with the bottom of floating ice shelves, increases its density and sinks to the sea floor as the Antarctic Bottom Water. The southward flow of the deep water is compensated by the northward flow of these two Antarctic water masses. The Antarctic Surface Water sinks under the subantarctic surface waters at the Antarctic Convergence (often called the Antarctic Polar Front), where it contributes to the production of Antarctic Intermediate Water, which spreads far to the north at depths near 1,000 m.

The upwelling deep water south of the polar front recycles nutrients into the Antarctic Surface Water, making it highly productive. Of special interest in the region discussed in this study is the large cyclonic gyre in the Weddell Sea (Gordon and Goldberg, 1970). This Weddell gyre is the largest of a number of cyclonic gyres ringing Antarctica. The waters under the gyre are colder and fresher than water to the north. The Weddell gyre extends eastward from the Antarctic Peninsula

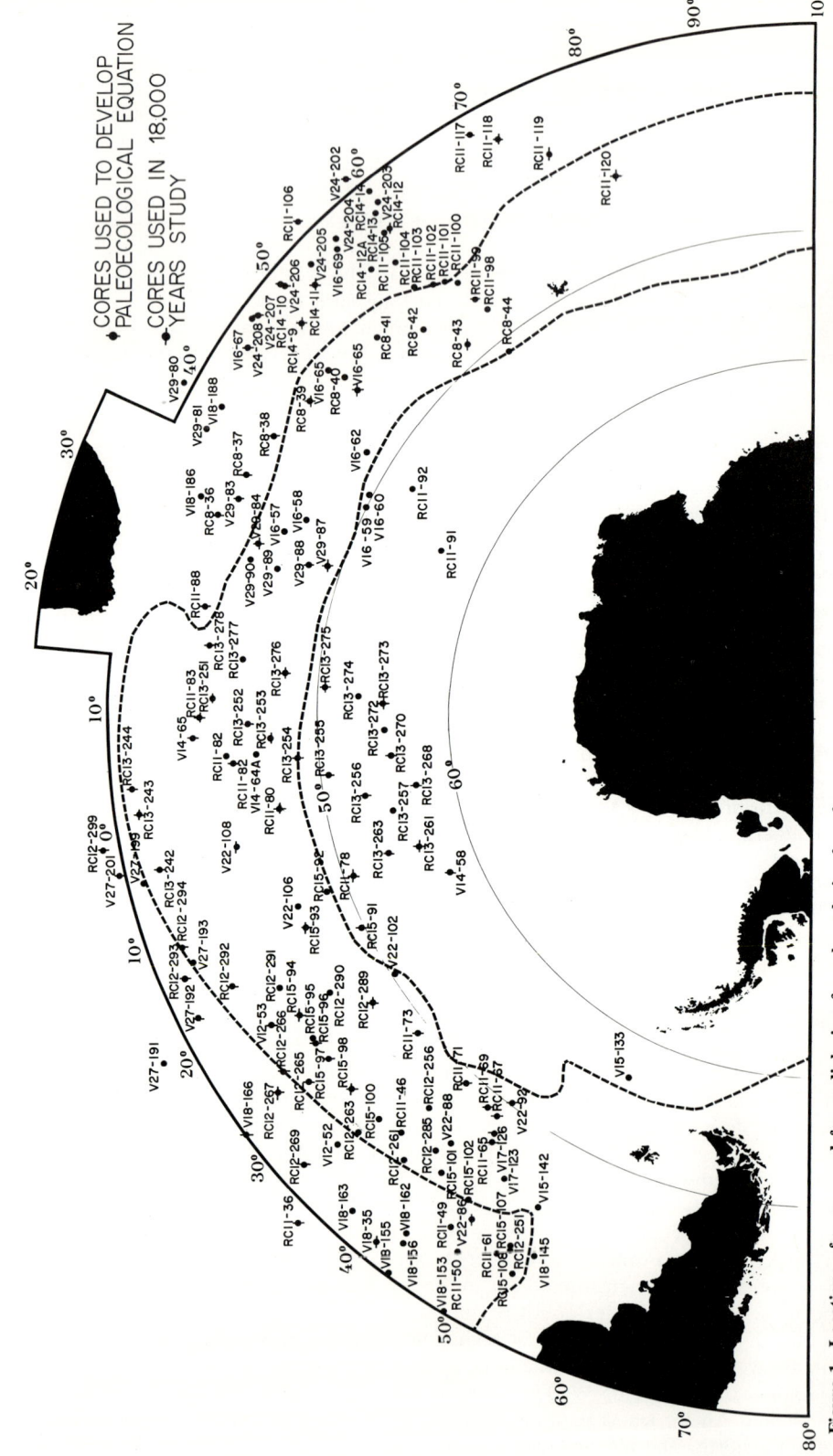

Figure 1. Locations of cores used for radiolarian faunal analysis of surface-sediment samples. Seventy-two of these were used to develop paleoecologic equations. Dashed lines represent average positions of subtropical convergence and Antarctic Polar Front.

TABLE 1. LOCATION OF CORES AND DEPTHS OF WATER

Core	Lat	Long	Water depth (m)	Core	Lat	Long	Water Depth (m)
RC8-36*	40°26.0′S	28°55.0′E	4,098	RC13-276*	47°41.9′S	14°42.3′E	5,015
RC8-37*	41°19.0′S	33°13.0′E	4,775	RC13-278*	42°03.7′S	16°43.8′E	4,790
RC8-38*	41°53.0′S	37°49.0′E	3,784	RC14-9*	39°01.0′S	47°53.0′E	2,692
RC8-39*	42°53.0′S	42°21.0′E	4,330	RC14-10*	36°15.0′S	49°28.0′E	4,017
RC8-40	43°47.0′S	46°12.0′E	2,540	RC14-11*	38°00.0′S	51°11.0′E	3,268
RC8-41	43°38.0′S	51°16.0′E	2,897	RC14-12	38°45.0′S	59°18.0′E	5,271
RC8-42	45°41.0′S	54°54.0′E	3,915	RC14-13	37°23.0′S	59°19.0′E	5,128
RC8-43	48°41.0′S	57°22.0′E	4,319	RC14-14	35°55.0′S	59°58.0′E	4,916
RC8-44	51°04.0′S	60°18.0′E	4,753	RC15-91*	49°53.3′S	15°34.1′W	3,775
RC11-36*	33°52.0′S	35°16.0′W	4,222	RC15-92*	48°29.0′S	10°19.9′W	3,378
RC11-46	43°21.5′S	36°06.0′W	5,179	RC15-93*	46°05.8′S	13°13.5′W	2,714
RC11-49	40°20.0′S	45°25.0′W	5,134	RC15-94*	42°53.9′S	20°51.3′W	3,762
RC11-50	39°08.4′S	47°12.0′W	5,282	RC15-95	42°57.3′S	23°30.1′W	4,264
RC11-61	40°30.0′S	50°10.0′W	5,438	RC15-96*	42°52.7′S	23°54.9′W	4,426
RC11-65*	47°02.0′S	43°41.0′W	5,435	RC15-97*	42°57.3′S	25°56.2′W	4,583
RC11-67*	48°40.0′S	42°38.0′W	3,150	RC15-98*	42°56.2′S	29°46.6′W	4,416
RC11-69*	48°54.0′S	41°00.0′W	5,492	RC15-100	42°58.1′S	33°51.3′W	4,998
RC11-71*	49°08.0′S	37°25.0′W	5,537	RC15-101	42°59.0′S	41°34.3′W	5,174
RC11-73*	49°04.0′S	29°14.0′W	5,321	RC15-102	42°37.8′S	45°19.4′W	5,000
RC11-78*	50°52.0′S	09°52.0′W	3,115	RC15-107	41°23.7′S	50°58.9′W	5,018
RC11-80*	46°45.0′S	00°03.0′W	3,656	RC15-108	39°52.4′S	52°28.8′W	5,082
RC11-81*	43°54.0′S	05°13.0′E	4,704	V12-52	39°35.6′S	32°52.5′W	5,015
RC11-82	43°29.0′S	05°57.0′E	4,609	V12-53*	40°54.3′S	20°22.9′W	3,797
RC11-83*	41°36.0′S	09°43.0′E	4,718	V14-58	57°37.0′S	13°36.0′W	3,543
RC11-88*	41°11.0′S	20°08.0′E	5,125	V14-64A	45°35.0′S	06°00.0′E	4,581
RC11-91	56°34.0′S	34°11.0′E	5,373	V14-65*	41°03.5′S	07°47.0′E	4,825
RC11-92	52°29.0′S	39°57.0′E	5,093	V15-133*	56°37.0′S	53°54.5′W	4,031
RC11-98	47°39.0′S	61°29.0′E	4,292	V15-142	44°53.7′S	51°32.0′W	5,856
RC11-99*	46°31.0′S	61°02.0′E	4,449	V16-57	45°14.0′S	29°29.0′E	5,289
RC11-100	44°50.0′S	60°52.0′E	4,742	V16-58	46°30.0′S	31°16.0′E	4,731
RC11-101	44°04.0′S	59°50.0′E	4,806	V16-59	50°03.0′S	35°11.0′E	4,868
RC11-102*	43°42.0′S	58°48.0′E	4,709	V16-60	49°59.5′S	36°45.5′E	4,574
RC11-103	43°02.0′S	57°21.0′E	4,673	V16-65*	45°00.0′S	45°46.0′E	1,618
RC11-104	40°55.0′S	57°39.0′E	4,885	V16-66	42°39.0′S	45°40.0′E	2,985
RC11-105	38°47.0′S	58°50.0′E	5,256	V16-67*	37°09.0′S	43°30.0′E	3,643
RC11-106*	34°20.0′S	54°13.0′E	4,212	V16-69	37°35.0′S	54°44.0′E	4,960
RC11-117*	36°29.0′S	69°33.0′E	4,367	V17-123	45°22.0′S	46°58.0′W	5,273
RC11-118*	37°48.0′S	71°32.0′E	4,354	V17-126*	47°35.0′S	43°21.0′W	5,497
RC11-119	40°18.0′S	74°34.0′E	3,709	V18-35	36°24.0′S	41°07.0′W	4,942
RC11-120*	43°31.0′S	79°52.0′E	3,193	V18-145	41°31.0′S	53°17.0′W	5,469
RC12-251	41°12.0′S	51°03.0′W	5,554	V18-153	35°02.0′S	48°51.0′W	4,401
RC12-256	45°59.0′S	36°24.0′W	5,455	V18-155	35°07.0′S	43°29.0′W	4,848
RC12-261	42°00.0′S	38°07.6′W	5,209	V18-156	37°27.0′S	42°50.0′W	5,044
RC12-263	41°03.1′S	33°17.7′W	5,020	V18-162	38°03.0′S	42°44.0′W	5,088
RC12-265*	40°54.3′S	26°51.0′W	4,395	V18-163	36°58.0′S	37°47.0′W	5,000
RC12-266*	39°49.2′S	24°47.5′W	3,939	V18-166*	34°59.0′S	27°07.0′W	4,257
RC12-267*	38°41.0′S	25°47.0′W	4,144	V18-186	38°59.0′S	29°56.0′E	4,204
RC12-269*	36°58.5′S	32°11.5′W	4,360	V18-188	37°38.0′S	37°52.0′E	4,960
RC12-285	43°58.5′S	39°57.0′W	5,114	V22-86	41°36.0′S	46°27.0′W	5,110
RC12-289*	47°54.2′S	23°41.5′W	4,484	V22-88	45°04.0′S	40°20.0′W	4,980
RC12-290	45°38.6′S	20°32.8′W	4,276	V22-92*	50°06.0′S	42°56.0′W	1,686
RC12-291	42°35.2′S	17°48.0′W	3,508	V22-102	50°11.0′S	22°24.0′W	4,319
RC12-292*	39°40.6′S	15°28.5′W	3,541	V22-106	46°08.0′S	10°54.0′W	3,037
RC12-293*	36°53.0′S	13°09.0′W	3,393	V22-108	43°11.0′S	03°15.0′W	4,171
RC12-294*	37°15.6′S	10°05.8′W	3,308	V24-202*	34°21.0′S	59°13.0′E	5,512
RC12-299*	34°04.8′S	01°00.0′W	4,296	V24-203	36°59.0′S	59°59.0′E	4,997
RC13-242*	37°32.3′S	03°35.2′W	4,266	V24-204	36°50.0′S	55°17.0′E	4,631
RC13-243*	36°54.0′S	01°19.8′E	4,790	V24-205	36°51.0′S	52°17.0′E	5,515
RC13-244*	36°30.0′S	03°32.7′E	5,222	V24-206*	36°30.0′S	49°34.0′E	3,365
RC13-251	42°30.7′S	11°40.3′E	4,341	V24-207*	36°18.0′S	46°13.0′E	3,449
RC13-252*	45°05.1′S	09°09.0′E	4,523	V24-208	36°13.0′S	45°39.0′E	3,231
RC13-253	46°36.0′S	07°37.5′E	2,494	V27-191	33°03.3′S	18°43.4′W	3,957
RC13-254*	48°34.2′S	05°07.6′E	3,636	V27-192*	36°33.9′S	16°52.0′W	3,087
RC13-255*	50°34.5′S	02°53.7′E	3,332	V27-193	37°52.2′S	12°04.2′W	3,354
RC13-256*	53°10.9′S	00°21.3′W	2,525	V27-199	36°16.1′S	04°24.4′W	4,071
RC13-257	55°00.2′S	03°00.1′W	2,836	V27-201*	34°46.2′S	03°24.5′W	4,186
RC13-261*	56°07.3′S	08°41.2′W	4,221	V29-80	34°40.0′S	38°16.0′E	5,264
RC13-263*	53°48.3′S	08°13.0′W	3,389	V29-81	37°22.0′S	35°31.0′E	4,834
RC13-268*	57°02.3′S	00°05.6′W	4,005	V29-83*	41°32.0′S	30°57.0′E	5,059
RC13-270*	55°28.8′S	04°38.2′E	3,160	V29-84*	43°51.0′S	27°36.0′E	5,451
RC13-272	55°05.1′S	08°00.0′E	2,538	V29-87*	49°06.0′S	27°23.0′E	5,314
RC13-273*	55°04.5′S	11°34.5′E	4,967	V29-88*	47°51.0′S	26°47.0′E	5,737
RC13-274	53°09.0′S	12°25.6′E	3,372	V29-89	45°44.0′S	25°39.0′E	5,945
RC13-275*	50°43.0′S	13°26.0′E	1,984	V29-90	43°42.0′S	25°44.0′E	5,148

Note: All cores are from Lamont-Doherty core library and were taken by Lamont research vessels *Conrad* (RC) and *Vema* (V).
*Sample was used to develop final paleoecologic equations.

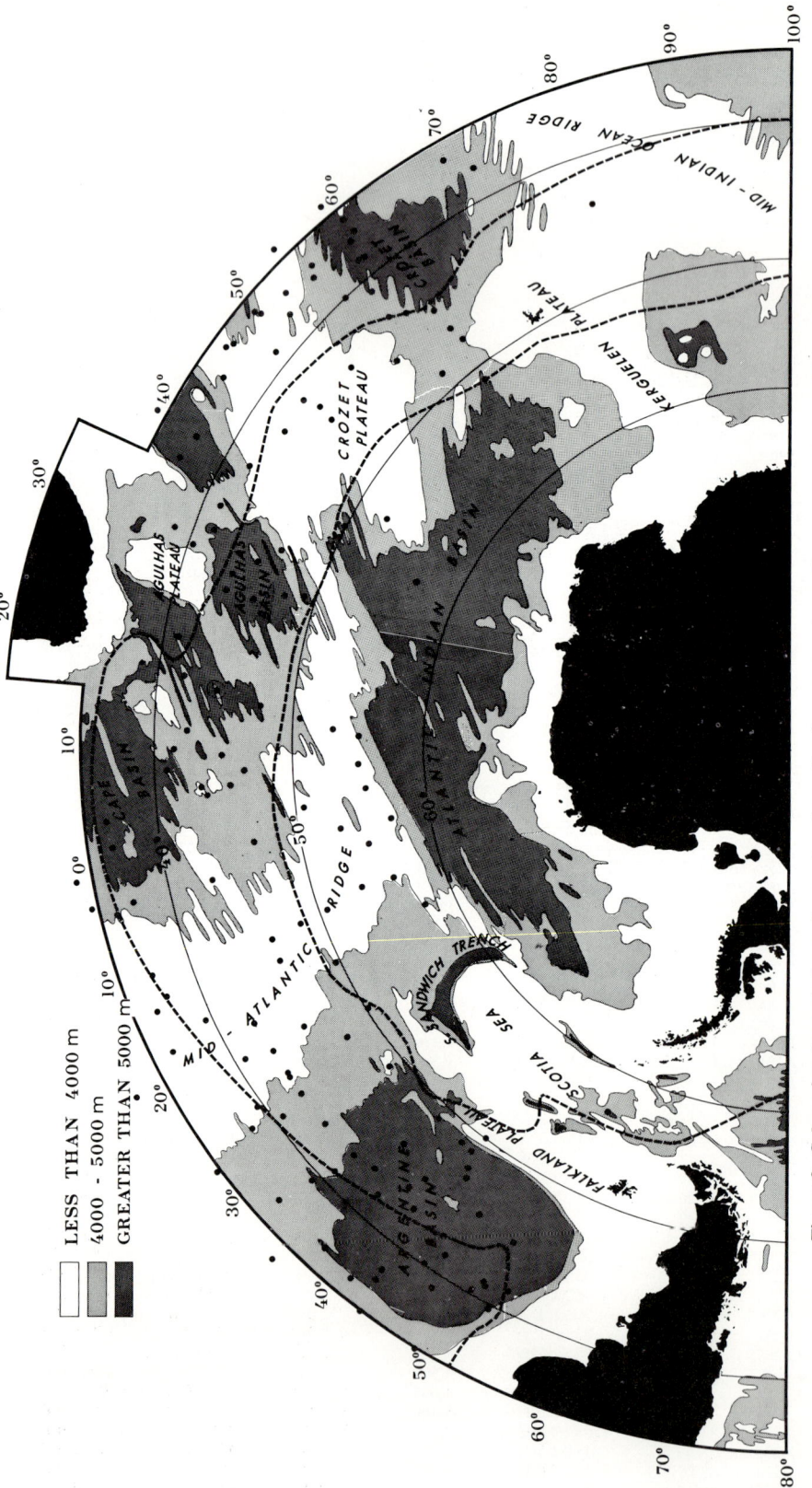

Figure 2. Submarine topography and core locations. Depth contours from Heezen and Tharp (1972). Dashed lines represent average positions of subtropical convergence and Antarctic Polar Front.

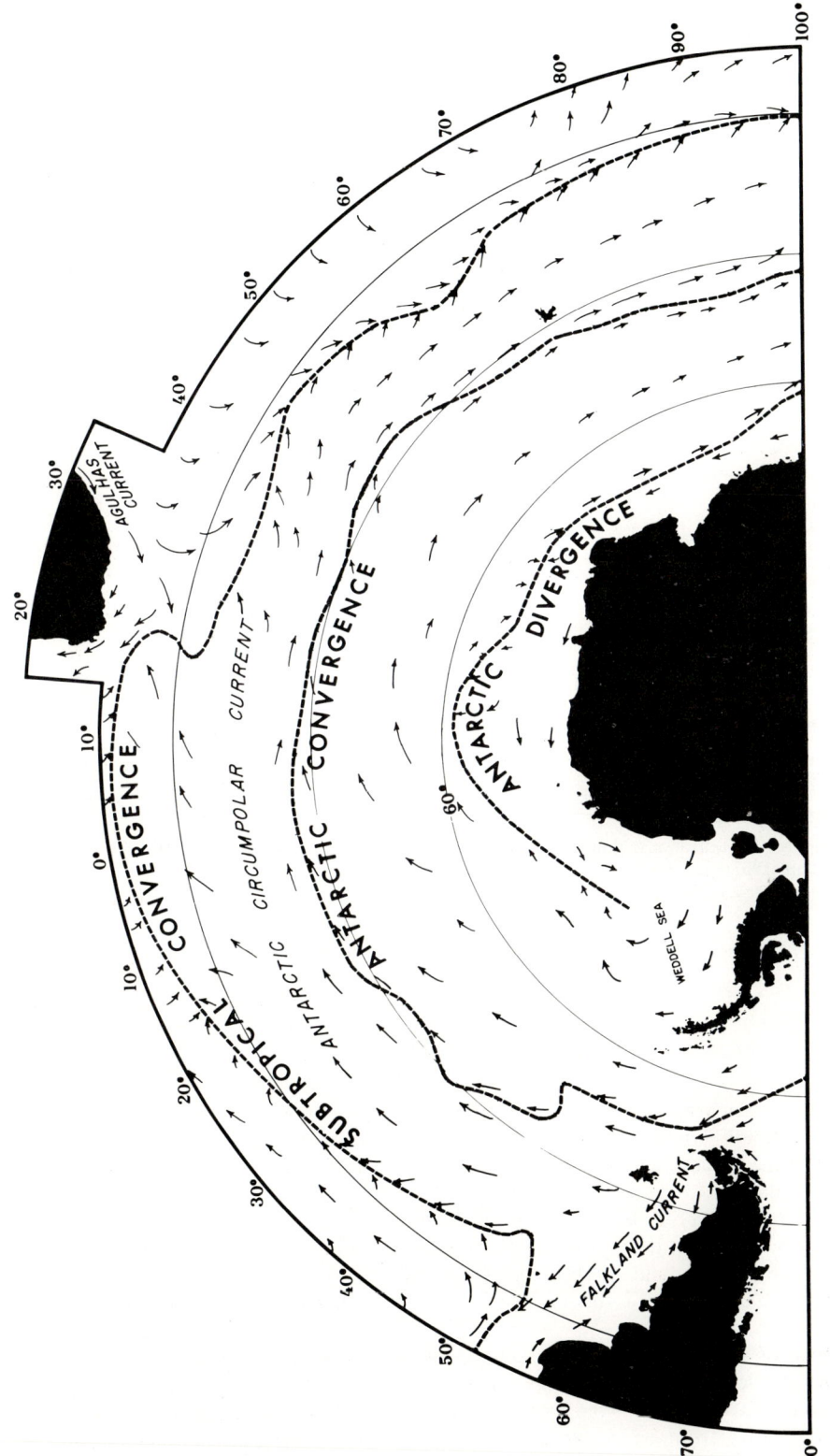

Figure 3. Generalized surface-water circulation (after Defant, 1961).

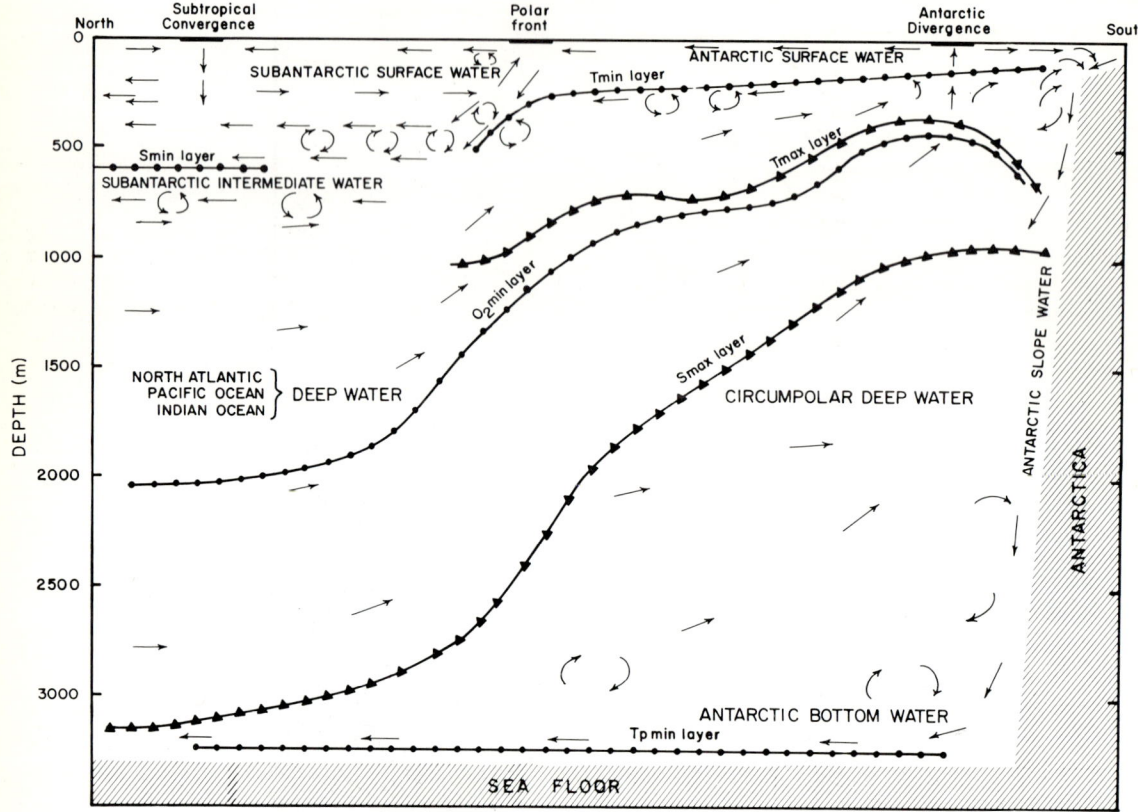

Figure 4. Vertical meridional circulation and water masses of the Antarctic Ocean (from Gordon, 1967).

to approximately long 30°E and northward from Antarctica to between lat 55°S to 60°S.

SEDIMENT DISTRIBUTION

The oceanographic properties of the water surrounding Antarctica largely control the sediment types accumulating on the ocean floor. Figure 5 is a generalized map of the predominant sediment types in the area, after Goodell (1973) and modified by our data.

North of the polar front, calcareous ooze predominates at depths shallower than about 4,500 m. In deeper water calcareous sediments give way to predominantly gray, olive, red, and brown clayey silt and silty clay, which sometimes have appreciable amounts of volcanic glass and become sandy adjacent to the continental shelves.

East of about long 25°W and slightly north of the polar front there is a narrow band of siliceous and calcareous ooze in which siliceous and calcareous tests together exceed 30% of the total sediment. Siliceous tests rapidly increase in number south of the polar front as calcareous ones decrease. An almost continuous zone of siliceous ooze (predominantly diatomaceous but with appreciable amounts of

Radiolaria) carpets the ocean floor beneath the polar waters. The southern limit of the siliceous-ooze deposits east of long 25°W corresponds with the mean positions of the ice edge for the months of January and April as given by Mackintosh (1972, Figs. 1, 2). As pointed out by Bunt (1968), the ice layer and associated epontic algae severely limit the available light for photosynthesis in the water column. The time of year of ice break-up and melting when the seas are ice free is of critical importance in controlling the total annual diatom production. South of this boundary the sea is covered with ice for more than eight months each year. Even though standing-crop values are highest at the Antarctic divergence and in neritic waters (El-Sayed, 1968, 1970), the annual total diatom production is drastically reduced by ice cover. As a result, south of a line marking the northern limit of eight months-old sea ice, the ocean floor is covered with siliceous silty clay. Besides distance to the source of terrigenous sediment, other important factors determining the southern boundary of siliceous diatom ooze may be the different diatom species produced in different latitudinal bands and related resistance to solution in the water column and preservation in the sediments. According to Koslova (1970), only those she included in the group of subantarctic diatoms are resistant to solution and are well preserved in bottom sediments. They do not live near the ice edge and display vegetative activity mostly in summer. They are distributed in waters of the West Wind Drift, with the polar front marking their northern limit. By contrast, the Antarctic diatom group in waters of the East Wind Drift south of the Antarctic divergence and the moderately warm-water diatoms that inhabit waters north of the Antarctic Polar Front are, for the most part, dissolved in the water column and therefore make a less important contribution to ocean-floor sediment.

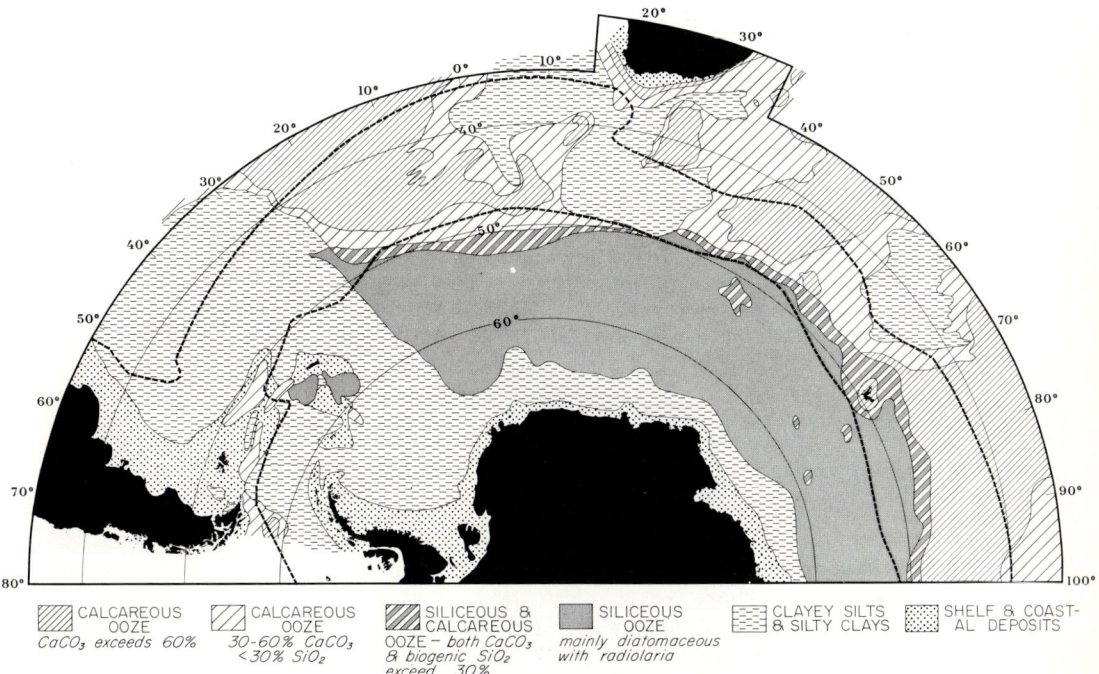

Figure 5. Distribution of sediment types (modified after Goodell, 1973, Fig.2, Pl. 2). Calcareous ooze is usually foraminiferal, sometimes calcilutaceous. Clayey silt and silty clay are calcareous adjacent to calcareous ooze, siliceous adjacent to siliceous ooze, and sandy adjacent to continental shelves.

TABLE 2. TAXONOMIC GROUPS CHOSEN AS REPRESENTATIVE
OF THE RADIOLARIAN FAUNA

Species name	Reference
Ommatodiscus sp.	Benson, 1966, Pl. 10, fig. 3 only
Theocorythium trachelium (Ehrenberg)	Nigrini, 1967, Pl. 8, figs. 1, 2
Lithelius minor (Jörgensen)	Benson, 1966, Pl. 18, figs. 1–4
Pterocanium praetextum (Ehrenberg)	Nigrini, 1967, Pl. 7, figs. 1, 2
Anthocyrtidium ophirense (Ehrenberg)	Nigrini, 1967, Pl. 6, fig. 3
Theoconus hertwigii (Haeckel)	Nigrini, 1967, Pl. 7, fig. 3
Lamprocyclas maritalis maritalis (Haeckel)	Nigrini, 1967, Pl. 7, fig. 5
Lamprocyclas maritalis Haeckel *polypora*	Nigrini, 1967, Pl. 7, fig. 6
Ommatartus tetrathalamus (Haeckel)	Nigrini, 1967, Pl. 2, figs. 4a–4d
Lithocampe sp.	Nigrini, 1967, Pl. 8, figs. 6a, 6b
Siphocampe sp.	Riedel, 1958, Pl. 4, figs. 9, 10
Tetrapyle octacantha Muller	Benson, 1966, Pl. 15, figs. 3–10; Pl. 16, fig. 1
Phorticium pylonium (Haeckel) Cleve	Benson, 1966, Pl. 16, figs. 5–9; Pl. 16, fig. 509; Pl. 17, figs. 1–3
Heliodiscus asteriscus Haeckel	Nigrini, 1967, Pl. 3, fig. 1
Eucyrtidium acuminatum (Ehrenberg)	Nigrini, 1967, Pl. 8, figs. 3a, 3b
Hexacontium cf. *heteracantha* (Popofsky)	Benson, 1966, Pl. 4, figs. 6, 7
Theocalyptra(?) *bicornis* (Popofsky)	Petrushevskaya, 1967, Fig. 71, p. 125
Spongotrochus glacialis Popofsky	Riedel, 1958, Pl. 2, figs. 1, 2 Includes *Spongodiscus*(?) *setosus* (Dreyer) Petrushevskaya, 1967, Fig. 20, p. 38
Antarctissa denticulata (Ehrenberg)	Petruskevskaya, 1967, Figs. 49, 50, p. 85, 87
Antarctissa strelkovi (Jörgensen)	Petruskevskaya, 1967, Fig. 51, *Helotholus histricosa* Jörgensen, Reidel, 1958, Pl. 3, fig. 8
Cycladophora davisiana Ehrenberg	Petruskevskaya, 1967, Fig. 69, p. 121
Triceraspyris(1) *antarctica* (Haeckel)	Petruskevskaya, 1967, Fig. 37, p. 63
Spongoplegma antarcticum Haeckel	Hays, 1965, Pl. 1, fig. 1, p. 166
Lithelius nautiloides Popofsky	Riedel, 1958, Pl. 2, figs. 3, 4
Spongurus pylomaticus Riedel	Riedel, 1958, Pl. 1, figs. 10, 11
Spongopyle osculosa Dreyer	Riedel, 1958, Pl. 1, fig. 12
Androcyclas gamphonycha Jörgensen	Hays, 1965, Pl. 3, fig. 2, p.174

The diatom ooze grades south into siliceous silty clay and sandy silt, and they in turn grade into the shelf and coastal deposits of Antarctica made by "glacial marine sediments" (Goodell, 1973) and submarine tills. The glacial marine sediments have 30% sand and larger grain sizes and are confined to the outer Antarctic continental shelf, slope, and rise. The shelf and coastal deposits of South America are silty sands that are often gravelly and calcareous.

RADIOLARIAN FAUNAL ANALYSIS[1]

Radiolaria retained on a 62-μm screen were prepared by the random-settling technique (Moore, 1973). All species were studied and counted in 30 trigger-weight core-top samples and spot samples from piston cores to determine their relative abundances. Those species that consistently composed less than 2% of the fauna were eliminated. Other species that may have represented more than 2% were also eliminated because of identification problems. From this preliminary survey,

---
[1]Faunal counts made on all samples are included in Appendix I on microfiche in pocket inside back cover.

27 taxonomic groups (Table 2) were chosen; these generally met the following requirements: (1) they could be identified with relative ease; (2) each of the 27 species has a preference for either Antarctic, subantarctic, or subtropical waters—as a group they represent the total range of sea surface conditions within the area; and (3) the 27 species accounted for a large percentage of the total fauna, ranging from more than 80% under polar water to 25% in the northernmost samples under subtropical water.

The following assumptions were made: (1) The chosen species have not evolved significantly within the last few hundred thousand years, and therefore the response of the species to oceanographic conditions has remained essentially constant through time. (2) The species either live in near-surface waters and therefore respond to change in surface-water conditions or the relationship between the nature of the water where they do live and the surface water has remained constant.

A minimum of 500 known specimens (that is, specimens belonging to any one of the 27 taxonomic groups) was counted in 145 trigger-weight core tops (Fig. 1, Table 1). The resulting paleontologic data were processed following the method of Imbrie and Kipp (1971).

Imbrie and Kipp (1971, p. 79) described their method in terms of five sequential procedures, which are explained in detail in their paper: (1) Paleontological core-top data are filtered to eliminate samples unrepresentative of Holocene conditions and rare (less than 2%) species. (2) These data are resolved into several assemblages by factor-analysis techniques. (3) A least-squares technique is used to write a set of paleoecologic equations relating the factor description of these assemblages to observed oceanographic parameters. (4) The paleontologic data for samples taken downcore are expressed in terms of the core-top assemblages. (5) The paleoecologic equations are used to estimate paleoenvironmental parameters.

The procedure employed to filter the data is different from that used by Imbrie and Kipp and is explained in the following section.

DATA ANALYSIS

Factor Analysis and Paleoecologic Equations

Because of strong bottom currents beneath Antarctic and subantarctic waters, the core tops are frequently not representative of Holocene conditions, since the most recent sediment has been removed or the contained fossil fauna is a mixture of several assemblages. Of the more than 200 cores studied in the area, we eliminated several dozen because sediments within the top metre were dated as pre-Ω zone (Hays, 1965). Another dozen were discarded because of their sparse radiolarian content. These were all in the northern reaches of the area. This left 145 cores that appeared to have Holocene core tops and sufficient Radiolaria for study.

An additional 36 samples located in the Argentine Basin west of long 30°W were deleted, because they were judged unreliable for climatic studies owing to the influence of strong bottom currents (Wüst, 1957; Le Pichon and others, 1971a, 1971b; Ewing and others, 1971). Most samples contained abundant rounded sand grains of larger than average size for Radiolaria; some of them also contained a few pre-Quaternary Radiolaria. This left 109 core-top samples located between long 80°E and 30°W (Fig. 1) that we judged suitable for processing by Imbrie and Kipp's (1971) techniques. The Q-mode factor analysis with varimax rotation resolved six assemblages (factors), which accounted for 96.8% of the variability of the data. Seasonal (March to September and October to February) oceanographic data published by Gordon and Goldberg (1970) were correlated with surface faunal distribution expressed as factors. A curvilinear fit produced a 27-term paleoecologic equation for each oceanographic parameter. Downcore radiolarian counts from four piston cores were used to test the reliability of the paleoecologic equations. Summer temperature estimates ranging from 15° to −10°C at the location of core RC8-39 indicated a major problem.

These clearly inaccurate results could be due to the following: (1) Past abundances of species are not represented in the Holocene sediment data base. (2) There are inaccuracies in the maps of sea-surface temperature used to estimate values at individual core sites. (3) Difficulties in Radiolaria identification have produced inconsistencies. (4) Certain radiolarian species have evolved different ecologic tolerances through time, or they do not live at the surface, and thus their past abundances reflect changes in deep rather than surface waters. (5) A significant number of core tops are either not Holocene or have been mixed.

To determine which of these possible errors affected our data, we studied each of them separately.

Past Species Abundances Not Reflected in Holocene Sediments. A fundamental prerequisite when paleoecologic equations are applied to older sediments is that no species have a past abundance that is greater than its abundance in any recent sediments. The percentage abundance of each of the 27 species was plotted down the four cores on which complete counts had been made at 20-cm intervals and then compared with the core-top data. *Cycladophora davisiana*, which consistently reaches levels of 10 to 30% at depth in the four cores, rarely reaches percentages greater than 5% in the core-top data, and the cores that have values of 5% at their tops are all in the Crozet Basin. For this reason, we eliminated it from the equation but continued to count it.

Temperature. Because of the paucity of oceanographic stations in our Antarctic Ocean sector, the temperature data used for the initial equations were subject to considerable error. Also, the most up-to-date map by Gordon and Goldberg

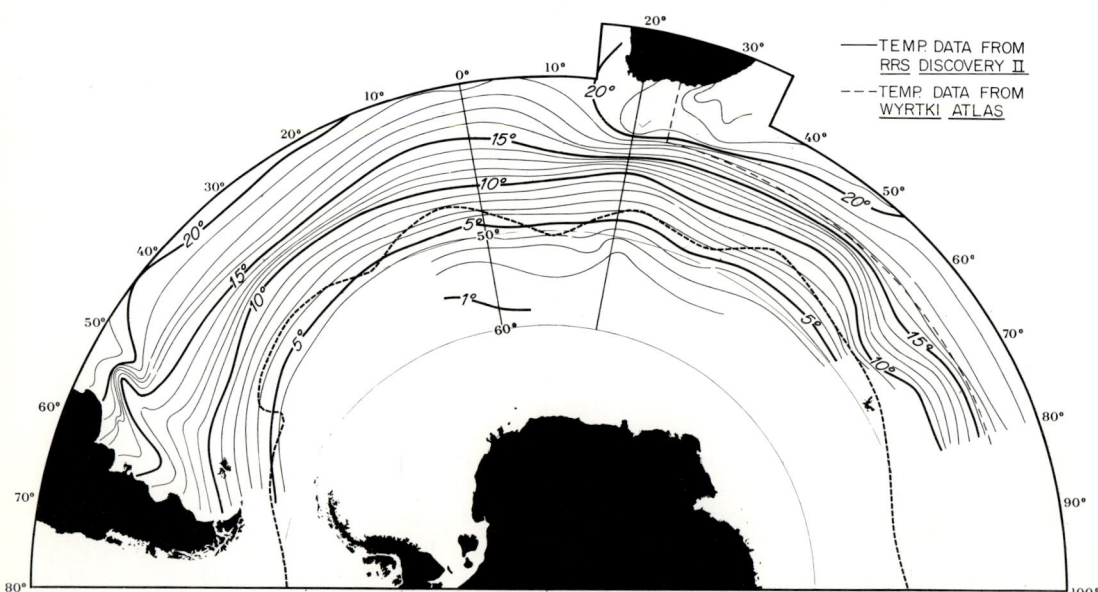

Figure 6. Mean February sea-surface temperature. Based on *World Atlas of Sea Surface Temperatures* (U.S. Navy Hydrographic Office, 1954), corrected according to data from Wyrtki (1971), Discovery Reports (1947), and NODC.

(1970) averages temperatures over two six-month intervals. The most detailed map covering the entire area and giving monthly averages of sea-surface temperature is the *World Atlas of Sea Surface Temperatures* (U.S. Navy Hydrographic Office, 1954). We corrected this map with more recent data from Wyrtki (1971), two *Discovery* traverses, and all available data from Naval Oceanographic Data Center (NODC) for August and February (Figs. 6, 7).

For our purposes, the best oceanographic data in the area are those gathered by RRS *Discovery* in traverses along long 0° and 20°E, which were repeated seven times between July 6, 1938 and March 18, 1939. Seven to 18 stations were reoccupied along each traverse, depending on ice conditions.

The *Discovery* temperature data were plotted in a plane of latitude versus day of the year similar to that used by Bowen and Stommel (1971) for the 2,000-m level. We contoured temperature values to obtain continuous temperature values for the 15th of August and February along long 0° and 20°E between lat 38°S and 65°S. Temperatures were then interpolated in the area between these two north-south lines. This provided an area within and immediately adjacent to which we have the best control of monthly sea-surface temperature (Fig. 8).

Inconsistencies in Radiolarian Identification. A third possible cause of large errors in our equation is inconsistency in identification of the 27 selected species or failure of these species to respond in a consistent way to surface water temperature changes. To test for these errors, 59 cores were selected in the vicinity of the *Discovery* lines, the area covered by Wyrtki's (1971) map, and a few cores that coincide or nearly coincide with oceanographic stations included in the NODC data (Fig. 8). The percentage of each species relative to the 27 species was plotted versus temperature for all 27 species at the 59 core sites. It was apparent from studying these plots that the species that were most clearly defined and contained the least morphologic variation showed the best correlation between relative abundance and temperature. Of the 27 taxonomic groups, six presented serious identification problems. *Phorticium pylonium* and *Tetrapyle octacantha* were elimi-

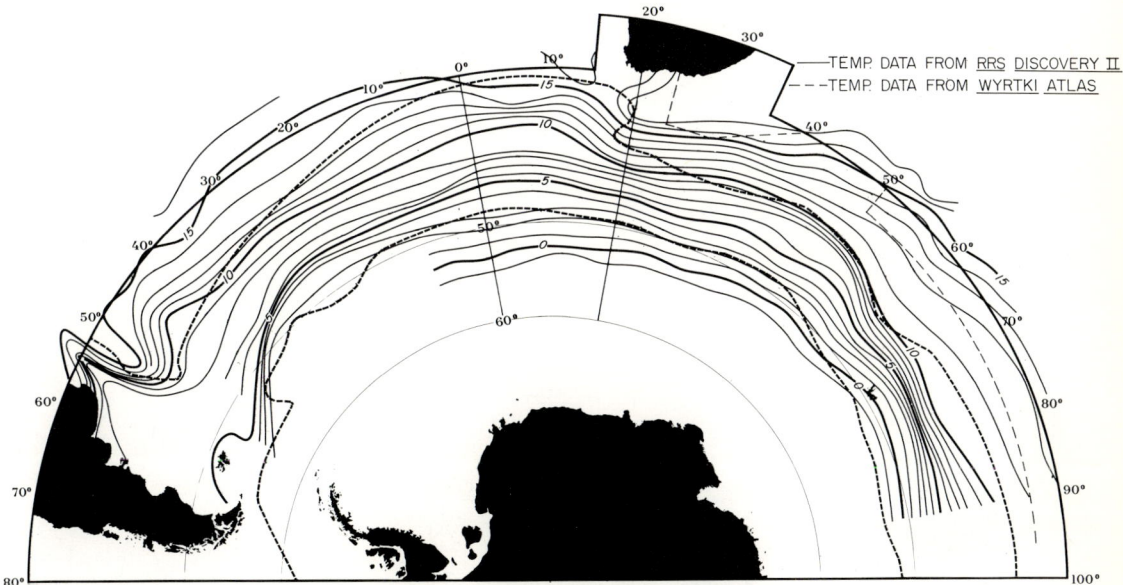

Figure 7. Mean August sea-surface temperature. Based on *World Atlas of Sea Surface Temperatures* (1954), corrected according to data from Wyrtki (1971), Discovery Reports (1947), and NODC.

nated because of their highly variable morphology and because they look very different, depending on their orientation on a slide. Together they represent a significant part of the total assemblage, but the high probability of their producing errors if they were included forced their elimination. Two additional taxonomic groups (*Siphocampium* sp. and *Hexacontium heteracantha*) were also eliminated because of identification problems. Three other pairs of taxonomic groups appear to intergrade: *Lamprocyclas maritalis polypora–Lamprocyclas maritalis maritalis*, *Antarctissa denticulata–Antarctissa strelkovi*, and *Spongotrochus glacialis–Spongopyle osculosa*. For this reason these taxonomic groups were combined under the headings *Lamprocyclas maritalis*, *Antarctissa strelkovi*, and *Spongotrochus glacialis*, respectively. While making counts we tried to separate them, but for the purpose of our equations they were combined (a detailed study of these forms is in preparation). *Theoconus hertwigii* was eliminated, because it was found in only 15 core-top samples and had a maximum relative abundance of only 0.8%.

The remaining 18 taxonomic groups all show a correlation between relative abundance and temperature. These 18 taxonomic groups account for more than 70% of the total fauna from Antarctic sediment and more than 20% from subtropical sediment. In six of the northernmost samples they account for just less than 20%. This northward decrease in the representation of our selected fauna reflects the disappearance of the Antarctic fauna and a large increase in diversity of subtropical fauna.

Evolutionary Change and Ecological Response. We have not observed any consistent morphologic change with time in any of the 18 species. The relative abundance of each one of the species was plotted against depth in cores RC11-120 and RC15-94. The consistency of the variations in relative abundance among the different species seems to confirm the validity of the assumption that the ecologic response of the species remained essentially constant during the time represented by the total length of these cores (more than 200,000 yr).

Test to Determine If Core Tops Reflect Recent Sea Surface Conditions. It was realized that many of the core-top samples might not be Holocene in age. The

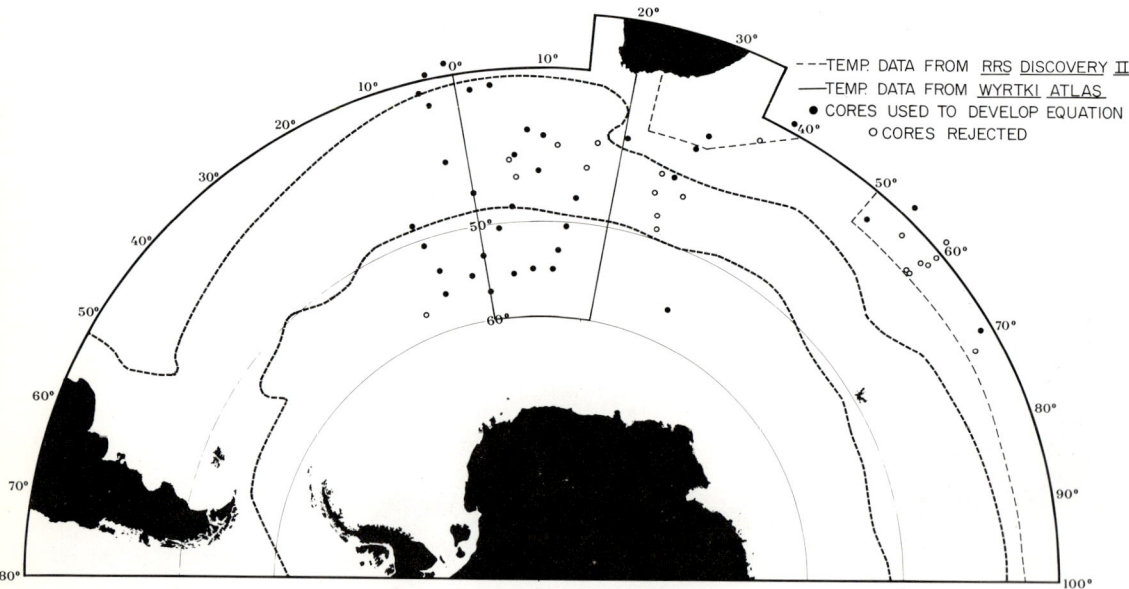

Figure 8. Location of 59 cores from which the 36 used to develop initial paleoecologic equations were chosen.

samples we initially eliminated for not being modern assemblages are older than 400,000 yr (the upper limit of *Stylatractus universus* Hays and the most recent datum we could identify routinely). Radiolarian transport by bottom currents was also expected to introduce errors.

Having chosen 50 cores with the best temperature control, we studied the graphs of percentage distribution against temperature for individual species. It was clear that some samples had percentages for one or several species, which were either too high or too low compared to their normal range of abundance relative to the other samples. A plot of *A. strelkovi* abundance versus summer sea-surface temperature gives an example of such a discrepancy (Fig. 9). From this and similar plots, it was apparent that the Crozet Basin samples have an anomalously cold fauna. The anomalously high values of *A. strelkovi*, an Antarctic species, in the northern Crozet Basin can be explained by northward transport of this species by bottom currents.

Based on a detailed analysis of these graphs, 36 surface-sediment samples were chosen showing the best correlation between sea-surface temperature and paleontologic data. The lack of correlation of other samples was interpreted as being caused

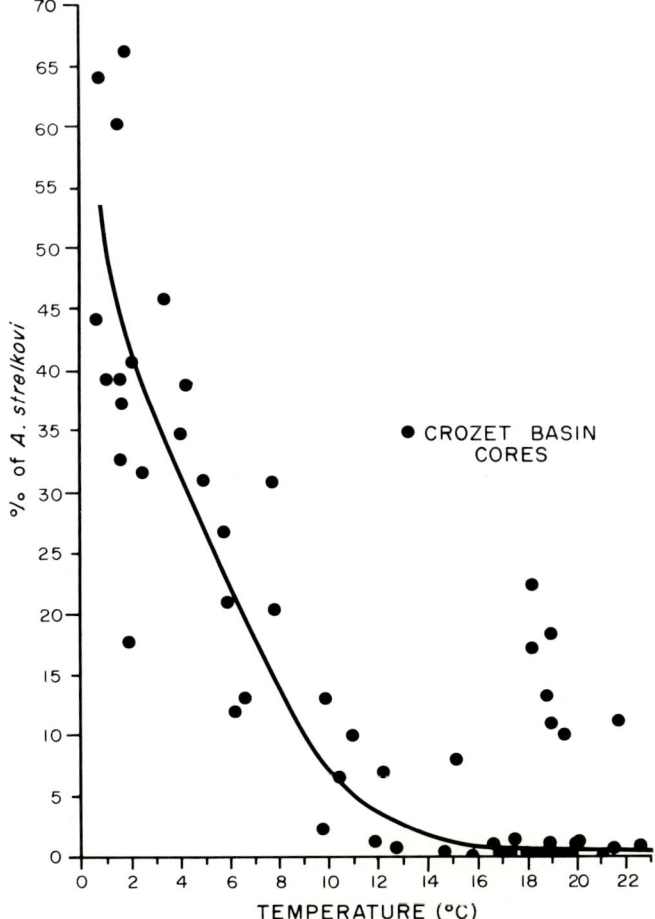

Figure 9. *A. strelkovi* **percentage of total number of identified Radiolaria versus February sea-surface temperature in 59 trigger-weight core tops. Figure 7 shows locations of samples.**

by bottom currents. Using only these 36 samples, we plotted percentage distributions of the two most abundant taxonomic groups against temperatures (Figs. 10, 11).

The faunal data for the 36 chosen samples were factor analyzed. Five factors account for 99.13% of the variance. Three factors with a cumulative variance of 97.18 and four factors with a variance of 98.39 were used to generate sets of linear and curvilinear correlation coefficients. Multiple correlation coefficients and standard errors of estimate for the equations are given in Table 3.

Using the linear equation with three factors, we obtained summer and winter (February and August) temperature estimates for 109 core locations. Using our February and August sea-surface-temperature maps, we were able to obtain residuals (differences between temperatures read from the map and the ones estimated using the formula). Seventy-four of these residuals have absolute values under 3°C, which is less than twice our standard error of estimate. Again, samples from the Crozet Basin have anomalously cold faunas. Residuals with positive values between 5.4° and 10.9°C were obtained for the 11 cores in this area.

Using the faunal data from only the 74 core tops with residuals under 3°C, we repeated the factor analysis, and the fauna was resolved into three assemblages (that is, factors). When the numerical values of the factors were contoured, nine samples did not fit into a coherent pattern and were eliminated, leaving 65 core top samples. Using three and four factors, linear and curvilinear equations were generated (see Table 4). Again, the linear equation with three factors was chosen.

Even though the results obtained with 65 cores were excellent (see Table 4), we made an effort to broaden the areal core-top coverage by using this equation on the paleontologic data for the original 145 surface-sediment samples, including

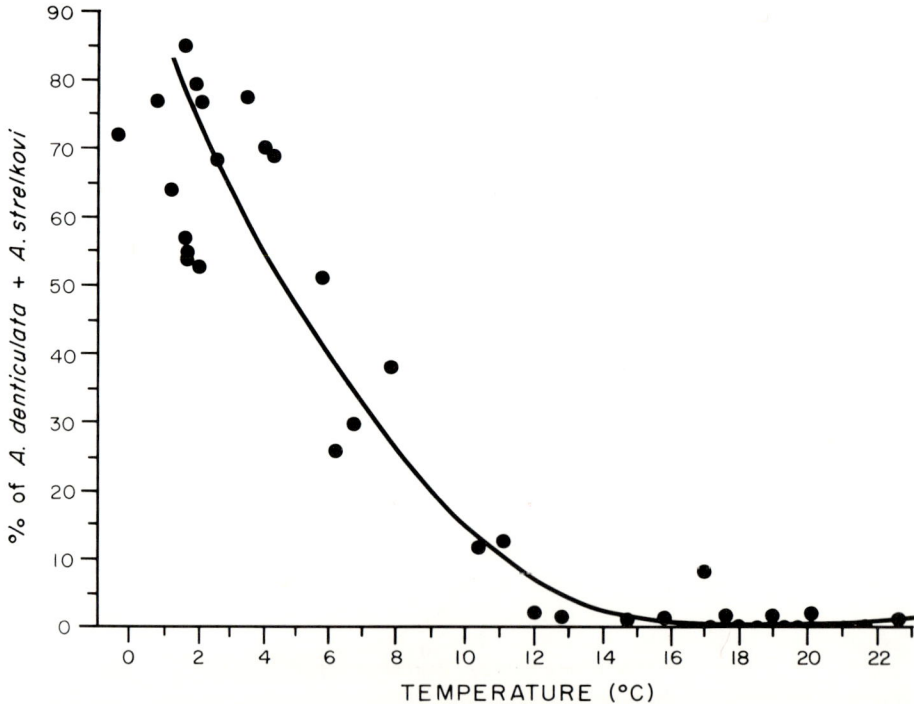

Figure 10. *A. denticulata* plus *A. strelkovi* percentage of total population of identified Radiolaria versus February sea-surface temperature in 36 trigger-weight cores used to develop initial paleoecologic equation. Figure 7 shows locations of samples.

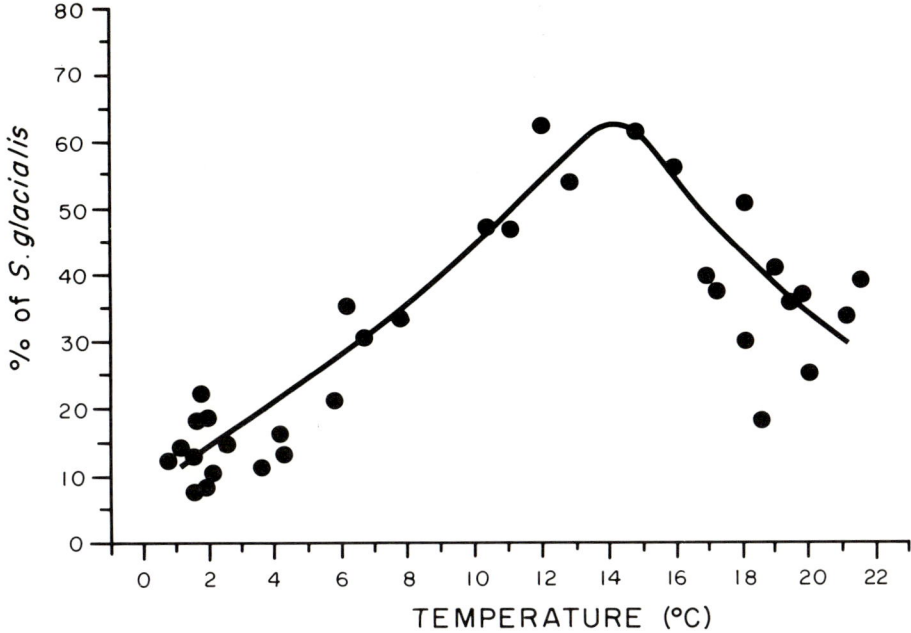

Figure 11. *S. glacialis* group percentage of total population of identified Radiolaria versus February sea-surface temperature in 36 trigger-weight cores used to develop initial paleoecologic equation. Figure 7 shows location of samples.

those from the Argentine Basin. A study of the results showed a concentration of high residuals (observed-estimated temperature) for those cores raised from depths of more than 5,000 m in the Argentine Basin. These residuals are of the same order and sign as those obtained for cores in the Crozet Basin (Figs. 12, 13). This finding supports our interpretation of the presence of cold faunas in the Crozet Basin due to the northward transport of Antarctic species by bottom currents.

Of the 36 core tops west of long 30°W, we were able to add only 9 cores located south and northeast of the Argentine Basin. The elimination of 2 core tops from the former 65 because of the presence of *C. davisiana* in excess of 5.5% of the total number of Radiolaria left us with a final total of 72 cores (Fig. 1, Table 1, and within the rectangle determined by lines at values of 0% and 5.5% *C. davisiana* and ±3°C residuals in Fig. 14). Figure 14 shows that cores in both the Crozet and Argentine Basins have high residual values, but the Crozet Basin cores contain consistently larger percentages of *C. davisiana*. This could be tentatively explained by different regimes of sedimentation existing in the two basins. The Argentine Basin is known to receive a large volume of sediment from the southern tip of South America or Antarctica (Biscaye, 1965; Le Pichon and others, 1971a, 1971b; Ewing and others, 1971). Thus, although erosion predominates at the axis of the bottom current, the Argentine Basin as a whole is a catchment for transported sediments. The influx of sediment into the Crozet Basin is probably less than that into the Argentine Basin. Bottom currents may scour the Crozet Basin more than the Argentine Basin and mix older sediments that have a high *C. davisiana* content with younger sediments (Hays and others, this volume).

In contrast with the Argentine Basin, the Atlantic Indian Basin samples have low residuals but high percentages of *C. davisiana*. As in the Crozet Basin, the

TABLE 3. MAIN PARAMETERS OF THE REGRESSION BASED ON 32 SAMPLES

	Linear 3 factor		Curvilinear 3 factor		Linear 4 factor		Curvilinear 4 factor	
	T_w	T_s	T_w	T_s	T_w	T_s	T_w	T_s
Multiple correlation coefficient	0.963	0.968	0.979	0.985	0.965	0.970	0.983	0.990
(Adjusted for degrees of freedom)	0.960	0.967	0.973	0.980	0.961	0.967	0.973	0.985
F value for analysis of variance	135.110	161.202	67.383	91.410	103.748	124.580	43.806	77.401
Standard error of estimate	1.895	1.989	1.576	1.550	1.877	1.963	1.574	1.359
(Adjusted for degrees of freedom)	1.952	2.048	1.794	1.765	1.963	2.053	1.985	1.714

Note: T_w, winter temperature (August); T_s, summer temperature (February).

scattered high values for *C. davisiana* seem to be better explained for the cores in the Atlantic Indian Basin by erosion of the most recent sediments or reworking and mixing with older sediments. The low residuals are probably due to the low temperature values in this area and corresponding small change in the assemblages through time.

Analysis of the 72 Samples

The faunal data for the 72 samples, expressed as percentages of 18 taxonomic groups, was factor analyzed. Three factors accounting for cumulative variances of 59.9, 95.0 and 97.8 were resolved.

The communalities are all larger than 0.9, except for two samples with values of 0.83 and 0.88. Fifty-one samples have values of 0.98, indicating that the three factors account for practically all the information used.

The statistics for the 18 taxonomic groups in the 72 samples are presented in Table 5. The relative importance of each taxonomic group in each of the three factors is given in Table 6.

Figures 15, 16, and 17 are contour maps of factor loadings for the 72 samples. Factor 1 is composed primarily of the *S. glacialis* group, with significant contributions from *Ommatodiscus* sp. and *Lithelius minor*. Values for factor 1 are highest in subtropical waters, but this factor is still dominant (with factor loadings of 0.9) as much as 7° of latitude south of the mean position of the subtropical convergence in the Atlantic Ocean. The importance of factor 1 decreases rapidly to the south. The contour of the 0.2 factor loading corresponds roughly to the Antarctic Polar Front. Factor 2 is dominated by the *Antarctissa denticulata* and *A. strelkovi* group. It is the dominant factor in Antarctic waters. North of the Antarctic Polar Front the loadings for factor 2 decrease as the values for factor 1 increase.

S. glacialis, Ommatodiscus sp., *Theocorythium trachelium, Lithelius minor*, and *Lamprocyclas maritalis* are the main constituents of factor 3. *S. glacialis* has a distribution pattern that is inverse to the patterns of other elements in the assemblage, as indicated by its negative sign in the varimax factor score matrix (Table 6; Imbrie and Kipp, 1971, p. 86). Factor 3 does not dominate at any location but attains maximum negative values under subantarctic waters. At a few locations north of the subtropical convergence, this factor has positive values of more than 0.20, corresponding to a relatively low percentage of *S. glacialis*, which is replaced in those samples by the other components in the factor.

TABLE 4. MAIN PARAMETERS OF THE REGRESSION
BASED ON 65 SAMPLES

	Linear 3 factor		Curvilinear 3 factor		Linear 4 factor		Curvilinear 4 factor	
	T_w	T_s	T_w	T_s	T_w	T_s	T_w	T_s
Multiple correlation coefficient	0.969	0.977	0.976	0.983	0.969	0.978	0.976	0.985
(Adjusted for degrees of freedom)	0.968	0.977	0.972	0.981	0.968	0.977	0.970	0.982
F value for analysis of variance	317.390	435.292	120.186	179.889	234.357	331.056	73.003	119.723
Standard error of estimate	1.429	1.455	1.350	1.315	1.441	1.446	1.390	1.295
(Adjusted for degrees of freedom)	1.452	1.479	1.443	1.406	1.476	1.481	1.557	1.451

Note: T_w, winter temperature (August); T_s, summer temperature (February).

The distribution of the factors agrees well with the distribution of surface water masses. The close correlation of the factors with temperature is clearly shown in Figure 18, which demonstrates that the assemblage distributions approximate the parabolic model of Imbrie and Kipp (1971). This figure also indicates that reliable temperature estimates for February are only possible when temperatures are above 3°C. This is also shown in Figure 19, a plot of February sea-surface temperatures obtained from the map versus those estimated using our final linear paleoecologic equation based on 72 core-top samples and three factors. In this figure, we also give the values for the multiple correlation coefficient and standard error of estimate for the paleoecologic equations. The cumulative proportion reduced is 0.932 for the winter (August) and 0.905 for the summer (February) surface-temperature regressions. The error of estimate for both equations is less than 10% of the observed temperature range, which is 17.7°C (−1.7° to 16.0°C) for August and 20.5°C (0.8° to 21.3°C) for February.

The equations for estimating sea-surface temperatures for August (T_w) and for February (T_s) in degrees Celsius are

$$T_w = 15.266A + 1.543B + 9.984C - 1.984$$

$$T_s = 13.966A - 2.904B + 10.545C + 5.103,$$

where A, B, and C stand for the loadings of factor 1 (subtropical assemblage), factor 2 (Antarctic assemblage), and factor 3 (subantarctic assemblage), respectively.

The procedure we used for choosing the 72 core tops provided a continuous check for the equations within the study area. When used on downcore radiolarian counts from four piston cores, the results were consistent and logical.

To check equation reliability further, August and February temperatures were estimated for seven Eltanin cores located outside the area of study, using radiolarian fauna from the trigger-weight top samples. The location of the cores and results are given in Table 7. The temperature data were obtained from the World Atlas of Sea Surface Temperatures (U.S. Navy Hydrographic Office, 1954). Only one of the fourteen residuals is larger than 3°C, about twice the standard error of estimate of the regression.

Three samples were processed twice in order to provide us with a check on the reproducibility of our data. The two sets were counted one month apart. The results presented in Table 8 are remarkably good.

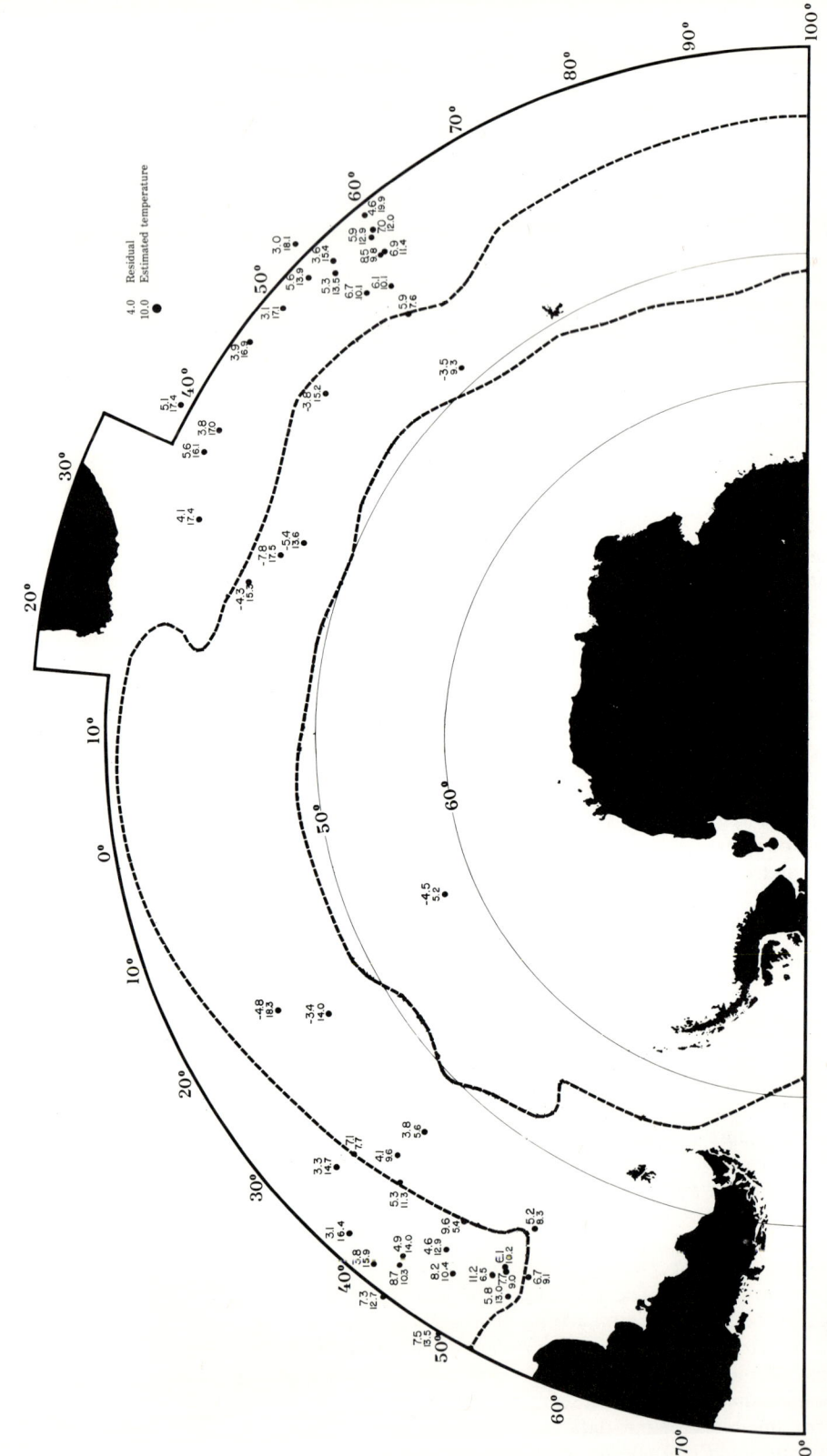

Figure 12. Locations of samples having February residuals greater than or equal to 3°C. Residuals are the differences between estimated values using the final paleoecologic equation and temperatures read from Figure 6. High positive values are concentrated in the Argentine and Crozet Basins.

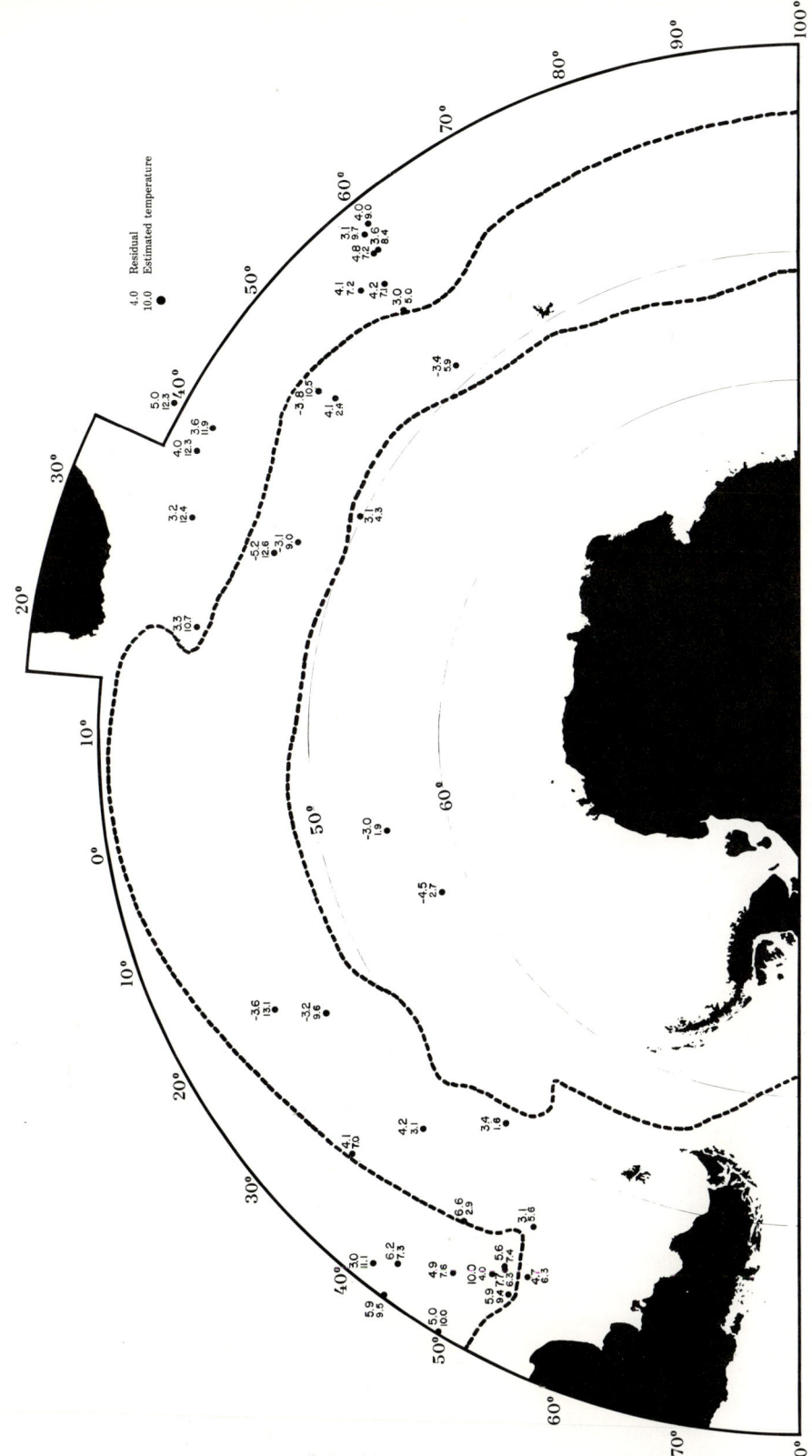

Figure 13. Locations of samples having August residuals greater than or equal to 3°C. Residuals are the differences between estimated values using the final paleoecologic equation and temperatures read from Figure 7. High positive values are concentrated in the Argentine and Crozet Basins.

TABLE 5. GENERAL STATISTICS FOR RAW DATA

Variable no.	Mean	Standard deviation	Minimum value	Maximum value
1	8.7488	7.5000	0.0	25.5578
2	3.3097	4.3195	0.0	25.8009
3	8.3377	6.8244	0.0	25.2772
4	0.3602	0.5394	0.0	2.4528
5	0.1013	0.4406	0.0	2.6316
6	1.1046	2.0614	0.0	8.1081
7	0.3128	0.5506	0.0	2.0089
8	0.5795	0.7960	0.0	3.3803
9	1.3181	1.3527	0.0	5.3191
10	3.2165	3.9700	0.0	23.0263
11	1.5154	1.7687	0.0	6.8349
12	2.2494	3.9520	0.0	21.8818
13	0.3175	0.4394	0.0	1.6103
14	1.7463	1.8120	0.0	8.0515
15	1.0371	1.4235	0.0	6.9971
16	2.3060	3.2794	0.0001	15.2710
17	39.6808	16.0197	7.7520	69.7531
18	23.6680	28.0575	0.0001	86.8218

Note: For species names, see Table 6.

CALCIUM CARBONATE

One hundred and forty-five trigger-weight core tops raised from depths ranging from 1,600 to 5,970 m (see Fig. 1 for location of cores) were analyzed using the Hülsemann (1966) gasometric technique. The generalized areal distribution of calcium carbonate is shown in Figure 5 and was briefly discussed under "Sediment Distribution."

Lisitzin (1971, Table 11.2) gives approximate average values of 5,000 m for the CCD between lat 40°S and 70°S in the southern Indian and South Atlantic Oceans. To study the relation between calcium carbonate content of bottom sediment and water depth in different areas, the percentage of calcium carbonate content

TABLE 6. VARIMAX FACTOR SCORE MATRIX

Variable no.	Species name	Factor 1	Factor 2	Factor 3
1	*Ommatodiscus* sp.	0.323	−0.024	0.666
2	*T. trachelium*	0.139	−0.015	0.359
3	*L. minor*	0.283	−0.031	0.356
4	*P. praetextum*	0.012	−0.002	0.011
5	*A. ophirense*	0.008	−0.001	0.012
6	*O. tetrathalamus*	0.042	−0.008	0.058
7	*Lithocampe* sp.	0.011	−0.002	0.010
8	*H. asteriscus*	0.020	−0.004	0.013
9	*E. acuminatum*	0.045	−0.007	0.044
10	*T. bicornis*	0.069	0.041	0.058
11	*T. antarctica*	−0.003	0.054	−0.040
12	*S. antarcticum*	−0.013	0.093	−0.046
13	*L. nautiloides*	−0.001	0.012	−0.008
14	*S. pylomaticus*	0.003	0.052	−0.038
15	*A. gamphonycha*	0.028	−0.004	−0.022
16	*L. maritalis*	0.100	−0.012	0.273
17	*S. glacialis*	0.870	0.178	−0.443
18	*A. strelkovi*	−0.139	0.975	0.125

is plotted against depth in Figure 20. In this figure the 145 cores are from three different regimes: (1) cores taken from Antarctic sediment (south of the Antarctic Polar Front); (2) cores taken west of the crest of the Mid-Atlantic Ridge and north of the Antarctic Polar Front; (3) cores taken east of the crest of the Mid-Atlantic Ridge (roughly long 10°W) and north of the Antarctic Polar Front. These three regimes have distinctive $CaCO_3$ values relative to depth. South of the Antarctic Polar Front no cores have significant (less than 5%) carbonate values in water more than 3,700 m deep. Even in the shallowest polar-water core (1,985 m) the $CaCO_3$ value is only 26%.

Under subtropical and subantarctic waters, two regimes can be recognized, with the boundary between them corresponding to the Mid-Atlantic Ridge. The CCD of the western regime is significantly shallower (4,900 m) than that of the eastern regime (5,200 m); this is similar to the findings of Ellis and Moore (1973). In order to examine the details of these two regimes on either flank of the Mid-Atlantic Ridge, Figure 21 was prepared. This diagram includes cores taken on either flank of the ridge and excludes cores taken in the Indian Ocean (east of long 25°E). It shows that the dissolution regimes vary between the western and eastern ridge flanks, with the CCD significantly deeper on the eastern flank.

Interpretation

Several factors control the calcium carbonate concentration in modern deep-sea sediments (for detailed examination of the factors, see Correns, 1939; Broecker, 1971; and Berger, 1971). There are three primary factors that control the $CaCO_3$ values measured in surficial Holocene deep-sea sediment: (1) rate of fixation of $CaCO_3$ in surface waters, (2) dilution by noncalcareous components, either biogenic silica or terrigenous minerals, and (3) dissolution at or near the sea floor. In the three regimes we have examined, all three factors play a role, but their importance differs from one region to another.

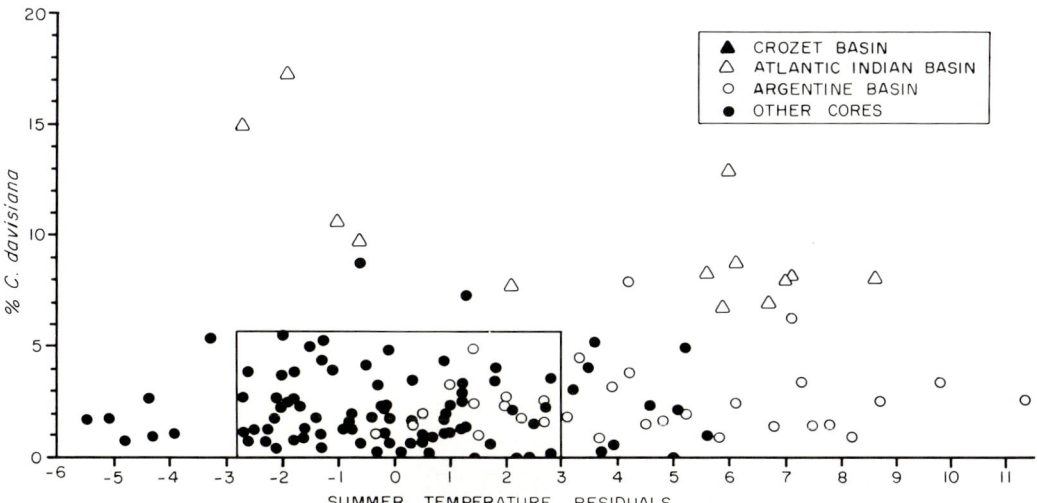

Figure 14. *C. davisiana* percentage of total Radiolaria versus summer (February) temperature residuals for 145 core tops. Residuals are obtained by subtracting the estimated value using the final paleoecologic equation from the temperature read from Figure 6. All 72 samples used in obtaining the paleoecologic equations fall within the rectangle.

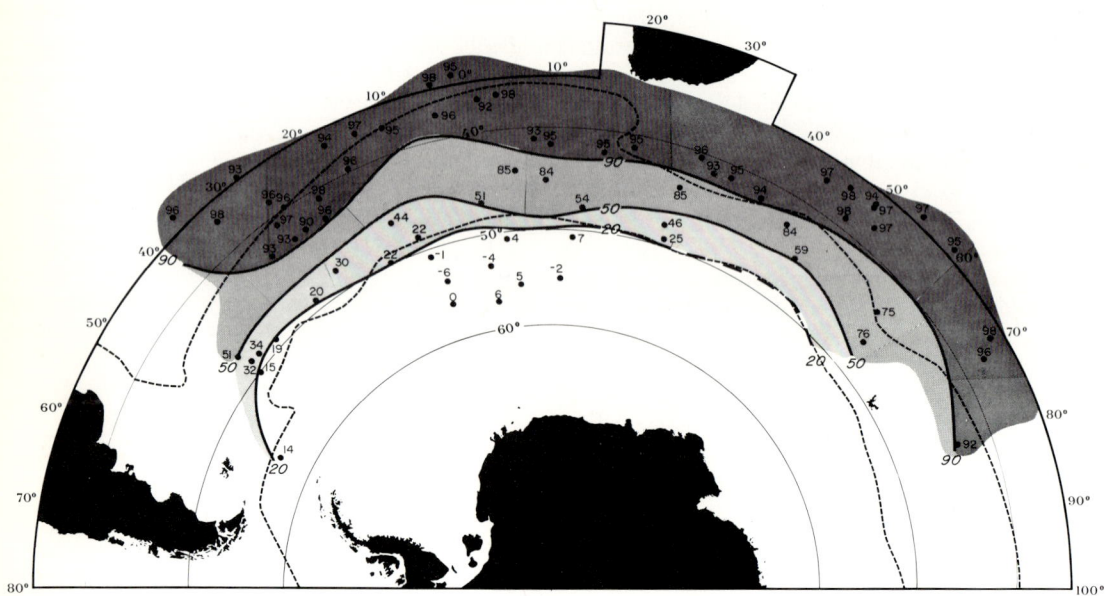

Figure 15. Distribution of factor 1 (subtropical assemblage) in surface sediments expressed as factor loadings ×100. The factor loading gives a measure of the importance of the factor at each station (Imbrie and Kipp, 1971).

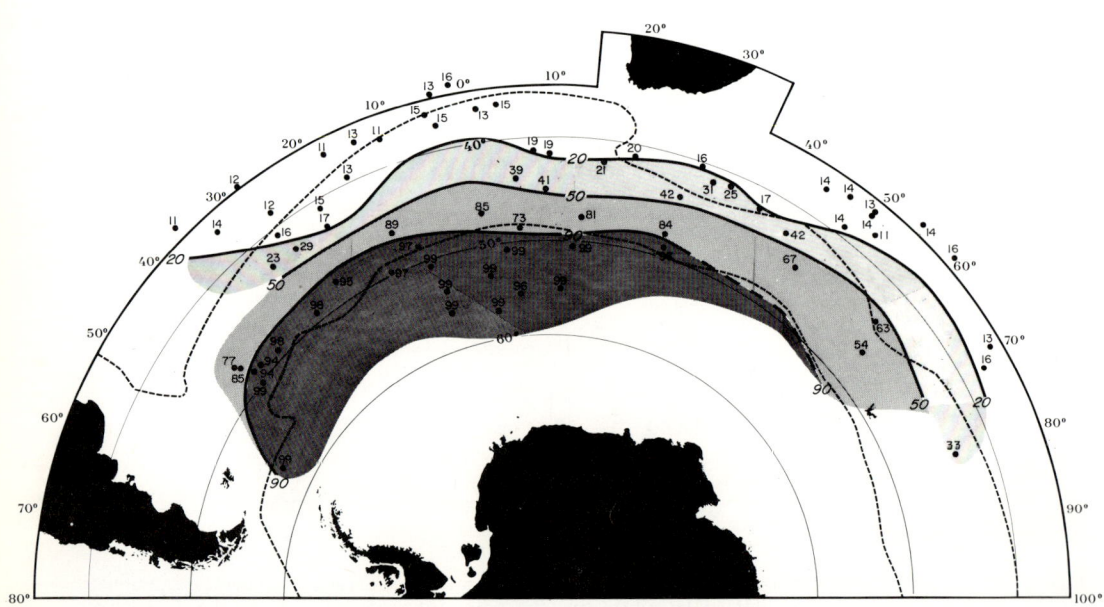

Figure 16. Distribution of factor 2 (Antarctic assemblage) in surface sediments expressed as factor loadings ×100. The factor loading gives a measure of the importance of the factor at each station (Imbrie and Kipp, 1971).

In Antarctic waters, calcite fixation is limited to one dominant species of Foraminifera, *Globigerina pachyderma*, although a few others (for example, *G. bulloides*) occur in minor amounts (Bé, 1969). No coccoliths live in polar waters (McIntyre and Bé, 1967). Biogenic silica is fixed in large amounts by diatoms and Radiolaria. Consequently, the CCD in Antarctic water is relatively shallow, only about 3,700 m. Belyaeva (1970) estimated that the CCD in the Antarctic Ocean is between 3,000 and 3,200 m. Kennett (1968) showed that the CCD in the Ross Sea is about 550 m.

In subantarctic and subtropical waters the CCD is significantly deeper than in polar water. $CaCO_3$ production in surface water is higher, with abundant coccoliths and various species of Foraminifera and significantly fewer diatoms and Radiolaria. Sverdrup and others (1942) and Bé and Tolderlund (1971) have shown that the total biomass of plankton in the upper 50 m is higher in water overlying the eastern flank of the Mid-Atlantic Ridge than over the western flank. If productivity is higher over the eastern flank, this would tend to depress the CCD and raise the lysocline (Berger, 1971). This suggests that the cause of the difference in CCD depth between the east and west flanks of the Mid-Atlantic Ridge is due to different bottom-water characteristics, as suggested by Ellis and Moore (1973).

Gordon and Goldberg (1970) illustrated a number of north-south profiles from Antarctica across the Antarctic Polar Front. Two of these in the Atlantic on the eastern and western flanks of the Mid-Atlantic Ridge show different thermal structures. Because of the blocking action of topographic features, the bottom waters are significantly warmer east of the Mid-Atlantic Ridge, where 0°C water does not penetrate north of the position of the Antarctic Polar Front (lat 50°S). In the western basin, however, 0°C water extends well beyond the position of the polar front. The characteristics associated with these colder bottom waters are probably the cause of the shallower CCD on the western flank of the Mid-Atlantic Ridge. A profile by Gordon and Goldberg (1970) in the western Indian Ocean

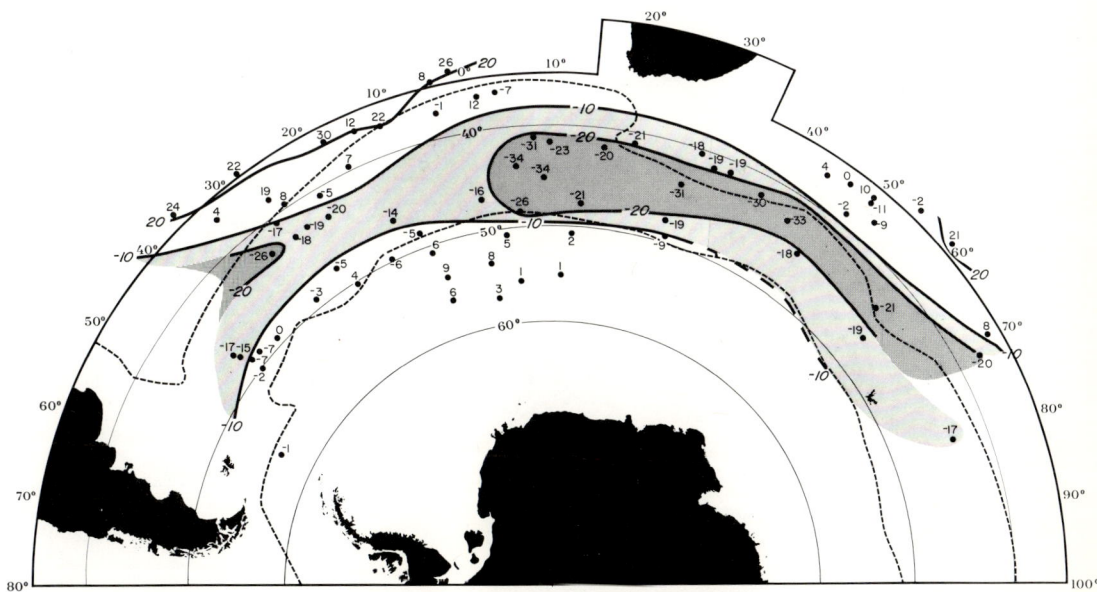

Figure 17. Distribution of factor 3 (subantarctic assemblage) in surface sediments expressed as factor loadings ×100. The factor loading gives a measure of the importance of the factor at each station (Imbrie and Kipp, 1971).

shows that at least in the location of the profile, bottom temperatures north of about lat 47°S are greater than 1°C. This may explain why the CCD in the western Indian Ocean sector of the subantarctic may be at least locally as deep as 5,500 m. Since all cores with high carbonate values raised from depths greater than 5,000 m occur in deep areas of the Mid-Atlantic Ridge or Crozet Basin, they were probably taken from deep fracture zones isolated from the main northward flow of Antarctic Bottom Water.

RADIOLARIA PER GRAM

When counting radiolarian species for temperature estimates, we also counted the total number of Radiolaria in the same area of the slide. Because the radiolarian slides were prepared using the technique described by Moore (1973), we were able to obtain the number of Radiolaria per gram on a $CaCO_3$-free basis. The area (A_c) on which the Radiolaria (T_r) were counted was obtained by multiplying the diameter of the microscope field by the length of the slide cover; this value was then multiplied by the number of traverses counted. The area of the beaker (A_b) in which the Radiolaria were allowed to settle on the slide was kept constant by always using the same type of 600-cc beakers. The bulk dry weight of the sample was corrected by using the $CaCO_3$ percentage values obtained using the Hülsemann (1966) gasometric technique to get the $CaCO_3$-free sample weight (W_c).

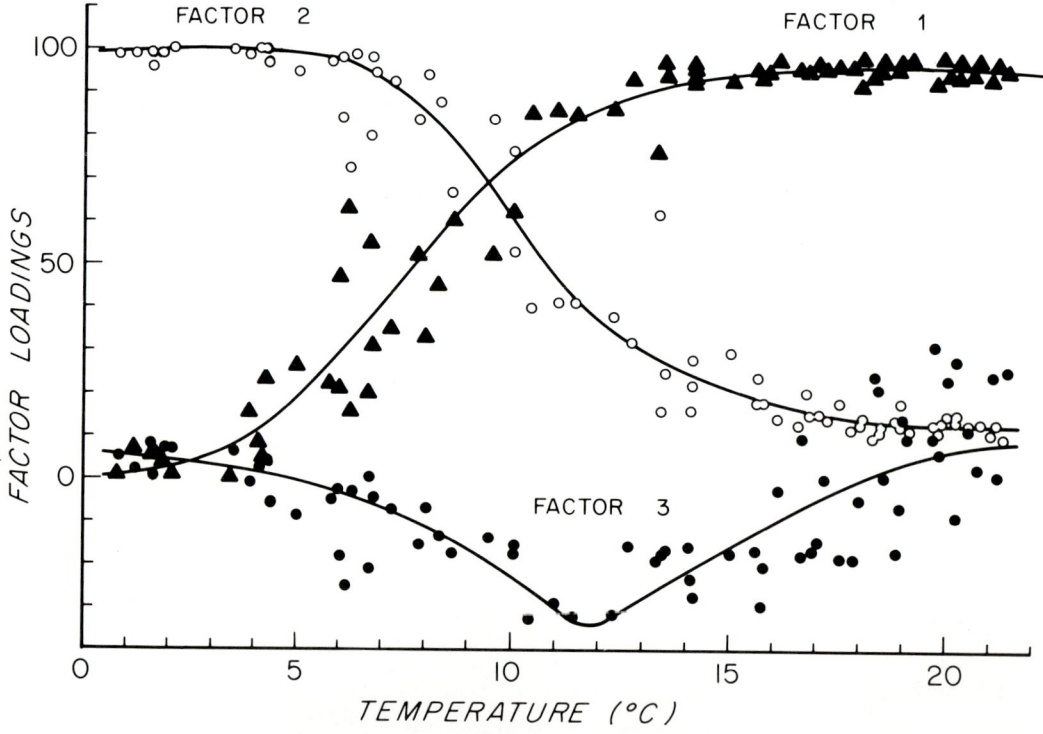

Figure 18. Dominance of assemblages versus February sea-surface temperature for the 72 samples used to obtain paleoecologic equations. Dominance is expressed as factor loadings ×100. Factor 1, subtropical assemblage; factor 2, Antarctic assemblage; factor 3, subantarctic assemblage.

Using the formula Radiolaria/gram = $(T_r \times A_b)/(A_c \times W_c)$, we obtained the number of Radiolaria retained on a 62-μm screen per gram of sediment on a $CaCO_3$-free basis.

Our values for total number of Radiolaria per gram of sediment are certainly lower than actual values in the sediment, because we lose an undetermined number of fragments and whole tests with a diameter smaller than 62 μm. Of the broken fragments still on the slides, we counted those that we judged to constitute 50% or more of the original shell.

The same samples that were used to check the reproducibility of paleotemperature estimates were used to check the reproducibility of our values of total Radiolaria per gram of sediment. The results are presented in Table 9, in which weight of

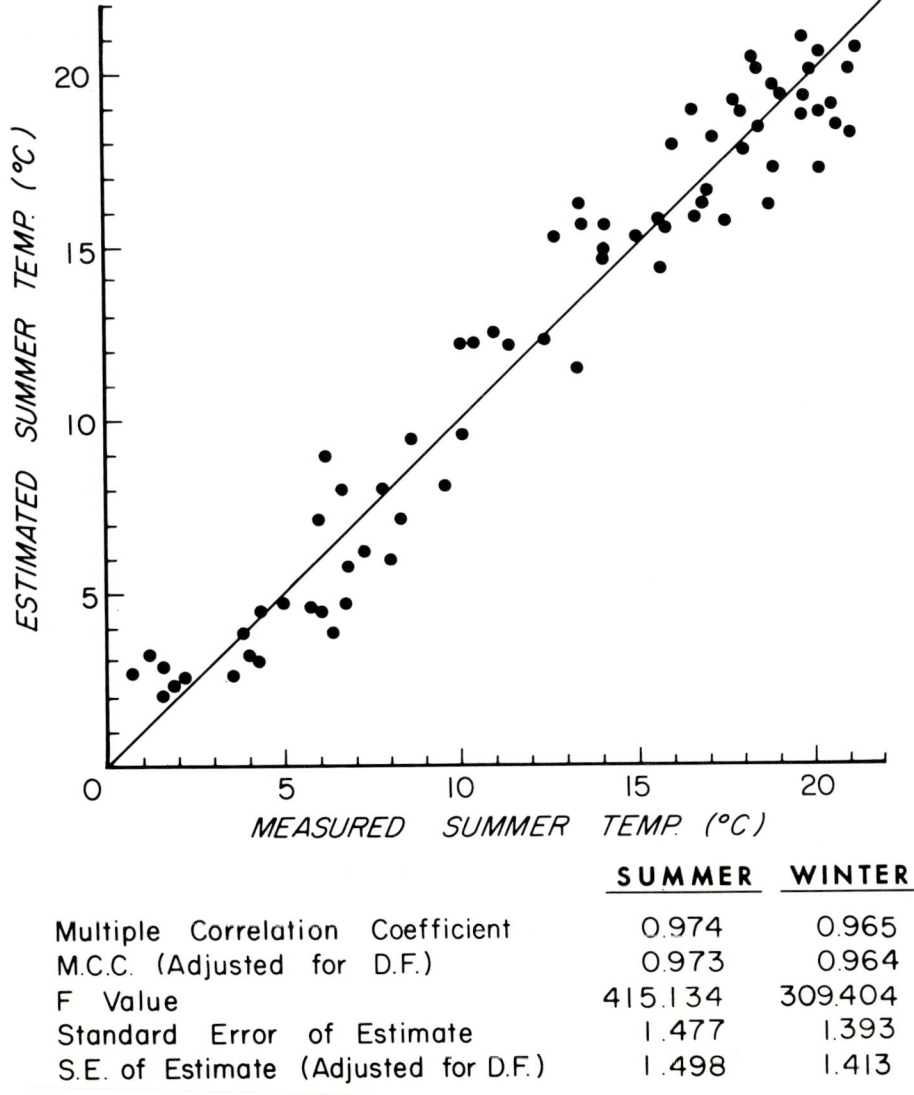

	SUMMER	WINTER
Multiple Correlation Coefficient	0.974	0.965
M.C.C. (Adjusted for D.F.)	0.973	0.964
F Value	415.134	309.404
Standard Error of Estimate	1.477	1.393
S.E. of Estimate (Adjusted for D.F.)	1.498	1.413

Figure 19. February sea-surface temperature in degrees Celsius obtained from Figure 6 versus estimated temperatures of 72 trigger-weight core-top samples.

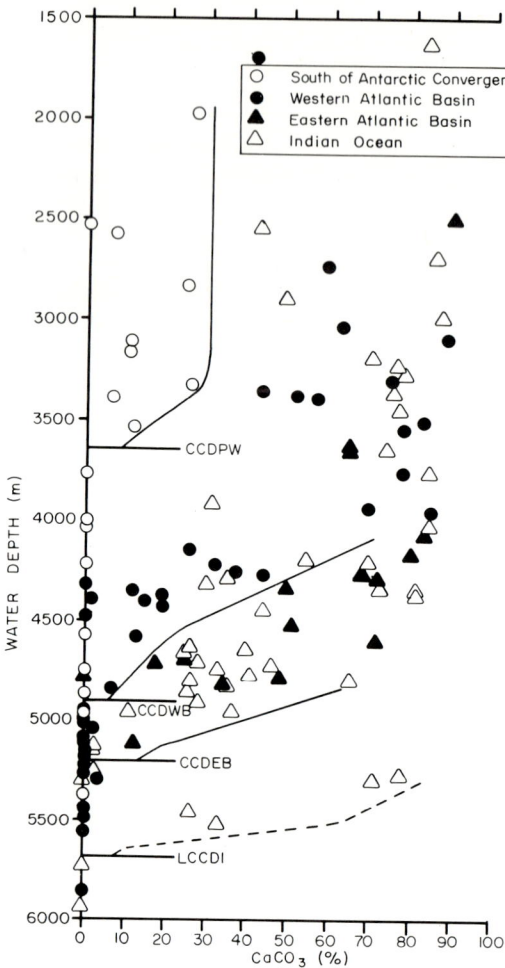

Figure 20. Calcium carbonate percentage of total sediment versus depth. CCDPW, calcium carbonate compensation depth under polar (Antarctic) waters; CCDWB, CCD for western Atlantic basin; CCDEB, CCD for eastern Atlantic basin; LCCDI, local CCD in the Indian Ocean.

the sediment is considered the total bulk weight. The table shows that our data are reproducible within 20%.

Goll and Bjørklund (1974) studied the distribution of Radiolaria in South Atlantic sediments, but they used a different sieve size (44 μm) and different counting techniques, and their values were not presented on a carbonate-free basis. Likewise, the data gathered by Petrushevskaya (1967) were obtained using different techniques, yet the general pattern of radiolarian abundance that we obtained is similar to that of these previous workers (Fig. 22). Although the numbers vary significantly between nearby cores, a definite pattern is present. A zone 5° to 10° of latitude wide between long 25°W and 20°E and with values higher than 50,000 is approximately bisected by the Antarctic Polar Front. In the Indian Ocean this zone narrows, except at long 60°E, where it is about 10° of latitude wide and lies parallel to and just north of the Antarctic Polar Front. Maximum values of over 100,000 are only present in a narrow belt between long 13°W and 27°E , which coincides with the polar front. This pattern is most easily explained by a gradual reduction in Radiolaria productivity from the Antarctic Polar Front northward. The decrease in values of Radiolaria per gram south of the polar front is probably due to an

TABLE 7. OBSERVED AND ESTIMATED TEMPERATURES AND RESIDUALS FOR SEVERAL *ELTANIN* CORES LOCATED OUTSIDE AREA OF STUDY

Core no.	Lat (S)	Long (E)	August temperature (°C)		August residual	February temperature (°C)		February residual
			Map	Estimated		Map	Estimated	
E35-7	49°58'	128°04'	4.7	6.5	−1.8	7.4	9.2	−1.8
E36-14	58°05'	150°14'	−0.8	0.4	−1.2	3.1	3.0	0.1
E39-13	45°00'	125°59'	8.6	10.8	−2.2	11.7	15.0	−3.3
E39-18	48°00'	126°05'	6.6	4.7	1.9	9.0	7.2	1.8
E39-21	48°51'	126°01'	5.8	7.2	−1.4	8.6	10.1	−1.5
E48-27	38°32'	79°54'	12.2	14.0	−1.8	16.6	18.9	−2.3
E50-17	63°00'	120°03'	−1.5	0.5	−2.0	1.4	3.1	−1.7

increase in the number of diatoms. The highest values of Radiolaria per gram generally occur under subantarctic water.

West of about lat 40°W lower values from subantarctic waters are probably due to an increase in clay and silt in Argentine Basin bottom sediment. The effects of dilution and changes of density could be eliminated if the rates of accumulation and the bulk densities of Holocene sediments were known, so that the values could be expressed in terms of Radiolaria/cm^2 × 1,000 yr. This problem is discussed by Hays and others (this volume).

CONCLUSIONS

Despite the increase in Antarctic Ocean exploration during the International Geophysical Year and the steady work carried out since then by a reduced but significant group of institutions and scientists, many areas around Antarctica remain unexplored. Very little is known about winter oceanographic conditions, owing to difficulties of exploration when the sea is covered by ice. Many of the processes taking place in the southern oceans are little understood. Nevertheless, from this study we conclude the following.

1. The composition of the radiolarian fauna at any location can be expressed in terms of the relative abundance of three assemblages (factors). The areal distribution of these assemblages correlates with the areal distribution of three main water masses.

The highest values for factor 1 come from under subtropical water. The isopleth

TABLE 8. TEMPERATURE ESTIMATES FOR DUPLICATE SAMPLES USED TO CHECK REPRODUCIBILITY OF DATA

Core no.	Sample depth (cm)	Factor 1	Factor 2	Factor 3	T_w	T_s
RC13-254	100	0.0102	0.9944	0.0542	0.25	2.93
RC13-254 A*	100	0.0166	0.9944	0.0470	0.27	2.94
V29-84	80	0.7271	0.6166	−0.2905	7.2	10.4
V29-84 A*	80	0.8206	0.4496	−0.3412	7.8	11.6
RC13-228	TW_{top}	0.9143	0.1272	−0.0708	11.5	16.8
RC13-228 A*	TW_{top}	0.9298	0.1344	−0.0783	11.6	16.9

Note: Factor 1, subtropical assemblage; factor 2, Antarctic assemblage; factor 3, subantarctic assemblage; T_w, winter temperature (August); T_s, summer temperature (February). TW_{top}, trigger-weight core top.

*Duplicate sample.

that corresponds to a factor loading of 0.9 for factor 1 also corresponds to the average position of the subtropical convergence in the Indian Ocean, but it is about 7° of latitude south of the subtropical convergence at lat 10°E and intersects it at about lat 38°W. The isopleth corresponding to a 0.2 factor loading correlates well with the Antarctic Polar Front.

Factor 2 reaches its maximum expression under Antarctic surface water. The isopleth for a factor loading of 0.9 corresponds to the Antarctic Polar Front, and the 0.2-factor loading contour for factor 2 nearly coincides with the corresponding contour of 0.9 for factor 1.

Factor 3 is never dominant, but it attains the highest negative factor loadings under subantarctic water.

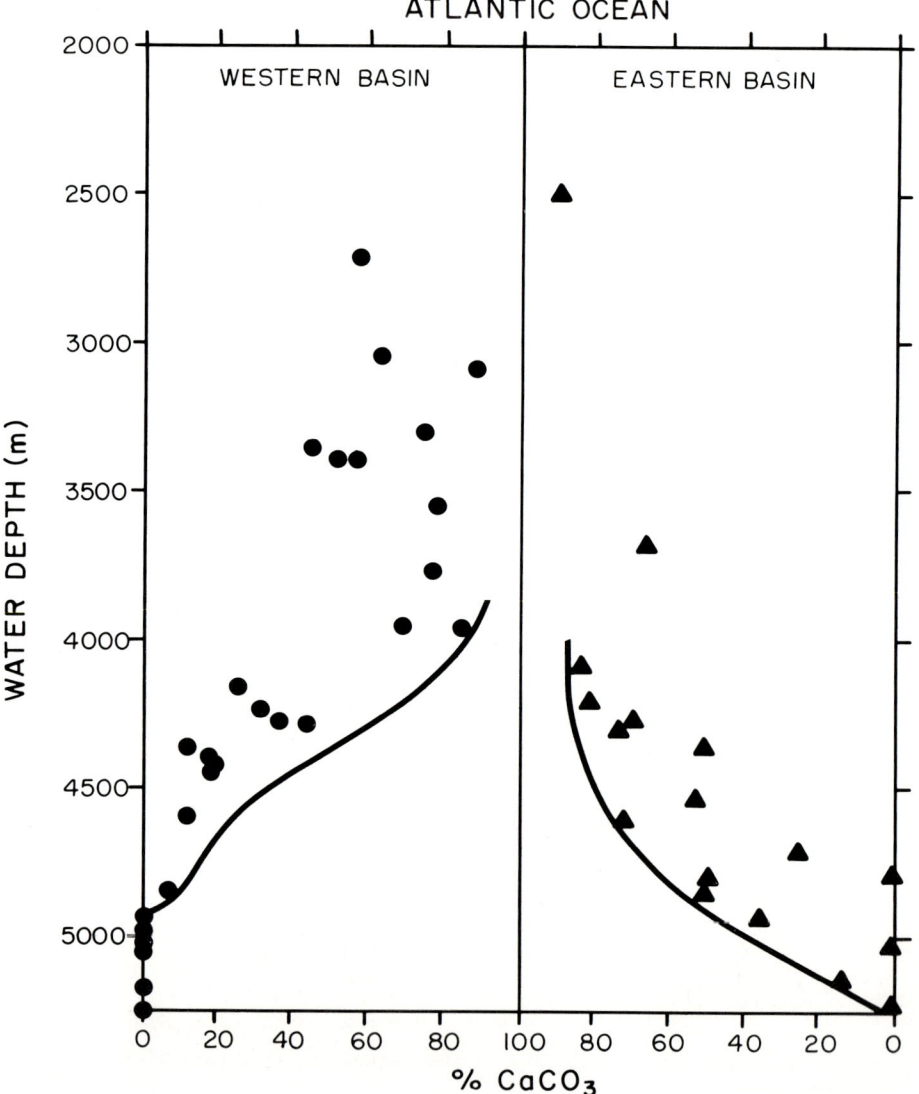

Figure 21. Calcium carbonate percentage of total sediment versus depth of water for the western and eastern basins of the Atlantic Ocean.

TABLE 9. VALUES OF RADIOLARIA/GRAM OBTAINED FOR DUPLICATE SAMPLES USED TO CHECK REPRODUCIBILITY OF DATA

Core no.	Sample depth (cm)	Total Radiolaria	Area counted (mm^2)	Area beaker (cm^2)	Bulk weight	Radiolaria/gm
RC13-254	100	703	300	50.26	0.79	14,908
RC13-254 A*	100	672	200	50.26	1.02	16,556
V29-84	80	1,296	120	50.26	2.10	25,848
V29-84 A*	80	1,115	100	50.26	1.90	29,494
RC13-228	TW$_{top}$	2,064	1,200	50.26	2.35	3,678
RC13-228 A*	TW$_{top}$	2,552	900	50.26	3.17	4,495

Note: TW$_{top}$, trigger-weight core top.
*Duplicate sample.

2. Using the regression technique of Imbrie and Kipp (1971), we established equations that permit us to calculate surface-water paleotemperatures for the months of February and August within the area of study. The standard error of estimate is less than 1.5°C.

One of the most difficult problems encountered was choosing samples truly representative of recent sediments, in which the thanatocoenosis corresponds to the biocoenosis. By establishing equations based on a small number of samples with good quality control, estimating temperatures for all samples, and including samples that gave better results and repetition of the process, we were able to establish that many samples located mainly in the Argentine and Crozet Basins have mixed faunas. These mixed faunas are produced by northward transport of Antarctic species by bottom currents.

The presence of the species *Cycladophora davisiana* in relative amounts over 5.5% is also considered to be a reliable indication of reworked or older than Holocene sediments. Out of the 145 trigger-weight core-top samples analyzed, only 72 were judged to be reliable for estimates of present sea-surface temperature.

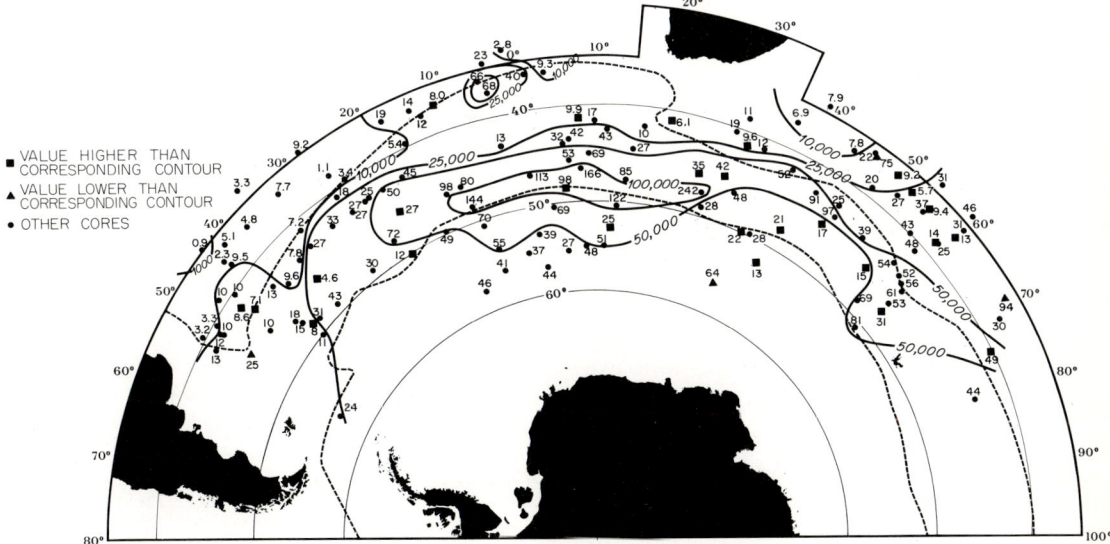

Figure 22. Radiolaria per gram of CaCO$_3$-free surface sediment. Values (in thousands) are the numbers of Radiolaria retained on a 62 μm sieve.

3. There is a strong correlation between oceanographic properties and underlying sediment types. The line bisecting the zone with highest values of Radiolaria per gram of $CaCO_3$-free sediment nearly coincides with the Antarctic Polar Front between long 25°W and 20°E. In the Indian Ocean sector this zone lies parallel to and just north of the Antarctic Polar Front.

Calcareous ooze is restricted to sediment deposited north of the polar front. The depth of the lysocline and the calcium carbonate compensation depth (CCD) are deeper in the eastern basin of the Atlantic Ocean. At least locally the CCD for the Indian Ocean north of the Antarctic Polar Front is deeper than 5,200 m. In the Atlantic Ocean the CCD is 5,200 m in the eastern basin and 4,800 m in the western basin. Differences in the CCD are mainly due to different degrees of dissolution produced by differences in deep-water characteristics.

South of the Antarctic Polar Front, biogenic siliceous, primarily diatomaceous, sediments predominate, and $CaCO_3$ productivity decreases. Waters are also colder and more corrosive at shallower depths, and the CCD is at about 3,700 m. The southern boundary of the diatom ooze seems to be determined primarily by reduction in the annual production of diatoms, mainly due to sea ice. The southern boundary of the diatom ooze corresponds to the average position of the ice edge for the months of January and April.

ACKNOWLEDGMENTS

A. L. Gordon, T. C. Moore, Jr., N. G. Kipp, and M. Roche critically reviewed the manuscript and made many valuable suggestions. Discussions with S. S. Streeter, J. V. Gardner, and M. Roche during the course of this study were helpful. We gratefully acknowledge the support provided by Grace Irving, who generated many of the radiolarian counts presented in this paper.

We gratefully acknowledge the efforts of the curatorial staff at Lamont-Doherty and the crews of RV *Vema* and RV *Conrad*, who made the basic core material available.

This research was supported by National Science Foundation Grants IDO71 04204 to the CLIMAP program, DES-72-01568 and Office of Naval Research Contract N00014-67-A-108-0004.

REFERENCES CITED

Bé, A.W.H., 1969, Planktonic foraminifera, *in* Distribution of selected groups of marine invertebrates in waters south of 35°S latitude: Am. Geog. Soc. Antarctic Map Folio Ser., folio 2, p. 9–12.

Bé, A.W.H., and Tolderlund, D. S., 1971, Distribution and ecology of living planktonic foraminifera in surface waters of the Atlantic and Indian Oceans, *in* Funnel, B. M., and Riedel, W. R., eds., The micropaleontology of oceans: London, University Press, p.105–149.

Belyaeva, N. V., 1970, Regularities in the distribution of planktonic foraminifera in the water and sediments of the Southern Ocean, *in* Holdgate, M. W., ed., Antarctic ecology, Vol. 1, The Southern Indian and Pacific Oceans: New York, Academic Press, p. 154–161.

Benson, R. M., 1966, Recent Radiolaria from the Gulf of California [Ph.D. dissert.]: Minneapolis, Univ. Minnesota, 577 p.

Berger, W. H., 1971, Sedimentation of planktonic foraminifera: Marine Geology, v. 11, p. 325–358.

Biscaye, P. E., 1965, Mineralogy and sedimentation of recent deep-sea clay in the Atlantic

Ocean and adjacent seas and oceans: Geol. Soc. America Bull., v. 76, p. 803-832.
Bowen, J. L., and Stommel, H., 1971, How variable is the Antarctic Circumpolar Current?, *in* Quam, L. O., and Porter, H. D., eds., Research in the Antarctic: Am. Assoc. Adv. Sci. Pub., no. 93, p. 645-650.
Broecker, W. S., 1971, Calcite accumulation rates and glacial to interglacial changes in oceanic mixing, *in* Turekian, K. K., ed., The late Cenozoic glacial ages: New Haven, Yale Univ. Press, p. 239-265.
Bunt, J. S., 1968, Microalgae of the Antarctic pack-ice zone, *in* Symposium on Antarctic oceanography: Cambridge, Scott Polar Research Institute, p. 198-219.
Correns, C. W., 1939, Pelagic sediments of the North Atlantic Ocean, *in* Trask, P. D., ed., Recent marine sediments: Tulsa, Am. Assoc. Petroleum Geologists, p. 373-395.
Deacon, G.E.R., 1933, A general account of the hydrology of the South Atlantic Ocean: Discovery Repts., v. 7, p. 171-238.
―――1937, The hydrology of Southern Ocean: Discovery Repts., v. 15, p. 1-124.
―――1963, The Southern Ocean, *in* Hill, M. N., ed., The sea: New York, Interscience Pubs., Inc., p. 281-296.
Defant, A., 1961, Physical oceanography, Vol. 1: London, Pergamon Press, 729 p.
Discovery Reports, 1947, Station list 1937-1939, v. 24: Cambridge Univ. Press, p. 197-242.
Ellis, D. B., and Moore, T. C., 1973, Calcium carbonate, opal, and quartz in Holocene pelagic sediments and the calcite compensation level in the South Atlantic Ocean: Jour. Marine Research, v. 31, p. 210-227.
El-Sayed, S. Z., 1968, On the productivity of the Southwest Atlantic Ocean and the waters of the Antarctic Peninsula, *in* Biology of the Antarctic seas, III: Antarctic Research Ser., v. 11, p. 15-47.
―――1970, On the productivity of the Southern Oceans, *in* Holdgate, M. W., ed., Antarctic ecology: New York, Academic Press, p. 119-135.
Ewing, M., Eittreim, S. L., Ewing, J. I., and Le Pichon, X., 1971, Sediment transport and distribution in the Argentine Basin. Pt. 3, Nepheloid layer and processes of sedimentation, *in* Physics and chemistry of the Earth, Vol. 8: London, Pergamon Press, p. 49-77.
Goll, R. M., and Bjørklund, K. R., 1974, Radiolaria in surface sediments of the South Atlantic: Micropaleontology, v. 20, p. 38-75.
Goodell, H. G., 1973, The sediments, *in* Marine sediments of the Southern Oceans: Am. Geog. Soc. Antarctic Map Folio Ser., folio 17.
Gordon, A. L., 1967, Structure of Antarctic waters between 20°W and 170°W: Am. Geog. Soc. Antarctic Map Folio Ser., folio 5.
―――1971, Antarctic Polar Front zone, *in* Reid, V. L., ed., Antarctic oceanology I: Am. Geophys. Union Antarctic Research Ser., v. 15, p. 205-221.
Gordon, A. L., and Goldberg, R. D., 1970, Circumpolar characteristics of Antarctic waters: Am. Geog. Soc. Antarctic Map Folio Ser., folio 13.
Hays, J. D., 1965, Radiolaria and Late Tertiary and Quaternary history of Antarctic seas: Am. Geophys. Union Antarctic Research Ser., v. 5, p. 125-184.
Hays, J. D., Lozano, J., Shackleton, N., and Irving, G., 1976, Reconstruction of the Atlantic and western Indian Ocean sectors of the 18,000 B.P. Antarctic Ocean, *in* Cline, R. M., and Hays, J. D., eds., Investigation of late Quaternary paleoceanography and paleoclimatology: Geol. Soc. America Mem. 145 (this volume).
Heezen, B. C., and Tharp, M., 1972, Submarine and subglacial topography, *in* Morphology of the Earth in the Antarctic and subantarctic: Am. Geog. Soc. Antarctic Map Folio Ser., folio 16.
Hülsemann, J., 1966, On the routine analysis of carbonates in unconsolidated sediments: Jour. Sed. Petrology, v. 36, p. 622-625.
Imbrie, J., and Kipp, N. G., 1971, A new micropaleontological method for quantitative paleoclimatology: Application to a late Pleistocene Caribbean core, *in* Turekian, K. K., ed., The late Cenozoic glacial ages: New Haven, Yale Univ. Press, p. 71-182.
Kennett, J. P., 1968, The fauna of the Ross Sea, Pt. 6, Ecology and distribution of foraminifera: New Zealand Dept. Scientific and Industrial Research, v. 186, 47 p.
Kort, V. G., 1962, The Antarctic Ocean: Sci. American, v. 207, no. 3, p. 113-128.
Koslova, O. G., 1970, Diatoms in suspension and in bottom sediments in the southern Indian

and Pacific Oceans, *in* Holdgate, M. W., ed., Antarctic ecology, Vol. 1, The southern Indian and Pacific Oceans: New York, Academic Press, p. 148–153.

Le Pichon, X., Eittreim, S. L., and Ludwig, W. J., 1971a, Sediment transport and distribution in the Argentine Basin. 1, Antarctic Bottom Current passage through the Falkland Fracture Zone, *in* Physics and chemistry of the Earth, Vol. 8: London, Pergamon Press, p. 1–28.

Le Pichon, X., Ewing, M., and Truchan, M., 1971b, Sediment transport and distribution in the Argentine Basin. 2, Antarctic bottom current into the Brazil basin: Physics and chemistry of the Earth, Vol. 8: London, Pergamon Press, p. 29–48.

Lisitzin, A. P., 1971, Distribution of siliceous microfossils in suspension and in bottom sediments, *in* Funnel, B. M., and Riedel, W. R., eds., Micropaleontology of the oceans: Cambridge, Cambridge Univ. Press, p. 73–195.

Mackintosh, N. A., 1972, Life cycle of Antarctic krill in relation to ice and water conditions: Discovery Repts., v. 36, p. 1–94.

McIntyre, A., and Bé, A.W.H., 1967, Modern Coccolithophoridae of the Atlantic Ocean—I, Placoliths and cyrtoliths: Deep-Sea Research, v. 14, p. 561–597.

Moore, T. C., Jr., 1973, Method of randomly distributing grains for microscopic examination: Jour. Sed. Petrology, v. 43, p. 904–906.

Murray, J., and Renard, A. F., 1891, Reports of the research voyage of the H.M.S. *Challenger*, Vol. 18, 1873–76: London, Majesty's Stationary Office.

Nigrini, C., 1967, Radiolaria in pelagic sediments from the Indian and Atlantic Oceans: Scripps Inst. Oceanography Bull., v. 11, p. 1–125.

Petrushevskaya, M. G., 1967, Radiolarians of orders Spumellaria and Nassellaria of the Antarctic region, *in* Andriyashev, A. P., and Ushakov, P. V., eds., Studies of marine fauna biological: Soviet Antarctic Exped. Repts. (1955–1958), v. 3, p. 2–186 (translated by M. Raveh, 1968, U.S. Dept. Commerce, Springfield, Va.).

Philippi, E., 1912, Die Grundproben, deutschen Sudpolar-Expedition, 1901–1903: Geographie and Geologie, v. 2, p. 415–616.

Riedel, W. R., 1958, Radiolaria in Antarctic sediments: British, Australian, and New Zealand Antarctic Research Exped. Repts., ser. B., v. 6, pt. 10, p. 217–255.

Sachs, H. M., 1973a, North Pacific radiolarian assemblages and their relationship to oceanographic parameters: Quaternary Research, v. 3, p. 73–88.

——1973b, Late Pleistocene history of the North Pacific: Evidence from a quantitative study of Radiolaria in core V21-173: Quaternary Research, v. 3, p. 89–98.

Schott, W., 1939, Deep-sea sediments of the Indian Ocean, *in* Trask, P. D., ed., Recent marine sediments, Pt. 5, Pelagic deposits: Tulsa, Okla., Am. Assoc. Petroleum Geologists, p. 396–408.

Sverdrup, H. U., Johnson, M. V., and Fleming, R. H., 1942, The oceans, their physics, chemistry, and general biology: Englewood Cliffs, N.J., Prentice Hall, 1,087 p.

U.S. Navy Hydrographic Office, 1954, World atlas of sea surface temperature (2nd ed.): U.S. Navy Hydrographic Office Pub. no. 225 (under the authority of the Secretary of the Navy).

Wüst, G., 1957, Quantitative Untersuchungen zur Statik und Dynamik des atlantischen Ozeans. Stromgeschwindigkeiten und Strommengen in den Tiefen des atlantischen Ozeans: Wiss. Ergeb. Deutsche Atlantische Expedition "Meteor" 1925–1927, v. 6.

Wyrtki, K., 1971, Oceanographic atlas of the International Indian Ocean Expedition: Washington, D.C., National Science Foundation, 53 p.

MANUSCRIPT RECEIVED BY THE SOCIETY JULY 1, 1974
REVISED MANUSCRIPT RECEIVED MAY 19, 1975
MANUSCRIPT ACCEPTED MAY 23, 1975
LAMONT-DOHERTY GEOLOGICAL OBSERVATORY CONTRIBUTION NO. 2274

Reconstruction of the Atlantic and Western Indian Ocean Sectors of the 18,000 B.P. Antarctic Ocean

James D. Hays
Jose A. Lozano*
*Lamont-Doherty Geological Observatory and Department of Geological Sciences
Columbia University
Palisades, New York 10964*

Nicholas Shackleton
*Sub-Department of Quaternary Research
Cambridge University
Cambridge, England*

AND

Grace Irving
*Lamont-Doherty Geological Observatory
Palisades, New York 10964*

ABSTRACT

Using relative-abundance changes of the radiolarian *Cycladophora davisiana* calibrated using O^{18}/O^{16} values and C^{14} ages, we have established a high-resolution biostratigraphy for the Antarctic and subantarctic ocean covering the past 150,000 yr. Extremes of this species' relative abundance have been estimated to have occurred approximately 18,000, 35,000, 60,000, 85,000, 105,000, and 120,000 yr ago. Crosscorrelation between O^{18}/O^{16} values and the most recent *C. davisiana* maximum shows that this maximum is isochronous over a broad area of the Antarctic and subantarctic sea floor.

Using diverse stratigraphic criteria, we selected 34 cores from the Atlantic and western Indian Ocean sectors of the Antarctic and subantarctic ocean containing a strong *C. davisiana* relative-abundance maximum near the core top, believing that this maximum occurred approximately 18,000 yr ago in the Antarctic and

---
*Present address: Departamento de Geologia, Universidad Nacional, Bogota, D.E., Colombia, South America.

subantarctic. Using techniques including radiolarian-based paleoecological equations, Radiolaria per gram of sediment, and percentage of calcium carbonate, we show that 18,000 yr ago the Antarctic Polar Front was displaced north of its present position by as much as 7° of latitude. The subtropical convergence was little changed from its present position; consequently, the width of subantarctic waters was reduced.

Using *C. davisiana* as a stratigraphic indicator, we estimate the sedimentation rate in one Antarctic core of the upper diatom-ooze layer (11.5 cm/1,000 yr) and the immediately underlying silty clay (2.7 cm/1,000 yr). The slower sedimentation rate of the silty clay strongly suggests that diatom production was inhibited during its deposition relative to production during deposition of the younger diatom ooze. We interpret this to indicate that summer ice cover extended nearly to lat 55°S 18,000 yr ago, whereas today it melts back to the Antarctic continent.

INTRODUCTION

In this paper we reconstruct paleoceanographic conditions for 18,000 B.P. in the Atlantic and western Indian Ocean sectors of the Antarctic and subantarctic oceans; this represents a small part of a global reconstruction for 18,000 B.P. undertaken by CLIMAP. Included in this volume is a reconstruction of the North Atlantic 18,000 yr ago (McIntyre and others, this volume). The North Atlantic and Antarctic Oceans show several marked differences that have affected their paleoceanographic histories. In the Southern Hemisphere the minimum percentage of land occurs between lat 40°S and 60°S, effectively, the Antarctic Ocean. This unbroken circumglobal ocean promotes a strong zonal flow. The maximum percentage of land area in the Northern Hemisphere occurs at comparable latitudes (40°N to 70°N). The North Atlantic Ocean, constricted by continents on both the east and west, has a strong meridional surface flow.

Because the radiation balance in high latitudes of both hemispheres is negative, the heat budget is balanced by advective heat transport conveyed by the atmosphere and oceans from lower latitudes. North Atlantic surface currents carry heat northward, but in the Antarctic Ocean (south of the polar front) heat is supplied by a southward-flowing and upwelling deep current.

During the last glacial period the continents on both sides of the North Atlantic were covered by thick ice sheets, most of which have melted. By contrast, the Southern Hemisphere land ice increased considerably less during the last glaciation. The ice increase was probably greatest in southern South America with the development of the Patagonian ice sheet (Flint, 1971). The Antarctic ice cap may have thickened owing to the grounding of large ice shelves in the Ross and Weddell Seas (Hollin, 1962). Other than these differences, the extent of continental ice in the Southern Hemisphere was apparently little different from today. South Georgia was probably covered by an ice cap (Clapperton, 1971), but the Falkland Islands were not (Greenway, 1972). Evidence from Kerguelen suggests that it was covered by an ice cap (Bellair, 1965), but Macquarie Island was not (Flint, 1971). The percentage change of ice volume was large in the Northern Hemisphere and small in the Southern Hemisphere between the last glacial stage and today.

The Antarctic Ocean represents the Southern Hemisphere equivalent of the Northern Hemisphere latitudes that supported the large build-up of ice that characterized the glacial epochs. The Antarctic Polar Front separates the cold nonstratified Antarctic water from the warmer, slightly saltier subantarctic surface water to the north. The position of the polar front shows little seasonal or secular change (Gordon, 1971; Bowen and Stommel, 1971; Botnikov, 1964), although irregular variations of 100 to 400 km are common (Gordon, 1971). South of the polar front, cold temperatures inhibit calcareous secreting organisms, but siliceous organisms (diatoms and Radiolaria) abound; thus, sediment blanketing the sea floor consists of thick layers of diatomaceous ooze south of the polar front and siliceous calcareous ooze underlying subantarctic waters. Because Radiolaria are ubiquitous in Antarctic and subantarctic sediments, our study focuses on them as a means for reconstructing the paleoceanography 18,000 yr ago.

In this paper we develop stratigraphic criteria that can be used to recognize the 18,000 B.P. level in Antarctic and subantarctic sediments. Through the use of paleoecological equations developed by Lozano and Hays (this volume), we estimate 18,000 B.P. paleotemperatures of the Southern Ocean south of lat 35°S and between Kerguelen and South America. Using these paleotemperature estimates and other criteria, we reconstruct the Antarctic Ocean as it was 18,000 yr ago.

PREVIOUS WORK

Study of Antarctic Radiolaria began with systematic studies of collections (surface sediment and plankton) made by the national oceanographic expeditions of the nineteenth and early twentieth centuries (Ehrenberg, 1844; Haeckel, 1887; Haecker, 1908; Popofsky, 1908). These studies were primarily systematic and paid little heed to geographic distributions. Modern work with Antarctic Radiolaria started with Riedel (1958), who described and emended earlier descriptions of Antarctic species and made some ecological inferences concerning the distribution of certain species. Lozano and Hays (this volume) reviewed the previous ecological studies; we confine ourselves here to previous stratigraphic work.

The first Antarctic stratigraphic studies were published by Hays (1965). He erected a biostratigraphic zonation with zones designated by Greek letters ϕ, X, Ψ, and Ω and based on the upward sequential disappearance of radiolarian species in sediment ranging in age from Pliocene (ϕ) through Pleistocene (X, Ψ, Ω). The Ψ-Ω boundary was dated by the ionium/thorium technique (Ku and Broecker, 1966) at about 400,000 B.P. With the development of paleomagnetic stratigraphic determinations on Antarctic cores (Opdyke and others, 1966), the older part of the radiolarian zonation was put in a time-stratigraphic framework (Table 1). This was later extended to lower Pliocene sediment (Hays and Opdyke, 1967) and correlated with other stratigraphic determinations for Pliocene-Pleistocene sediment from other parts of the world (Hays and others, 1969; Hays, 1969; Hays and Berggren, 1971).

This Antarctic stratigraphy has been checked and largely confirmed by subsequent workers (Goodell and Watkins, 1968; Bandy and others, 1971). Since then, further subdivisions have been made using Foraminifera (Kennett, 1970), diatoms (Donahue,

TABLE 1. ANTARCTIC RADIOLARIAN ZONATION

Epoch	Zone	Age (m.y.)[6]
	Ω	0.0–0.4
	Ψ	0.4–0.7
Pleistocene	X	0.7–1.8
Pliocene	ϕ	1.8–2.4
	γ	2.4–4.2?
	τ	4.2–5.0

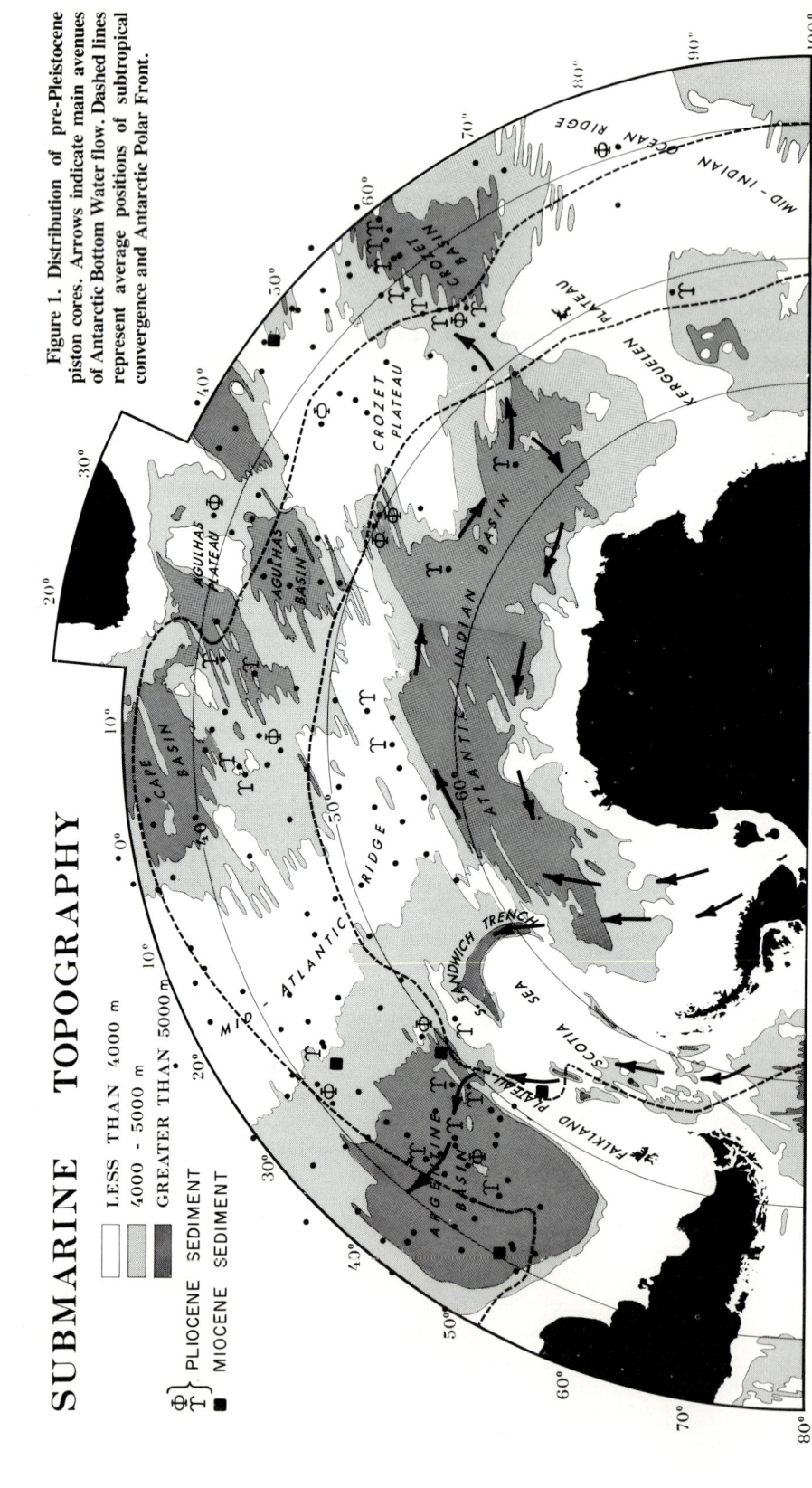

Figure 1. Distribution of pre-Pleistocene piston cores. Arrows indicate main avenues of Antarctic Bottom Water flow. Dashed lines represent average positions of subtropical convergence and Antarctic Polar Front.

1970), calcareous nannofossils (Geitzenauer, 1971), and Radiolaria (Bandy and others, 1971).

To determine the gross stratigraphy of most cores in the area, bottom samples from piston cores were examined and dated using the radiolarian stratigraphy. Of 257 cores, 51 penetrated pre-Pleistocene sediment. Most, if not all, of these cores contain hiatuses. Although hiatuses might be encountered in any region of rough topography or strong bottom currents, in this area, cores penetrating older sediment are not concentrated along the rugged topography of the mid-oceanic ridge, but rather in two deep basins, the Crozet and the Argentine (Fig. 1). Cores penetrating pre-Pleistocene sediment are also concentrated on relatively shallow plateaus such as the Agulhas (Saito and Fray, 1965) and Falkland Plateaus (Burckle and Hays, 1966).

Watkins and Kennett (1971) provided evidence of large regional hiatuses south of Australia in the southern Tasman Sea; they attributed these hiatuses to a late Cenozoic increase in the speed of the Antarctic Bottom Water and consequent sediment erosion. Gordon (1972) showed that this region of strong bottom scour is along the path of northward-flowing bottom water which breaks through the mid-oceanic ridge at approximately long 145° to 155°E.

The Argentine and Crozet Basins represent similar situations. Northward-flowing water passing through a gap in the Scotia Ridge has produced large sedimentary ripples on the Argentine Basin floor to the north (Ewing and others, 1971). Our core data show both high sedimentation rates and many hiatuses in this area, indicating that sediment is being scoured and rapidly redeposited. We attempted to relate cores with hiatuses to the geometry of these megaripples but found our core coverage insufficient to show a clear relationship. The Crozet Basin appears to be another major avenue of northward flow of Antarctic Bottom Water. A string of cores (Fig. 1) taken along the southern axis of the Crozet Basin shows hiatuses. In the central part of the basin, neighboring cores show high deposition rates and have many hiatuses. An inspection of seismic-profiler records in this region revealed large ripplelike sedimentary structures similar to those observed by Ewing and others (1971) in the Argentine Basin.

A number of cores in both the Argentine and Crozet Basins penetrate sediment of the γ zone (Gauss or Gilbert age), which is similar in age to sediment beneath hiatuses in the south Tasman Basin reported by Watkins and Kennett (1971). Watkins and Kennett (1972) speculated that erosion was initiated by an increase in Antarctic Bottom Water production associated with the development of extensive ice shelves around Antarctica in post-Gilbert or Gauss time. Because erosion in the Brunhes magnetic epoch could easily remove sediment as old as lower Gilbert (γ or τ), it is not possible to know if the erosion occurred prior to the Brunhes epoch. Watkins and Kennett (1971, 1972) have shown that most cores containing hiatuses have a thin Brunhes section at the top, suggesting that erosion is intermittent and that there has been a recent interval with less bottom scour than in the more distant past. Many of our cores in the Crozet and Argentine Basins show a similar sequence. A regime of less-active erosion could have occurred repeatedly, but its record would have been destroyed by periods of subsequently increased erosion. Gordon (1972) suggested that strengthening the Antarctic Circumpolar Current would increase near-surface salinities and could lead to production of larger volumes of Antarctic Bottom Water. Therefore, we suspect that repeatedly during the Brunhes epoch (past 700,000 yr), and perhaps before, increased wind stress engendered increased bottom-water production and caused erosion along the paths of northward-flowing streams of Antarctic Bottom Water, as in the Argentine, Crozet, and southern Tasman Basins. When the first episode of erosion occurred remains a question.

STRATIGRAPHY

In this study we are dealing with a short interval of time (upper part of the Ω zone), where the previous stratigraphy helps to provide only a gross picture of ages. It is necessary, therefore, to develop a detailed stratigraphy that can be used within the Ω zone, particularly within the past 50,000 yr.

To develop a detailed stratigraphic framework for the upper part of the Ω zone, faunal criteria other than evolutionary changes (faunal appearances, extinctions, and so forth) must be used. Various stratigraphic schemes have been established based on nonevolutionary changes, such as the stratigraphic work by Ericson and others (1961) in the Atlantic based principally on the presence and absence of *Globorotalia menardii*. Changes in coiling direction of a number of foraminiferal species have also been used (Bandy, 1959; Saito in Hays and others, 1969; and others). To determine if any Antarctic radiolarian species had stratigraphically valuable abundance or morphologic changes, we studied more than 40 species in several cores that we judged to have spanned several hundred thousand years. *Cycladophora davisiana* was found to occur at depth in a number of cores in significantly greater abundance than it ever occurred in sea-floor surface sediment. It is therefore inappropriate for paleotemperature estimates, but it may have considerable stratigraphic value. To test the stratigraphic value, its abundance relative to all other Radiolaria was determined down four cores (RC11-120, RC8-39, V29-84, and RC15-94; Fig. 2). Although the amplitude of abundance changes differ from core to core, the patterns have a number of similarities. All core tops have a low abundance of *C. davisiana* [less than 5%; (a) in Fig. 2]. Below this low-abundance zone, there is an abrupt increase to the maximum percentage of *C. davisiana* obtained at any level in the cores [(b) in Fig. 2]. Below this peak at (b) its abundance fluctuates but generally declines to (e_3), where the abundance of *C. davisiana* reaches a minimum comparable to minimum (a) at the top of the core. Not only is the sequence of peaks and troughs in the cores comparable through this interval

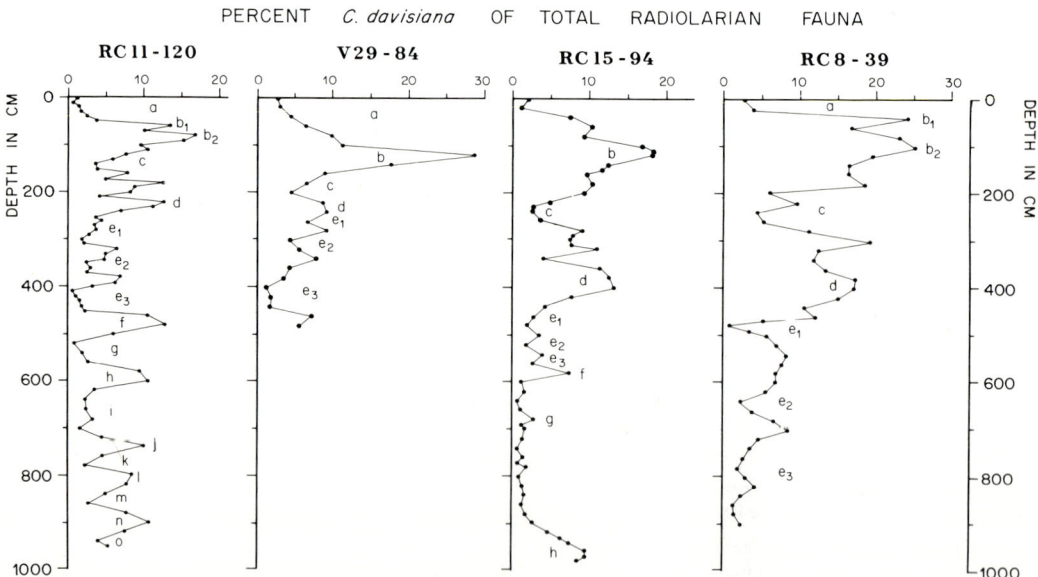

Figure 2. *C. davisiana* stratigraphy down the length of piston cores RC11-120, V29-84, RC15-94, and RC8-39. For location of cores see Lozano and Hays (this volume, Fig. 1, Table 1).

[(a) to (e₃)], but in general, the relative abundances of maximums and minimums are similar. For example, the maximum (b) is usually greater than 20%, minimum (c) is on the order of 5%, and maximum (d) is approximately 10 to 15%.

Lozano and Hays (this volume, their Fig. 14) have shown that the percentage of *C. davisiana* in the vast majority of 145 core tops from this area is less than 5%. Cores having greater than 5% are principally found in basins where bottom-water scour has probably removed the topmost layer of sediment or mixed it with underlying sediment containing higher proportions of *C. davisiana*. There is no correlation of *C. davisiana* abundance in core tops with either latitude or longitude. No Holocene core tops contain high *C. davisiana* percentages, so the cause of the older maximums cannot be easily related to shifts of surface water masses.

K. Bjørklund (1973, oral commun.) examined a series of six plankton tows taken in the Norwegian Sea. Four were gathered from depths of 500 m, and two were taken at depths of 2,000 m. *C. davisiana* was not present in the samples taken from the tows raised through less than 500 m of water. However, *C. davisiana* was among the five most abundant species identified from the tows raised from 2,000 m. Petrushevskaya (1967) noted that *C. davisiana* was only found in deep-water Antarctic sediment (1,000 m). In a review of this species occurrence as reported in the literature, she noted that it is cosmopolitan and occurs in most ocean basins. These observations suggest that *C. davisiana* is a deep-living species; thus, its relative-abundance variations may be in response to changes in deep-water characteristics, not near-surface water changes. If so, it may be a good stratigraphic indicator for Antarctic sediment, and it has potential as a stratigraphic indicator for other parts of the ocean.

Calibration of the *C. davisiana* Curve

The *C. davisiana* curves (Fig. 2) show consistency in pattern from core to core, but to determine the time significance of the maximums and minimums requires independent time calibration. Shackleton and Opdyke (1973) have shown that a large part of the variations of δO^{18} in the foraminiferal shells extracted from deep-sea cores through a glacial cycle is due to the formation and melting of glacial ice. The mixing time of the ocean is not more than a few thousand years, so changes in δO^{18} due to changing ice volume are nearly synchronous on a global scale. The last δO^{18} maximum has been dated in core V12-122 (Ku and Broecker, 1966) at between 16,000 and 20,000 B.P. Thus, this last δO^{18} maximum can be used to approximate this time interval.

Core RC11-120 has a calcium carbonate content that ranges between 30 and 80%. O^{18}/O^{16} values were determined on *G. builoides* at 20-cm intervals and were plotted on the same scale as the *C. davisiana* percentage (Fig. 3). The percentage scale for *C. davisiana* in Figure 3 has been reversed to allow for easier comparison with the O^{18}/O^{16} data. The similarity between these patterns is remarkable. Isotopic maximums and minimums coincide with *C. davisiana* maximums and minimums in a very direct way.

Assuming that *C. davisiana* is a deep-living species, the fluctuations in its relative abundance may be caused by either a general decrease in near-surface radiolarian productivity during glacial periods or by an increase in *C. davisiana* during these periods, or both. Although it is not possible to rule out categorically either alternative, it seems more likely that the change in relative abundance of *C. davisiana* is caused by changes in its absolute abundance. This possibility is examined below in the discussion of the absolute accumulation rates of Radiolaria.

The relationship between the O^{18}/O^{16} curve and the abundance of *C. davisiana*

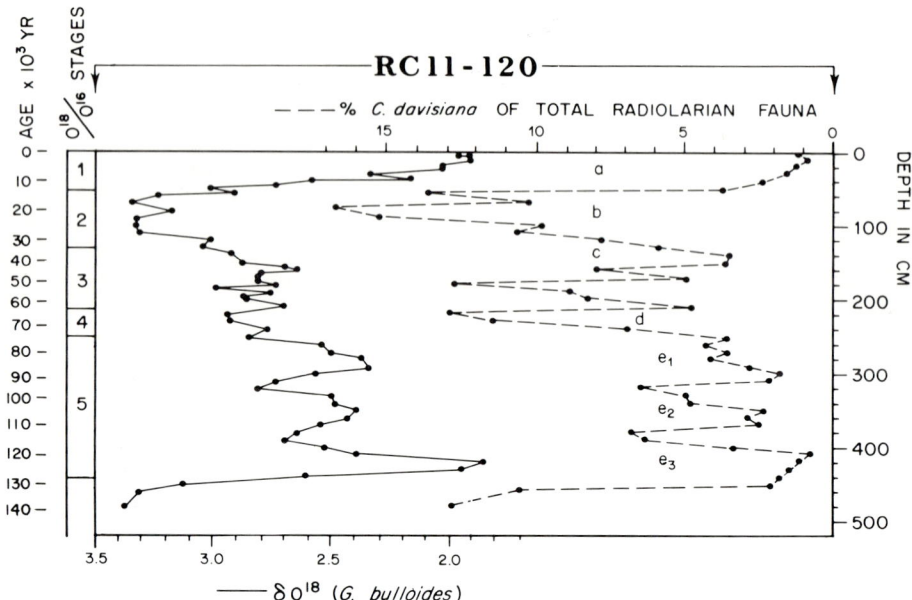

Figure 3. Correlation of stratigraphies determined by the oxygen isotope method and *C. davisiana* for the top 500 cm of core RC11-120. Oxygen isotopic composition expressed as a deviation per mil from the PDB standard. Stages after Emiliani (1955); estimated ages after Shackleton and Opdyke (1973). For location of core, see Figure 4.

presents an intriguing ecological problem and provides an important stratigraphic tool. Because the O^{18}/O^{16} curve is now well dated and has global validity, the *C. davisiana* curve can be calibrated by comparison with the isotopic record (Fig. 3, Table 2). The 18,000 B.P. level falls in isotopic stage 2, and the maximum peak of *C. davisiana* [interval (b)] falls very close to the O^{18}/O^{16} value considered to show the peak ice volume during the last glaciation, which can be used as a first approximation of the 18,000 B.P. level. However, to use the *C. davisiana* curve independently of the isotopic curve, it must be shown that the *C. davisiana* curve is not regionally time transgressive relative to the isotopic curve.

To test this, cores at extreme points in the area were selected (Fig. 4, Table 3) to see if either changes in latitude or longitude might cause time-transgressive changes in the relationship between the O^{18}/O^{16} curve and the pattern of the *C. davisiana* curve. No O^{18}/O^{16} data were available for cores taken south of

TABLE 2. CORRELATION AND AGE OF INTERVALS IN THE *C. DAVISIANA* CURVE

C. davisiana interval	Isotopic stage	Barbados high sea-level stands	Age* ($\times 10^3$ yr)
a	1		0–13
b	2		13–32
c	3		32–64
d	4		64–75
e	5		75–128
e_1		I	82
e_2		II	103
e_3		III	125

*Ages after Shackleton and Opdyke (1973).

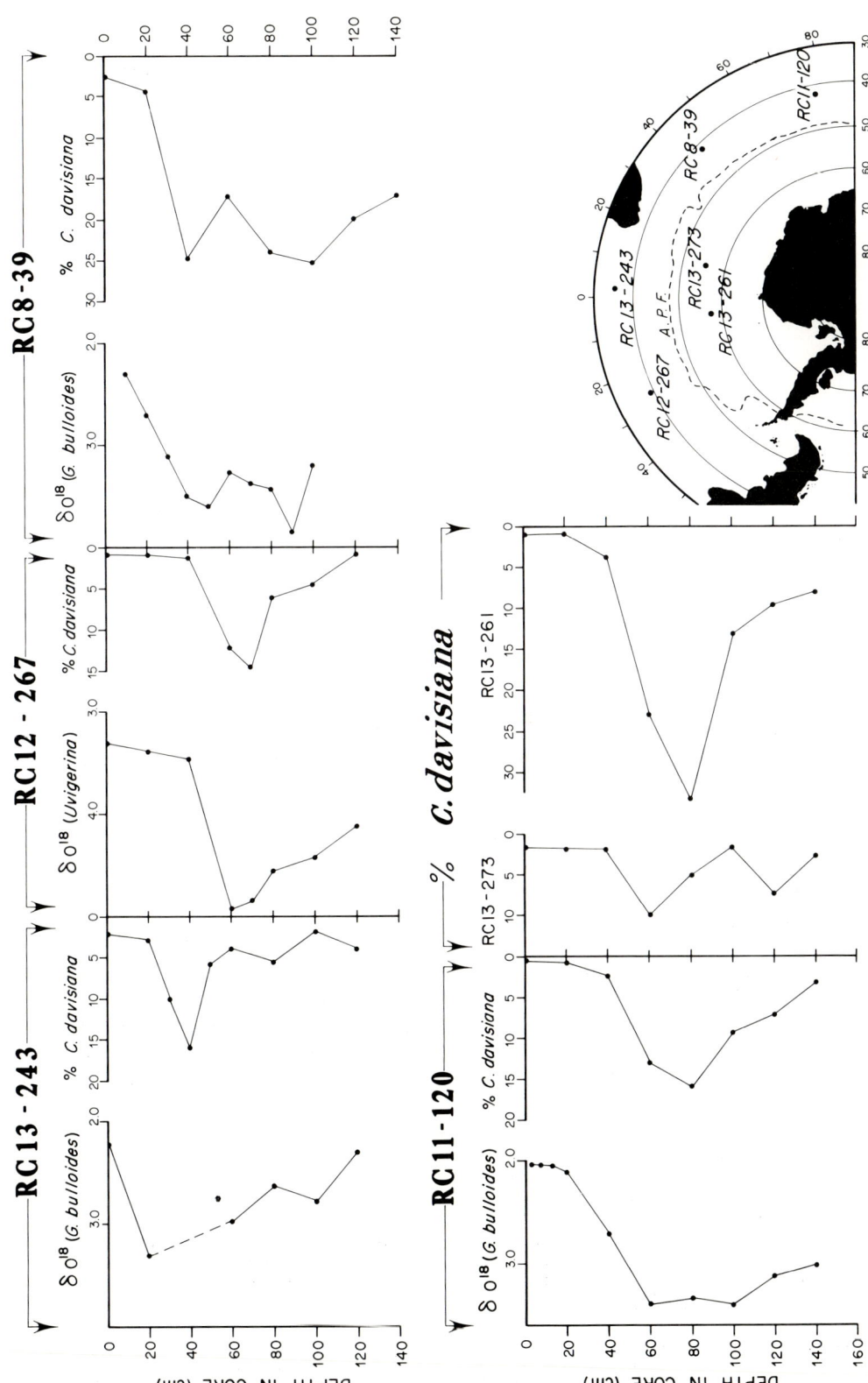

Figure 4. Correlation of oxygen-isotope and *C. davisiana* curves for the upper part of four cores at extreme positions within the area of study. Oxygen isotopic composition expressed as a deviation per mil from the PDB standard; *C. davisiana* expressed as percentage of total Radiolaria.

TABLE 3. LOCATION AND WATER DEPTH FOR CORES WITH OXYGEN ISOTOPE AND *C. DAVISIANA* INFORMATION

Core	Long	Lat	Water depth (m)
RC13-243	01°20'E	36°54'S	4,790
RC12-267	25°47'W	38°41'S	4,144
RC8-39	42°21'E	42°53'S	4,330
RC11-120	79°21'E	43°31'S	3,193
RC13-273	11°34'E	55°04'S	4,967
RC13-261	08°41'W	56°07'S	4,221

Note: From Figure 4.

the Antarctic Polar Front (very little carbonate is present in this region at any depth). The *C. davisiana* curves for two cores from this region are included in Figure 4 to show that the *C. davisiana* curve has the same general shape as it does north of the polar front. Figure 4 shows that although the amplitude of the *C. davisiana* peak varies from core to core, its position remains constant relative to the O^{18}/O^{16} curve throughout the area.

We have assumed on the basis of Northern Hemisphere data that the δO^{18} maximum corresponds to 18,000 yr ago. To check this, we obtained C^{14} dates of two cores with significantly different sedimentation rates (Table 4). Extrapolation from the two dates for core RC11-83 gives a 2,500-yr age for the core top. Radiocarbon dates on a number of cores taken from various depths in the Atlantic give extrapolated core-top ages of 1,000 to 5,800 yr, with an average of about 2,900 yr (Kulp and others, 1951; Suess, 1954; Rubin and Suess, 1955; Broecker and Kulp, 1957). Emery (1960) found the average age of five sea-floor surface sediments in various basins off southern California extrapolated from C^{14} dates to be 2400 B.P. Table 4 shows that when we assume a 2500 B.P. radiocarbon date for the top of core RC11-120, the extrapolated age of the *C. davisiana* maximum and the δO^{18} maximum approximates 18,000 B.P. in both cores.

The most recent *C. davisiana* maximum therefore appears to be a reliable stratigraphic indicator for the 18,000 B.P. level south of lat 35°S.

18,000 B.P. Level

Having established a criterion for recognizing the 18,000 B.P. level in Antarctic and subantarctic sediments, counts of *C. davisiana* were made on 68 cores that, on the basis of our preliminary stratigraphic work, contain thick Ω sections. Of these, 37 (Table 5) show a characteristic *C. davisiana* curve, 33 of which are shown in Figure 5. The remaining 31 cores either had high *C. davisiana* abundance values at the top, which suggests that the tops were missing, or the curve showed

TABLE 4. EXTRAPOLATED C^{14} AGES FOR δO^{18} AND *C. DAVISIANA* MAXIMUMS IN CORES RC11-83 AND RC11-120

Core	Depth of C^{14} sample (cm)	C^{14} age	Depth of *C. davisiana* maximums (cm)	Depth of δO^{18} maximums (cm)	Extrapolated age of *C. davisiana* maximums*	Extrapolated age of δO^{18} maximums*
RC11-83	94–100	6,590 ± 200	360	400	~17,000 ± 1500	~18,500 ± 1,000
	185–188	9,950 ± 900				
RC11-120	36–39	9,400 ± 600	80	80	~17,300	~17,300

*Assumes that core top has a finite age of approximately 2,500 yr.

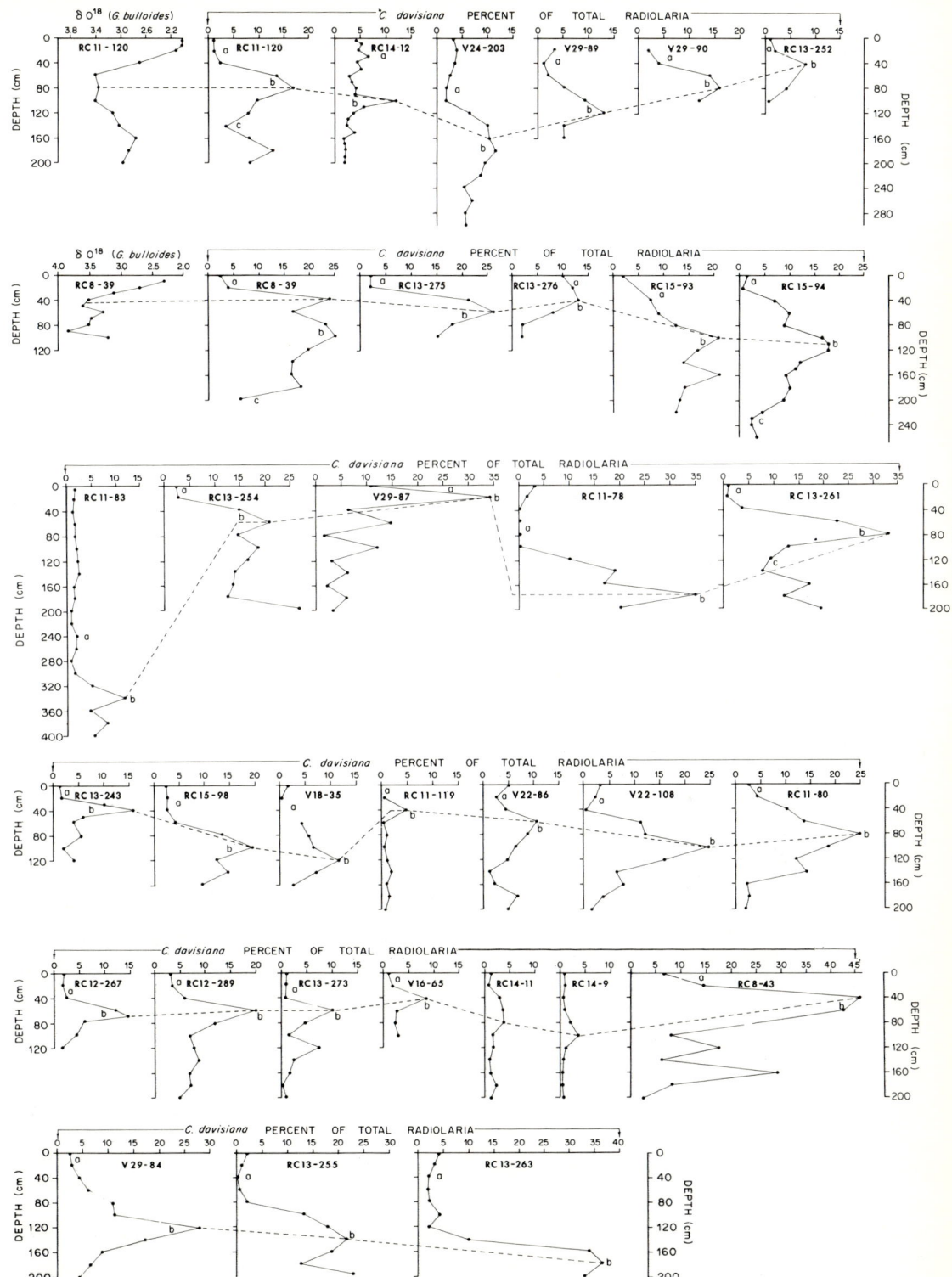

Figure 5. Percentage of *C. davisiana* versus depth in 33 cores, showing depth at which the 18,000 B.P. level was discerned. For location of cores, see Table 5 and Figure 1 in Lozano and Hays (this volume).

poorly defined maximums (less than 5%). This latter type occurred well north of the Antarctic Polar Front. Only those cores containing a significant peak (greater than 5%) just below the topmost minimum were judged to be reliable for the 18,000 B.P. estimates. The *C. davisiana* curves and the 18,000 B.P. level are shown in Figure 5. More than 300 of the 18 temperature-sensitive species discussed by Lozano and Hays (this volume) were counted in the 18,000 B.P. level sample. The faunal data were expressed in terms of the three factors resolved in the surface sediments.[1] Factors 1 and 2 have factor loadings ranging from 0.2 to 1.0 (Figs. 6, 7), but factor 3 (Fig. 8) is transitional between the two major factors, with negative factor loadings ranging to −0.35. The difference between the 18,000 B.P. factor maps and the present distribution of the factors lies primarily in the northward advance of the Antarctic assemblage (factor 2) 18,000 yr ago at the expense of the transitional subantarctic assemblage (factor 3; Figs. 7, 8). The subtropical assemblage (factor 1) shows little displacement between 18,000 B.P. and today; however, the 0.2 contour is somewhat displaced to the north (Fig. 6). The maximum northward displacement of the Antarctic assemblage occurs in the western Atlantic (long 0° to 40°W) and the western Indian Ocean (long 30°E to 60°E); minimal displacement occurs in the easternmost Atlantic and south of Africa.

Values of 0.9 for factor 2 today approximate the mean position of the Antarctic Polar Front. The pattern of this factor in Figure 7 suggests that the polar front shifted north significantly in the western basin of the South Atlantic and the westernmost part of the Indian Ocean but showed only a small minor shift south of Africa and a similar small shift in the region of Kerguelen.

COMPARISON OF CALCIUM CARBONATE AT THE 18,000 B.P. LEVEL AND IN SURFACE SEDIMENT

Lozano and Hays (this volume) studied the distribution of $CaCO_3$ in surface sediment in the study area. Our purpose is to examine the differences in $CaCO_3$ content in recent sediments and those of 18,000 B.P. Figure 9 illustrates the changes in $CaCO_3$ content with depth between today and 18,000 B.P. in 34 cores. It is clear that some cores show great changes in $CaCO_3$ content (75%) between 18,000 B.P. and today but that others show very little change (less than 10%).

In most cases, $CaCO_3$ content was lower 18,000 yr ago than today. However, there are two exceptions—cores RC11-83 and RC14-9—where the dominant pattern is reversed. Core RC14-9 was raised from beneath subtropical water, and core RC11-83 from under subantarctic water. We have no explanation for the departure of these cores from the normal pattern.

From Figure 9 it is clear that there is no relationship between core depth and degree of change of $CaCO_3$ content from 18,000 B.P. to today. If depth does not control the degree of change, then it may be related to the deep-water characteristics of specific ocean basins. Core RC15-93 from the western Atlantic and core RC11-80 from the eastern Atlantic were raised from two different water depths, but they both show similarly large changes in $CaCO_3$ content between today and 18,000 B.P. Consequently, neither water depth nor present bottom-water characteristics can explain the total-carbonate variation between today and 18,000 B.P.

---

[1] Faunal counts made on all samples are included in Appendix I on microfiche in pocket inside back cover.

Therefore, the pattern observed may be related to changes in surface-water characteristics between today and 18,000 yr ago. In Figure 10 we have replotted the carbonate data of Figure 9 in three different fields (Antarctic, subantarctic, and subtropical cores). In each of these fields the cores were plotted against water depth. It is clear from Figure 10 that the cores that show the maximum change in $CaCO_3$ content between today and 18,000 yr ago were raised from under subantarctic water. Cores raised from under subtropical and polar water show consistently less change. This phenomenon can be explained by a northward shift of the Antarctic Polar Front. An examination of the factor maps shows that the

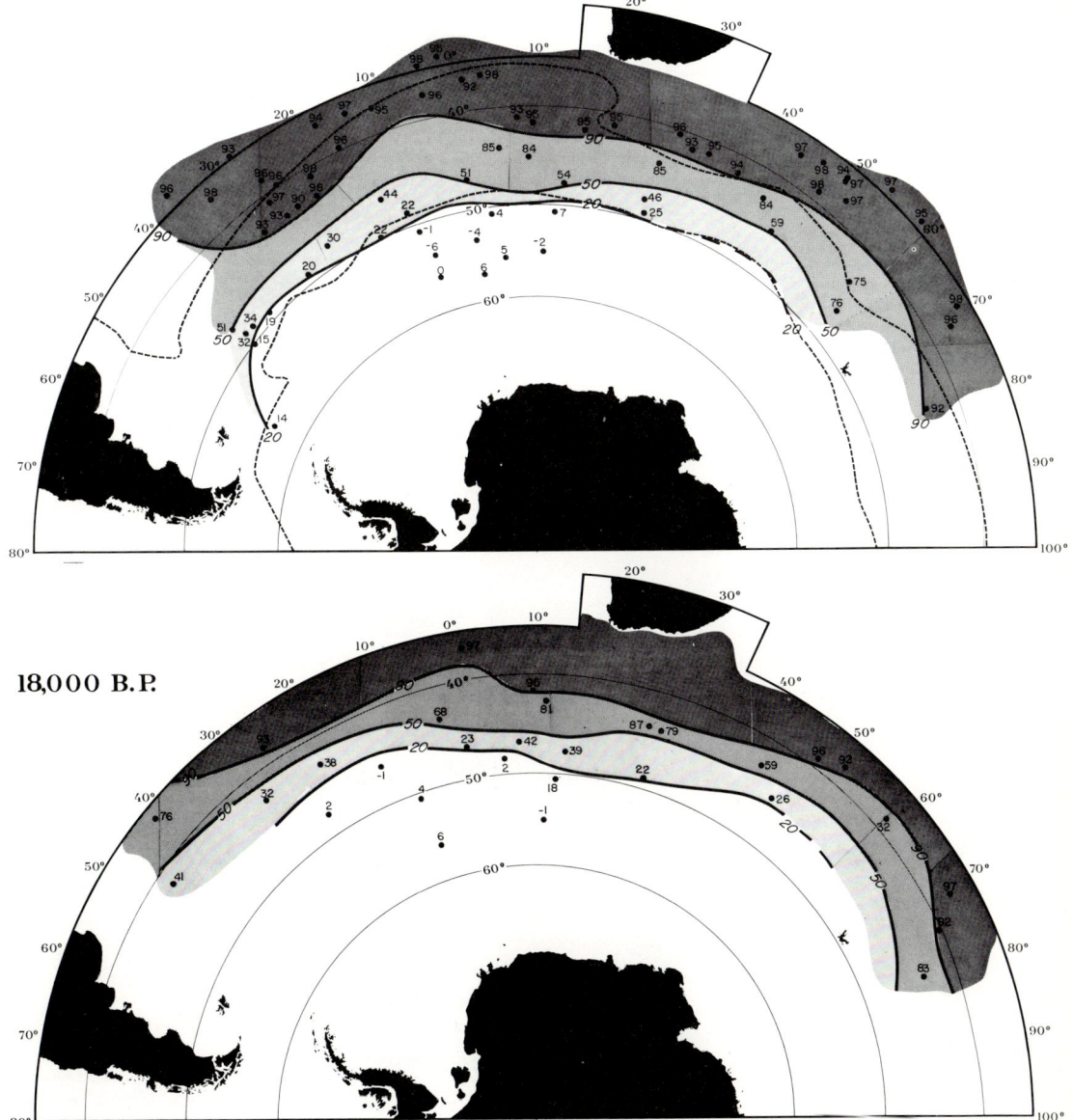

Figure 6. Distribution of factor 1 (subtropical assemblage) in surface sediment today and 18,000 yr ago. Values are factor loadings × 100.

0.9 factor loading of factor 2 today closely follows the present mean position of the Antarctic Polar Front, as we have already pointed out. The 0.9 factor loading 18,000 yr ago was shifted northward, particularly in the South Atlantic western basin (Fig. 7). Consequently, it overrides a number of subantarctic cores (RC13-254, RC11-80, RC15-93, RC15-94, V16-65, and RC15-98). All of these except core V16-65 show a sharp decline of $CaCO_3$ at 18,000 B.P. compared with today. This would have the immediate effect of reducing $CaCO_3$ fixation in the surface water overlying these cores because of the absence of coccolithophores in polar water. Core V16-65 shows a relatively small $CaCO_3$ drop compared with the other

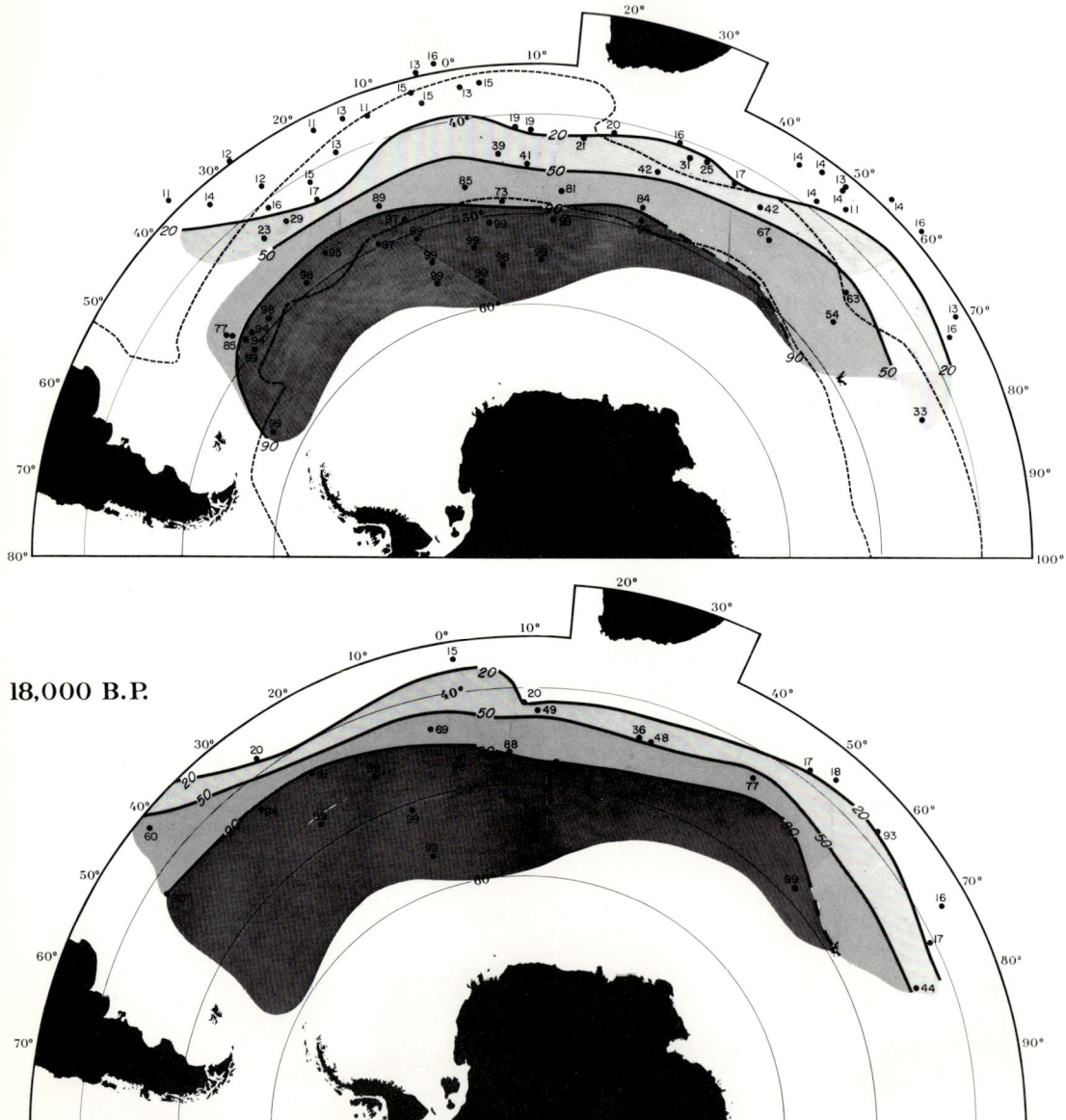

Figure 7. Distribution of factor 2 (Antarctic assemblage) in surface sediment today and 18,000 yr ago. Values are factor loadings × 100.

cores, which must be due to its very shallow depth (1,618 m). Most likely, $CaCO_3$ change in this core is produced only by changes of surface-water productivity, because it has remained above the calcium carbonate compensation depth (CCD). Goodell (1973) showed Holocene $CaCO_3$ values of greater than 60% south of the polar front at comparably shallow depths. Other cores within the subantarctic water mass also show sharp declines, but the factor maps indicate that they were not overrun by the Antarctic Polar Front (cores V22-108 and RC13-243). These cores were raised from depths between 4,100 and 4,800 m in the eastern basin of the South Atlantic.

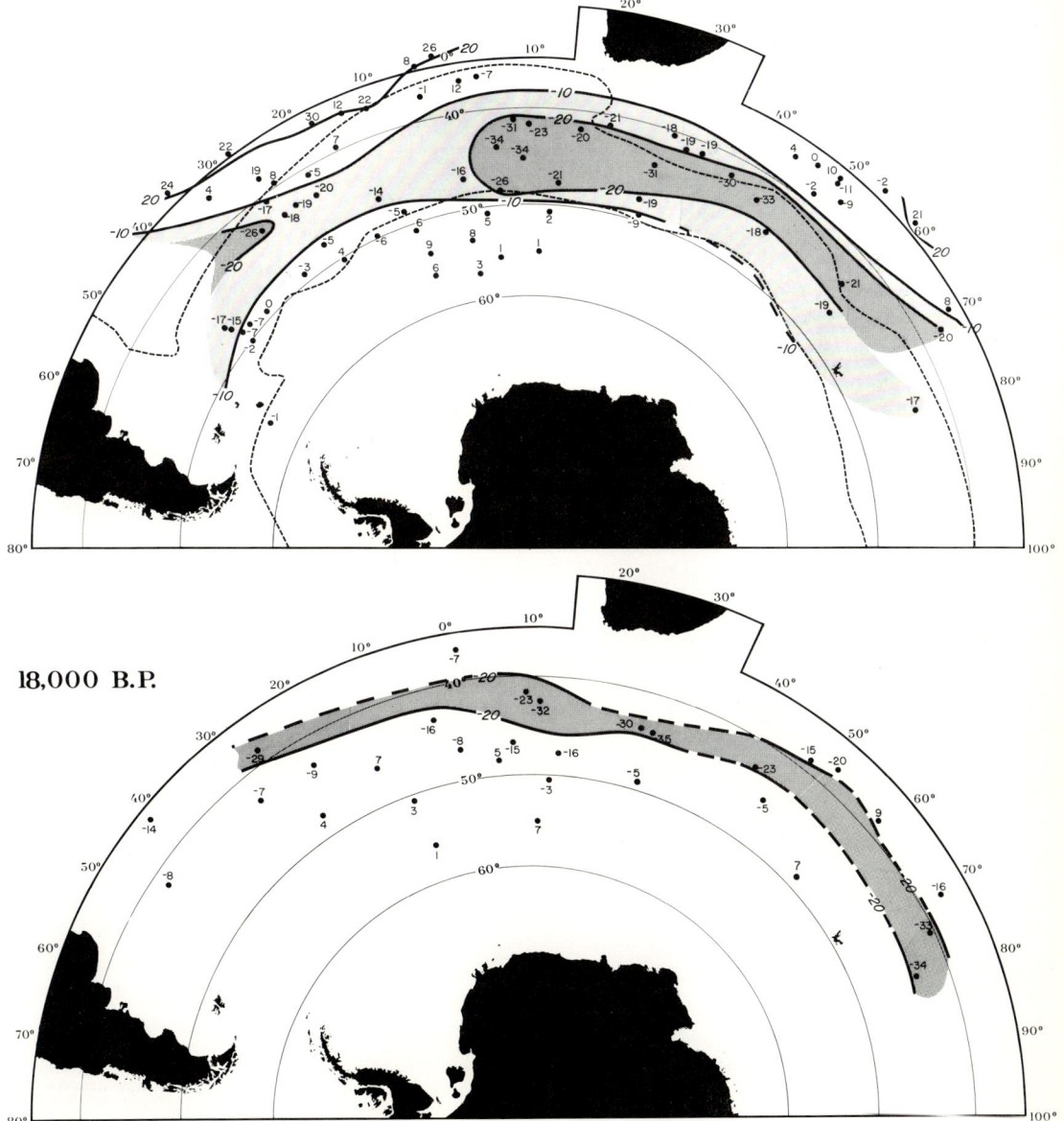

Figure 8. Distribution of factor 3 (subantarctic assemblage) in surface sediment today and 18,000 yr ago. Values are factor loadings × 100.

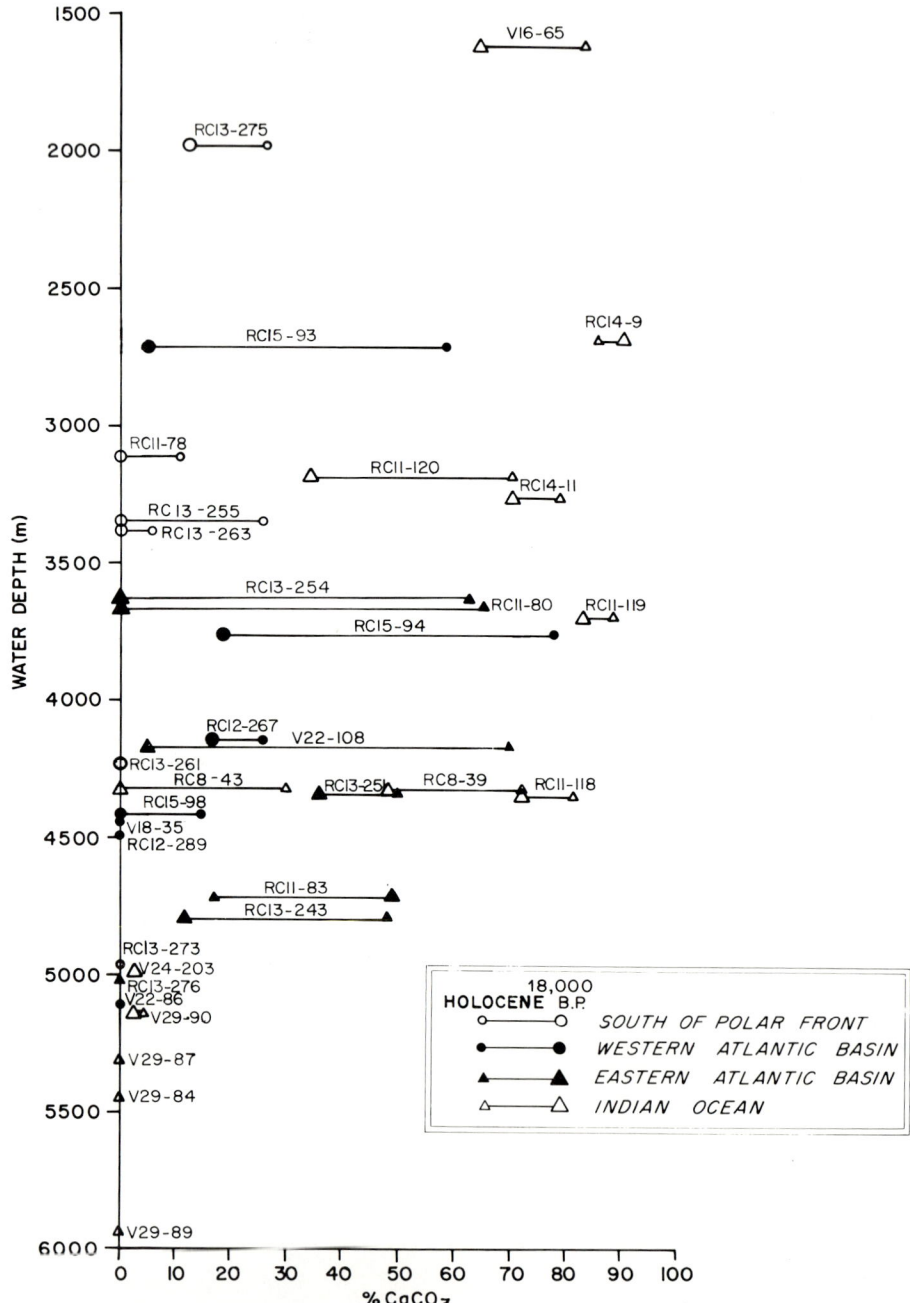

Figure 9. Calcium carbonate percentage of total sediment today and 18,000 yr ago versus water depth. Bar length represents difference in $CaCO_3$ percentage between today and 18,000 B.P. level for each core location.

TABLE 5. LOCATION, DEPTH OF WATER, AND DEPTH IN CORE
OF THE 18,000 B.P. LEVEL FOR 37 CORES

Core	Lat (S)	Long	Water depth (m)	18,000 B.P. level depth in core (cm)	Core	Lat (S)	Long	Water depth (m)	18,000 B.P. level depth in core (cm)
RC8-39	42°53.0'	42°21.0'E	4,330	40	RC13-273	55°04.5'	11°34.5'E	4,967	60
RC8-43	48°41.0'	57°22.0'E	4,319	40	RC13-275	50°43.0'	13°26.0'E	1,984	60
RC11-78	50°52.0'	09°52.0'W	3,115	180	RC13-276	47°41.9'	14°42.3'E	5,015	40
RC11-80	46°45.0'	00°03.0'W	3,656	80	RC14-9	39°01.0'	47°53.0'E	2,692	100
RC11-83	41°36.0'	09°43.0'E	4,718	340	RC14-11	38°00.0'	51°11.0'E	3,268	80
RC11-118	37°48.0'	71°32.0'E	4,354	40	RC14-12	38°45.0'	59°18.0'E	5,271	100
RC11-119	40°18.0'	74°34.0'E	3,709	40	RC15-93	46°05.8'	13°13.5'W	2,714	100
RC11-120	43°31.0'	79°52.0'E	3,193	80	RC15-94	42°53.9'	20°51.3'W	3,762	110
RC12-267	38°41.0'	25°47.0'W	4,144	60	RC15-98	42°56.2'	29°46.6'W	4,416	100
RC12-289	47°54.2'	23°41.5'W	4,484	60	V16-65	45°00.0'	45°46.0'E	1,618	40
RC13-243	36°54.0'	01°19.8'E	4,790	40	V18-35	36°24.0'	41°07.0'W	4,942	120
RC13-251	42°30.7'	11°40.3'E	4,341	20	V22-86	41°36.0'	46°27.0'W	5,110	60
RC13-253	46°36.0'	07°37.5'E	2,494	40	V22-108	43°11.0'	03°15.0'W	4,171	100
RC13-254	48°34.2'	05°07.6'E	3,636	60	V24-203	36°59.0'	59°59.0'E	4,997	180
RC13-255	50°34.5'	02°53.7'E	3,332	140	V29-84	43°51.0'	27°36.0'E	5,451	120
RC13-256	53°10.9'	00°21.3'W	2,525	400	V29-87	49°06.0'	27°23.0'E	5,314	20
RC13-257	55°00.2'	03°00.1'W	2,836	20	V29-89	45°44.0'	25°39.0'E	5,945	120
RC13-261	56°07.3'	08°41.2'W	4,221	80	V29-90	43°42.0'	25°44.0'E	5,148	60
RC13-263	53°48.3'	08°13.0'W	3,389	180					

We have sufficient data from the eastern basin of the South Atlantic to study the change between the present CCD and its level 18,000 yr ago. We have plotted eastern basin South Atlantic cores and Antarctic cores in a depth-latitude plane, indicating $CaCO_3$ content of the core tops as well as the 18,000 B.P. level where it is known (Fig. 11). With these data we can determine the meridional slope of the CCD today and 18,000 yr ago. It is clear that the CCD was significantly steeper within this latitudinal range 18,000 yr ago than it is today. We believe, therefore, that the southward increase in thickness of the bottom water 18,000 yr ago caused significantly greater corrosion in cores RC13-243 and V22-108 than occurs today.

In Figure 11 we plotted the slope of the 1°C isotherm along a traverse at long 0° (Bowen and Stommel, 1971). This 1°C isotherm closely parallels the slope of the present CCD between lat 45°S and 35°S, where we have good control. We have dashed in an estimated position of the CCD where we have less firm control so that it conforms to our data but also parallels the 1°C isotherm. The parallelism between the CCD and the 1° isotherm suggests that the CCD follows an isotherm colder than 1°C. Assuming that a similar relationship between isotherms and the CCD existed 18,000 yr ago, this suggests that the northward slope of the wedge of cold water that is the Antarctic Polar Front moved north 18,000 yr ago and that the northward shift of factor 2 was accompanied by a northward shift of the deep thermal structure.

RADIOLARIA CONCENTRATION 18,000 YR AGO

Lozano and Hays (this volume) mapped the number of Radiolaria per gram in Antarctic surface sediment. Their map is shown in Figure 12 along with a map of Radiolaria per gram at the 18,000 B.P. level, which also gives the percentage of change from Holocene concentrations. In general, there are fewer Radiolaria per gram at the 18,000 B.P. level than in the surface sediment, with exceptions occurring in the northern part. The maximum radiolarian concentration (30,000/g) occurred in the region of the paleo-Antarctic Polar Front in the Atlantic sector (Figs. 12, 13). One possible explanation for reduction in radiolarian concentration in the area between the present polar front and its paleoposition is that the highest radiolarian concentration today is primarily under subantarctic water, and the northward advance of polar water with its high production of diatoms would cause dilution of Radiolaria by diatoms. This explanation is supported by the fact that in this region the sediment was generally more diatomaceous 18,000 yr ago than

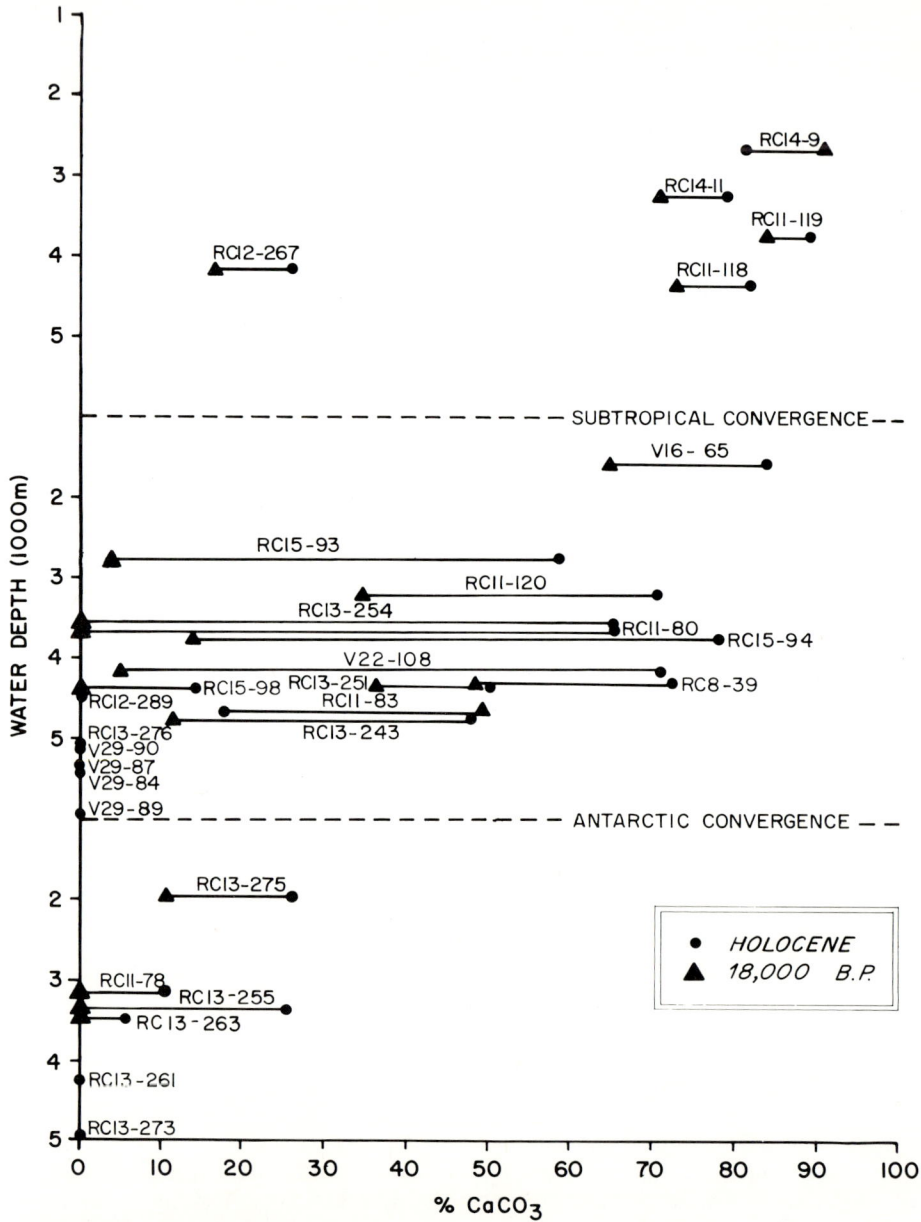

Figure 10. Calcium carbonate percentage of total sediment versus water depth under subtropical, subantarctic and Antarctic waters. Bar length represents difference in CaCO$_3$ between today and 18,000 B.P. level for each core location.

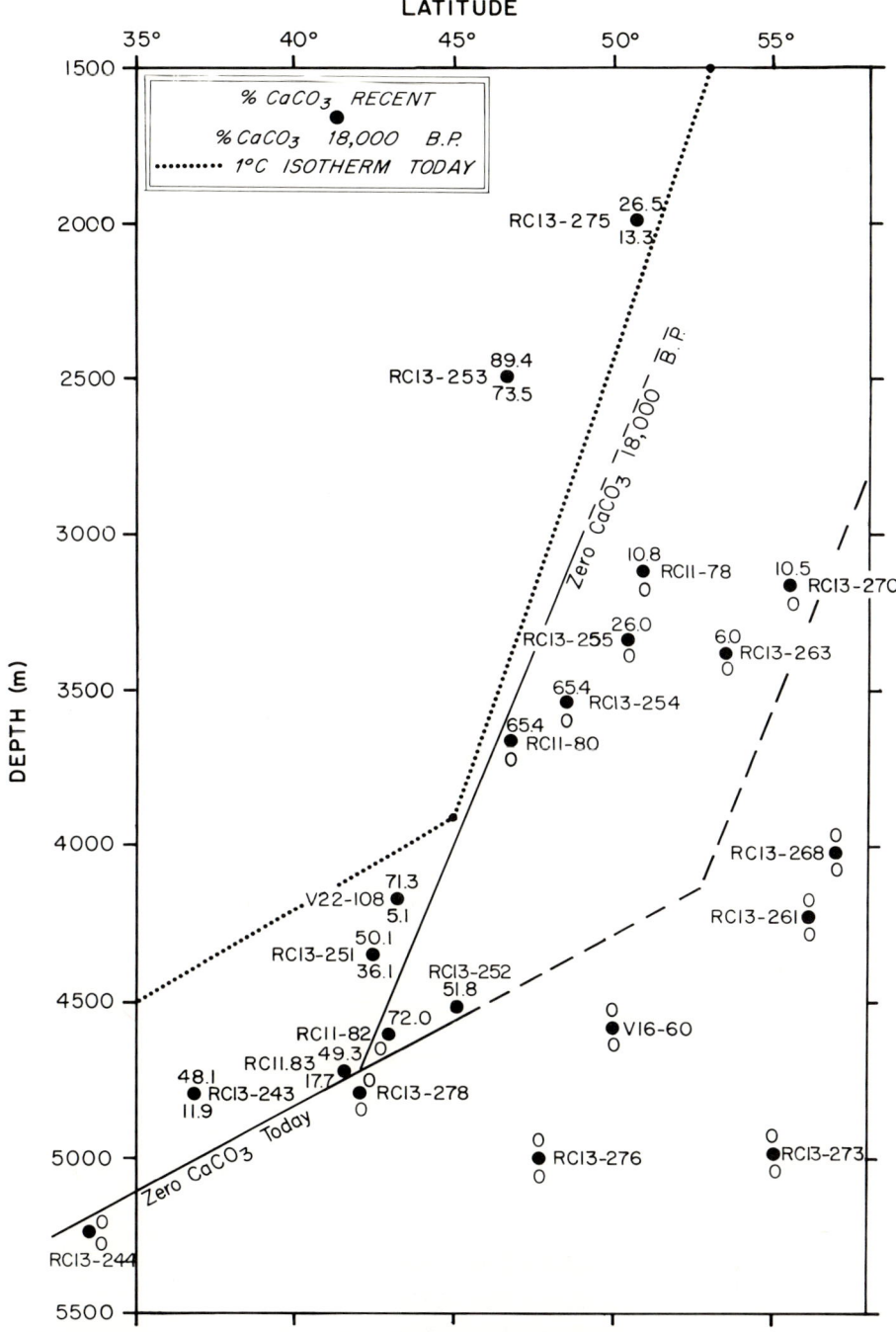

Figure 11. Calcium carbonate percentage of total sediment today and 18,000 yr ago plotted on a latitude-depth plane for cores south of polar front and in the eastern Atlantic basin. Isotherm of 1°C along long 0° from Bowen and Stommel (1971). Movement of the CCD from its position 18,000 B.P. to the position it has today represents retreat of the polar front thermohaline structure to the south.

it is today. This, however, does not explain the reduction in the number of Radiolaria per gram 18,000 yr ago in cores that are south of the present position of the Antarctic Polar Front. This could be caused by a number of factors such as increased diatom production 18,000 yr ago in polar water, increased terrigenous input, or the northward advance of the mean position of the pack ice, which would have reduced overall productivity of both diatoms and Radiolaria. For the southernmost cores in the area, this latter suggestion is probable.

The values of radiolarian concentration are affected by changes in sedimentation rate from core to core as well as changes in bulk density, and thus they do not

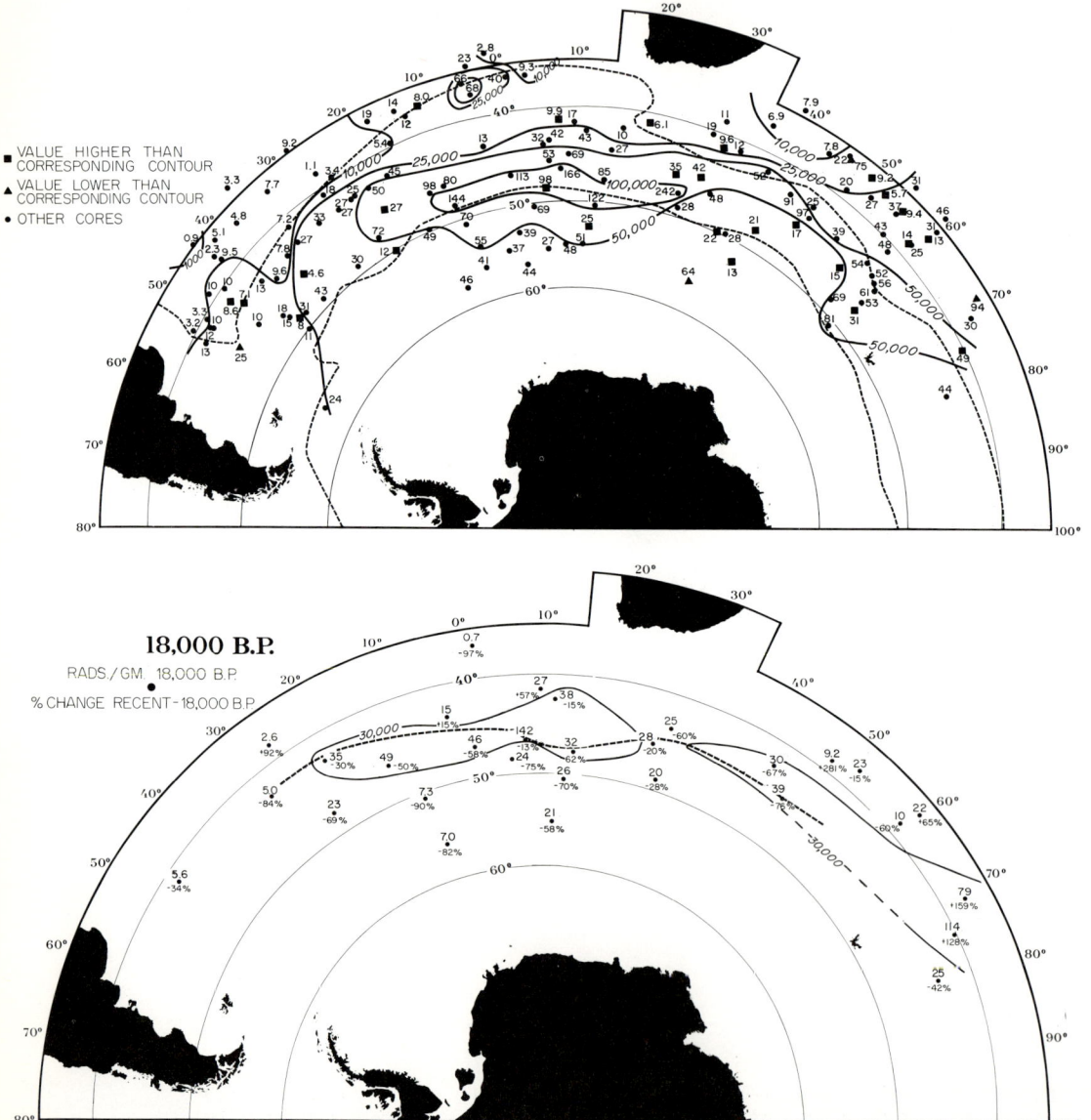

Figure 12. Radiolaria per gram of surface sediment today and 18,000 yr ago. The 18,000 B.P. map also shows percentage of change from value at present. Negative sign indicates a decrease relative to present value.

give a measure of the absolute rates of accumulation of Radiolaria. To correct this, we expressed absolute rates of accumulation of Radiolaria as number of Radiolaria/cm^2/1,000 yr. Of the 37 cores used to determine the 18,000 B.P. level, 16 were considered to offer the best possibility for determining sedimentation rates for both Holocene and glacial sediments.

Rates for Holocene sedimentation were calculated for the interval from the piston core top (considered to represent the present) to the (a)-(b) boundary of the *C. davisiana* curve (Fig. 5), to which an age of 13,000 yr was assigned. Rates for glacial sedimentation were calculated for the interval between boundaries (a)-(b) and (b)-(c) (13,000 to 32,000 yr). All cores used are now dry, and densities (dry weight per unit wet volume) had to be estimated. For those cores in which the main change is from calcareous ooze to calcareous clay, values were obtained from a plot of $CaCO_3$ against D (dry weight in grams per cm^3 of wet sediment) on the basis of 41 measurements made on samples from cores RC13-229, RC13-232, and RC13-244 in the Cape and Angola Basins; these cores varied in $CaCO_3$ content from 89 to 0% (R. Embley, 1974, oral commun). For cores in which the major lithologic change is from diatom ooze to diatomaceous clay, the D values were estimated from average values for similar types of sediment in the Pacific Ocean given by Horn and others (1975).

The results (Fig. 13, Table 6) are in general agreement with the interpretation given above for the values of Radiolaria per gram, but they display in a clearer way the drastic reduction of Radiolaria/cm^2/1,000 yr for those cores south of the present average position of the Antarctic Polar Front. It should be noted that any sediment lost during the coring process from the core top would exaggerate the values for the Holocene deposits.

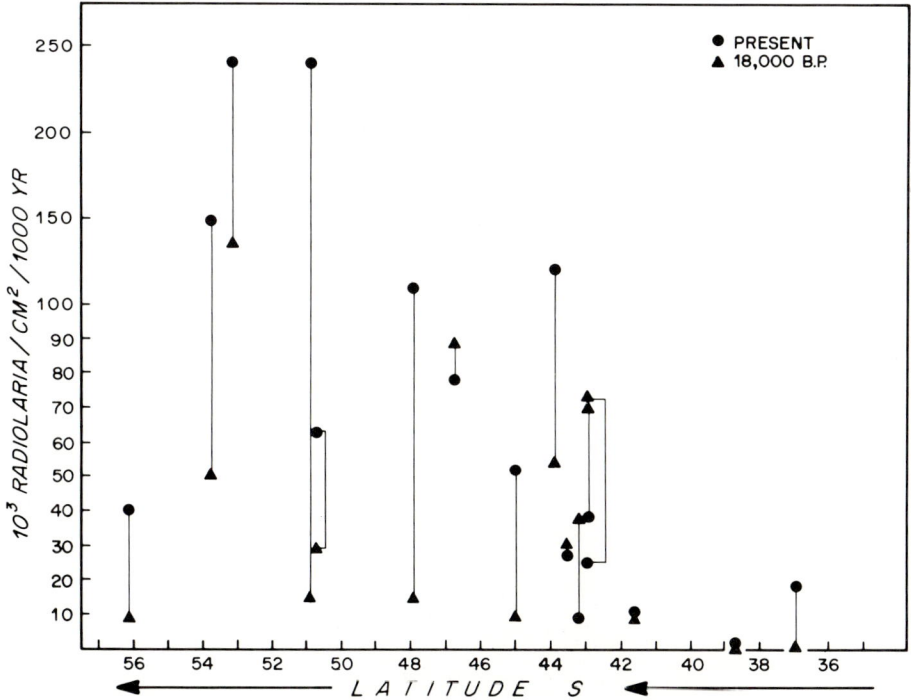

Figure 13. Radiolaria/cm^2/1,000 yr versus latitude today and 18,000 yr ago. Notice change of scale for values over 100,000. Core numbers and locations are given in Table 6.

TABLE 6. ABSOLUTE RATES OF ACCUMULATION OF RADIOLARIA AND C. DAVISIANA TODAY AND 18,000 B.P.

Core	Lat (S)	Long	Density		Sedimentation rate (cm/1,000 yr)		Radiolaria/cm²/1,000 yr		C. davisiana/cm²/1,000 yr	
			Top	18,000 B.P. level	Holocene	Glacial	Top	18,000 B.P. level	Top	18,000 B.P. level
RC13-243	36°54'	1°20'E	0.54	0.43	1.9	1.2	21,344	372	297	136
RC12-267	38°41'	25°47'W	0.46	0.43	3.8	1.1	1,407	1,024	33	124
V22-86	41°36'	46°27'W	0.40	0.40	3.1	4.2	10,652	9,485	278	1,024
RC8-39	42°53'	42°21'E	0.68	0.54	2.3	8.4	39,383	70,452	1,063	17,084
RC15-94	42°54'	20°51'W	0.74	0.44	3.1	8.9	25,580	73,434	297	13,181
V22-108	43°11'	3°15'W	0.67	0.41	3.8	6.3	9,904	37,968	111	9,378
RC11-120	43°31'	79°52'W	0.67	0.49	3.1	4.2	27,751	34,748	339	5,772
V29-84	43°51'	27°36'E	0.40	0.40	4.6	5.3	121,887	54,295	6,192	15,583
V16-65	45°00'	45°46'E	0.79	0.62	2.3	1.1	52,511	9,485	866	988
RC11-80	46°45'	0°03'W	0.63	0.40	3.1	4.7	78,846	88,266	1,908	21,978
RC12-289	47°54'	23°42'W	0.40	0.40	3.8	1.6	110,299	14,970	2,647	2,994
RC13-275	50°43'	13°26'E	0.30	0.30	2.3	4.2	63,139	29,034	1,282	7,276
RC11-78	50°52'	9°52'W	0.42	0.40	9.2	5.3	241,597	15,298	4,494	5,339
RC13-256	53°11'	0°21'W	0.25	0.40	24.6	8.4	241,996	135,420	8,010	49,550
RC13-263	53°48'	8°13'W	0.25	0.5	11.5	3.2	149,416	50,491	6,380	18,439
RC13-261	56°07'	8°41'W	0.25	0.5	3.9	2.6	40,329	9,184	1,024	3,049

Note: See text for estimation of densities and rates of sedimentation; see Figure 13.

In Table 6 we also show the rates of accumulation of *C. davisiana*/cm^2/1,000 yr. Although there are exceptions, in general the absolute rates of accumulation of *C. davisiana* were higher 18,000 yr ago than they were during the Holocene Epoch. However, changes in absolute accumulation rates of *C. davisiana* cannot completely explain the changes in relative percentage of *C. davisiana* (Fig. 5, Table 6).

MOVEMENT OF THE POLAR FRONT

Studies of seasonal and secular changes in the position of the Antarctic Polar Front have largely led to the conclusion that there is little or no change (Bowen and Stommel, 1971; Botnikov, 1963, 1964). Botnikov (1964) showed that deviations from the mean position of the Antarctic Polar Front between 1901 and 1960 are rarely more than 2° of latitude. We have presented evidence that the polar front was significantly farther north 18,000 yr ago (some 7° of latitude in the western Atlantic and about 4° in the eastern Atlantic and western Indian Ocean) than it is today throughout most of the area, with one exception—the region directly south of Africa. Gordon (1967) presented strong arguments suggesting that along much of its length, the position of the polar front is controlled by bottom topography. It is beyond the scope of this paper to go into a detailed analysis of why the polar front moved north in some areas and remained stationary in others. South of Africa, where the polar front did not move, the ridge axis trends east, forming a zonal topographic barrier. To the west, the ridge trends north with numerous east-trending fracture zones cutting across it. The Antarctic Polar Front apparently could move north in this region and in the western basin of the South Atlantic to reach a new equilibrium position over one or several more-northerly fracture zones. To the east, the front moved north over the broad surface of the Crozet Plateau.

PALEOTEMPERATURE ESTIMATES

Choosing samples located at the maximum abundance of *C. davisiana*, we estimated temperatures for 18,000 B.P. using the paleoecological equations developed by Lozano and Hays (this volume), who followed the techniques of Imbrie and Kipp (1971). From these data we have developed a set of maps (Figs. 14 through 17) that were drawn by making corrections to the temperature data and making certain assumptions about isotherm and isoanomaly trends, including the following:

1. The anomaly at any point was considered the difference between sea-surface temperature today over the core site, as read from the sea-surface temperature map of Lozano and Hays, and the estimated temperature 18,000 yr ago. An exception to this rule was made for three cores—the two westernmost South Atlantic cores, V18-35 and V22-86, and core RC14-12 from the Crozet Basin. The temperature difference used for these cores in the anomaly map is the difference between the estimated temperature for the core top and the estimated temperature at the 18,000 B.P. level. This was done because data for sea-surface temperatures are highly variable in the area of cores V18-35 and V22-86. In the area of core RC14-12, bottom mixing is strong. Because it is not possible to correct the 18,000 B.P. level for mixing effects, the difference between the estimate for 18,000 B.P. and today's estimate may be more representative of the true difference than the difference between today's temperature and the estimate for 18,000 B.P.

360   Hays and Others

2. Today, a factor-2 loading of 0.9 corresponds approximately with the Antarctic Polar Front. We also assume this was true 18,000 yr ago. Because the polar front today represents a zone of steep thermal gradient, we constructed, wherever possible, the temperature anomaly between 18,000 B.P. and today so that the factor-2 loading of 0.9 cuts the anomaly contours approximately at the point of maximum curvature. When our data precluded this, we used our difference data to construct the anomaly in the most conservative way.

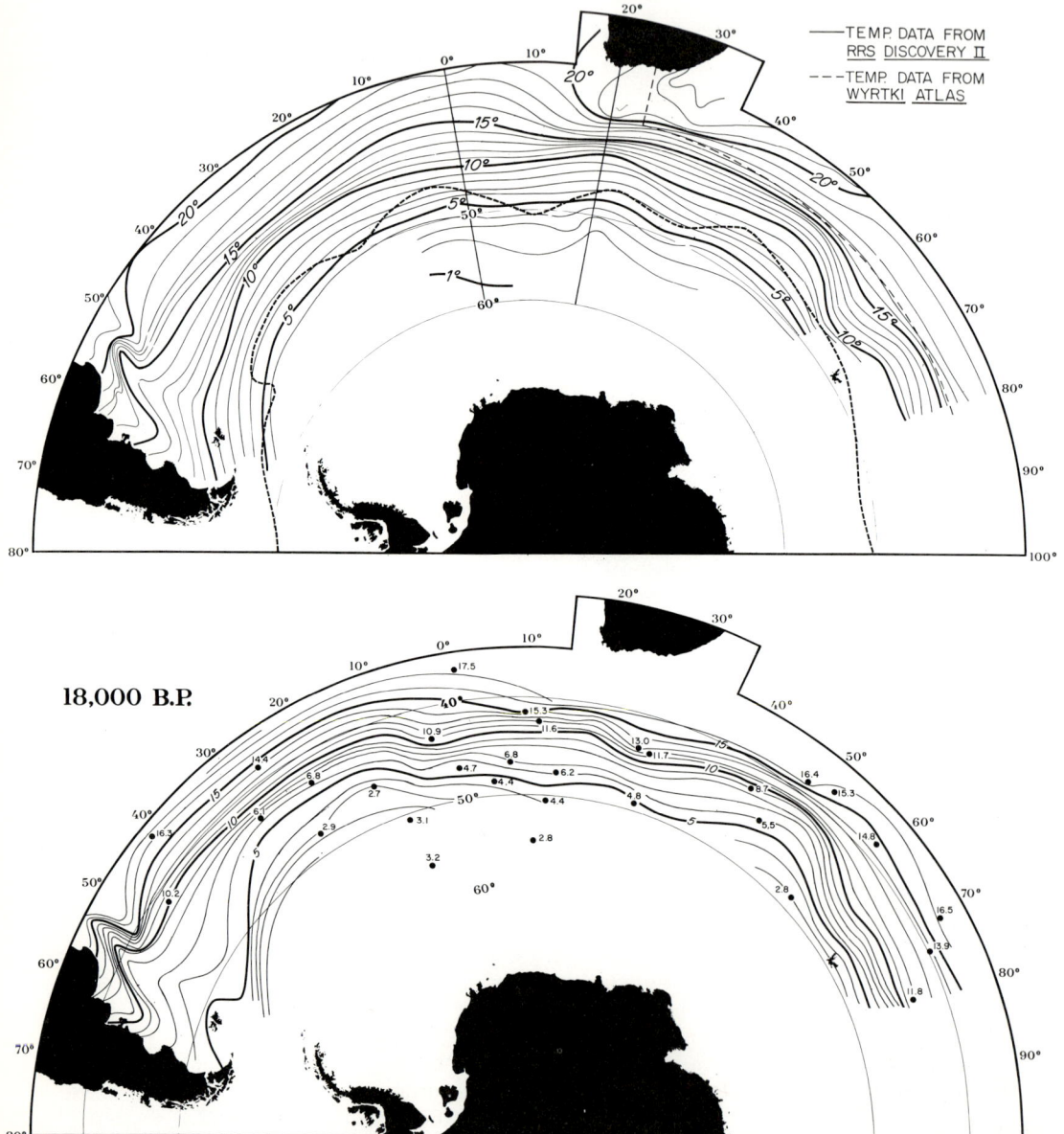

Figure 14. February sea-surface temperature today and 18,000 yr ago. The 18,000 B.P. values were obtained through the use of the paleoecological equation developed by Lozano and Hays (this volume). Today's temperature data are based on a compilation from various sources (Lozano and Hays, this volume).

The corrected anomaly maps (Figs. 16, 17) show a strong temperature anomaly in the western South Atlantic (8°C in summer, 5°C in winter) and a weaker anomaly in the western Indian Ocean. These anomalies would be expected from the above discussion of the movement of the polar front. There is effectively no anomaly south of Africa or south of the present position of the polar front. In this latter area we are hampered, because our techniques cannot estimate temperatures colder than 3°C in the summer and 0°C in the winter. However, because temperatures

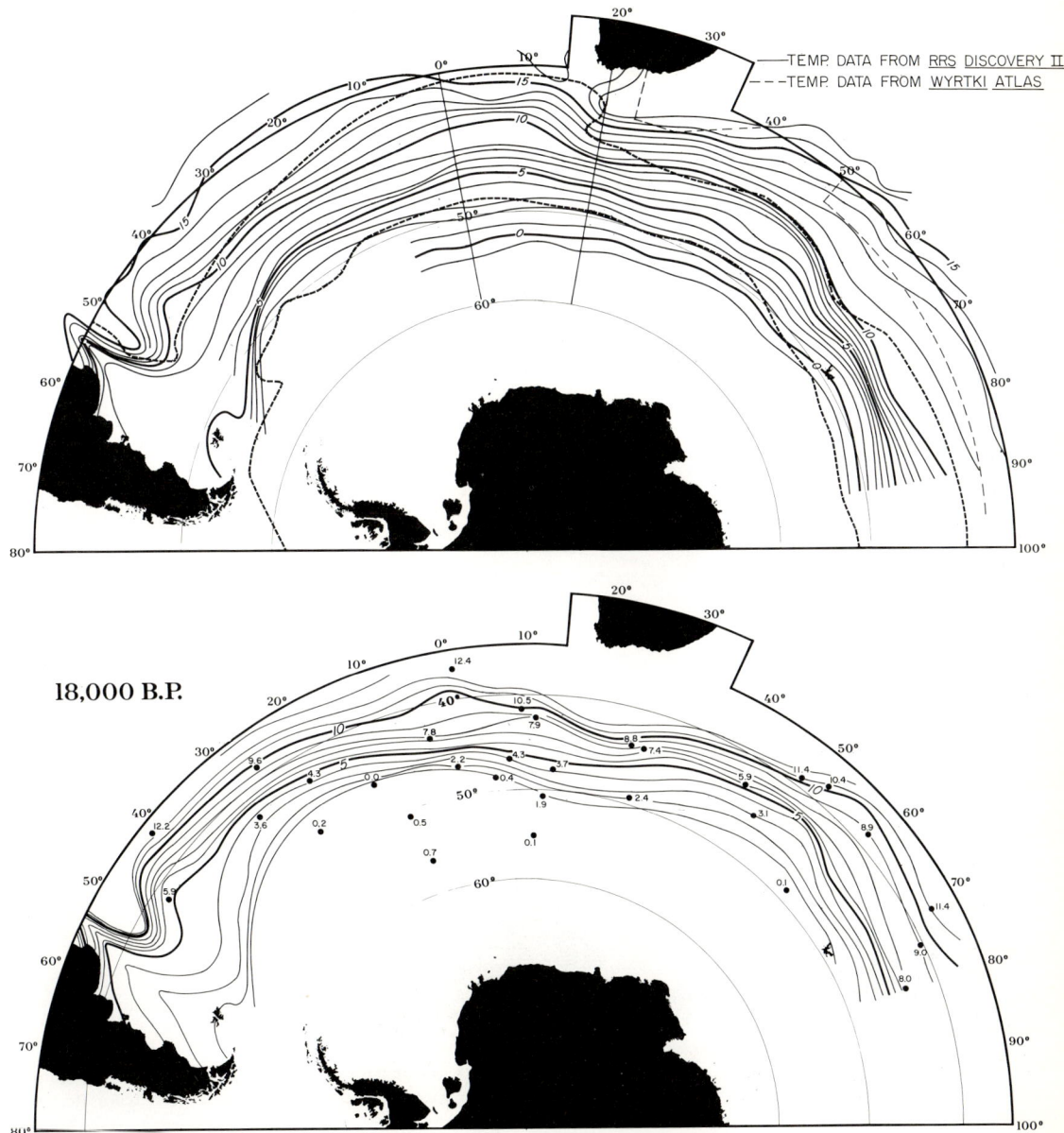

Figure 15. August sea-surface temperature today and 18,000 yr ago. The 18,000 B.P. values were obtained through use of the paleoecological equation developed by Lozano and Hays (this volume). Today's temperature data are based on a compilation from various sources (Lozano and Hays, this volume).

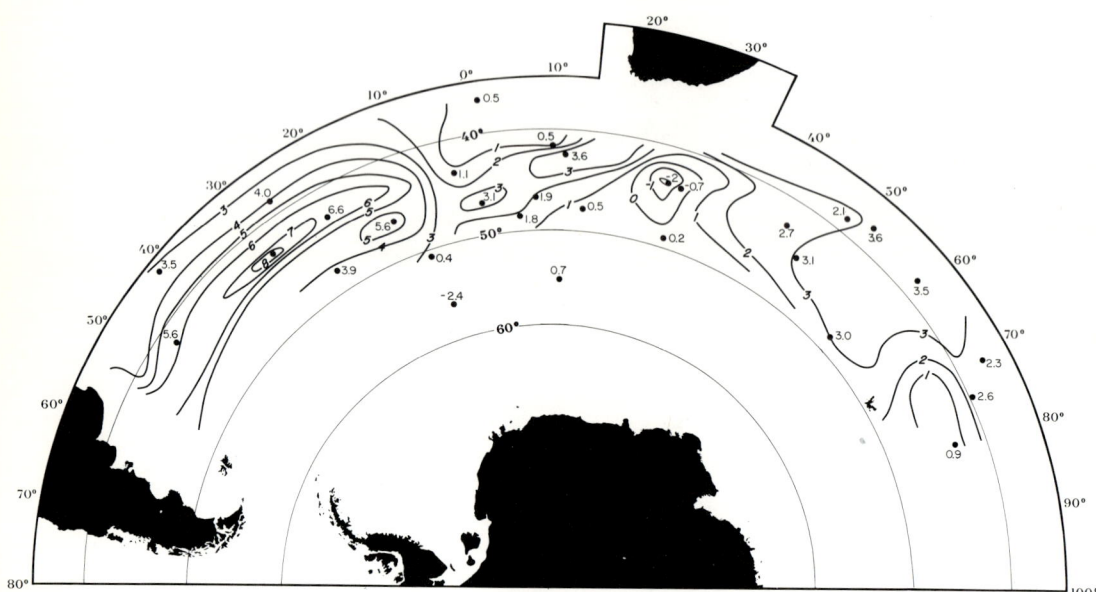

Figure 16. February sea-surface temperature anomaly (today and 18,000 B.P.). Value for each location was obtained by subtracting the estimated paleotemperature for the 18,000 B.P. level sample from the value read from map in Figure 6 of Lozano and Hays (this volume).

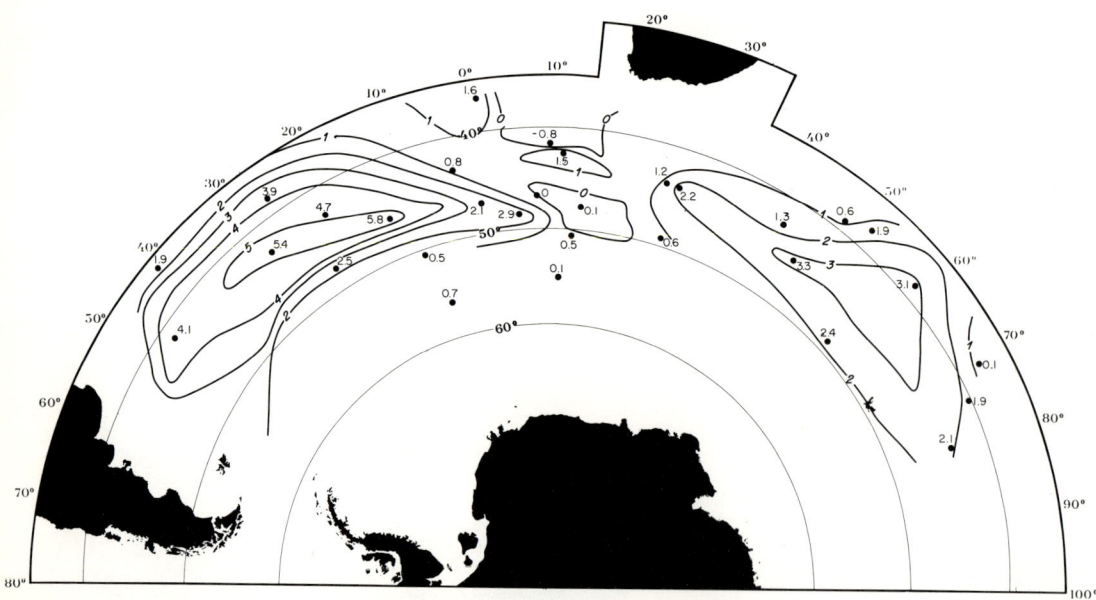

Figure 17. August sea-surface temperature anomaly (today and 18,000 B.P.). Value for each location was obtained by subtracting the estimated paleotemperature for the 18,000 B.P. level sample from the value read from the map in Figure 7 of Lozano and Hays (this volume).

today are very low south of the polar front, the real anomaly, if it exists, must be small.

The difference between the summer and winter anomalies cannot be explained by our inability to estimate very low temperatures, and we conclude therefore that it is real. These maps clearly show that the main differences between today and 18,000 yr ago occurred in the subantarctic region; subtropical waters show little temperature change.

The temperature maps are largely self-explanatory. The northward movement of the polar water 18,000 yr ago steepened the temperature gradients in the subantarctic region compared with today. In regions where the isotherms are closely spaced today, 18,000 yr ago they were also. In essence, most of the temperature data generated in this study can be explained by a northward movement of the polar water 18,000 yr ago.

SEA ICE AROUND ANTARCTICA, 18, 000 B.P.

The extent of pack ice 18,000 yr ago around Antarctica is critical to an understanding of Southern Hemisphere climate. Yet it is difficult to find criteria for recognizing the ice extent in the past. The lithology of cores south of the present position of the Antarctic Polar Front is striking because of the fluffy nature of the diatom ooze that predominates. Our data indicate that this diatom ooze had sedimentation rates as much as 40 cm/1,000 yr, making it the fastest accumulating open-ocean pelagic sediment. A pattern repeated in a number of cores is a surficial layer of nearly pure diatom ooze overlying less diatomaceous silty radiolarian clay. Philippi (1912), the first to observe this lithologic sequence, interpreted the underlying less diatomaceous sediment to be glacial marine because of its similarity to sediment he had observed close to the Antarctic continent. Later workers (Schott, 1939; Hough, 1950; Litsitzin, 1960, 1962, 1972) have also observed this sequence, but it has been difficult to date. Using our *C. davisiana* stratigraphy, it is possible to estimate the age of the change between the two lithologies. The data in Table 7 indicate that the change from sandy radiolarian clay to nearly pure diatom ooze occurred about 18,000 yr ago or somewhat later. This sedimentary sequence can be explained either by an increase in ice-rafted detritus, which effectively diluted the diatoms, or a reduction in diatom production or both. To test which of these possibilities is correct, we have determined the relative abundance of *C. davisiana* in the upper 6.5 m of core RC13-263 and compared it with the relative abundance

TABLE 7. DEPTH OF *C. DAVISIANA* PEAK CORRESPONDING TO 18,000 B.P. LEVEL AND UPPERMOST LITHOLOGIC CHANGE FROM DIATOM OOZE TO DIATOMACEOUS CLAY IN 9 CORES

Core	Depth of *C. Davisiana* peak (cm)	Depth lithologic change (cm)
RC15-92	not done	100?, core disturbed
RC11-78	180	140?, not striking
RC13-263	180	160–210
RC13-261	80	50
RC13-255	320	270
RC13-275	60	50
RC13-273	60	50
RC11-96	100	90
RC11-97	120	100

of *C. davisiana* in core RC11-120 (Fig. 18), a core with O^{18}/O^{16} data. The lithology of core RC13-263 through this interval can be divided into a topmost layer of nearly pure diatom ooze about 160 cm thick, an underlying layer of diatomaceous clay between 160 and 460 cm, and a lower interval of diatom ooze between 460 cm and about 620 cm (Fig. 18).

The *C. davisiana* abundance peaks (a) and (d) (Fig. 18) are clearly defined in core RC13-263, indicating that the upper diatom ooze interval corresponds approximately to the Holocene Epoch and has a sedimentation rate of 11.5 cm/1,000 yr. The interval between *C. davisiana* abundance peaks (b) and (d) has a sedimentation rate of about 2.7 cm/1,000 yr. The lower diatomaceous ooze corresponds to either *C. davisiana* minimums e_2 or e_3 or both. Whichever interpretation is used for the duration of the lower diatom-ooze layer, this layer has a higher sedimentation rate than the interval between *C. davisiana* abundance peaks (b) and (d). These data strongly suggest that the lithologic changes in the core are due in large part to diatom productivity variations. If the more clayey layer was caused by terrigenous material influx, its sedimentation rate should be much higher than the diatom-ooze layers. We conclude therefore that increased ice cover in the summer months was responsible for inhibiting diatom production between 90,000 to 110,000 and 15,000 yr ago. We cannot at this point determine whether this ice was glacial

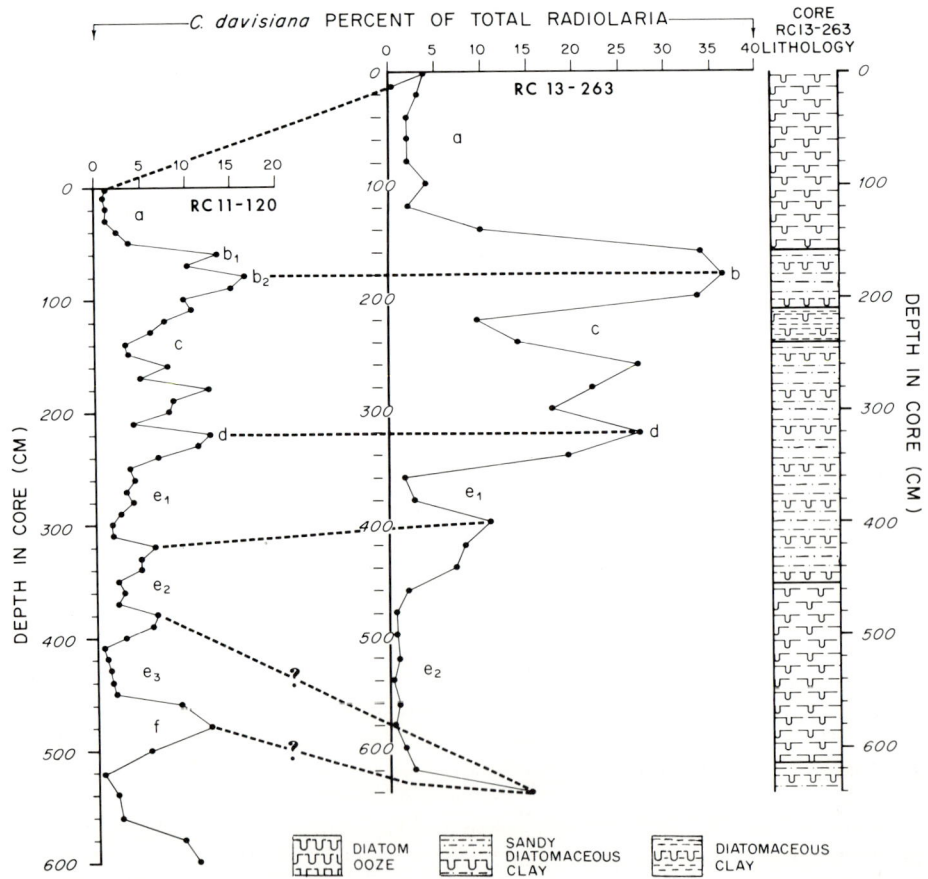

Figure 18. Correlation of *C. davisiana* stratigraphies in cores RC11-120 and RC13-263.

or sea ice. The less diatomaceous layer does contain some sand and silt, which might be interpreted as support for Wilson's (1964) surge theory so strongly defended by Hollin (1965, 1969). If the ice cover was at least initially glacial, then we might expect the highest concentration of ice-rafted detritus in the lower part of the layer. The petrology of the less diatomaceous layer is now being studied.

None of the cores in our area is far enough south to be covered by sea ice through most of the year. By studying USS *Eltanin* cores from the Pacific and Indian Oceans, Goodell (1973) has been able to map sediments close to the continent. A thin band of siliceous silty clay extends to north of lat 65°N in the Indian Ocean and to nearly lat 60°S in parts of the Pacific (Fig. 19). Today this corresponds approximately to the mean position of sea ice in April and January (Mackintosh, 1972).

Gordon (1971) made a detailed study of the Antarctic Ocean thermal structure along long 120°W. He illustrated (his Fig. 3) stably stratified water extending from lat 60°S to 63°S and strongly mixed water with little thermal stratification between lat 63°S and the Antarctic Polar Front at about lat 57°S. Today, siliceous clay is accumulating under the stably stratified water, and diatom ooze is accumulating under the mixed water. The stably stratified region is probably caused by the presence of ice throughout most of the year.

We have mapped sediment types from the 18,000 B.P. level in the cores (Fig. 19). The most striking difference between glacial sediment distribution patterns and those of today is the great northward displacement of the southern boundary of the diatom-ooze belt 18,000 yr ago (Fig. 19). The diatom-ooze belt is reduced in area, but its northern limit is extended north a distance comparable to the northward advance of the Antarctic Polar Front.

We interpret this sediment pattern to indicate a much greater extent of sea ice in the summer months 18,000 yr ago, assuming that the boundary between siliceous clay and silt and diatom ooze roughly corresponded to the mean annual extent of the ice as it does today.

Figure 20 gives our best estimate of conditions around Antarctica in the summer and winter 18,000 yr ago. The greatest difference between today and 18,000 yr ago occurs in the summer when the sea ice was extended well beyond its present summer limits. As we pointed out above, the Antarctic Polar Front was displaced northward, and the subantarctic waters were compressed.

CONCLUSIONS AND QUESTIONS

From this study we draw the following conclusions.

Major changes in the relative abundance of the radiolarian species *Cycladophora davisiana* are time dependent through the past 35,000 yr and probably longer, thus constituting good stratigraphic indicators. The synchroneity of relative-abundance changes of *C. davisiana* with changes in O^{18}/O^{16} values throughout a broad area of the Antarctic and subantarctic indicates that these changes are time dependent and thus constitute a powerful high-resolution stratigraphic tool for the Antarctic Ocean. Because *C. davisiana* is a cosmopolitan species, similar abundance changes in other parts of the world may also prove to be useful as stratigraphic guides for upper Quaternary deposits.

The Antarctic Polar Front was north of its present position 18,000 yr ago approximately 7° of latitude in the western basin of the South Atlantic and somewhat less in the eastern basin of the South Atlantic and the western Indian Ocean. Much of the data presented in this paper support this conclusion; this includes

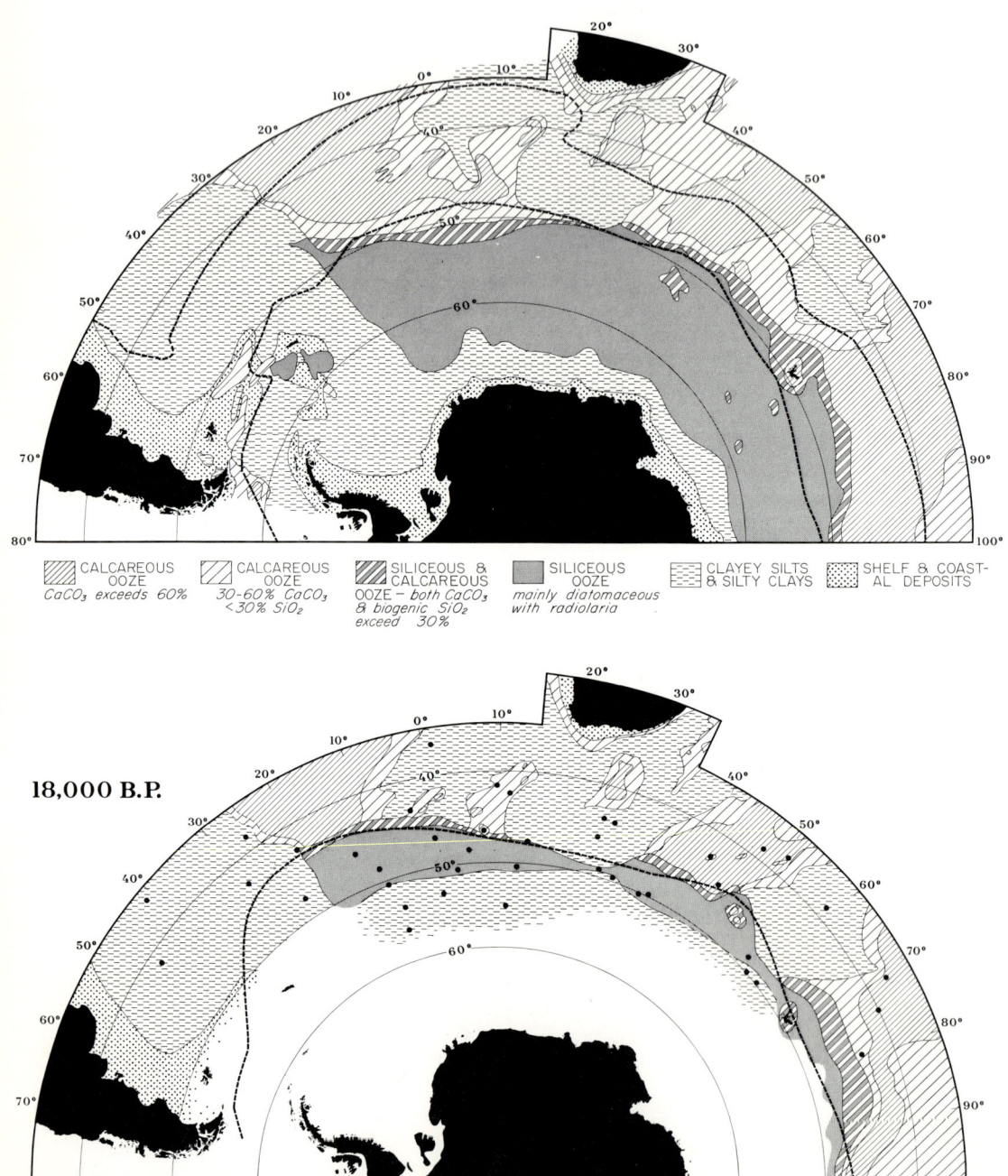

Figure 19. Generalized distribution of sediments today and 18,000 B.P. Sediment types after Goodell (1973). Calcareous ooze is usually foraminiferal, sometimes calcilutite. The clayey silt and silty clay are calcareous near calcareous ooze, siliceous adjacent to siliceous ooze, and sandy adjacent to continental shelves.

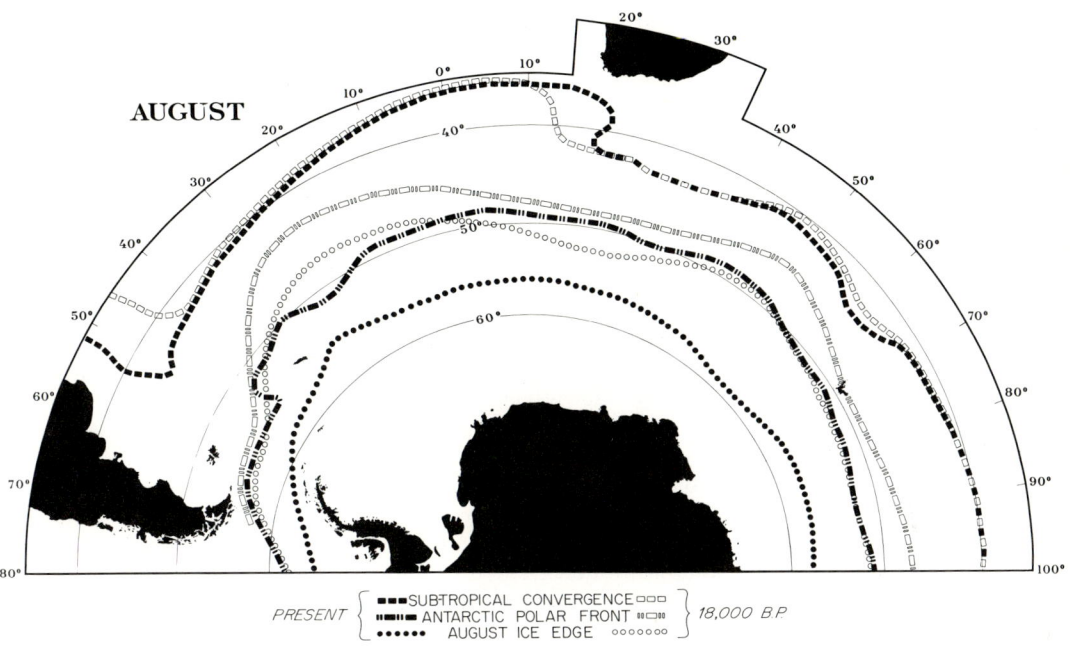

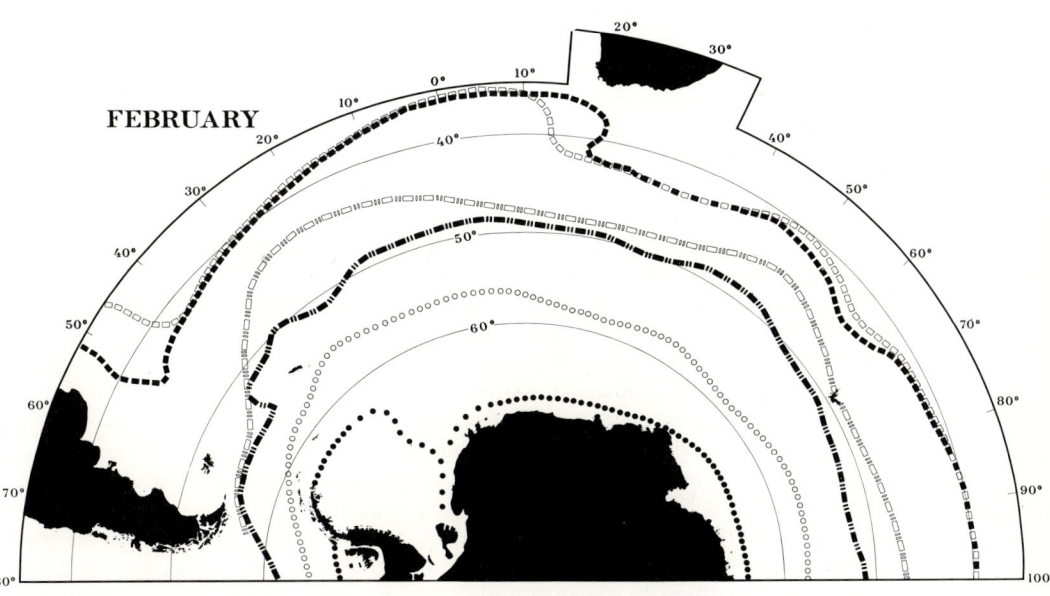

Figure 20. Main oceanographic features around Antarctica today and 18,000 yr ago for August (winter) and February (summer). Ice limits for today from Mackintosh (1972). Present ice coverage within limits shown averages about 50% (Neumann and Pierson, 1966).

a northward shift of the maximum concentration zone of Radiolaria (today this boundary approximates the position of the Antarctic Polar Front) and a northward shift of the boundary between low and high calcareous sediments (today this boundary also corresponds approximately with the position of the polar front).

The boundary between the subtropical and subantarctic water masses was approximately the same as it is today. The factor analysis of the radiolarian assemblages indicates that the boundary between the subtropical assemblage (factor 1) and the subantarctic assemblage (factor 3) was approximately the same 18,000 yr ago as today. Consequently, the expansion of the polar water resulted in a compression of the subantarctic water mass and a steepening of the thermal gradient between the Antarctic Polar Front and subtropical convergence.

The extent of summer sea ice on the Antarctic Ocean was much greater 18,000 yr ago than it is today, although winter sea ice may have been only moderately greater. A lithologic change in Antarctic cores from a topmost sequence of fluffy diatom ooze with a high deposition rate to an underlying radiolarian diatomaceous silty clay with a lower deposition rate occurred between 15,000 and 20,000 yr ago, the exact time varying from one core to another. This change is most probably caused by significantly greater summer coverage of sea ice during the deposition of radiolarian diatomaceous clay than during the more recent deposition of fluffy diatomaceous ooze. The lower sedimentation rate during the deposition of the less diatomaceous sediment precludes the possibility that the lithologic change was caused by an upward increase in diatom productivity.

These observed changes in the Antarctic between 18,000 yr ago and today would have had significant consequences for the rest of the world. Below we list several of the consequences:

1. Increased sea ice in the Antarctic would intensify global atmospheric and oceanic circulation. Fletcher (1969) made a comprehensive study of the effect of increased Antarctic sea ice on global circulation. Because sea ice insulates the atmosphere from oceanic heat, increased sea-ice cover in the Antarctic will greatly increase atmospheric cooling in the winter. This will increase the thermal gradient to the Equator and consequently increase the vigor of atmospheric circulation. The increased atmospheric circulation would in turn increase the oceanic surface circulation. Evidence of this is available from the equatorial regions of both the Atlantic and Pacific. Gardner and Hays (this volume) have shown that the south equatorial current system was significantly stronger than it is today. Various authors have suggested that the Pacific equatorial current system was significantly stronger during glacial periods than during interglacial periods (Arrhenius, 1952; Hays and others, 1969; Luz, 1973). How increased vigor of Southern Hemisphere circulation may have influenced the Northern Hemisphere is difficult to assess, but it seems likely that it did.

2. A rise in the calcium carbonate compensation depth (CCD) in the Antarctic would have caused a depression of the CCD in lower latitudes. The expansion of polar waters into lower latitudes in the Southern Hemisphere, as well as in the North Atlantic as shown by McIntyre and others (this volume), reduced the area of calcium carbonate deposition 18,000 yr ago as compared with today. Maintenance of the calcium carbonate budget would require increased calcium carbonate deposition in lower latitudes and depression of the CCD. Work in the equatorial Pacific (Arrhenius, 1952; Hays and others, 1969) has shown that the CCD was depressed in this area during glacial but not during interglacial times.

The position of the Antarctic Polar Front controlled the glaciation of subantarctic islands. Examination of the data from subantarctic islands has shown that the islands positioned north of the Antarctic Polar Front were not significantly glaciated

18,000 yr ago (for example, Falkland Islands), whereas those lying to the south of the polar front were covered by extensive ice caps (for example, South Georgia). Thus, the area of potential continental low-altitude glaciation in the Southern Hemisphere (the area south of the 18,000 B.P. position of the polar front) extended to latitudes comparable to the southern limits of Northern Hemisphere ice sheets.

This study has raised more questions than it has answered, and some of the more important questions deserve to be stated.

Why did the Antarctic sea ice not melt back to the continent in the summer 18,000 yr ago as it does today? If our conclusions about the extent of summer sea ice in the Antarctic are correct, this is the most important question raised by our study; and its answer will probably have to wait for a definitive explanation of the causes of glacial epochs. However, because Antarctic sea ice seems to have been the Southern Hemisphere counterpart of Northern Hemisphere midlatitude ice sheets, it is important to define the problem. Southern Hemisphere continental ice was not much greater 18,000 yr ago than it is today, and its influence cannot adequately explain the great expansion of summer sea ice. There are three main sources of heat for the Antarctic Ocean: the sun, upwelling circumpolar deep water (primarily North Atlantic Deep Water), and atmospheric heat flux—any or all of which may have varied to affect the sea-ice cover.

Important areas of future research to resolve this question will be detailed studies of deep-sea cores to determine precisely when oceanic ice cover waxed and waned theoretical studies of how atmospheric heat transport into the Antarctic may have differed with varying amounts of sea ice, and global modeling experiments of deep-ocean circulation to determine how net transport of the deep water may have varied with changing surface conditions.

Why was the Antarctic Polar Front farther north 18,000 yr ago than it is today? Because the reason for the present position of the polar front is unclear, it is difficult to guess why it was farther north 18,000 yr ago. Two possibilities are the influence of increased sea ice and increased wind stress effectively expanding the polar gyre. However, it is unclear how effective the wind stress would be in increasing net transport of the ocean water if much of the surface were covered with sea ice.

ACKNOWLEDGMENTS

We gratefully acknowledge encouragement and useful suggestions from A. Gordon, S. Streeter, and J. Imbrie during the preparation of the manuscript. A. McIntyre, T. Moore, G. Kukla, and B. Luz read the manuscript and made several helpful suggestions. M. Perry, who is responsible for the drafting, offered a number of creative ideas.

The cores used in this study were taken and curated through National Science Foundation Grant DES-72-01568 and Office of Naval Research Contract N00014-67-A-0108-0004. CLIMAP research was supported by National Science Foundation Grant IDO71 04204. N. Shackleton was a 1974–1975 Lamont-Doherty Senior Post-Doctoral Fellow.

REFERENCES CITED

Arrhenius, G., 1952, Sediment cores from the East Pacific: Swedish Deep-Sea Exped. Rept., 1947–1948, v. 5, p. 1–228.

Bandy, O. L., 1959, The geologic significance of coiling ratios in the foraminifer *Globigerina pachyderma* (Ehrenberg) [abs]: Geol. Soc. America Bull., v. 70, p. 1708.

Bandy, O. L., Casey, R. E., and Wright, R. C., 1971, Late Neogene planktonic zonation, magnetic reversals, and radiometric dates, Antarctic to the Tropics, *in* Reid, J., ed., Antarctic oceanology I: Am. Geophys. Union Antarctic Research Ser., v. 15, p. 1-26.

Bellair, P., 1965, Un exemple de glaciation aberrante. Les Îles Kerguelen: Com. Nat. Français des Recherches Antarctiques, Pub. 11, p. 1-27.

Botnikov, V. N., 1963, Geographical position of the Antarctic Convergence Zone in the Antarctic Ocean: Soviet Antarctic Exped. Inf. Bull., no. 41 (English ed.), v. 4, no. 6, p. 324-327.

——1964, Seasonal and long term fluctuations of the Antarctic Convergence Zone: Soviet Antarctic Exped. Inf. Bull., no. 45 (English ed.), v. 5, no. 2, p. 92-95.

Bowen, J. L., and Stommel, H., 1971, How variable is the Antarctic Circumpolar Current?, *in* Quam, L. O., and Porter, H. D., eds, Research in the Antarctic: Am. Assoc. Adv. Sci. Pub. 93, p. 645-650.

Broecker, W. S., and Kulp, J. L., 1957, Lamont natural radio-carbon measurements, IV: Science, v. 126, p. 1324-1334.

Burckle, L. H., and Hays, J. D., 1966, Tertiary sediments of Falkland platform and Argentine continental slope [abs.]: Am. Assoc. Petroleum Geologists Bull., v. 50, no. 3, p. 607.

Clapperton, C. M., 1971, Geomorphology of the Stromness Bay-Cumberland Bay area, South Georgia: British Antarctic Survey Sci. Rep., no. 70, 25 p.

Donahue, J. G., 1970, Diatoms as Quaternary biostratigraphic and paleoclimatic indicators in high latitudes of the Pacific Ocean [Ph.D. dissert.]: New York, Columbia Univ., 230 p.

Ehrenberg, C. G., 1844, Einige vorläufige Resultate seiner Untersuchungen der ihm von der Südpolreise des Capitan Ross, so wie von den Herren Schayer und Darwin qugekommenen Materialien, über das Verhalten des kleinsten Levens in den Oceanen und den grössten bisher zugänglichen Tiefen des Weltmeeres: Berlin, Akad. Wiss. K. Preuss., p. 182-207.

Emery, K. O., 1960, The sea of southern California: New York, John Wiley & Sons, Inc., 366 p.

Emiliani, C., 1955, Pleistocene temperatures: Jour. Geology, v. 63, p. 538-578.

Ericson, D. B., Ewing, M., Wollin, G., and Heezen, B. C., 1961, Atlantic deep-sea sediment cores: Geol. Soc. America Bull., v. 72, p. 193-286.

Ewing, M., Eittreim, S. L., Ewing, J. I., and Le Pichon, X., 1971, Sediment transport and distribution in the Argentine Basin. Pt. 3, Nepheloid layer and processes of sedimentation: Physics and Chemistry of the Earth, v. 8, p. 49-77.

Fletcher, J. O., 1969, Ice extent on the Southern Ocean and its relation to world climate, *in* Natl. Sci. Found. Memorandum R17-3793: Santa Monica, Calif., The Rand Corp.

Flint, R. F., 1971, Glacial and Quaternary geology: New York, John Wiley & Sons, 892 p.

Gardner, J. V., and Hays, J. D., 1976, The eastern equatorial Atlantic: Sea-surface temperature and circulation response to global climatic change during the past 200,000 years, *in* Cline, R. M., and Hays, J. D., eds., Investigation of late Quaternary paleoceanography and paleoclimatology: Geol. Soc. America Memoir 145 (this volume).

Geitzenauer, K. R., 1971, The Pleistocene calcareous nannoplankton of the subantarctic Pacific Ocean: Deep-Sea Research, v. 19, p. 45-60.

Goodell, H. G., 1973, The sediments, *in* Marine sediments of the southern oceans: Am. Geog. Soc. Antarctic Map Folio Ser., folio 17.

Goodell, H. G., and Watkins, N. D., 1968, The paleomagnetic stratigraphy of the Southern Ocean: 20° West to 160° East longitude: Deep-Sea Research, v. 15, p. 89-112.

Gordon, A. L., 1967, Structure of Antarctic waters between 20°W and 170°W: Am. Geog. Soc. Antarctic Map Folio Ser., folio 5.

——1971, Recent physical oceanographic studies of Antarctic waters, *in* Quam, L. O., and Porter, H. D., eds., Research in the Antarctic: Am. Assoc. Adv. Sci. Pub. 93, p. 609-629.

——1972, Physical oceanography of the southeast Indian Ocean, *in* Hays, D. E., ed., Antarctic oceanography II: Am. Geophys. Union Antarctic Research Ser., v. 19, p. 5-9.

Greenway, M. E., 1972, The geology of the Falkland Islands: British Antarctic Survey Sci. Repts., no. 76, 42 p.

Haeckel, E., 1887, Report of the radiolarian collected by H.M.S. *Challenger* during the years 1873-1876, *in* Challenger expedition reports, Zool., vol. 18: New York, Johnson Reprint Co.

Haecker, V., 1908, Die Tripyleen, Collodarien und Mikroradiolarien der Tiefsee: Wiss. Ergebnisse Deutsch Tiefsee-Expedition 1898-1899, v. 14, 476 p.

Hays, J. D., 1965, Radiolaria and Late Tertiary and Quaternary history of Antarctic seas: Am. Geophys. Union Antarctic Research Ser., v. 5, p. 125-184.

——1969, Climatic record of late Cenozoic Antarctic sediments related to the record of world climate: Paleoecology of Africa and Antarctica, Scientific Comm. on Antarctic Research 1968: Cape Town, South Africa, A. A. Balkena, p. 139-163.

Hays, J. D., and Berggren, W. A., 1971, Quaternary boundaries and correlations, *in* Funnel, B. M., and Riedel, W. R., eds., Micropaleontology of the oceans: Cambridge, England, Cambridge Univ. Press, p. 669-691.

Hays, J. D., and Opdyke, N. D., 1967, Antarctic Radiolaria, magnetic reversals and climatic change: Science, v. 158, p. 1001-1011.

Hays, J. D., Saito, T., Opdyke, N. D., and Burckle, L. H., 1969, Pliocene-Pleistocene sediments of the equatorial Pacific: Their paleomagnetic, biostratigraphic and climatic record: Geol. Soc. America Bull., v. 80, p. 1481-1514.

Hollin, J. T., 1962, On the glacial history of the Antarctic: Jour. Glaciology, v. 4, p. 173-195.

——1965, Wilson's theory of ice ages: Nature, v. 208, p. 12-16.

——1969, Ice-sheet surges and the geological record: Canadian Jour. Earth Sci., v. 6, p. 903.

Horn, D. R., Delach, M. N., and Horn, B. M., 1975, Physical properties of sedimentary provinces, North Pacific and North Atlantic Oceans, *in* Inderbitzen, A. L., ed., Deep sea sediments: Physical and mechanical properties: New York, Plenum Pub. Corp. (in press).

Hough, J. L., 1950, Pleistocene lithology of Antarctic bottom sediments: Jour. Geology, v. 58, p. 254-260.

Imbrie, J., and Kipp, N. G., 1971, A new micropaleontological method for quantitative paleoclimatology: Application to a late Pleistocene Caribbean core, *in* Turekian, K. K., ed., The late Cenozoic glacial ages: New Haven, Conn., Yale Univ. Press, p. 71-79.

Kennett, J. P., 1970, Pleistocene paleoclimates and foraminiferal biostratigraphy in subantarctic deep-sea cores: Deep-Sea Research, v. 17, p. 125-140.

Ku, T. L., and Broecker, W. S., 1966, Atlantic deep-sea stratigraphy: Extension of absolute chronology to 320,000 years: Science, v. 151, p. 448-450.

Kulp, J. L., Feely, H. W., and Tryon, L. E., 1951, Lamont natural radiocarbon measurements, I: Science, v. 114, p. 565-568.

Litsitzin, A. P., 1960, Bottom sediments of the eastern Antarctic and the southern Indian Ocean: Deep-Sea Research, v. 7, p. 89-99.

——1962, Bottom sediments of the Antarctic: Am. Geophys. Union Geophys. Mon., v. 7, p. 81-88.

——1972, Sedimentation in the World Ocean: Soc. Econ. Paleontologists and Mineralogists Spec. Pub. 17.

Lozano, J. A., and Hays, J. D., 1976, Relationship of radiolarian assemblages to sediment types and physical oceanography in the Atlantic and western Indian Ocean sectors of the Antarctic Ocean, in Cline, R. M., and Hays, J. D., eds., Investigation of late Quaternary paleoceanography and paleoclimatology: Geol. Soc. America Mem. 145 (this volume).

Luz, B., 1973, Stratigraphic and paleoclimatic analysis of late Pleistocene tropical southeast Pacific cores (appendix by N. J. Shackleton): Quaternary Research, v. 3, p. 56-72.

Mackintosh, N. A., 1972, Life cycle of Antarctic krill in relation to ice and water conditions: Discovery Repts., v. 36, p. 1-94.

McIntyre, A., Kipp, N., and others, 1976, Glacial North Atlantic 18,000 years ago: A CLIMAP reconstruction, *in* Cline, R. M., and Hays, J. D., eds., Investigation of late Quaternary paleoceanography and paleoclimatology: Geol. Soc. America Mem. 145 (this volume).

Neumann, G., and Pierson, W. J., Jr., 1966, Principles of physical oceanography: Englewood Cliffs, N.J., Prentice-Hall, Inc., 545 p.

Opdyke, N. D., Glass, B., Hays, J. D., and Foster, J., 1966, Paleomagnetic studies of Antarctic deep-sea cores: Science, v. 154, p. 349-357.

Petrushevskaya, N. G., 1967, Radiolarians of orders Spumellaria and Nassellaria of the antarctic region, *in* Andriyashev, A. P., and Ushakov, P. V., eds., Studies of marine fauna: Biological Repts. Soviet Antarctic Exped. (1955-1958) v. 3, p. 2-186 [Translated by Raveh, M., 1968, Springfield, Va., U.S. Dept. Commerce].

Philippi, E., 1912, Die Grundproben der deutschen Südpolar-Expedition 1901-1903, *in* Deutsche Südpolar-Expedition 1901-1903, vol. 2: Berlin, Geographie und Geologie, p. 415-616.

Popofsky, A., 1908, Die Radiolarien der Antarktic (mit Ausnahme der Tripyleen): Deutsch Sudpolar-Expedition, v. 10, no. 2, p. 185-305.

Riedel, W. R., 1958, Radiolaria in Antarctic sediments: British, Australian, New Zealand Antarctic Research Exped. Repts., ser. B, v. 6, pt. 10, p. 217-255.

Rubin, M., and Suess, H. E., 1955, U.S. Geological Survey radiocarbon dates II: Science, v. 121, p. 481-488.

Saito, T., and Fray, C., 1965, Cretaceous and Tertiary sediments from the southwestern Indian Ocean [abs.]: Geol. Soc. America, Abs. for 1964, Spec. Paper 82, p. 171-172.

Schott, W., 1939, Deep-sea sediments of the Indian Ocean, *in* Trask, P. D., ed., Recent marine sediments: Soc. Econ. Paleontologists and Mineralogists Spec. Pub. 4, p. 398-407.

Shackleton, N. J., and Opdyke, N. D., 1973, Oxygen isotope and paleomagnetic stratigraphy of equatorial Pacific core V28-238: Oxygen isotope temperatures and ice volumes on a 10^5 year and 10^6 year scale: Quaternary Research, v. 3, p. 39-55.

Suess, H. E., 1954, U.S. Geological Survey radiocarbon dates I: Science, v. 120, p. 167-173.

Watkins, N. D., and Kennett, J. P., 1971, Antarctic Bottom Water: Major changes in velocity during the late Cenozoic between Australia and Antarctica: Science, v. 173, p. 813-818.

——1972, Regional sedimentary disconformities and upper Cenozoic changes in bottom water velocities between Australia and Antarctica: Am. Geophys. Union Antarctic Research Ser., v. 19, p. 273-293.

Wilson, A. T., 1964, Origin of ice ages: An ice shelf theory for Pleistocene glaciation: Nature, v. 210, p. 147-149.

MANUSCRIPT RECEIVED BY THE SOCIETY AUGUST 9, 1974
REVISED MANUSCRIPT RECEIVED MARCH 20, 1975
MANUSCRIPT ACCEPTED APRIL 9, 1975
LAMONT-DOHERTY GEOLOGICAL OBSERVATORY CONTRIBUTION NO. 2275

PACIFIC

Geological Society of America
Memoir 145
© 1976

Late Quaternary Sediment of the Panama Basin: Sedimentation Rates, Periodicities, and Controls of Carbonate and Opal Accumulation

NICKLAS G. PISIAS*
School of Oceanography
Oregon State University
Corvallis, Oregon 97331

ABSTRACT

Assuming a constant rate of quartz accumulation for deep-sea sediment core Y69-106P, from the Panama Basin, I have estimated the age of samples from the core and have constructed a curve for sedimentation-rate versus time. Stratigraphic controls for the calculated time scale include three C^{14} measurements, the extinction of the radiolarian *Stylatractus universus*, correlation with oxygen-isotope curves from other dated cores, and an Ar^{40}/Ar^{39} age determination.

The model sedimentation rates, when combined with data on mineralogic composition, allow the determination of accumulation rates for $CaCO_3$, opaline SiO_2, and remaining "detritus." Fluctuations in $CaCO_3$ accumulation rates over time correlate with variations in oxygen-isotope ratios in biogenous carbonate from two other equatorial Pacific cores; high oxygen-isotope ratios correlate with high $CaCO_3$ accumulation rates. Opaline SiO_2 accumulation rates reflect changes in the dominance of different radiolarian fossil groups, changes which can be related to surface circulation in the Panama Basin.

Spectral analysis of the accumulation rate of $CaCO_3$ in core Y69-106P and of the oxygen-isotope record of core V28-238 indicates a 23,000-yr periodicity. Spectral analysis of the opal accumulation rate suggests a 100,000-yr periodicity, which is an important frequency in the oxygen-isotope record. Both these periodicities are close to those found in fluctuations in the Earth's orbital parameters. The eccentricity of the orbit fluctuates with an average period of 98,000 yr, and the precession of the equinoxes changes with a period of 22,000 yr.

*Present address: University of Rhode Island, Graduate School of Oceanography, Kingston, Rhode Island 02881.

INTRODUCTION

A core of deep-sea sediment contains a detailed record of the ocean's response to various climatic conditions. Such a record, unlike that obtained from continental deposits, gives a relatively continuous history of climatic change over long periods of time. Resolution of the climatic record determined from the analysis of deep-sea sediment varies with the rate of sediment accumulation. Cores that reflect normal pelagic accumulation rates (1 to 3 cm/1,000 yr) give a picture of climatic change over the last 1 m.y. (Shackleton and Opdyke, 1973) with a resolution of no more than one sample per 1,000 yr. Cores that reflect high sedimentation rates, taken near ocean margins, yield records of climatic change for the past 50,000 yr with a resolution of less than one sample per 200 yr (Pisias and others, 1973; Soutar, 1971).

Interpretation of the climatic record preserved in deep-sea sediment cores is dependent on the establishment of an accurate time scale for the cores being studied. An accurate time scale is necessary for quantitative descriptions of time-transgressive events such as the movement of oceanographic fronts or of major current systems (McIntyre and others, 1972) or any sequence of changes in oceanic processes that reflects climatic change (Pisias and others, 1975). Statistical analysis of climatic fluctuations (for example, spectral analysis) demands the best possible time scale. A time scale is usually developed by dating levels in a core and assuming continuous and constant sedimentation between these levels. Sediment samples can be dated either directly, by radiometric analyses, or indirectly, by recognition of previously dated paleomagnetic or stratigraphic events. Sedimentation rates can also be estimated from measurements of the activity of radiogenic elements such as Th^{230} and Pa^{231} (Ku, 1966). Again a constant sedimentation rate between dated horizons is usually assumed.

All these methods give sedimentation rates averaged over long periods of geologic time; only if many data points are determined can fluctuations in rates be observed. Where such fluctuations are observed, an abrupt change from one average rate to another is usually assumed. Rarely is sufficient information obtained to enable recognition of a continuously fluctuating sediment accumulation rate in a deep-sea sediment core.

I have developed a sedimentation-rate model to overcome the limitation imposed by a finite number of age determinations and to examine the relationship between climate and accumulation rates. In this model, one component of the sediment is assumed to accumulate at a constant rate.

Similar models were used by Wiseman (1956) and Arrenhius (1952) in the absence of adequate radiometric dating methods. Wiseman, in his study of equatorial Atlantic cores, assumed that deep-sea sediment consisted of only two components: $CaCO_3$ accumulating at a variable rate and fine-grained silicate detritus accumulating at a constant rate. This assumption of a constant accumulation rate for fine-grained silicate detritus has subsequently been shown to be inadequate for the Atlantic (Broecker and others, 1958). Arrenhius assumed that fine-grained silicate detritus in equatorial Pacific cores accumulated at a constant rate and that TiO_2 concentration could be used as a measure of its abundance. Arrenhius' time scale compares well with other equatorial cores dated by paleomagnetic stratigraphy (Hays and others, 1969).

In the sedimentation-rate model developed here, a sediment sample is considered to be a combination of four components: $CaCO_3$ (determined by LECO induction furnace), opaline SiO_2 and crystalline quartz (both determined by x-ray diffraction; Ellis, 1972), and "detritus," where detritus percentage equals 100−(carbonate +

opal + quartz percentage). Of these four components it is assumed that quartz has a constant accumulation rate.[1]

SEDIMENTATION-RATE MODEL

Assumptions Used in the Model

The sedimentation-rate model of core Y69-106P (taken at 2°59′N, 86°33′W, 2,870-m water depth; Fig. 1) is based on the assumption that quartz accumulation has remained constant throughout the late Pleistocene and Holocene Epochs. This assumption is consistent with the known pattern of terrigenous sedimentation in the Panama Basin: eolian input is small (Prospero and Bonatti, 1969) and unlikely to vary from glacial to interglacial times, fluviatile input is largely trapped in the eastern part of the Panama Basin (Heath and others, 1974), and the deep bottom-water circulation responsible for carrying most of the quartz entering the western part of the basin (Heath and others, 1974; Plank and others, 1973) is controlled by geothermal heating (Laird, 1971). The consistency of calculated average rates of quartz accumulation between stratigraphic datums (0.0186, 0.0188, and 0.0182 g/cm^2/1,000 yr for the intervals 0 to 12, 12 to 35 and 35 to 341 cm, respectively), together with the coherent picture of temporal changes in sedimentation rates indicated by the model, support the validity of the basic assumption.

There are additional assumptions used in the sedimentation-rate model that are usually made in deep-sea sediment studies but not explicitly stated: (a) Both the sample interval and the use of linear interpolation between sample points provide an adequate characterization of the real fluctuations in any parameter of the sediment in a core, and (b) All maximum and minimum values of a parameter are sampled.

Age Determinations for Core Y69-106P

The low average sedimentation rate (about 2 cm/1,000 yr; Table 1) recorded in core Y69-106P and the length of the core allowed the recovery of sediment deposited throughout the last three glacial-interglacial cycles. From light-scattering data Plank and others (1973) concluded that there has been little erosion of material in the 2 to 10-μm grain size they studied at this site. The core consists mostly of pale-olive (Munsell code 10Y6/2) foraminiferal clay, with two distinct layers of volcanic ash at 120 cm and 506 cm below the core top. The mean carbonate content is 65%, with the most carbonate-rich sediment at the base of the core.

A summary of the sedimentation rates determined from radiometric procedures (Pisias, 1974) is given in Table 1. The radiometric data give average rates for this core that range from 0.5 to 6.6 cm/1,000 yr, with no two procedures in agreement.

The extinction of *Stylatractus universus* in the Pacific Ocean, the Indian Ocean, and the Antarctic region occurred about 400,000 yr ago, according to Hays (1970). This age is based on interpolation, assuming constant sedimentation rates, of the extinction level between the sediment surface and the Brunhes-Matuyama paleomagnetic reversal boundary in 11 North Pacific cores (Hays and Ninkovich, 1970). The mean extinction age in these cores is 400,000 yr, with a standard deviation of 20,000 yr. I did not use two North Pacific cores, V20-109 and V20-119, because of extremely young ages for the extinction level. Such extreme values probably

---

[1] Appendixes II and III are on microfiche in pocket inside back cover.

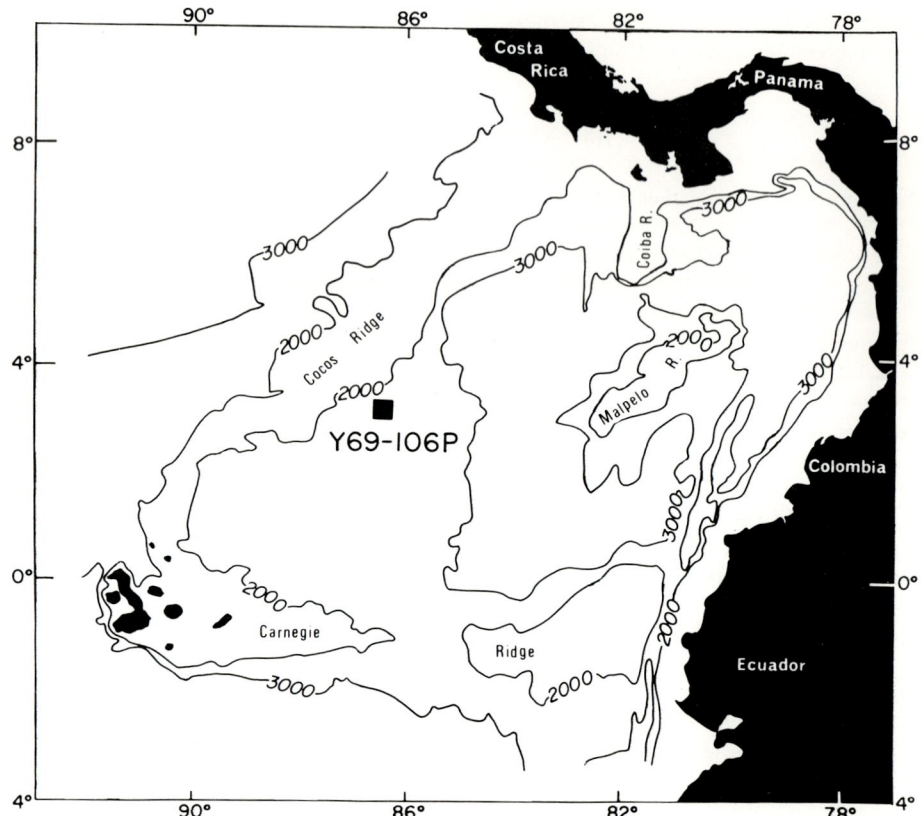

Figure 1. Panama Basin and location of core Y69-106P.

reflect loss of sediment from the core tops during coring operations. Since all age estimates of this extinction could be too young because of a systematic loss of the upper layers of sediment from the piston cores, 400,000 yr is probably a minimum estimate for the age of this stratigraphic level. Four cores taken in the equatorial Pacific (Hays and others, 1969) give extinction ages younger than 400,000 yr. This difference between the interpolated age of *S. universus* extinction determined in the equatorial Pacific and in the North Pacific may be real, or again it may result from missing tops in these few cores. The four equatorial cores are more calcareous than the North Pacific cores, and if the fluctuations of their $CaCO_3$ contents correspond to varying sedimentation rates, a simple interpolation from the Brunhes-Matuyama paleomagnetic boundary could be systematically in error. If data for all the cores with *S. universus* extinctions and magnetic stratigraphy are averaged, excluding the two North Pacific and the two equatorial values discussed above, the mean age is 396,000 yr. Thus, a 400,000-yr age for the extinction of *S. universus* seems to be a good estimate for this stratigraphic level.

Oxygen-isotope ratios in the calcareous tests of foraminifers reflect the temperature and the isotopic composition of the sea water in which the foraminifer deposited its test. The major control over the oxygen-isotope ratio in sea water is the amount of ice stored in continental ice masses (Shackleton and Opdyke, 1973). Thus the youngest oxygen-isotope maximum found in deep-sea sediment corresponds to the last maximum ice advance of the Wisconsin Glaciation. This maximum occurrence should be nearly synchronous in all oceans. Any variations in the timing of the event at different localities should be related only to the mixing time of the oceans.

Figure 2 shows oxygen-isotope curves obtained from the upper parts of nine cores for which analyses have been made. C^{14} measurements are also available for all the cores except V28-238 and P6304-9. The time scale for core V28-238 is based on the identification of a coccolith datum of which the 73,000-yr age has been determined using cores with C^{14} dates (Geitzenauer, 1973, written commun.). The time scale for core P6304-9, taken from Broecker and van Donk (1970), is based on the identification of "termination II" from the oxygen-isotope profiles and on an assumed age for this level of 125,000 yr. The ages of the youngest oxygen-isotope maxima identified in each of these cores are given in Table 2. A180-73 gives an age much younger than any of the other cores. When correlated with a neighboring core (A180-74) for which eight C^{14} ages are available, core A180-73 consistently gives younger ages for all stratigraphic datums (Ericson and others, 1961). If core A180-73 is excluded from the calculation, the mean age of the most recent oxygen-isotope maximum is 17,900 yr, with a standard deviation of 1,300 yr. This age is similar to the time of the last maximum ice advance measured in Europe and North America (G. Denton, 1973, written commun.), and the 1,300-yr standard deviation is of the same magnitude as the mixing time of the oceans.

For core Y69-106P, the 400,000-yr age of *S. universus* extinction and the 17,900-yr age for the youngest oxygen-isotope maximum give an average sedimentation rate of about 2 cm/1,000 yr. This value agrees well with the value obtained from C^{14} measurements on the core. The average sedimentation rate determined from Th^{230} data (Pisias, 1974), however, is about three times this value. Such a discrepancy may result from enrichment of uranium in the sediment of Y69-106P due to reducing conditions of the sediment (Veeh, 1967). No correction was made for the Th^{230} generated by the decay of such an excess.

If one accepts a 400,000-yr age for the extinction of *S. universus* at 841 cm, the 1-m.y. K-Ar age of the volcanic ash at 506 cm is too old to be a depositional age. The extreme age of this sample may reflect incomplete outgassing of Ar during ash solidification. On the other hand, the 1-m.y. date could be the real age of formation of the ash, and its high stratigraphic position in Y69-106P could

TABLE 1. TIME SCALE FOR CORE Y69-106P

Dating method	Depth of samples in core (cm)	Age (B.P.)	Average accumulation rate (cm/1,000 yr)
		Radiometric	
C^{14}	0–6	4450 ± 110	2.07
	10–15	9030 ± 155	1.93
	29–36	19,400 ± 800	
K-Ar	506	1,000,000 ± 20,000	0.506
Ar^{40}/Ar^{39}	506	260,000	2.0
Th^{230}*			6.6 ± 1.0
		Stratigraphic	
O^{18} max	35	17,900 ± 1,300	2.3
O^{18} min	238	122,000 ± 7,000†	2.0
S. universus	841	400,000 ± 20,000	2.2

*Pisias (1974).
†See Ku and others (1974).

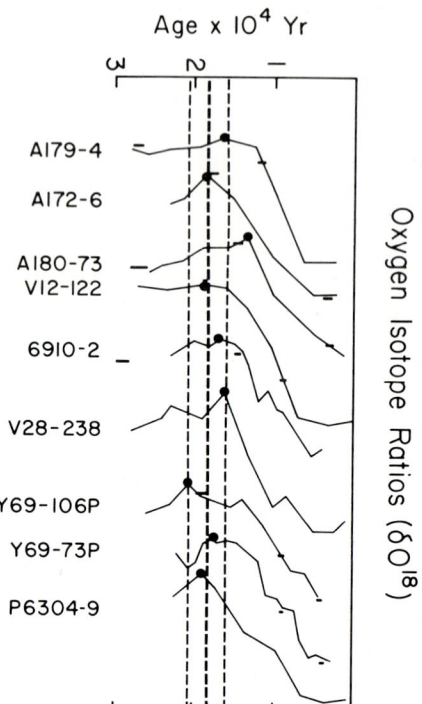

Figure 2. Oxygen-isotope curves showing youngest oxygen-isotope maximum. Bars indicate C^{14} dates of cores. Also shown is mean age plus and minus two standard deviations from ages in Table 2. For references see Table 2.

be a result of subsequent displacement. Ar^{40}/Ar^{39} measurements (Brereton, 1972) were made to determine if excess radiogenic Ar could be detected, thereby confirming the hypothesis that the 1-m.y. age is indeed too old because of incomplete Ar outgassing. The Ar^{40}/Ar^{39} stepwise degassing technique (Brereton, 1972) gave an age of about 260,000 yr for this ash, and the measurements indicated that excess radiogenic Ar was present (J. Dymond, 1973, oral commun.).

Use of the Model

Assuming a constant rate of quartz accumulation through time, the age at any core level is linearly related to the total amount of quartz between that level and the surface of the sediment. The quartz mass versus depth in the core is calculated from the quartz concentration per unit dry weight and the bulk density of the samples in the core (where bulk density is defined as dry weight/wet volume, in g/cm^3). Integration of the quartz mass versus depth in the core by use of linear interpolation yields total accumulated quartz mass versus depth in the core. This curve can be converted to age versus depth by calibration using the 400,000-yr radiolarian datum, the C^{14} date at 12.5 cm (9,030 yr) and the accumulated quartz masses at these levels. The model yields an age at a 2-cm depth in the core that is within the error limits of the C^{14} age for the interval 0 to 6 cm. It gives the age of the oxygen-isotope maximum found at 35 cm as 18,400 yr, well within the previously discussed limits for this datum age. The calculated average age of the oxygen-isotope maximum (Table 2) and the model time scale suggest that the C^{14} date for the interval 29 to 36 cm is slightly too old. This C^{14} measurement has a large error associated with it (Table 1).

The model age for the volcanic ash at 506 cm is within 3% of the age determined by the Ar^{40}/Ar^{39} measurement. The model age for the last oxygen-isotope minimum at 112,000 yr is within the 90% confidence interval for the age determined by Ku and others (1974) for the associated high sea-level stand.

From the age-versus-depth relationship determined by the model, sedimentation rates were calculated for each point sampled along the core. The rate assigned to a sample point is the average of the rates calculated for the intervals preceding and succeeding each point. The accumulation rates of individual sediment components are determined from the model sedimentation rates and the data on sediment composition. The accumulation rate of any sedimentary component at a given level is equal to the product of the total rate of sedimentation (cm/1,000 yr), the component concentration (wt %), and the bulk density (g/cm^3). The rate of each component is expressed in the units g/cm^2/1,000 yr.

Results and Discussion

The accumulation rate of each sedimentary component as a function of age is given in Figure 3. In general, the sedimentation rates fluctuate smoothly in the younger part of the record, with higher rates and more variation in the samples older than 350,000 model-estimated years (e.y.). The sedimentation rates for detritus, which clearly show the ash deposition at 51,000 and 267,000 e.y., imply that increased deposition of volcanic ash may have occurred prior to 350,000 e.y. Samples from the lower section of the core also have a higher average CaCO$_3$ concentration than those from the upper part of the core (75% versus 65%); bulk density values show a similar increase. Clearly, there are indications of an important change in the sedimentation history of this core site at about 350,000 e.y. Either the model is not valid for the lowest section of the core, or the intermittent addition of reworked material added a distinct and different character to the sedimentation record prior to 350,000 e.y. Kowsmann (1973) suggested that foraminiferal sand and volcanic ash are both concentrated by winnowing. This process of reworking would explain the observed correlations of CaCO$_3$ with volcanic ash in the oldest part of the sedimentation record of core Y69-106P. The peaks in the lower section of the detritus curve correspond to maxima in the CaCO$_3$ sedimentation rate and

TABLE 2. AGE OF LAST OXYGEN-ISOTOPE MAXIMUM IN DEEP-SEA CORES

Core	Age (B.P.)	Dating method	References
A179-14	16,400	C^{14}	Emiliani (1955)
A172-6	18,400	C^{14}	Emiliani (1955)
A180-73	13,200	C^{14}	Emiliani (1955)
V12-122	18,500	C^{14}	Imbrie and others (1973)
6910-2	16,800	C^{14}	Shackleton (1973, written commun.) Phipps (1974)
V23-238	17,000	Biostratigraphy dating at 75,000 yr	Shackleton and Opdyke (1973) McIntyre (1973, written commun.)
Y69-106P	20,400	C^{14}	Appendix I, this paper
Y69-73P	17,200	C^{14}	Shackleton (1973, written commun.)
P6304-9	18,800	Based on the assumption of a 125,000-yr age for "termination" II	Broecker and van Donk (1970)
Mean	17,900 ± 1,300, excluding A180-73		

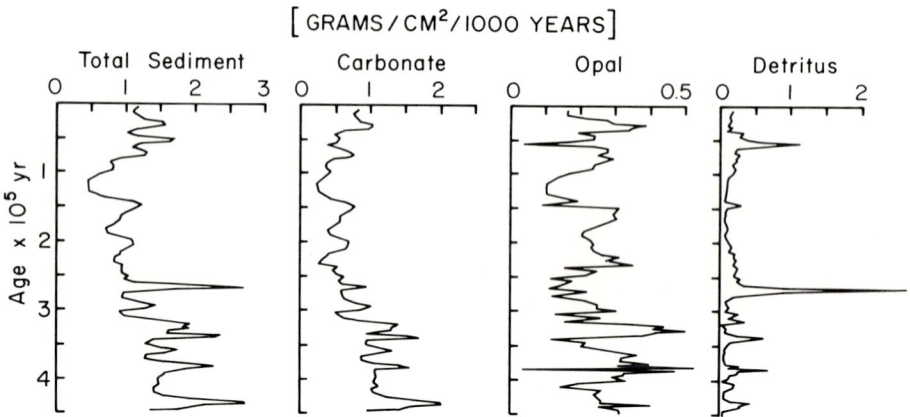

Figure 3. Model accumulation rates versus ages for core Y69-106P.

to minima in the opaline SiO_2 sedimentation rate (Fig. 3).

Because of the difficulty in reconstructing temporal changes in sedimentation rates, some fundamental questions on how climate may affect $CaCO_3$ and opaline SiO_2 sedimentation are still in debate. In Atlantic sediment, for example, high carbonate concentrations are correlated with interglacial periods, whereas in the Pacific, they are correlated with glacial periods. Wiseman (1954) concluded that the high $CaCO_3$ concentrations in the Atlantic during interglacial periods resulted from high productivity due to higher temperatures. Arrenhius (1952) hypothesized that high $CaCO_3$ concentrations found in sections of equatorial Pacific cores resulted from increased productivity in the surface waters due to more vigorous circulation of ocean water during glacial times. However, Berger (1973) suggested that productivity changes in surface waters of the ocean have a minor effect on carbonate deposition and that changes in the corrosiveness of bottom water are the primary control of the accumulation rate of $CaCO_3$.

To help resolve the climatic information contained in core Y69-106P, $CaCO_3$ sedimentation rates are compared with oxygen-isotope variations in core V28-238, from the western equatorial Pacific (Fig. 4; Shackleton and Opdyke, 1973), and in core V19-55, from the eastern tropical Pacific (Luz, 1973). The time scale for V28-238 was determined by linear interpolation between five known extinction-age levels (Pisias, 1974). According to Luz (1973) the sedimentation rate for V19-55 was about 0.98 cm/1,000 yr. Correlation with the isotope curve V28-238, however, indicates a slightly lower rate of about 0.80 cm/1,000 yr.

The correlation of the carbonate sedimentation rates of core Y69-106P with the oxygen-isotope curves is very good above the level that corresponds to 350,000 yr B.P., but poor at greater depths (Fig. 4). When ice caps were large and sea level low (large δo^{18}), $CaCO_3$ accumulation rates in the Panama Basin were high, and during periods of high sea level and low ice volumes, the rates of accumulation were low. This general relationship has held for the past 350,000 yr.

Opaline SiO_2, like $CaCO_3$, is a product of biological activity. Its rate of accumulation in sediment is a function both of the rate of supply of opaline tests from surface waters and of the rate of dissolution. The rate of opal dissolution is highest in the relatively warm, SiO_2-depleted near-surface water (<1,000 m depth); however, leaching of SiO_2 from deep-sea sediment is an important process that contributes to the geochemical budget of SiO_2 in the ocean (Heath, 1974). In areas of high fertility, Pacific surface sediment is characteristically rich in siliceous microfossils. In the Atlantic, however, surface sediment is relatively SiO_2-poor, even in areas of high productivity. The difference in SiO_2 preservation between the Atlantic and Pacific reflects differences in dissolved SiO_2 distribution in the two oceans. The Pacific is relatively rich in dissolved SiO_2 compared to the Atlantic and therefore is less corrosive to opal. The difference in SiO_2 distribution between the major oceans can be explained by the fractionation model discussed by Berger (1970) to describe differences in $CaCO_3$ deposition patterns. The Pacific exchanges SiO_2-depleted surface water for deep water that is rich in SiO_2, thus favoring opal production and preservation. The Atlantic exchanges deep water for shallow water, which sinks to form young, nutrient-depleted bottom waters corrosive to opaline SiO_2.

Because few cores have been analyzed for opal, fluctuations in opaline SiO_2 concentrations in equatorial cores have not been related to climatic conditions. The presence of opal-rich sediment in the Pacific beneath highly productive surface waters suggests that Arrenhius' argument for enhanced deposition due to increased upwelling during glacial periods should apply to opal as well as to carbonate concentrations. Again, the rate curves of Figure 3 do not allow us to distinguish enhanced productivity from better preservation.

To distinguish the effects of productivity and preservation, it is necessary to examine the faunal and floral makeup of the opaline sediment. Dinkelman (1974) studied the radiolarian assemblages in surface sediment of the Panama Basin and related them to productivity and circulation in the surface waters. Using Q-mode factor analysis (Klovan and Imbrie, 1971) of the radiolarian thanatocoenoses, Dinkelman identified four fossil assemblages, which he called "tropical," "solution," "Peru Current," and "equatorial undercurrent" (Cromwell Current) factors.

The two alternative interpretations of the oceanographic significance of these factors given by Dinkelman are: (1) Dominance of the Peru and equatorial undercurrent factors in a sample reflects higher productivity associated with intensified circulation and upwelling due to these currents; and (2) The Peru and equatorial undercurrent factors are well-preserved fossil groups, and the tropical and solution factors are highly dissolved radiolarian fossil assemblages derived from the same biocoenoses.

Using matrix algebra (Imbrie and Kipp, 1971), radiolarian populations in samples from Y69-106P have been resolved into these four radiolarian factors (Dinkelman, 1974). The relative abundances of the Peru and equatorial factors from Y69-106P samples correlate well with the opal accumulation rates (Figs. 5, 6A). Such correlation can be explained by either of Dinkelman's interpretations: Either minimal solution resulted in high opal accumulation rates and a dominant equatorial undercurrent factor, or increased upwelling related to acceleration of the equatorial undercurrent system led to increased opal accumulation rates. If the first explanation is correct, the magnitude of the solution factor in the core samples should be negatively correlated with the total sedimentation rate. This assumes that low sedimentation rates increase solution of opaline tests by long exposure to bottom water and that SiO_2 diffusion from deep-sea sediment is restricted to the upper layer (4 cm) of the sediment (Heath, 1974). These relationships do not appear to exist in this core. Neither the magnitude of the solution factor nor of the equatorial undercurrent factor is significantly correlated with the total sedimentation rate of Y69-106P (Figs. 6C, 6D).

The equatorial undercurrent factor shows a positive correlation with opal accumulation but shows no correlation with the total sedimentation rate (Figs. 6A, 6C). The solution factor is negatively correlated with opal accumulation and not

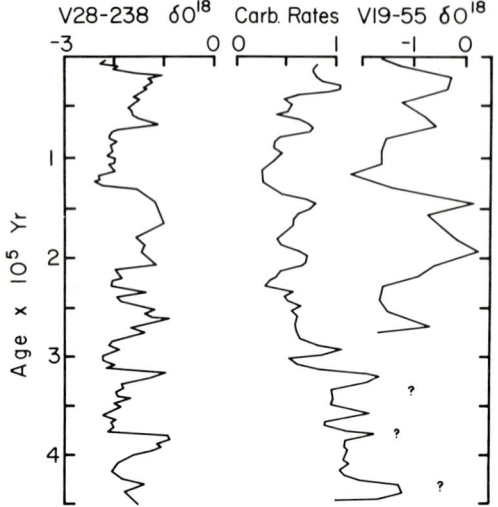

Figure 4. Comparison of Y69-106P CaCO$_3$ accumulation rates and oxygen-isotope records. Question marks indicate where peaks in CaCO$_3$ accumulation rate curve that are believed to be related to reworking have been removed.

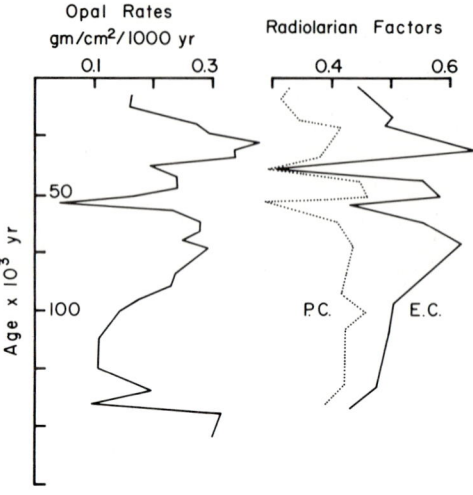

Figure 5. Model opal accumulation rates for Y69-106P versus radiolarian fossil assemblage factors. P.C., "Peru current" factor; E.C., "equatorial undercurrent" factor.

correlated with the total sedimentation rate (Figs. 6B, 6D). Therefore, for this core either (1) the rate of opal supply is relatively uniform and independent of carbonate supply and dissolution, and dissolution of SiO$_2$ is variable; or (2) the supply of opal is variable, and dissolution of SiO$_2$ is relatively uniform. At present, I consider the second alternative to apply to Y69-106P. This conclusion is supported by an observational estimate of dissolution in the radiolarian assemblages from this core, which shows no significant differences between samples (Dinkelman, 1974).

In general, accumulation rates of CaCO$_3$ show very good correlation with oxygen-isotope data, an indicator of global climatic change. Opal accumulation rates show good correlation with indicators of surface circulation, suggesting that biologic productivity is the primary controlling factor of these rates.

A time lag between changes in carbonate and opal accumulation rates and changes in the oxygen-isotope data for Y69-106 (Pisias and others, 1975; Fig. 7) indicates that the following sequence of events occurred during the transition from the last glacial to the present interglacial period: (1) Increased corrosiveness of bottom water to CaCO$_3$ in the equatorial Pacific (indicated by reduced carbonate accumulation) preceded (2) a reduction in the intensity of surface circulation in the Panama Basin, which in turn preceded (3) recession of the continental glaciers (indicated by a reduction in O^{18}/O^{16} ratios).

SPECTRAL ANALYSIS OF Y69-106P AND V28-238 CLIMATIC RECORDS

Of major interest in the study and prediction of climatic change through time is the question of whether observed fluctuations are random or have periodic characteristics. If periodicities do occur in climatic change, can they be related to known periodic environmental changes? Climatic fluctuations have been measured on scales ranging from a year to hundreds of thousands of years. Major climatic cycles (glacial and interglacial stages) have been attributed to periodic changes in the orbital parameters of the Earth relative to the Sun (Milankovitch, 1938). Shorter term climatic changes measured on a time scale of centuries, such as the "little" ice age of the fifteenth and sixteenth centuries, have been attributed to changes in solar activity (Bray, 1968). According to Suess (1970), the variations in solar activity during the last 7,000 yr are also periodic. Because the sun is the ultimate source of energy for the Earth's oceans and atmosphere, the periodic characteristics of solar activity and of the Earth's orbit should be evident in climatic records.

Spectral analyses of the oxygen-isotope curves of V28-238 and the sedimentation-rate curves of Y69-106P were made to test for periodicities. The analyses were performed using the procedures described by Pisias and others (1973).

Numerical techniques used to analyze a time series require that the series be sampled at equal time intervals. Cores are usually sampled at constant depth intervals, and only later is a time scale assigned to these samples. If the sedimentation rate varies with time, as it does here, then samples at constant depth intervals do not represent constant time intervals. If the time scale is reliable, either the core must be resampled or, more realistically, interpolated data values must be computed at constant time intervals. The data sets from cores V28-238 and Y69-106P have been interpolated to a 5,000-yr sample interval. It has been demonstrated that different numerical interpolation procedures do not affect the sample spectrum obtained from the data set of V28-238 (Pisias, 1974).

The spectrum of the $CaCO_3$ sedimentation-rate curves of Y69-106P is shown in Figure 8A. The older parts (>350,000 e.y.) of the sedimentation-rate curves indicate sedimentation conditions that were different from those for the younger sections and thus represent a non-"stationarity" in the series. $CaCO_3$ rates of the past 350,000 yr are considered to be less disturbed by winnowing and possible

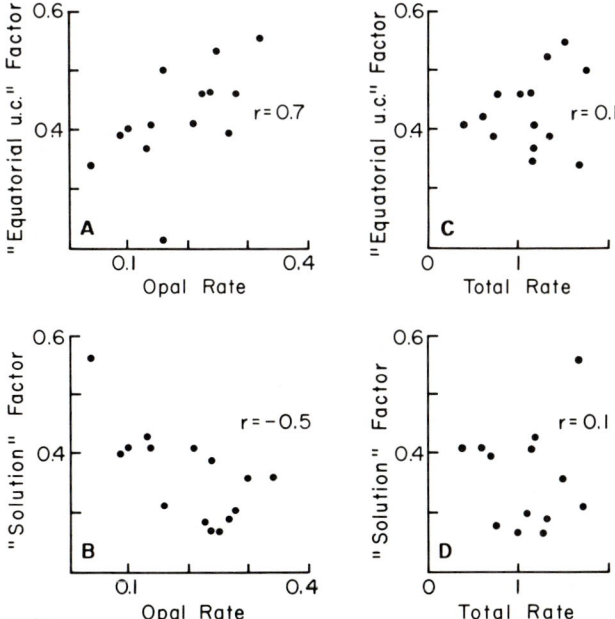

Figure 6. Opal and total accumulation rates versus "equatorial undercurrent" and "solution" radiolarian assemblage facors.

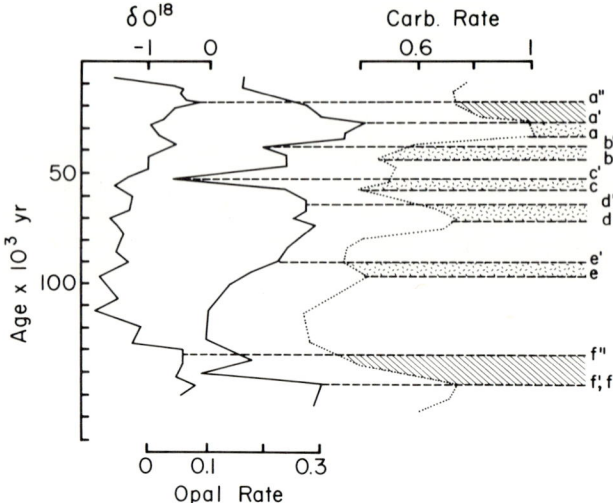

Figure 7. Lag time between CaCO₃ and opal accumulation rates and oxygen-isotope ratios of core Y69-106P (from Pisias and others, 1975).

slumps. The difference between the spectrum of the younger part of the core and that of the total carbonate record is clear (Fig. 8A). The spectrum of the shortened record of $CaCO_3$ has a distinct peak at the frequency corresponding to a 23,000-yr period.

Spectra of $CaCO_3$ accumulation rates versus model age, carbonate concentrations versus model age, and carbonate concentrations versus age estimated by linear interpolation between dated samples are compared in Figure 8B. The concentration spectra are similar for the two time scales, with spectra based on the model ages shifted slightly toward lower frequencies. Neither concentration spectrum shows the 23,000-yr period observed in the $CaCO_3$ accumulation-rate spectrum. For core Y69-106P, fluctuations in the $CaCO_3$ accumulation rate are a better guide to the climatic control of the oceanic carbonate system than are fluctuations in $CaCO_3$ concentration through time.

The time scale for V28-238 is not corrected for variations in sedimentation rates. However, this core is rich in $CaCO_3$ and shows little variation in concentration. Thus its sedimentation rate may be fairly uniform. As shown in Figure 8B, the spectrum of variations in the concentration of a sedimentary component calculated using a time scale that has been interpolated from stratigraphic data does not differ in form from that obtained using a time scale corrected for variations in sedimentation rates. This suggests that the spectrum of the V28-238 oxygen-isotope data probably would not be significantly changed by the construction of a sedimentation-rate model equivalent to that used for Y69-106P. The spectrum obtained for the V28-238 oxygen-isotope data (Fig. 9) has a peak corresponding to a period of 23,000 yr, the same peak found in the $CaCO_3$ sedimentation-rate data of Y69-106P.

For Y69-106P, the spectral estimate of a shortened opal accumulation record (0 to 350,000 e.y.) does not indicate the presence of an important periodicity at 23,000 yr (Fig. 9). The opal and oxygen-isotope spectra do have isolated low-frequency maxima centered at about 100,000 yr.

Spectral analysis has identified the importance of two periodicities, 23,000 and 100,000 yr, in climatically influenced sedimentary properties. These periodicities correspond to two of the three periodic fluctuations of the Earth's orbital parameters about the Sun: precession of the equinoxes (22,000-yr period; Imbrie and Kipp, 1971), and changes in the eccentricity of the orbit (98,000-yr period). (The third fluctuation is tilt of the rotational axis relative to the orbital plane [44,000-yr period].)

Precession of the equinoxes changes the distance from the Earth to the Sun during particular seasons. Winters will be colder when precession occurs at aphelion (farthest from the Sun) than when it occurs at perihelion. When winter occurs at aphelion, the following summer occurs at perihelion, and the seasonal contrast is a maximum. The opposite condition occurs if winter coincides with perihelion. The effects of precession of the equinoxes are opposite in the northern and southern hemispheres: when one hemisphere has maximal seasonal contrast, the other has minimal contrast.

Variations in the eccentricity of the Earth's orbit, which have an average period of 98,000 yr, modulate the effect of equinox precession. The closer the orbit is to a circle, the less precession affects seasonal contrast; the more elliptical the orbit, the more important precession becomes.

Therefore, the fluctuations of oxygen-isotope ratios and $CaCO_3$ accumulation rates, controlled respectively by the high-latitude processes of ice-cap melting and build-up and bottom-water circulation, may be related to precession of the equinoxes. Changes in the opal accumulation rate are controlled by changes in low-latitude surface circulation that appear to be related to the eccentricity of the Earth's orbit and may be related to changes in the distribution of incoming radiation throughout the year.

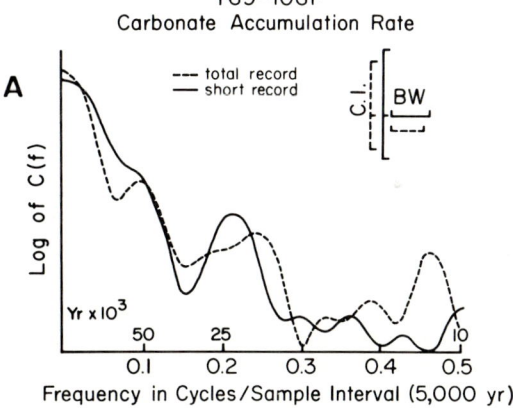

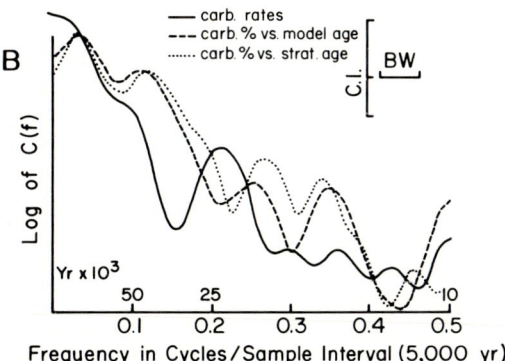

Figure 8. A. Spectral estimates for $CaCO_3$ accumulation rates from Y69-106P for entire and shortened record. B. Spectral estimates of $CaCO_3$ accumulation rate, $CaCO_3$ concentration versus model age, and $CaCO_3$ concentration versus stratigraphic age. C.I., 80% confidence interval; BW, band width; C(f), spectral estimates at frequency f. Peaks that are larger in magnitude than C.I. are considered to represent important periodicities in the record.

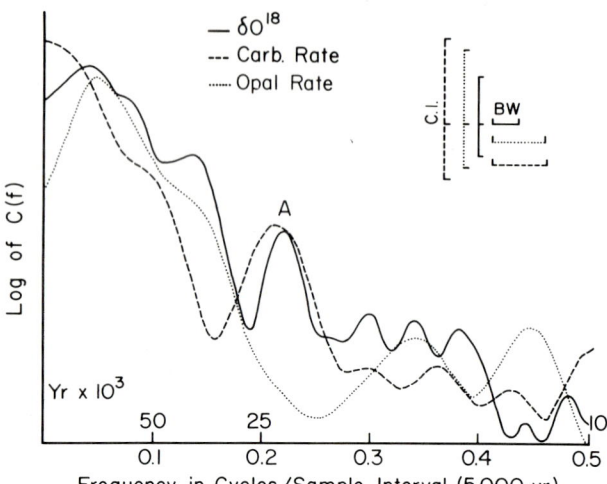

Figure 9. Spectral estimates of shortened opal and CaCO₃ accumulation rates from Y69-106P and oxygen isotopes of V28-238. See Figure 8 caption for explanation of terms.

CONCLUSIONS

The assumption of a constant rate of quartz accumulation during the past 350,000 yr in the Panama Basin allows detailed profiles of $CaCO_3$ and opal accumulation rates to be calculated. The validity of such a model is confirmed by the uniform rates of accumulation of quartz in three independently dated intervals in core Y69-106P.

Fluctuations in $CaCO_3$ accumulation rates from Y69-106P show a very close correlation with climatic indicators during the past 350,000 yr. Correlation with oxygen-isotope curves from other deep-sea cores indicates that $CaCO_3$ accumulation rates were higher during glacial advances than they were during glacial retreat. Fluctuations in opaline SiO_2 rates parallel faunal changes that are related to local surface circulation. During periods of increased upwelling associated with the equatorial undercurrent, opal accumulation rates are high. Correlations between the presence of statistically derived radiolarian assemblages and the accumulation of opaline SiO_2 suggest that changes in the rate of opal supply from surface waters are the primary factors controlling opal accumulation rates.

Periodicities in the fluctuations of the Earth's orbital parameters are reflected in climatic records from cores Y69-106P and V28-238. Spectral analysis indicates the presence of a 23,000-yr period in oxygen-isotope data from V28-238 and in the carbonate accumulation rates of Y69-106P; opal accumulation rates from Y69-106P suggest the presence of a 100,000-yr period. The 23,000-yr period is close to the periodicity of the precession of the equinoxes; 100,000 yr is the period of fluctuations in the eccentricity of the Earth's orbit.

ACKNOWLEDGMENTS

This research was supported by the National Science Foundation/International Decade of Ocean Exploration Grant GX-28673 to the CLIMAP project. I acknowledge G. R. Heath and T. C. Moore for their helpful suggestions and for critically reviewing the manuscript. N. J. Shackleton wrote Appendix I. I thank him for the oxygen-isotope analyses, which greatly added to the results of this paper. Many helpful discussions with B. Malfait and M. Dinkelman are gratefully acknowledged.

N. J. Shackleton's work was supported by the Natural Environmental Research Council under Grant GR3/1762. The mass spectrometer was operated with meticulous care by M. A. Hall.

APPENDIX I. OXYGEN-ISOTOPE DETERMINATIONS FOR Y69-106P

Core Y69-106P was sampled at 10-cm intervals to 280 cm, with closer sampling intervals in the upper part, for oxygen-isotope analysis. Analytical procedure was as described by Shackleton and Opdyke (1973), except that the isotope analysis was performed in a new VG Micromass 602C mass spectrometer with an improved overall analytical precision of ± 0.05 per mil (1 σ). Analyses are referred to the PDB standard on the assumption that the Emiliani B-1 standard has an oxygen-isotope composition of ± 0.29 per mil on the PDB scale. Each sample analyzed comprised 12 individuals of the species *Globigerinoides sacculifera*.

The boundary between stages 1 and 2 lies at about 15 cm; the boundary between stages 4 and 5 at about 115 cm, and the boundary between stages 5 and 6 at about 245 cm. The sedimentation rate in stages 2 to 4 is not sufficiently high to permit reliable identification of the boundaries; on the basis of the oxygen-isotope data it is not possible to say whether this section of the core is complete. The substages of stage 5 (Shackleton, 1969) are, however, clearly resolved; according to current thinking, substage 5a is at about 170 cm with Barbados II (105,000 B.P.) and 5e is at about 238 cm with Barbados II (125,000 B.P.) (Matthews, 1973).

REFERENCES CITED

Arrenhius, G., 1952, Sediment cores from the East Pacific: Swedish Deep-Sea Expedition Reports (1947–1948), v. 5, p. 1–89.

Berger, W. H., 1970, Biogenous deep-sea sediments: Fractionation by deep-sea circulation: Geol. Soc. America Bull., v. 81, no. 5, p. 1385–1402.

———1973, Deep-sea carbonates: Pleistocene dissolution cycles: Jour. Foram. Research, v. 3, no. 4, p. 187–195.

Brereton, N. R., 1972, A reappraisal of ^{40}Ar/^{39}Ar stepwise degassing technique: Royal Astron. Soc. Geophys. Jour., v. 27, p. 449–478.

Bray, J. R., 1968, Glaciation and solar activity since the fifth century BC and the solar cycle: Nature, v. 220, p. 672–674.

Broecker, W. S., and van Donk, J., 1970, Insolation changes, ice volumes, and O^{18} record in deep-sea cores: Rev. Geophysics and Space Physics, v. 8, no. 1, p. 169–198.

Broecker, W. S., Turekian, K. K., and Heezen, B. C., 1958, The relation of deep-sea sedimentation rates to variations in climate: Am. Jour. Sci., v. 256, p. 503–517.

Dinkelman, M. G., 1974, Late Quaternary radiolarian paleo-oceanography of the Panama Basin, eastern equatorial Pacific [Ph.D. thesis]: Corvallis, Oregon State Univ., 123 p.

Ellis, D. B., 1972, Holocene sediment of the South Atlantic Ocean: The calcite compensation depth and concentration of calcite, opal, and quartz [M.S.thesis]: Corvallis, Oregon State Univ., 77 p.

Emiliani, C., 1955, Pleistocene temperatures: Jour. Geology, v. 63, no. 6, p. 539–578.

Ericson, D. B., Ewing, M., Wollin, G., and Heezen, B. C., 1961, Atlantic deep-sea sediment cores: Geol. Soc. America Bull., v. 72, no. 2, p. 193–286.

Hays, J. D., 1970, Stratigraphy and evolutionary trends of radiolaria in North Pacific deep-sea sediments: Geol. Soc. America Mem. 126, p. 185–218.

Hays, J. D., and Ninkovich, D., 1970, North Pacific deep-sea ash chronology and age of present Aleutian underthrusting: Geol. Soc. America Mem. 126, p. 263-290.

Hays, J. D., Saito, T., Opdyke, N. D., and Burckle, L. H., 1969, Pliocene-Pleistocene sediments of the equatorial Pacific: Their paleomagnetic, biostratigraphic and climatic record: Geol. Soc. America Bull., v. 80, no. 8, p. 1481-1514.

Heath, G. R., 1974, Dissolved silica and deep-sea sediments, in Hay, W. W., ed., Studies in paleo-oceanography: Soc. Econ. Paleontologists and Mineralogists Spec. Pub. 20, p. 77-93.

Heath, G. R., Moore, T. C., and Roberts, G. L., 1974, Mineralogy of surface sediments from the Panama Basin, eastern equatorial Pacific: Jour. Geology, v. 81, p. 145-160.

Imbrie, J., and Kipp, N. G., 1971, A new micropaleontological method for quantitative paleoclimatology: Application to a late Pleistocene Caribbean core, in Turekian, K.K., ed., The late Cenozoic glacial ages: New Haven, Conn., Yale Univ. Press, p. 71-181.

Imbrie, J., van Donk, J., and Kipp, N. G., 1973, Paleoclimatic investigation of late Pleistocene Caribbean deep-sea core: Comparison of isotopic and faunal methods: Quaternary Research, v. 3, no. 2, p. 10-18.

Klovan, J. E., and Imbrie, J., 1971, An algorithm and FORTRAN-4 program for large-scale Q-mode factor analysis and calculation of factor scores: Math. Geology, v. 3, no. 1, p. 61-77.

Kowsmann, R. O., 1973, Panama Basin surface sediments: Coarse components [M.S. thesis]: Corvallis, Oregon State Univ., 73 p.

Ku, T.-L., 1966, Uranium disequilibrium in deep-sea sediments [Ph.D. thesis]: New York, Columbia Univ., 157 p.

Ku, T.-L., Kimmel, M. A., Easter, W. H., and O'Neil, T. J., 1974, Eustatic sea level 120,000 years ago on Oahu, Hawaii: Science, v. 183, p. 959-962.

Laird, N. P., 1971, Panama Basin deep water properties and circulation: Jour. Marine Research, v. 29, p. 226-234.

Luz, B., 1973, Stratigraphic and paleoclimate analysis of late Pleistocene tropical southeast Pacific cores: Quaternary Research, v. 3, no. 1, p. 56-72.

Matthews, R. K., 1973, Relative elevation of late Pleistocene high sea level stands: Barbados uplift rates and their implications: Quaternary Research, v. 3, p. 147-153.

McIntyre, A., Ruddiman, W. F., and Jantzen, R., 1972, Southward penetration of the North Atlantic polar front: Faunal and floral evidence of large-scale surface water mass movements over the last 225,000 years: Deep-Sea Research, v. 19, p. 61-77.

Milankovitch, M., 1938, Die chronologie des Pleistocene: Acad. Sci. Math. Nat. Bull., Belgrad, v. 4, p. 49.

Phipps, J. B., 1974, Sediments and tectonics of the Gorda-Juan de Fuca plate [Ph.D. thesis]: Corvallis, Oregon State Univ., 118 p.

Pisias, N. G., 1974, Model of late Pleistocene-Holocene variations in rate of sediment accumulation: Panama Basin, eastern equatorial Pacific [M.S. thesis]: Corvallis, Oregon State Univ., 77 p.

Pisias, N. G., Dauphin, J. P., and Sancetta, C. S., 1973, Spectral analysis in late Pleistocene-Holocene sediments: Quaternary Research, v. 3, no. 1, p. 3-9.

Pisias, N. G., Heath, G. R., and Moore, T. C., Jr., 1975, Lag times for oceanic responses to climate change: Nature, v. 256, p. 716-717.

Plank, W. S., Zaneveld, J. V., and Pak, H., 1973, Distribution of suspended matter in the Panama Basin: Jour. Geophys. Research, v. 78, no. 30, p. 7113-7121.

Prospero, J. M., and Bonatti, E., 1969, Continental dust in the atmosphere of the eastern equatorial Pacific: Jour. Geophys. Research, v. 74, p. 3362-3371.

Shackleton, N. J., 1969, The last interglacial in the marine and terrestrial records: Royal Soc. [London] Proc., ser. B, v. 174, p. 123-154.

Shackleton, N. J., and Opdyke, N. D., 1973, Oxygen isotope and paleomagnetic stratigraphy of equatorial Pacific core V23-238: Oxygen isotope temperatures and ice volumes on 10^5 year scale: Quaternary Research, v. 3, no. 1, p. 39-55.

Soutar, A., 1971, Micropaleontology of anaerobic sediments and the California Current, in Funnell, B. M., and Riedel, W. R., eds., Micropaleontology of the oceans: Cambridge, Cambridge Univ. Press, p. 223-230.

Suess, H. E., 1970, The three causes of secular C^{14} fluctuations, their amplitudes and time constants, *in* Olsson, I.U., ed., Radiocarbon variations and absolute chronology: Nobel Symp., 12th, Proc., Uppsala 1969, p. 595–605.

Veeh, H. H., 1967, Deposition of uranium from the ocean: Earth and Planetary Sci. Letters, v. 3, no. 2, p. 145–150.

Wiseman, J.D.H., 1956, The rates of accumulation of nitrogen and calcium carbonate on the equatorial Atlantic floor: Adv. Sci., v. 12, p. 579.

MANUSCRIPT RECEIVED BY THE SOCIETY SEPTEMBER 23, 1974
REVISED MANUSCRIPT RECEIVED MARCH 31, 1975
MANUSCRIPT ACCEPTED APRIL 9, 1975

Printed in U.S.A.

Geological Society of America
Memoir 145
© 1976

Late Quaternary Accumulation Rates of Opal, Quartz, Organic Carbon, and Calcium Carbonate in the Cascadia Basin Area, Northeast Pacific

G. Ross Heath*
Ted C. Moore, Jr.*
and
J. Paul Dauphin*
School of Oceanography
Oregon State University
Corvallis, Oregon 97331

ABSTRACT

Accumulation rates of terrigenous (quartz, organic carbon) and biogenic (calcium carbonate, opaline silica) components of two cores from southern Cascadia Basin are highly correlated with total-sediment accumulation rate and each other throughout late Pleistocene and Holocene time. In contrast, the correlations of the biogenic and terrigenous rates in a core on the east flank of Gorda Ridge are much lower (nonsignificant in the case of carbonate). In Cascadia Basin, sedimentation during late Pleistocene time was controlled by turbidite deposition. Apparently, the turbidity currents entrained biogenic debris produced by intense upwelling close to shore; this has resulted in the high correlations between the accumulation rates of the two sediment types. The only climatic signals evident in the Cascadia Basin cores are (1) pulses of accelerated deposition that correspond to glacial advances and catastrophic Columbia River floods at 32,000 and 18,000 B.P. and (2) marked reductions in sedimentation rates due to the sea-level rise at the end of the Wisconsin glaciation.

Accumulation rates of terrigenous components on the east flank of Gorda Ridge decreased fairly uniformly from 35,000 B.P. to the present. In contrast, the rate of carbonate accumulation was markedly lower prior to 32,000 B.P., from 29,000 to 27,000 B.P., and since, 15,000 B.P., apparently because of increased corrosiveness of bottom waters. The marked increase in carbonate dissolution in all cores in

---

*Present address: Graduate School of Oceanography, University of Rhode Island, Kingston, Rhode Island 02881

the area during Holocene time points to a regional increase in corrosiveness of bottom waters at the end of Pleistocene time.

Our study suggests that the increased resolution of late Quaternary oceanographic changes, which is theoretically possible in rapidly deposited hemipelagic sediments, is readily degraded by turbidite deposition, even where classic coarse-grained turbidites are absent. Cores from hemipelagic areas free of sediments carried in autosuspension are most likely to preserve detailed climatic and oceanographic records of late Quaternary time.

INTRODUCTION

Sedimentation in the northeast Pacific at about lat 45°N adjacent to North America has been intensely studied for the past decade. The area is of interest because of (1) turbidites that were trapped in Cascadia Basin and are spilling out onto nearby Tufts Abyssal Plain (Duncan and Kulm, 1970; Horn and others, 1971); (2) the shallow pelagic deposits preserved on the actively spreading Gorda and Juan de Fuca Ridges (Phipps, 1974); (3) the length of the Cascadia deep-sea channel and the complex sediment types deposited in it (Griggs and Kulm, 1970); (4) the influence of plate tectonics on sedimentation on the Gorda-Juan de Fuca plate (Kulm and others, 1973); (5) the marked variations in sediment properties that record the change from glacial Pleistocene to Holocene conditions (Duncan and others, 1970; Dauphin, 1972); and (6) the sensitivity of biogenic deposits of the area to changes in the influence of Transitional Zone, Alaskan Gyre, and Central Subarctic water masses during glacial and interglacial periods (Moore, 1973).

On the basis of these earlier works and the accumulation rates of key sediment components, we have attempted to reconstruct in detail the sedimentation history of the area for the past 35,000 yr. Although numerous deep-sea cores have been analyzed for their biogenic, mineralogic, and elemental compositions, the results have rarely been converted from concentrations to accumulation rates. As a result, it is seldom possible to determine whether a component changes in abundance because of variations in its input rate, variations in the supply of one or more other sedimentary components, or some combination of these. The few studies that considered rate changes (Broecker and others, 1958; Emery, 1960) have occasionally come to conclusions diametrically opposed to those based purely on concentration changes.

Despite the obvious benefit of working with rate data down cores, the requirements that the bulk density of the sediment be determined in order to convert linear deposition rates to mass deposition rates and that the absolute age of each sample be known make such data difficult and expensive to obtain. This study is an attempt to determine the value of such data in interpreting late Pleistocene and Holocene deep-sea deposits.

METHODOLOGY

Three cores were selected for detailed study. The sediments of these cores (Table 1, Fig. 1) were deposited fairly rapidly, appear to be free of major hiatuses for late Pleistocene-Holocene time, are relatively free of sand-size turbidite material (compared to other Cascadia Basin cores), and have detailed radiolarian stratigraphies available (Moore, 1973). Foraminiferal-radiolarian curves (Fig. 2) and C^{14} dates (Fig. 3) provide the basic time framework for the study. Detailed correlations

TABLE 1. CASCADIA AREA CORES USED FOR THE STUDY

Core	Lat (N)	Long (W)	Water depth (m)	Age of sediments in this study (yr)	Mean accumulation rate for interval (g/cm^2/1,000 yr)
6604-10	43°16'	126°24'	3,002	0–36,000	17.6
6609-5	43°34'	126°28'	2,978	0–28,000	12.4
6910-2	41°16'	127°01'	2,615	0–37,000	6.6

of the cores were carried out using profiles of biostratigraphic data. Only the most reliable correlations between core 6910-2 and the other cores are indicated on Figure 3. Correlated points were dated by assuming uniform sedimentation in core 6910-2 and interpolating between C^{14}-dated intervals in that core. Thus, the time scales for the three cores (Fig. 3) should be internally consistent, even though some absolute errors may be present.

Bulk-density measurements were made by weighing a fixed volume of sediment before and after drying overnight at 110°C. The apparent bulk densities thus obtained were corrected for core shrinkage (all cores are stored in sealed tubes at 4°C, so shrinkage is minimal) to yield data on the salt-free dry mass per unit volume used in the rate calculation.

The variables considered in this study are calcium carbonate, organic carbon, opaline silica, quartz (together forming 20 to 25% of the total sediment), and proportions of three radiolarian assemblages determined from detailed faunal analyses (Moore, 1973).

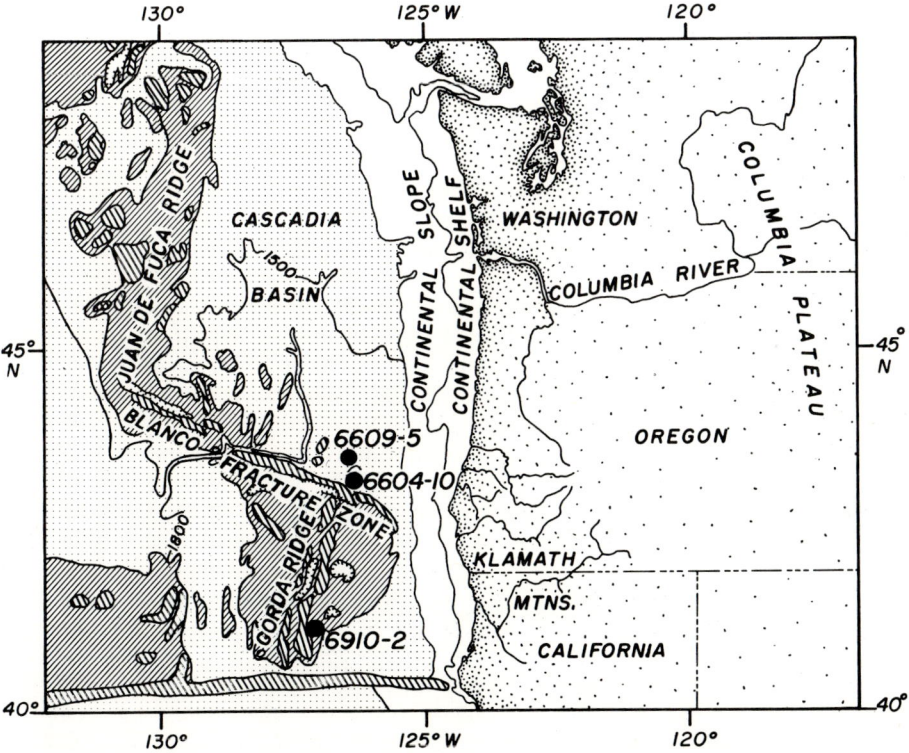

Figure 1. Location of cores in relation to nearby morphologic features (McManus, 1967). Stippled area, largely covered by Pleistocene turbidites; fine cross hatching, turbidite-free sediments; coarse hatching, steep hills and ridges.

The carbonate, determined by a LECO analyzer, records only biogenic calcite in the sediment, because the major rivers of the area do not carry detectable calcite (Glenn, 1973). The organic carbon, also determined by a LECO analyzer, is largely terrigenous plant debris. A small part may be of oceanic origin, but this has never been positively determined. The opaline silica, determined by x-ray diffractometry after conversion to cristobalite, is a measure of biogenic input from diatoms and radiolarians. The quartz, also determined by x-ray diffractometry (Till and Spears, 1969), is derived from general continental weathering and is a rough index of terrigenous input to the area.

The sequence of operations necessary to convert the raw concentration, bulk density, and age data to rates of accumulation are outlined in Figure 4. The bulk-density values are first fitted by a bicubic spline routine to convert the discrete observations to a function varying continuously with core depth. This function is then integrated to give cumulative mass per square centimeter as a function of depth below the core top. The integrated function is used to convert the dated levels in the core from depths to cumulative masses. The ages at these levels are then fitted by a second bicubic spline expression to give ages as a function of cumulative mass down the core. Differentiation of this function gives the sedimentation rate at any position in the core. The analytical data are then reduced by converting their depths to cumulative masses and then to ages; the concentrations are then multiplied by the appropriate sedimentation rates to yield accumulation rates of each component as a function of sample age.

RESULTS

Bulk Sediment

Figure 5 shows the total-sediment accumulation rates in the three cores as a function of age. The striking differences reflect the complex sedimentation pattern that prevailed in the area during Pleistocene time. Appendix 1 lists analytical data for the three cores, and sample depths and bulk density for the Cascadia Basin cores are given in Appendix II.[1]

At present, all three sites are receiving hemipelagic sediments. These fine-grained, olive to gray deposits are not well understood, but they appear to be derived from the continental shelf and upper slope and dispersed in intermediate and deep waters (Moore, 1972; Harlett and Kulm, 1972; Heath and others, 1974). Such deposits lack the grading of turbidites and are laid down over all types of terrain (Heath and others, 1974). This type of sedimentation has prevailed throughout the analyzed interval of core 6910-2, a period of about 37,000 yr. Thus, this core site on the west flank of the Gorda Ridge has never been covered by the turbidity currents that spread over much of the area during Pleistocene time (Duncan and Kulm, 1970).

The other cores, 6604-10 and 6609-5, are close together on a broad ridge that lies between Cascadia and Astoria channels (Fig. 6). The ridge is above the influence of the minor turbidite flows that have passed down the channels since the rise of sea level in Holocene time, but it was intermittently flooded by sand-bearing turbidity currents during the last glaciation (Duncan, 1968). Thus, the complex nature of the sedimentation rate curves for cores 6604-10 and 6609-5 relative to that of core 6910-2 largely results from turbidite deposition. The difference between

---

[1] Appendixes I and II are on microfiche in pocket inside back cover.

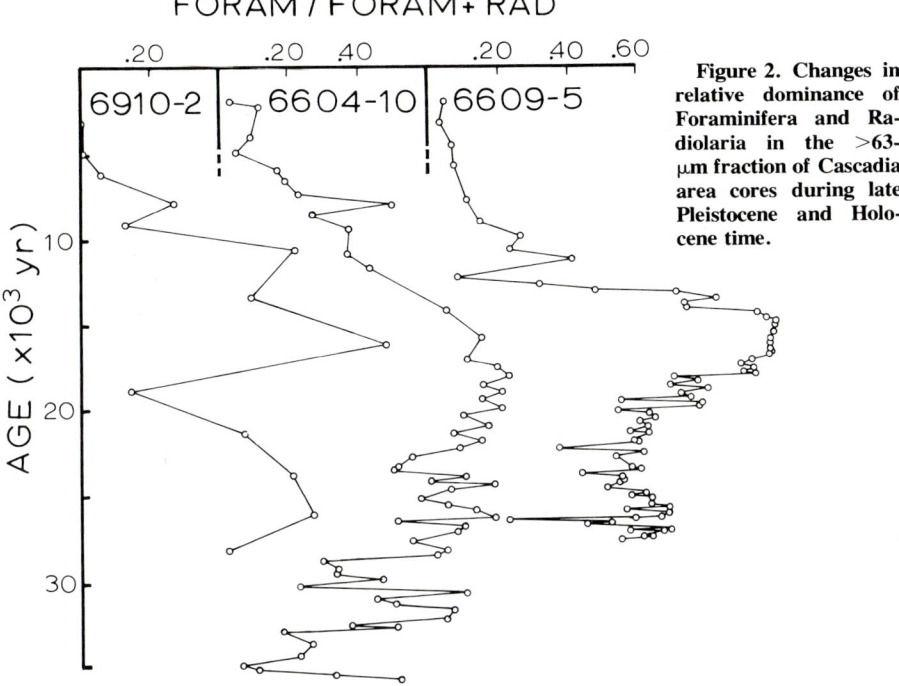

Figure 2. Changes in relative dominance of Foraminifera and Radiolaria in the >63-μm fraction of Cascadia area cores during late Pleistocene and Holocene time.

the curves for cores 6604-10 and 6609-5 (Fig. 5) probably results from differences in their positions relative to topographic highs in the area and from variations in the relative importance of Cascadia and Astoria turbidites at the two sites. Thus, from Figure 6 it appears that core 6604-10 received sediment from Astoria channel flows but was fairly well protected from Cascadia channel turbidity currents. Core 6609-5 may have received detritus from both sources, a hypothesis supported

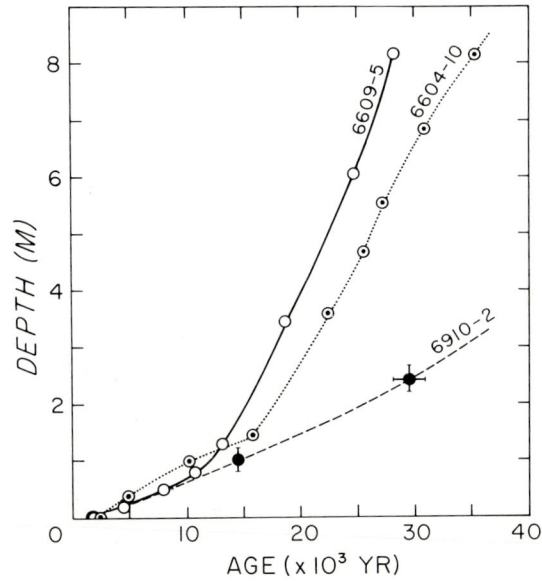

Figure 3. Age-depth relations in the three cores. Circles on core 6910-2 are C^{14} dates (bars give dated interval and age uncertainty). Other circles are points correlated with core 6910-2 on the basis of radiolarian stratigraphy.

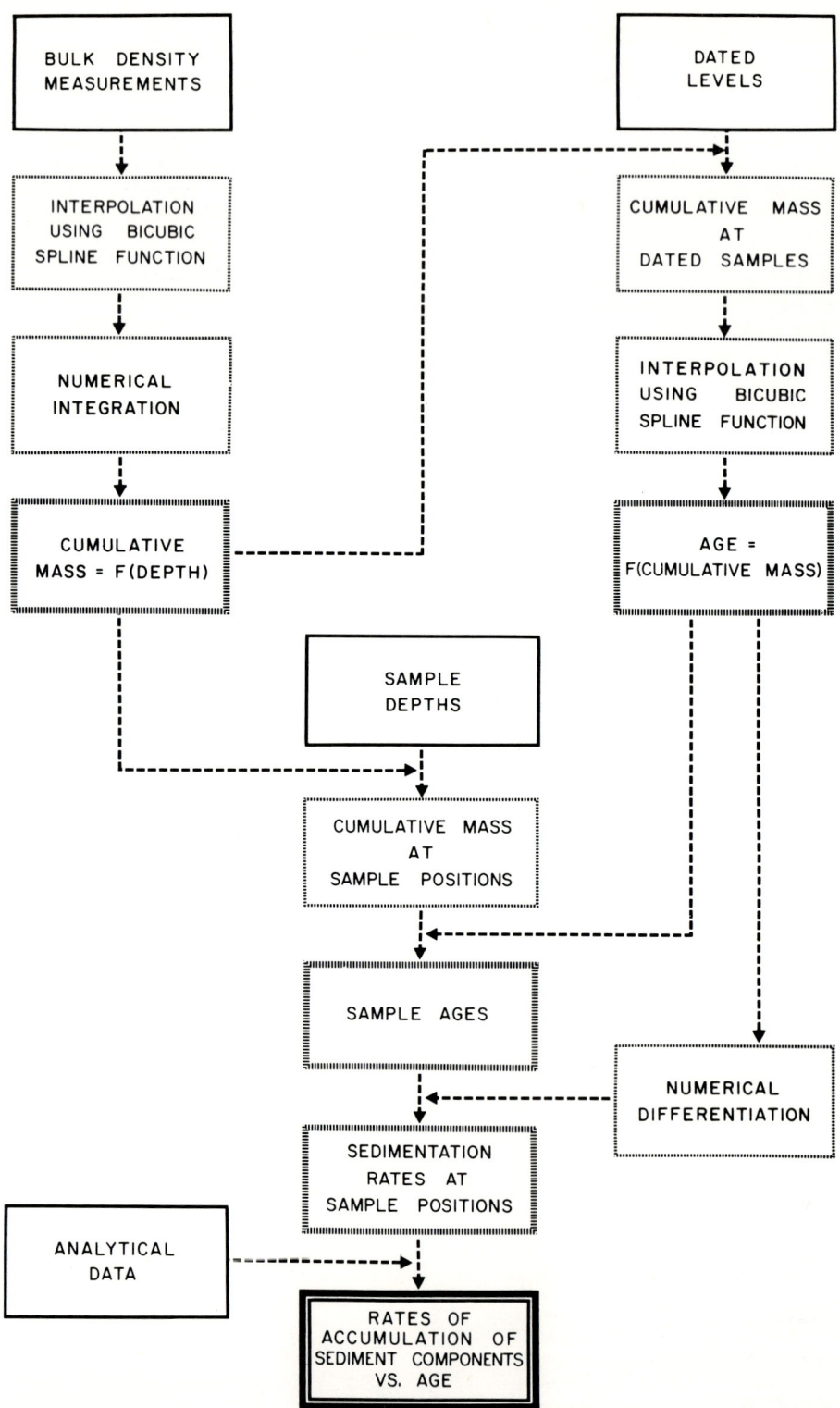

Figure 4. Procedure used to combine analytical data with dated levels in the cores to yield rates of accumulation of sediment components as a function of age.

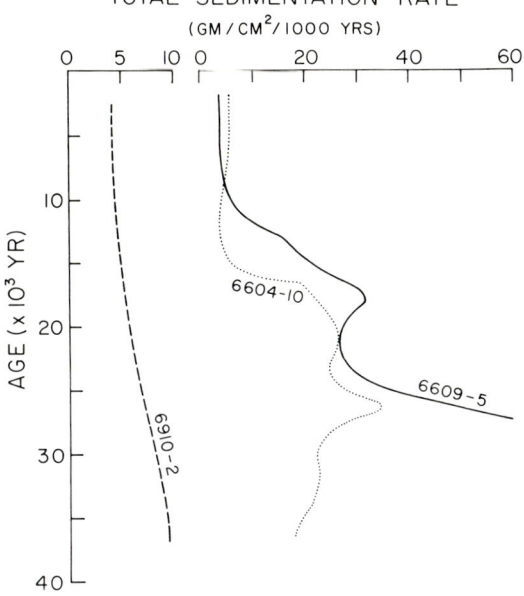

Figure 5. Instantaneous total sedimentation rates in the three cores during late Quaternary time.

by the higher ratio of late Pleistocene to Holocene sedimentation rates in this core than in core 6604-10 (Table 2). During Pleistocene time, flow down Astoria channel extended to the south end of Gorda Ridge and even turned north into its median valley (McManus and others, 1970). It appears that these flows also influenced hemipelagic deposition on the east flank of Gorda Ridge, inasmuch as Pleistocene sediments in core 6910-2 and nearby areas have mineralogic characteristics more reminiscent of the Columbia than the nearby Rogue or Klamath Rivers (Duncan and others, 1970; Dauphin, 1972).

Dauphin (1972) recorded a change in the grain-size distribution of quartz in core 6910-2 at 15,000 to 16,000 B.P. If this marks the end of major turbidite flows down Astoria channel, it appears to coincide with the decrease in sedimentation rate recorded in core 6604-10 (Fig. 5). Major turbidity-current flows continued in Cascadia channel after Astoria channel had become relatively inactive (Griggs, 1969). This may explain the persistence of high sedimentation rates in core 6609-5 beyond the time of their decrease in core 6604-10.

Terrigenous Components

As discussed above, quartz and organic carbon are considered to be dominantly, if not entirely, of terrigenous origin. The general shapes of the accumulation-rate curves for quartz, organic carbon, and total sediment in each of the three cores are very similar (Fig. 7). As would be expected, correlation coefficients between the three rates are very high ($p < 0.01$ in all cases; Table 3). Thus, even in core 6910-2, which lacks discernible turbidites, terrigenous material seems to overwhelm other sedimentary components.

In the two northern cores, pulses of accelerated quartz deposition occurred at about 34,000 B.P. (core 6604-10 only), 26,000 B.P., and 18,000 to 21,000 B.P. The youngest of these pulses corresponds to the last major glacial advance in North America (Kukla, 1972), as well as to the youngest "glacial" peaks in oxygen-isotope curves for the Greenland ice cap (Dansgaard and others, 1971)

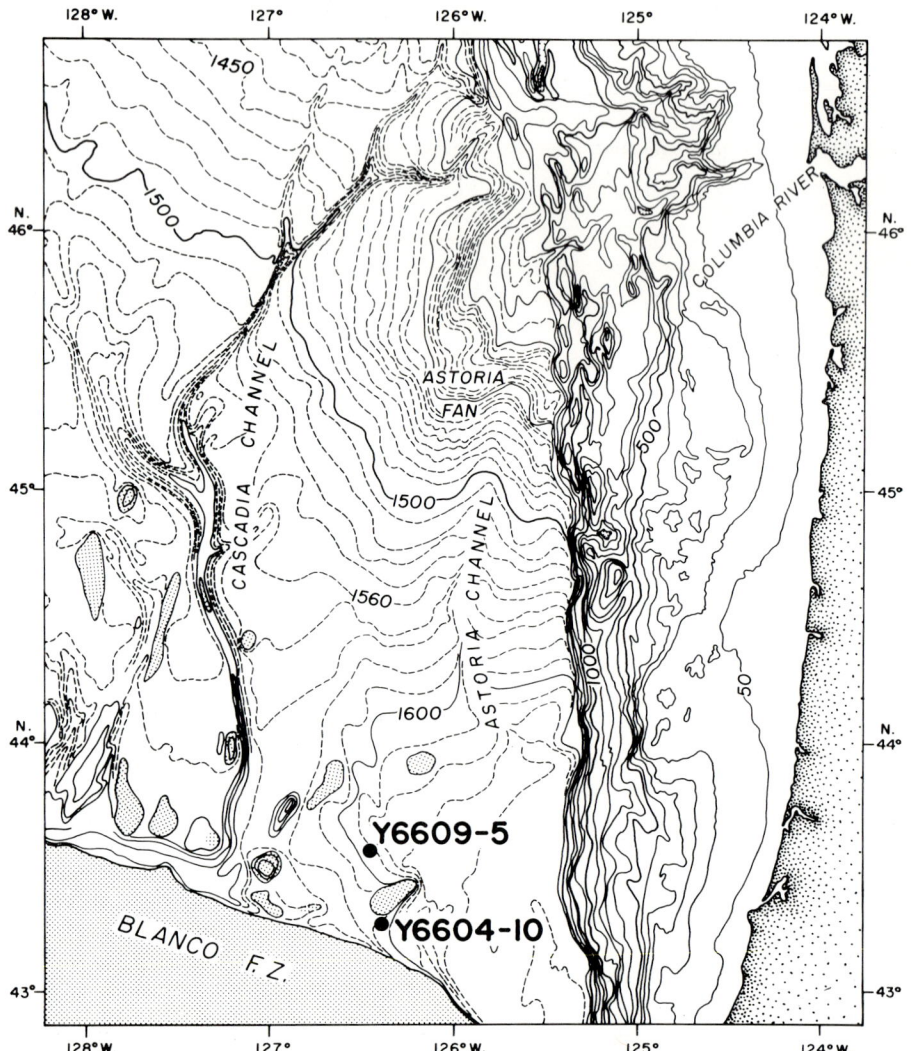

Figure 6. Position of Cascadia Basin cores 6604-10 and 6609-5 relative to the principal topographic features of the Cascadia Basin floor (McManus, 1967) and to hills and areas of rough topography (stippled).

TABLE 2. RATIO OF ACCUMULATION RATES OF TOTAL SEDIMENT AND ANALYZED COMPONENTS IN PLEISTOCENE AND HOLOCENE SECTIONS OF CASCADIA AREA CORES

Accumulation rate	Core 6604-10	Core 6609-5	Core 6910-2
Total	4.5	7.5	1.8
Quartz	4.7	7.7	2.0
Organic carbon	4.7	6.9	1.6
Carbonate	8.0	12.0	8.2
Opal	3.0	5.0	1.9

Note: Values are ratios of Pleistocene rates to Holocene rates.

and for planktonic and benthic foraminifers in deep-sea sediments (Emiliani, 1966; Shackleton and Opdyke, 1973). The 34,000-B.P. pulse corresponds to glacial advances of both the Erie lobe of the Laurentide ice sheet and the Frazer-Puget segments of the Cordilleran ice sheet. Both the 18,000- to 21,000- and 34,000-B.P. dates correspond to times of catastrophic flooding of the Columbia River basin due to breaking of ice dams on the headwaters of the river (Richmond and others, 1965). These floods cut channels in the loess deposits of the basin and carried enormous volumes of quartz-rich detritus out to sea (Phipps, 1974). It is not surprising that they triggered major turbidity currents which carried coarse debris far out into Cascadia Basin (Griggs, 1969).

The 26,000-B.P. pulse does not correspond to a documented glacial event in North America. In fact, the estimated temperature record in core V23-82 from the North Atlantic (Sancetta and others, 1973) suggests that this pulse corresponds to a slightly warmer interval in the late Wisconsin glaciation rather than to a glacial advance.

The accumulation rates of organic carbon in the three cores generally parallel the quartz rates but differ in detail. No systematic pattern can be recognized

TABLE 3. CORRELATION MATRIX FOR SEDIMENTATION RATES AND RADIOLARIAN FACTOR LOADINGS (MOORE, 1973) IN THE THREE CASCADIA AREA CORES

	Total sedimentation rate	Quartz rate	Organic carbon rate	Carbonate rate	Opal rate	F1 (Transitional)	F2 (Alaskan Gyre)
Quartz rate	+0.96						
	+0.98						
	+0.95						
Organic carbon rate	+0.90	+0.87					
	+1.00	+0.98					
	+0.86	+0.84					
Carbonate rate	+0.88	+0.86	+0.76				
	+0.93	+0.94	+0.92				
	. .	. .	. .				
Opal rate	+0.90	+0.87	+0.83	+0.78			
	+0.76	+0.70	+0.77	+0.68			
	+0.69	+0.72	+0.66	. .			
F1 (Transitional)	−0.63	−0.62	−0.55	−0.48	−0.52		
	−0.60	−0.55	−0.59	−0.66	−0.56		
	. .	. .	. .	. .	. .		
F2 (Alaskan Gyre)	+0.81	+0.58	+0.76	+0.72	+0.80	−0.63	
	+0.52	+0.80	+0.51	+0.63	+0.40	. .	
	+0.83	+0.74	+0.59	. .	. .	. .	
F3 (Central subarctic)	−0.35	−0.35	−0.41	−0.34	−0.46	. .	−0.58
	. .	. .	. .	. .	. .	. .	−0.77
	−0.54	. .	. .	. .	. .	. .	−0.83

Note: Number of samples per core: 81 (6604-10), 65 (6609-5), and 29 (6910-2). Values of r for $p = 0.05$: 0.34 (6604-10), 0.37 (6609-5), and 0.54 (6910-2). Values of r for $p = 0.01$: 0.38 (6604-10), 0.42 (6609-5), and 0.60 (6910-2). Ellipsis points denote a value not significant at $p = 0.05$.

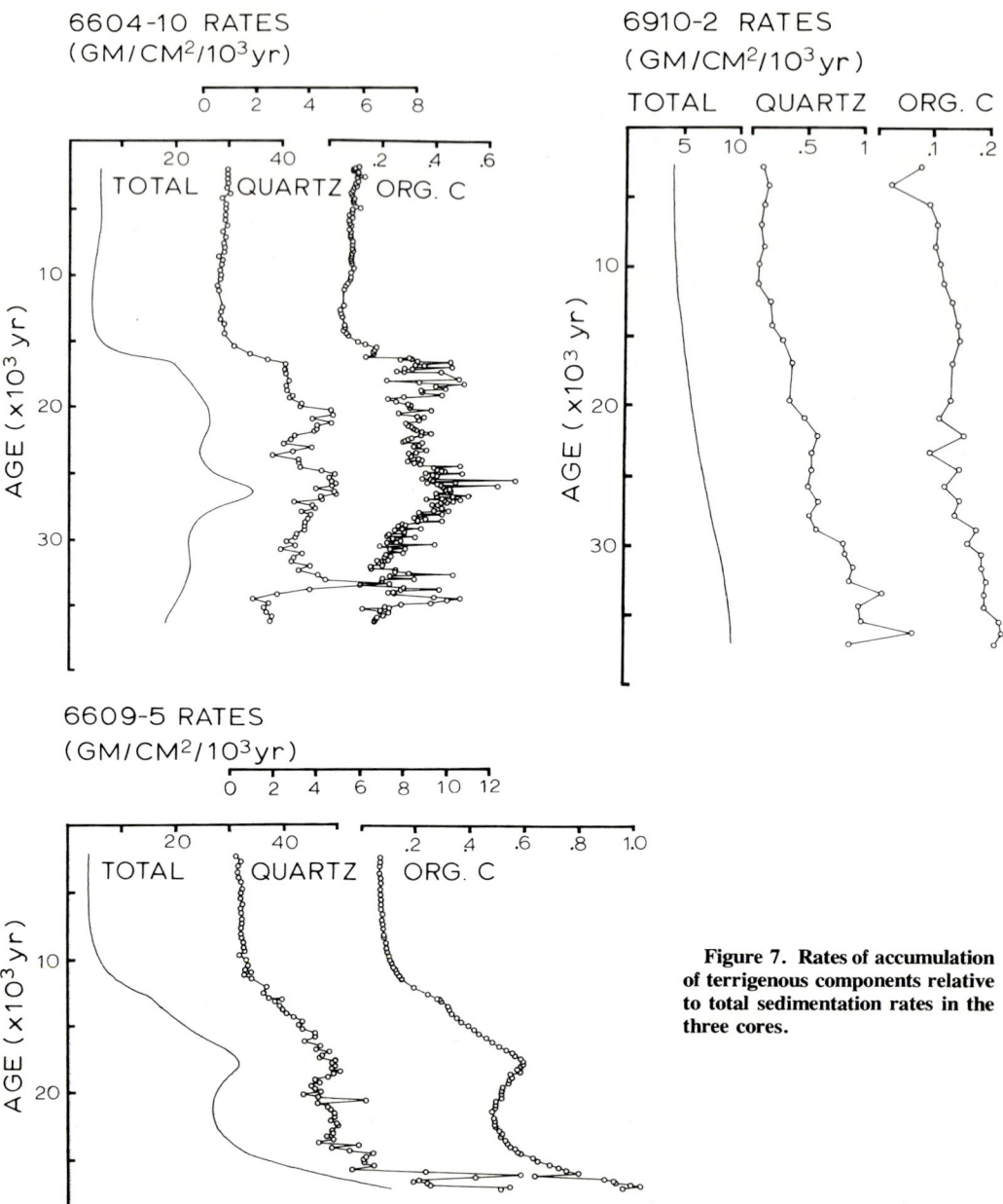

Figure 7. Rates of accumulation of terrigenous components relative to total sedimentation rates in the three cores.

in the differences, which suggests that they may result from slight variations in the relative proportions of clay (richer in organic carbon) and silt (richer in quartz) rather than from more fundamental changes in sediment supply.

Comparison of quartz and organic-carbon accumulation rates with changes in radiolarian assemblages measured in the cores (Moore, 1973) shows some correlation between variations in sedimentation conditions and oceanographic conditions in the overlying near-surface waters. The oceanographic conditions are estimated by counting radiolarian assemblages in northeastern Pacific surface sediments, factoring

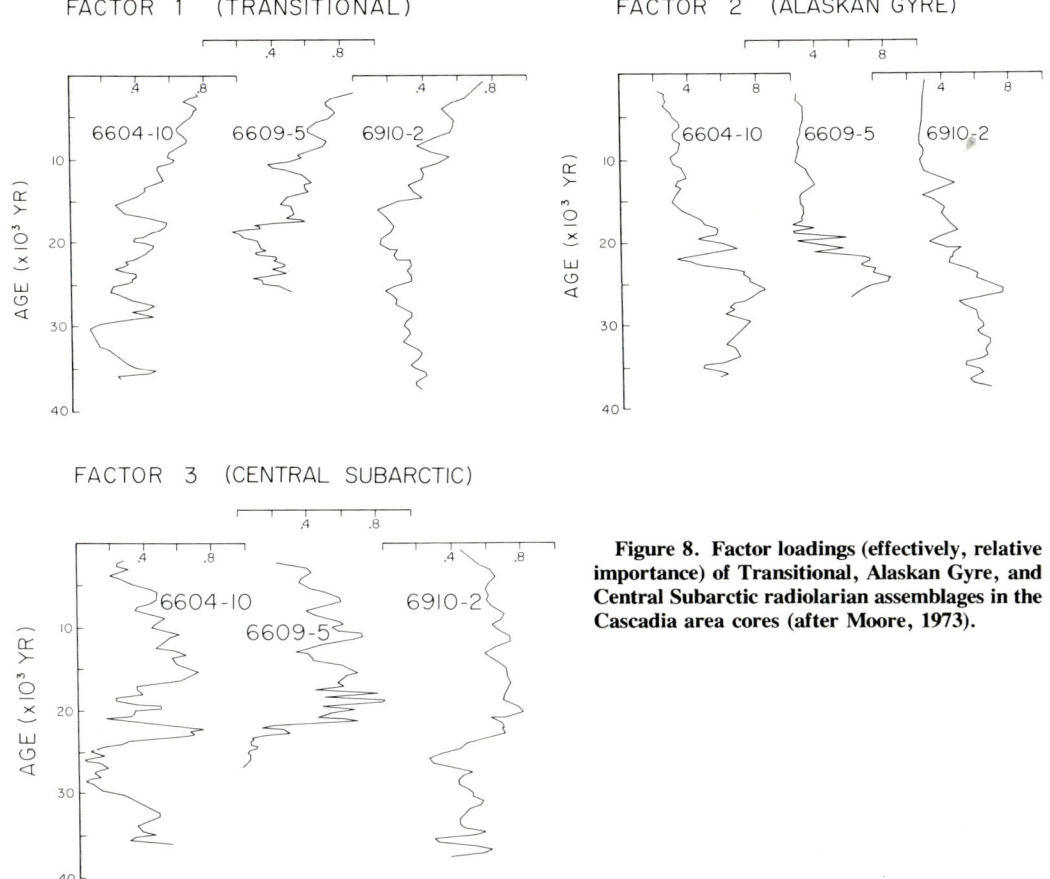

Figure 8. Factor loadings (effectively, relative importance) of Transitional, Alaskan Gyre, and Central Subarctic radiolarian assemblages in the Cascadia area cores (after Moore, 1973).

the population data (see Klovan and Imbrie, 1971), regressing the factors on modern sea-surface parameters [temperature, salinity, and so forth (Imbrie and Kipp, 1971)], and applying the regression equations to radiolarian populations sampled down the cores (Moore, 1973). Three factors account for most of the variance in surface radiolarian assemblages. From their geographic distributions relative to present-day surface water masses, these have been identified as Transitional Zone (factor 1), Alaskan Gyre (factor 2), and Central Subarctic (factor 3) assemblages. The three cores of this study now lie beneath the Transitional Zone, with Central Subarctic and Alaskan Gyre Waters lying farther to the north.

For the downcore data (Fig. 8), the Alaskan Gyre factor (F2) is most highly correlated with the total-sediment accumulation rate (Table 3), as well as with the quartz and organic-carbon rates. This is not surprising, because increased influx of terrigenous material, due in part to lowered sea level, and southward shift of northeastern Pacific water-mass boundaries are typical glacial phenomena in this area.

In the northern cores (6604-10 and 6609-5) the Transitional Zone factor (F1) is weakly negatively correlated with the rates of accumulation of total sediment, organic carbon, and quartz. In the hemipelagic sediments of core 6910-2, however, such correlations are lacking.

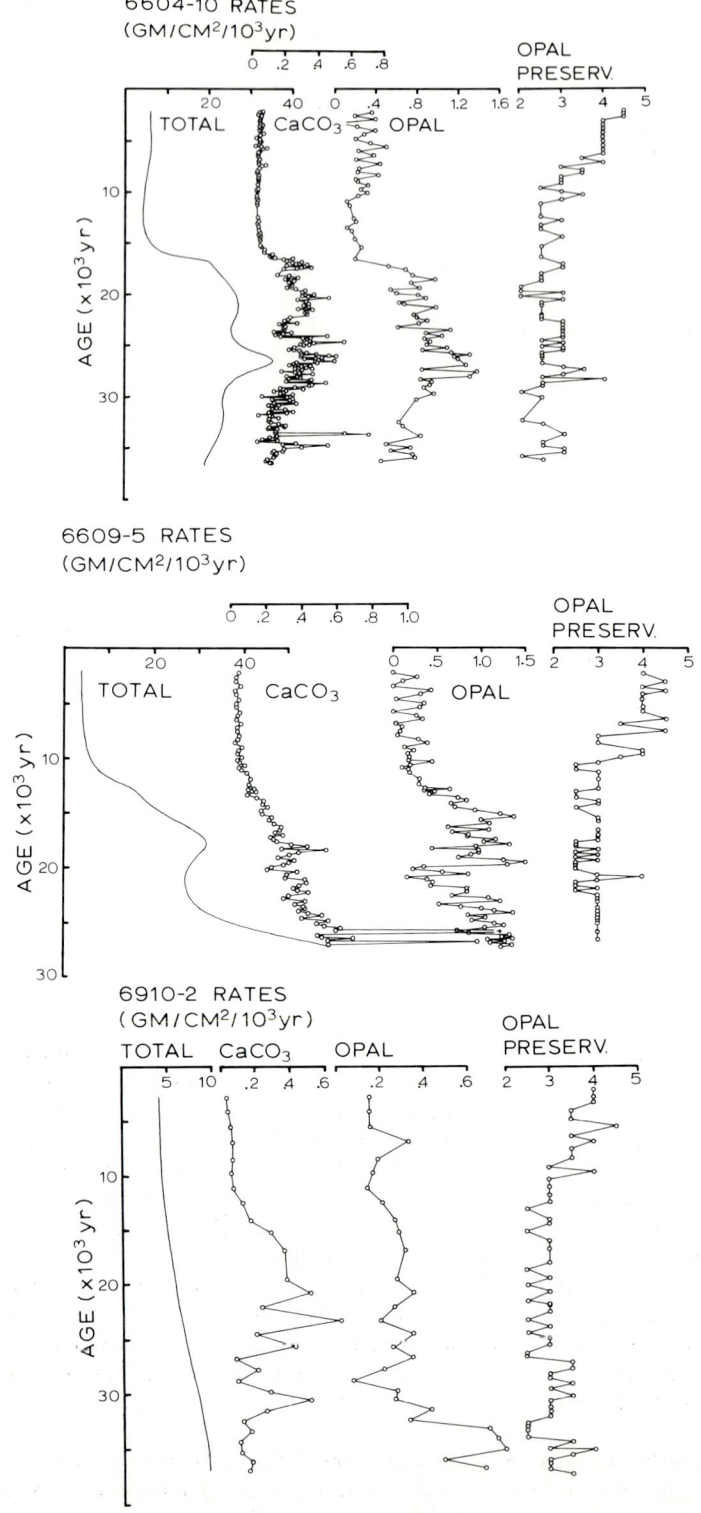

Figure 9. Rates of accumulation of biogenous components relative to total sedimentation rates in the three cores, compared with degree of preservation of radiolarian tests larger than 63 μm. Well-preserved samples have higher preservation indices.

Biogenic Components

The accumulation rates of the biogenic components (opaline silica and calcium carbonate) in the three cores are shown in Figure 9. In the two northern cores, the opal and carbonate rates closely parallel the total sedimentation rate, a trend also apparent in the correlation coefficients (Table 3). In the more hemipelagic core 6910-2, opal is somewhat correlated with total sedimentation rate (Table 3), but carbonate is not correlated with any other parameter.

In the two northern cores, both carbonate and opal rates are positively correlated with the Alaskan Gyre radiolarian factor (F2) and negatively correlated with the Transitional Zone factor (F1). Because of strong correlations between bulk sedimentation rate and both faunal factors and carbonate and opal, however, it is difficult to say whether there is a direct relationship between surface oceanographic parameters and biogenic sedimentation in this area rather than independent control of both by an additional factor, climatic change. This conclusion is supported by the poor core-to-core correlation of the biogenic rates of accumulation (Fig.9).

DISCUSSION

Perhaps the most puzzling aspect of the results presented here is the high correlation between the total-sediment accumulation rate and the rates of accumulation of all the sediment components analyzed. As already discussed, the covariation of quartz and organic carbon with total accumulation rate is reasonable in view of the dominantly terrigenous origin of the sediment in the area. The reasons for the correlation of opal and carbonate rates with total-sediment accumulation rate in the two northern cores, and of opal with total-sedimentation rate in core 6910-2, however, are much less clear. The following two possible explanations appear plausible.

1. Changing oceanographic conditions in near-surface waters have changed the rate of biogenic detrital supply to the sea floor, but variable preservation largely dependent on the accumulation rate of the bulk sediment masks such changes. Such a phenomenon is consistent with suggestions that most calcareous (Berger, 1974) and siliceous (Heath, 1974) tests secreted in the photic zone are dissolved before they can be incorporated into the sedimentary record. If the accumulation rate does largely reflect variable dissolution of the microfossil assemblages reaching the sea floor, the preservation state of the tests should be better in sections of the core with high opal or carbonate accumulation rates than in low-rate sections. Figure 9 shows no such opal pattern in any of the three cores. In all cases, a subjective index of the preservation of radiolarian assemblages tends to be negatively rather than positively correlated with opal accumulation rates. In the case of carbonate, however, the Holocene section contains only badly preserved foraminifers. Furthermore, the increased rate of carbonate accumulation in upper Pleistocene relative to Holocene sections of all three cores (Table 3) is greater than the corresponding increases for other sediment components. Such a pattern strongly implies variable dissolution. The fact that the Pleistocene/Holocene ratio is relatively independent of total deposition rates (Table 3) suggests that variations in bottom-water aggressiveness, rather than variations in the rate of burial, exert the greatest influence on carbonate preservation in the Cascadia area. In either case, the foraminiferal-radiolarian abundance curves used for stratigraphic correlation in Cascadia Basin and surrounding areas (Griggs and others, 1970; Phipps, 1974; Barnard and McManus, 1973) reflect variable dissolution rather than changing productivity in near-surface waters.

2. Large quantities of biogenic opal and carbonate were deposited along the continental margin and then entrained in the turbidity currents that carried most of the terrigenous sediment to the core sites. If the amount of biogenic debris picked up was related to the size of the turbidity current, biogenic and bulk sediment rates of accumulation would covary.

Such an explanation appears somewhat forced, but it is supported by recent observations that the intense upwelling and associated biological productivity that occurs off the Pacific Northwest is largely confined to a zone within 10 km of the coast (L. Small, 1973, oral commun.). Since the sediments beneath the upwelling areas are not dominated by microfossils, the carbonate and opal produced in these areas must either dissolve rapidly or be carried into deeper water by near-bottom currents. Direct current measurements (Harlett and Kulm, 1972) and evidence from cores such as those described here suggest that the latter explanation is responsible for at least part of the missing biogenic sediment. The preservation patterns of Figure 9 are readily explained if the entrainment and lateral transport of fine-grained sediment accelerates the rate of opal dissolution. Such an explanation is supported by the observation that the Pleistocene/Holocene ratios of opal accumulation are less than the ratios for other sedimentary components in the basin cores (Table 3).

The situation in the southern core, 6910-2, is more complex. As discussed previously, this core has obviously received fine debris from the north, but it lacks sedimentary structures typical of turbidites. Whether the currents that carried the sediment were true turbidity currents that barely impinged on the site or dilute autosuspensions that carried no coarse debris (Moore, 1972) is unclear. In either case, the explanation presented for the correlation of biogenic and terrigenous accumulation rates in the northern cores could also account for the correlation of opal and total sedimentation rates for core 6910-2. The lower correlation coefficient at this site (Table 2), however, together with the independent behavior of the carbonate accumulation rate (Fig. 9C, Table 2), suggests that core 6910-2 more closely resembles a typical deep-sea pelagic core, in which the abundance of biogenic material is determined more by production and dissolution than by lateral transport. The differences between the opal and carbonate curves (Fig. 9C) could reflect either differences in the relative rates of dissolution and lateral influx of the two phases (opal dissolving more slowly) or inability of the autosuspensions reaching the site to carry the denser carbonate particles.

The absence of significant correlations between the carbonate accumulation rate in core 6910-2 and Moore's (1973) radiolarian factors (Table 2) suggests that dissolution rather than production is the primary factor controlling carbonate accumulation. If so, the bottom waters at site 6910-2 were most corrosive prior to 32,000 B.P., from about 29,000 to 27,000 B.P., and since about 15,000 B.P. Such variations could result either from increases in the amount of oxidizable organic matter supplied to the site or from changes in the character of the bottom water over the site. Rate data for too few cores are available to decide between the two possibilities at the present time.

CONCLUSIONS

The rapid yet fairly continuous sedimentation that characterizes hemipelagic areas close to continental margins makes cores from such areas attractive to students of the late Pleistocene. The present study has made clear, however, that variations in sedimentation due to channelized bottom currents can mask changes due to variations in oceanographic conditions or continental climate. Thus, fluctuations in the concentration of organic carbon in terrigenous sediments of Cascadia Basin are periodic and may reflect global climatic variations (Pisias and others, 1973), but changes in the deposition rate of organic carbon in the Cascadia Basin are largely determined by patterns of turbidity-current flow that have little areal coherency.

Of the three cores studied, two are so strongly influenced by turbidite deposition that they record little beyond the sea-level rise marking the end of Pleistocene time. Pulses of accelerated terrigenous deposition at 18,000 to 21,000 B.P. and 34,000 B.P. may reflect glacial advances and periods of catastrophic flooding in the Columbia watershed, but a pulse of equal intensity at 26,000 B.P. cannot be explained in this way. The strong covariation of biogenic and terrigenous accumulation rates in Cascadia Basin suggests that fluctuations in biogenic sedimentation due to changing oceanographic conditions are largely masked by variations in the rate of influx of turbidity current-transported opal and carbonate, at least during late Pleistocene time.

Only core 6910-2, which sampled the east flank of the Gorda Ridge above the depth of major turbidity-current influence, appears to contain a sedimentary record in which oceanographic factors play an important role in determining the accumulation rate of biogenic sediments. Here the carbonate record shows markedly lower accumulation rates prior to 32,000 B.P., 29,000 to 27,000 B.P., and since 15,000 B.P. Lack of correlation between those lows and surface oceanographic conditions recorded in radiolarian faunas suggests that solution variations, rather than productivity variations, are the cause. Further core studies are needed to confirm this suggestion and to determine whether the reduced rate of opal deposition typical of the Holocene sections of all the cores reflects regional changes in productivity or simply a reduction in the supply of turbidite-entrained siliceous tests originally deposited close to shore. The parallel but more extreme reduction in the carbonate accumulation rate during Holocene time appears to result from dissolution of calcite tests. This may result partly from slow overall deposition rates and consequent long exposure of particles at the sea floor, but most of the dissolution appears to reflect increased corrosiveness of bottom waters since the end of Pleistocene time.

ACKNOWLEDGMENTS

We are grateful to L. Kulm and J. Phipps for counsel when we were selecting cores for study and for helpful discussions of results. P. Price and C. Rathbun made most of the sediment analyses; J. Imlah and K. Torvik assisted with the preparation of the manuscript. Comments from N. Pisias, P. Biscaye, W. Berger, and H. Poelchau have been most helpful in our interpretations. This research was largely supported by the International Decade of Ocean Exploration CLIMAP project (National Science Foundation Grant GX-28673).

REFERENCES CITED

Barnard, W. D., and McManus, D. A., 1973, Planktonic foraminiferan-radiolarian stratigraphy and the Pleistocene-Holocene boundary in the northeast Pacific: Geol. Soc. America Bull., v. 84, p. 2097–2100.

Berger, W. H., 1974, Biogenous deep sea sediments: Production, preservation, and interpretation, *in* Riley, J. P., and Chester, R., eds., Chemical oceanography; Vol. 3: London, Academic Press (in press).

Broecker, W. S., Turekian, K. K., and Heezen, B. C., 1958, The relation of deep sea sedimentation rates to variations in climate: Am. Jour. Sci., v. 256, p. 503–517.

Dansgaard, W., Johnsen, S. J., Clausen, H. B., and Langway, C. C., 1971, Climatic record revealed by the Camp Century ice core, *in* Turekian, K. K., ed., The late Cenozoic glacial ages: New Haven, Yale Univ. Press, p. 37–56.

Dauphin, J. P., 1972, Size distribution of chemically extracted quartz used to characterize fine grained sediments [M.S. thesis]: Corvallis, Oregon State Univ., 63 p.

Duncan, J. R., 1968, Late Pleistocene and postglacial sedimentation and stratigraphy of deep-sea environments off Oregon [Ph.D. thesis]: Corvallis, Oregon State Univ., 222 p.

Duncan, J. R., and Kulm, L. D., 1970, Mineralogy, provenance, and dispersal history of late Quaternary deep-sea sands in Cascadia Basin and Blanco Fracture Zone off Oregon: Jour. Sed. Petrology, v. 40, p. 874–887.

Duncan, J. R., Kulm, L. D., and Griggs, G. B., 1970, Clay mineral composition of late Pleistocene and Holocene sediments of Cascadia Basin, northeastern Pacific Ocean: Jour. Geology, v. 78, p. 213–221.

Emery, K. O., 1960, The sea off southern California: New York, John Wiley & Sons, 366 p.

Emiliani, C., 1966, Paleotemperature analysis of Caribbean cores P6304-8 and P6304-9 and a generalized temperature curve for the last 425,000 years: Jour. Geology, v. 74, p. 109–126.

Glenn, J. L., 1973, Relations among radionuclide content and physical, chemical, and mineral characteristics of Columbia River sediments: U.S. Geol. Survey Prof. Paper 433-M, p. M1–M52.

Griggs, G. B., 1969, Cascadia channel: The anatomy of a deep-sea channel [Ph.D. thesis]: Corvallis, Oregon State Univ., 183 p.

Griggs, G. B., and Kulm, L. D., 1970, Sedimentation in Cascadia deep-sea channel: Geol. Soc. America Bull., v. 81, p. 1361–1384.

Griggs, G. B., Kulm, L. D., Duncan, J. R., and Fowler, G. A., 1970, Holocene faunal stratigraphy and paleoclimatic implications of deep-sea sediments in Cascadia Basin: Paleogeography, Paleoclimatology, Paleoecology, v. 7, p. 5–12.

Harlett, J. C., and Kulm, L. D., 1972, Some observations of near-bottom currents in deep-sea channels: Jour. Geophys. Research, v. 77, p. 499–504.

Heath, G. R., 1974, Dissolved silica and deep-sea sediments, *in* Hay, W. W., ed., Studies in paleo-oceanography: Soc. Econ. Paleontologists and Mineralogists Spec. Pub. 20, p. 77–93.

Heath, G. R., Moore, T. C., Jr., and Roberts, G. L., 1974, Mineralogy of surface sediments from the Panama Basin, eastern equatorial Pacific: Jour. Geology, v. 82, p. 145–160.

Horn, D. B., Ewing, M., Delach, M. N., and Horn, B. M., 1971, Turbidites of the northeast Pacific: Sedimentology, v. 16, p. 55–69.

Imbrie, J., and Kipp, N. G., 1971, A new micropaleontological method of quantitative paleoclimatology: Application to a late Pleistocene Caribbean core, *in* Turekian, K. K., ed., The late Cenozoic glacial ages: New Haven, Yale Univ. Press, p. 71–181.

Klovan, J. E., and Imbrie, J., 1971, An algorithm and FORTRAN-IV program for large-scale Q-mode factor analysis and calculation of factor scores: Internat. Assoc. Math. Geology Jour., v. 3, p. 61–77.

Kukla, G. J., 1972, Insolation and glacials: Boreas, v. 1, p. 63–96.

Kulm, L. D., von Huene, R., and others, 1973, Initial reports of the Deep Sea Drilling Project, Vol. 18: Washington, D.C., U.S. Govt. Printing Office, 1,077 p.

McManus, D. A., 1967, Physiography of Cobb and Gorda Rises, northeast Pacific Ocean: Geol. Soc. America Bull., v. 78, p. 527–546.

McManus, D. A., Burns, R. E., and others, 1970, Initial reports of the Deep Sea Drilling Project, Vol. V: Washington, D.C., U.S. Govt. Printing Office, 827 p.

Moore, D. G., 1972, Reflection profiling studies of the California continental borderland: Structure and Quaternary turbidite basins: Geol. Soc. America Spec. Paper 107, 142 p.

Moore, T. C., Jr., 1973, Late Pleistocene–Holocene oceanographic changes in the northeastern Pacific: Quaternary Research, v. 3, p. 99–109.

Phipps, J. B., 1974, Sediments and tectonics of the Gorda–Juan de Fuca plate [Ph.D. thesis]: Corvallis, Oregon State Univ., 115 p.

Pisias, N. G., Dauphin, J. P., and Sancetta, C., 1973, Spectral analysis of late Pleistocene–Holocene sediments: Quaternary Research, v. 3, p. 3–9.

Richmond, G. M., Fryxell, R., Neff, G. E., and Weis, P. L., 1965, The Cordilleran ice sheet of the northern Rocky Mountains and related Quaternary history of the Columbia plateau, in Wright, H. E., and Frey, D. G., eds., The Quaternary of the United States: Princeton, N.J., Princeton Univ. Press, p. 231–241.

Sancetta, C., Imbrie, J., and Kipp, N. G., 1973, Climatic record of the past 130,000 years in North Atlantic deep-sea core V23-82: Correlation with the terrestrial record: Quaternary Research, v. 3, p. 110–116.

Shackleton, N. J., and Opdyke, N. D., 1973, Oxygen isotope and paleomagnetic stratigraphy of equatorial Pacific core V28-238: Oxygen isotope temperatures and ice volumes on a 10^5 year and 10^6 year scale: Quaternary Research, v. 3, p. 39–55.

Till, R., and Spears, D. A., 1969, The determination of quartz in sedimentary rocks using an X-ray diffraction method: Clays and Clay Minerals, v. 17, p. 323–327.

Manuscript Received by the Society September 23, 1974
Revised Manuscript Received March 31, 1975
Manuscript Accepted April 11, 1975

Printed in U.S.A.

Geological Society of America
Memoir 145
© 1976

Glacial Advance in the Gulf of Alaska Area Implied by Ice-Rafted Material

ROLAND VON HUENE
U.S. Geological Survey
Menlo Park, California 94025

JIM CROUCH
U.S. Geological Survey
Menlo Park, California 94025

AND

EDWIN LARSON
Department of Geological Sciences
University of Colorado
Boulder, Colorado 80302

ABSTRACT

An abundance curve for ice-rafted detritus based on data from cores from the Gulf of Alaska depicts the main continental glacial events, on which curves of a series of shorter events are superimposed. Interpretation of the shorter curves is limited by the difficulty of distinguishing between glacial events and the "noise level" of the data. However, the correspondence is remarkable between the curves developed here and curves of Pleistocene ocean water and ice temperatures, glacial advances, and lake levels. The curves suggest cycles of glacial advance lasting 12,000 to 15,000 yr; if more cores were available, the cyclicity would probably be established with greater confidence. Our studies of ice-rafted detritus in the Gulf of Alaska revealed that under certain conditions, a rapidly deposited deep-sea marine section, which characteristically has temporal continuity, can record variations in the intensity of alpine glaciation, which characteristically has high sensitivity to climatic change.

INTRODUCTION

Various investigators have studied cores of North Pacific marine sediment containing ice-rafted detritus in order to identify periods of glaciation. In an early study, Conolly and Ewing (1970) examined 22 cores from the northwest Pacific and found six zones of ice-rafted detritus deposited during the Brunhes normal epoch in a core near Kamchatka. In general, Conolly and Ewing found that the cores close to Kamchatka, a presumed source area, contained the most zones. Kent and others (1971) also found six zones of ice-rafted sand in cores from the northeast Pacific. A closer sampling interval per unit of time, enabled them to identify zones of ice-rafted material in greater detail than did Conolly and Ewing. This detail, and careful correlation between eight cores, allowed them to relate the local abundance data for ice-rafted material to broad terrestrial glacial sequences. As in the study by Conolly and Ewing (1970), the resolution of periods of ice-rafting generally improved in cores closest to a source area, but the cores of Kent and others recovered a shorter time interval because of the greater rates of deep-ocean sedimentation near a source area.

We reported data on abundances of ice-rafted material in a preliminary paper (von Huene and others, 1973). At site 178 of Leg 18 of the Deep Sea Drilling Project (DSDP), we sampled a Pleistocene core (1.8 m.y. old) about 190 m long at intervals representing 1,700 to 2,600 yr. The most expanded sections in the piston cores of Kent and others (1971) could only be sampled at intervals representing 8,000 to 13,000 yr if the cores were not to be disturbed by the analysis. Also, limitations on depth of penetration with a piston corer did not apply on the R/V *Glomar Challenger*. The upper parts of most holes near a source area in the Gulf of Alaska were continuously cored; however, the *Glomar Challenger* core drilling recovered less than 50% of the section, which imposed severe limitations on our ability to interpret the glacial chronology.

Despite limitations, the results of greater sampling density obtained by us (von

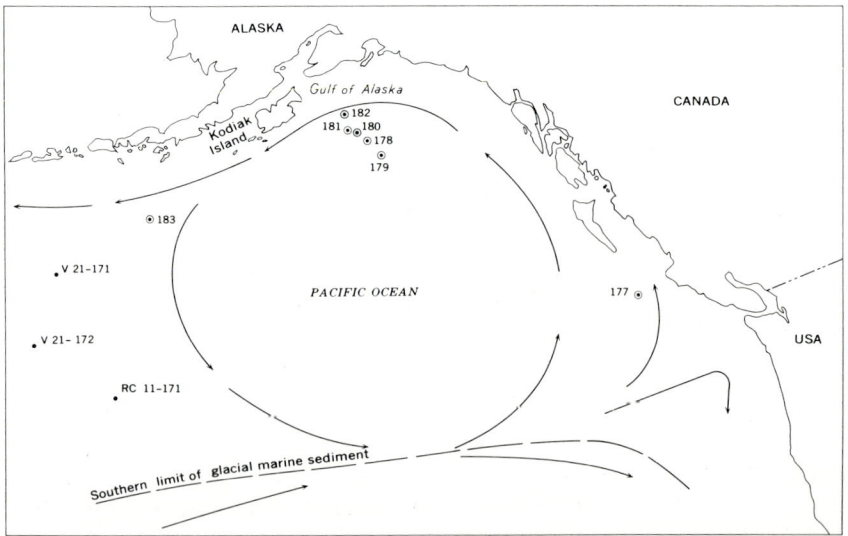

Figure 1. Gulf of Alaska showing location of Lamont-Doherty Geological Observatory (·) and Deep Sea Drilling Project cores (⊙), direction of currents in the Alaska current gyre (arrows), and southern limit of glacial marine sediment (from Kent and others, 1971).

Huene and others, 1973) and by Kent and others (1971) suggest that resolution of glacial advance and retreat during periods of less than 10,000 yr is possible by measuring the abundance of ice-rafted detritus in cores near the Gulf of Alaska sources.

In this paper we report a more complete dating of core 178 and a further analysis of Lamont-Doherty Geological Observatory piston core V21-171, a core used by Kent and others (1971) which was selected as optimum for this study. The additional material from V21-171 is sufficient to establish a general Gulf of Alaska chronology for the last 300,000 yr. We also point out a program designed to define cycles of glacial advance and retreat lasting 10,000 yr or less.

ANALYSIS OF ICE-RAFTED DETRITUS

Material Used

Measurements were made of the coarse-sand content of cores from four DSDP sites in the Gulf of Alaska (178, 179, 180, 183; Fig. 1). Analytical procedures were described in our earlier paper (von Huene and others, 1973). All the cores contain many ice-rafted pebbles and even cobbles. Graphs of detritus abundance show much detail; however, missing sections break the record, and the gaps are so large that the data cannot provide a general chronology (von Huene and others, 1973). Detritus-abundance graphs of cores from site 178 (Fig. 2) (the most complete cores), illustrate the type of detail possible despite a low relative sand content (generally 1% or less). Ages were established using diatoms, Radiolaria, and paleomagnetic reversals (Fig. 3). High and low events in the graph, based on three or more stepwise related measurements rather than just one or two, give assurance of data validity.

In an attempt to complete part of the record from DSDP drill sites, samples were run from core V21-171, a core analyzed by Kent and others (1971) that was taken from a site closest to a Gulf of Alaska source area. Additional sampling and analysis decreased the average time between sample midpoints from Kent's interval of 10,000 yr to about 3,000 yr. The resulting coarse-sand abundance graph is shown in Figure 4. Despite analysis in two laboratories where different grain-size limits and analytical procedures were used, the combined data still show single data point peaks of less than 1% of the dry sample weight. These peaks may be "noise," and the only criterion for deciding between "signal" and "noise" is the stepwise succession of data points. A peak defined by three to five data points probably represents a glacial advance, but peaks based on single points are questionable, as will be discussed later. A possible correlation of V21-171 with the first three cores from DSDP site 178 (Fig. 4) is shown for comparison with less "noisy looking" data.

Limits on Interpretation

It is difficult to interpret the significance of single data points that seem to suggest short ice advances or retreats. Small departures from a direct correlation between climate and the abundance of ice-rafted detritus are likely. The relation implied in curves showing abundance of ice-rafted material is that relative abundance of sand is proportional to abundance of icebergs, which is in turn related to rate of glacial advance. None of these relations can be established in a quantitative way, but for a gross qualitative delineation of glacial epochs, the implied relation

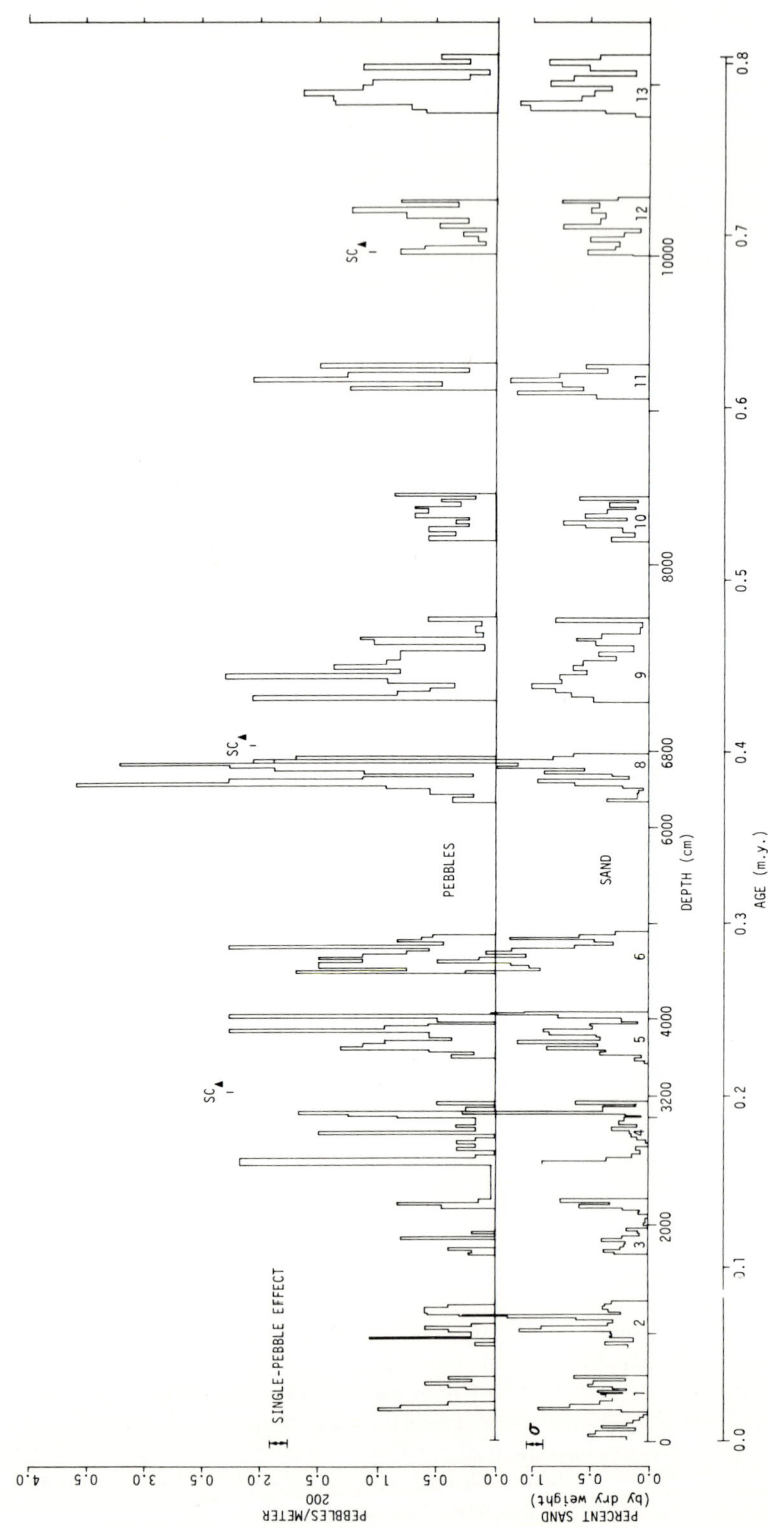

Figure 2. Graph of dry weight of sand and pebbles as a function of age. Width of each bar indicates length of channel sample from center of core shown at left margin. SC△ indicates point where age control is available. Age between control points is interpreted linearly with respect to depth. Bold numbers along sand curve are site 178 core numbers. Standard deviation (σ) and general effect of single pebble from DSDP site 178.

is probably reasonable. However, this may not hold in dealing with discrete events within an epoch that are at the threshold of resolution (or where the desired signal amplitude is near the noise).

One uncertainty in attempts to correlate glacial advance and abundance of rafted material is caused by the question of the uniformity with which glaciers and icebergs

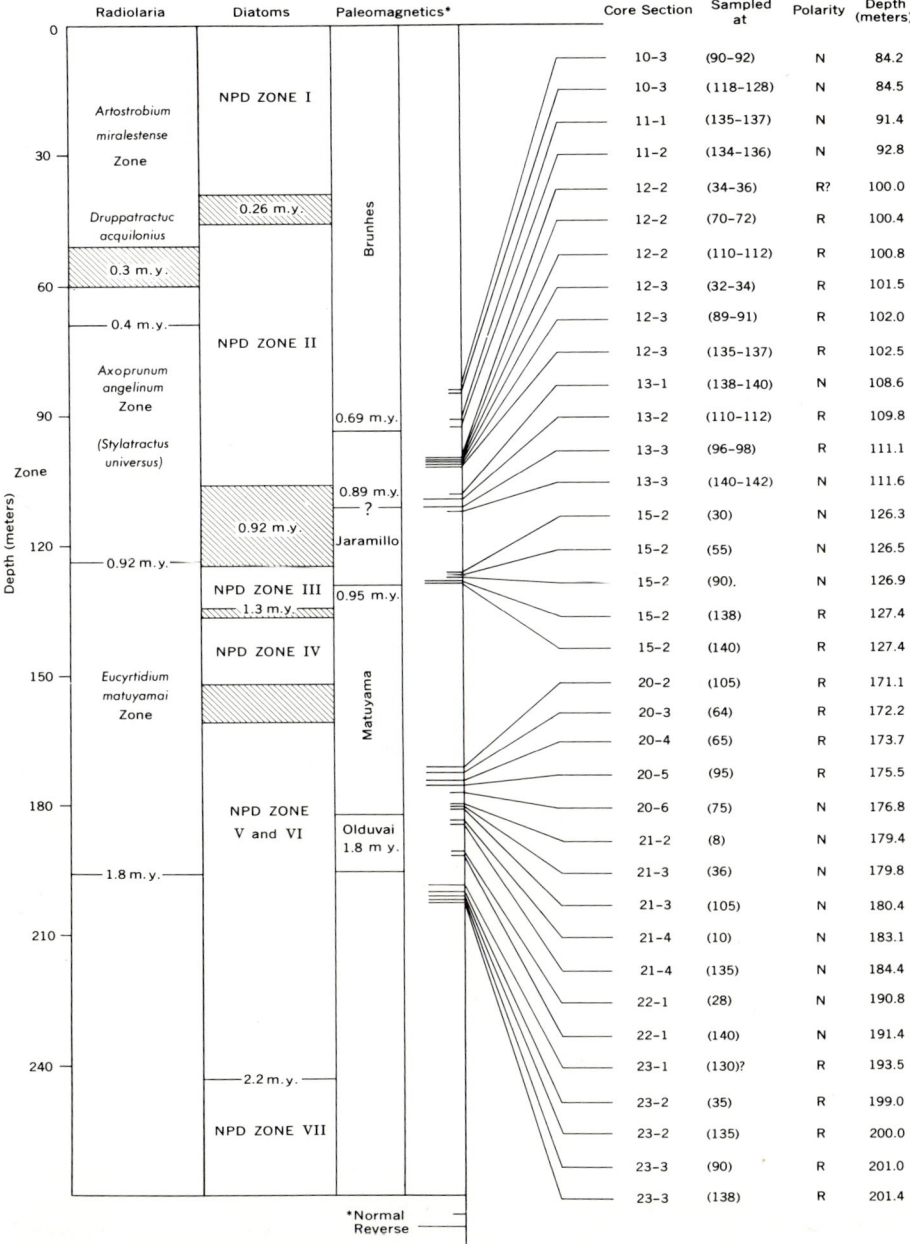

Figure 3. Age determinations on cores from site 178. Uncertainty in position of biostratigraphic boundaries shown by extent of hachured zones. Queried paleomagnetic polarities indicate nearly horizontal inclinations of remanent magnetization.

are loaded with detritus. Information bearing on this question came from the study of ice-rafted material. Fifty percent or more of the large clasts found in DSDP cores 178, 179, 180, and 182 showed no signs of abrasion and intergranular collision, as would be found in a beach or fluvial environment, and only 5% or less showed any glacial facets. This indicates that rockfalls from the walls of the glacial valley are the largest source of detritus. Therefore, the quantity of glacial debris in an iceberg will probably vary with the thickness of the glacier or the relative amount of valley wall exposed, with the speed at which the glacier moves, and with the availability of loose detritus from the previous interglacial period.

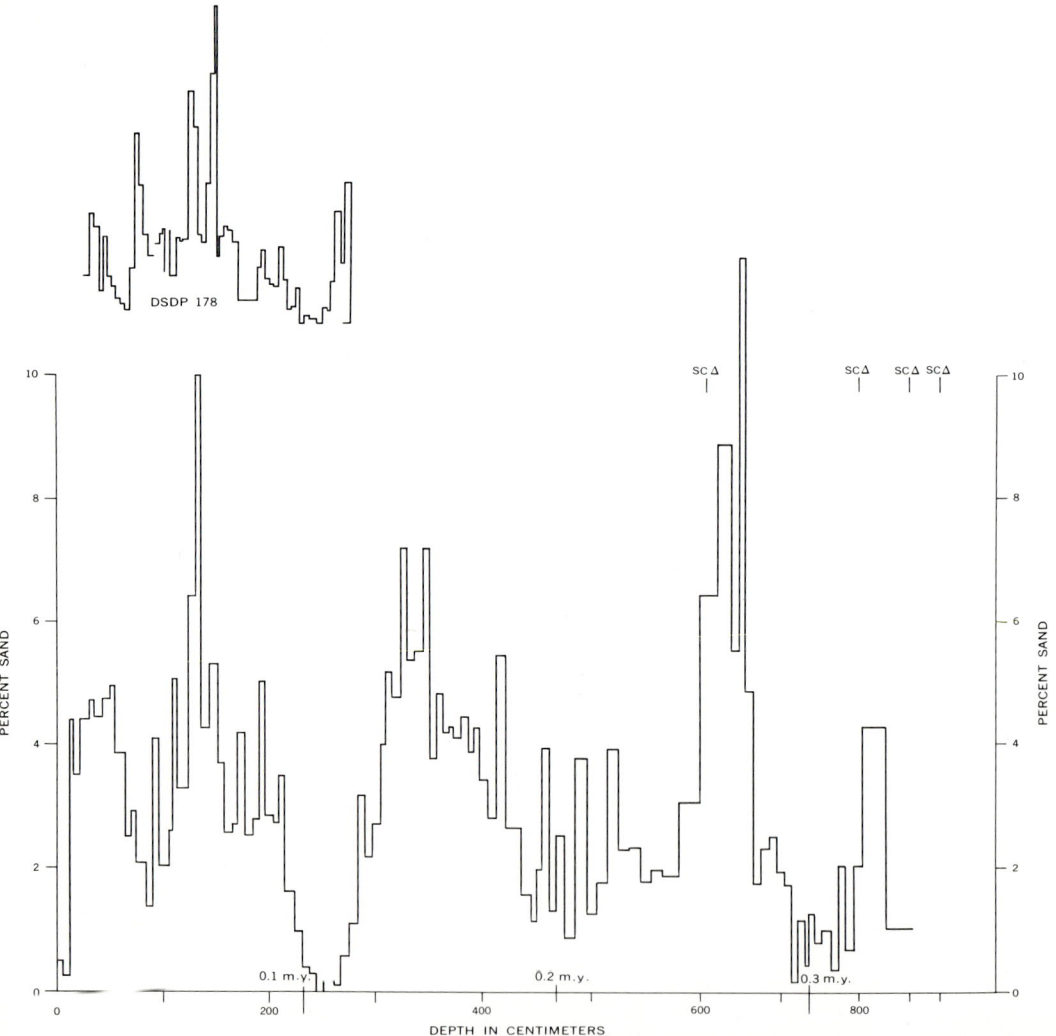

Figure 4. Graph of dry weight percent of coarse sand in core V21-171 as function of age. Midpoint of each bar corresponds to midpoint depth of sample. SC△ indicates where age control is available. Percentage of sand was adjusted at these points to reflect abundance as function of age rather than depth. Possible correlation with first three cores from DSDP site 178 shown above first 100,000 year segment. Curves have been joined, eliminating indicated gaps between cores. Top of core 178 was missing and absence of ramp at beginning of V21-171 suggests that top of this core was also not recovered or is badly disturbed.

A second uncertainty arises from the way detritus is released from icebergs. Icebergs become progressively more unstable as they melt, and they usually overturn, causing large accumulations of detritus exposed by ablation to be suddenly dumped. The dumped material spreads out during its fall through 3 to 5 km of water and settles on the ocean floor in a pattern that is a function of grain size and current.

Both these considerations would introduce uncertainty or "noise" in the simple concept of the relations between abundance of ice-rafted material and glacial advance, because they cannot be quantitatively determined. Less significant effects might also be introduced by changes in current, water temperature, and the position of the ice front from which bergs break away (for example, at the edge of the shelf or near the present coastline).

Certain other basic assumptions contain inherent uncertainties. The assumption of an ice-rafted origin for coarse sand has been validated (Kent and others, 1971; von Huene and others, 1973), but some sampling problems have not been resolved, such as the critical size of a sample with respect to grain size and with respect to time span. Probably the greatest uncertainty is caused by the degree of variability in matrix deposition rates. The deep-ocean mud containing rafted debris was deposited at differing rates that appear to correlate with intensity of glaciation (von Huene and others, 1973). Therefore it is probably inaccurate to interpolate time linearly from depth across glacial and interglacial periods.

Since none of these factors is quantitatively known, the level of resolution in the curves of ice-rafted detritus, and particularly the significance of events defined by single data points, is uncertain. The only criterion for evaluation available without well-correlated replicate cores is the character of the graphs.

Significance of Ice-Rafted Abundance Data from the Gulf of Alaska

The major periods of abundant icebergs indicated by the V21-171 curve are in general agreement with data obtained from studies of paleo-ocean temperature, oxygen isotope fluctuations, lake-level and sea-level fluctuations, and positions of moraines in other areas. The Gulf of Mexico water-temperature curve developed by Kennett and Huddlestun (1972) and the Caribbean oxygen isotope fluctuation curve developed by Emiliani (1971; based on ages proposed by Broecker and van Donk 1970), are shown for comparison with the V21-171 curve (Fig. 5). The major difference between these curves lies in the amplitude of the warmer period 25,000 to 50,000 yr ago. A striking agreement in main features, however, is seen in a comparison of our detritus-abundance curve with the northern European ice-advance curve developed by Mörner (1972). Other interesting similarities are seen in a comparison with the southeast California lake-level curve by Smith (1968).

There are two advantages in using the Gulf of Alaska for the study of glacial advances as reflected by ice-rafted material. First, the Alaska current gyre probably remained constant during the Pleistocene, regardless of climatic change. This is indicated in the detrital fraction of core samples by the absence of rock types from places other than the adjacent coast (von Huene and others, 1973). Second, because of the northern latitude and the high glacier-covered coastal mountains circumjacent to the gulf, iceberg abundance is probably a sensitive indicator of climatic changes. Large mountain valley glaciers and ice fields in the coastal ranges are at present nearly within tidal range, and small glacial advances would produce icebergs in the open ocean. In this setting, the sensitivity of alpine glaciation to climatic change is reflected in the marine record, and the record is preserved with the characteristic temporal continuity of deep-ocean sedimentation.

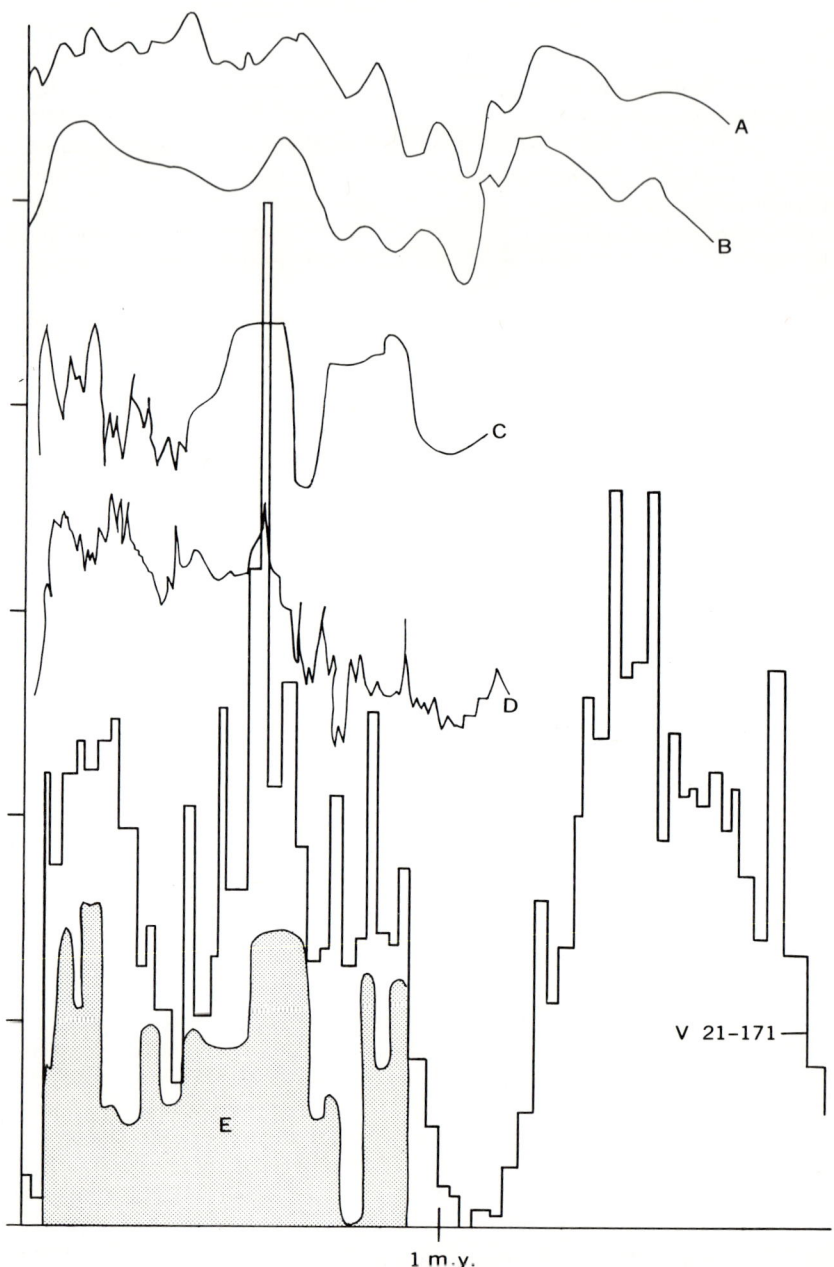

Figure 5. Comparison of graph for core V21-171 with other curves. A, from Kennett and Huddlestun (1972); Gulf of Mexico, water temperature inferred from microfossil assemblages. B, from Emiliani (1971), using age scale of Broecker and van Donk (1970); Caribbean, oxygen isotope fluctuations. C, from Smith (1968); southeast California, lake levels (based on Searles Lake). D, Dansgaard and others (1971); Greenland, oxygen isotopes. E, from Mörner (1972); northern Europe, glacial advances.

It has been pointed out to us by D. Hopkins (1973, written commun.) that the mountains bordering the Gulf of Alaska are remarkable for the perversity of their glaciers, many of which advance at times when glaciers in other regions are retreating. This apparently happens because the level of maximum snowfall fluctuates to some favored altitude, where the supply for a particular glacier then increases. Hopkins further notes, however, that the local effects do not appear to have influenced the major glacial events in the data of core V21-171. This is shown by the good agreement with Mörner's (1972) time-distance curve. The good agreement with the Greenland oxygen isotope fluctuations (Dansgaard and others, 1971) as well, suggests that when averaged over the whole gulf, ice advances reflect increased northern hemisphere ice accumulation.

The significance of short-period fluctuations in the abundance curves for ice-rafted detritus is indefinite without comparison with other cores from nearby sites. Some confidence in the present data is gained from the fact that most peaks with a 10,000-yr period are formed by three or more data points in V21-171. Other northern-latitude data indicate minor cold periods or ice advances of short duration (Dansgaard and others, 1971; Mörner, 1972), but they are also based on single data sets without the opportunity to validate by comparison. Because the 200-yr spacing of points in the data developed by Dansgaard and others (1971) shows isotopic fluctuations 1,000 yr long, the 3,000-yr interval peaks in V21-171 could be significant. Also, analytical errors for data from DSDP sites (von Huene and others, 1973) that should apply to the V21-171 data are smaller than the amplitude of single-point peaks in the V21-171 curve. Therefore, although the significance level cannot be set quantitatively, there is some basis for confidence at a 10,000-yr interval when three or four data points define a peak by an orderly stepwise progression.

Cyclicity in Gulf of Alaska Curves

One objective of the study of past climates is to develop the ability to forecast major climatic change, and particularly to distinguish between natural cyclicity and the effects of man's activities.

Inspection of the detritus-abundance graph for site 178 (Fig. 2) suggests a possible periodicity of glacial advance of 12,000 to 15,000 yr. Verification of this observation was tried using a Fourier spectral analysis. The relative frequency content of the data, which can also be interpreted as the number of times a cycle of specific length occurs, is shown in Figure 6. Although suggestive, these data are not very convincing.

Future Research

The study of detritus abundance is a promising way to establish a detailed glacial chronology for the Gulf of Alaska. Sampling along the Kodiak-Bowie seamount chain looks especially worthwhile for the following reasons: (1) the seamounts project above the calcium carbonate compensation level, assuring recovery of foraminifera and coccoliths as well as diatoms and Radiolaria; (2) many ash layers in the Gulf of Alaska (Pratt and others, 1973) provide the opportunity for correlation between cores; (3) cores can be recovered close enough to source areas so that minor glacial advances can be detected. The cores recovered to date are relatively far from source areas. Correlations between several cores at each sampling site would permit recognition of anomalous peaks due to iceberg dumps and other undesired effects. Three sampling sites at different distances from the major source

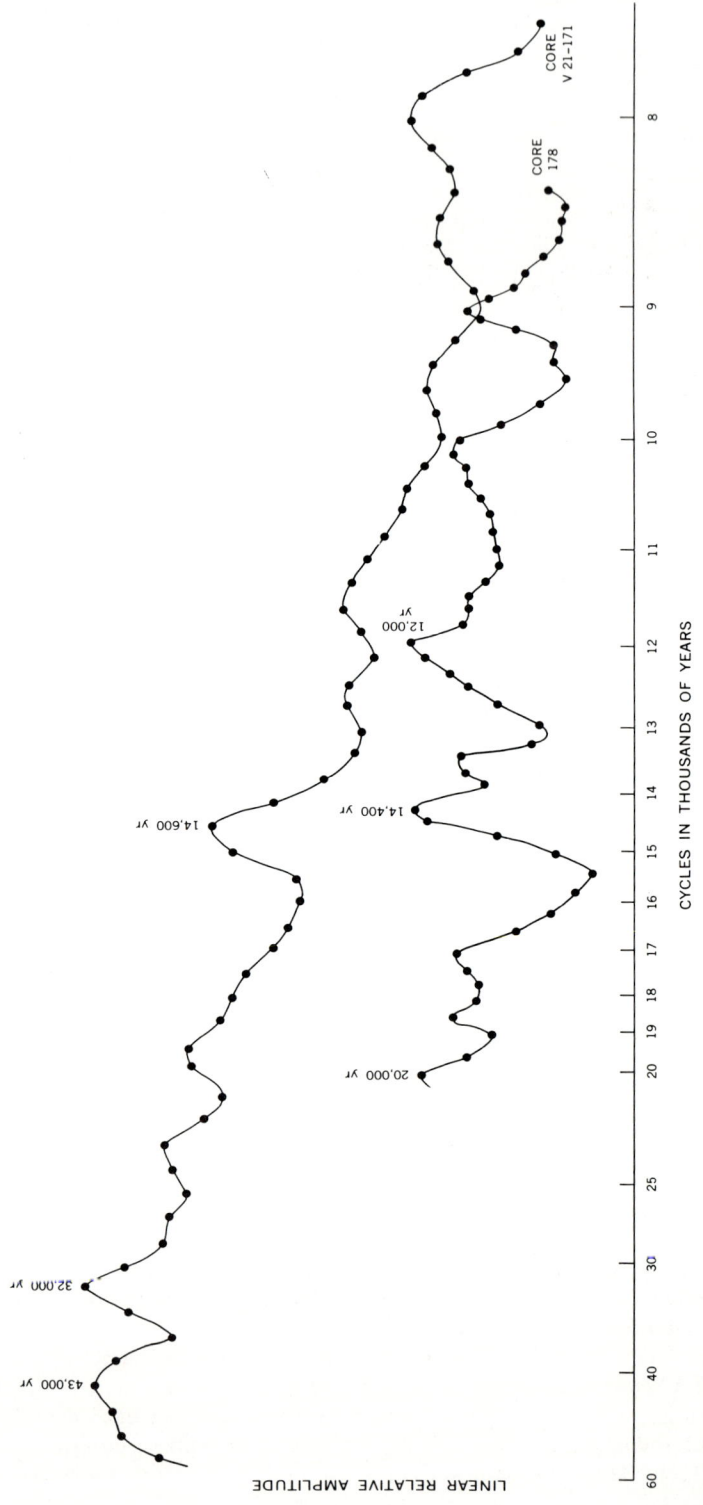

Figure 6. Amplitude spectra of coarse sand percent versus age curves for cores V21-171 and 178. Vertical axis is linear and gives relative frequency content of specific time periods in core data. For discussion of this type of analysis, see Dansgaard and others (1971).

areas would provide a statistical basis for establishing the correlation between intensity of advance and detritus abundance. In this regard, it would be particularly helpful to measure oxygen isotope fluctuations on the same cores.

CONCLUSIONS

Abundances for ice-rafted detritus appear to provide a good indication of glacial advances in the Gulf of Alaska. Major fluctuations are in agreement with those seen in other climatically influenced data, especially the northern European ice-advance data of Mörner (1972). The resolution of this method is not quantitatively determined, but in this study some confidence is derived from the consistency of the data in 10,000-yr-long events. The level of uncertainty could be greatly decreased by a field and laboratory program designed specifically to study past climatic changes. Resolution could be improved. Abundance studies of ice-rafted detritus could define short-period cyclicity (less than 10,000 yr) over a long period of time, possibly 1 million yr or more. A mild 10,000- to 15,000-yr cyclicity of glacial advance is suggested in the curves, but further verification is needed.

ACKNOWLEDGMENTS

We were greatly helped in this study by Alan Cooper, who made the Fourier spectral analysis; by Dennis Kent, who provided us with original data; by Jim Hays and Roy Capo, who arranged to provide us with samples from core V21-171; and by Paula Worstel, Dennis Bohrer, Trudy Wood, and Victor Sotelo, who assisted in sampling DSDP cores. The reviews of our manuscript by David Hopkins and Robert Rowland were very helpful and improved the paper. This study was partially supported by National Science Foundation Grants GA-11050 and GA-10635 and by Office of Naval Research Grant N00014-67-A-0108-0004.

REFERENCES CITED

Broecker, W. S., and van Donk, Jan, 1970, Insolation changes, ice volumes, and the O^{18} record in deep-sea cores: Rev. Geophysics and Space Physics, v. 8, no. 1, p. 169–198.

Conolly, J. R., and Ewing, Maurice, 1970, Ice-rafted detritus in North Pacific deep-sea sediments, *in* Hays, J. D., ed., Geological investigations of the North Pacific: Geol. Soc. America Mem. 126, p. 219–231.

Dansgaard, W., Johnsen, S. J., Clausen, H. B., and Langway, C. C., 1971, Climatic record revealed by the Camp Century ice core, *in* The late Cenozoic glacial ages: New Haven, Yale Univ. Press, p. 37–56.

Emiliani, Cesari, 1971, The last interglacial; paleotemperatures and chronology: Science, v. 171, no. 3971, p. 571–573.

Kennett, J. P., and Huddlestun, P., 1972, Late Pleistocene paleoclimatology, foraminiferal biostratigraphy and tephrochronology, western Gulf of Mexico: Quaternary Research, v. 2, no. 1, p. 38–69.

Kent, D., Opdyke, N. D., and Ewing, Maurice, 1971, Climate change in the North Pacific using ice-rafted detritus as a climatic indicator: Geol. Soc. America Bull., v. 82, no. 10, p. 2741–2754.

Mörner, N.-A., 1972, The cold/warm changes during the last ice age with special reference to the stratigraphy at Dösenbacka and Ellesbo in southwest Sweden: Stockholm Contr. Geology, v. 24, no. 4, p. 51–77.

Pratt, R. M., Scheidegger, K. F., and Kulm, L. D., 1973, Volcanic ash from DSDP site 178, Gulf of Alaska, *in* Kulm, L. D., von Huene, Roland, and others, eds., Initial reports of the Deep Sea Drilling Project, Vol. XVIII: Washington, U.S. Govt. Printing Office, p. 833–834.

Smith, G. I., 1968, Late-Quaternary geologic and climatic history of Searles Lake, southeastern California, *in* Means of correlation of Quaternary successions: Internat. Assoc. Quaternary Research, Cong., 7th, U.S.A. 1965, Proc., v. 8, Salt Lake City, Univ. Utah Press, p. 293–310.

von Huene, R., Larson, E., and Crouch, J., 1973, Preliminary study of ice-rafted erratics as indicators of glacial advances in the Gulf of Alaska, *in* Kulm, L. D., von Huene, Roland, and others, eds., Initial reports of the Deep Sea Drilling Project, Vol. XVIII: Washington, U.S. Govt. Printing Office, p. 835–842.

MANUSCRIPT RECEIVED BY THE SOCIETY MARCH 28, 1974
REVISED MANUSCRIPT RECEIVED SEPTEMBER 23, 1974
MANUSCRIPT ACCEPTED OCTOBER 7, 1974

Geological Society of America
Memoir 145
© 1976

Modern Pacific Coccolith Assemblages: Derivation and Application to Late Pleistocene Paleotemperature Analysis

Kurt R. Geitzenauer
Lamont-Doherty Geological Observatory
Columbia University, Palisades, New York 10964
and
Department of Geology and Geography
Herbert H. Lehman College
City University of New York, New York, New York 10468

Michael B. Roche*
Lamont-Doherty Geological Observatory
Columbia University, Palisades, New York 10964

AND

Andrew McIntyre
Lamont-Doherty Geological Observatory
Columbia University, Palisades, New York 10964
and
Department of Earth and Environmental Sciences
Queens College
City University of New York, Flushing, New York 11367

ABSTRACT

Factor analyses of relative abundance counts of coccoliths from selected Pacific Ocean core tops yield six assemblages that coincide with the following surface water masses and regions: (1) the Kuroshio system and Pacific Equatorial Water Mass; (2) the southern and northern central Pacific regions (both occupied by two assemblages), (3) the Western South Pacific Central Water Mass, and (4)

---

*Present address: Jersey Central Power and Light Company, Morristown, New Jersey 07960.

the Subarctic Pacific and Intermediate Water Masses. The sixth assemblage represents a flora that is transitional between the subpolar and central Pacific regions in both hemispheres and that occurs as well within the Peru Current system.

Transfer functions relating the assemblages to sea-surface temperatures were applied to coccolith populations that existed over the past 200,000 yr and that were taken from cores in the western equatorial Pacific (cores V28-238 and V28-239) and the eastern Pacific (core V19-55). The values produced are floral indices calibrated on seabed samples to be unbiased estimates of summer sea-surface temperatures (T_s), winter sea-surface temperatures (T_w), and seasonality ($T_s - T_w$). T_s in the western equatorial Pacific varies within relatively narrow limits, with a maximum range of 5.7° and 3.9°C. T_w shows relatively greater fluctuation, with a maximum range of 8.9° and 6.3°C. T_s and T_w in the eastern Pacific show much less variation, with a maximum range of 1.9° and 3.3°C, respectively. There appears to be no correlation between T_s and glacial events. A seasonality ($T_s - T_w$) maximum occurs in midcycle just before glaciation in "glacial cycles" (Kukla, 1961) B, D, and E. The seasonality change is greater in the western equatorial cores.

The foregoing adds evidence to support the theory that sea-surface temperature over the past 200,000 yr has not globally varied sinusoidally but is subject to complex regional dynamics.

INTRODUCTION

Changes in relative abundances and species migration of Coccolithophoridae and Foraminifera have been used as semiquantitative paleoclimatic indicators of Pleistocene climatic changes (coccoliths: McIntyre, 1967; Geitzenauer, 1972; McIntyre and others, 1972; foraminiferids: Schott, 1935; Phleger and others, 1953; Ericson and Wollin, 1968; McIntyre and others, 1972; Ruddiman and McIntyre, this volume). A quantitative method for the treatment of foraminiferal population data has recently been developed (Imbrie and Kipp, 1971). This method relates surface-sediment foraminiferal assemblages to sea-surface ecological parameters such as temperature and salinity and thus allows absolute estimates of these parameters through time by determination of foraminiferal assemblages downcore. The Imbrie and Kipp technique has been successfully adapted and applied to the Atlantic Coccolithophoridae (Roche and others, 1975). To date, the cores studied by this quantitative method have been from high to middle latitudes where Pleistocene temperature changes have been relatively large.

Phase 1 of the research design for this study involves two steps: first, ecologically valid coccolith assemblages from the Pacific Ocean are determined by factor analysis of coccolith populations from surface sediments, and second, transfer functions capable of estimating absolute winter and summer sea-surface temperatures from coccolith populations are derived by means of multivariate stepwise regression. This provides the basis for paleoenvironmental analysis in phase 2, in which we apply the transfer functions to coccolith floras of three deep-sea Pleistocene cores from the Pacific in order to estimate absolute sea-surface paleotemperatures over the past 275,000 yr.

This design is applied to the Pacific Ocean to elicit answers to the following questions: (a) Can the Pacific sediments, in which carbonate is less extensively preserved than in the Atlantic sediments, still yield ecologically valid assemblages with the Q-mode factor analysis; (b) can the Imbrie and Kipp technique be applied to coccolith floras of the Pacific; and (c) can the derived paleoenvironmental equations be used to determine the thermal history of the low-latitude Pacific?

MODERN FLORAS

Samples

Fifty-one surface-sediment samples were taken from the tops of trigger-weight cores (Fig. 1; Table 1). The samples were prepared as described by McIntyre and others (1967), with ultrasonic agitation omitted.

Random counts of at least 350 coccoliths per sample were made with a transmission electron microscope.[1] This number provides a 95% probability that species occurring with relative abundances of 1% will be counted (Shaw, 1964). The samples were selected on the basis of their high carbonate content, depth (all <3,900 m), a qualitative estimate of preservation of the coccolith flora, and finally, a quantitative comparison of assemblages during data treatment.

The Pacific, especially the North Pacific, is often considered a "noncarbonate" ocean. Since much of the Pacific Basin is deeper than the calcite compensation depth (CCD), the dominant sediment is red clay. However, extensive coring by Lamont-Doherty Geological Observatory ships in areas of the western and central North Pacific has recovered carbonate-rich sediments showing little or no effects of solution on the coccolith flora. We have attempted to maintain a somewhat uniform sample density throughout the Pacific, so little emphasis is placed on any particular area during the data treatment, but some areas have been sampled more densely than others (Fig. 1).

A number of carbonate coccolith samples were taken in the North Pacific from the Emperor Seamounts, Mid-Pacific Mountains, and just west of the Mariana Trench. Certain areas, notably the Northeast Pacific Basin from about lat 10°S to the northern edge, lacked suitable samples, except for one area in the vicinity of the Cobb Seamount. Other broad areas lacking suitable samples are the Northwest Pacific Basin, bounded by lats 10°N and 30°N and longs 150°E to 180°E, and the northern part of the Southwest Pacific Basin.

Although rich in carbonate, several samples from the eastern edge of the South Pacific, including some from the Nazca Ridge, proved unsuitable for this study. The coccoliths were poorly preserved and probably not representative of the biocoenoses even though they were well above the CCD.

Coccolith Species

Every coccolith observed during the counting procedure was tallied, but only 16 species (Table 2) were used for paleoceanographic analysis. These 16 species were chosen because of their abundance in deep-sea sediments and their relatively narrow ranges of temperature preference (McIntyre and Bé, 1967; McIntyre and others, 1970). Two important and sometimes dominant coccolith species, *Emiliani huxleyi* (Lohman) Hay and Mohler and *Gephyrocapsa caribbeanica* Hay and Mohler, are not included in the equations. *E. huxleyi* evolved during the time period under study, and *G. caribbeanica* has undergone a recent marked decline in abundance. Because of evolutionary and perhaps ecological variations, these two species were considered unreliable climatic paleoindicators and have been excluded. Certain species, such as *Gephyrocapsa ericsonii* McIntyre and Bé and a number of *Syracosphaera* sp., occur too infrequently in the surface sediments to be significant.

---

[1] For the raw count data, see Appendix I on microfiche in pocket inside back cover.

Treatment of Recent Coccolith Data

Klovan and Imbrie (1971) described the Q-mode factor analysis program (CAB-FAC), which was applied to the surface-sediment coccolith population samples in this study. The purpose of factor analysis is to account for the maximum amount of population variance in a sample group with a minimum of factors. Six assemblages have been produced from our analysis of 51 samples and they account for 97.6% of the data variance. The relationship of the 16 species to the six assemblages used in the analysis is given in the factor score matrix (F, Table 3). This matrix indicates the importance of each species to each of the assemblages. The higher the F value, the greater the importance of that species to the assemblage. Five of the factors are dominated by one species; the sixth is an assemblage of three species.

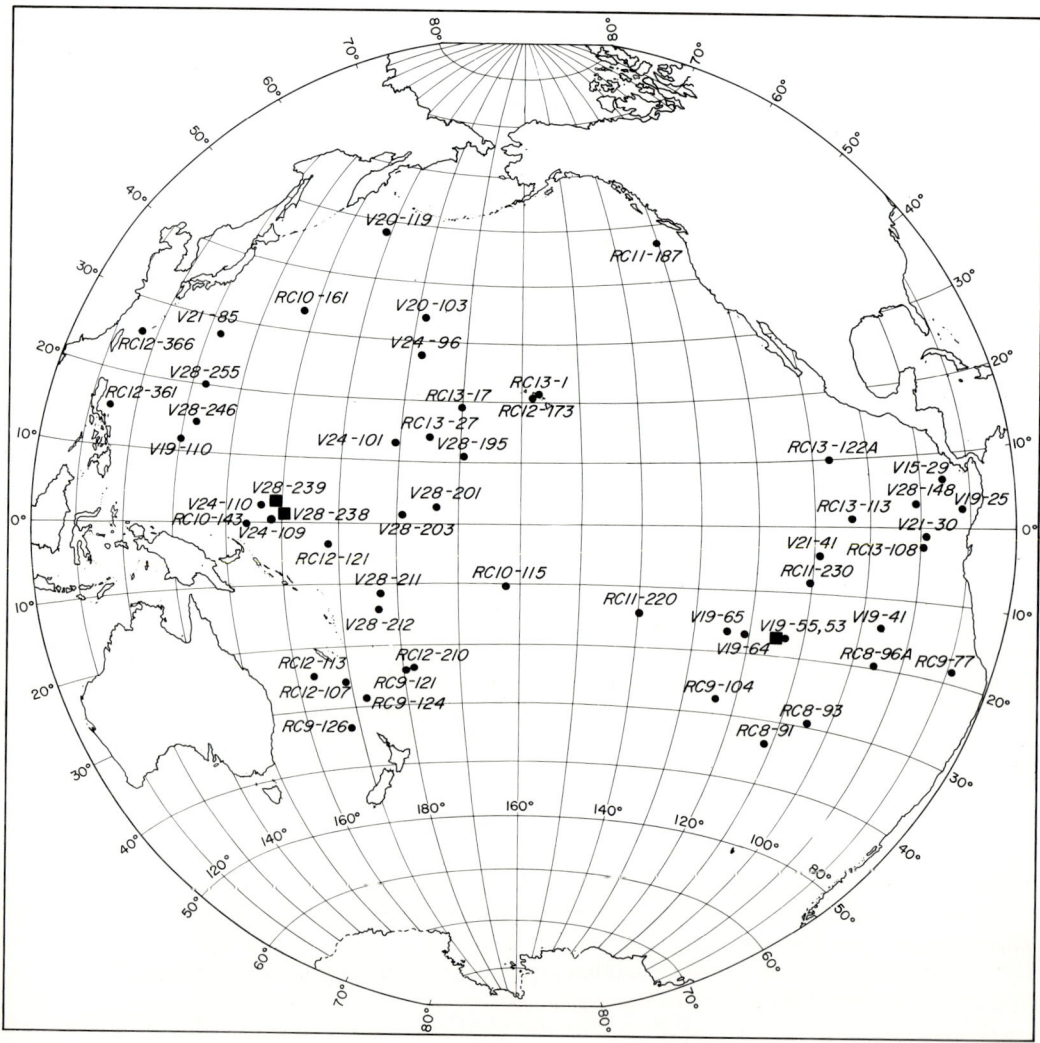

Figure 1. Map showing trigger-weight core top (circles) and piston core (squares) locations.

TABLE 1. TRIGGER-WEIGHT CORE DATA

Trigger-weight core	Lat	Long	Depth (m)	Trigger-weight core	Lat	Long	Depth (m)
RC8-91	33°25′S	111°54′W	2,723	V15-29	06°21′N	85°17′W	1,889
RC8-93	29°22′S	105°14′W	3,157	V19-25	02°28′N	81°42′W	2,404
RC9-104	27°26′S	123°26′W	3,632	V19-41	14°06′S	96°12′W	3,243
RC9-121	24°46′S	179°41′W	2,153	V19-53	17°01′S	113°31′W	3,058
RC9-124	28°45′S	172°36′E	2,540	V19-55	17°00′S	114°11′W	3,177
RC9-126	33°14′S	168°44′E	2,060	V19-64	16°56′S	121°12′W	3,570
RC10-115	10°35′S	163°18′W	3,135	V19-65	16°39′S	124°23′W	3,867
RC10-143	0°20′S	153°59′E	3,074	V19-110	11°51′N	140°03′E	3,532
RC10-161	33°05′N	158°00′E	3,587	V20-103	35°59′N	177°50′W	3,442
RC11-187	47°09′N	130°07′W	2,670	V20-119	47°57′N	168°47′E	2,739
RC11-220	14°49′S	139°58′W	2,950	V21-30	01°13′S	89°41′W	617
RC11-230	08°48′S	110°48′W	3,259	V21-41	04°43′S	109°25′W	3,435
RC12-107	26°00′S	169°12′E	3,115	V21-85	27°58′N	140°30′E	1,684
RC12-113	24°53′S	163°31′E	2,454	V24-96	27°40′N	177°59′W	3,305
RC12-121	03°44′S	168°23′E	3,519	V24-101	13°10′N	178°53′E	3,336
RC12-173	20°47′N	157°56′W	1,476	V24-109	00°26′N	158°48′E	2,367
RC12-210	24°14′S	177°36′W	1,529	V24-110	02°21′N	156°42′E	2,613
RC12-361	15°06′N	124°08′E	3,528	V28-148	03°42′N	91°32′W	2,834
RC12-366	26°35′N	126°20′E	1,644	V28-195	10°39′N	169°48′W	2,439
RC13-1	21°28′N	157°14′W	2,114	V28-201	02°46′N	173°56′W	3,217
RC13-17	19°05′N	170°04′W	3,374	V28-203	00°57′N	179°25′W	3,248
RC13-27	13°52′N	175°16′W	3,555	V28-211	11°46′S	177°00′E	2,644
RC13-108	03°07′S	89°25′W	3,308	V28-212	14°03′S	176°26′E	2,972
RC13-113	01°39′S	103°38′W	3,195	V28-239	03°15′N	159°11′E	3,490
RC13-122A	10°03′N	107°00′W	3,275	V28-246	14°22′N	142°43′E	2,745
				V28-255	20°06′N	142°27′E	3,261

The relationship of the six assemblages to each of the surface-sediment samples is given by a varimax matrix (B_{ct}).[2] The communality for each sample is the sum of squares of that row in the B_{ct} or the amount of population variation that can be accounted for by the six assemblages with a value of 0.6, considered acceptable (Imbrie and Kipp, 1971). The mean communality is 0.9755. In most cases, a sample can be defined in terms of one assemblage, although some show relatively high values for several assemblages.

The biogeographic distribution of phytoplankton is determined by a number of ecological parameters such as light, temperature, salinity, fertility, and currents (Smayda, 1958). Since several of these parameters also dictate the uniqueness of water masses, the distribution of phytoplankton is often correlative with water-mass distribution. The Pacific water masses and general current directions, as taken from Sverdrup and others (1942), are shown in Figure 2 and can be compared to plots of varimax values for the six assemblages (Figs. 3 through 8).

Assemblage A, dominated by *Gephyrocapsa oceanica* Kamptner, accounts for 32.1% of the data variance. *G. oceanica* is a large species most abundant in the Pacific Equatorial Water Mass (McIntyre and others, 1970; Okada and Honjo, 1973) and in the marginal seas of the western Pacific (Okada, 1973).

Our data show *G. oceanica* to be a strong equatorial species. It is also associated with the Kuroshio Current system, resulting in strong varimax values not only across the Pacific within the Pacific Equatorial Water Mass but also along the western edge of the tropical North Pacific (Philippine Sea) and along the Kuroshio Current and Kuroshio extension into the subtropical zone of the Western North

---

[2] See Appendix II on microfiche in pocket inside back cover.

TABLE 2. COCCOLITH SPECIES

Syracosphaera pulchra Lohmann
Syracosphaera variabilis
Umbellosphaera irregularis Paasche
Umbellosphaera tenuis (Kamptner) Paasche
Helicopontosphaera kamptneri Hay and Mohler
Calciosolenia sinuosa Schlauder
Discosphaera tubifera (Murray and Blackman) Lohmann
Rhabdosphaera clavigera Murray and Blackman
Cyclococcolithina fragilis (Lohmann) Wilcoxon
Cyclococcolithina leptopora (var. B) (Murray and Blackman) Wilcoxon
Cyclococcolithina leptopora (var. C)
Coccolithus pelagicus (Wallich) Schiller
Gephyrocapsa oceanica Kamptner
Umbilicosphaera hulbertiana Gaarder
Umbilicosphaera mirabilis (cold var.) Lohmann
Umbilicosphaera mirabilis (warm var.)

Pacific Central Water Mass (WNPC; Fig. 3). This distributional pattern is very similar to that shown by certain Pacific Equatorial Water Mass zooplankton (McGowan, 1971). McIntyre and others (1970) found *G. oceanica* to be confined to waters with temperatures of 19°C or greater. Our surface-sediment population study corroborates this distribution.

Assemblage B is dominated by *Cyclococcolithina leptopora* (Murray and Blackman) Wilcoxon (var. C) and accounts for 27% of the data variance. This form has been shown by McIntyre and others (1970) to have a wide biogeographic distribution in Pacific plankton with a maximum concentration in the central South Pacific between lats 40°S and 55°S, an area not covered by this study. Figure 4 shows the symmetrical distributional pattern of high varimax values for this assemblage. In the North Pacific, the highest values occur in the northern part of the WNPC and in the northeastern Pacific Transition Region. In the Southern Hemisphere, the highest assemblage B varimax values occur in the southern part of the Eastern South Pacific Central Water Mass (ESPC) and within the Pacific Equatorial Water Mass between longs 100°W and 111°W. This latter occurrence might be explained by the surface current patterns in that sector. The Transition

TABLE 3. FACTOR SCORE MATRIX (F MATRIX)

Species	Assemblages (factors)					
	A	B	C	D	E	F
S. pulchra	−0.0360	0.0930	0.2380	−0.0170	−0.0430	−0.0280
S. variabilis	−0.0040	−0.0030	0.0170	−0.0020	0.0110	0.0010
U. irregularis	−0.0260	−0.0440	0.3120	0.0690	0.1970	0.0390
U. tenuis	−0.0240	−0.0490	0.6710	0.1050	−0.1480	0.0540
H. kamptneri	0.0320	0.0250	0.0930	0.0050	−0.0120	0.0110
C. sinuosa	0.0380	0.0560	0.1840	0.0740	0.2400	−0.0500
D. tubifera	−0.0030	−0.0030	0.0200	−0.0010	0.0110	0.0020
R. clavigera	−0.0330	0.1040	0.5570	0.0730	−0.1870	−0.0770
C. fragilis	0.0940	−0.0160	0.0900	0.2050	0.8910	0.0010
C. leptopora (var. B.)	0.0560	0.0470	−0.0010	0.0090	−0.0540	−0.0030
C. leptopora (var. C.)	0.0720	0.9810	−0.0490	0.0250	0.0160	−0.0070
C. pelagicus	0.0150	0.0240	0.0090	0.0090	−0.0070	0.9930
G. oceanica	0.9810	−0.0730	0.0420	−0.0520	−0.0900	−0.0150
U. hulbertiana	0.0000	0.0290	0.0590	0.0050	0.0250	−0.0200
U. mirabilis (cold var.)	0.1140	0.0180	0.0070	−0.0160	−0.0100	0.0080
U. mirabilis (warm var.)	−0.0350	0.0280	0.1640	−0.9630	0.1980	0.0100

Region to the east of the ESPC results from the northward-flowing Peru Current system. Unfortunately, no well-preserved coccolith assemblages were found along the eastern margin of the South Pacific. However, it is possible that the Peru Current system does carry *C. leptoporus* (var. C) from a known center of abundance along the northern part of the Subantarctic Water Mass (McIntyre and others, 1970) into the lower latitudes. McGowan (1971) shows distribution boundaries for 12 species of transition-zone planktonic organisms that are quite similar to our plot of high varimax values for assemblage B even to the northern extension along the Peru Current system.

Another water-mass distribution map, as summarized by Knox (1970, Fig. 7),

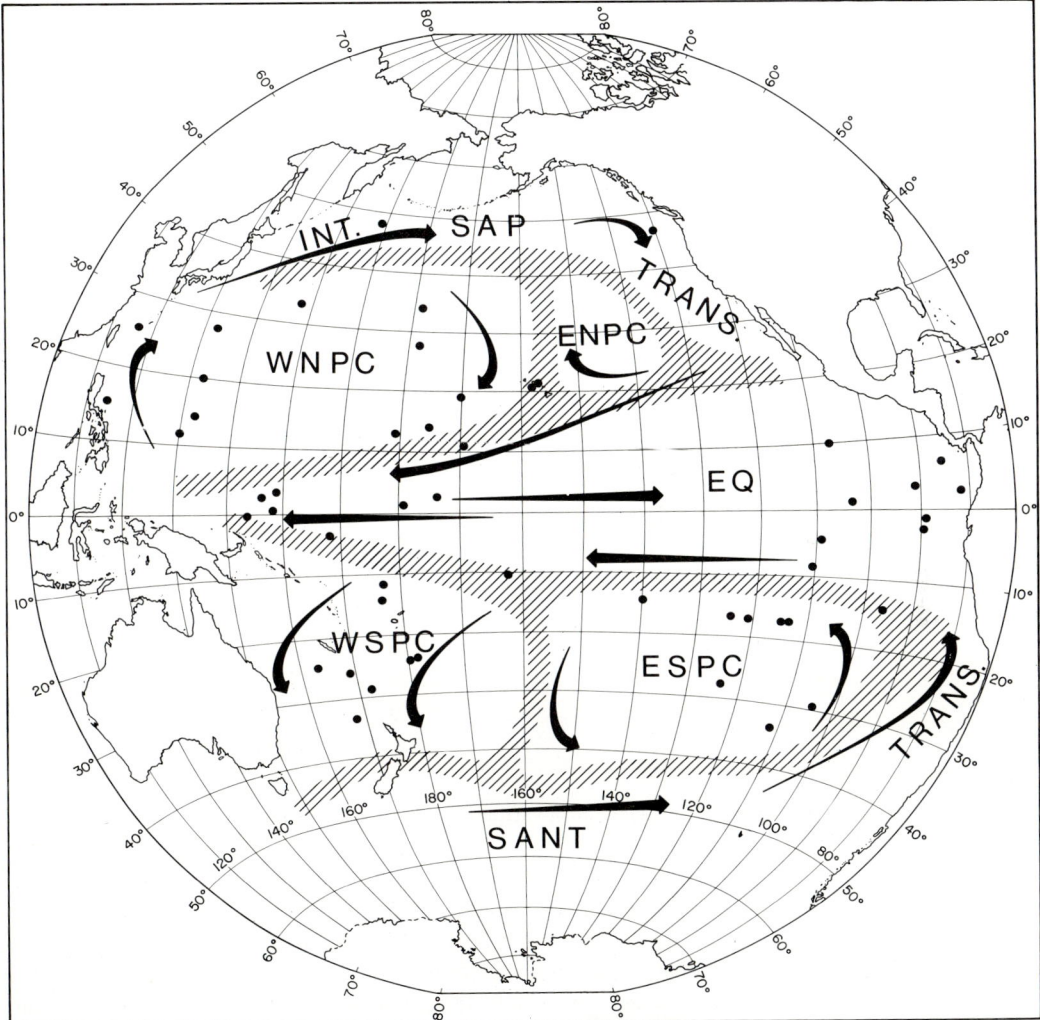

Figure 2. Generalized diagram of surface currents and water masses in the Pacific. Locations of trigger-weight core samples used in this study are indicated by dots. EQ = Pacific Equatorial Water; SAP = Subarctic Pacific Water; WNPC = Western North Pacific Central Water; ENPC = Eastern North Pacific Central Water; WSPC = Western South Pacific Central Water; ESPC = Eastern South Pacific Central Water; SANT = Subantarctic Water; INT. = Intermediate Water; TRANS. = Transition Region (from Sverdrup and others, 1942).

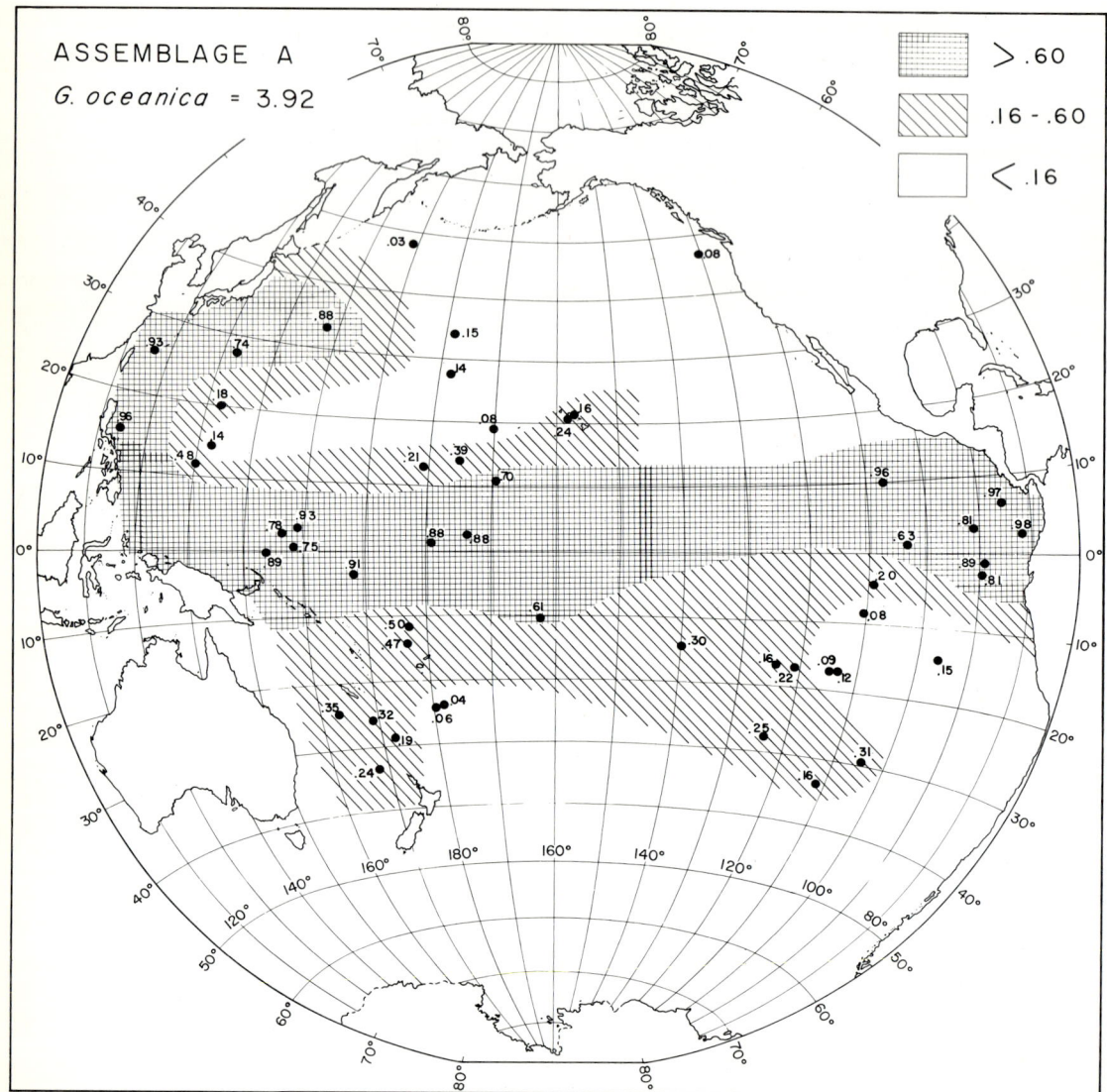

Figure 3. Geographic distribution of varimax values for assemblage A (*Gephyrocapsa oceanica*) in core tops.

shows a southern subtropical gyral system that is equivalent to the WSPC and ESPC of Sverdrup and others (1942) but that extends about 10° farther north in the central and eastern Pacific, resulting in a narrower equatorial zone. This gyral system would include all of the high assemblage B varimax values in the southern Pacific except one. Assemblage B is, as far as our study can confirm, an assemblage of the higher latitude part of the central regions. In addition, this assemblage is apparently carried into lower latitudes by the Peru Current system and South Pacific Central Water circulation in the eastern part of the basin.

Assemblage C, composed of *Umbellosphaera tenuis* (Kamptner) Paasche, *Umbellosphaera irregularis* Paasche, and *Rhabdosphaera clavigera* Murray and Blackman,

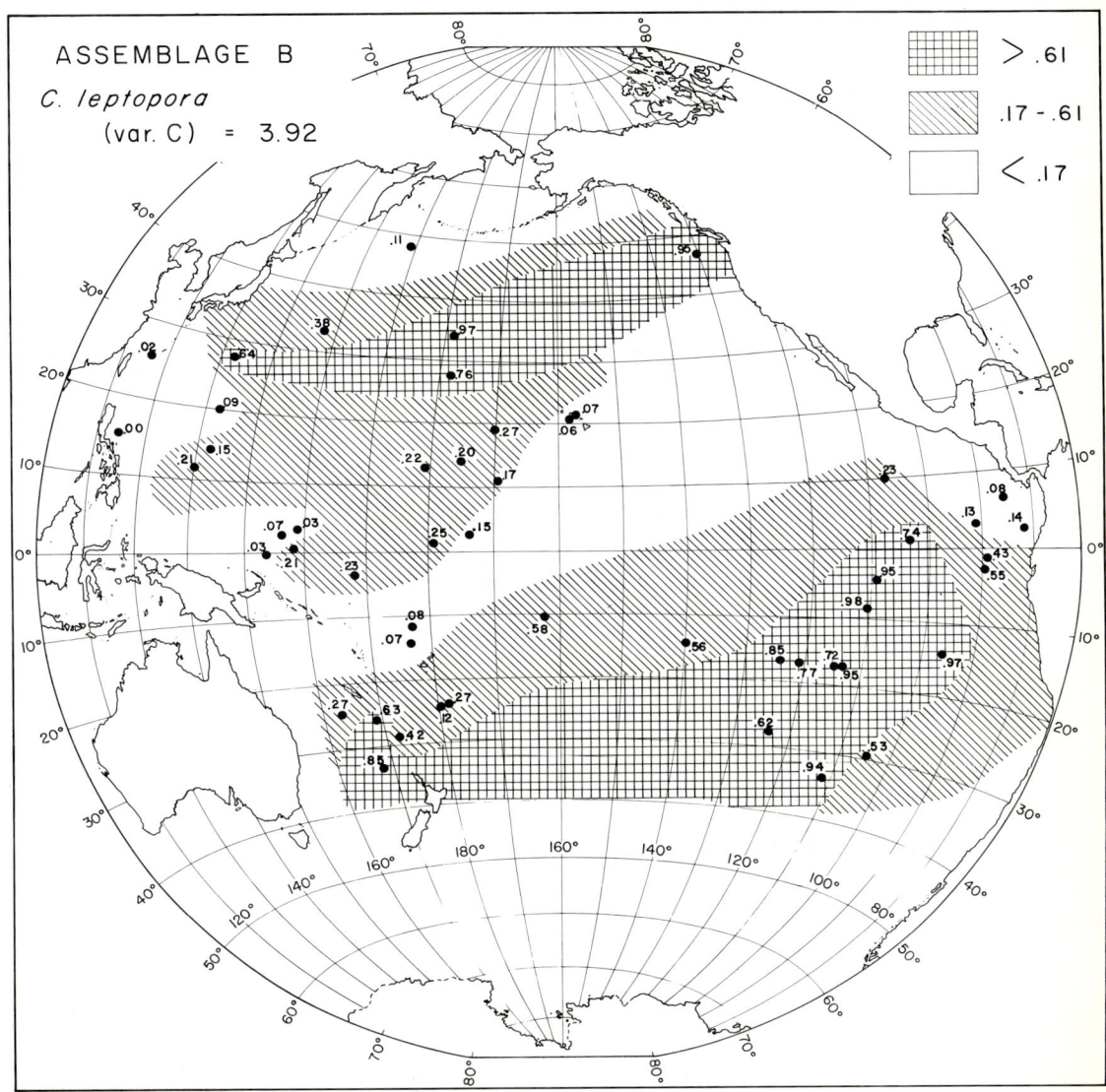

Figure 4. Geographic distribution of varimax values for assemblage B (*Cyclococcolithina leptopora* [var. C]) in core tops.

and assemblage D, dominated by *Umbilicosphaera mirabilis* Lohmann (warm variety), account for 16.1% and 13.1%, respectively, of the data variance. Our data show both these assemblages to be indicative of the central water masses, but their highest varimax values generally occur south of the high assemblage B values in the North Pacific and north of the high B values in the western South Pacific (Figs. 5, 6). The varimax distribution map for assemblage C in particular shows a geographic pattern very similar to that of the central water masses shown in Figure 2. Although they have emerged as separate factors on the basis of their biogeographic distribution, assemblages C and D might best be described as subassemblages.

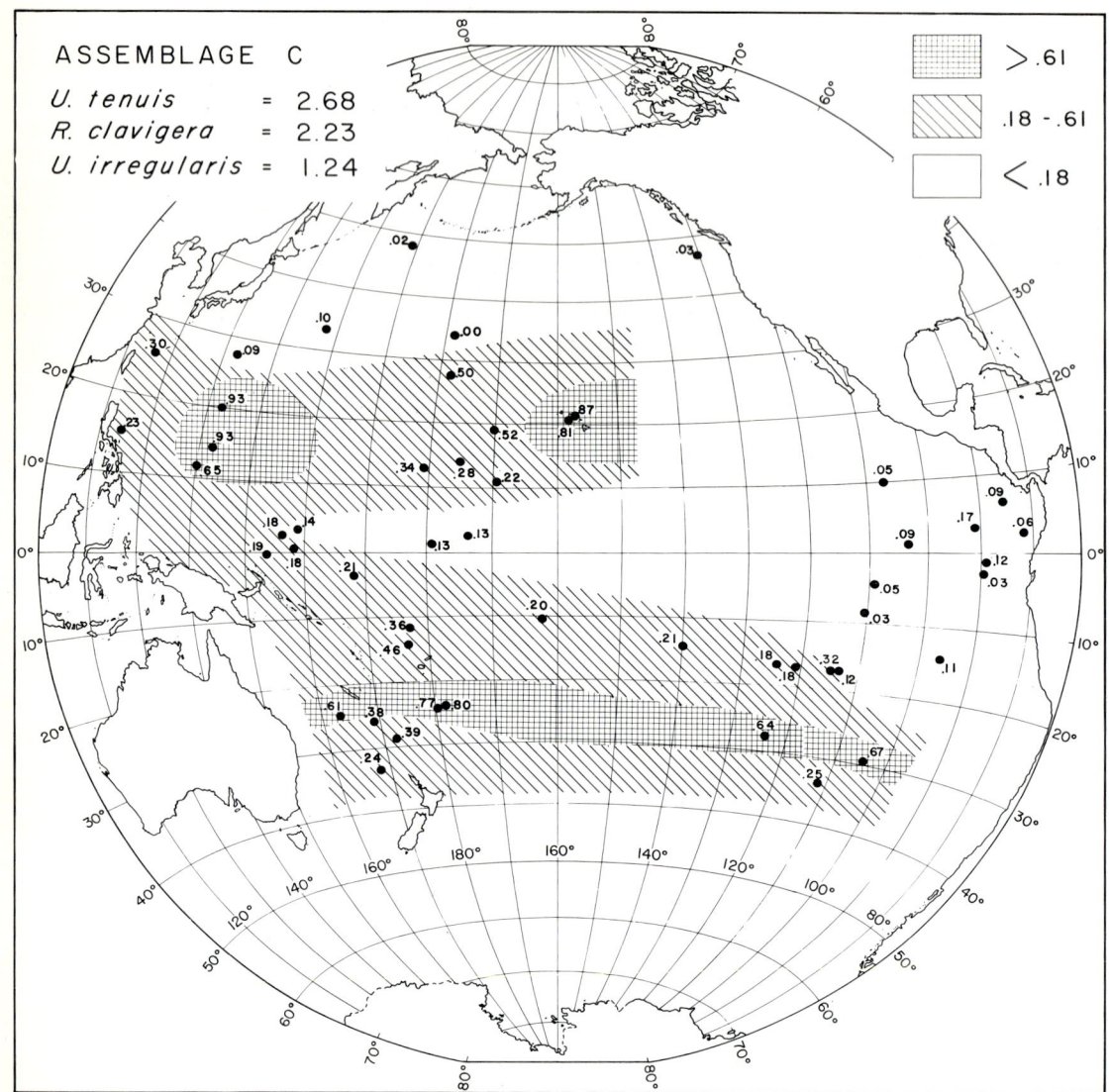

Figure 5. Geographic distribution of varimax values for assemblage C (*Umbellosphaera tenuis*, *Rhabdosphaera clavigera* and *Umbellosphaera irregularis*) in core tops.

McIntyre and others (1970), in their study of Pacific coccolithophorida from water samples, found both *U. tenuis* and *R. clavigera* to have their maximum concentration within the central water masses. *U. irregularis*, the least important of the three in our assemblage C, was found by them to be an equatorial species, although it occurs in the central water masses as well. Okada and Honjo (1973) found both *U. tenuis* and *U. irregularis* to be most important in the central water masses, with *U. tenuis* generally occurring at deeper levels. *R. clavigera* was also important in the northern part of the WNPC. Neither of the authors mentioned above discussed the distribution of *Umbilicosphaera mirabilis* (warm variety), so no comparison can be made for assemblage D.

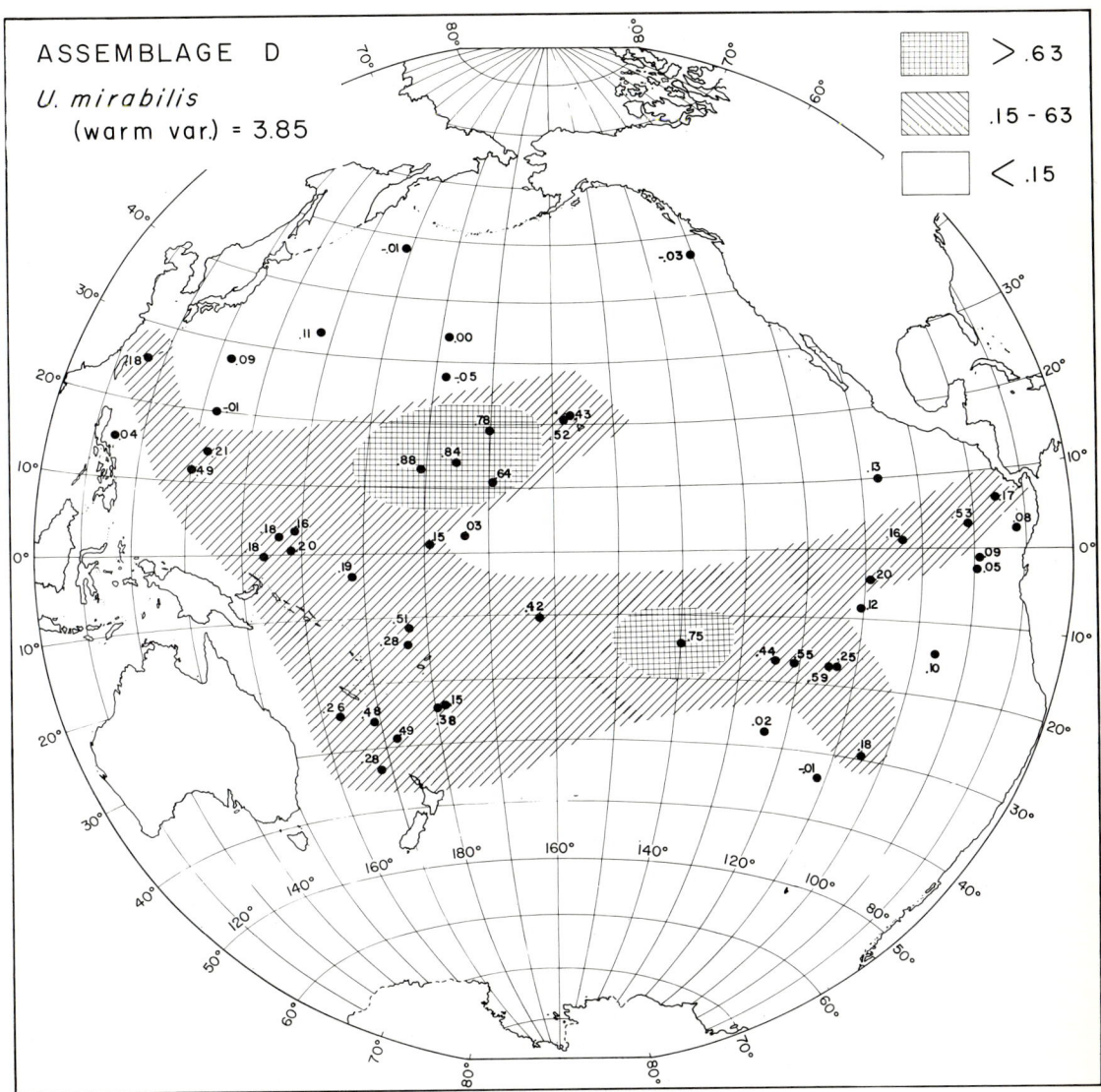

Figure 6. Geographic distribution of varimax values for assemblage D (*Umbilicosphaera mirabilis* [warm var.]) in core tops.

Assemblage E, dominated by *Cyclococcolithina fragilis* (Lohmann) Wilcoxon, is a weak factor accounting for 7.1% of the data variance and having a high varimax value of only 0.68. Significantly, high values for this assemblage occur in only four cores, all of them situated within the WSPC, although the assemblage does show a wide geographic distribution in the western Pacific, apparently centered about the equatorial region (Fig. 7). McIntyre and others (1970) did not consider *C. fragilis* in their study, but Okada and Honjo (1973) found this species to occur over a wide area, with maximum concentration in the equatorial and northern WNPC mass. The latter authors sampled the South Pacific Central Water Mass only in the central Pacific to about lat 15°S, so little is known about the living

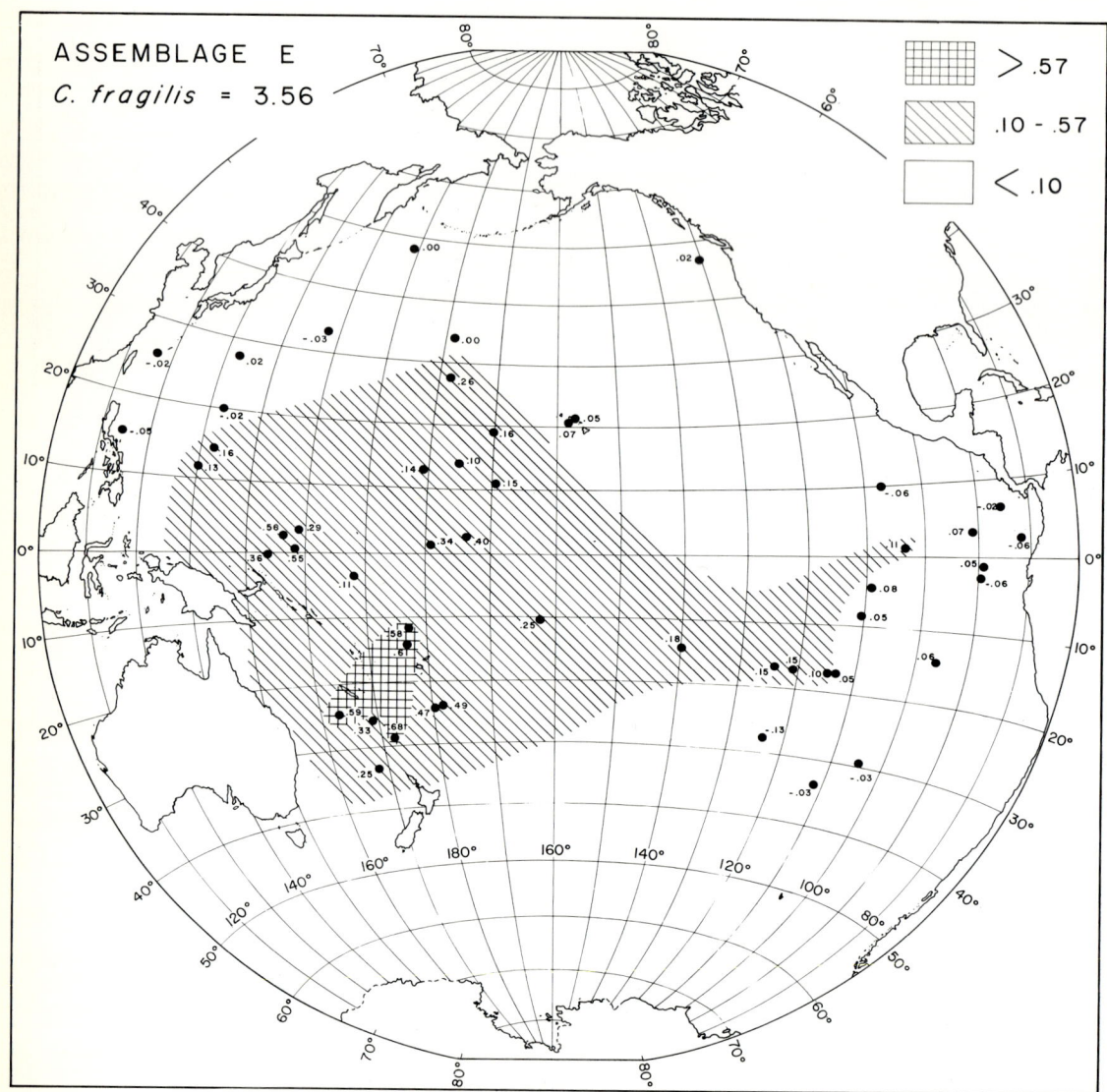

Figure 7. Geographic distribution of varimax values for assemblage E (*Cyclococcolithina fragilis*) in core tops.

occurrence of *C. fragilis* in our area of maximum assemblage E varimax.

Assemblage F, dominated by *Coccolithus pelagicus* (Wallich) Schiller, is represented in only three samples and accounts for only 2.2% of the data variance. This, of course, is because of our unequal sampling pattern, which is controlled by sediment distribution. Our limited data show this assemblage to be subarctic in character (Fig. 8). The highest varimax value occurs within the North Pacific Intermediate Water of Sverdrup and others (1942; the subarctic biogeographical zone of Knox, 1971). The only other two occurrences are in the northern WNPC and the northeastern Pacific Transition Region (Fig. 2). McIntyre and Bé (1967) and McIntyre and others (1970) described *C. pelagicus* as a stenothermal subarctic

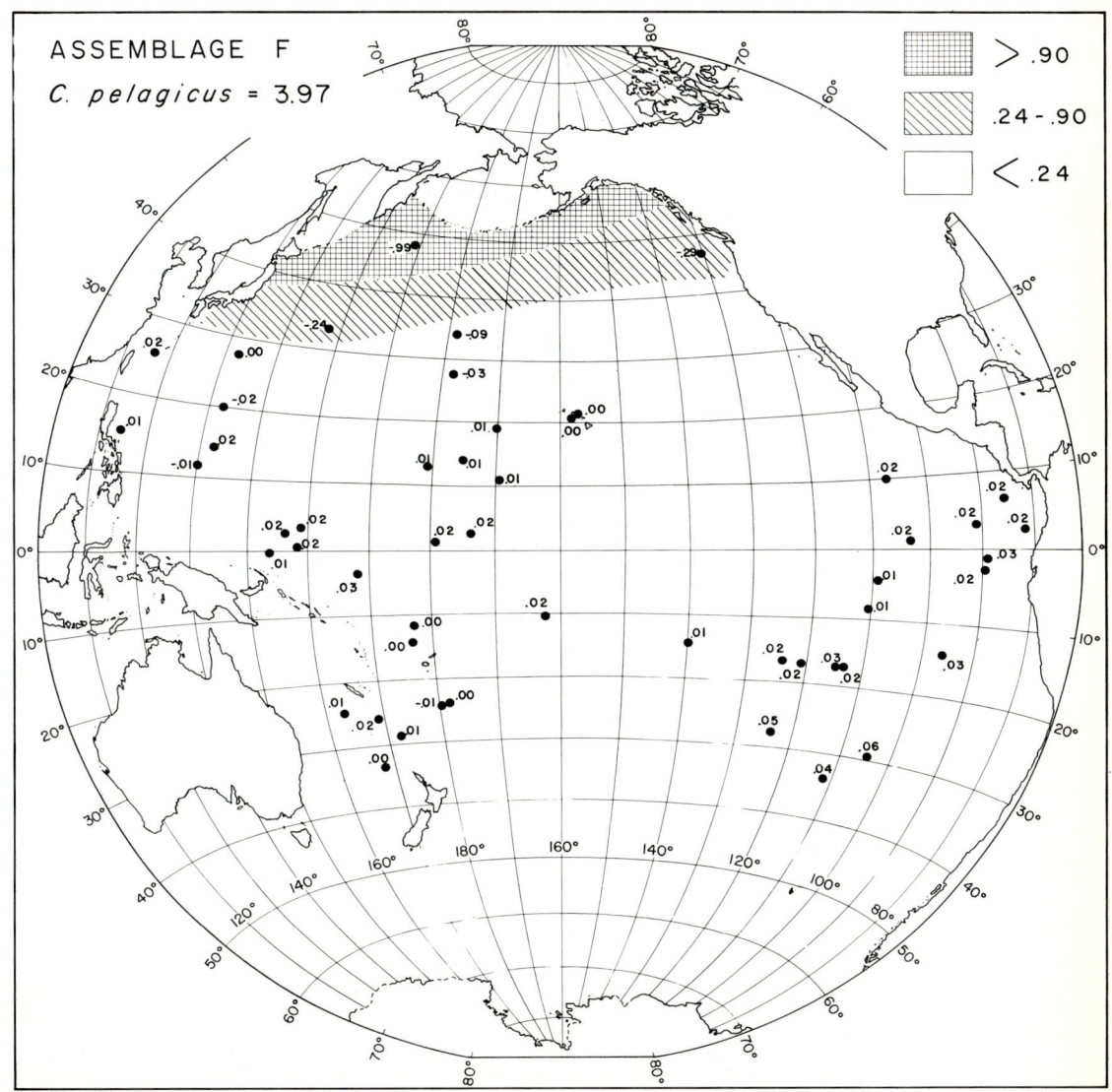

Figure 8. Geographic distribution of varimax values for assemblage F (*Coccolithus pelagicus*) in core tops.

species with a temperature range of 6° through 14°C.

Berger (1968, 1970, 1973a, 1973b) has shown quite convincingly that any attempt to quantify deep-sea paleontological studies must include a consideration of the effects of selective dissolution on the organism studied. Because calcite solution begins well above the CCD, sample selection is complicated. Samples found above the CCD can be affected by solution, but if only very shallow samples are used the data set becomes too limited geographically. The data must be tested to ensure that the results are due to biogeographic distributions and not to solution. If solution effects are significant, then fragile species such as *U. tenuis* and *U. irregularis* will be removed, and robust species such as *G. oceanica* and *C. leptopora* will be preferentially preserved. Varimax values would, therefore, react in a similar way, with assemblages A and B increasing and C decreasing as solution of the samples increases.

Varimax values for all six assemblages in the samples used in this study have been plotted against depth (Fig. 9). These plots show a random distribution with no observable trends. We feel confident, therefore, that the observed species-abundance variation in our surface-sediment samples is primarily due to biogeographic distribution and not to selective dissolution.

Regression Analysis

The relationships between the assemblages (as described by the six factors) and the winter and summer sea-surface temperatures are quantified by producing two regression equations, with the assemblages (B_{ct}) as independent variables and temperatures as dependent variables. Surface-water temperature values at each core location were taken from Defant (1961), Wyrtki (1964), and La Violette and Seim (1969) (see Table 4). Curvilinear regression was used because it produced a slightly better fit to the data. The associative relationship between the assemblages and temperatures can be described by multiple correlation coefficients (MCC) and standard error of estimates (SEE; both adjusted for the degrees of freedom). These statistics (MCC, 0.915 and 0.934, and SEE, 2.4° and 1.4°, for winter and summer, respectively) confirm the high degree of correlation between these data sets and establish the standard error of estimate for downcore parameter reconstruction. The SEE values are equivalent to 9% and 7.5% of the winter and summer temperature ranges of the data, respectively.

DOWNCORE COCCOLITH STUDY

Samples

In 1971, R/V *Vema* obtained two long piston cores (V28-238 and V28-239) on the Solomon Rise in the western equatorial Pacific (Table 5) beneath the present westward-flowing South Equatorial Current. The cores are composed of well-preserved foraminifer-coccolith ooze exhibiting almost no lithologic changes and no evidence of mixing or discontinuity. Both cores contain a record of the entire Brunhes epoch and have relatively high sedimentation rates for the Pacific. Because of these factors, total floral-faunal, magnetic, and isotopic analyses of these cores have been or are being made to delineate the climatic history of this area (Shackleton and Opdyke, 1973; Saito and Thompson, 1973). Core V19-55 (Table 5) from the East Pacific Rise was examined because (1) the foraminiferal (Luz, 1973) and oxygen-isotope data (Shackleton, 1973) are already available; (2) coccolith preservation is relatively good; (3) its geographic position is within the tropical zone on the side of the Pacific opposite the location of the other two piston cores; and (4) the core is surrounded by a relatively high density of surface-sediment samples. Core V19-55 lies beneath the eastern South Pacific Central Water Mass (Sverdrup and others, 1942) and is composed primarily of a well-preserved foraminifer-coccolith ooze with several thin interbedded foraminiferal sands. Luz (1973) reported two zones of foraminiferal dissolution in core V19-55, but these are not reflected in the coccolith assemblages.

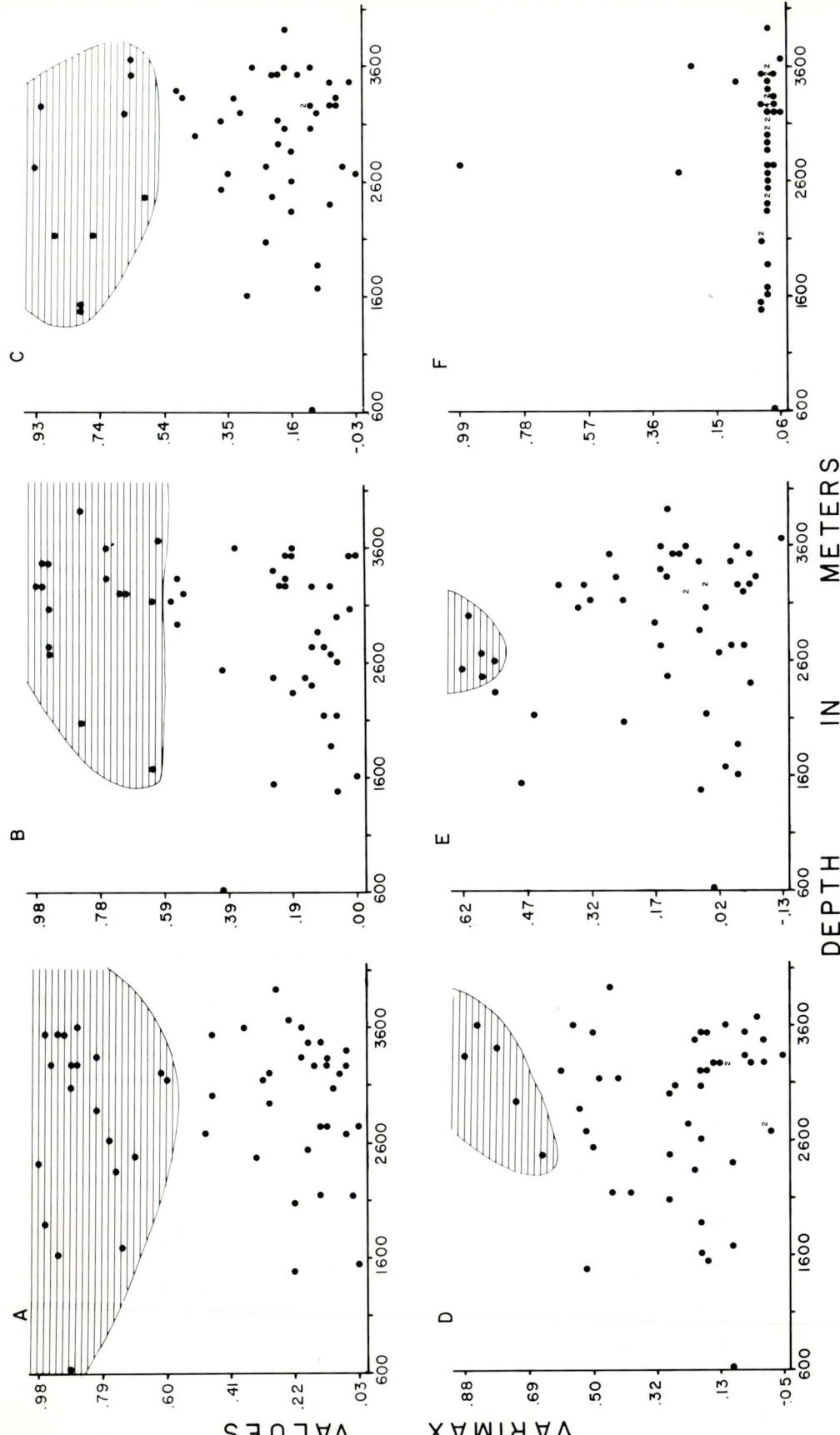

Figure 9. Varimax values for each of the assemblages plotted against depth. Samples showing highest varimax values, as indicated in distribution maps (Figs. 3 through 8), are included within shaded areas.

TABLE 4. OBSERVED AND ESTIMATED SEA-SURFACE TEMPERATURES FROM SURFACE-SEDIMENT SAMPLES

Core	Summer		Winter		Source*
	Observed	Estimated	Observed	Estimated	
RC8-91	21.8	23.8	15.6	19.0	2
RC8-93	23.7	25.4	18.1	19.5	2
RC9-104	25.2	24.1	20.2	18.8	2
RC9-121	24.5	25.1	21.3	22.4	3
RC9-124	23.3	23.8	19.5	20.9	3
RC9-126	21.0	23.0	16.0	19.5	3
RC10-115	28.8	27.4	28.4	24.6	1
RC10-143	30.0	30.1	28.8	29.6	1
RC10-161	26.7	27.7	17.3	17.3	1
RC11-187	16.8	17.3	9.1	8.3	1
RC11-220	28.2	28.2	26.2	27.0	2
RC11-230	25.9	25.2	24.1	21.6	1
RC12-107	24.5	25.0	20.5	21.7	3
RC12-113	25.1	26.6	21.0	22.7	3
RC12-121	29.0	29.1	28.8	26.1	1
RC12-173	26.7	27.3	24.4	25.0	1
RC12-210	25.0	24.6	22.0	20.4	3
RC12-361	28.8	27.9	26.3	25.2	1
RC12-366	28.5	28.3	20.5	25.7	1
RC13-1	26.3	27.2	23.8	25.5	1
RC13-17	27.7	26.1	25.4	24.2	1
RC13-27	27.8	28.4	26.8	26.5	1
RC13-108	26.2	26.7	21.1	21.7	1
RC13-113	27.1	26.5	21.7	22.0	1
RC13-122	28.2	26.7	28.0	24.0	1
V15-29	27.7	27.1	26.5	25.5	1
V19-25	26.6	26.6	25.7	24.4	1
V19-41	23.6	24.8	21.1	20.6	2
V19-53	25.6	25.2	23.0	22.4	2
V19-64	26.4	26.7	23.9	24.8	2
V19-65	26.7	25.7	24.2	23.3	2
V19-110	30.0	28.1	27.7	24.8	1
V20-103	25.1	23.4	16.2	17.5	1
V20-119	11.3	11.1	2.7	2.8	1
V21-30	26.3	27.8	21.9	23.7	1
V21-41	25.7	25.6	23.2	22.3	1
V21-85	28.0	26.6	20.0	21.1	1
V24-96	26.7	24.4	20.8	18.4	1
V24-101	28.3	27.9	27.2	25.9	1
V24-109	29.9	28.8	28.8	27.9	1
V24-110	30.0	30.2	28.8	30.4	1
V28-148	27.7	28.3	25.5	27.2	1
V28-195	27.7	28.9	26.3	27.5	1
V28-201	28.9	29.8	28.2	28.4	1
V28-203	28.9	29.2	28.4	27.3	1
V28-211	28.8	27.6	28.2	26.4	1
V28-212	28.8	27.8	27.5	25.5	1
V28-239	30.0	30.1	28.8	29.7	1
V28-246	30.0	29.3	27.3	25.7	1
V28-255	29.5	30.1	25.8	27.0	1

*The observed sea-surface temperatures have been obtained from the following sources: 1, La Violette and Seim (1969); 2, Wyrtki (1964); 3, Defant (1961).

TABLE 5. PISTON-CORE DATA

Core	Latitude	Longitude	Depth (m)	Core length (cm)
V19-55	17°00'S	114°11'W	3,177	405
V28-238	01°01'N	160°29'E	3,120	1,602
V28-239	03°15'N	159°11'E	3,490	2,102

Chronology

To discuss Pleistocene paleotemperatures and compare results with other investigations, it is imperative that a good chronostratigraphy be established for each core. The following four primary datum levels were used (Fig. 10; Table 6): (1) The Brunhes-Matuyama paleomagnetic boundary at about 690,000 B.P. (Cox, 1969) and three coccolith events. (2) The extinction of *Pseudoemiliania lacunosa* (Kamptner) Gartner, estimated by us to have occurred at approximately 380,000 B.P. This date is based on our examination of other faunally, isotopically, or paleomagnetically dated cores from the Atlantic (V12-18, V16-205, V22-230, and V22-163), the Caribbean (V12-122), and the Pacific (RC11-209). The mean value of the *P. lacunosa* extinction levels interpolated from the Brunhes-Matuyama paleomagnetic boundary for the five Pacific cores is 380,000 B.P. Gartner (1973) believed the date should be somewhat younger, about 350,000 B.P., but because these dates differ by less than 10% and are not especially crucial to this study, we will continue to use our slightly older date. (3) The first appearance of *E. huxleyi* Lohmann (Hay and Mohler), estimated by us to have occurred at about 210,000 B.P. This is an approximate age based on our observations in the Atlantic, but it is probably correct to within ±10% of that age. (4) The relative abundance

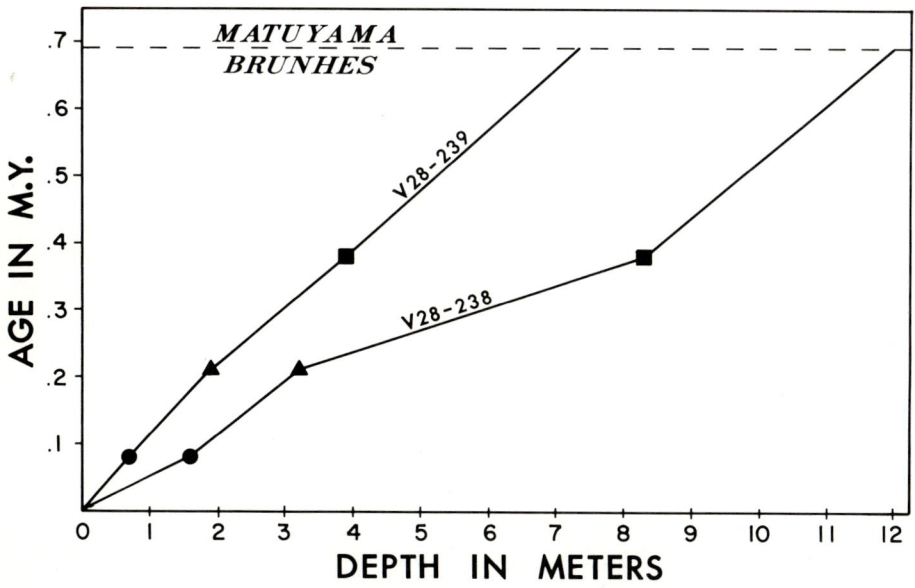

■ Last occurrence of *P. lacunosa* (380,000 y.)
▲ First appearance of *E. huxleyi* (210,000 y.)
● *G. caribbeanica*/*E. huxleyi* switchover (73,000 y.)

Figure 10. Time-depth plot of cores V28-238 and V28-239.

TABLE 6. CORE TIME-DATUM LEVELS AND SEDIMENTATION RATES

Datum	Approximate age (B.P.)	V28-238 a*	V28-238 b†	V28-239 a*	V28-239 b†	V19-55 a*	V19-55 b†
			1.8		1.1		0.9
Initial dominance of E. huxleyi over G. caribbeanica	73,000	135		80		65	
			1.5		0.8		1.0
First appearance of E. huxleyi	210,000	335		190		200	
			2.9		1.2		
Last occurrence of P. lacunosa	380,000	825		390			
			1.2		1.1		
Brunhes-Matuyama boundary	690,000	1,200		735			

*a = depth of datum level in core (cm).
†b = sedimentation rate as interpolated between datum levels (cm/10^3 yr).

of *E. huxleyi*, which has been found to increase suddenly during carbonate Minimum 2 (~72,000 to 73,000 B.P. in the Atlantic; McIntyre and others, 1972), becoming dominant over the previously more abundant *G. caribbeanica* Boudreaux and Hay. This reversal in dominance is also recorded in the western equatorial and eastern Pacific cores and has been tentatively assumed to be isochronous.

The oxygen-isotope stratigraphy of cores V28-238 and V19-55 has been determined (Figs. 11, 13; Shackleton, 1973; Shackleton and Opdyke, 1973). Identification of Termination II (127,000 B.P.) in core V19-55 at about 115 cm (Luz, 1973) confirms the reliability of the coccolith datum levels in this core, indicating an almost linear sedimentation rate of approximately 0.9 cm/1,000 yr for the past 250,000 yr.

The last occurrence of *P. lacunosa* in core V28-238 occurs approximately at the boundary between isotope stages 12 and 13, previously dated at 380,000 B.P. (Broecker and van Donk, 1970). Other cores in which the extinction level of *P. lacunosa* and oxygen-isotope stratigraphy can be compared show a similar correlation (Gartner, 1973; van Donk, this volume). Shackleton and Opdyke (1973), however, assumed a uniform rate of sedimentation for core V28-238 from the Brunhes-Matuyama boundary to the present and calculated the age of the boundary between stages 12 and 13 at about 472,000 B.P. The age discrepancy can be attributed to a nonlinear sedimentation rate. Specifically, the sedimentation rate increases from about 1.2 cm/1,000 yr, between the Brunhes-Matuyama boundary and the extinction level of *P. lacunosa*, to about 2.9 cm/1,000 yr, between the *P. lacunosa* datum and the first appearance of *E. huxleyi*; it then decreases to about 1.5 to 1.8 cm/1,000 yr (Fig. 10; Table 6). The time-depth plot for core V28-239, based on coccolith datum levels, yields calculated sedimentation rates of 0.8 to 1.2 cm/1,000 yr (Table 6).

For the eastern equatorial Pacific and the North Atlantic, total calcium carbonate curves have proved to be excellent climatic and time indicators (Hays and others, 1969; McIntyre and others, 1972). The total carbonate curve for the western Pacific cores yields less definitive carbonate maximums and minimums (Figs. 11 through 13), but a few observations can be made. A carbonate minimum occurs in core V28-238 at 39 cm and a subdued minimum in core V28-239 at 21 to 30 cm. Assuming sedimentation rates of 1.8 cm and 1.1 cm/1,000 yr for cores V28-238 and V28-239, respectively, we arrive at an age of about 19,000 through 27,000 yr for this minimum, comparable to the calculated age of 15,000 through 30,000 yr for Minimum 1 in

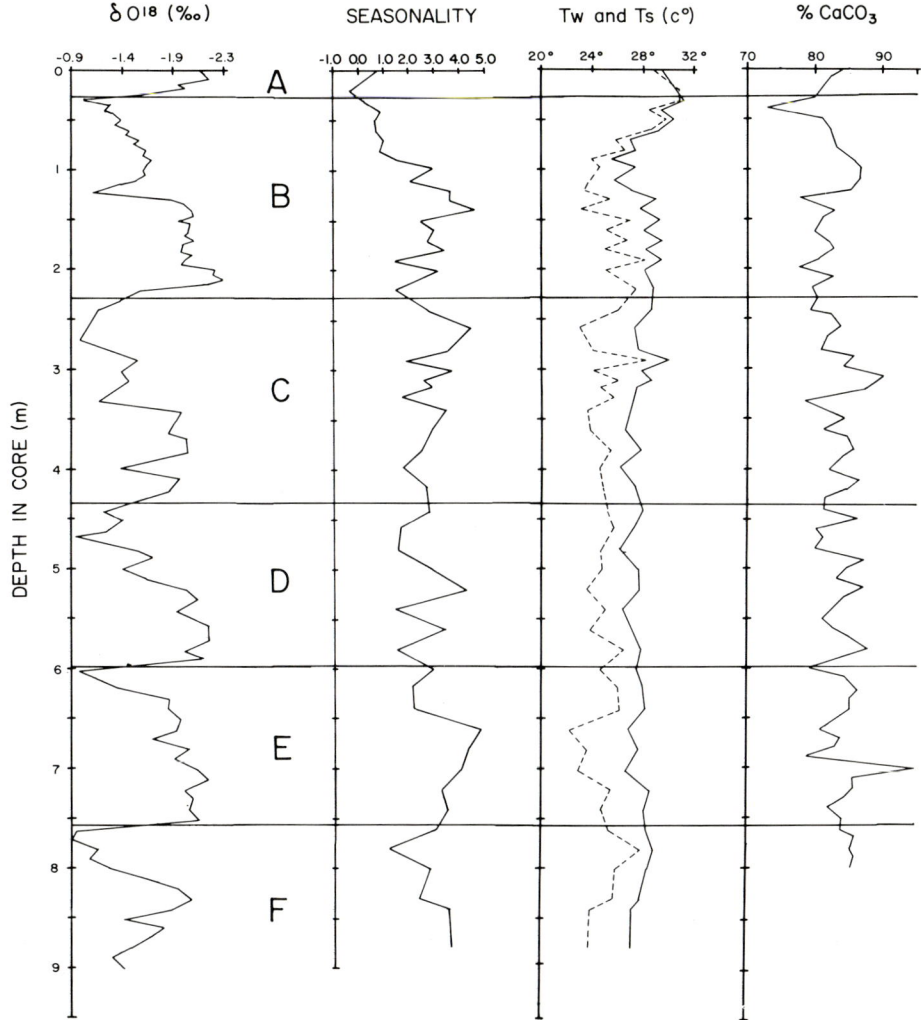

Figure 11. Core V28-238. δO^{18} (Shackleton and Opdyke, 1973), glacial stages A through F (Kukla, 1961), seasonality ($T_s - T_w$), summer (T_s = solid line) and winter temperatures (T_w = dashed line) as estimated during this study, and relative percent calcium carbonate plotted to depth.

the North Atlantic (McIntyre and others, 1972). A single-sample carbonate minimum occurs in core V28-238 at 128 cm at approximately 71,000 B.P. This compares favorably with the age of 72,000 yr for Minimum 2 in the North Atlantic (McIntyre and others, 1972). These carbonate minimums correlate closely with but slightly precede the O^{18} minimums at 20 and 123 cm in core V28-238 (Fig. 11). There is no recognizable second carbonate minimum in core V28-239. Below the above-mentioned minimums, small carbonate fluctuations preclude further development of definitive carbonate stratigraphies.

Whereas the upper 1.5 m of core V28-238 yields a typical "North Atlantic" carbonate curve (carbonate minima corresponding to glacial maxima; McIntyre and others, 1972), core V19-55 shows a more typical "Pacific" curve (Arrhenius, 1952), with carbonate minimums at 0, 70 to 110, 130, and 180 to 210 cm correlating

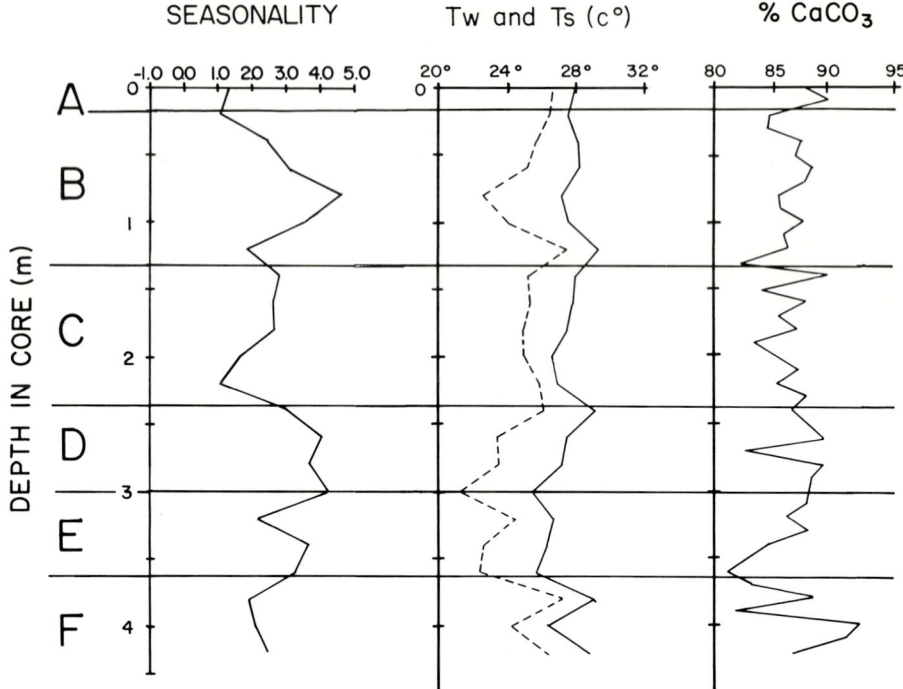

Figure 12. Core V28-239. Glacial stages A through F (Kukla, 1961) are correlated on the basis of the estimated sedimentation rates shown in Table 6. Seasonality ($T_s - T_w$) and summer (T_s = solid line) and winter temperatures (T_w = dashed line) as estimated during this study, and relative percent calcium carbonate plotted to depth.

with δO^{18} maximums (glacial minimums; Fig. 13).

The "Pacific" pattern of carbonate fluctuations has been attributed to an intensification of oceanic circulation and therefore to an increase in biogenic carbonate production during glacial maximums (Arrhenius, 1952; Hays and others, 1969). Berger (1973b), however, believed that dissolution during interglacial periods is probably more important than increased productivity during glacial periods in formation of the carbonate cycles.

Data Treatment

Once factor and regression analyses have been completed, downcore species abundance data[3] are converted to assemblages identical in composition to those found in the surface sediments. The conversion is made by post-multiplication of the normalized species data for the core by the transposed factor score matrix (Table 3). The resulting varimax matrix[4] expresses the importance of each of the assemblages to each of the samples from the core. The communalities[4] are a measure of the amount of population variation that can be accounted for by the six assemblages; that is, the higher the communality, the greater the amount of species variance incorporated. All communalities derived in cores V28-238, V28-239, and V19-55 are above 0.61, which means that the least amount of original data

---

[3] See Appendix III on microfiche in pocket inside back cover.
[4] See Appendix IV on microfiche in pocket inside back cover.

included in any sample is $0.61^{1/2}$ or 78%. The mean communalities (core V28-238, 0.941; V28-239, 0.927; V19-55, 0.988; trigger weights, 0.976) are higher than those reported by Imbrie and Kipp (1971) and Roche and others (1975).

Paleotemperature estimates (Table 7) are calculated by deriving the squares and cross-products of the six assemblages (producing 27 variables), summing the products of the variables and the previously derived regression coefficients, and adding this sum to a regression constant. The values produced are floral indices calibrated on seabed samples to be unbiased estimates of summer sea-surface temperature (T_s), winter sea-surface temperature (T_w), and seasonality ($T_s - T_w$).

Results

Figures 11 through 14 show the downcore paleotemperature curves plotted against depth and time. Paleotemperature estimates indicate that for the time period 0 to 400,000 B.P. the response of sea-surface temperature to climatic fluctuation was moderate to minimal in these regions of the world ocean. The maximum ranges of variation of T_s in cores V28-238 and V28-239 are 5.7° and 3.9°C, respectively; of T_w, 8.9° and 6.3°C, respectively. T_s and T_w in core V19-55 show much less variation, with maximum ranges of 1.9° and 3.3°C, respectively. The lesser variation of temperature in core V19-55 probably reflects the position of that core in the South Pacific Central Water Mass. There appears to be no correlation between low communalities (<0.9) and high or low temperature estimates.

There is little or no indication of cyclic temperature fluctuation that correlates with known glacial-interglacial history as recorded in the isotopic variations within these cores; however, a number of observations can be made. Paleotemperature plots for core V28-238 indicate little seasonal difference or seasonality ($T_s - T_w$) over the past 50,000 yr. This is not clearly reflected by core V28-239, although the upper two samples do have two of the lowest seasonality values for the entire core.

The lower sedimentation rate and wider sampling interval in core V28-239 result in a smoothing of the curve and may be somewhat analogous to making a running mean of the values in core V28-238. A feature shown by the V28-238 curve but not by core V28-239 is the relatively well pronounced temperature peak at about 17,000 B.P. (31 cm). There is a less well developed, possibly analogous peak

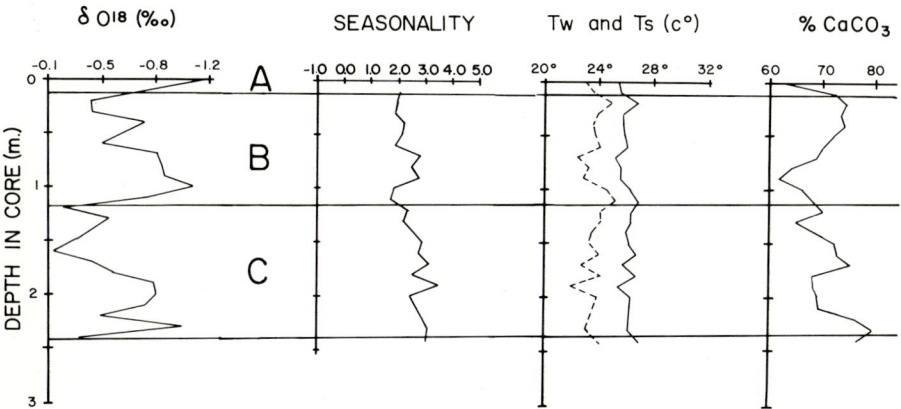

Figure 13. Core V19-55. δO^{18} (Shackleton, 1973), glacial stages A through C (Kukla, 1961), seasonality ($T_s - T_w$), summer (T_s = solid line) and winter temperatures (T_w = dashed line) as estimated during this study, and relative percent calcium carbonate plotted to depth.

in core V19-55 at an interpolated age of about 22,000 B.P. (20 cm). The O^{18} data (Fig. 11) clearly show an increase in ice volume at these times.

Of considerable interest is the apparent seasonality increase at four levels in the cores (Figs. 11 through 13). It is only at these levels that the seasonality values exceed those of the combined standard errors of estimate (SEE) for T_s and T_w ($\pm 1.4°$ and $\pm 2.4°$, respectively). Thus, only when the seasonality value

TABLE 7. DOWN CORE PALEOTEMPERATURE ESTIMATES

Core	Depth (cm)	T_s	T_w	Core	Depth (cm)	T_s	T_w
V28-238	1	29.5	28.8	V28-238	761	28.3	25.3
V28-238	20	30.6	30.9	V28-238	781	28.9	27.8
V28-238	31	31.2	31.0	V28-238	801	28.5	25.7
V28-238	41	29.4	28.5	V28-238	830	27.9	25.5
V28-238	50	30.4	29.7	V28-238	841	27.3	23.7
V28-238	61	29.3	28.6	V28-238	880	27.2	23.6
V28-238	71	26.9	25.9	V28-239	0	28.0	26.7
V28-238	81	27.4	26.5	V28-239	20	27.6	26.5
V28-238	90	25.5	23.9	V28-239	40	28.2	25.7
V28-238	98	27.5	24.5	V28-239	60	28.3	25.2
V28-238	111	25.9	23.8	V28-239	80	27.2	22.6
V28-238	121	27.1	23.4	V28-239	100	27.7	24.1
V28-238	130	29.2	25.5	V28-239	120	29.4	27.5
V28-238	140	27.8	23.2	V28-239	140	28.1	25.2
V28-238	151	29.4	26.9	V28-239	160	27.9	25.3
V28-238	161	28.1	25.1	V28-239	180	27.6	24.9
V28-238	171	29.5	26.7	V28-239	200	26.6	24.9
V28-238	180	28.3	24.9	V28-239	220	27.0	25.9
V28-238	191	29.6	28.1	V28-239	240	29.2	26.2
V28-238	201	28.2	25.0	V28-239	260	27.5	23.4
V28-238	220	28.9	27.4	V28-239	280	27.3	23.5
V28-238	241	28.7	26.0	V28-239	300	25.5	21.2
V28-238	258	27.4	23.0	V28-239	320	26.7	24.5
V28-238	281	27.7	24.1	V28-239	340	26.3	22.7
V28-238	291	30.0	28.2	V28-239	360	25.6	22.3
V28-238	301	27.9	24.1	V28-239	380	29.2	27.2
V28-238	311	28.7	26.1	V28-239	400	26.4	24.3
V28-238	318	27.6	24.6	V28-239	420	28.9	26.4
V28-238	328	27.4	25.7	V19-55	10	25.6	23.5
V28-238	341	27.1	23.6	V19-55	20	26.8	24.9
V28-238	361	26.7	23.8	V19-55	30	25.8	23.9
V28-238	381	28.0	25.4	V19-55	40	25.8	23.5
V28-238	398	26.3	24.5	V19-55	50	25.9	23.7
V28-238	418	27.5	24.8	V19-55	60	26.0	24.1
V28-238	441	28.1	25.2	V19-55	70	25.1	22.4
V28-238	458	27.4	25.7	V19-55	80	25.6	23.2
V28-238	481	26.2	24.7	V19-55	90	25.5	22.8
V28-238	501	27.7	24.8	V19-55	100	26.3	24.5
V28-238	521	27.8	23.5	V19-55	110	26.9	25.2
V28-238	540	26.5	25.0	V19-55	120	26.4	24.1
V28-238	561	27.2	23.8	V19-55	130	26.3	24.2
V28-238	581	28.0	26.5	V19-55	140	26.0	23.5
V28-238	601	27.6	24.6	V19-55	150	26.2	23.4
V28-238	618	28.0	25.9	V19-55	160	26.8	24.0
V28-238	641	28.3	26.1	V19-55	170	25.8	22.7
V28-238	661	26.9	22.1	V19-55	180	26.7	24.2
V28-238	680	27.7	23.4	V19-55	190	25.4	21.9
V28-238	701	26.7	22.7	V19-55	200	26.3	23.9
V28-238	721	28.7	25.4	V19-55	230	26.1	23.1
V28-238	741	28.1	24.6	V19-55	242	27.0	24.0

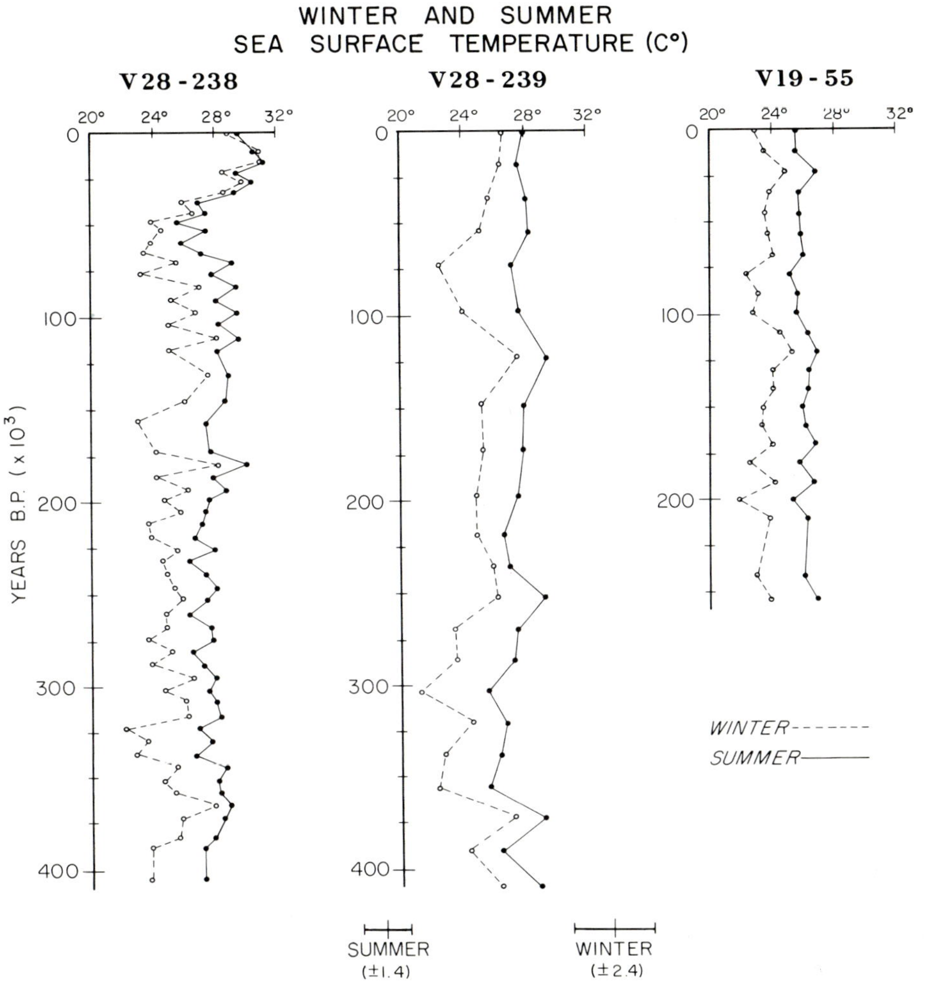

Figure 14. Estimated summer (solid lines) and winter temperatures (dashed lines) for cores V28-238, V28-239, and V19-55 plotted to time. Calculated standard error of estimate indicated at bottom.

exceeds 3.8° may we consider there to have been a significant difference between summer and winter sea-surface temperatures. The most recent occurrence of significant seasonality occurs at 77,000 B.P. (140 cm) and 73,000 B.P. (80 cm) in cores V28-238 and V28-239, respectively. Although seasonality does not exceed the combined standard error of estimate in any V19-55 sample, the most recent seasonality peak occurs at 78,000 B.P. (70 cm).

Previous significant seasonality events in these cores are given in Table 8.

With the exception of the event at 155,000 B.P. (258 cm) in core V28-238, the seasonality maximums occur near the end of an O^{18} stage. The approximate amounts of time by which the seasonality maximums precede each termination are 16,000 yr for stage 11, 8,300 yr for stage 9, and 9,300 yr for stage 5. The seasonality maximum at 155,000 B.P. in core V28-238 occurs in cycle B, during glacial stage 6. (The climatic cycle concept as defined by Kukla (1961) divides the Pleistocene record into interglacial-glacial cycles, with the Holocene being the early interglacial

TABLE 8. APPROXIMATE AGES (M.Y.) OF SEASONALITY PEAKS AND CYCLES DURING WHICH THEY OCCUR

Core	Glacial Cycles*			
	A	B	C	D
V28-238	77,000	155,000	272,000	320,000 to 336,000
V28-239	73,000		268,000 to 305,000	337,000
V19-55	78,000	299,000		

*See Kukla (1961).

stage of cycle A; it is a convenient way of discussing these climatic events.) Cycle B has been described by Ruddiman and McIntyre (this volume) as an atypical cycle for the North Atlantic, and this may be reflected in the displacement of the seasonality maximum in this cycle. In core V19-55 the seasonality maximums occur during stages 5 and 7, preceding the ends of these stages by approximately 10,000 to 20,000 yr and 10,000 yr, respectively.

The occurrence of seasonality peaks in midcycle or just before glacial events is intriguing, but the significance that these data might have for a better understanding of the mechanisms responsible for glacial-interglacial cycles is not clear. However, some thoughts and observations can be discussed. It is clear from the curves that the seasonality peaks are due primarily to a decrease in T_w, with T_s remaining relatively constant. We must, therefore, present a model which could account for a decrease in T_w of 1.5° to 2.0°C in the western equatorial area and 1.0°C in the southeastern subtropical area just before a glacial event.

Surface-temperature maps of the Pacific show a tongue of cooler water extending along the equator from the eastern to the central Pacific as a result of upwelling. If trade wind circulation were seasonally intensified, thereby intensifying the westward-flowing South Equatorial Current and the associated equatorial upwelling, this cooler surface water might extend into the western equatorial region. The eastern South Pacific Central Water Mass, where core V19-55 was taken, might also undergo an increase in circulation, but, not being directly affected by the cooler upwelled surface waters, would show less of a temperature change, thus explaining the lesser decrease in T_w. What remains to be explained is why the intensification of trade winds would occur at midcycle rather than during a midglacial or mid-interglacial period.

CONCLUSIONS

1. Factor analyses of surface-sediment coccolith populations have resulted in the definition of six coccolith assemblages or factors that are related to surface water-mass distribution.

2. Mathematical manipulation of the data has shown a significant correlation between the six assemblages and summer and winter sea-surface temperatures.

3. A carbonate minimum at about 22,000 B.P. and 18,000 to 27,000 B.P. in cores V28-238 and V28-239, respectively, roughly correlates with an O^{18} minimum (glacial maximum) in core V28-238 and is the only carbonate event recognized with certainty from the western equatorial Pacific for the past 400,000 yr.

4. A calcium carbonate curve from an eastern tropical Pacific core (V19-55) shows a clear negative correlation with O^{18} values from the same core for the past 225,000 yr. Carbonate minimums correlate with O^{18} maximums (glacial minimums).

5. Paleotemperature estimates for the past 400,000 yr in two cores from the

western equatorial Pacific and for the past 250,000 yr in one core from the eastern tropical Pacific indicate that sea-surface temperature has varied within a relatively greater range during this period in the western Pacific (maximum possible range: $T_s = 5.7°C$, $T_w = 8.9°C$) than in the eastern Pacific ($T_s = 1.9°C$, $T_w = 3.3°C$). Glacial and interglacial stages, however, are not correlative with sea-surface temperature change. Instead, there appears to have been a trend toward increasing seasonality ($T_s - T_w$) slightly before the end of O^{18} interglacial stages 5, 9, and 11.

ACKNOWLEDGMENTS

We thank the many people involved in the CLIMAP program who have donated their time and data—in particular, N. Opdyke, N. Shackleton, B. Luz and B. Molfino. The manuscript was reviewed by T. Moore, A. Bé, L. Burckle, and N. Kipp, and their pertinent comments were appreciated. The research was supported by National Science Foundation IDOE Grant IDO71-04204. Cores were collected and processed under Office of Naval Research Grant N00014-67A-0108-0004 and National Science Foundation Grant DES-72-01568.

REFERENCES CITED

Arrhenius, G., 1952, Sediment cores from the East Pacific: Swedish Deep-Sea Exped. (1947–1948) Repts.,: v. 5, fasc. 1, 89 p.

Berger, W. H., 1968, Planktonic foraminifera: Selective solution and paleoclimatic interpretation: Deep-Sea Research, v. 15, p. 31–43.

―――1970, Planktonic foraminifera: Selective solution and the lysocline: Marine geology, v. 8, p. 11–138.

―――1973a, Deep-sea carbonates: Evidence for a coccolith lysocline: Deep-Sea Research, v. 20, p. 917–921.

―――1973b, Deep-sea carbonates: Pleistocene dissolution cycles: Jour. Foram, Research, v. 3, p. 187–195.

Broecker, W. S., and van Donk, J., 1970, Insolation changes, ice volumes, and the O^{18} record in deep-sea cores: Rev. Geophysics and Space Physics, v. 8, p. 169–198.

Cox, A., 1969, Geomagnetic reversals: Science, v. 163, p. 237–245.

Defant, A., 1961, Physical oceanography (Vol. 1): New York, Pergamon Press, 729 p.

Ericson, D. B., and Wollin, G., 1968, Pleistocene climates and chronology in deep-sea sediments: Science, v. 162, p. 1227–1234.

Gartner, S., 1973, Late Pleistocene calcareous nannofossils in the Caribbean and their interoceanic correlation: Palaeogeography, Palaeoclimatology, Palaeoecology, v. 12, p. 169–191.

Geitzenauer, K. R., 1972, The Pleistocene calcareous nannoplankton of the subantarctic Pacific Ocean: Deep-Sea Research, v. 19, p. 45–60.

Hays, J. D., Saito, T., Opdyke, N. D., and Burckle, L. H., 1969, Pliocene-Pleistocene sediments of the equatorial Pacific; Their paleomagnetic, biostratigraphic and climatic record: Geol. Soc. America Bull., v. 80, p. 1481–1514.

Imbrie, J., and Kipp, N. G., 1971, A new micropaleontological method for quantitative paleoclimatology: Application to a late Pleistocene Caribbean core, in Turekian, K., ed., The late Cenozoic glacial ages: New Haven, Yale Univ. Press, p. 71–181.

Klovan, J. E., and Imbrie, J., 1971, An algorithm and FORTRAN IV program for large-scale Q-mode factor scores: Jour. Internat. Assoc. Math. Geol., v. 3, p. 61–77.

Knox, G. A., 1970, Biological oceanography of the South Pacific, in Wooster, W. S., ed, Scientific exploration of the South Pacific: Natl. Acad. Sci., p. 155–182.

Kukla, J., 1961, Quaternary sedimentation cycle. Survey of Czechoslovak Quaternary: Institut Geologiczny, Prace, Tom XXXIV, Warszawa, p. 145–154.

La Violette, P. E., and Seim, S. E., 1969, Monthly reports of mean, minimum, and maximum sea surface temperature in the North Pacific Ocean: U.S. Naval Oceano. Office Spec. Pub. 123, 62 p.

Luz, B., 1973, Stratigraphic and paleoclimatic analysis of late Pleistocene tropical southeast Pacific cores: Quaternary Research, v. 3, p. 56–72.

McGowan, J. A., 1971, Oceanic biogeography of the Pacific, in Funnell, B. M., and Riedel, W. R., ed., The micropaleontology of Oceans: Cambridge, Cambridge Univ. Press, p. 3–74.

McIntyre, A., 1967, Coccoliths as paleoclimatic indicators of Pleistocene glaciations: Science, v. 158, p. 1314–1317.

McIntyre, A., and Bé, A. W. H., 1967, Modern Coccolithophoridae of the Atlantic Ocean. I. Placoliths and cyrtoliths: Deep-Sea Research, v. 14, p. 561–597.

McIntyre, A., Bé, A. W. H., and Preikstas, R., 1967, Coccoliths and the Plio-Pleistocene boundary, in Sears, M., ed., Progress in oceanography. Vol. 4, The Quaternary history of the ocean basins: New York, Pergamon Press, p. 3–24.

McIntyre, A., Bé, A. W. H., and Roche, M. B., 1970, Modern Pacific Coccolithophorida: A paleontological thermometer: New York Acad. Sci. Trans., v. 32, p. 720–731.

McIntyre, A., Ruddiman, W. F., and Jantzen, R., 1972, Southward penetration of the North Atlantic polar front: Faunal and floral evidence of large scale surface water mass movements over the last 225,000 years: Deep-Sea Research, v. 19, p. 61–77.

Okada, H., 1973, Modern coccolithophorids in neritic environment of western Pacific Ocean: Geol. Soc. America Abs. with Programs, v. 5, p. 757–758.

Okada, H., and Honjo, S., 1973, The distribution of oceanic coccolithophorids in the Pacific: Deep-Sea Research, v. 20, p. 355–374.

Phleger, F. B., Parker, F. L., and Pierson, J. F., 1953, North Atlantic Foraminifera: Swedish Deep-Sea Exped. Repts., v. 7, 122 p.

Roche, M. B., McIntyre, A., and Imbrie, J., 1975, Quantitative paleo-oceanography of the late Pleistocene-Holocene North Atlantic: Coccolith evidence, in Saito, T. and Burckle, L., eds., Neogene boundaries: New York, Micropaleontology Press (in press).

Ruddiman, W., and McIntyre, A., 1976 Northeast Atlantic paleoclimatic changes over the past 600,000 years, in Cline, R. M., and Hays, J. D., eds., Investigation of Late Quaternary paleoceanography and paleoclimatology: Geol. Soc. America Mem. 145 (this volume).

Saito, T., and Thompson, P., 1973, Equatorial Pacific climatic fluctuations during the last 500,000 years: Geol. Soc. America Abs. with Programs, v. 5, p. 840–841.

Schott, W., 1935, Die Foraminiferan in dem aquatorialen Teil des Atlantischen Ozeans: Deutsche Atlantische Exped. auf "Meteor" 1925-1927, v. 3, p. 43–131.

Shackleton, N. J., 1973, Appendix, in Luz, B., Stratigraphic and paleoclimatic analysis of late Pleistocene tropical southeast Pacific cores: Quaternary Research, v. 3, p. 70.

Shackleton, N. J., and Opdyke, N. D., 1973, Oxygen isotope and paleomagnetic stratigraphy of equatorial Pacific core V28-238: Oxygen isotope temperatures and ice volumes on a 10^5 year and 10^6 year scale: Quaternary Research, v. 3, p. 39–55.

Shaw, A. B., 1964, Time in stratigraphy: New York, McGraw-Hill, 365 p.

Smayda, T. J., 1958, Biogeographical studies of marine phytoplankton: Oikos, v. 9, p. 158–191.

Sverdrup, H. U., Johnson, M. W., and Fleming, R. H., 1942, The oceans: Their physics, chemistry and general biology: New York, Prentice-Hall, 1060 p.

van Donk, J., 1976, O^{18} record of the Atlantic Ocean for the entire Pleistocene Epoch, in Cline, R. M., and Hays, J. D., Investigation of Late Quaternary paleoceanography and paleoclimatology: Geol. Soc. America Mem. 145 (this volume).

Wyrtki, K., 1964, The thermal structure of the Eastern Pacific Ocean: Deutschen Hydrographischen Zeitschrift, Egaenzungsheft Reihe A (8°), v. 6, p. 4–84.

MANUSCRIPT RECEIVED BY THE SOCIETY NOVEMBER 11, 1974
REVISED MANUSCRIPT RECEIVED JUNE 24, 1975
MANUSCRIPT ACCEPTED JULY 14, 1975
LAMONT-DOHERTY GEOLOGICAL OBSERVATORY CONTRIBUTION NO. 2276

Printed in U.S.A.

Geological Society of America
Memoir 145
© 1976

Oxygen-Isotope and Paleomagnetic Stratigraphy of Pacific Core V28-239 Late Pliocene to Latest Pleistocene

N. J. Shackleton
Sub-department of Quaternary Research
University of Cambridge,
5 Salisbury Villas, Station Road
Cambridge, England CB1 2JF

AND

N. D. Opdyke
Lamont-Doherty Geological Observatory
Columbia University
Palisades, New York 10964
and
Department of Geological Sciences
Columbia University
New York, New York 10027

ABSTRACT

V28-239 core from cruise 28 of R/V *Vema* preserves a detailed oxygen-isotope and paleomagnetic record for all of the Pleistocene Epoch. The entire 21-m-long core has been analyzed at 5-cm intervals. Glacial stage 22, above the Jaramillo magnetic event, may represent the first major Northern Hemisphere continental glaciation of middle Pleistocene character. Prior to this, higher frequency glacial events extend to near the level of the Olduvai magnetic event. Glacial events of less regular frequency extend to the bottom of the core, which represents late Pliocene time. Fluctuations in carbonate dissolution intensity occur throughout the core with a similar frequency to the oxygen-isotope fluctuations.

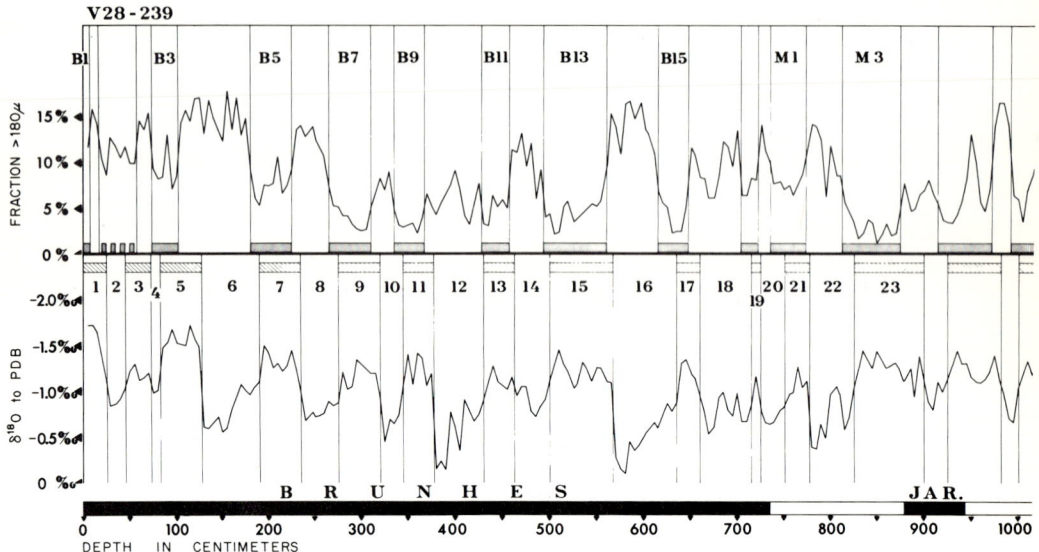

Figure 1. Coarse-fraction record (above), oxygen-isotope record (below), and paleomagnetic record in core V28-239. Dissolution zones in the coarse-fraction record are numbered after

INTRODUCTION

The combination of oxygen-isotope and paleomagnetic stratigraphy in core V28-238 (Shackleton and Opdyke, 1973) has provided an excellent framework within which to investigate the history of events in the western equatorial Pacific during the past 800,000 yr and to correlate this history with events elsewhere. We have now extended this study to about 2.1 m.y. by analyzing a stratigraphically longer core, V28-239, taken relatively close to core V28-238.

In the upper part of the record, comparison between the two cores provides valuable insight into the effects of postdepositional solution and mixing at the sea floor on the oxygen-isotope record and accumulation-rate variations. The lower part of the core provides, for the first time, detailed information on early Pleistocene climates.

ANALYTICAL RESULTS

Core V28-239 was raised from the Solomon Rise at lat 3°15′N, long 159°11′E from a depth of 3,490 m during cruise 28 of the R/V *Vema*. Preliminary magnetic stratigraphy of the core has already been reported (Shackleton and Opdyke, 1973). Analyses have been performed continuously across magnetic reversal boundaries in samples approximately 3 cm across, thus sharply constraining the magnetic boundaries. The Brunhes-Matuyama boundary is located at 726 cm, the Jaramillo event between 877 and 940 cm, and the Olduvai event between 1,553 and 1,781 cm.

Samples for oxygen-isotope analysis were taken at 5-cm intervals throughout the core. Magnetic stratigraphy indicates an average accumulation rate of 1 cm/10^3 yr, giving one sample every 5,000 yr; by comparison, the sampling interval of 10 cm in core V28-238 is one sample every 6,000 yr.

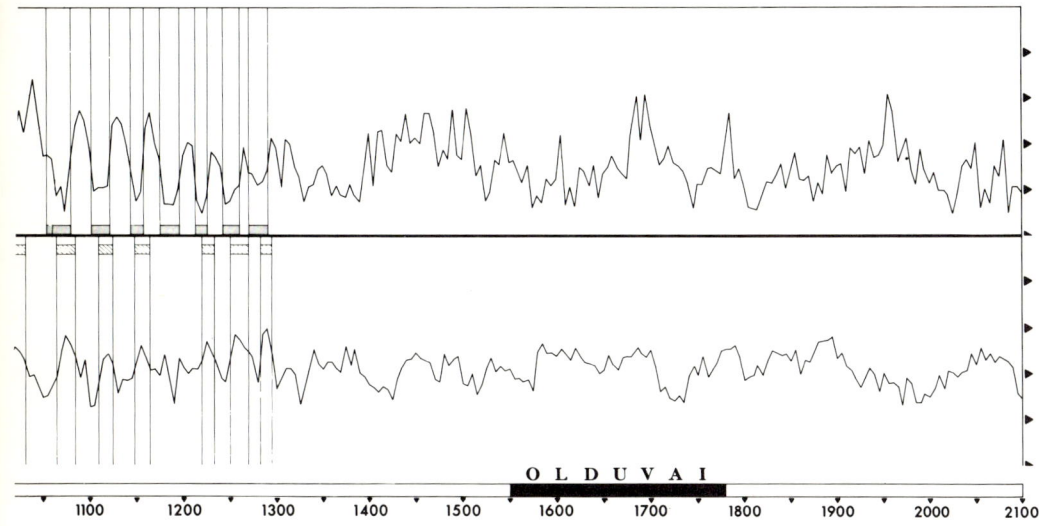

Hays and others (1969); stages in the oxygen-isotope record are numbered after Emiliani (1955, 1966) and Shackleton and Opdyke (1973).

Sediment samples were disaggregated in distilled water; foraminifers were selected for analysis from the >180-μm fraction after sieving and ultrasonic cleaning. Sample pretreatment and chemical processing were identical to those used for core V28-238 (Shackleton and Opdyke, 1973). Isotope analysis was performed in a new V.G. Micromass 602C mass spectrometer. Analyses are referred to the PDB standard (Epstein and others, 1951, 1953) using a value of +0.29‰ for the Emiliani B-1 standard (Shackleton, 1974). This calibration is accurate to better than ±0.05‰. Analytical results in Shackleton and Opdyke (1973) were referred to the B-1 standard and must be corrected by +0.29‰ before comparison with the data presented in this paper.

A single analysis has been made at each level in the core. For each analysis, 15 specimens of *Globigerinoides sacculifer* were selected (in the lower part of the core *G. fistulosus* was used in some samples, three samples contained insufficient specimens for analysis, and a few contained less than 15). Analytical precision is estimated to be ± 0.05‰, the standard deviation for 100 analyses of a standard carbonate performed during the first six months of instrument operation. However, the uncertainty in analysis of a single sample from the sediment is ±0.11‰ (Shackleton and Opdyke, 1973). Isotopic variability among the specimens and analytical precision are combined in this figure. Analytical results are given in Table 1. Figure 1 shows the percentage by weight retained on the 180-μm sieve for each sample, the oxygen-isotope record, and the paleomagnetic record.

Character of the Isotopic Record

Jaramillo Magnetic Event to Present. Figure 1 suggests that the oxygen-isotope record may be divided into three episodes of differing character. The upper part, all of which is represented in core V28-238 (Fig. 2) as well as core V28-239, contains glacial stages at approximately 100,000-yr intervals. Apparently, the isotopic composition of the ocean changed by almost the same extent in every glaciation

TABLE 1. OXYGEN-ISOTOPIC COMPOSITION OF *GLOBIGERINOIDES SACCULIFER* IN CORE V28-239

Depth	$\delta^{18}O$	Depth	$\delta^{18}O$	Depth	$\delta^{18}O$	Depth	$\delta^{18}O$	Depth	$\delta^{18}O$	Depth	$\delta^{18}O$	Depth	$\delta^{18}O$
5	−1.72	305	. .	605	−0.58	905	−0.88	1205	−0.99	1505	−0.86	1805	−0.95
10	−1.72	310	−1.20	611	−0.66	910	−0.79	1210	−1.05	1510	−0.99	1810	−0.97
15	−1.64	315	−1.20	615	−0.60	915	−1.10	1215	−1.04	1515	−0.76	1814	−1.08
20	−1.36	320	−0.92	620	−0.73	920	−0.99	1220	−1.13	1520	−0.87	1820	−0.97
25	−1.10	325	−0.45	625	−0.87	925	−1.12	1225	−1.35	1525	−1.04	1825	−1.13
29	−0.84	330	−0.70	630*	−0.78	930	. .	1230	−1.23	1530	−1.17	1830	−1.24
35	−0.86	335	−0.65	635	−0.87	935	−1.44	1235	−1.11	1535	−1.01	1835	−1.22
40	−0.92	340	−0.74	640	−1.31	940	−1.30	1240	−0.95	1540	−1.07	1840	−1.19
45	−1.03	345*	−1.06	645	−1.35	945	−1.31	1245	−0.90	1545	−1.02	1845	−1.21
50	−1.21	350*	−1.40	651	−1.18	950	−1.15	1250	−1.19	1550	−1.04	1850	−1.13
55	−1.30	355*	−1.07	655	−1.14	955	−1.10	1255	−1.42	1555	−0.93	1855	−1.25
60	−1.12	360*	−1.42	660	−0.94	960	−1.09	1260	−1.37	1560	−0.94	1860	−0.98
65	−1.15	365	−1.36	665	−0.78	965	−1.13	1265	−1.28	1564	−0.90	1865	−1.14
70	−1.20	370	−1.06	669	−0.53	970	−1.21	1270	−1.24	1570	−0.92	1870	−1.12
75	−0.98	375	−1.19	675	−0.61	975	−1.39	1274	−1.18	1575	−0.80	1875	−1.23
80*	−1.01	380	−0.15	680	−0.93	980	−1.13	1280	−0.90	1580	−1.23	1880	−1.34
85	−1.48	385	−0.24	685	−0.99	985	−0.96	1285	−1.42	1585	−1.32	1885	−1.35
90	−1.53	390	−0.14	690	−0.79	990	−0.70	1290	−1.48	1590	−1.21	1890	−1.36
95	−1.68	395	−0.77	695	−0.73	995	−0.66	1295	−1.21	1595	−1.22	1895	−1.39
100*	−1.53	400	−0.61	700	−0.98	1000	−1.02	1300	−0.82	1600	−1.18	1900	−1.15
105	−1.51	405	−0.35	705	−0.67	1005	. .	1306	−0.95	1605	−1.26	1905	−1.21
110	−1.50	410	−0.91	710	−0.67	1010	−1.33	1310	−1.05	1610	−1.20	1910	−1.07
115	−1.72	415	. .	715	−0.84	1015	−1.17	1315	−1.05	1615	−1.14	1915	−1.03
120	−1.57	420	−0.67	720	−1.16	1020	−1.29	1320	−0.98	1620	−1.27	1920	−0.93
125	−0.98	425	−0.78	726	−0.77	1025	−1.26	1325	−0.65	1625	−1.18	1925	−0.84
130	−0.61	430	−0.90	730	−0.66	1030	−1.14	1330	−0.84	1630	−1.11	1930	−0.95
135	−0.59	435	. .	735	−0.64	1035	−0.96	1335	−1.05	1635	−1.10	1935	−1.05
140	−0.65	440	−1.28	739	−0.66	1040	−0.98	1340	−1.26	1640	−1.08	1940	−0.97
146	−0.72	445	−1.11	746	−0.79	1045	−0.87	1345	−1.10	1645	−0.99	1945	−0.79
150	−0.56	450	−1.06	750	−0.82	1050	−0.73	1350	−1.04	1650	−1.03	1950	−0.99
155	−0.60	455	−1.02	756	−0.97	1055	−0.75	1355	−1.12	1655	−1.19	1955	−0.88
160	−0.80	460	−1.15	760	−0.98	1060	−0.86	1360	−1.12	1659	−1.09	1960	−0.89
165	−0.94	465	−0.95	765	−1.27	1065	−0.96	1365	−1.04	1665	−1.03	1966	−0.84
170	−1.07	470	−1.05	770	−1.04	1070	−1.25	1369	−1.01	1670	−1.17	1970	−0.64
175	−1.01	475	−1.05	775	−1.11	1074*	−1.41	1375	−1.29	1675	−1.18	1975	−0.94
180	−0.96	480	−0.78	780	−0.39	1080	−1.30	1380	−1.08	1679	−1.18	1980	−0.88
185	−1.05	485	−0.72	785	−0.37	1085	−1.17	1384	−1.26	1686	−1.28	1985	−0.67
190	−1.10	490	−0.83	790	−0.64	1090	−0.95	1390	−0.99	1690	−1.21	1990	−0.67
195	−1.50	495	−0.90	795	−0.49	1095	−1.14	1395	−0.96	1695	−1.17	1994	−0.77
200	−1.42	500	−1.11	800	−0.96	1100	−0.63	1400	−0.87	1700	−1.25	2000	−0.73
205	−1.26	505	. .	807	−1.05	1105	−0.64	1405	−0.84	1705	−1.06	2005	−0.82
210	−1.31	510*	−1.45	810	−0.97	1110	−0.93	1410	−0.79	1710	−0.79	2010	−0.97
215	−1.22	515	−1.29	815	−0.58	1115	−1.15	1415	−0.85	1715	−0.84	2016	−0.82
220	−1.28	519	−1.22	820	−0.71	1120	−1.21	1420	−0.82	1720	−0.72	2020	−1.02
225	−1.40	526	−1.03	825	−1.04	1125	−1.09	1425	−0.71	1725	−0.69	2025	−1.00
230	−1.24	529*	−1.07	830	. .	1130	−0.77	1430	−0.93	1730	−0.75	2030	−0.95
235	−0.99	535	−1.32	835	−1.45	1135	−0.93	1435	−1.06	1735	−0.67	2035	−1.01
240	−0.68	540	−1.24	840	−1.34	1140	−0.92	1440	−1.15	1740	−0.95	2040	−1.03
248	−0.77	545	−1.11	845	−1.25	1145	−0.94	1445	−1.16	1746	−1.07	2045	−1.21
251	−0.72	550	−1.26	850	−1.44	1150	−1.13	1450	−1.22	1750	−0.97	2050	−1.09
255	−0.73	554	−1.25	855	—	1155	−1.30	1455	−1.16	1755	−1.18	2055	−1.27
260	−0.76	561	−1.10	860	−1.25	1160	−1.16	1460	−1.14	1761	−0.95	2060	−1.14
265	−0.89	565	−1.09	865	. .	1165	−1.03	1465	−1.11	1765	−1.02	2065	−1.21
270	−0.84	570	−0.27	870	−1.31	1170	. .	1470	−0.93	1770	−1.11	2070	−1.11
275	−0.87	575	−0.15	874	−1.25	1175	. .	1475	−0.90	1775	−1.24	2075	−1.18
280	−1.21	580	−0.10	879	−1.10	1180	−1.20	1480	−1.21	1780	−1.26	2080	−1.15
285*	−1.02	585	−0.45	886	−1.24	1185*	−0.94	1485	−1.08	1785	−1.26	2085	−1.12
290	−1.05	590	−0.35	890	−0.84	1190*	−0.66	1490	−1.15	1790	−1.30	2090	−1.06
295*	−1.35	595	−0.41	895	−1.38	1195	−1.17	1495	−1.18	1795	−1.17	2095	−0.77
300	. .	602	−0.54	900	−1.13	1200	−1.06	1500	−0.88	1800	−0.91	2100	−0.71

Note: Composition is expressed as a deviation per mil from the PDB standard. Samples contained 15 individuals.
*Less than 15.

during this interval. The rather large variability among glacial extreme isotopic values in core V28-239 is an artifact of sedimentation processes. This is evident from the fact that the extreme isotopic values in successive glaciations are both less variable and more distant from the Holocene value in cores with higher accumulation rates. In core V28-239 (1.0 cm/10^3 yr) the extreme isotopic values in glacial stages 2 and 6 to 22 differ from the Holocene value by 1.22 ± 0.24‰. In core V28-238 (1.7 cm/10^3 yr) the same ten glacial extreme values differ from the Holocene value by 1.04 ± 0.14‰. In core V19-28 (4.0 cm/10^3 yr) the last five glacial extreme values differ from the Holocene value by 1.62 ± 0.11‰. (Ninkovich and Shackleton, 1975).

Figure 1 shows cyclic changes in the percentage of sediment that is greater than 180 μm as well as in oxygen-isotope composition. Thompson and Saito (1974) documented correlative cyclic variations in dissolution intensity in cores V28-238, V28-239, and RC11-210. The latter core is in the region where Hays and others (1969) defined dissolution zones on the basis of changing carbonate percentage. Thus, we may confidently ascribe the observed variations in coarse-fraction percentage to changes in dissolution intensity and assign them to zones according to the definition of Hays and others (1969). Figure 1 shows these dissolution zones. The delay between the climatic change recorded in the oxygen-isotope record and the dissolution change, noted by Luz and Shackleton (1975) and by Ninkovich and Shackleton (1975), is preserved throughout the sequence.

Characteristically, the transition from glacial to interglacial extreme occurred very rapidly (Broecker and van Donk, 1970); indeed, 12,000 yr ago deglaciation took place so fast that its record in sediment cores is almost invariably determined by the sediment-mixing depth rather than by the actual rate of change in the isotopic composition of the ocean (at least 0.3‰/10^3 yr). Glacial stages 2, 6, 10, 12, 20, and 22 terminated in this manner. Stages 4, 8, and 14 probably did

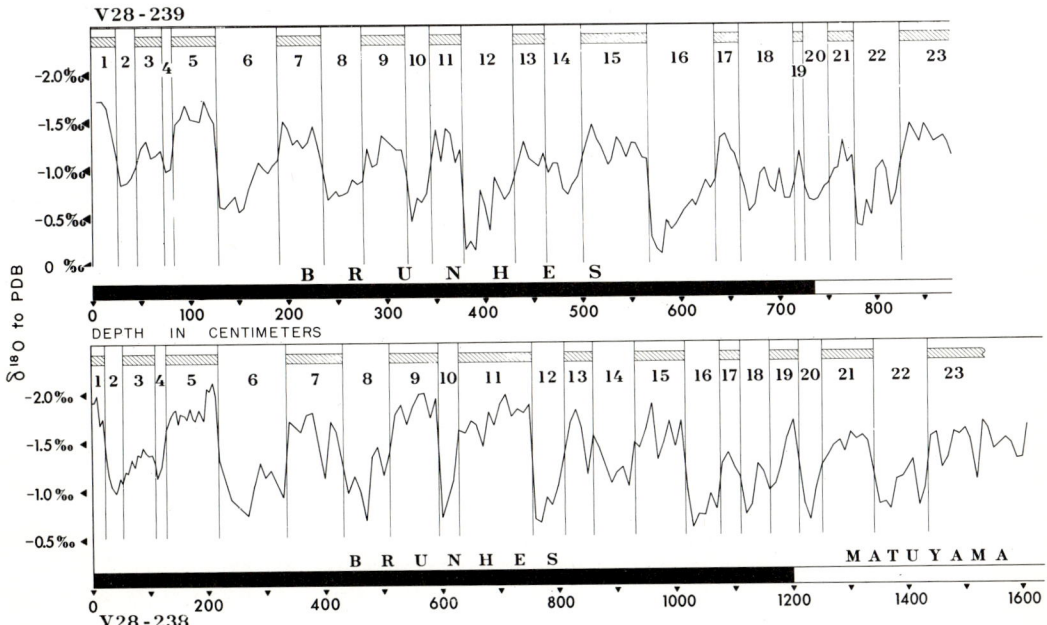

Figure 2. Oxygen-isotope and paleomagnetic record in upper 880 cm of cores V28-239 (above) and V28-238 (below).

not, whereas the evidence from stage 18 is inconclusive. The character of the record in other respects does not seem susceptible to generalization. There is a significant amount of fine structure that can be reliably appreciated only in cores with a higher sedimentation rate. Much potential information for marine-terrestial correlation is contained in this detail.

In core V28-239 glacial stage 22 is the first glaciation of duration and intensity similar to the glacial stages of the Brunhes epoch.

Mid-Matuyama Region. Between 0.8 to 1.4 m.y. ago (800 and 1,400 cm) isotopic fluctuations occurred at about 40,000-yr intervals. Because of the effects of mixing, it is not possible to say whether the true amplitude of glacial-interglacial change was smaller in mid-Matuyama time. The record of changing coarse fraction (Fig. 1) has a similar frequency, and again, changes in coarse fraction appear about 5 cm above oxygen-isotope changes, representing a lag of about 5,000 yr.

Among the stages so far defined, the isotopic difference between adjacent stages generally exceeds the isotopic variations within a single stage. In order to extend this principle to the section between 800 and 1,400 cm, one would need to define shorter stages; yet, with present coring techniques this degree of refinement could seldom be utilized. For the present we defer extending the scheme for numbering stages.

The pattern of coarse-fraction percentage may be zoned easily by reference to Hays and others (1969) as far back as their M3 zone (Fig. 1). Their M11 zone probably falls between 1,330 and 1,390 cm. Between M3 and M11, the records they described lack the resolution to display events of the frequency that we demonstrate.

Olduvai Region. The lower third of the core, from 1,400 to 2,100 cm, records lower frequency oxygen-isotopic changes with an amplitude no greater than in the middle section. The sediment recording the Olduvai magnetic event contains a well-marked "glaciation" lasting about 25,000 yr near the base of the event and another at the top of the event. A glaciation of similar magnitude is found at 2,000 cm (about 2 m.y. B.P., late Pliocene). Assuming that the Pliocene-Pleistocene boundary is in or near the Olduvai magnetic event, there is no associated climatic event that would enable late Pliocene to be distinguished from earliest Pleistocene time on the basis of climate. However, there is sufficient information in the isotopic record to facilitate detailed long-distance correlation in the region of this boundary. The relationship between the oxygen-isotope record and the record of changing coarse-fraction percentage is less obvious in this part of the core.

Comparison with Atlantic Core V16-205

The only other core extending to the base of the Pleistocene record that has been analyzed using the oxygen-isotope technique is North Atlantic core V16-205 (van Donk, this volume). This core has a sedimentation rate of only about 0.55 cm/10^3 yr, so that the distortion in the core V28-239 isotopic record, discussed below, is present to a considerably greater extent in core V16-205. In the Brunhes section of the core the major climatic cycles can still be distinguished with confidence, but in the section below the Jaramillo magnetic event, core V16-205 shows scarcely any change over an interval interpreted to represent 300,000 yr. The frequency of isotopic changes that we observe in core V28-239 is about one cycle per 40,000 yr, corresponding to a wavelength in core V16-205 of only 22 cm. It would be expected that mixing would largely obscure the record of these fluctuations, although high-precision analyses at closer sampling intervals might possibly detect a residual record. An additional impediment to correlation between the two cores is that

V16-205 seems to contain even greater changes in accumulation rate than V28-239. The rate between the Jaramillo and the top of the Brunhes in core V16-205 is reported to be 0.25 cm/10^3 yr, although the average rate through the entire core is 0.55 cm/10^3 yr. It is to be hoped that interoceanic correlations for the Matuyama epoch will become more reliable with the analysis of more cores in both the oceans.

ISOTOPE STRATIGRAPHY AND ITS LIMITATIONS

Oxygen-Isotope Stages: Terminology

Emiliani (1955, 1966) used numbers 1 to 16 to designate stages that he recognized in oxygen-isotope records he obtained in sediment cores from the Caribbean Sea and Atlantic Ocean. We (Shackleton and Opdyke, 1973) recognized 22 stages in core V28-238, the first 16 coinciding with those used by Emiliani. As a step toward formalizing this nomenclature, stage boundaries were defined by the depth at which they were located in core V28-238 (Fig. 2). For core V28-239, 23 stages are shown in Figure 1; the depths of stage boundaries, placed by correlation with core V28-238, are given in Table 2.

Before considering the extension of this terminology, it is important to consider the assumptions on which use of the oxygen-isotope record as a stratigraphic tool is based and the limitations of its usefulness. It is universally agreed that at the

TABLE 2. STAGE BOUNDARIES IN CORE V28-239 AS DETERMINED BY CORRELATION WITH CORE V28-238

Boundary	Depth in core* (cm)	Age† (B.P.)	Termination§
1-2	25	13,000	I
2-3	45	32,000	
3-4	72	64,000	
4-5	82	75,000	
5-6	127	128,000	II
6-7	190	195,000	
7-8	235	251,000	III
8-9	275	297,000	
9-10	320	347,000	IV
10-11	345	367,000	
11-12	377	440,000	V
12-13	430	472,000	
13-14	462	502,000	
14-15	500	542,000	
15-16	567	592,000	VI
16-17	635	627,000	
17-18	660	647,000	
18-19	715	688,000	
19-20	725		
20-21	750		
21-22	777		
22-23	825		

*Determined by correlation with core V28-238.

†Ages are those estimated by Shackleton and Opdyke (1973) by linear interpolation in core V28-238 using a rate of 1.7 cm/10^3 yr.

§Terminations from Broecker and van Donk (1969). They defined terminations on the basis of their interpretation of the saw-toothed character of the oxygen-isotope record. Owing to a possible hiatus in core V12-122, it appears that the event labeled termination VI by them is the stage 16-15 boundary.

maximum of the most recent glaciation, between 20,000 and 15,000 yr ago, sea level was lowered about 100 m as the result of ice storage on the continents. The same situation prevailed at each glacial maximum. This ice was certainly depleted in O^{18} relative to ocean water, and the result of its removal on the isotopic composition of the remaining ocean water has been discussed many times since its effect on the isotope record in fossil Foraminifera was first considered (Emiliani, 1955; Olausson, 1965; Shackleton, 1967; Dansgaard and Tauber, 1969; Shackleton and Opdyke, 1973). The resemblance between the actual (at present unknown) record of mean isotopic composition of oceans as a function of time and any particular measured record of isotopic composition in calcareous fossils from a sediment core depends on a number of factors which are discussed below.

Ocean Mixing and the Oxygen-Isotope Record

When isotopically light water is removed from or added to a particular part of the ocean, the resulting change will be noticed in another part of the ocean only after a delay resulting from the finite mixing time of the oceans. This has been variously estimated, but it is probably under 1,000 yr (Gordon, 1975). Thus, the effect of a sudden change in continental ice volume could be a transient spike in isotopic composition of the Atlantic, followed a few hundred years later by an isotopic change in the Pacific.

In reality this effect is limited by the finite response time of an ice sheet (Weertman, 1964). Nevertheless, transient effects can be detected in restricted areas such as the Mediterranean (Olausson, 1965) and the Gulf of Mexico (Kennett and Shackleton, 1975). This is the ultimate limit to the precision of oxygen-isotope stratigraphy.

Deep-Ocean Temperatures and the Ideal Oxygen-Isotope Record of Benthic Foraminifera

The isotopic composition of the calcite tests of foraminifera depends on both the isotopic composition and the temperature of the water they inhabited (Urey, 1947). In the ocean, surface-temperature changes probably were not synchronous over the entire globe. However, deep-ocean temperature changes during Pleistocene time were probably small, and the stability requirements of the ocean suggest that they should have occurred essentially synchronously. Thus, a record of the isotopic composition of the tests of suitable benthic foraminifers in a core from a depth of 3,000 m (and with a sufficiently high sedimentation rate to provide a stratigraphic resolution of about 1,000 yr) would approach the ideal with which the oxygen-isotope record in other cores might be correlated. Consideration of the available sections of record approaching this ideal suggests that in favorable circumstances such a record would contain enough information to permit a correlation accuracy of better than 2,000 yr throughout the past 900,000 yr.

Because the ideal isotopic record for benthic foraminifers has not yet been obtained, it is not possible to say in what manner the isotopic records of planktonic foraminifers in cores V28-238 and V28-239 differ from the ideal. We consider below the effects of mixing and burrowing, postdepositional solution, and surface temperature change on such isotopic records.

Sediment Mixing and Burrowing

Table 3 compares the peak-to-peak amplitude of isotopic changes in cores V28-238 and V28-239. The 5-cm sampling interval in core V28-239 approaches the mixing depth (Berger and Heath, 1968; Ruddiman and Glover, 1972), and the extremes analyzed must therefore approximate the extremes present in the core. The mean amplitude of isotopic fluctuations is less in core V28-239 than in V28-238, probably because the accumulation rate is lower in comparison with the mixing depth.

We have argued that the observed peak-to-peak amplitude of oxygen-isotope changes in core V28-238 was reduced by mixing (Shackleton and Opdyke, 1973). Thus, the full range of isotopic variation is attenuated in the sediment in core V28-238 and even more so in core V28-239. However, both cores preserve sufficient record that successive stages can be unambiguously recognized.

Carbonate Dissolution and the Oxygen-Isotope Record

Core V28-239 was taken at a depth of 3,490 m, compared to 3,120 m for core V28-238. This accounts for more intense dissolution occurring in core V28-239. Savin and Douglas (1973) pointed out that increasing dissolution not only progressively removes the more solution-susceptible species (often those that lived in shallower water), but it also selectively removes from the population of a single species those members that lived closer to the surface. Thus, the fossil population that has suffered more dissolution registers a lower isotopic temperature as a consequence of that dissolution.

TABLE 3. STAGE-BY-STAGE ISOTOPE EXTREMES FOR CORES V28-239 AND V28-238

Stage	Interval	V28-239 A (δ, ‰)	V28-239 B (range)	V28-238 C (δ, ‰)	V28-238 D (range)	Between-core Difference (A − C)*
1	1-2	−1.72	0.88	−1.98	1.01	0.26
2	2-3	−0.84	0.46	−0.97	0.47	0.13
3	3-4	−1.30	0.32	−1.44	0.59	0.14
4	4-5	−0.98	0.74	−0.85	1.26	−0.13
5	5-6	−1.72	1.16	−2.11	1.37	0.39
6	6-7	−0.56	0.94	−0.74	1.05	0.18
7	7-8	−1.50	0.82	−1.79	1.10	0.29
8	8-9	−0.68	0.67	−0.69	1.30	0.01
9	9-10	−1.35	0.90	−1.99	1.28	0.64
10	10-11	−0.45	0.97	−0.71	1.26	0.26
11	11-12	−1.42	1.28	−1.97	1.32	0.55
12	12-13	−0.14	1.14	−0.65	1.16	0.51
13	13-14	−1.28	0.56	−1.81	0.76	0.53
14	14-15	−0.72	0.73	−1.05	0.82	0.33
15	15-16	−1.45	1.35	−1.87	1.28	0.42
16	16-17	−0.10	1.25	−0.59	0.77	0.49
17	17-18	−1.35	0.82	−1.36	0.64	0.01
18	18-19	−0.53	0.63	−0.72	0.97	0.19
19	19-20	−1.16	0.52	−1.69	1.03	0.53
20	20-21	−0.64	0.63	−0.66	0.87	0.02
21	21-22	−1.27	0.90	−1.53	0.76	0.26
22		−0.37		−0.77		0.40

Note: Column A, extreme oxygen-isotopic composition in each stage in core V28-239, from Table 1. Column B, isotopic difference between adjacent stages in core V28-239. Mean 0.84 ± 0.28. Column C, extreme oxygen-isotopic composition in each stage in core V28-238, from Shackleton and Opdyke (1973, Table 1), corrected to PDB standard. Column D, isotopic difference between adjacent stages in core V28-238. Mean 1.00 ± 0.27

*Difference between the extreme reached in cores V28-239 and V28-238 for each stage. Mean 0.29 ± 0.21.

Samples from the two core tops differ isotopically by about 0.2‰. The isotopic extremes in each stage differ between the cores by 0.29 ± 0.21‰ (Table 3). This is probably due to greater solution in core V28-239, which is evident in the degree of preservation of the foraminifers. Part of the between-stage variation in core V28-239 may be due to the effect of fluctuating carbonate dissolution on the isotopic composition, and this clearly limits the veracity of a record of ocean oxygen-isotope composition derived from measurements in a deep-water core.

Sediment Accumulation-Rate Changes and Their Effect on the Oxygen-Isotope Record

Despite the distortions in the shape of the record of oxygen-isotopic changes in core V28-239 caused by dissolution and by sediment mixing discussed above, the position of the "terminations" (glacial-interglacial stage boundaries) of Broecker and van Donk (1970) may be placed confidently. Thus, the sediment thickness deposited during each climatic cycle may be measured and compared with the thickness deposited over the same time intervals in other cores.

Figure 3 compares the thickness of sedimentary deposits representing these major climatic cycles in cores V28-238 and V28-239 (data from Table 4). If the accumulation had remained constant in both cores, the points in Figure 3 would lie on a single straight line. The fact that they do not shows that in one or both cores the accumulation rate varied. In Figures 4 and 5, the thicknesses of sedimentary deposits representing the climatic cycles in cores V28-239 and V28-238 are compared with the average thicknesses of the same cycles in a suite of four cores analyzed by Emiliani (1966, 1972). Core V28-238 (Fig. 5) displays better agreement with the Caribbean suite. Therefore, the scatter in Figure 3 is probably due to stage-to-stage variations in accumulation rate in core V28-239.

In Figure 5 the greatest discrepancy between the Caribbean cores and core V28-238 is in stages 2 to 5. This suggests that in one of these areas there was a systematic change in accumulation rate about 130,000 yr ago (end of stage 6), as suggested by Emiliani and Shackleton (1974). The existence of a regional change in accumulation rate for any area at the end of stage 6 serves as a warning that linear extrapolation

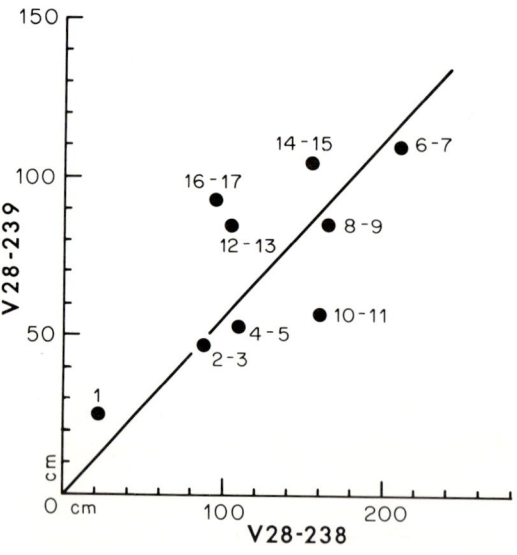

Figure 3. Comparison of thickness of climatic cycles (odd- and succeeding even-numbered stages) in cores V28-239 and V28-238. Data from Table 4.

TABLE 4. THICKNESS OF CLIMATIC CYCLES IN A SUITE OF CARIBBEAN
CORES AND IN PACIFIC CORES V28-238 AND V28-239

Stages	Cores						
	P6304-4	P6304-7	P6304-8	P6304-9	Mean P6304 suite	V28-238	V28-239
1	25	30	30	30	29	22	25
2-3	130	145	150	165	148	88	47
4-5	175	185	170	145	169	110	53
6-7	240	275	280	240	259	210	110
8-9	160	195	210	200	191	165	85
10-11	145	180	190	180	174	160	57
12-13	155	150	. .	130	144	105	85
14-15	85	80?	. .	200	200*	155	105
16-17						95	93

Note: Data for Caribbean cores from Emiliani (1955, 1972); for Pacific cores from Shackleton and Opdyke (1973); for core V28-239 from this paper.
*Data for stages 14-15 from P6304 suite is inconsistent. Value for 6304-9 has been adopted, because it is more consistent with the Pacific cores (Figs. 3, 4).

performed on the basis of an assumed uniform accumulation rate may be in error even if the extrapolation is based on average accumulation rates from numerous cores.

Effect of Varying Surface Temperature

In the Caribbean, Emiliani (1966) has shown that the oxygen-isotope composition of *G. sacculifer* in recent sediment implies deposition at or near surface temperature. Vincent and Shackleton (1975) have shown that this is also true in the Indian Ocean. While this situation holds, changes in surface temperature during Pleistocene time, if present, should affect the isotopic composition of *G. sacculifer* populations in Pleistocene sediment. Hence, it is generally assumed that changes in surface temperature may be estimated by subtracting the component that is ascribed to glacially induced changes in ocean isotopic composition from the total record of oxygen-isotopic change (Imbrie and others, 1973).

In the Pacific a different situation prevails: The oxygen-isotope composition of *G. sacculifer* corresponds to a temperature several degrees below sea-surface temperature in many core-top samples (Savin and Douglas, 1973). To what extent this figure is an indication of a difference in the depth distribution of calcification in the species and to what extent it is a function of selective dissolution of the individuals from shallower depths in the water column (Savin and Douglas, 1973) remain to be evaluated. However, we (Shackleton and Opdyke, 1973) argued that the close similarity between the isotopic records of *G. sacculifer* and benthic species in core V28-238 implies that both records depict the history of the isotopic composition of the ocean, and that changes in surface temperature, temperature-depth structure, depth distribution of *G. sacculifer*, and selective dissolution all play minor roles. Discrepancies between the planktonic and benthic records in core V28-238 were ascribed by us (Shackleton and Opdyke, 1973) to the effects of postdepositional sediment mixing by burrowing organisms rather than to the factors mentioned above.

Long-Term Trends in the Oxygen-Isotope Record

There is no general agreement regarding any long-term trends in climate during Pleistocene time; trends that emerge from studies based on the present-day ecological requirements of any organisms may be due to the effect of evolution or of evolutionary adaptation. It might be hoped that oxygen-isotope analysis would provide a means of discerning real trends. Unfortunately, in the case of core V28-239 it seems certain that the trends observed are due to factors that alter the amount of smoothing the record has undergone, and not to trends in isotopic composition of the ocean. A careful comparison between the records of cores V28-238 and V28-239 shows that there is a relationship between trends in isotopic composition and changes in accumulation rate. For example, in core V28-239 the most extreme isotopic values are found in glacial stages 12 and 16. This might be taken to imply that these were the stages of greatest glacial advance. However, Table 2 shows that these two glacial stages are represented by almost the same sediment thickness in core V28-239 as in V28-238, whereas most units in V28-239 are about two-thirds of the thickness found in V28-238. Thus, these two glaciations appear more extreme only because sediment mixing has had a smaller blurring effect. It may be the case that the accumulation rate changed in response to climate, in which case the trend would have some second-order relationship to climate. In the lower part of the core, there may be long-term changes in dissolution intensity and perhaps evolutionary changes in calcite secretion depth. The *fistulosus* form of *G. sacculifer* is isotopically indistinguishable from the usual form, but it may prove to be related to a change in calcite secretion depth.

We conclude that it would be unwise to deduce from available oxygen-isotope measurements any trend either in surface temperature in the equatorial Pacific or in the amplitude of glacial advances in the Northern Hemisphere. In more favorable circumstances (higher accumulation rate, shallower water) oxygen-isotope analysis should be capable of yielding information on such trends.

CALCIUM CARBONATE DISSOLUTION STRATIGRAPHY

During Pleistocene time the extent of solution on the Pacific floor varied cyclically (Arrhenius, 1952; Hays and others, 1969) as climate changed. Luz and Shackleton (1975) have shown that during stage 5, dissolution systematically lagged behind

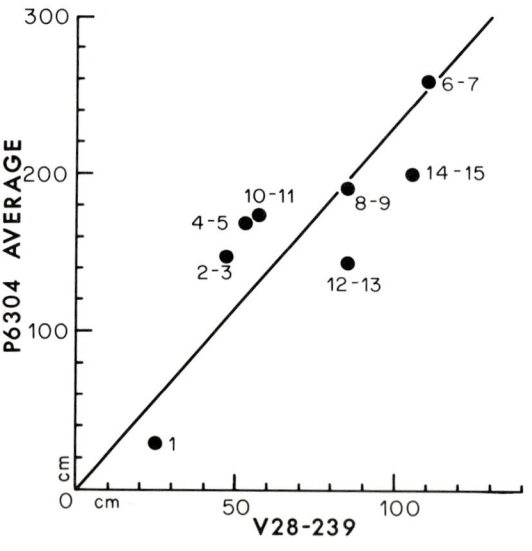

Figure 4. Comparison of thickness of climatic cycles (odd- and succeeding even-numbered stages) in core V28-239 and in suite of cores P6304-4, P6304-7, P6304-8, and P6304-9 from Emiliani (1966, 1972). Data from Table 4.

the climatic record. The intensity of dissolution rose not at the boundary between stages 6 and 5 (termination II of Broecker and van Donk, 1970) but a few thousand years later. We now show that this relationship has held through the past 1.5 m.y.

Figure 1 indicates the boundaries of the dissolution zones, numbered according to the scheme of Hays and others (1969), and their relation to the oxygen-isotope stages. The dissolution zones are not manifested as carbonate fluctuations, and the carbonate content is high throughout the core (Thompson, 1976). However, Thompson and Saito (1974) have shown that change in dissolution intensity is the dominant factor in determining the downcore changes in foraminiferal faunal composition in this area.

Figure 1 clearly indicates that Hays and others (1969) were correct in their assertion that changes in carbonate content in eastern equatorial Pacific sediments could be correlated with the Northern Hemisphere climatic record and with oxygen-isotope records from the Caribbean. However, use of carbonate cycles as a precise stratigraphic tool may be misleading. Figure 1 shows that the base of each dissolution zone in the sediment is not found at the same position as the glacial-to-interglacial isotopic transition, but rather some 5 to 20 cm above. This represents a delay of a few thousand years between the climatic change and its effect on bottom-water chemistry in the equatorial Pacific. This delay may not be constant from one latitude to another or from one climatic cycle to another.

CHRONOLOGY

The record of changes in the oxygen-isotope composition of the world oceans may be readily used as a stratigraphic tool in Pleistocene deep-sea sediments of all oceans (the Atlantic and Caribbean, Emiliani, 1955; the Arctic, van Donk and Mathieu, 1969; the Indian Ocean, Oba, 1969; the Pacific Ocean, Shackleton and Opdyke, 1973; the sub-Antarctic regions, Hays and others, this volume). Moreover, since the primary mechanism giving rise to these changes is the growth and retreat of continental ice sheets in the Northern Hemisphere, the record is of considerable

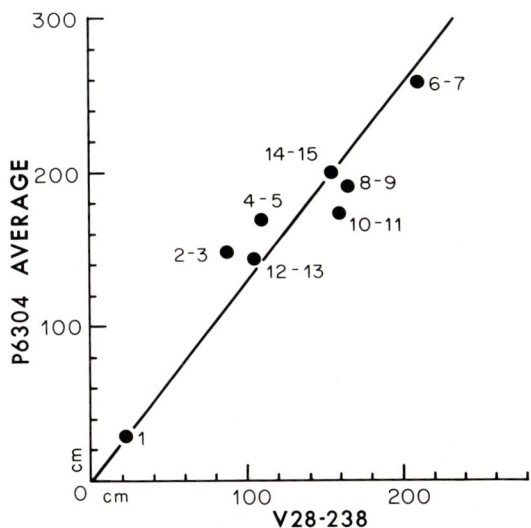

Figure 5. Comparison of thickness of climatic cycles (odd- and succeeding even-numbered stages) in core V28-238 and in suite of cores P6304-4, P6304-7, P6304-8, and P6304-9 from Emiliani (1966, 1972). Data from Table 4.

value as a basic Pleistocene stratigraphic tool to which the less continuous fragments of the Pleistocene record from the continents may be correlated (for example, the Netherlands, van der Hammen and others, 1967; Central Europe, Kukla, 1970).

Hence, the absolute dating of the oxygen-isotope record is of wide interest. Moreover, accurate dating is essential if oxygen-isotope records are to be used to test the hypothesis that variations in the Earth's orbital parameters caused climatic changes.

Isotopic stage boundaries in core V28-238 were first dated by assuming a constant rate of sediment accumulation above the Brunhes-Matuyama magnetic reversal boundary at 1,200 cm (Shackleton and Opdyke, 1973). Although more accurate ages will be obtained in the future by making use of several cores, it is important that this be done critically. We have argued that the accumulation rate within the Brunhes magnetic epoch was more uniform in core V28-238 than in V28-239. Thus, ages derived by assuming uniform accumulation rates may be more reliable if based on core V28-238 alone than they would be if based either on V28-239 alone or on a combination of the two cores.

For this reason, we do not offer new estimates of the ages of stage boundaries but consider those estimates made earlier (Shackleton and Opdyke, 1973) as the best available (Table 2). We emphasize again that these cannot be regarded as definitive.

CONCLUSIONS

Oxygen-isotope analysis provides an excellent stratigraphic tool for the past 900,000 yr. Oxygen-isotope stratigraphy has been used to locate the timing of carbonate dissolution zones relative to major climatic changes. Dissolution increased a few thousand years after each deglaciation.

Oxygen-isotope stratigraphy enables us to compare sediment accumulation rates. Cores V28-238 and V28-239 and a suite from the Caribbean analyzed by Emiliani have been compared in detail. Core V28-238 had a more constant accumulation rate from stage to stage than the deeper core V28-239.

In lower Pleistocene sediment, oxygen-isotope stratigraphy will not be useful in cores with an accumulation rate lower than $1 \text{ cm}/10^3 \text{ yr}$, because the isotopic changes were more frequent than in the Brunhes epoch and therefore more readily obscured by mixing in the sediment.

ACKNOWLEDGMENTS

Coring was supported by Office of Naval Research Grant N00014-67-A-0108-0004 and National Science Foundation Grant GA-29460 to Lamont-Doherty Geological Observatory. The research was supported by National Science Foundation IDOE Grant IDO71-04204. Oxygen-isotope analysis was supported by Natural Environment Research Council Grant GR3/1762 to N. J. Shackleton at Cambridge. We gratefully thank M. A. Hall for his operation of the mass spectrometer with consistent care. We are grateful to W. A. Berggren, K. King, and G. J. Kukla for making instructive comments on an early version of this paper, and to S. M. Savin for a very constructive final review.

REFERENCES CITED

Arrhenius, G., 1952, Sediment cores from the East Pacific: Swedish Deep-Sea Exped. Repts., v. 5, p. 6-227.

Berger, W. H., and Heath, G. R., 1968, Vertical mixing in pelagic sediments: Jour. Marine Research, v. 26, p. 135-143.

Broecker, W. S., and van Donk, J., 1970, Insolation changes, ice volumes and the O^{18} record in deep-sea sediments: Rev. Geophysics and Space Physics, v. 8., p. 169-198.

Dansgaard, W., and Tauber, H., 1969, Glacier oxygen-18 content and Pleistocene ocean temperatures: Science, v. 166, p. 499-502.

Emiliani, C., 1955, Pleistocene temperatures: Jour. Geology, v. 63, p. 538-578.

——1966, Palaeotemperature analysis of Caribbean cores P6304-8 and P6304-9 and a generalized temperature curve for the past 425,000 years: Jour. Geology, v. 74, p. 109-126.

——1972, Quaternary paleotemperatures and the duration of the high-temperature intervals: Science, v. 178, p. 398-401.

Emiliani, C., and Shackleton, N. J., 1974, The Brunhes epoch: Palaeotemperatures and geochronology: Science, v. 183, p. 511-514.

Epstein, S., Buchsbaum, R., Lowenstam, H.A., and Urey, H. C., 1951, Carbonate-water isotopic temperature scale: Geol. Soc. America Bull., v. 62, p. 417-426.

——1953, Revised carbonate-water isotopic temperature scale: Geol. Soc. America Bull., v. 64, p. 1315-1326.

Gordon, A. L., 1975, General ocean circulation, in Numerical models of ocean circulation: Washington, D.C., Natl. Acad. Sci., p. 39-53.

Hays, J. D., Saito, T., Opdyke, N. D., and Burckle, L. H., 1969, Pliocene-Pleistocene sediments of the equatorial Pacific: Their paleomagnetic, biostratigraphic, and climatic record: Geol. Soc. America Bull., v. 80, p. 1481-1514.

Hays, J. D., Lozano, J., Shackleton, N., and Irving, G., 1976, Reconstruction of the Atlantic Ocean and western Indian Ocean sectors of the 18,000 B.P. Antarctic Ocean, in Cline, R. M., and Hays, J. D., eds., Investigation of late Quaternary paleoceanography and paleoclimatology: Geol. Soc. America Mem. 145 (this volume).

Imbrie, J., van Donk, J., and Kipp, N. G., 1973, Paleoclimatic investigation of a Late Pleistocene deep-sea core: Comparison of isotopic and faunal methods: Quaternary Research, v. 3, p. 10-38.

Kennett, J. P., and Shackleton, N. J., 1975, Latest Pleistocene melting of the Laurentide ice sheet recorded in deep-sea cores from the Gulf of Mexico: Science, v. 188, p. 147-150.

Kukla, J., 1970, Correlations between loesses and deep-sea sediments: Geol. Fören. Stockholm Förh., v. 92, p. 148-180.

Luz, B., and Shackleton, N. J., 1975, $CaCO_3$ solution in the tropical East Pacific during the past 130,000 years: Cushman Found. Foram. Research Spec. Pub. 13, p. 142-150.

Ninkovich, D., and Shackleton, N. J., 1975, Distribution, stratigraphic position and age of ash layer "L," in the Panama Basin region: Earth and Planetary Sci. Letters, v. 27, p. 20-34.

Oba, T., 1969, Biostratigraphy and isotopic paleotemperatures of some deep-sea cores from the Indian Ocean: Tohoku Univ. Sci. Repts., 2nd ser. (Geology), v. 41, p. 129-195.

Olausson, E., 1965, Evidence of climatic changes in North Atlantic deep-sea cores, with remarks on isotopic palaeotemperature analysis: Prog. Oceanography, v. 3, p. 221-252.

Ruddiman, W. F., and Glover, L. K., 1972, Vertical mixing of ice-rafted volcanic ash in North Atlantic sediments: Geol. Soc. America Bull., v. 83, p. 2817-2836.

Savin, S. M., and Douglas, R. G., 1973, Stable isotope and magnesium geochemistry of recent planktonic Foraminifera from the South Pacific: Geol. Soc. America Bull., v. 84, p. 2327-2342.

Shackleton, N. J., 1967, Oxygen isotope analyses and paleotemperatures re-assessed: Nature, v. 215, p. 15-17.

——1974, Attainment of isotopic equilibrium between ocean water and the benthonic foraminifera genus *Uvigerina*: Isotopic changes in the ocean during the last glacial: Paris, Centre National de la Recherche Scientifique, Colloquium 219.

Shackleton, N. J., and Opdyke, N. D., 1973, Oxygen isotope and palaeomagnetic stratigraphy of equatorial Pacific core V28-238: Oxygen isotope temperatures and ice volumes on a 10^5 and 10^6 year scale: Quaternary Research, v. 3, p. 39-55.

Thompson, P.R., 1976, Planktonic foraminiferal dissolution and the progress towards a Pleistocene equatorial Pacific transfer function: Jour. Foram. Research (in press).

Thompson, P. R., and Saito, T., 1974, Pacific Pleistocene sediments: Planktonic foraminifera dissolution cycles and geochronology: Geology, v. 2, p. 333-335.

Urey, H. C., 1947, The thermodynamic properties of isotopic substances: Chem. Soc. Jour., p. 562-581.

van der Hammen, T., Maarleveld, G. C., Vogel, J. C., and Zagwijn, W. H., 1967, Stratigraphy, climatic succession and radiocarbon dating of the last glacial in the Netherlands: Geologie en Mijnbouw, v. 46e, p. 79-95.

van Donk, J., 1976, An O^{18} record of the Atlantic Ocean for the entire Pleistocene, *in* Cline, R. M., and Hays, J. D., eds., Investigation of late Quaternary paleoceanography and paleoclimatology: Geol. Soc. America Mem. 145 (this volume).

van Donk, J., and Mathieu, G., 1969, Oxygen isotope compositions of foraminifera and water samples from the Arctic Ocean: Jour. Geophys. Research, v. 74, p. 3396-3407.

Vincent, E., and Shackleton, N. J., 1975, Oxygen and carbon isotope composition of recent planktonic foraminifera from the Southwest Indian Ocean: Geol. Soc. America Abs. with Programs, v. 7, no. 7, p. 1308.

Weertman, J., 1964, Rate of growth or shrinkage of nonequilibrium ice sheets: Jour. Glaciology, v. 5, p. 145-158.

MANUSCRIPT RECEIVED BY THE SOCIETY DECEMBER 18, 1974
REVISED MANUSCRIPT RECEIVED JUNE 11, 1975
MANUSCRIPT ACCEPTED JUNE 25, 1975
LAMONT-DOHERTY GEOLOGICAL OBSERVATORY CONTRIBUTION NO. 2277